Student Solutions Manual

for
Kaufmann and Schwitters's

Algebra for College Students

Eighth Edition

Karen Schwitters
Seminole Community College

Laurel Fischer

Jesse Turner
Seminole Community College

Australia • Brazil • Canada • Mexico • Singapore • Spain • United Kingdom • United States

Printed in the United States of America
1 2 3 4 5 6 7 10 09 08 07 06

Printer: Thomson/West

0-495-10515-5
Cover Image Credit: Doug Smock/Getty Images

Thomson Higher Education
10 Davis Drive
Belmont, CA 94002-3098
USA

Contents

Chapter 1 Basic Concepts and Properties

PROBLEM SET 1.1 Sets, Real Numbers, and Numerical Expressions

1. True
3. False
5. True
7. False
9. True
11. 0 and 14
13. $0, 14, \frac{2}{3}, -\frac{11}{14}, 2.34, 3.2\overline{1}, \frac{55}{8}, -19$, and -2.6
15. 0 and 14
17. All of them
19. $R \not\subseteq N$
21. $I \subseteq Q$
23. $Q \not\subseteq H$
25. $N \subseteq W$
27. $I \not\subseteq N$
29. -8 is real, rational, an integer, and negative.
31. $-\sqrt{2}$ is real, irrational, and negative.
33. $\{1, 2\}$
35. $\{0, 1, 2, 3, 4, 5\}$
37. $\{..., -1, 0, 1, 2\}$
39. $\varnothing$
41. $\{0, 1, 2, 3, 4\}$
43. -6
45. 2
47. $3x + 1$
49. $5x$

51. $16 + 9 - 4 - 2 + 8 - 1$
$25 - 4 - 2 + 8 - 1$
$21 - 2 + 8 - 1$
$19 + 8 - 1$
$27 - 1$
26

53. $9 \div 3 \bullet 4 \div 2 \bullet 14$
$3 \bullet 4 \div 2 \bullet 14$
$12 \div 2 \bullet 14$
$6 \bullet 14$
84

55. $7 + 8 \bullet 2$
$7 + 16$
23

57. $9 \bullet 7 - 4 \bullet 5 - 3 \bullet 2 + 4 \bullet 7$
$63 - 20 - 6 + 28$
$43 - 6 + 28$
$37 + 28$
65

59. $(17 - 12)(13 - 9)(7 - 4)$
$(5)(4)(3)$
$(20)(3)$
60

61. $13 + (7 - 2)(5 - 1)$
$13 + (5)(4)$
$13 + 20$
33

63. $(5 \bullet 9 - 3 \bullet 4)(6 \bullet 9 - 2 \bullet 7)$
$(45 - 12)(54 - 14)$
$(33)(40)$
1320

65. $7[3(6 - 2)] - 64$
$7[3(4)] - 64$
$7(12) - 64$
$84 - 64$
20

67. $[3 + 2(4 \bullet 1 - 2)][18 - (2 \bullet 4 - 7 \bullet 1)]$
$[3 + 2(4 - 2)][18 - (8 - 7)]$
$[3 + 2(2)][18 - 1]$
$(3 + 4)(17)$
$(7)(17)$
119

69. $14 + 4\left(\frac{8 - 2}{12 - 9}\right) - 2\left(\frac{9 - 1}{19 - 15}\right)$
$14 + 4\left(\frac{6}{3}\right) - 2\left(\frac{8}{4}\right)$
$14 + 4(2) - 2(2)$

$14 + 8 - 4$
$22 - 4$
18

71. $[7 + 2 \bullet 3 \bullet 5 - 5] \div 8$
$[7 + 6 \bullet 5 - 5] \div 8$
$[7 + 30 - 5] \div 8$
$(37 - 5) \div 8$
$(32) \div 8$
4

73. $\dfrac{3 \bullet 8 - 4 \bullet 3}{5 \bullet 7 - 34} + 19$

$\dfrac{24 - 12}{35 - 34} + 19$

$\dfrac{12}{1} + 19$

$12 + 19$
31

PROBLEM SET 1.2 Operations with Real Numbers

1. $8 + (-15)$
-7

3. $(-12) + (-7)$
-19

5. $-8 - 14$
-22

7. $9 - 16$
-7

9. $(-9)(-12)$
108

11. $(5)(-14)$
-70

13. $(-56) \div (-4)$
14

15. $\dfrac{-112}{16}$

-7

17. $-2\frac{3}{8} + 5\frac{7}{8} = \left(\left|5\frac{7}{8}\right| - \left|-2\frac{3}{8}\right|\right) =$
$\left(5\frac{7}{8} - 2\frac{3}{8}\right) = 3\frac{4}{8} = 3\frac{1}{2}$

19. $4\frac{1}{3} - \left(-1\frac{1}{6}\right) = 4\frac{1}{3} + 1\frac{1}{6} = 4\frac{2}{6} + 1\frac{1}{6} =$
$5\frac{3}{6} = 5\frac{1}{2}$

21. $\left(-\frac{1}{3}\right)\left(\frac{2}{5}\right) = -\left(\left|-\frac{1}{3}\right| \cdot \left|\frac{2}{5}\right|\right) =$
$-\left(\frac{1}{3} \cdot \frac{2}{5}\right) = -\frac{2}{15}$

23. $\frac{1}{2} \div \left(-\frac{1}{8}\right) = -\left(\left|\frac{1}{2}\right| \div \left|-\frac{1}{8}\right|\right) =$
$-\left(\frac{1}{2} \div \frac{1}{8}\right) = -\left(\frac{1}{2} \cdot 8\right) = -4$

25. $0 \div (-14)$
0

27. $(-21) \div 0$
Undefined

29. $-21 - 39 = -60$

31. $-17.3 + 12.5 = -4.8$

33. $21.42 - 7.29 = 14.13$

35. $-21.4 - (-14.9) =$
$-21.4 + 14.9 = -6.5$

37. $(5.4)(-7.2) = -38.88$

39. $\dfrac{-1.2}{-6} = 0.2$

41. $\left(-\frac{1}{3}\right) + \left(-\frac{3}{4}\right) =$

$\left(-\frac{4}{12}\right) + \left(-\frac{9}{12}\right) = -\frac{13}{12}$

43. $-\frac{3}{2} - \left(-\frac{3}{4}\right) = -\frac{3}{2} + \frac{3}{4} =$

$-\frac{6}{4} + \frac{3}{4} = -\frac{3}{4}$

45. $-\frac{2}{3} - \frac{7}{9} = -\frac{6}{9} - \frac{7}{9} = -\frac{13}{9}$

47. $\left(-\frac{3}{4}\right)\left(\frac{4}{5}\right) = -\frac{3}{5}$

49. $\frac{3}{4} \div \left(-\frac{1}{2}\right) = \frac{3}{4}\left(-\frac{2}{1}\right) = -\frac{3}{2}$

51. $9 - 12 - 8 + 5 - 6$
$-3 - 8 + 5 - 6$
$-11 + 5 - 6$
$-6 - 6$
-12

53. $-21 + (-17) - 11 + 15 - (-10)$
$-38 - 11 + 15 + (+10)$
$-49 + 15 + (+10)$
$-34 + (+10)$
-24

55. $7\frac{1}{8} - \left(2\frac{1}{4} - 3\frac{7}{8}\right) = 7\frac{1}{8} - \left(2\frac{2}{8} - 3\frac{7}{8}\right) =$

$7\frac{1}{8} - \left(-1\frac{5}{8}\right) = 7\frac{1}{8} + 1\frac{5}{8} = 8\frac{6}{8} =$

$8\frac{3}{4} = \frac{35}{4}$

57. $16 - 18 + 19 - [14 - 22 - (31 - 41)]$
$16 - 18 + 19 - [14 - 22 - (-10)]$
$16 - 18 + 19 - (14 - 22 + 10)$
$16 - 18 + 19 - (-8 + 10)$
$16 - 18 + 19 - (2)$
$16 - 18 + 19 - 2$
$-2 + 19 - 2$
$17 - 2$
15

59. $[14 - (16 - 18)] - [32 - (8 - 9)]$
$[14 - (-2)] - [32 - (-1)]$
$(14 + 2) - (32 + 1)$
$16 - 33$
-17

61. $4\frac{1}{12} - \frac{1}{2}\left(\frac{1}{3}\right) = 4\frac{1}{12} - \frac{1}{6} =$

$4\frac{1}{12} - \frac{2}{12} = 3\frac{11}{12} = \frac{47}{12}$

63. $-5 + (-2)(7) - (-3)(8)$
$-5 + (-14) - (-24)$
$-5 + (-14) + 24$
$-19 + 24$
5

65. $\frac{2}{5}\left(-\frac{3}{4}\right) - \left(-\frac{1}{2}\right)\left(\frac{3}{5}\right) =$

$-\frac{3}{10} - \left(-\frac{3}{10}\right) =$

$-\frac{3}{10} + \frac{3}{10} = 0$

67. $(-6)(-9) + (-7)(4)$
$54 - 28$
26

69. $3(5 - 9) - 3(-6)$
$3(-4) + 18$
$-12 + 18$
6

71. $(6 - 11)(4 - 9)$
$(-5)(-5)$
25

73. $-6(-3 - 9 - 1)$
$-6(-12 - 1)$
$-6(-13) = 78$

75. $56 \div (-8) - (-6) \div (-2)$
$-7 - (3)$
$-7 - 3$
-10

77. $-3[5 - (-2)] - 2(-4 - 9)$
$-3(5 + 2) - 2(-13)$
$-3(7) + 26$
$-21 + 26$
5

79. $\frac{-6 + 24}{-3} + \frac{-7}{-6 - 1}$
$\frac{18}{-3} + \frac{-7}{-7}$
$-6 + 1$
-5

81. $14.1 - (17.2 - 13.6) =$
$14.1 - (3.6) = 14.1 - 3.6 = 10.5$

83. $3(2.1) - 4(3.2) - 2(-1.6) =$
$6.3 - 12.8 + 3.2 =$
$-6.5 + 3.2 = -3.3$

85. $7(6.2 - 7.1) - 6(-1.4 - 2.9) =$
$7(-0.9) - 6(-4.3) =$
$-6.3 + 25.8 = 19.5$

87. $\frac{2}{3} - \left(\frac{3}{4} - \frac{5}{6}\right) = \frac{8}{12} - \left(\frac{9}{12} - \frac{10}{12}\right) =$
$\frac{8}{12} - \left(-\frac{1}{12}\right) = \frac{8}{12} + \frac{1}{12} = \frac{9}{12} = \frac{3}{4}$

89. $3\left(\frac{1}{2}\right) + 4\left(\frac{2}{3}\right) - 2\left(\frac{5}{6}\right) =$
$\frac{3}{2} + \frac{8}{3} - \frac{10}{6} = \frac{9}{6} + \frac{16}{6} - \frac{10}{6} =$
$\frac{25}{6} - \frac{10}{6} = \frac{15}{6} = \frac{5}{2}$

93. $6(-1) + 4(2) + 1(3) + 2(0) + 5(1) =$
$-6 + 8 + 3 + 0 + 5 = 10$
Jeff was 10 over par.

95. $9(-5) + 28.50 =$
$-45 + 28.50 = -16.50$
He lost \$16.50 for the day.

97. $1.25 + 0.88 - 0.50 - 1.13 + 0.38 =$
$2.51 + (-1.63) = 0.88$
The net gain was 0.88 dollars.

99. $2(-15.6) - 4.8 - 23.7 + 10.6 =$
$-31.2 - 4.8 - 23.7 + 10.6 = -49.1$
No, they did not meet the goal.

PROBLEM SET 1.3 Properties of Real Numbers and Use of Exponents

1. Associative property of addition
3. Commutative property of addition
5. Additive inverse property
7. Multiplicative property of negative one
9. Commutative property of multiplication
11. Distributive property
13. Associative property of multiplication

15. $36 + (-14) + (-12) + 21 + (-9) - 4$
$36 + 21 + (-14) + (-12) + (-9) + (-4)$
$57 + (-39)$
18

17. $[83 + (-99)] + 18$
$-16 + 18$
2

19. $(25)(-13)(4)$
$(25)(4)(-13)$
$(100)(-13)$
-1300

21. $17(97) + 17(3)$
$17(97 + 3)$
$17(100)$
1700

23. $14 - 12 - 21 - 14 + 17 - 18 + 19 - 32$
$14 - 14 - 21 - 12 - 18 - 32 + 17 + 19$
$0 - 83 + 36$
-47

25. $(-50)(15)(-2)-(-4)(17)(25)$
$15(-50)(-2)-(17)(-4)(25)$
$15(100)-17(-100)$
$1500+1700$
3200

27. 2^3-3^3
$8-27$
-19

29. -5^2-4^2
$-25-16$
-41

31. $(-2)^3-3^2$
$-8-9$
-17

33. $3(-1)^3-4(3)^2$
$3(-1)-4(9)$
$-3-36$
-39

35. $7(2)^3+4(-2)^3$
$7(8)+4(-8)$
$56-32$
24

37. $-3(-2)^3+4(-1)^5$
$-3(-8)+4(-1)$
$24-4$
20

39. $(-3)^2-3(-2)(5)+4^2$
$9-3(-2)(5)+16$
$9+6(5)+16$
$9+30+16$
$39+16$
55

41. $2^3+3(-1)^3(-2)^2-5(-1)(2)^2$
$8+3(-1)(4)-5(-1)(4)$
$8-3(4)+5(4)$
$8-12+20$
$-4+20$
16

43. $(3+4)^2=(7)^2=49$

45. $[3(-2)^2-2(-3)^2]^3$
$[3(4)-2(9)]^3$
$(12-18)^3$
$(-6)^3$
-216

47. $2(-1)^3-3(-1)^2+4(-1)-5$
$2(-1)-3(1)+4(-1)-5$
$-2-3-4-5$
-14

49. $2^4-2(2)^3-3(2)^2+7(2)-10$
$16-2(8)-3(4)+7(2)-10$
$16-16-12+14-10$
$0-12+14-10$
$-12+14-10$
$2-10$
-8

51. $3\left(\frac{1}{2}\right)^4-2\left(\frac{1}{2}\right)^3+5\left(\frac{1}{2}\right)^2-4\left(\frac{1}{2}\right)+1$
$3\left(\frac{1}{16}\right)-2\left(\frac{1}{8}\right)+5\left(\frac{1}{4}\right)-2+1$
$\frac{3}{16}-\frac{1}{4}+\frac{5}{4}-2+1=\frac{3}{16}$

53. $-\left(\frac{2}{3}\right)^2+5\left(\frac{2}{3}\right)-4$
$-\left(\frac{4}{9}\right)+\frac{10}{3}-4$
$-\frac{4}{9}+\frac{30}{9}-\frac{36}{9}=-\frac{10}{9}$

57. 3^7
2187

59. $(-2)^{11}$
-2048

61. -5^6
$-1\bullet 5^6$
$-1\bullet 15625$
$-15{,}625$

63. $(1.41)^4$
3.95254161

PROBLEM SET 1.4 Algebraic Expressions

1. $-7x+11x$
$(-7+11)x$
$4x$

3. $5a^2-6a^2$
$(5-6)a^2$
$-a^2$

5. $4n-9n-n$
$(4-9-1)n$
$-6n$

7. $4x-9x+2y$
$(4-9)x+2y$
$-5x+2y$

9. $-3a^2+7b^2+9a^2-2b^2$
$-3a^2+9a^2+7b^2-2b^2$
$(-3+9)a^2+(7-2)b^2$
$6a^2+5b^2$

11. $15x-4+6x-9$
$15x+6x-4-9$
$(15+6)x-13$
$21x-13$

13. $5a^2b-ab^2-7a^2b$
$5a^2b-7a^2b-ab^2$
$(5-7)a^2b-ab^2$
$-2a^2b-ab^2$

15. $3(x+2)+5(x+3)$
$3x+6+5x+15$
$3x+5x+6+15$
$8x+21$

17. $-2(a-4)-3(a+2)$
$-2a+8-3a-6$
$-2a-3a+8-6$
$-5a+2$

19. $3(n^2+1)-8(n^2-1)$
$3n^2+3-8n^2+8$
$3n^2-8n^2+3+8$
$-5n^2+11$

21. $-6(x^2-5)-(x^2-2)$
$-6x^2+30-x^2+2$
$-6x^2-x^2+30+2$
$-7x^2+32$

23. $5(2x+1)+4(3x-2)$
$10x+5+12x-8$
$10x+12x+5-8$
$22x-3$

25. $3(2x-5)-4(5x-2)$
$6x-15-20x+8$
$6x-20x-15+8$
$-14x-7$

27. $-2(n^2-4)-4(2n^2+1)$
$-2n^2+8-8n^2-4$
$-2n^2-8n^2+8-4$
$-10n^2+4$

29. $3(2x-4y)-2(x+9y)$
$6x-12y-2x-18y$
$6x-2x-12y-18y$
$4x-30y$

31. $3(2x-1)-4(x+2)-5(3x+4)$
$6x-3-4x-8-15x-20$
$6x-4x-15x-3-8-20$
$-13x-31$

33. $-(3x-1)-2(5x-1)+4(-2x-3)$
$-3x+1-10x+2-8x-12$
$-3x-10x-8x+1+2-12$
$-21x-9$

35. $3x+7y$
For $x=-1$ and $y=-2$,
$3(-1)+7(-2)$
$-3-14$
-17

37. $4x^2-y^2$
For $x=2$ and $y=-2$,
$4(2)^2-(-2)^2$
$4(4)-(4)$
$16-4=12$

39. $2a^2 - ab + b^2$
For $a = -1$ and $b = -2$,
$2(-1)^2 - (-1)(-2) + (-2)^2$
$2(1) - (-1)(-2) + 4$
$2 - 2 + 4$
4

41. $2x^2 - 4xy - 3y^2$
For $x = 1$ and $y = -1$,
$2(1)^2 - 4(1)(-1) - 3(-1)^2$
$2(1) - 4(1)(-1) - 3(1)$
$2 + 4 - 3$
$6 - 3$
3

43. $3xy - x^2y^2 + 2y^2$
For $x = 5$ and $y = -1$,
$3(5)(-1) - (5)^2(-1)^2 + 2(-1)^2$
$3(5)(-1) - (25)(1) + 2(1)$
$-15 - 25 + 2$
$-40 + 2$
-38

45. $7a - 2b - 9a + 3b$
$7a - 9a - 2b + 3b$
$-2a + b$
For $a = 4$ and $b = -6$
$-2(4) + (-6)$
$-8 - 6$
-14

47. $(x - y)^2$
For $x = 5$ and $y = -3$,
$[(5) - (-3)]^2$
$[5 + 3]^2$
$(8)^2$
64

49. $-2a - 3a + 7b - b$
$-5a + 6b$
For $a = -10$ and $b = 9$,
$-5(-10) + 6(9)$
$50 + 54$
104

51. $-2(x + 4) - (2x - 1)$
$-2x - 8 - 2x + 1$
$-2x - 2x - 8 + 1$
$-4x - 7$, for $x = -3$
$-4(-3) - 7$
$12 - 7$
5

53. $2(x - 1) - (x + 2) - 3(2x - 1)$
$2x - 2 - x - 2 - 6x + 3$
$2x - x - 6x - 2 - 2 + 3$
$-5x - 1$, for $x = -1$
$-5(-1) - 1$
$5 - 1$
4

55. $3(x^2 - 1) - 4(x^2 + 1) - (2x^2 - 1)$
$3x^2 - 3 - 4x^2 - 4 - 2x^2 + 1$
$-3x^2 - 6$, for $x = \frac{2}{3}$
$-3\left(\frac{2}{3}\right)^2 - 6$
$-3\left(\frac{4}{9}\right) - 6 =$
$-\frac{4}{3} - \frac{18}{3} =$
$-\frac{22}{3}$

57. $5(x - 2y) - 3(2x + y) - 2(x - y)$
$5x - 10y - 6x - 3y - 2x + 2y$
$-3x - 11y$, for $x = \frac{1}{3}; y = -\frac{3}{4}$
$-3\left(\frac{1}{3}\right) - 11\left(-\frac{3}{4}\right)$
$-1 + \frac{33}{4} =$
$-\frac{4}{4} + \frac{33}{4} = \frac{29}{4}$

59. πr^2
$(3.14)(8.4)^2$
$(3.14)(70.56)$
221.5584
221.6

61. $\pi r^2 h$
$(3.14)(4.8)^2(15.1)$
$(3.14)(23.04)(15.1)$
$(72.3456)(15.1)$
1092.41856
1092.4

63. $2\pi r^2 + 2\pi rh$
$2(3.14)(7.8)^2 + 2(3.14)(7.8)(21.2)$
$2(3.14)(60.84) + (6.28)(7.8)(21.2)$
$(6.28)(60.84) + (48.984)(21.2)$
$382.0752 + 1038.4608$
1420.536
1420.5

65. $n + 12$

67. $n - 5$

69. $n(50)$
$50n$

71. $\frac{1}{2}n - 4$

73. $\frac{n}{8}$

75. $2n - 9$

77. $10(n - 6)$

79. $n + 20$

81. $2t - 3$

83. $n + 47$

85. $8y$

87. $25cm$

89. $\frac{c}{25}$

91. $n + 2$

93. $\frac{c}{5}$

95. $12d$

97. $3y + f$

99. $5280m$

CHAPTER 1 Review Problem Set

1a. 67
b. 0, -8, and 67
c. 0 and 67
d. 0, $\frac{3}{4}$, $-\frac{5}{6}$, $\frac{25}{3}$, -8, 0.34, $0.2\overline{3}$, 67, $\frac{9}{7}$
e. $\sqrt{2}$ and $-\sqrt{3}$
2. Associative property of addition
3. Substitution property for equality
4. Multiplication property of negative one
5. Distributive property
6. Associative property of multiplication
7. Commutative property of addition
8. Distributive property
9. Multiplicative inverse property
10. Symmetric property of equality

11. $-8\frac{1}{4} + \left(-4\frac{5}{8}\right) - \left(-6\frac{3}{8}\right)$
$-8\frac{2}{8} + \left(-4\frac{5}{8}\right) + 6\frac{3}{8}$
$-12\frac{7}{8} + 6\frac{3}{8} = -6\frac{4}{8} = -6\frac{1}{2}$

12. $9\frac{1}{3} - 12\frac{1}{2} + \left(-4\frac{1}{6}\right) - \left(-1\frac{1}{6}\right)$
$9\frac{2}{6} - 12\frac{3}{6} + \left(-4\frac{1}{6}\right) + 1\frac{1}{6}$
$9\frac{2}{6} + 1\frac{1}{6} - 12\frac{3}{6} + \left(-4\frac{1}{6}\right)$
$10\frac{3}{6} + \left(-16\frac{4}{6}\right) = -6\frac{1}{6}$

13. $-8(2)-16\div(-4)+(-2)(-2)$
$-16-16\div(-4)+(-2)(-2)$
$-16+4+4$
$-12+4$
-8

14. $4(-3)-12\div(-4)+(-2)(-1)-8$
$-12-(-3)+(+2)-8$
$-12+(+3)+(+2)-8$
$-9+(+2)-8$
$-7-8$
-15

15. $-3(2-4)-4(7-9)+6$
$-3(-2)-4(-2)+6$
$6+8+6$
$14+6$
20

16. $[48+(-73)]+74$
$48+[-73+74]$
$48+1$
49

17. $[5(-2)-3(-1)][-2(-1)+3(2)]$
$(-10+3)(2+6)$
$(-7)(8)$
-56

18. -4^2-2^3
$-16-8$
-24

19. $(-2)^4+(-1)^3-3^2$
$16+(-1)-9$
$15-9$
6

20. $2(-1)^2-3(-1)(2)-2^2$
$2(1)+6-4$
$2+6-4$
4

21. $[4(-1)-2(3)]^2$
$(-4-6)^2$
$(-10)^2$
100

22. $3-[-2(3-4)]+7$
$3-[-2(-1)]+7$
$3-(2)+7$
$3+(-2)+7$
$1+7$
8

23. $3a^2-2b^2-7a^2-3b^2$
$3a^2-7a^2-2b^2-3b^2$
$-4a^2-5b^2$

24. $4x-6-2x-8+x+12$
$4x-2x+x-6-8+12$
$3x-2$

25. $\frac{1}{5}ab^2-\frac{3}{10}ab^2+\frac{2}{5}ab^2+\frac{7}{10}ab^2$
$\left(\frac{1}{5}-\frac{3}{10}+\frac{2}{5}+\frac{7}{10}\right)ab^2$
$\left(\frac{2}{10}-\frac{3}{10}+\frac{4}{10}+\frac{7}{10}\right)ab^2$
$\left(1\right)ab^2=ab^2$

26. $-\frac{2}{3}x^2y-\left(-\frac{3}{4}x^2y\right)-\frac{5}{12}x^2y-2x^2y$
$-\frac{2}{3}x^2y+\frac{3}{4}x^2y-\frac{5}{12}x^2y-2x^2y$
$\left(-\frac{2}{3}+\frac{3}{4}-\frac{5}{12}-2\right)x^2y$
$\left(-\frac{8}{12}+\frac{9}{12}-\frac{5}{12}-\frac{24}{12}\right)x^2y$
$\left(-\frac{28}{12}\right)x^2y=-\frac{7}{3}x^2y$

27. $3(2n^2+1)+4(n^2-5)$
$6n^2+3+4n^2-20$
$6n^2+4n^2+3-20$
$10n^2-17$

28. $-2(3a-1)+4(2a+3)-5(3a+2)$
$-6a+2+8a+12-15a-10$
$-6a+8a-15a+2+12-10$
$-13a+4$

29. $-(n-1)-(n+2)+3$
$-n+1-n-2+3$
$-n-n+1-2+3$
$-2n+2$

30. $3(2x-3y)-4(3x+5y)-x$
$6x-9y-12x-20y-x$
$6x-12x-x-9y-20y$
$-7x-29y$

31. $4(a-6)-(3a-1)-2(4a-7)$
$4a-24-3a+1-8a+14$
$4a-3a-8a-24+1+14$
$-7a-9$

32. $-5(x^2-4)-2(3x^2+6)+(2x^2-1)$
$-5x^2+20-6x^2-12+2x^2-1$
$-5x^2-6x^2+2x^2+20-12-1$
$-9x^2+7$

33. $-5x+4y$
For $x=\frac{1}{2}$ and $y=-1$,
$-5\left(\frac{1}{2}\right)+4\left(-1\right)$
$-\frac{5}{2}-4$
$-\frac{5}{2}-\frac{8}{2}$
$-\frac{13}{2}=-6\frac{1}{2}$

34. $3x^2-2y^2$
For $x=\frac{1}{4}$ and $y=-\frac{1}{2}$,
$3\left(\frac{1}{4}\right)^2-2\left(-\frac{1}{2}\right)^2$
$3\left(\frac{1}{16}\right)-2\left(\frac{1}{4}\right)$
$\frac{3}{16}-\frac{1}{2}$
$\frac{3}{16}-\frac{8}{16}$
$-\frac{5}{16}$

35. $-5(2x-3y)$
For $x=1$ and $y=-3$,
$-5[2(1)-3(-3)]$
$-5(2+9)$
$-5(11)$
-55

36. $(3a-2b)^2$
For $a=-2$ and $b=3$,
$[3(-2)-2(3)]^2$
$(-6-6)^2$
$(-12)^2$
144

37. $a^2+3ab-2b^2$
For $a=2$ and $b=-2$,
$(2)^2+3(2)(-2)-2(-2)^2$
$4+3(2)(-2)-2(4)$
$4-12-8$
$-8-8$
-16

38. $3n^2-4-4n^2+9$
$-n^2+5$
$-(7)^2+5$, for $n=7$
$-49+5$
-44

39. $3(2x-1)+2(3x+4)$
$6x-3+6x+8$
$12x+5$
$12(1.2)+5$, for $x=1.2$
$14.4+5=19.4$

40. $-4(3x-1)-5(2x-1)$
$-12x+4-10x+5$
$-22x+9$
$-22(-2.3)+9$, for $x=-2.3$
$50.6+9=59.6$

41. $2(n^2+3)-3(n^2+1)+4(n^2-6)$
$2n^2+6-3n^2-3+4n^2-24$
$3n^2-21$
$3\left(-\frac{2}{3}\right)^2-21$, for $n=-\frac{2}{3}$
$3\left(\frac{4}{9}\right)-21=\frac{4}{3}-\frac{63}{3}=-\frac{59}{3}$

42. $5(3n-1)-7(-2n+1)+4(3n-1)$
$15n-5+14n-7+12n-4$
$41n-16$
$41\left(\frac{1}{2}\right)-16$, for $n=\frac{1}{2}$
$\frac{41}{2}-\frac{32}{2}=\frac{9}{2}$

43. $4+2n$

44. $3n-50$

45. $\frac{2}{3}n-6$

46. $10(n-14)$

47. $5n-8$

48. $\frac{n}{n-3}$

49. $5(n+2)-3$

50. $\frac{3}{4}(n+12)$

51. $37-n$

52. w words/1 hour $=w$ words/60 min.
$\frac{w}{60}$

53. Brother's age = 2(Harry's age) − 7
$2y-7$

54. $n+3$

55. $p+5n+25q$

56. Perimeter $=i$ inches.
side $=\frac{i}{4}$ inches
side $=\frac{1}{12}\left(\frac{i}{4}\right)=\frac{i}{48}$ feet

57. length $=y$ yards $=36y$ inches
width $=f$ feet $=12f$ inches
$P=2l+2w$
$P=2(36y)+2(12f)$
$P=(72y+24f)$ inches

58. 10 cm = 1 decimeter
$10d$

59. 1 foot = 12 inches
$12f+i$

60. $P=2l+2w$
$50=2c+2w$
$25=c+w$
$25-c=w$

CHAPTER 1 Test

1. Symmetric Property
2. Distributive Property

3. $-4-(-3)+(-5)-7+10$
$-4+(+3)+(-5)-7+10$
$-1+(-5)-7+10$
$-6-7+10$
$-13+10$
-3

4. $7-8-3+4-9-4+2-12$
$7+4+2-8-3-9-4-12$
$13-36$
-23

5. $5\left(-\frac{1}{3}\right)-3\left(-\frac{1}{2}\right)+7\left(-\frac{2}{3}\right)+1$
$-\frac{5}{3}+\frac{3}{2}-\frac{14}{3}+1$
$-\frac{10}{6}+\frac{9}{6}-\frac{28}{6}+\frac{6}{6}$
$-\frac{23}{6}$

6. $(-6) \bullet 3 \div (-2) - 8 \div (-4)$
$-18 \div (-2) - (-2)$
$9 - (-2)$
$9 + 2 = 11$

7. $-\frac{1}{2}(3-7) - \frac{2}{5}(2-17)$
$-\frac{1}{2}(-4) - \frac{2}{5}(-15)$
$2 + 6$
8

8. $[48 + (-93)] + (-49)$
$[-93 + 48)] + (-49)$
$-93 + [48 + (-49)]$
$-93 + (-1) = -94$

9. $3(-2)^3 + 4(-2)^2 - 9(-2) - 14$
$3(-8) + 4(4) - 9(-2) - 14$
$-24 + 16 + 18 - 14$
$-8 + 18 - 14$
$10 - 14$
-4

10. $[2(-6) + 5(-4)][-3(-4) - 7(6)]$
$(-12 - 20)(12 - 42)$
$(-32)(-30)$
960

11. $[-2(-3) - 4(2)]^5$
$(6 - 8)^5$
$(-2)^5$
-32

12. $6x^2 - 3x - 7x^2 - 5x - 2$
$6x^2 - 7x^2 - 3x - 5x - 2$
$-x^2 - 8x - 2$

13. $3(3n - 1) - 4(2n + 3) + 5(-4n - 1)$
$9n - 3 - 8n - 12 - 20n - 5$
$9n - 8n - 20n - 3 - 12 - 5$
$-19n - 20$

14. $-7x - 3y$
For $x = -6$ and $y = 5$,
$-7(-6) - 3(5)$
$42 - 15$
27

15. $3a^2 - 4b^2$
For $a = -\frac{3}{4}$ and $b = \frac{1}{2}$,
$3\left(-\frac{3}{4}\right)^2 - 4\left(\frac{1}{2}\right)^2$
$3\left(\frac{9}{16}\right) - 4\left(\frac{1}{4}\right)$
$\frac{27}{16} - \frac{16}{16}$
$\frac{11}{16}$

16. $6x - 9y - 8x + 4y$
$-2x - 5y$
For $x = \frac{1}{2}$ and $y = -\frac{1}{3}$,
$-2\left(\frac{1}{2}\right) - 5\left(-\frac{1}{3}\right)$
$-1 + \frac{5}{3}$
$\frac{2}{3}$

17. $-5n^2 - 6n + 7n^2 + 5n - 1$
$2n^2 - n - 1$
For $n = -6$,
$2(-6)^2 - (-6) - 1$
$2(36) - (-6) - 1$
$72 + 6 - 1$
$78 - 1$
77

18. $-7(x - 2) + 6(x - 1) - 4(x + 3)$
$-7x + 14 + 6x - 6 - 4x - 12$
$-5x - 4$
$-5(3.7) - 4$, for $x = 3.7$
$-18.5 - 4 = -22.5$

19. $-2xy - x + 4y$
For $x = -3$ and $y = 9$,
$-2(-3)(9) - (-3) + 4(9)$
$54 + 3 + 36$
$57 + 36$
93

20. $4(n^2+1)-(2n^2+3)-2(n^2+3)$
$4n^2+4-2n^2-3-2n^2-6$
-5

21. $6n-30$

22. $3(n+8)+4$

23. $\frac{72}{n}$

24. $5n+10d+25q$

25. length $= x$ yards $= 3x$ feet
width $= y$ feet

Perimeter $= 2l+2w$
Perimeter $= 2(3x)+2(y)$
Perimeter $= (6x+2y)$ feet

Chapter 2 Equations, Inequalities, and Problem Solving

PROBLEM SET 2.1 Solving First-Degree Equations

1. $3x+4=16$
$3x+4-4=16-4$
$3x=12$
$\frac{1}{3}(3x)=\frac{1}{3}(12)$
$x=4$
The solution set is $\{4\}$.

3. $5x+1=-14$
$5x+1-1=-14-1$
$5x=-15$
$\frac{1}{5}(5x)=\frac{1}{5}(-15)$
$x=-3$
The solution set is $\{-3\}$.

5. $-x-6=8$
$-x-6+6=8+6$
$-x=14$
$-1(-x)=-1(14)$
$x=-14$
The solution set is $\{-14\}$.

7. $4y-3=21$
$4y-3+3=21+3$
$4y=24$
$\frac{1}{4}(4y)=\frac{1}{4}(24)$
$y=6$
The solution set is $\{6\}$.

9. $3x-4=15$
$3x-4+4=15+4$
$3x=19$
$\frac{1}{3}(3x)=\frac{1}{3}(19)$
$x=\frac{19}{3}$
The solution set is $\left\{\frac{19}{3}\right\}$.

11. $-4=2x-6$
$-4+6=2x-6+6$
$2=2x$
$\frac{1}{2}(2)=\frac{1}{2}(2x)$
$1=x$
The solution set is $\{1\}$.

13. $-6y-4=16$
$-6y-4+4=16+4$
$-6y=20$
$-\frac{1}{6}(-6y)=-\frac{1}{6}(20)$
$y=-\frac{20}{6}$
$y=-\frac{10}{3}$
The solution set is $\left\{-\frac{10}{3}\right\}$.

15. $4x-1=2x+7$
$4x-1+1=2x+7+1$
$4x=2x+8$
$4x-2x=2x-2x+8$
$2x=8$
$\frac{1}{2}(2x)=\frac{1}{2}(8)$
$x=4$
The solution set is $\{4\}$.

17. $5y+2=2y-11$
$5y+2-2=2y-11-2$
$5y=2y-13$
$5y-2y=2y-2y-13$
$3y=-13$
$\frac{1}{3}(3y)=\frac{1}{3}(-13)$
$y=-\frac{13}{3}$
The solution set is $\left\{-\frac{13}{3}\right\}$.

19. $3x+4=5x-2$
$3x+4-4=5x-2-4$
$3x=5x-6$
$3x-5x=5x-5x-6$
$-2x=-6$

$-\frac{1}{2}(-2x) = -\frac{1}{2}(-6)$
$x = 3$
The solution set is $\{3\}$.

21. $-7a + 6 = -8a + 14$
$-7a + 6 - 6 = -8a + 14 - 6$
$-7a = -8a + 8$
$-7a + 8a = -8a + 8a + 8$
$a = 8$
The solution set is $\{8\}$.

23. $5x + 3 - 2x = x - 15$
$3x + 3 = x - 15$
$3x + 3 - 3 = x - 15 - 3$
$3x = x - 18$
$3x - x = x - x - 18$
$2x = -18$
$\frac{1}{2}(2x) = \frac{1}{2}(-18)$
$x = -9$
The solution set is $\{-9\}$.

25. $6y + 18 + y = 2y + 3$
$7y + 18 = 2y + 3$
$7y + 18 - 18 = 2y + 3 - 18$
$7y = 2y - 15$
$7y - 2y = 2y - 2y - 15$
$5y = -15$
$\frac{1}{5}(5y) = \frac{1}{5}(-15)$
$y = -3$
The solution set is $\{-3\}$.

27. $4x - 3 + 2x = 8x - 3 - x$
$6x - 3 = 7x - 3$
$6x - 3 + 3 = 7x - 3 + 3$
$6x = 7x$
$6x - 6x = 7x - 6x$
$0 = x$
The solution set is $\{0\}$.

29. $6n - 4 - 3n = 3n + 10 + 4n$
$3n - 4 = 7n + 10$
$3n - 4 + 4 = 7n + 10 + 4$
$3n = 7n + 14$
$3n - 7n = 7n - 7n + 14$
$-4n = 14$
$-\frac{1}{4}(-4n) = -\frac{1}{4}(14)$
$n = -\frac{14}{4}$
$n = -\frac{7}{2}$
The solution set is $\left\{-\frac{7}{2}\right\}$.

31. $4(x - 3) = -20$
$4x - 12 = -20$
$4x - 12 + 12 = -20 + 12$
$4x = -8$
$\frac{1}{4}(4x) = \frac{1}{4}(-8)$
$x = -2$
The solution set is $\{-2\}$.

33. $-3(x - 2) = 11$
$-3x + 6 = 11$
$-3x + 6 - 6 = 11 - 6$
$-3x = 5$
$-\frac{1}{3}(-3x) = -\frac{1}{3}(5)$
$x = -\frac{5}{3}$
The solution set is $\left\{-\frac{5}{3}\right\}$.

35. $5(2x + 1) = 4(3x - 7)$
$10x + 5 = 12x - 28$
$10x + 5 - 5 = 12x - 28 - 5$
$10x = 12x - 33$
$10x - 12x = 12x - 12x - 33$
$-2x = -33$
$-\frac{1}{2}(-2x) = -\frac{1}{2}(-33)$
$x = \frac{33}{2}$
The solution set is $\left\{\frac{33}{2}\right\}$.

37. $5x - 4(x - 6) = -11$
$5x - 4x + 24 = -11$
$x + 24 = -11$
$x + 24 - 24 = -11 - 24$
$x = -35$
The solution set is $\{-35\}$.

39. $-2(3x-1)-3=-4$
$-6x+2-3=-4$
$-6x-1=-4$
$-6x-1+1=-4+1$
$-6x=-3$
$-\frac{1}{6}(-6x)=-\frac{1}{6}(-3)$
$x=\frac{3}{6}$
$x=\frac{1}{2}$
The solution set is $\left\{\frac{1}{2}\right\}$.

41. $-2(3x+5)=-3(4x+3)$
$-6x-10=-12x-9$
$-6x-10+10=-12x-9+10$
$-6x=-12x+1$
$-6x+12x=-12x+12x+1$
$6x=1$
$\frac{1}{6}(6x)=\frac{1}{6}(1)$
$x=\frac{1}{6}$
The solution set is $\left\{\frac{1}{6}\right\}$.

43. $3(x-4)-7(x+2)=-2(x+18)$
$3x-12-7x-14=-2x-36$
$-4x-26=-2x-36$
$-4x-26+26=-2x-36+26$
$-4x=-2x-10$
$-4x+2x=-2x+2x-10$
$-2x=-10$
$-\frac{1}{2}(-2x)=-\frac{1}{2}(-10)$
$x=5$
The solution set is $\{5\}$.

45. $-2(3n-1)+3(n+5)=-4(n-4)$
$-6n+2+3n+15=-4n+16$
$-3n+17=-4n+16$
$-3n+17-17=-4n+16-17$
$-3n=-4n-1$
$-3n+4n=-4n+4n-1$
$n=-1$
The solution set is $\{-1\}$.

47. $3(2a-1)-2(5a+1)=4(3a+4)$
$6a-3-10a-2=12a+16$
$-4a-5=12a+16$
$-4a-5+5=12a+16+5$
$-4a=12a+21$
$-4a-12a=12a-12a+21$
$-16a=21$
$-\frac{1}{16}(-16a)=-\frac{1}{16}(21)$
$a=-\frac{21}{16}$
The solution set is $\left\{-\frac{21}{16}\right\}$.

49. $-2(n-4)-(3n-1)=-2+(2n-1)$
$-2n+8-3n+1=-2+2n-1$
$-5n+9=2n-3$
$-5n+9-9=2n-3-9$
$-5n=2n-12$
$-5n-2n=2n-2n-12$
$-7n=-12$
$-\frac{1}{7}(-7n)=-\frac{1}{7}(-12)$
$n=\frac{12}{7}$
The solution set is $\left\{\frac{12}{7}\right\}$.

51. Let $x=$ number.
$3x-15=27$
$3x-15+15=27+15$
$3x=42$
$\frac{1}{3}(3x)=\frac{1}{3}(42)$
$x=14$
The number is 14.

53. Let $n=$ 1st integer,
$n+1=$ 2nd integer and
$n+2=$ 3rd integer.

$n+n+1+n+2=42$
$3n+3=42$
$3n+3-3=42-3$
$3n=39$
$\frac{1}{3}(3n)=\frac{1}{3}(39)$
$n=13$
The integers are 13, 14, and 15.

55. Let $n =$ 1st odd integer,
$n+2 =$ 2nd odd integer and
$n+4 =$ 3rd odd integer.

$3(n+2)-(n+4)=n+11$
$3n+6-n-4=n+11$
$2n+2=n+11$
$2n+2-2=n+11-2$
$2n=n+9$
$2n-n=n-n+9$
$n=9$
The integers are 9, 11, and 13.

57. Let $x =$ smaller number and
$6x-3 =$ larger number.

$6x-3-x=67$
$5x-3=67$
$5x-3+3=67+3$
$5x=70$
$\frac{1}{5}(5x)=\frac{1}{5}(70)$
$x=14$
smaller number $= 14$
larger number $= 6(14)-3=84-3=81$
The numbers are 14 and 81.

59. Let $x =$ normal rate of pay.
Angelo worked 40 hours at regular pay and 6 hours at double pay.

$40x+6(2x)=572$
$40x+12x=572$
$52x=572$
$\frac{1}{52}(52x)=\frac{1}{52}(572)$
$x=11.00$
Angelo's normal hourly rate is \$11.00 per hour.

61. Let $x =$ number of pennies,
$2x-10 =$ number of nickels and
$3x-20 =$ number of dimes.

$x+2x-10+3x-20=150$
$6x-30=150$
$6x-30+30=150+30$
$6x=180$
$\frac{1}{6}(6x)=\frac{1}{6}(180)$
$x=30$
number of pennies $= 30$
number of nickels $= 2(30)-10$
$= 60-10=50$
number of dimes $= 3(30)-20$
$= 90-20=70$
There are 30 pennies, 50 nickels, and 70 dimes.

63. Let $x =$ cost of the ring.
$3x-150=750$
$3x-150+150=750+150$
$3x=900$
$\frac{1}{3}(3x)=\frac{1}{3}(900)$
$x=300$
The cost of the ring is \$300.

65. Let $x =$number of 3-bedroom apartments,
$3x+10 =$ number of 2-bedroom apartments and
$2(3x+10) =$ number of 1-bedroom apartments.

$x+3x+10+2(3x+10)=230$
$x+3x+10+6x+20=230$
$10x+30=230$
$10x+30-30=230-30$
$10x=200$
$\frac{1}{10}(10x)=\frac{1}{10}(200)$
$x=20$
number of 3-bedroom apartments
$= 20$
number of 2-bedroom apartments
$= 3(20)+10=20+10=70$
number of 1-bedroom apartments
$= 2(70)=140$
There are 20 three-bedroom apartments, 70 two-bedroom apartments, and 140 one-bedroom apartments.

73a) $5x+7=5x-4$
$5x-5x+7=5x-5x-4$
$7=-4$ contradiction
The solution set is $\emptyset$.

b) $4(x-1)=4x-4$
$4x-4=4x-4$
$4x-4x-4=4x-4x-4$

$-4 = -4$ identity
The solution set is all real numbers.

c) $3(x-4) = 2(x-6)$
$3x - 12 = 2x - 12$
$3x - 12 + 12 = 2x - 12 + 12$
$3x = 2x$
$3x - 2x = 2x - 2x$
$x = 0$
The solution set is $\{0\}$.

d) $7x - 2 = -7x + 4$
$7x - 2 + 2 = -7x + 4 + 2$
$7x = -7x + 6$
$7x + 7x = -7x + 7x + 6$
$14x = 6$
$\frac{1}{14}(14x) = \frac{1}{14}(6)$
$x = \frac{6}{14} = \frac{3}{7}$
The solution set is $\left\{\frac{3}{7}\right\}$.

e) $2(x-1) + 3(x+2) = 5(x-7)$
$2x - 2 + 3x + 6 = 5x - 35$
$5x + 4 = 5x - 35$
$5x - 5x + 4 = 5x - 5x - 35$
$4 = -35$ contradiction
The solution set is $\emptyset$.

f) $-4(x-7) = -2(2x+1)$
$-4x + 28 = -4x - 2$
$-4x + 4x + 28 = -4x + 4x - 2$
$28 = -2$ contradiction
The solution set is $\emptyset$.

75. Let x = 1st integer,
$x + 1$ = 2nd integer,
$x + 2$ = 3rd integer and
$x + 3$ = 4th integer.

$x(x+3) = (x+2)(x+3)$
$x^2 + 3x = x^2 + 3x + 2x + 6$
$x^2 + 3x = x^2 + 5x + 6$
$x^2 - x^2 + 3x = x^2 - x^2 + 5x + 6$
$3x = 5x + 6$
$3x - 5x = 5x - 5x + 6$
$-2x = 6$
$x = -3$
The only potential solution is the integers -3, -2, -1, and 0. However, these do not check because $-3(0) \neq -2(-1)$
$0 \neq 2$
Therefore, there are no such integers.

PROBLEM SET 2.2 Equations Involving Fractional Forms

1. $\frac{3}{4}x = 9$
$\frac{4}{3}\left(\frac{3}{4}x\right) = \frac{4}{3}(9)$
$x = 12$
The solution set is $\{12\}$.

3. $-\frac{2x}{3} = \frac{2}{5}$
$-\frac{3}{2}\left(-\frac{2x}{3}\right) = -\frac{3}{2}\left(\frac{2}{5}\right)$
$x = -\frac{3}{5}$
The solution set is $\left\{-\frac{3}{5}\right\}$.

5. $\frac{n}{2} - \frac{2}{3} = \frac{5}{6}$
$6\left(\frac{n}{2} - \frac{2}{3}\right) = 6\left(\frac{5}{6}\right)$
$6\left(\frac{n}{2}\right) + 6\left(-\frac{2}{3}\right) = 6\left(\frac{5}{6}\right)$
$3n - 4 = 5$
$3n - 4 + 4 = 5 + 4$
$3n = 9$
$\frac{1}{3}(3n) = \frac{1}{3}(9)$
$n = 3$
The solution set is $\{3\}$.

7. $\frac{5n}{6} - \frac{n}{8} = \frac{-17}{12}$

$24\left(\frac{5n}{6} - \frac{n}{8}\right) = 24\left(-\frac{17}{12}\right)$

$24\left(\frac{5n}{6}\right) + 24\left(-\frac{n}{8}\right) = 24\left(-\frac{17}{12}\right)$

$20n - 3n = -34$

$\frac{1}{17}(17n) = \frac{1}{17}(-34)$

$n = -2$

The solution set is $\{-2\}$.

9. $\frac{a}{4} - 1 = \frac{a}{3} + 2$

$12\left(\frac{a}{4} - 1\right) = 12\left(\frac{a}{3} + 2\right)$

$12\left(\frac{a}{4}\right) + 12(-1) = 12\left(\frac{a}{3}\right) + 12(2)$

$3a - 12 = 4a + 24$

$3a - 12 + 12 = 4a + 24 + 12$

$3a = 4a + 36$

$3a - 4a = 4a - 4a + 36$

$-a = 36$

$-1(-a) = -1(36)$

$a = -36$

The solution set is $\{-36\}$.

11. $\frac{h}{4} + \frac{h}{5} = 1$

$20\left(\frac{h}{4} + \frac{h}{5}\right) = 20(1)$

$20\left(\frac{h}{4}\right) + 20\left(\frac{h}{5}\right) = 20(1)$

$5h + 4h = 20$

$9h = 20$

$\frac{1}{9}(9h) = \frac{1}{9}(20)$

$h = \frac{20}{9}$

The solution set is $\left\{\frac{20}{9}\right\}$.

13. $\frac{h}{2} - \frac{h}{3} + \frac{h}{6} = 1$

$6\left(\frac{h}{2} - \frac{h}{3} + \frac{h}{6}\right) = 6(1)$

$6\left(\frac{h}{2}\right) + 6\left(-\frac{h}{3}\right) + 6\left(\frac{h}{6}\right) = 6(1)$

$3h - 2h + h = 6$

$2h = 6$

$\frac{1}{2}(2h) = \frac{1}{2}(6)$

$h = 3$

The solution set is $\{3\}$.

15. $\frac{x-2}{3} + \frac{x+3}{4} = \frac{11}{6}$

$12\left(\frac{x-2}{3} + \frac{x+3}{4}\right) = 12\left(\frac{11}{6}\right)$

$12\left(\frac{x-2}{3}\right) + 12\left(\frac{x+3}{4}\right) = 12\left(\frac{11}{6}\right)$

$4(x-2) + 3(x+3) = 2(11)$

$4x - 8 + 3x + 9 = 22$

$7x + 1 = 22$

$7x + 1 - 1 = 22 - 1$

$7x = 21$

$\frac{1}{7}(7x) = \frac{1}{7}(21)$

$x = 3$

The solution set is $\{3\}$.

17. $\frac{x+2}{2} - \frac{x-1}{5} = \frac{3}{5}$

$10\left(\frac{x+2}{2} - \frac{x-1}{5}\right) = 10\left(\frac{3}{5}\right)$

$10\left(\frac{x+2}{2}\right) - 10\left(\frac{x-1}{5}\right) = 10\left(\frac{3}{5}\right)$

$5(x+2) - 2(x-1) = 2(3)$

$5x + 10 - 2x + 2 = 6$

$3x + 12 = 6$

$3x + 12 - 12 = 6 - 12$

$3x = -6$

$\frac{1}{3}(3x) = \frac{1}{3}(-6)$

$x = -2$

The solution set is $\{-2\}$.

19. $\frac{n+2}{4} - \frac{2n-1}{3} = \frac{1}{6}$

$12\left(\frac{n+2}{4} - \frac{2n-1}{3}\right) = 12\left(\frac{1}{6}\right)$

$12\left(\frac{n+2}{4}\right) - 12\left(\frac{2n-1}{3}\right) = 12\left(\frac{1}{6}\right)$

$3(n+2) - 4(2n-1) = 2$

$3n + 6 - 8n + 4 = 2$

$-5n + 10 = 2$

$-5n + 10 - 10 = 2 - 10$

$-5n = -8$

$-\frac{1}{5}(-5n) = -\frac{1}{5}(-8)$

$n = \frac{8}{5}$

The solution set is $\left\{\frac{8}{5}\right\}$.

21. $\frac{y}{3} + \frac{y-5}{10} = \frac{4y+3}{5}$

$30\left(\frac{y}{3} + \frac{y-5}{10}\right) = 30\left(\frac{4y+3}{5}\right)$

$30\left(\frac{y}{3}\right) + 30\left(\frac{y-5}{10}\right) = 30\left(\frac{4y+3}{5}\right)$

$10y + 3(y-5) = 6(4y+3)$

$10y + 3y - 15 = 24y + 18$

$13y - 15 = 24y + 18$

$13y - 15 + 15 = 24y + 18 + 15$

$13y = 24y + 33$

$13y - 24y = 24y - 24y + 33$

$-11y = 33$

$-\frac{1}{11}(-11y) = -\frac{1}{11}(33)$

$y = -3$

The solution set is $\{-3\}$.

23. $\frac{4x-1}{10} - \frac{5x+2}{4} = -3$

$20\left(\frac{4x-1}{10} - \frac{5x+2}{4}\right) = 20(-3)$

$20\left(\frac{4x-1}{10}\right) - 20\left(\frac{5x+2}{4}\right) = 20(-3)$

$2(4x-1) - 5(5x+2) = 20(-3)$

$8x - 2 - 25x - 10 = -60$

$-17x - 12 = -60$

$-17x - 12 + 12 = -60 + 12$

$-17x = -48$

$-\frac{1}{17}(-17x) = -\frac{1}{17}(-48)$

$x = \frac{48}{17}$

The solution set is $\left\{\frac{48}{17}\right\}$.

25. $\frac{2x-1}{8} - 1 = \frac{x+5}{7}$

$56\left(\frac{2x-1}{8} - 1\right) = 56\left(\frac{x+5}{7}\right)$

$56\left(\frac{2x-1}{8}\right) - 56(1) = 56\left(\frac{x+5}{7}\right)$

$7(2x-1) - 56 = 8(x+5)$

$14x - 7 - 56 = 8x + 40$

$14x - 63 = 8x + 40$

$14x - 63 + 63 = 8x + 40 + 63$

$14x = 8x + 103$

$14x - 8x = 8x - 8x + 103$

$6x = 103$

$\frac{1}{6}(6x) = \frac{1}{6}(103)$

$x = \frac{103}{6}$

The solution set is $\left\{\frac{103}{6}\right\}$.

27. $\frac{2a-3}{6} + \frac{3a-2}{4} + \frac{5a+6}{12} = 4$

$12\left(\frac{2a-3}{6} + \frac{3a-2}{4} + \frac{5a+6}{12}\right) = 12(4)$

$$12\left(\frac{2a-3}{6}\right)+12\left(\frac{3a-2}{4}\right)+12\left(\frac{5a+6}{12}\right)=12(4)$$
$2(2a-3)+3(3a-2)+1(5a+6)=12(4)$
$4a-6+9a-6+5a+6=48$
$18a-6=48$
$18a-6+6=48+6$
$18a=54$
$\frac{1}{18}\left(18a\right)=\frac{1}{18}\left(54\right)$
$a=3$
The solution set is $\{3\}$.

29. $x+\frac{3x-1}{9}-4=\frac{3x+1}{3}$

$9\left(x+\frac{3x-1}{9}-4\right)=9\left(\frac{3x+1}{3}\right)$

$9(x)+9\left(\frac{3x-1}{9}\right)-9(4)=3(3x+1)$

$9x+3x-1-36=9x+3$
$12x-37=9x+3$
$12x-37+37=9x+3+37$
$12x=9x+40$
$12x-9x=9x-9x+40$
$3x=40$
$\frac{1}{3}\left(3x\right)=\frac{1}{3}\left(40\right)$
$x=\frac{40}{3}$
The solution set is $\left\{\frac{40}{3}\right\}$.

31. $\frac{x+3}{2}+\frac{x+4}{5}=\frac{3}{10}$

$10\left(\frac{x+3}{2}+\frac{x+4}{5}\right)=10\left(\frac{3}{10}\right)$

$10\left(\frac{x+3}{2}\right)+10\left(\frac{x+4}{5}\right)=3$

$5(x+3)+2(x+4)=3$
$5x+15+2x+8=3$
$7x+23=3$
$7x+23-23=3-23$
$7x=-20$
$\frac{1}{7}\left(7x\right)=\frac{1}{7}\left(-20\right)$
$x=-\frac{20}{7}$
The solution set is $\left\{-\frac{20}{7}\right\}$.

33. $n+\frac{2n-3}{9}-2=\frac{2n+1}{3}$

$9\left(n+\frac{2n-3}{9}-2\right)=9\left(\frac{2n+1}{3}\right)$

$9n+9\left(\frac{2n-3}{9}\right)-9(2)=3(2n+1)$

$9n+2n-3-18=6n+3$
$11n-21=6n+3$
$11n-21+21=6n+3+21$
$11n=6n+24$
$11n-6n=6n-6n+24$
$5n=24$
$\frac{1}{5}\left(5n\right)=\frac{1}{5}\left(24\right)$
$n=\frac{24}{5}$
The solution set is $\left\{\frac{24}{5}\right\}$.

35. $\frac{3}{4}(t-2)-\frac{2}{5}(2t-3)=\frac{1}{5}$

$20\left[\frac{3}{4}(t-2)-\frac{2}{5}(2t-3)\right]=20\left(\frac{1}{5}\right)$

$20\left[\frac{3}{4}(t-2)\right]-20\left[\frac{2}{5}(2t-3)\right]=4$

$15(t-2)-8(2t-3)=4$
$15t-30-16t+24=4$
$-t-6=4$
$-t-6+6=4+6$
$-t=10$
$-1(-t)=-1(10)$
$t=-10$
The solution set is $\{-10\}$.

37. $\frac{1}{2}\left(2x-1\right)-\frac{1}{3}\left(5x+2\right)=3$

$6\left[\frac{1}{2}(2x-1)-\frac{1}{3}(5x+2)\right]=3(6)$

$6\left[\frac{1}{2}(2x-1)\right]-6\left[\frac{1}{3}(5x+2)\right]=18$

$3(2x-1)-2(5x+2)=18$

$6x-3-10x-4=18$

$-4x-7=18$

$-4x-7+7=18+7$

$-4x=25$

$-\frac{1}{4}\left(-4x\right)=-\frac{1}{4}\left(25\right)$

$x=-\frac{25}{4}$

The solution set is $\left\{-\frac{25}{4}\right\}$.

39. $3x-1+\frac{2}{7}\left(7x-2\right)=-\frac{11}{7}$

$7\left[3x-1+\frac{2}{7}(7x-2)\right]=7\left(-\frac{11}{7}\right)$

$7(3x)+7(-1)+7\left[\frac{2}{7}(7x-2)\right]=-11$

$21x-7+2(7x-2)=-11$

$21x-7+14x-4=-11$

$35x-11=-11$

$35x-11+11=-11+11$

$35x=0$

$\frac{1}{35}\left(35x\right)=\frac{1}{35}\left(0\right)$

$x=0$

The solution set is $\{0\}$.

41. Let $x=$ number.

$\frac{1}{2}x=\frac{2}{3}x-3$

$6\left(\frac{1}{2}x\right)=6\left(\frac{2}{3}x-3\right)$

$3x=6\left(\frac{2}{3}x\right)-6(3)$

$3x=4x-18$

$3x-4x=4x-4x-18$

$-x=-18$

$x=18$

The number is 18.

43. Let $x=$ length and

$\frac{1}{4}x+1=$ width.

$P=2l+2w$

$42=2(x)+2\left(\frac{1}{4}x+1\right)$

$42=2x+\frac{2}{4}x+2$

$42=2x+\frac{1}{2}x+2$

$2(42)=2\left(2x+\frac{1}{2}x+2\right)$

$84=2(2x)+2\left(\frac{1}{2}x\right)+2(2)$

$84=4x+x+4$

$84=5x+4$

$84-4=5x+4-4$

$80=5x$

$\frac{1}{5}\left(80\right)=\frac{1}{5}\left(5x\right)$

$16=x$

length $=16$ inches

width $=\frac{1}{4}(16)+1=4+1=5$ inches

The length is 16 inches. The width is 5 inches.

45. Let $x=$ 1st integer,

$x+1=$ 2nd integer and

$x+2=$ 3rd integer.

$x+\frac{1}{3}\left(x+1\right)+\frac{3}{8}\left(x+2\right)=25$

$24\left[x+\frac{1}{3}(x+1)+\frac{3}{8}(x+2)\right]=24(25)$

$24(x)+24\left[\frac{1}{3}(x+1)\right]+24\left[\frac{3}{8}(x+2)\right]=600$

$24x+8(x+1)+9(x+2)=600$

$24x + 8x + 8 + 9x + 18 = 600$
$41x + 26 = 600$
$41x + 26 - 26 = 600 - 26$
$41x = 574$
$\frac{1}{41}\left(41x\right) = \frac{1}{41}\left(574\right)$
$x = 14.$
The integers are 14, 15, and 16.

47. Let x = one piece and
$\frac{2}{3}x$ = other piece.

$x + \frac{2}{3}x = 20$

$3\left(x + \frac{2}{3}x\right) = 3(20)$

$3(x) + 3\left(\frac{2}{3}x\right) = 60$

$3x + 2x = 60$
$5x = 60$
$\frac{1}{5}\left(5x\right) = \frac{1}{5}\left(60\right)$
$x = 12$
One piece = 12 ft.
The other piece = $\frac{2}{3}(12) = 8$ ft.
The shorter piece is 8 ft.

49. Let x = present age of Angie and
$64 - x$ = present age of mother.
$x + 8$ = Angie's age in eight years.
$64 - x + 8 = 72 - x$ = mother's age in eight years.

$\frac{3}{5}\left(72 - x\right) = x + 8$

$5\left[\frac{3}{5}(72 - x)\right] = 5(x + 8)$

$3(72 - x) = 5x + 40$
$216 - 3x = 5x + 40$
$216 - 3x + 3x = 5x + 3x + 40$
$216 = 8x + 40$
$216 - 40 = 8x + 40 - 40$
$176 = 8x$
$\frac{1}{8}\left(176\right) = \frac{1}{8}\left(8x\right)$
$22 = x$

Angie's present age = 22.
Mother's present age = $64 - 22 = 42$.
At the present time Angie is 22 years old and her mother is 42 years old.

51. Let x = Marcus' present age.
Let $\frac{1}{2}x$ = Sydney's present age.

Also $x + 12$ = Marcus' age in 12 years,
and $\frac{1}{2}x + 12$ = Sydney's age in 12 years.

Sydney's age in 12 years will be the same as $\frac{5}{8}$ Marcus' age in 12 years.

Therefore, $\frac{1}{2}x + 12 = \frac{5}{8}\left(x + 12\right)$
$8\left[\frac{1}{2}x + 12\right] = 8\left[\frac{5}{8}\left(x + 12\right)\right]$
$4x + 96 = 5(x + 12)$
$4x + 96 = 5x + 60$
$36 = x$
Marcus is 36 years old and Sydney is 18 years old.

53. Let x = Aura's first exam score,
$x + 10$ = Aura's second exam score and
$x + 10 + 4$ = Aura's third exam score.

$\frac{x + x + 10 + x + 14}{3} = 88$

$\frac{3x + 24}{3} = 88$

$3x + 24 = 264$
$3x = 240$
$x = 80$

Aura's first exam score is 80,
the second exam score is 90, and
the third exam score is 94.

55. Let $x =$ one supplementary angle and

$\frac{1}{3}x + 4 =$ other supplementary angle.

$$x + \frac{1}{3}x + 4 = 180$$
$$3\left(x + \frac{1}{3}x + 4\right) = 3(180)$$
$$3x + 3\left(\frac{1}{3}x\right) + 3(4) = 540$$
$$3x + x + 12 = 540$$
$$4x + 12 = 540$$
$$4x + 12 - 12 = 540 - 12$$
$$4x = 528$$
$$\frac{1}{4}\left(4x\right) = \frac{1}{4}\left(528\right)$$
$$x = 132$$

One angle is 132°, the

other angle $= \frac{1}{3}(132) + 4 = 44 + 4 = 48°$.

The angles are 132° and 48°.

57. Let $x =$ angle,
$90 - x =$ complement of the angle and
$180 - x =$ supplement of the angle.

$$90 - x = \frac{1}{6}\left(180 - x\right) - 5$$
$$6(90 - x) = 6\left[\frac{1}{6}(180 - x) - 5\right]$$
$$6(90) - 6x = 6\left[\frac{1}{6}(180 - x)\right] - 6(5)$$
$$540 - 6x = 180 - x - 30$$
$$540 - 6x = 150 - x$$
$$540 - 6x + 6x = 150 - x + 6x$$
$$540 = 150 + 5x$$
$$540 - 150 = 150 - 150 + 5x$$
$$390 = 5x$$
$$\frac{1}{5}\left(390\right) = \frac{1}{5}\left(5x\right)$$
$$78 = x$$

The angle is 78°.

PROBLEM SET 2.3 Equations Involving Decimals and Problem Solving

1. $0.14x = 2.8$
$100(0.14x) = 100(2.8)$
$14x = 280$
$x = 20$
The solution set is $\{20\}$.

3. $0.09y = 4.5$
$100(0.09y) = 100(4.5)$
$9y = 450$
$y = 50$
The solution set is $\{50\}$.

5. $n + 0.4n = 56$
$10(n + 0.4n) = 10(56)$
$10n + 4n = 560$
$14n = 560$
$n = 40$
The solution set is $\{40\}$.

7. $s = 9 + 0.25s$
$100(s) = 100(9 + 0.25s)$
$100s = 900 + 25s$
$100s - 25s = 900 + 25s - 25s$
$75s = 900$
$s = 12$
The solution set is $\{12\}$.

9. $s = 3.3 + 0.45s$
$100(s) = 100(3.3 + 0.45s)$
$100s = 330 + 45s$
$100s - 45s = 330 + 45s - 45s$
$55s = 330$
$s = 6$
The solution set is $\{6\}$.

11. $0.11x + 0.12(900 - x) = 104$
$100[0.11x + 0.12(900 - x)] = 100(104)$
$100(0.11x) + 100[0.12(900 - x)] = 10400$
$11x + 12(900 - x) = 10400$

$11x + 10800 - 12x = 10400$
$-x + 10800 = 10400$
$-x + 10800 - 10800 = 10400 - 10800$
$-x = -400$
$x = 400$
The solution set is $\{400\}$.

13. $0.08(x + 200) = 0.07x + 20$
$100[0.08(x + 200)] = 100(0.07x + 20)$
$8(x + 200) = 100(0.07x) + 100(20)$
$8x + 1600 = 7x + 2000$
$x + 1600 = 2000$
$x = 400$
The solution set is $\{400\}$.

15. $0.12t - 2.1 = 0.07t - 0.2$
$100(0.12t - 2.1) = 100(0.07t - 0.2)$
$12t - 210 = 7t - 20$
$12t = 7t + 190$
$5t = 190$
$t = 38$
The solution set is $\{38\}$.

17. $0.92 + 0.9(x - 0.3) = 2x - 5.95$
$100[0.92 + 0.9(x - 0.3)] = 100(2x - 5.95)$
$92 + 90(x - 0.3) = 200x - 595$
$92 + 90x - 27 = 200x - 595$
$90x + 65 = 200x - 595$
$90x = 200x - 660$
$-110x = -660$
$x = 6$
The solution set is $\{6\}$.

19. $0.1d + 0.11(d + 1500) = 795$
$100[0.1d + 0.11(d + 1500)] = 100(795)$
$10d + 11(d + 1500) = 79500$
$10d + 11d + 16500 = 79500$
$21d + 16500 = 79500$
$21d = 63000$
$d = 3000$
The solution set is $\{3000\}$.

21. $0.12x + 0.1(5000 - x) = 560$
$100[0.12x + 0.1(5000 - x)] = 100(560)$
$12x + 10(5000 - x) = 56000$
$12x + 50000 - 10x = 56000$
$2x + 50000 = 56000$
$2x = 6000$
$x = 3000$
The solution set is $\{3000\}$.

23. $0.09(x + 200) = 0.08x + 22$
$100[0.09(x + 200)] = 100(0.08x + 22)$
$9(x + 200) = 8x + 2200$
$9x + 1800 = 8x + 2200$
$9x = 8x + 400$
$x = 400$
The solution set is $\{400\}$.

25. $0.3(2t + 0.1) = 8.43$
$100[0.3(2t + 0.1)] = 100(8.43)$
$30(2t + 0.1) = 843$
$60t + 3 = 843$
$60t = 840$
$t = 14$
The solution set is $\{14\}$.

27. $0.1(x - 0.1) - 0.4(x + 2) = -5.31$
$100[0.1(x - 0.1) - 0.4(x + 2)] = 100(-5.31)$
$10(x - 0.1) - 40(x + 2) = -531$
$10x - 1 - 40x - 80 = -531$
$-30x - 81 = -531$
$-30x = -450$
$x = 15$
The solution set is $\{15\}$.

29. Let $p =$ original price.
$(100\%)p - (20\%)p = 72$
$1.00p - 0.20p = 72$
$0.80p = 72$
$100(0.80p) = 100(72)$
$80p = 7200$
$p = 90$
The original price is \$90.

31. Let $s =$ sale price.
$s = 64 - (15\%)(64)$
$s = 64 - 0.15(64)$
$s = 64 - 9.60$
$s = 54.40$
The sales price is \$54.40.

33. Let $s =$ selling price.
$s = 30 + (60\%)(30)$
$s = 30 + 0.60(30)$
$s = 30 + 18$

$s = 48$
The selling price should be \$48.

35. Let $s =$ selling price.
$s = 200 + 50\%(s)$
$s = 200 + 0.50s$
$100(s) = 100(200 + 0.50s)$
$100s = 20000 + 50s$
$50s = 20000$
$s = 400$
The selling price is \$400.

37. Amount of profit $= 39.60 - 24 = 15.60$

Rate of profit $= \dfrac{15.60}{24.00} = 0.65 = 65\%$

39. Let x represent the rate of profit.

selling price $=$ cost $+$ profit
$800 = 300 + x(800)$

$500 = 800x$

$\dfrac{500}{800} = x$

$62.5\% = x$
The rate of profit based on selling price is 62.5%.

41. Let $x =$ this year's salary.

$34775 = x + x(0.07)$
$100(34775) = 100[x + x(0.07)]$
$3477500 = 100x + 7x$
$3477500 = 107x$
$32500 = x$
The present salary is \$32,500.

43. Let $x =$ amount invested at 10% and $x + 1500 =$ amount invested at 11%.

$(10\%)(x) + (11\%)(x + 1500) = 795$
$0.10x + 0.11(x + 1500) = 795$
$100[0.10x + 0.11(x + 1500)] = 100(795)$
$10x + 11(x + 1500) = 79500$
$10x + 11(x + 1500) = 79500$
$10x + 11x + 16500 = 79500$
$21x + 16500 = 79500$
$21x = 63000$

$x = 3000$
amount invested at 10% $=$ \$3000
amount invested at 11% $=$
$3000 + 1500 =$ \$4500

45. Let x represent the amount invested at 6%, and $95000 - x$ represent the amount invested at 9%.

$$\begin{pmatrix}\text{Interest from}\\ \text{6\% dollars}\end{pmatrix} + \begin{pmatrix}\text{Interest from}\\ \text{9\% dollars}\end{pmatrix} = \begin{pmatrix}\text{Total}\\ \text{interest}\end{pmatrix}$$

$(6\%)(x) + (9\%)(95000 - x) = 7290$
$0.06x + 0.09(95000 - x) = 7290$
$6x + 9(95000 - x) = 729000$
$6x + 855000 - 9x = 729000$
$-3x = -126000$
$x = 42000$
Therefore, $95000 - x = 53000$
There is \$53,000 invested at 9%.

47. Let $x =$ number of pennies,
$2x - 1 =$ number of nickels and
$2x - 1 + 3 = 2x + 2 =$ number of dimes.

$0.01x + 0.05(2x - 1) + 0.10(2x + 2) = 2.63$
$100[0.01x+0.05(2x - 1)+0.10(2x+2)] = 100(2.63)$
$1x + 5(2x - 1) + 10(2x + 2) = 263$
$1x + 10x - 5 + 20x + 20 = 263$
$31x + 15 = 263$
$31x = 248$
$x = 8$
number of pennies $= 8$
number of nickels $= 2(8) - 1 = 16 - 1 = 15$
number of dimes $= 2(8) + 2 = 16 + 2 = 18$
There are 8 pennies, 15 nickels, and 18 dimes.

49. Let $x =$ number of dimes,
$3x =$ number of quarters and
$70 - (x + 3x) = 70 - 4x$
$=$ number of half-dollars.

$0.10x + 0.25(3x) + 0.50(70 - 4x) = 17.75$
$100[0.10x+0.25(3x)+0.50(70 - 4x)] = 100(17.75)$
$10x + 25(3x) + 50(70 - 4x) = 1775$
$10x + 75x + 3500 - 200x = 1775$
$-115x + 3500 = 1775$

$-115x = -1725$
$x = 15$
numbers of dimes = 15
number of quarters = 3(15) = 45
number of half-dollars
= 70 − 4(15) = 70 − 60 = 10
There are 15 dimes, 45 quarters, and 10 half-dollars.

55. $0.12x - 0.24 = 0.66$
$0.12x = 0.9$
$x = 7.5$
checking
$0.12(7.5) - 0.24 = 0.66$
$0.9 - 0.24 = 0.66$
$0.66 = 0.66$
The solution set is {7.5}.

57. $0.14t + 0.13(890 - t) = 67.95$
$0.14t + 115.7 - 0.13t = 67.95$
$0.01t + 115.7 = 67.95$
$0.01t = -47.75$
$t = -4775$
checking
$0.14(-4775) + 0.13[890 - (-4775)] = 67.95$
$-668.5 + 0.13(5665) = 67.95$
$-668.5 + 736.45 = 67.95$
$67.95 = 67.95$
The solution set is {−4775}.

59. $0.14n - 0.26 = 0.958$
$0.14n = 1.218$
$n = 8.7$
checking
$0.14(8.7) - 0.26 = 0.958$
$1.218 - 0.26 = 0.958$
$0.958 = 0.958$
The solution set is {8.7}.

61. $0.6(d - 4.8) = 7.38$
$0.6d - 2.88 = 7.38$
$0.6d = 10.26$
$d = 17.1$
checking
$0.6(17.1 - 4.8) = 7.38$
$0.6(12.3) = 7.38$
$7.38 = 7.38$
The solution set is {17.1}.

63. $0.5(3x + 0.7) = 20.6$
$1.5x + 0.35 = 20.6$
$1.5x = 20.25$
$x = 13.5$
checking
$0.5[3(13.5) + 0.7] = 20.6$
$0.5(40.5 + 0.7) = 20.6$
$0.5(41.2) = 20.6$
$20.6 = 20.6$
The solution set is {13.5}.

65. amount of profit = 100 − 90 = 10
rate of profit $= \frac{10}{100} = 0.10 = 10\%$
His claim is correct based on selling price.

PROBLEM SET 2.4 Formulas

1. $i = Prt$
$i = 300(8\%)(5)$
$i = 300(0.08)(5)$
$i = 120$
The interest is $120.

3. $i = Prt$
$132 = 400(11\%)t$
$132 = 400(0.11)t$
$132 = 44t$
$3 = t$
The time is 3 years.

5. $i = Prt$
$90 = 600(r)\left(2\frac{1}{2}\right)$
$90 = 600(r)(2.5)$
$90 = 1500r$
$.06 = r$
$6\% = r$
The rate is 6%.

7. $i = Prt$
$216 = P(9\%)(3)$
$216 = P(0.09)(3)$

$216 = P(.27)$
$800 = P$
The principal is \$800.

9. $A = P + Prt$
$A = 1000 + 1000(12\%)(5)$
$A = 1000 + 1000(0.12)(5)$
$A = 1000 + 600$
$A = 1600$
The amount is \$1600.

11. $A = P + Prt$
$1372 = 700 + 700(r)(12)$
$1372 = 700 + 8400r$
$672 = 8400r$
$0.08 = r$
$8\% = r$
The rate is 8%.

13. $A = P + Prt$
$326 = P + P(7\%)(9)$
$326 = P + P(0.07)(9)$
$326 = P + 0.63P$
$326 = 1.63P$
$200 = P$
The principal is \$200.

15. $A = \frac{1}{2}h(b_1 + b_2)$

$2(A) = 2\left[\frac{1}{2}h(b_1 + b_2)\right]$

$2A = h(b_1 + b_2)$

$\frac{2A}{h} = b_1 + b_2$

$\frac{2A}{h} - b_1 = b_2$

$b_2 = \frac{2(98)}{14} - 8$ $\quad b_2 = \frac{2(104)}{8} - 12$ $\quad b_2 = \frac{2(49)}{7} - 4$

$b_2 = 14 - 8$ $\quad b_2 = 26 - 12$ $\quad b_2 = 14 - 4$

$b_2 = 6$ $\quad b_2 = 14$ $\quad b_2 = 10$

$b_2 = \frac{2(162)}{9} - 16$ $\quad b_2 = \frac{2(16\frac{1}{2})}{3} - 4$ $\quad b_2 = \frac{2(38\frac{1}{2})}{11} - 5$

$b_2 = 36 - 16$ $\quad b_2 = 11 - 4$ $\quad b_2 = 7 - 5$

$b_2 = 20$ $\quad b_2 = 7$ $\quad b_2 = 2$

Area(A)	98	104	49	162	$16\frac{1}{2}$	$38\frac{1}{2}$
Height(h	14	8	7	9	3	11
One base(b_1)	8	12	4	16	4	5
Other base(b_2)	6	14	10	20	7	2

17. $V = Bh$ for h

$\frac{1}{B}(V) = \frac{1}{B}(Bh)$

$\frac{V}{B} = h$

19. $V = \pi r^2 h$ for h

$\frac{1}{\pi r^2}(V) = \frac{1}{\pi r^2}(\pi r^2 h)$

$\frac{V}{\pi r^2} = h$

21. $C = 2\pi r$ for r

$\frac{1}{2\pi}(C) = \frac{1}{2\pi}(2\pi r)$

$\frac{C}{2\pi} = r$

23. $I = \frac{100M}{C}$ for C

$C(I) = C\left(\frac{100M}{C}\right)$

$CI = 100M$

$\frac{I}{I}(CI) = \frac{I}{I}(100M)$

$C = \frac{100M}{I}$

25. $F = \frac{9}{5}C + 32$ for C

$F - 32 = \frac{9}{5}C$

$\frac{5}{9}(F - 32) = \frac{5}{9}\left(\frac{9}{5}C\right)$

$\frac{5}{9}(F - 32) = C$

27. $y = mx + b$ for x

$y - b = mx$

$\frac{1}{m}(y - b) = \frac{1}{m}(mx)$

$\frac{y - b}{m} = x$

29. $y - y_1 = m(x - x_1)$ for x

$y - y_1 = mx - mx_1$

$y - y_1 + mx_1 = mx$

$\frac{1}{m}(y - y_1 + mx_1) = \frac{1}{m}(mx)$

$\frac{y - y_1 + mx_1}{m} = x$

31. $a(x + b) = b(x - c)$ for x

$ax + ab = bx - bc$

$ax = bx - bc - ab$

$ax - bx = -bc - ab$

$x(a - b) = -bc - ab$

$\frac{1}{(a - b)}[x(a - b)] = \frac{1}{(a - b)}[-bc - ab]$

$x = \frac{-bc - ab}{a - b}$

$x = \frac{-1(-bc - ab)}{-1(a - b)}$

$x = \frac{bc + ab}{-a + b}$

$x = \frac{ab + bc}{b - a}$

33. $\frac{x - a}{b} = c$ for x

$b\left(\frac{x - a}{b}\right) = b(c)$

$x - a = bc$

$x = bc + a$

35. $\frac{1}{3}x + a = \frac{1}{2}b$ for x

$6\left(\frac{1}{3}x + a\right) = 6\left(\frac{1}{2}b\right)$

$2x + 6a = 3b$

$2x = 3b - 6a$

$\frac{1}{2}(2x) = \frac{1}{2}(3b - 6a)$

$x = \frac{3b - 6a}{2}$

37. $2x - 5y = 7$ for x

$2x = 5y + 7$

$\frac{1}{2}(2x) = \frac{1}{2}(5y + 7)$

$x = \frac{5y + 7}{2}$

39. $-7x - y = 4$ for y

$-y = 4 + 7x$

$-1(-y) = -1(4 + 7x)$

$y = -4 - 7x$

$y = -7x - 4$

41. $3(x - 2y) = 4$ for x

$3x - 6y = 4$

$3x = 6y + 4$

$x = \frac{6y + 4}{3}$

43. $\frac{y - a}{b} = \frac{x + b}{c}$ for x

$bc\left(\frac{y - a}{b}\right) = bc\left(\frac{x + b}{c}\right)$

$c(y - a) = b(x + b)$

$cy - ac = bx + b^2$

$cy - ac - b^2 = bx$

$\frac{1}{b}(cy - ac - b^2) = \frac{1}{b}(bx)$

$$\frac{cy - ac - b^2}{b} = x$$

45. $(y+1)(a-3) = x - 2$ for y

$ay - 3y + a - 3 = x - 2$

$ay - 3y = x - 2 - a + 3$

$(a-3)y = x - a + 1$

$\frac{1}{(a-3)}[(a-3)y] = \frac{1}{(a-3)}(x - a + 1)$

$y = \frac{x - a + 1}{a - 3}$

47. Let x = width and.
$4x - 2$ = length.
$P = 2l + 2w$

$56 = 2(4x - 2) + 2x$
$56 = 8x - 4 + 2x$
$56 = 10x - 4$
$60 = 10x$
$6 = x$

width = 6 meters
length = 4(6) − 2 = 24 − 2 = 22 meters

The width is 6 meters and the length is 22 meters.

49. Let 500 = principal.
Then 1000 = double the principal.

$A = P + Prt$
$1000 = 500 + 500(9\%)t$
$1000 = 500 + 500(0.09)t$
$1000 = 500 + 45t$
$500 = 45t$

$\frac{500}{45} = t$
$11\frac{1}{9} = t$

The principal will double in $11\frac{1}{9}$ years at 9%.

51. Let P = principal and
$2P$ = double the principal.

$A = P + Prt$
$2P = P + P(9\%)t$

$2P = P + P(0.09)t$
$2P = P + 0.09Pt$

$P = 0.09Pt$

$\frac{1}{0.09P}(P) = \frac{1}{0.09P}(0.09Pt)$
$11.\bar{1} = t$

$11\frac{1}{9} = t$

The principal will double in $11\frac{1}{9}$ years at 9%.

53. Let t = time traveling.
$d = rt$
$d_1 = 450t$
$d_2 = 550t$

$450t + 550t = 4000$
$1000t = 4000$
$t = 4$ hours

It will take 4 hours.

55. Let t = Time traveling for Juan and.
$t - \frac{3}{2}$ = Time traveling for Cathy.
$d = rt$

Distance for Juan = Distance for Cathy

$4t = 6\left(t - \frac{3}{2}\right)$

$4t = 6t - 9$

$-2t = -9$

$t = \frac{9}{2} = 4.5$

Time traveling for Juan
= 4.5 hours
Time traveling for Cathy
= 4.5 − 1.5 = 3.0 hours
It takes Cathy 3 hours to catch up with Juan.

57. Let t = time traveling at 20 mph and.
$4.5 - t$ = time traveling at 12 mph.

$d = rt$
$20t + 12(4.5 - t) = 70$
$20t + 54 - 12t = 70$
$8t + 54 = 70$

$8t = 16$
$t = 2$
distance $= 20(2) = 40$ miles
Bret has traveled for 40 miles
before he slowed down.

59. Let $x =$ quarts of 30% solution and
$20 - x =$ quarts of 70% solution.

$(30\%)(x) + (70\%)(20 - x) = (40\%)(20)$
$0.30x + 0.70(20 - x) = 0.40(20)$
$100[0.30x + 0.70(20 - x)] = 100[0.40(20)]$
$30x + 70(20 - x) = 40(20)$
$30x + 1400 - 70x = 800$
$-40x + 1400 = 800$
$-40x = -600$
$x = 15$

quarts of 30% solution $= 15$
quarts of 70% solution $= 20 - 15 = 5$
There should be 15 quarts of
the 30% solution and 5 quarts
of the 70% solution.

61. Let $x =$ amount of pure acid.

$(100\%)(x) + (30\%)(150) = (40\%)(x + 150)$
$1x + 0.30(150) = 0.40(x + 150)$
$100[1x + 0.30(150)] = [0.40(x + 150)]$
$100x + 30(150) = 40(x + 150)$
$100x + 4500 = 40x + 6000$
$100x = 40x + 1500$
$60x = 1500$
$x = 25$

There must be 25 ml of pure acid.

67. $i = Prt$
$i = 1125\left(13\frac{1}{4}\%\right)(4)$
$i = 1125(0.1325)(4)$
$i = 596.25$
The interest is \$596.25.

69. $i = Prt$
$243.75 = 1250(13\%)t$
$243.75 = 1250(0.13)t$
$243.75 = 162.5t$
$1.5 = t$
It will take 1.5 years.

71. $i = Prt$
$159.50 = 2200(r)(0.5)$
$159.50 = 1100r$
$0.145 = r$
$14.5\% = r$
The rate is 14.5%.

73. $A = P + Prt$
$2173.75 = P + P\left(8\frac{3}{4}\%\right)(2)$
$2173.75 = P + P(0.0875)(2)$
$100(2173.75) = 100[P + P(0.0875)(2)]$
$217375 = 100P + 17.5P$
$217375 = 117.5P$
$1850 = P$
The principal is \$1850.

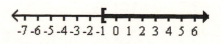

PROBLEM SET 2.5 Inequalities

1. $x > 1$
$(1, \infty)$

3. $x \geq -1$
$[-1, \infty)$

-7 -6 -5 -4 -3 -2 -1 0 1 2 3 4 5 6

5. $x < -2$
$(-\infty, -2)$

Problem Set 2.5

7. $x \leq 2$
$(-\infty, 2]$

9. $(-\infty, 4)$
$x < 4$

11. $(-\infty, -7]$
$x \leq -7$

13. $(8, \infty)$
$x > 8$

15. $[-7, \infty)$
$x \geq -7$

17. $x - 3 > -2$
$x - 3 + 3 > -2 + 3$
$x > 1$

19. $-2x \geq 8$
$-\frac{1}{2}\left(-2x\right) \leq -\frac{1}{2}\left(8\right)$
$x \leq -4$

21. $5x \leq -10$
$\frac{1}{5}\left(5x\right) \leq \frac{1}{5}\left(-10\right)$
$x \leq -2$

23. $2x + 1 < 5$
$2x < 4$
$\frac{1}{2}\left(2x\right) < \frac{1}{2}\left(4\right)$
$x < 2$

25. $3x - 2 > -5$
$3x - 2 + 2 > -5 + 2$
$3x > -3$
$\frac{1}{3}\left(3x\right) > \frac{1}{3}\left(-3\right)$
$x > -1$

27. $-7x - 3 \leq 4$
$-7x \leq 7$
$-\frac{1}{7}\left(-7x\right) \geq -\frac{1}{7}\left(7\right)$
$x \geq -1$

29. $2 + 6x > -10$
$6x > -12$
$\frac{1}{6}\left(6x\right) > \frac{1}{6}\left(-12\right)$
$x > -2$

31. $5 - 3x < 11$
$-3x < 6$
$-\frac{1}{3}\left(-3x\right) > -\frac{1}{3}\left(6\right)$
$x > -2$

33. $15 < 1 - 7x$
$14 < -7x$
$-\frac{1}{7}(14) > -\frac{1}{7}(-7x)$
$-2 > x$
$x < -2$

-7 -6 -5 -4 -3 -2 -1 0 1 2 3 4 5 6

35. $-10 \le 2 + 4x$
$-12 \le 4x$
$\frac{1}{4}(-12) \le \frac{1}{4}(4x)$
$-3 \le x$
$x \ge -3$

-7 -6 -5 -4 -3 -2 -1 0 1 2 3 4 5 6

37. $3(x+2) > 6$
$3x + 6 > 6$
$3x > 0$
$\frac{1}{3}(3x) > \frac{1}{3}(0)$
$x > 0$

-7 -6 -5 -4 -3 -2 -1 0 1 2 3 4 5 6

39. $5x + 2 \ge 4x + 6$
$5x \ge 4x + 4$
$x \ge 4$

-7 -6 -5 -4 -3 -2 -1 0 1 2 3 4 5 6

41. $2x - 1 > 6$
$2x > 7$
$x > \frac{7}{2}$
The solution set is $\left(\frac{7}{2}, \infty\right)$.

43. $-5x - 2 < -14$
$-5x < -12$
$-\frac{1}{5}(-5) > -\frac{1}{5}(-12)$
$x > \frac{12}{5}$
The solution set is $\left(\frac{12}{5}, \infty\right)$.

45. $-3(2x+1) \ge 12$
$-6x - 3 \ge 12$
$-6x \ge 15$
$-\frac{1}{6}(-6x) \le -\frac{1}{6}(15)$
$x \le -\frac{15}{6}$
$x \le -\frac{5}{2}$
The solution set is $\left(-\infty, -\frac{5}{2}\right]$.

47. $4(3x-2) \ge -3$
$12x - 8 \ge -3$
$12x \ge 5$
$\frac{1}{12}(12x) \ge \left(\frac{1}{12}5\right)$
$x \ge \frac{5}{12}$
The solution set is $\left[\frac{5}{12}, \infty\right)$.

49. $6x - 2 > 4x - 14$
$6x > 4x - 12$
$2x > -12$
$\frac{1}{2}(2x) > \frac{1}{2}(-12)$
$x > -6$
The solution set is $(-6, \infty)$.

51. $2x - 7 < 6x + 13$
$2x < 6x + 20$
$-4x < 20$

$-\frac{1}{4}\left(-4x\right) > -\frac{1}{4}\left(20\right)$
$x > -5$
The solution set is $(-5, \infty)$.

53. $4(x-3) \leq -2(x+1)$
$4x - 12 \leq -2x - 2$
$4x \leq -2x + 10$
$6x \leq 10$
$\frac{1}{6}\left(6x\right) \leq \frac{1}{6}\left(10\right)$
$x \leq \frac{10}{6}$
$x \leq \frac{5}{3}$
The solution set is $\left(-\infty, \frac{5}{3}\right]$.

55. $5(x-4) - 6(x+2) < 4$
$5x - 20 - 6x - 12 < 4$
$-x - 32 < 4$
$-x < 36$
$-1(-x) > -1(36)$
$x > -36$
The solution set is $(-36, \infty)$.

57. $-3(3x+2) - 2(4x+1) \geq 0$
$-9x - 6 - 8x - 2 \geq 0$
$-17x - 8 \geq 0$
$-17x \geq 8$
$-\frac{1}{17}\left(-17x\right) \leq -\frac{1}{17}\left(8\right)$
$x \leq -\frac{8}{17}$
The solution set is $\left(-\infty, -\frac{8}{17}\right]$.

59. $-(x-3) + 2(x-1) < 3(x+4)$
$-x + 3 + 2x - 2 < 3x + 12$
$x + 1 < 3x + 12$
$x < 3x + 11$
$-2x < 11$
$-\frac{1}{2}\left(-2x\right) > -\frac{1}{2}\left(11\right)$
$x > -\frac{11}{2}$
The solution set is $\left(-\frac{11}{2}, \infty\right)$.

61. $7(x+1) - 8(x-2) < 0$
$7x + 7 - 8x + 16 < 0$
$-x + 23 < 0$
$-x < -23$
$-1(-x) > -1(-23)$
$x > 23$
The solution set is $(23, \infty)$.

63. $-5(x-1) + 3 > 3x - 4 - 4x$
$-5x + 5 + 3 > -x - 4$
$-5x + 8 > -\text{x} - 4$
$-5x > -x - 12$
$-4x > -12$
$-\frac{1}{4}\left(-4x\right) < -\frac{1}{4}\left(-12\right)$
$x < 3$
The solution set is $(-\infty, 3)$.

65. $3(x-2) - 5(2x-1) \geq 0$
$3x - 6 - 10x + 5 \geq 0$
$-7x - 1 \geq 0$
$-7x \geq 1$
$-\frac{1}{7}\left(-7x\right) \leq -\frac{1}{7}\left(1\right)$
$x \leq -\frac{1}{7}$
The solution set is $\left(-\infty, -\frac{1}{7}\right]$.

67. $-5(3x+4) < -2(7x-1)$
$-15x - 20 < -14x + 2$
$-15x < -14x + 22$
$-x < 22$
$-1(-x) > -1(22)$
$x > -22$
The solution set is $(-22, \infty)$.

69. $-3(x+2) > 2(x-6)$
$-3x - 6 > 2x - 12$
$-3x > 2x - 6$
$-5x > -6$
$-\frac{1}{5}\left(-5x\right) < -\frac{1}{5}\left(-6\right)$
$x < \frac{6}{5}$
The solution set is $\left(-\infty, \frac{6}{5}\right)$.

PROBLEM SET 2.6 More on Inequalities and Problem Solving

1. $\frac{2}{5}x+\frac{1}{3}x>\frac{44}{15}$

$15\left(\frac{2}{5}x+\frac{1}{3}x\right)>15\left(\frac{44}{15}\right)$

$15\left(\frac{2}{5}x\right)+15\left(\frac{1}{3}x\right)>44$

$6x+5x>44$

$11x>44$

$x>4$

The solution set is $(4, \infty)$.

3. $x-\frac{5}{6}<\frac{x}{2}+3$

$6\left(x-\frac{5}{6}\right)<6\left(\frac{x}{2}+3\right)$

$6(x)-6\left(\frac{5}{6}\right)<6\left(\frac{x}{2}\right)+6(3)$

$6x-5<3x+18$

$6x<3x+23$

$3x<23$

$x<\frac{23}{3}$

The solution set is $\left(-\infty, \frac{23}{3}\right)$.

5. $\frac{x-2}{3}+\frac{x+1}{4}\geq\frac{5}{2}$

$12\left(\frac{x-2}{3}+\frac{x+1}{4}\right)\geq 12\left(\frac{5}{2}\right)$

$12\left(\frac{x-2}{3}\right)+12\left(\frac{x+1}{4}\right)\geq 30$

$4(x-2)+3(x+1)\geq 30$

$4x-8+3x+3\geq 30$

$7x-5\geq 30$

$7x\geq 35$

$x\geq 5$

The solution set is $[5, \infty)$.

7. $\frac{3-x}{6}+\frac{x+2}{7}\leq 1$

$42\left(\frac{3-x}{6}+\frac{x+2}{7}\right)\leq 42(1)$

$42\left(\frac{3-x}{6}\right)+42\left(\frac{x+2}{7}\right)\leq 42$

$7(3-x)+6(x+2)\leq 42$

$21-7x+6x+12\leq 42$

$-x+33\leq 42$

$-x\leq 9$

$-1(-x)\geq -1(9)$

$x\geq -9$

The solution set is $[-9, \infty)$.

9. $\frac{x+3}{8}-\frac{x+5}{5}\geq\frac{3}{10}$

$40\left(\frac{x+3}{8}-\frac{x+5}{5}\right)\geq 40\left(\frac{3}{10}\right)$

$40\left(\frac{x+3}{8}\right)-40\left(\frac{x+5}{5}\right)\geq 12$

$5(x+3)-8(x+5)\geq 12$

$5x+15-8x-40\geq 12$

$-3x-25\geq 12$

$-3x\geq 37$

$-\frac{1}{3}(-3x)\leq -\frac{1}{3}(37)$

$x\leq -\frac{37}{3}$

The solution set is $\left(-\infty, -\frac{37}{3}\right]$.

11. $\frac{4x-3}{6}-\frac{2x-1}{12}<-2$

$12\left(\frac{4x-3}{6}-\frac{2x-1}{12}\right)<12(-2)$

$12\left(\frac{4x-3}{6}\right)-12\left(\frac{2x-1}{12}\right)<-24$

$2(4x-3)-(2x-1)<-24$

$8x-6-2x+1<-24$

$6x-5<-24$

$6x<-19$

$x<-\frac{19}{6}$

The solution set is $\left(-\infty, -\frac{19}{6}\right)$.

Problem Set 2.6

13. $0.06x + 0.08(250 - x) \geq 19$

$100[0.06x + 0.08(250 - x)] \geq 100(19)$

$100(0.06x) + 100[0.08(250 - x)] \geq 1900$

$6x + 8(250 - x) \geq 1900$

$6x + 2000 - 8x \geq 1900$

$-2x + 2000 \geq 1900$

$-2x \geq -100$

$-\frac{1}{2}(-2x) \leq -\frac{1}{2}(-100)$

$x \leq 50$

The solution set is $(-\infty, 50]$.

15. $0.09x + 0.1(x + 200) > 77$

$100[0.09x + 0.1(x + 200)] > 100(77)$

$100(0.09x) + 100[0.1(x + 200)] > 7700$

$9x + 10(x + 200) > 7700$

$9x + 10x + 2000 > 7700$

$19x + 2000 > 7700$

$19x > 5700$

$x > 300$

The solution set is $(300, \infty)$.

17. $x \geq 3.4 + 0.15x$

$100(x) \geq 100(3.4 + 0.15x)$

$100x \geq 100(3.4) + 100(0.15x)$

$100x \geq 340 + 15x$

$85x \geq 340$

$x \geq 4$

The solution set is $[4, \infty)$.

19. $x > -1$ and $x < 2$

The solution set is $(-1, 2)$.

21. $x \leq 2$ and $x > -1$

The solution set is $(-1, 2]$.

23. $x > 2$ or $x < -1$

The solution set is $(-\infty, -1) \cup (2, \infty)$.

25. $x \leq 1$ or $x > 3$

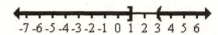

The solution set is $(-\infty, 1] \cup (3, \infty)$.

27. $x > 0$ and $x > -1$

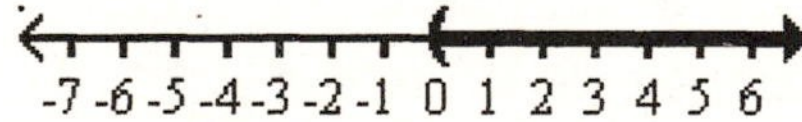

The solutions set is $(0, \infty)$.

29. $x < 0$ and $x > 4$

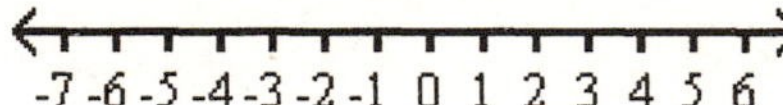

The solution set is $\emptyset$.

31. $x > -2$ or $x < 3$

The solution set is $(-\infty, \infty)$.

33. $x > -1$ or $x > 2$

-7 -6 -5 -4 -3 -2 -1 0 1 2 3 4 5 6

The solution set is $(-1, \infty)$.

35. $x - 2 > -1$ and $x - 2 < 1$
$x > 1$ and $x < 3$

-7 -6 -5 -4 -3 -2 -1 0 1 2 3 4 5 6

The solution set is $(1, 3)$.

37. $x + 2 < -3$ or $x + 2 > 3$
$x < -5$ or $x > 1$

-7 -6 -5 -4 -3 -2 -1 0 1 2 3 4 5 6

The solution set is $(-\infty, -5) \cup (1, \infty)$.

39. $2x - 1 \geq 5$ and $x > 0$
$2x \geq 6$ and $x > 0$
$x \geq 3$ and $x > 0$

-7 -6 -5 -4 -3 -2 -1 0 1 2 3 4 5 6

The solution set is $[3, \infty)$.

41. $5x - 2 < 0$ and $3x - 1 > 0$
$5x < 2$ and $3x > 1$
$x < \frac{2}{5}$ and $x > \frac{1}{3}$

-2 -1 0 1 2

The solution set is $\left(\frac{1}{3}, \frac{2}{5}\right)$.

43. $3x + 2 < -1$ or $3x + 2 > 1$
$3x < -3$ or $3x > -1$
$x < -1$ or $x > -\frac{1}{3}$

-2 -1 0 1 2

The solution set is $(-\infty, -1) \cup \left(-\frac{1}{3}, \infty\right)$.

45. $-3 < 2x + 1 < 5$
$-4 < 2x < 4$
$-2 < x < 2$
The solution set is $(-2, 2)$.

47. $-17 \leq 3x - 2 \leq 10$
$-15 \leq 3x \leq 12$
$-5 \leq x \leq 4$
The solution set is $[-5, 4]$.

49. $1 < 4x + 3 < 9$
$-2 < 4x < 6$
$-\frac{2}{4} < x < \frac{6}{4}$
$-\frac{1}{2} < x < \frac{3}{2}$
The solution set is $\left(-\frac{1}{2}, \frac{3}{2}\right)$.

51. $-6 < 4x - 5 < 6$
$-1 < 4x < 11$
$-\frac{1}{4} < x < \frac{11}{4}$
The solution set is $\left(-\frac{1}{4}, \frac{11}{4}\right)$.

53. $-4 \leq \frac{x-1}{3} \leq 4$
$3(-4) \leq 3\left(\frac{x-1}{3}\right) \leq 3(4)$
$-12 \leq x - 1 \leq 12$
$-11 \leq x \leq 13$
The solution set is $[-11, 13]$.

55. $-3 < 2 - x < 3$
$-5 < -x < 1$
$-1(-5) > -1(-x) > -1(1)$
$5 > x > -1$
$-1 < x < 5$
The solution set is $(-1, 5)$.

57. Let r = rate.
$(9\%)(300) + r(200) > 47$
$0.09(300) + r(200) > 47$
$100[0.09(300) + r(200)] > 100(47)$
$100[0.09(300)] + 100r(200) > 4700$
$9(300) + 20000r > 4700$
$2700 + 20000r > 4700$
$20000r > 2000$
$r > \frac{2000}{20000}$
$r > 0.1$
$r > 10\%$
The rate must be more than 10%.

59. Let x = average height of the two guards.
6 ft. 8 in. $= 6\frac{8}{12}$ feet $= 6\frac{2}{3}$ feet
6 ft. 4in. $= 6\frac{4}{12}$ feet $= 6\frac{1}{3}$ feet
$\frac{3(6\frac{2}{3}) + 2(x)}{5} \geq 6\frac{1}{3}$
$\frac{20 + 2x}{5} \geq \frac{19}{3}$
$15\left(\frac{20 + 2x}{5}\right) \geq 15\left(\frac{19}{3}\right)$
$3(20 + 2x) \geq 5(19)$
$60 + 6x \geq 95$
$6x \geq 35$
$x \geq \frac{35}{6}$
$x \geq 5\frac{5}{6}$
$x \geq 5\frac{10}{12}$
The average height should be 5 ft. 10 inches or better.

61. Let x = score for the third game.
$\frac{142 + 170 + x}{3} \geq 160$
$\frac{312 + x}{3} \geq 160$
$3\left(\frac{312 + x}{3}\right) \geq 3(160)$
$312 + x \geq 480$
$x \geq 168$
The score should be 168 or better.

63. Let x = score on the fifth day.
$\frac{82 + 84 + 78 + 79 + x}{5} \leq 80$
$\frac{323 + x}{5} \leq 80$
$5\left(\frac{323 + x}{5}\right) \leq 5(80)$
$323 + x \leq 400$
$x \leq 77$
The score should be 77 or less.

65. $325 \leq \frac{9}{5}C + 32 \leq 425$
$293 \leq \frac{9}{5}C \leq 393$
$\frac{5}{9}(293) \leq \frac{5}{9}\left(\frac{9}{5}C\right) \leq \frac{5}{9}(393)$
$163 \leq C \leq 218$
The temperature would be between 163°C and 218°C inclusive.

67. $70 \leq \frac{100M}{C} \leq 125$
$70 \leq \frac{100M}{9} \leq 125$
$\frac{9}{100}(70) \leq \frac{9}{100}\left(\frac{100M}{9}\right) \leq \frac{9}{100}(125)$
$6.3 \leq M \leq 11.25$
The mental age ranges from 6.3 to 11.25 years inclusive.

PROBLEM SET 2.7 Equations and Inequalities Involving Absolute Value

1. $|x| < 5$
$-5 < x < 5$
The solution set is $(-5, 5)$.

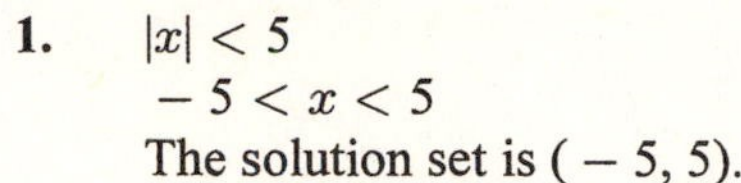

3. $|x| \leq 2$
$-2 \leq x \leq 2$
The solution set is $[-2, 2]$.

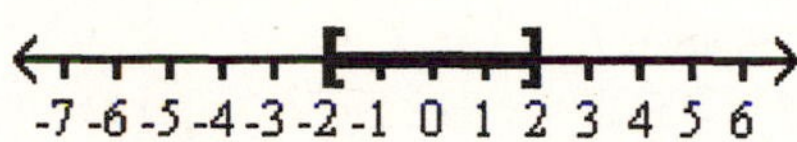

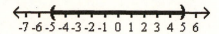

5. $|x| > 2$
$x < -2$ or $x > 2$
The solution set is $(-\infty, -2) \cup (2, \infty)$.

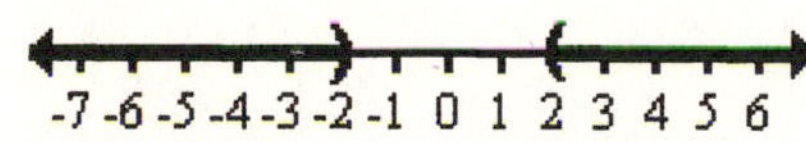

7. $|x - 1| < 2$
$-2 < x - 1 < 2$
$-1 < x < 3$
The solution set is $(-1, 3)$.

9. $|x + 2| \leq 4$
$-4 \leq x + 2 \leq 4$
$-6 \leq x \leq 2$
The solution set is $[-6, 2]$.

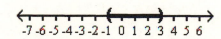

11. $|x + 2| > 1$
$x + 2 < -1$ or $x + 2 > 1$
$x < -3$ or $x > -1$
The solution set is $(-\infty, -3) \cup (-1, \infty)$.

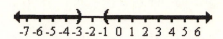

13. $|x - 3| \geq 2$
$x - 3 \leq -2$ or $x - 3 \geq 2$
$x \leq 1$ or $x \geq 5$
The solution set is $(-\infty, 1] \cup [5, \infty)$.

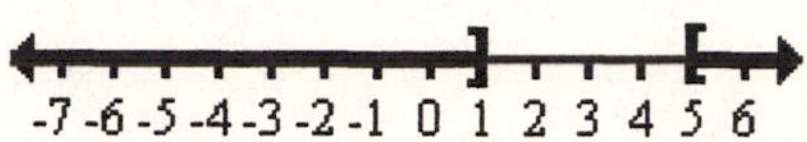

15. $|x - 1| = 8$
$x - 1 = 8$ or $x - 1 = -8$
$x = 9$ or $x = -7$
The solution set is $\{-7, 9\}$.

17. $|x - 2| > 6$
$x - 2 < -6$ or $x - 2 > 6$
$x < -4$ or $x > 8$
The solution set is $(-\infty, -4) \cup (8, \infty)$.

19. $|x + 3| < 5$
$-5 < x + 3 < 5$
$-8 < x < 2$
The solution set is $(-8, 2)$.

21. $|2x - 4| = 6$
$2x - 4 = 6$ or $2x - 4 = -6$
$2x = 10$ or $2x = -2$
$x = 5$ or $x = -1$
The solution set is $\{-1, 5\}$.

23. $|2x - 1| \leq 9$
$-9 \leq 2x - 1 \leq 9$
$-8 \leq 2x \leq 10$
$-4 \leq x \leq 5$
The solution set is $[-4, 5]$.

25. $|4x + 2| \geq 12$
$4x + 2 \leq -12$ or $4x + 2 \geq 12$
$4x \leq -14$ or $4x \geq 10$
$x \leq -\frac{14}{4}$ or $x \geq \frac{10}{4}$

$x \leq -\frac{7}{2}$ or $x \geq \frac{5}{2}$

The solution set is $\left(-\infty, -\frac{7}{2}\right] \cup \left[\frac{5}{2}, \infty\right)$.

27. $|3x+4| = 11$

$3x + 4 = 11$ or $3x + 4 = -11$

$3x = 7$ or $3x = -15$

$x = \frac{7}{3}$ or $x = -5$

The solution set is $\left\{-5, \frac{7}{3}\right\}$.

29. $|4 - 2x| = 6$

$4 - 2x = 6$ or $4 - 2x = -6$

$-2x = 2$ or $-2x = -10$

$x = -1$ or $x = 5$

The solution set is $\{-1, 5\}$.

31. $|2 - x| > 4$

$2 - x < -4$ or $2 - x > 4$

$-x < -6$ or $-x > 2$

$-1(-x) > -1(-6)$ or $-1(-x) < -1(2)$

$x > 6$ or $x < -2$

The solution set is $(-\infty, -2) \cup (6, \infty)$.

33. $|1 - 2x| < 2$

$-2 < 1 - 2x < 2$

$-3 < -2x < 1$

$-\frac{1}{2}(-3) > -\frac{1}{2}(-2x) > -\frac{1}{2}(1)$

$\frac{3}{2} > x > -\frac{1}{2}$

$-\frac{1}{2} < x < \frac{3}{2}$

The solution set is $\left(-\frac{1}{2}, \frac{3}{2}\right)$.

35. $|5x + 9| \leq 16$

$-16 \leq 5x + 9 \leq 16$

$-25 \leq 5x \leq 7$

$-5 \leq x \leq \frac{7}{5}$

The solution set is $\left[-5, \frac{7}{5}\right]$.

37. $\left|x - \frac{3}{4}\right| = \frac{2}{3}$

$x - \frac{3}{4} = \frac{2}{3}$ or $x - \frac{3}{4} = -\frac{2}{3}$

$x = \frac{2}{3} + \frac{3}{4}$ or $x = -\frac{2}{3} + \frac{3}{4}$

$x = \frac{8}{12} + \frac{9}{12}$ or $x = -\frac{8}{12} + \frac{9}{12}$

$x = \frac{17}{12}$ or $x = \frac{1}{12}$

The solution set is $\left\{\frac{1}{12}, \frac{17}{12}\right\}$.

39. $|-2x + 7| \leq 13$

$-13 \leq -2x + 7 \leq 13$

$-20 \leq -2x \leq 6$

$-\frac{1}{2}(-20) \geq -\frac{1}{2}(-2x) \geq -\frac{1}{2}(6)$

$10 \geq x \geq -3$

$-3 \leq x \leq 10$

The solution set is $[-3, 10]$.

41. $\left|\frac{x-3}{4}\right| < 2$

$-2 < \frac{x-4}{4} < 2$

$4(-2) < 4\left(\frac{x-3}{4}\right) < 4(2)$

$-8 < x - 3 < 8$

$-5 < x < 11$

The solution set is $(-5, 11)$.

43. $\left|\frac{2x+1}{2}\right| > 1$

$\frac{2x+1}{2} < -1$ or $\frac{2x+1}{2} > 1$

$2\left(\frac{2x+1}{2}\right) < -1(2)$ or $2\left(\frac{2x+1}{2}\right) \geq 1(2)$

$2x + 1 < -2$ or $2x + 1 > 2$

$2x < -3$ or $2x > 1$

$x < -\frac{3}{2}$ or $x > \frac{1}{2}$

The solution set is $\left(-\infty, -\frac{3}{2}\right) \cup \left(\frac{1}{2}, \infty\right)$.

45. $|2x-3|+2=5$
$|2x-3|=3$
$2x-3=3$ or $2x-3=-3$
$2x=6$ or $2x=0$
$x=3$ or $x=0$
The solution set is $\{0, 3\}$.

47. $|x+2|-6=-2$
$|x+2|=4$
$x+2=4$ or $x+2=-4$
$x=2$ or $x=-6$
The solution set is $\{-6, 2\}$.

49. $|4x-3|+2=2$
$|4x-3|=0$
$4x-3=0$
$4x=3$
$x=\frac{3}{4}$
The solution set is $\left\{\frac{3}{4}\right\}$.

51. $|x+7|-3\geq 4$
$|x+7|\geq 7$
$x+7\leq -7$ or $x+7\geq 7$
$x\leq -14$ or $x\geq 0$
The solution set is $(-\infty, -14]\cup[0, \infty)$.

53. $|2x-1|+1\leq 6$
$|2x-1|\leq 5$
$-5\leq 2x-1\leq 5$
$-4\leq 2x\leq 6$
$-2\leq x\leq 3$
The solution set is $[-2, 3]$.

55. $|2x+1|=-4$
The solution set is $\emptyset$.
Absolute value is never negative.

57. $|3x-1|>-2$
The solution set is $(-\infty, \infty)$.
Absolute value is always greater than a negative value.

59. $|5x-2|=0$
$5x-2=0$
$5x=2$
$x=\frac{2}{5}$
The solution set is $\left\{\frac{2}{5}\right\}$.

61. $|4x-1|<-1$
The solution set is $\emptyset$.
Absolute value is never less than a negative number.

63. $|x+4|<0$
The solution set is $\emptyset$.
Absolute value is never less than zero.

69. $|-2x-3|=|x+1|$
$-2x-3=x+1$ or $-2x-3=-(x+1)$
$-2x=x+4$ or $-2x-3=-x-1$
$-3x=4$ or $-2x=-x+2$
$x=-\frac{4}{3}$ or $-x=2$
$x=-\frac{4}{3}$ or $x=-2$
The solution set is $\left\{-2, -\frac{4}{3}\right\}$.

71. $|x-2|=|x+6|$
$x-2=x+6$ or $x-2=-(x+6)$
$x=x+8$ or $x-2=-x-6$
$0=8$ or $x=-x-4$
contradiction $2x=-4$
$x=-2$
The solution set is $\{-2\}$.

73. $|x+1|=|x-1|$
$x+1=x-1$ or $x+1=-(x-1)$
$x=x-2$ or $x+1=-x+1$
$0=-2$ or $x=-x+0$
contradiction $2x=0$
$x=0$
The solution set is $\{0\}$.

75. $|ax+b|<k$
If $ax+b\geq 0$, then $|ax+b|=ax+b$
so, $ax+b<k$

Problem Set 2.7

If $ax+b<0$, then $|ax+b| = -(ax+b)$
so, $-(ax+b) < k$
$-1[-(ax+b)] > -1(k)$
$ax+b > -k$
Therefore $ax+b<k$ and $ax+b > -k$
is $-k < ax+b < -k$.

CHAPTER 2 Review Problem Set

1. $5(x-6) = 3(x+2)$
$5x-30 = 3x+6$
$5x-30+30 = 3x+6+30$
$5x = 3x+36$
$5x-3x = 3x-3x+36$
$2x = 36$
$\frac{1}{2}(2x) = \frac{1}{2}(36)$
$x = 18$
The solution set is $\{18\}$.

2. $2(2x+1)-(x-4) = 4(x+5)$
$4x+2-x+4 = 4x+20$
$3x+6 = 4x+20$
$3x = 4x+14$
$-x = 14$
$x = -14$
The solution set is $\{-14\}$.

3. $-(2n-1)+3(n+2) = 7$
$-2n+1+3n+6 = 7$
$n+7 = 7$
$n+7-7 = 7-7$
$n = 0$
The solution set is $\{0\}$.

4. $2(3n-4)+3(2n-3) = -2(n+5)$
$6n-8+6n-9 = -2n-10$
$12n-17 = -2n-10$
$12n = -2n+7$
$14n = 7$
$n = \frac{7}{14} = \frac{1}{2}$
The solution set is $\left\{\frac{1}{2}\right\}$.

5. $\frac{3t-2}{4} = \frac{2t+1}{3}$
$12\left(\frac{3t-2}{4}\right) = 12\left(\frac{2t+1}{3}\right)$
$3(3t-2) = 4(2t+1)$
$9t-6 = 8t+4$
$9t = 8t+10$
$t = 10$
The solution set is $\{10\}$.

6. $\frac{x+6}{5} + \frac{x-1}{4} = 2$
$20\left(\frac{x+6}{5} + \frac{x-1}{4}\right) = 20(2)$
$20\left(\frac{x+6}{5}\right) + 20\left(\frac{x-1}{4}\right) = 40$
$4(x+6)+5(x-1) = 40$
$4x+24+5x-5 = 40$
$9x+19 = 40$
$9x = 21$
$x = \frac{21}{9} = \frac{7}{3}$
The solution set is $\left\{\frac{7}{3}\right\}$.

7. $1 - \frac{2x-1}{6} = \frac{3x}{8}$
$24\left(1 - \frac{2x-1}{6}\right) = 24\left(\frac{3x}{8}\right)$
$24(1) - 24\left(\frac{2x-1}{6}\right) = 3(3x)$
$24 - 4(2x-1) = 9x$
$24 - 8x + 4 = 9x$
$28 - 8x = 9x$
$28 = 17x$
$\frac{28}{17} = x$
The solution set is $\left\{\frac{28}{17}\right\}$.

8. $\dfrac{2x+1}{3} + \dfrac{3x-1}{5} = \dfrac{1}{10}$

$30\left(\dfrac{2x+1}{3} + \dfrac{3x-1}{5}\right) = 30\left(\dfrac{1}{10}\right)$

$30\left(\dfrac{2x+1}{3}\right) + 30\left(\dfrac{3x-1}{5}\right) = 3$

$10(2x+1) + 6(3x-1) = 3$

$20x + 10 + 18x - 6 = 3$

$38x + 4 = 3$

$38x = -1$

$x = -\dfrac{1}{38}$

The solution set is $\left\{-\dfrac{1}{38}\right\}$.

9. $\dfrac{3n-1}{2} - \dfrac{2n+3}{7} = 1$

$14\left(\dfrac{3n-1}{2} - \dfrac{2n+3}{7}\right) = 14(1)$

$14\left(\dfrac{3n-1}{2}\right) - 14\left(\dfrac{2n+3}{7}\right) = 14$

$7(3n-1) - 2(2n+3) = 14$

$21n - 7 - 4n - 6 = 14$

$17n - 13 = 14$

$17n = 27$

$n = \dfrac{27}{17}$

The solution set is $\left\{\dfrac{27}{17}\right\}$.

10. $|3x-1| = 11$

$3x - 1 = 11$ or $3x - 1 = -11$

$3x = 12$ or $3x = -10$

$x = 4$ or $x = -\dfrac{10}{3}$

The solution set is $\left\{-\dfrac{10}{3}, 4\right\}$.

11. $0.06x + 0.08(x+100) = 15$

$100[0.06x + 0.08(x+100)] = 100(15)$

$100(0.06x) + 100[0.08(x+100)] = 1500$

$6x + 8(x+100) = 1500$

$6x + 8x + 800 = 1500$

$14x + 800 = 1500$

$14x = 700$

$x = 50$

The solution set is $\{50\}$.

12. $0.4(t-6) = 0.3(2t+5)$

$10[0.4(t-6)] = 10[0.3(2t+5)]$

$4(t-6) = 3(2t+5)$

$4t - 24 = 6t + 15$

$4t = 6t + 39$

$-2t = 39$

$t = -\dfrac{39}{2}$

The solution set is $\left\{-\dfrac{39}{2}\right\}$.

13. $0.1(n+300) = 0.09n + 32$

$100[0.1(n+300)] = 100[0.09n + 32]$

$10(n+300) = 100(0.09n) + 100(32)$

$10n + 3000 = 9n + 3200$

$10n = 9n + 200$

$n = 200$

The solution set is $\{200\}$.

14. $0.2(x-0.5) - 0.3(x+1) = .4$

$10[0.2(x-0.5) - 0.3(x+1)] = 10(0.4)$

$10[0.2(x-0.5)] - 10[0.3(x+1)] = 4$

$2(x-0.5) - 3(x+1) = 4$

$2x - 1 - 3x - 3 = 4$

$-x - 4 = 4$

$-x = 8$

$x = -8$

The solution set is $\{-8\}$.

15. $|2n+3| = 4$

$2n + 3 = 4$ or $2n + 3 = -4$

$2n = 1$ or $2n = -7$

$n = \dfrac{1}{2}$ or $n = -\dfrac{7}{2}$

The solution set is $\left\{-\dfrac{7}{2}, \dfrac{1}{2}\right\}$.

16. $ax - b = b + 2$

$ax = 2b + 2$

$x = \dfrac{2b+2}{a}$

17. $ax = bx + c$
$ax - bx = c$
$x(a - b) = c$
$x = \frac{c}{a - b}$

18. $m(x + a) = p(x + b)$
$mx + ma = px + pb$
$mx = px + pb - ma$
$mx - px = pb - ma$
$x(m - p) = pb - ma$
$x = \frac{pb - ma}{m - p}$

19. $5x - 7y = 11$
$5x = 11 + 7y$
$x = \frac{11 + 7y}{5}$

20. $\frac{x - a}{b} = \frac{y + 1}{c}$
$bc\left(\frac{x - a}{b}\right) = bc\left(\frac{y + 1}{c}\right)$
$c(x - a) = b(y + 1)$
$cx - ac = by + b$
$cx = by + b + ac$
$x = \frac{by + b + ac}{c}$

21. $A = \pi r^2 + \pi rs$ for s
$A - \pi r^2 = \pi rs$
$\frac{A - \pi r^2}{\pi r} = s$

22. $A = \frac{1}{2}h(b_1 + b_2)$ for b_2
$2(A) = 2[\frac{1}{2}h(b_1 + b_2)]$
$2A = h(b_1 + b_2)$
$2A = hb_1 + hb_2$
$2A - hb_1 = hb_2$
$\frac{2A - hb_1}{h} = b_2$

23. $S_n = \frac{n(a_1 + a_2)}{2}$ for n
$2S_n = n(a_1 + a_2)$
$\frac{2S_n}{a_1 + a_2} = n$

24. $\frac{1}{R} = \frac{1}{R_1} + \frac{1}{R_2}$ for R
$RR_1R_2\left(\frac{1}{R}\right) = RR_1R_2\left(\frac{1}{R_1} + \frac{1}{R_2}\right)$
$R_1R_2 = RR_1R_2\left(\frac{1}{R_1}\right) + RR_1R_2\left(\frac{1}{R_2}\right)$
$R_1R_2 = RR_2 + RR_1$
$R_1R_2 = R(R_2 + R_1)$
$\frac{R_1R_2}{R_2 + R_1} = R$

25. $5x - 2 \geq 4x - 7$
$5x \geq 4x - 5$
$x \geq -5$
The solution set is $[-5, \infty)$.

26. $3 - 2x < -5$
$-2x < -8$
$-\frac{1}{2}(-2x) > -\frac{1}{2}(-8)$
$x > 4$
The solution set is $(4, \infty)$.

27. $2(3x - 1) - 3(x - 3) > 0$
$6x - 2 - 3x + 9 > 0$
$3x + 7 > 0$
$3x > -7$
$x > -\frac{7}{3}$
The solution set is $\left(-\frac{7}{3}, \infty\right)$.

28. $3(x + 4) \leq 5(x - 1)$
$3x + 12 \leq 5x - 5$
$3x \leq 5x - 17$

$-2x \le -17$
$-\frac{1}{2}(-2x) \ge -\frac{1}{2}(-17)$
$x \ge \frac{17}{2}$
The solution set is $\left[\frac{17}{2}, \infty\right)$.

29. $\frac{5}{6}n - \frac{1}{3}n < \frac{1}{6}$
$6\left(\frac{5}{6}n - \frac{1}{3}n\right) < 6\left(\frac{1}{6}\right)$
$6\left(\frac{5}{6}n\right) - 6\left(\frac{1}{3}n\right) < 1$
$5n - 2n < 1$
$3n < 1$
$n < \frac{1}{3}$
The solution set is $\left(-\infty, \frac{1}{3}\right)$.

30. $\frac{n-4}{5} + \frac{n-3}{6} > \frac{7}{15}$
$30\left(\frac{n-4}{5} + \frac{n-3}{6}\right) > 30\left(\frac{7}{15}\right)$
$30\left(\frac{n-4}{5}\right) + 30\left(\frac{n-3}{6}\right) > 14$
$6(n-4) + 5(n-3) > 14$
$6n - 24 + 5n - 15 > 14$
$11n - 39 > 14$
$11n > 53$
$n > \frac{53}{11}$
The solution set is $\left(\frac{53}{11}, \infty\right)$.

31. $s \ge 4.5 + 0.25s$
$100(s) \ge 100(4.5 + 0.25s)$
$100s \ge 100(4.5) + 100(0.25s)$
$100s \ge 450 + 25s$
$75s \ge 450$
$s \ge 6$
The solution set is $[6, \infty)$.

32. $0.07x + 0.09(500 - x) \ge 43$
$100[0.07x + 0.09(500 - x)] \ge 100(43)$
$7x + 9(500 - x) \ge 4300$
$7x + 4500 - 9x \ge 4300$
$-2x + 4500 \ge 4300$
$-2x \ge -200$
$-\frac{1}{2}(-2x) \le -\frac{1}{2}(-200)$
$x \le 100$
The solution set is $(-\infty, 100]$.

33. $|2x - 1| < 11$
$-11 < 2x - 1 < 11$
$-10 < 2x < 12$
$-5 < x < 6$
The solution set is $(-5, 6)$.

34. $|3x + 1| > 10$
$3x + 1 < -10$ or $3x + 1 > 10$
$3x < -11$ or $3x > 9$
$x < -\frac{11}{3}$ or $x > 3$
The solution set is $\left(-\infty, -\frac{11}{3}\right) \cup (3, \infty)$.

35. $-3(2t - 1) - (t + 2) > -6(t - 3)$
$-6t + 3 - t - 2 > -6t + 18$
$-7t + 1 > -6t + 18$
$-7t > -6t + 17$
$-t > 17$
$-1(-t) < -1(17)$
$t < -17$
The solution set is $(-\infty, -17)$.

36. $\frac{2}{3}(x - 1) + \frac{1}{4}(2x + 1) < \frac{5}{6}(x - 2)$
$12[\frac{2}{3}(x - 1) + \frac{1}{4}(2x + 1)] < 12[\frac{5}{6}(x - 2)]$
$12[\frac{2}{3}(x - 1)] + 12[\frac{1}{4}(2x + 1)] < 10(x - 2)$
$8(x - 1) + 3(2x + 1) < 10x - 20$
$8x - 8 + 6x + 3 < 10x - 20$
$14x - 5 < 10x - 20$
$14x < 10x - 15$
$4x < -15$
$x < -\frac{15}{4}$
The solution set is $\left(-\infty, -\frac{15}{4}\right)$.

37. $x > -1$ and $x < 1$

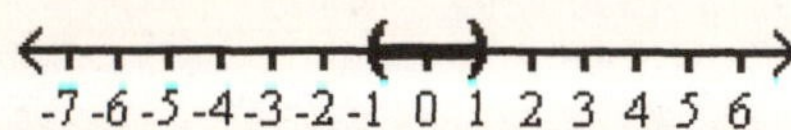

38. $x > 2$ or $x \leq -3$

39. $x > 2$ and $x > 3$

40. $x < 2$ or $x > -1$

41.

$2x + 1 > 3$	or	$2x + 1 < -3$
$2x > 2$	or	$2x < -4$
$x > 1$	or	$x < -2$

-7 -6 -5 -4 -3 -2 -1 0 1 2 3 4 5 6

42. $2 \leq x + 4 \leq 5$
$-2 \leq x \leq 1$

-7 -6 -5 -4 -3 -2 -1 0 1 2 3 4 5 6

43. $-1 < 4x - 3 \leq 9$
$2 < 4x \leq 12$
$\frac{2}{4} < x \leq 3$
$\frac{1}{2} < x \leq 3$

-7 -6 -5 -4 -3 -2 -1 0 1 2 3 4 5 6

44. $x + 1 > 3$ and $x - 3 < -5$
$x > 2$ and $x < -2$
$\emptyset$

45. Let $x =$ length and
$\frac{1}{3}x + 2 =$ width.

$P = 2l + 2w$

$44 = 2(x) + 2\left(\frac{1}{3}x + 2\right)$

$44 = 2x + \frac{2}{3}x + 4$

$40 = 2x + \frac{2}{3}x$

$3(40) = 3\left(2x + \frac{2}{3}x\right)$

$120 = 3(2x) + 3\left(\frac{2}{3}x\right)$

$120 = 6x + 2x$
$120 = 8x$
$15 = x$

length $= 15$ meters

width $= \frac{1}{3}(15) + 2 = 7$ meters

The length is 15 meters and the width is 7 meters.

46. Let $x =$ amount invested at 7% and
$500 - x =$ amount invested at 8%.

$(7\%)(x) + (8\%)(500 - x) = 38$

$0.07x + 0.08(500 - x) = 38$

$100[0.07x + 0.08(500 - x)] = 100(38)$

$100(0.07x) + 100[0.08(500 - x)] = 3800$
$7x + 8(500 - x) = 3800$
$7x + 4000 - 8x = 3800$
$-x + 4000 = 3800$
$-x = -200$
$x = 200$

amount invested at 7% = \$200
amount invested at 8% = 500 − 200 = \$300
There is \$200 invested at 7% and \$300 invested at 8%.

47. Let x = score on the 4th exam.

$\frac{3(84)+x}{4} \geq 85$

$\frac{252+x}{4} \geq 85$

$4\left(\frac{252+x}{4}\right) \geq 4(85)$

$252 + x \geq 340$

$x \geq 88$

The score should be 88 or better.

48. Let x = 1st integer,
$x + 1$ = 2nd integer and
$x + 2$ = 3rd integer.

$\frac{1}{2}x + \frac{1}{3}(x+2) = x + 1 - 1$

$\frac{1}{2}x + \frac{1}{3}(x+2) = x$

$6[\frac{1}{2}x + \frac{1}{3}(x+2)] = 6(x)$

$6\left(\frac{1}{2}x\right) + 6[\frac{1}{3}(x+2)] = 6x$

$3x + 2(x+2) = 6x$

$5x + 4 = 6x$

$4 = x$

The 1st integer = 4,
the 2nd integer = 5 and
the 3rd integer = 6.

49. Let x = normal hourly rate
and $\frac{3}{2}x$ = over 36 hours rate.

Pat worked 36 hours at the normal hourly rate and 6 hours at the overtime rate.

$36x + 6\left(\frac{3}{2}x\right) = 472.50$

$36x + 9x = 472.50$

$45x = 472.50$

$x = 10.50$

The normal hourly rate is \$10.50.

50. Let x = number of nickels,
$2x + 10$ = number of dimes and
$2x + 10 + 25$ = number of quarters.

$0.05(x)+0.10(2x+10)+.25(2x+10+25) = 24.75$

$0.05(x) + 0.10(2x+10) + 0.25(2x+35) = 24.75$

$100[.05 + .10(2x+10) + .25(2x+35)]=100(24.75)$

$100(.05x)+100[.10(2x+10)]+100[.25(2x+35)]=2475$

$5x + 10(2x+10) + 25(2x+35) = 2475$

$5x + 20x + 100 + 50x + 875 = 2475$

$75x + 975 = 2475$

$75x = 1500$

$x = 20$

number of nickels = 20
number of dimes = 2(20) + 10 = 50
number of quarters = 2(20) + 10 + 25 = 75

There are 20 nickels, 50 dimes, and 75 quarters.

51. Let x = angle,
$90 - x$ = complement of the angle and
$180 - x$ = supplement of the angle.

$90 - x = \frac{1}{10}(180 - x)$

$10(90 - x) = 10\left[\frac{1}{10}(180 - x)\right]$

$900 - 10x = 1(180 - x)$

$900 - 10x = 180 - x$

$-10x = -720 - x$

$-9x = -720$

$x = 80$

The angle is 80°.

52. Price = cost + 20%(cost)

Price = 38 + (20%)(38)
Price = 38 + 0.20(38)
Price = 38 + 7.6
Price = 45.60
The selling price should be \$45.60.

53. Let x represent pints of 1% solution, and let $10 - x$ represent pints of 4% solution.

$$\begin{pmatrix}\text{pure peroxide}\\ \text{in 1\% solution}\end{pmatrix} + \begin{pmatrix}\text{pure peroxide}\\ \text{in 4\% solution}\end{pmatrix} = \begin{pmatrix}\text{pure peroxide}\\ \text{in 2\% solution}\end{pmatrix}$$

$(1\%)(x) + (4\%)(10 - x) = (2\%)(10)$
$0.01x + 0.04(10 - x) = 0.02(10)$
$x + 4(10 - x) = 2(10)$
$x + 40 - 4x = 20$
$-3x + 40 = 20$
$-3x = -20$
$x = \frac{20}{3}$
The amount of 1% hydrogen peroxide solution to be mixed is $6\frac{2}{3}$ pints.

54. Reena's time = 5 hours 20 minutes = $5\frac{1}{3}$ hr

Gladys' time = $5\frac{1}{3} + 2 = 7\frac{1}{3}$ hr

Let $x =$ Reena's rate.

Reena's distance = $(5\frac{1}{3})(x) = \frac{16}{3}x$

Gladys' distance = $(7\frac{1}{3})(40) = \frac{22}{3}(40)$

$\frac{16}{3}x = \frac{22}{3}(40)$

$3[\frac{16}{3}x] = 3[\frac{22}{3}(40)]$

$16x = 22(40)$
$16x = 880$
$x = 55$
Reena's rate is 55 miles per hour.

55. Let t = time Sonya rode.
$t + \frac{5}{4}$ = time Rita rode.

Sonya's distance = 16t
Rita's distance = $12\left(t + \frac{5}{4}\right)$

Rita's distance = Sonya's distance + 2

$12\left(t + \frac{5}{4}\right) = 16t + 2$

$12t + 12\left(\frac{5}{4}\right) = 16t + 2$

$12t + 15 = 16t + 2$
$12t = 16t - 13$
$-4t = -13$
$t = \frac{-13}{-4} = 3\frac{1}{4}$

Sonya's time = $3\frac{1}{4}$ hours

Rita's time = $\frac{13}{4} + \frac{5}{4} = \frac{18}{4} = 4\frac{1}{2}$ hours.

Sonya rides for $3\frac{1}{4}$ hours and

Rita rides for $4\frac{1}{2}$ hours.

56. Let x = number of cups of orange juice.
$(100\%)(x) + (10\%)(50) = (20\%)(x + 50)$
$1.00x + 0.10(50) = 0.20(x + 50)$
$100[1x + 0.10(50)] = 100[0.20(x + 50)]$
$100(1x) + 100[0.10(50)] = 20(x + 50)$
$100x + 10(50) = 20x + 1000$
$100x + 500 = 20x + 1000$
$100x = 20x + 500$
$80x = 500$
$x = 6.25$
$x = 6\frac{1}{4}$
There should be $6\frac{1}{4}$ cups of orange juice.

CHAPTER 2 Test

1. $5x - 2 = 2x - 11$
$5x = 2x - 9$
$3x = -9$
$x = -3$
The solution set is $\{-3\}$.

2. $6(n - 2) - 4(n + 3) = -14$
$6n - 12 - 4n - 12 = -14$
$2n - 24 = -14$
$2n = 10$
$n = 5$
The solution set is $\{5\}$.

3. $-3(x+4)=3(x-5)$
$-3x-12=3x-15$
$-3x=3x-3$
$-6x=-3$
$x=\frac{-3}{-6}=\frac{1}{2}$
The solution set is $\left\{\frac{1}{2}\right\}$.

4. $3(2x-1)-2(x+5)=-(x-3)$
$6x-3-2x-10=-x+3$
$4x-13=-x+3$
$4x=-x+16$
$5x=16$
$x=\frac{16}{5}$
The solution set is $\left\{\frac{16}{5}\right\}$.

5. $\frac{3t-2}{4}=\frac{5t+1}{5}$
$20\left(\frac{3t-2}{4}\right)=20\left(\frac{5t+1}{5}\right)$
$5(3t-2)=4(5t+1)$
$15t-10=20t+4$
$15t=20t+14$
$-5t=14$
$t=-\frac{14}{5}$
The solution is $\left\{-\frac{14}{5}\right\}$.

6. $\frac{5x+2}{3}-\frac{2x+4}{6}=-\frac{4}{3}$
$6\left(\frac{5x+2}{3}-\frac{2x+4}{6}\right)=6\left(-\frac{4}{3}\right)$
$6\left(\frac{5x+2}{3}\right)-6\left(\frac{2x+4}{6}\right)=-8$
$2(5x+2)-1(2x+4)=-8$
$10x+4-2x-4=-8$
$8x=-8$
$x=-1$
The solution set is $\{-1\}$.

7. $|4x-3|=9$
$4x-3=9$ or $4x-3=-9$
$4x=12$ or $4x=-6$
$x=3$ or $x=-\frac{6}{4}=-\frac{3}{2}$
The solution set is $\left\{-\frac{3}{2}, 3\right\}$.

8. $\frac{1-3x}{4}+\frac{2x+3}{3}=1$
$12\left(\frac{1-3x}{4}+\frac{2x+3}{3}\right)=12(1)$
$12\left(\frac{1-3x}{4}\right)+12\left(\frac{2x+3}{3}\right)=12(1)$
$3(1-3x)+4(2x+3)=12$
$3-9x+8x+12=12$
$-x+15=12$
$-x=-3$
$x=3$
The solution set is $\{3\}$.

9. $2-\frac{3x-1}{5}=-4$
$5\left(2-\frac{3x-1}{5}\right)=5(-4)$
$5(2)-5\left(\frac{3x-1}{5}\right)=-20$
$10-(3x-1)=-20$
$10-3x+1=-20$
$-3x+11=-20$
$-3x=-31$
$x=\frac{-31}{-3}=\frac{31}{3}$
The solution set is $\left\{\frac{31}{3}\right\}$.

10. $0.05x+0.06(1500-x)=83.5$
$100[0.05x+0.06(1500-x)]=100(83.5)$
$100(0.05x)+100[0.06(1500-x)]=8350$
$5x+6(1500-x)=8350$
$5x+9000-6x=8350$
$-x+9000=8350$
$-x=-650$
$x=650$
The solution set is $\{650\}$.

11. $\frac{2}{3}x - \frac{3}{4}y = 2$ for y

$12\left(\frac{2}{3}x - \frac{3}{4}y\right) = 12(2)$

$12\left(\frac{2}{3}x\right) - 12\left(\frac{3}{4}y\right) = 24$

$8x - 9y = 24$

$-9y = -8x + 24$

$y = \frac{-8x + 24}{-9}$

$y = \frac{-1(-8x + 24)}{-1(-9)}$

$y = \frac{8x - 24}{9}$

12. $S = 2\pi r(r + h)$ for h

$S = 2\pi r^2 + 2\pi rh$

$S - 2\pi r^2 = 2\pi rh$

$\frac{S - 2\pi r^2}{2\pi r} = h$

13. $7x - 4 > 5x - 8$
$7x > 5x - 4$
$2x > -4$
$x > -2$
The solution set is $(-2, \infty)$.

14. $-3x - 4 \leq x + 12$
$-3x \leq x + 16$
$-4x \leq 16$
$-\frac{1}{4}(-4x) \geq -\frac{1}{4}(16)$
$x \geq -4$
The solution set is $[-4, \infty)$.

15. $2(x - 1) - 3(3x + 1) \geq -6(x - 5)$
$2x - 2 - 9x - 3 \geq -6x + 30$
$-7x - 5 \geq -6x + 30$
$-7x \geq -6x + 35$
$-x \geq 35$
$-1(-x) \leq -1(35)$
$x \leq -35$
The solution set is $(-\infty, -35]$.

16. $\frac{3}{5}x - \frac{1}{2}x < 1$
$10\left(\frac{3}{5}x - \frac{1}{2}x\right) < 10(1)$
$10\left(\frac{3}{5}x\right) - 10\left(\frac{1}{2}x\right) < 10$
$6x - 5x < 10$
$x < 10$
The solution set is $(-\infty, 10)$.

17. $\frac{x - 2}{6} - \frac{x + 3}{9} > -\frac{1}{2}$
$18\left(\frac{x - 2}{6} - \frac{x + 3}{9}\right) > 18\left(-\frac{1}{2}\right)$
$18\left(\frac{x - 2}{6}\right) - 18\left(\frac{x + 3}{9}\right) > -9$
$3(x - 2) - 2(x + 3) > -9$
$3x - 6 - 2x - 6 > -9$
$x - 12 > -9$
$x > 3$
The solution set is $(3, \infty)$.

18. $0.05x + 0.07(800 - x) \geq 52$
$100[0.05x + 0.07(800 - x)] \geq 100(52)$
$100(0.05x) + 100[0.07(800 - x)] \geq 5200$
$5x + 7(800 - x) \geq 5200$
$5x + 5600 - 7x \geq 5200$
$-2x + 5600 \geq 5200$
$-2x \geq -400$
$-\frac{1}{2}(-2x) \leq -\frac{1}{2}(-400)$
$x \leq 200$
The solution set is $(-\infty, 200]$.

19. $|6x - 4| < 10$
$-10 < 6x - 4 < 10$
$-6 < 6x < 14$
$-1 < x < \frac{14}{6}$
$-1 < x < \frac{7}{3}$
The solution set is $\left(-1, \frac{7}{3}\right)$.

20. $|4x + 5| \geq 6$

$4x + 5 \leq -6$ or $4x + 5 \geq 6$

$4x \leq -11$ or $4x \geq 1$

$x \leq -\frac{11}{4}$ or $x \geq \frac{1}{4}$

The solution set is

$\left(-\infty, -\frac{11}{4}\right] \cup \left[\frac{1}{4}, \infty\right).$

21. Let $x =$ original price.

$(100\%)(x) - (20\%)(x) = 57.60$

$1.00x - 0.20x = 57.60$

$100[1.00x - 0.20x] = 100[57.60]$

$100x - 20x = 5760$

$80x = 5760$

$x = 72$

The original price is \$72.

22. Let $x =$ width and

$3x + 1 =$ length.

$P = 2L + 2W$

$50 = 2(3x + 1) + 2(x)$

$50 = 6x + 2 + 2x$

$50 = 8x + 2$

$48 = 8x$

$6 = x$

width $=$ 6 centimeters

length $= 3(6) + 1 = 19$ centimeters

The width is 6 centimeters and the length is 19 centimeters.

23. Let $x =$ number of cups of grapefruit juice.

$100\%(x) + (8\%)(30) = (10\%)(x + 30)$

$1.00x + 0.08(30) = 0.10(x + 30)$

$100[1.00x + 0.08(30)] = 100[0.10(x + 30)]$

$100(1.00x) + 100[0.08(30)] = 10(x + 30)$

$100x + 8(30) = 10x + 300$

$100x + 240 = 10x + 300$

$100x = 10x + 60$

$90x = 60$

$x = \frac{60}{90} = \frac{2}{3}$

There should be $\frac{2}{3}$ cup of grapefruit juice added.

24. Let $x =$ score on 6th exam

$\frac{85 + 92 + 87 + 88 + 91 + x}{6} \geq 90$

$\frac{443 + x}{6} \geq 90$

$6\left(\frac{443 + x}{6}\right) \geq 6(90)$

$443 + x \geq 540$

$x \geq 97$

The score should be 97 or better.

25. Let $x =$ angle,

$90 - x =$ complement of the angle and

$180 - x =$ supplement of the angle.

$90 - x = \frac{2}{11}(180 - x)$

$11(90 - x) = 11\left[\frac{2}{11}(180 - x)\right]$

$11(90) - 11(x) = 2(180 - x)$

$990 - 11x = 360 - 2x$

$-11x = -630 - 2x$

$-9x = -630$

$x = 70$

The angle is 70°.

Chapter 3 Polynomials

PROBLEM SET 3.1 Polynomials: Sums and Differences

1. $7xy + 6y$
The degree is the sum of the exponents of the $7xy$ term.
The degree is 2.

3. $-x^2y + 2xy^2 - xy$
The degree is the sum of the exponents of either the $-x^2y$ or $2x^2y$ term. The degree is 3.

5. $5x^2 - 7x - 2$
The degree is the exponent of the $5x^2$ term. The degree is 2.

7. $8x^6 + 9$
The degree is the exponent of the $8x^6$ term. The degree is 6.

9. $-12 = -12x^0$
The degree is the exponent.
The degree is 0.

11. $(3x - 7) + (7x + 4)$
$3x - 7 + 7x + 4$
$3x + 7x - 7 + 4$
$10x - 3$

13. $(-5t - 4) + (-6t + 9)$
$-5t - 4 - 6t + 9$
$-5t - 6t - 4 + 9$
$-11t + 5$

15. $(3x^2 - 5x - 1) + (-4x^2 + 7x - 1)$
$3x^2 - 5x - 1 - 4x^2 + 7x - 1$
$3x^2 - 4x^2 - 5x + 7x - 1 - 1$
$-x^2 + 2x - 2$

17. $(12a^2b^2 - 9ab) + (5a^2b^2 + 4ab)$
$12a^2b^2 - 9ab + 5a^2b^2 + 4ab$
$12a^2b^2 + 5a^2b^2 - 9ab + 4ab$
$17a^2b^2 - 5ab$

19. $(2x - 4) + (-7x + 2) + (-4x + 9)$
$2x - 4 - 7x + 2 - 4x + 9$
$2x - 7x - 4x - 4 + 2 + 9$
$-9x + 7$

21. $(3x + 4) - (5x - 2)$
$3x + 4 - 5x + 2$
$3x - 5x + 4 + 2$
$-2x + 6$

23. $(6a + 2) - (-4a - 5)$
$6a + 2 + 4a + 5$
$6a + 4a + 2 + 5$
$10a + 7$

25. $(7x^2 + 9x + 8) - (3x^2 - x + 2)$
$7x^2 + 9x + 8 - 3x^2 + x - 2$
$7x^2 - 3x^2 + 9x + x + 8 - 2$
$4x^2 + 10x + 6$

27. $(-4a^2 + 6a + 10) - (2a^2 - 6a - 4)$
$-4a^2 + 6a + 10 - 2a^2 + 6a + 4$
$-4a^2 - 2a^2 + 6a + 6a + 10 + 4$
$-6a^2 + 12a + 14$

29. $(5x^3 + 2x^2 + 6x - 13) - (2x^3 + x^2 - 7x - 2)$
$5x^3 + 2x^2 + 6x - 13 - 2x^3 - x^2 + 7x + 2$
$5x^3 - 2x^3 + 2x^2 - x^2 + 6x + 7x - 13 + 2$
$3x^3 + x^2 + 13x - 11$

31.
$$\begin{array}{r} 12x + 6 \\ -(5x - 2) \\ \hline 7x + 8 \end{array}$$

33.
$$\begin{array}{r} -7x - 9 \\ -(-4x + 7) \\ \hline -3x - 16 \end{array}$$

35.
$$\begin{array}{r} 4x^2 - x - 2 \\ -(2x^2 + x + 6) \\ \hline 2x^2 - 2x - 8 \end{array}$$

37.
$$\begin{array}{r} -2x^3+6x^2-3x+8 \\ -(x^3+\ \ x^2-\ \ x-1) \\ \hline -3x^3+5x^2-2x+9 \end{array}$$

39.
$$\begin{array}{r} 2x-1 \\ -(-5x^2+6x-12) \\ \hline 5x^2-4x+11 \end{array}$$

41. $(x^2+9x-4)+(-5x^2-7x+10)$
$-(2x^2-7x-1)$
$x^2+9x-4-5x^2-7x+10-2x^2+7x+1$
$x^2-5x^2-2x^2+9x-7x+7x-4+10+1$
$-6x^2+9x+7$

43. $(4x^2+3)+(-7x^2+2x)$
$-(-x^2-7x-1)$
$4x^2+3-7x^2+2x+x^2+7x+1$
$4x^2-7x^2+x^2+2x+7x+3+1$
$-2x^2+9x+4$

45. $(-12n^2-n+9)-[(5n^2-3n-2)$
$+(-7n^2+n+2)]$
$-12n^2-n+9-[5n^2-3n-2-7n^2+n+2]$
$-12n^2-n+9-(-2n^2-2n)$
$-12n^2-n+9+2n^2+2n$
$-12n^2+2n^2-n+2n+9$
$-10n^2+n+9$

47. $(5x+2)+(7x-1)+(-4x-3)$
$5x+2+7x-1-4x-3$
$5x+7x-4x+2-1-3$
$8x-2$

49. $(12x-9)-(-3x+4)-(7x+1)$
$12x-9+3x-4-7x-1$
$12x+3x-7x-9-4-1$
$8x-14$

51. $(2x^2-7x-1)+(-4x^2-x+6)$
$+(-7x^2-4x-1)$
$2x^2-7x-1-4x^2-x+6-7x^2-4x-1$
$2x^2-4x^2-7x^2-7x-x-4x-1+6-1$
$-9x^2-12x+4$

53. $(7x^2-x-4)-(9x^2-10x+8)$
$+(12x^2+4x-6)$
$7x^2-x-4-9x^2+10x-8+12x^2+4x-6$
$7x^2-9x^2+12x^2-x+10x+4x-4-8-6$
$10x^2+13x-18$

55. $(n^2-7n-9)-(-3n+4)-(2n^2-9)$
$n^2-7n-9+3n-4-2n^2+9$
$n^2-2n^2-7n+3n-9-4+9$
$-n^2-4n-4$

57. $3x-[5x-(x+6)]$
$3x-[5x-x-6]$
$3x-[4x-6]$
$3x-4x+6$
$-x+6$

59. $2x^2-[-3x^2-(x^2-4)]$
$2x^2-[-3x^2-x^2+4]$
$2x^2-[-4x^2+4]$
$2x^2+4x^2-4$
$6x^2-4$

61. $-2n^2-[n^2-(-4n^2+n+6)]$
$-2n^2-[n^2+4n^2-n-6]$
$-2n^2-[5n^2-n-6]$
$-2n^2-5n^2+n+6$
$-7n^2+n+6$

63. $[4t^2-(2t+1)+3]-[3t^2+(2t-1)-5]$
$[4t^2-2t-1+3]-[3t^2+2t-1-5]$
$4t^2-2t+2-(3t^2+2t-6)$
$4t^2-2t+2-3t^2-2t+6$
$4t^2-3t^2-2t-2t+2+6$
t^2-4t+8

65. $[2n^2-(2n^2-n+5)]+[3n^2+(n^2-2n-7)]$
$[2n^2-2n^2+n-5]+[3n^2+n^2-2n-7]$
$(n-5)+(4n^2-2n-7)$
$n-5+4n^2-2n-7$
$4n^2+n-2n-5-7$
$4n^2-n-12$

67. $[7xy-(2x-3xy+y)]-[3x-(x-10xy-y)]$
$[7xy-2x+3xy-y]-[3x-x+10xy+y]$
$(10xy-2x-y)-(2x+10xy+y)$
$10xy-2x-y-2x-10xy-y$
$10xy-10xy-2x-2x-y-y$
$-4x-2y$

69. $[4x^3-(2x^2-x-1)]-[5x^3-(x^2+2x-1)]$
$[4x^3-2x^2+x+1]-[5x^3-x^2-2x+1]$
$4x^3-2x^2+x+1-5x^3+x^2+2x-1$
$4x^3-5x^3-2x^2+x^2+x+2x+1-1$
$-x^3-x^2+3x$

71a. Rectangle
$P=2l+2w$
$P=2(3x-2)+2(x+4)$
$P=6x-4+2x+8$
$P=8x+4$

b. $P=3x+x+3+x+x+1+2x+x+2+4$
$P=9x+10$

c. Triangle
$P=\mathrm{s}+\mathrm{s}+\mathrm{s}$
$P=4x+2+4x+2+4x+2$
$P=12x+6$

73. $S=2\pi(4)^2+2\pi(4)h$
$S=32\pi+8\pi h$

a. $h=5$
$S=32(3.14)+8(3.14)(5)$
$S=100.48+125.6$
$S=226.08\approx 226.1$

b. $h=7$
$S=32(3.14)+8(3.14)(7)$
$S=100.48+175.84$
$S=276.32\approx 276.3$

c. $h=14$
$S=32(3.14)+8(3.14)(14)$
$S=100.48+351.68$
$S=452.16\approx 452.2$

d. $h=18$
$S=32(3.14)+8(3.14)(18)$
$S=100.48+452.16$
$S=552.64\approx 552.6$

PROBLEM SET 3.2 Products and Quotients of Monomials

1. $(4x^3)(9x)$
$36x^{3+1}$
$36x^4$

3. $(-2x^2)(6x^3)$
$-12x^{2+3}$
$-12x^5$

5. $(-a^2b)(-4ab^3)$
$4a^{2+1}b^{1+3}$
$4a^3b^4$

7. $(x^2yz^2)(-3xyz^4)$
$-3x^{2+1}y^{1+1}z^{2+4}$
$-3x^3y^2z^6$

9. $(5xy)(-6y^3)$
$-30xy^{1+3}$
$-30xy^4$

11. $(3a^2b)(9a^2b^4)$
$27a^{2+2}b^{1+4}$
$27a^4b^5$

13. $(m^2n)(-mn^2)$
$-m^{2+1}n^{1+2}$
$-m^3m^3$

15. $\left(\frac{2}{5}xy^2\right)\left(\frac{3}{4}x^2y^4\right)$
$\frac{6}{20}x^{1+2}y^{2+4}$
$\frac{3}{10}x^3y^6$

17. $\left(-\frac{3}{4}ab\right)\left(\frac{1}{5}a^2b^3\right)$
$-\frac{3}{20}a^{1+2}b^{1+3}$
$-\frac{3}{20}a^3b^4$

19. $\left(-\frac{1}{2}xy\right)\left(\frac{1}{3}x^2y^3\right)$
$-\frac{1}{6}x^{1+2}y^{1+3}$
$-\frac{1}{6}x^3y^4$

21. $(3x)(-2x^2)(-5x^3)$
$30x^{1+2+3}$
$30x^6$

23. $(-6x^2)(3x^3)(x^4)$
$-18x^{2+3+4}$
$-18x^9$

25. $(x^2y)(-3xy^2)(x^3y^3)$
$-3x^{2+1+3}y^{1+2+3}$
$-3x^6y^6$

27. $(-3y^2)(-2y^2)(-4y^5)$
$-24y^{2+2+5}$
$-24y^9$

29. $(4ab)(-2a^2b)(7a)$
$-56a^{1+2+1}b^{1+1}$
$-56a^4b^2$

31. $(-ab)(-3ab)(-6ab)$
$-18a^{1+1+1}b^{1+1+1}$
$-18a^3b^3$

33. $\left(\frac{2}{3}xy\right)(-3x^2y)(5x^4y^5)$
$-10x^{1+2+4}y^{1+1+5}$
$-10x^7y^7$

35. $(12y)(-5x)\left(-\frac{5}{6}x^4y\right)$
$50x^{1+4}y^{1+1}$
$50x^5y^2$

37. $(3xy^2)^3$
$3^3x^{1(3)}y^{2(3)}$
$27x^3y^6$

39. $(-2x^2y)^5$
$(-2)^5x^{2(5)}y^{1(5)}$
$-32x^{10}y^5$

41. $(-x^4y^5)^4$
$(-1)^4x^{4(4)}y^{5(4)}$
$x^{16}y^{20}$

43. $(ab^2c^3)^6$
$a^{1(6)}b^{2(6)}c^{3(6)}$
$a^6b^{12}c^{18}$

45. $(2a^2b^3)^6$
$2^{1(6)}a^{2(6)}b^{3(6)}$
$2^6a^{12}b^{18}$
$64a^{12}b^{18}$

47. $(9xy^4)^2$
$9^2x^{1(2)}y^{4(2)}$
$81x^2y^8$

49. $(-3ab^3)^4$
$(-3)^4a^{1(4)}b^{3(4)}$
$81a^4b^{12}$

51. $-(2ab)^4$
$-[2^4a^{1(4)}b^{1(4)}]$
$-(16a^4b^4)$
$-16a^4b^4$

53. $-(xy^2z^3)^6$
$-[x^6y^{2(6)}z^{3(6)}]$
$-[x^6y^{12}z^{18}]$
$-x^6y^{12}z^{18}$

55. $(-5a^2b^2c)^3$
$(-5)^3a^{2(3)}b^{2(3)}c^{1(3)}$
$-125a^6b^6c^3$

57. $(-xy^4z^2)^7$
$(-1)^7x^{1(7)}y^{4(7)}z^{2(7)}$
$-x^7y^{28}z^{14}$

59. $\frac{9x^4y^5}{3xy^2}$
$3x^{4-1}y^{5-2}$
$3x^3y^3$

Problem Set 3.2

61. $\dfrac{25x^5y^6}{-5x^2y^4}$

$-5x^{5-2}y^{6-4}$

$-5x^3y^2$

63. $\dfrac{-54ab^2c^3}{-6abc}$

$9a^{1-1}b^{2-1}c^{3-1}$

$9a^0bc^2$

$9(1)bc^2$

$9bc^2$

65. $\dfrac{-18x^2y^2z^6}{xyz^2}$

$-18x^{2-1}y^{2-1}z^{6-2}$

$-18xyz^4$

67. $\dfrac{a^3b^4c^7}{-abc^5}$

$-a^{3-1}b^{4-1}c^{7-5}$

$-a^2b^3c^2$

69. $\dfrac{-72x^2y^4}{-8x^2y^4}$

$9x^{2-2}y^{4-4}$

$9x^0y^0$

$9(1)(1)$

9

71. $\dfrac{14ab^3}{-14ab}$

$-1a^{1-1}b^{3-1}$

$-a^0b^2$

$-(1)(b^2)$

$-b^2$

73. $\dfrac{-36x^3y^5}{2y^5}$

$-18x^3y^{5-5}$

$-18x^3y^0$

$-18x^3(1)$

$-18x^3$

75. $(2x^n)(3x^{2n})$

$6x^{n+2n}$

$6x^{3n}$

77. $(a^{2n-1})(a^{3n+4})$

$a^{2n-1+3n+4}$

a^{5n+3}

79. $(x^{3n-2})(x^{n+2})$

$x^{3n-2+n+2}$

x^{4n}

81. $(a^{5n-2})(a^3)$

a^{5n-2+3}

a^{5n+1}

83. $(2x^n)(-5x^n)$

$-10x^{n+n}$

$-10x^{2n}$

85. $(-3a^2)(-4a^{n+2})$

$12a^{2+n+2}$

$12a^{n+4}$

87. $(x^n)(2x^{2n})(3x^2)$

$6x^{n+2n+2}$

$6x^{3n+2}$

89. $(3x^{n-1})(x^{n+1})(4x^{2-n})$

$12x^{n-1+n+1+2-n}$

$12x^{n+2}$

91. length $= 2x$

width $= 3x$

height $= x$

$SA = 2lw + 2lh + 2wh$

$SA = 2(2x)(3x) + 2(2x)(x) + 2(3x)(x)$

$SA = 12x^2 + 4x^2 + 6x^2$

$SA = 22x^2$

$V = lwh$

$V = (2x)(3x)(x)$

$V = 6x^3$

93. Area of larger circle $= \pi r^2$

Area of smaller circle $= \pi(6)^2 = 36\pi$

Area of shaded region $= \pi r^2 - 36\pi$

PROBLEM SET 3.3 Multiplying Polynomials

1. $2xy(5xy^2 + 3x^2y^3)$
$2xy(5xy^2) + 2xy(3x^2y^3)$
$10x^2y^3 + 6x^3y^4$

3. $-3a^2b(4ab^2 - 5a^3)$
$-3a^2b(4ab^2) - 3a^2b(-5a^3)$
$-12a^3b^3 + 15a^5b$

5. $8a^3b^4(3ab - 2ab^2 + 4a^2b^2)$
$8a^3b^4(3ab) + 8a^3b^4(-2ab^2) + 8a^3b^4(4a^2b^2)$
$24a^4b^5 - 16a^4b^6 + 32a^5b^6$

7. $-x^2y(6xy^2 + 3x^2y^3 - x^3y)$
$-x^2y(6xy^2)+(-x^2y)(3x^2y^3)+(-x^2y)(-x^3y)$
$-6x^3y^3 - 3x^4y^4 + x^5y^2$

9. $(a + 2b)(x + y)$
$a(x + y) + 2b(x + y)$
$ax + ay + 2bx + 2by$

11. $(a - 3b)(c + 4d)$
$a(c + 4d) - 3b(c + 4d)$
$ac + 4ad - 3bc - 12bd$

13. $(x + 6)(x + 10)$
$x(x + 10) + 6(x + 10)$
$x^2 + 10x + 6x + 60$
$x^2 + 16x + 60$

15. $(y - 5)(y + 11)$
$y(y + 11) - 5(y + 11)$
$y^2 + 11y - 5y - 55$
$y^2 + 6y - 55$

17. $(n + 2)(n - 7)$
$n(n - 7) + 2(n - 7)$
$n^2 - 7n + 2n - 14$
$n^2 - 5n - 14$

19. $(x + 6)(x - 6)$
$x(x - 6) + 6(x - 6)$
$x^2 - 6x + 6x - 36$
$x^2 - 36$

21. $(x - 6)^2$
$(x)^2 + 2(x)(-6) + (-6)^2$
$x^2 - 12x + 36$

23. $(x - 6)(x - 8)$
$x(x - 8) - 6(x - 8)$
$x^2 - 8x - 6x + 48$
$x^2 - 14x + 48$

25. $(x + 1)(x - 2)(x - 3)$
$(x + 1)[x(x - 3) - 2(x - 3)]$
$(x + 1)[x^2 - 3x - 2x + 6]$
$(x + 1)(x^2 - 5x + 6)$
$x(x^2 - 5x + 6) + 1(x^2 - 5x + 6)$
$x^3 - 5x^2 + 6x + x^2 - 5x + 6$
$x^3 - 4x^2 + x + 6$

27. $(x - 3)(x + 3)(x - 1)$
$(x - 3)[x(x - 1) + 3(x - 1)]$
$(x - 3)[x^2 - x + 3x - 3]$
$(x - 3)(x^2 + 2x - 3)$
$x(x^2 + 2x - 3) - 3(x^2 + 2x - 3)$
$x^3 + 2x^2 - 3x - 3x^2 - 6x + 9$
$x^3 - x^2 - 9x + 9$

29. $(t + 9)^2$
$(t)^2 + 2(t)(9) + (9)^2$
$t^2 + 18t + 81$

31. $(y - 7)^2$
$(y)^2 + 2(y)(-7) + (-7)^2$
$y^2 - 14y + 49$

33. $(4x + 5)(x + 7)$
$4x(x + 7) + 5(x + 7)$
$4x^2 + 28x + 5x + 35$
$4x^2 + 33x + 35$

35. $(3y - 1)(3y + 1)$
$3y(3y + 1) - 1(3y + 1)$
$9y^2 + 3y - 3y - 1$
$9y^2 - 1$

37. $(7x - 2)(2x + 1)$
$7x(2x + 1) - 2(2x + 1)$
$14x^2 + 7x - 4x - 2$
$14x^2 + 3x - 2$

Problem Set 3.3

39. $(1+t)(5-2t)$
$1(5-2t)+t(5-2t)$
$5-2t+5t-2t^2$
$5+3t-2t^2$

41. $(3t+7)^2$
$(3t)^2+2(3t)(7)+(7)^2$
$9t^2+42t+49$

43. $(2-5x)(2+5x)$
$2(2+5x)-5x(2+5x)$
$4+10x-10x-25x^2$
$4-25x^2$

45. $(7x-4)^2$
$(7x)^2+2(7x)(-4)+(-4)^2$
$49x^2-56x+16$

47. $(6x+7)(3x-10)$
$6x(3x-10)+7(3x-10)$
$18x^2-60x+21x-70$
$18x^2-39x-70$

49. $(2x-5y)(x+3y)$
$2x(x+3y)-5y(x+3y)$
$2x^2+6xy-5xy-15y^2$
$2x^2+xy-15y^2$

51. $(5x-2a)(5x+2a)$
$5x(5x+2a)-2a(5x+2a)$
$25x^2+10ax-10ax-4a^2$
$25x^2-4a^2$

53. $(t+3)(t^2-3t-5)$
$t(t^2-3t-5)+3(t^2-3t-5)$
$t^3-3t^2-5t+3t^2-9t-15$
$t^3-14t-15$

55. $(x-4)(x^2+5x-4)$
$x(x^2+5x-4)-4(x^2+5x-4)$
$x^3+5x^2-4x-4x^2-20x+16$
$x^3+x^2-24x+16$

57. $(2x-3)(x^2+6x+10)$
$2x(x^2+6x+10)-3(x^2+6x+10)$
$2x^3+12x^2+20x-3x^2-18x-30$
$2x^3+9x^2+2x-30$

59. $(4x-1)(3x^2-x+6)$
$4x(3x^2-x+6)-1(3x^2-x+6)$
$12x^3-4x^2+24x-3x^2+x-6$
$12x^3-7x^2+25x-6$

61. $(x^2+2x+1)(x^2+3x+4)$
$x^2(x^2+3x+4)+2x(x^2+3x+4)$
$\quad+1(x^2+3x+4)$
$x^4+3x^3+4x^2+2x^3+6x^2+8x$
$\quad+x^2+3x+4$
$x^4+5x^3+11x^2+11x+4$

63. $(2x^2+3x-4)(x^2-2x-1)$
$2x^2(x^2-2x-1)+3x(x^2-2x-1)$
$\quad-4(x^2-2x-1)$
$2x^4-4x^3-2x^2+3x^3-6x^2-3x$
$\quad-4x^2+8x+4$
$2x^4-x^3-12x^2+5x+4$

65. $(x+2)^3$
$(x)^3+3(x)^2(2)+3(x)(2)^2+(2)^3$
$x^3+6x^2+12x+8$

67. $(x-4)^3$
$(x)^3+3(x)^2(-4)+3(x)(-4)^2+(-4)^3$
$x^3-12x^2+48x-64$

69. $(2x+3)^3$
$(2x)^3+3(2x)^2(3)+3(2x)(3)^2+(3)^3$
$8x^3+3(4x^2)(3)+3(2x)(9)+27$
$8x^3+36x^2+54x+27$

71. $(4x-1)^3$
$(4x)^3+3(4x)^2(-1)+3(4x)(-1)^2+(-1)^3$
$64x^3+3(16x^2)(-1)+3(4x)(1)-1$
$64x^3-48x^2+12x-1$

73. $(5x+2)^3$
$(5x)^3+3(5x)^2(2)+3(5x)(2)^2+(2)^3$
$125x^3+3(25x^2)(2)+3(5x)(4)+8$
$125x^3+150x^2+60x+8$

75. $(x^n-4)(x^n+4)$
$x^n(x^n+4)-4(x^n+4)$
$x^{2n}+4x^n-4x^n-16$
$x^{2n}-16$

77. $(x^a + 6)(x^a - 2)$
$x^a(x^a - 2) + 6(x^a - 2)$
$x^{2a} - 2x^a + 6x^a - 12$
$x^{2a} + 4x^a - 12$

79. $(2x^n + 5)(3x^n - 7)$
$2x^n(3x^n - 7) + 5(3x^n - 7)$
$6x^{2n} - 14x^n + 15x^n - 35$
$6x^{2n} + x^n - 35$

81. $(x^{2a} - 7)(x^{2a} - 3)$
$x^{2a}(x^{2a} - 3) - 7(x^{2a} - 3)$
$x^{4a} - 3x^{2a} - 7x^{2a} + 21$
$x^{4a} - 10x^{2a} + 21$

83. $(2x^n + 5)^2$
$(2x^n)^2 + 2(2x^n)(5) + (5)^2$
$4x^{2n} + 20x^n + 25$

85. length $= x + 2$
width $= x + 6$

$A = lw$
$A = (x + 2)(x + 6)$

There are four rectangles with the following dimensions and areas.

x by x	$A = x^2$
x by 6	$A = 6x$
2 by x	$A = 2x$
2 by 6	$A = 12$

The sum of the four rectangles equals the area of the figure.
$A = x^2 + 6x + 2x + 12$

From above A $= (x + 2)(x + 6)$, therefore
$(x + 2)(x + 6) = x^2 + 6x + 2x + 12$
$(x + 2)(x + 6) = x^2 + 8x + 12$

87. Area of larger rectangle
$A = lw$
$A = (2x + 3)(x)$
$A = 2x^2 + 3x$
Area of smaller rectangle
$A = lw$
$A = 3(x - 2)$
$A = 3x - 6$
Area of shaded portion
$A = (2x^2 + 3x) - (3x - 6)$
$A = 2x^2 + 3x - 3x + 6$
$A = 2x^2 + 6$

89. Dimensions of the box
length $= 16 - 2x$
width $= 16 - 2x$
height $= x$

$V = lwh$
$V = (16 - 2x)(16 - 2x)(x)$
$V = (16 - 2x)(16x - 2x^2)$
$V = 16(16x - 2x^2) - 2x(16x - 2x^2)$
$V = 256x - 32x^2 - 32x^2 + 4x^3$
$V = 4x^3 - 64x^2 + 256x$

Surface Area = area of the bottom plus the area of the four sides.

Dimensions of the bottom
length $= 16 - 2x$
width $= 16 - 2x$
Area $= (16 - 2x)(16 - 2x)$
Area $= 256 - 64x + 4x^2$

Dimensions of the sides
length $= 16 - 2x$
width $= x$
Area $= (16 - 2x)(x)$
Area $= 16x - 2x^2$

Surface Area
$SA = (256 - 64x + 4x^2) + 4(16x - 2x^2)$
$SA = 256 - 64x + 4x^2 + 64x - 8x^2$
$SA = 256 - 4x^2$

93a) $(a + b)^6$

$= 1a^6b^0 + 6a^5b^1 + 15a^4b^2 + 20a^3b^3 + 15a^2b^4 + 6a^1b^5 + 1a^0b^6$

$= a^6 + 6a^5b + 15a^4b^2 + 20a^3b^3 + 15a^2b^4 + 6ab^5 + b^6$

b) $(a+b)^7$

$= 1a^7b^0 + 7a^6b^1 + 21a^5b^2 + 35a^4b^3 + 35a^3b^4 + 21a^2b^5 + 7a^1b^6 + 1a^0b^7$

$= a^7 + 7a^6b + 21a^5b^2 + 35a^4b^3 + 35a^3b^4 + 21a^2b^5 + 7ab^6 + b^7$

c) $(a+b)^8$

$= 1a^8b^0 + 8a^7b^1 + 28a^6b^2 + 56a^5b^3 + 70a^4b^4 + 56a^3b^5 + 28a^2b^6 + 8a^1b^7 + 1a^0b^8$

$= a^8 + 8a^7b + 28a^6b^2 + 56a^5b^3 + 70a^4b^4 + 56a^3b^5 + 28a^2b^6 + 8ab^7 + b^8$

d) $(a+b)^9$

$= 1a^9b^0 + 9a^8b^1 + 36a^7b^2 + 84a^6b^3 + 126a^5b^4 + 126a^4b^5 + 84a^3b^6 + 36a^2b^7 + 9a^1b^8 + 1a^0b^9$

$= a^9 + 9a^8b + 36a^7b^2 + 84a^6b^3 + 126a^5b^4 + 126a^4b^5 + 84a^3b^6 + 36a^2b^7 + 9ab^8 + b^9$

95a) $21^2 = (20+1)^2 = 400 + 40 + 1 = 441$
b) $41^2 = (40+1)^2 = 1600 + 80 + 1 = 1681$
c) $71^2 = (70+1)^2 = 4900 + 140 + 1 = 5041$
d) $32^2 = (30+2)^2 = 900 + 120 + 4 = 1024$
e) $52^2 = (50+2)^2 = 2500 + 200 + 4 = 2704$
f) $82^2 = (80+2)^2 = 6400 + 320 + 4 = 6724$

97a) $15^2 = [10(1)+5]^2 = 100(1)(1+1) + 25 = 225$

b) $25^2 = [10(2)+5]^2 = 100(2)(2+1) + 25 = 625$

c) $45^2 = [10(4)+5]^2 = 100(4)(4+1) + 25 = 2025$

d) $55^2 = [10(5)+5]^2 = 100(5)(5+1) + 25 = 3025$

e) $65^2 = [10(6)+5]^2 = 100(6)(6+1) + 25 = 4225$

f) $75^2 = [10(7)+5]^2 = 100(7)(7+1) + 25 = 5625$

g) $85^2 = [10(8)+5]^2 = 100(8)(8+1) + 25 = 7225$

h) $95^2 = [10(9)+5]^2 = 100(9)(9+1) + 25 = 9025$

i) $105^2 = [10(10)+5]^2 = 100(10)(10+1) + 25 = 11025$

PROBLEM SET 3.4 Factoring: Use of the Distributive Property

1. $63 = 9 \bullet 7$ composite
3. 59 prime
5. $51 = 3 \bullet 17$ composite
7. $91 = 7 \bullet 13$ composite
9. 71 prime

11. 28
$4 \bullet 7$
$2 \bullet 2 \bullet 7$
$2^2 \bullet 7$

13. 44
$4 \bullet 11$
$2 \bullet 2 \bullet 11$
$2^2 \bullet 11$

15. 56
$8 \bullet 7$
$2 \bullet 4 \bullet 7$
$2 \bullet 2 \bullet 2 \bullet 7$
$2^3 \bullet 7$

17. 72
$8 \bullet 9$
$2 \bullet 4 \bullet 3 \bullet 3$
$2 \bullet 2 \bullet 2 \bullet 3 \bullet 3$
$2^3 \bullet 3^2$

19. 87
$3 \bullet 29$

21. $6x + 3y$
$3(2x) + 3(y)$
$3(2x + y)$

23. $6x^2 + 14x$
$2x(3x) + 2x(7)$
$2x(3x + 7)$

25. $28y^2 - 4y$
$4y(7y) + 4y(-1)$
$4y(7y - 1)$

27. $20xy - 15x$
$5x(4y) + 5x(-3)$
$5x(4y - 3)$

29. $7x^3 + 10x^2$
$x^2(7x) + x^2(10)$
$x^2(7x + 10)$

31. $18a^2b + 27ab^2$
$9ab(2a) + 9ab(3b)$
$9ab(2a + 3b)$

33. $12x^3y^4 - 39x^4y^3$
$3x^3y^3(4y) + 3x^3y^3(-13x)$
$3x^3y^3(4y - 13x)$

35. $8x^4 + 12x^3 - 24x^2$
$4x^2(2x^2) + 4x^2(3x) + 4x^2(-6)$
$4x^2(2x^2 + 3x - 6)$

37. $5x + 7x^2 + 9x^4$
$x(5) + x(7x) + x(9x^3)$
$x(5 + 7x + 9x^3)$

39. $15x^2y^3 + 20xy^2 + 35x^3y^4$
$5xy^2(3xy) + 5xy^2(4) + 5xy^2(7x^2y^2)$
$5xy^2(3xy + 4 + 7x^2y^2)$

41. $x(y + 2) + 3(y + 2)$
$(y + 2)(x + 3)$

43. $3x(2a + b) - 2y(2a + b)$
$(2a + b)(3x - 2y)$

45. $x(x + 2) + 5(x + 2)$
$(x + 2)(x + 5)$

47. $ax + 4x + ay + 4y$
$x(a + 4) + y(a + 4)$
$(a + 4)(x + y)$

49. $ax - 2bx + ay - 2by$
$x(a - 2b) + y(a - 2b)$
$(a - 2b)(x + y)$

51. $3ax - 3bx - ay + by$
$3x(a - b) - y(a - b)$
$(a - b)(3x - y)$

53. $2ax + 2x + ay + y$
$2x(a + 1) + y(a + 1)$
$(a + 1)(2x + y)$

55. $ax^2 - x^2 + 2a - 2$
$x^2(a - 1) + 2(a - 1)$
$(a - 1)(x^2 + 2)$

57. $2ac + 3bd + 2bc + 3ad$
$2ac + 2bc + 3bd + 3ad$
$2c(a + b) + 3d(b + a)$
$(a + b)(2c + 3d)$

59. $ax - by + bx - ay$
$ax + bx - ay - by$
$x(a + b) - y(a + b)$
$(a + b)(x - y)$

61. $x^2 + 9x + 6x + 54$
$x(x + 9) + 6(x + 9)$
$(x + 9)(x + 6)$

63. $2x^2 + 8x + x + 4$
$2x(x + 4) + 1(x + 4)$
$(x + 4)(2x + 1)$

65. $x^2 + 7x = 0$
$x(x + 7) = 0$
$x = 0$ or $x + 7 = 0$
$x = 0$ or $x = -7$
The solution set is $\{-7, 0\}$.

67. $x^2 - x = 0$
$x(x - 1) = 0$
$x = 0$ or $x - 1 = 0$
$x = 0$ or $x = 1$

The solution set is $\{0, 1\}$.

69. $a^2 = 5a$
$a^2 - 5a = 0$
$a(a-5) = 0$
$a = 0$ or $a - 5 = 0$
$a = 0$ or $a = 5$
The solution set is $\{0, 5\}$.

71. $-2y = 4y^2$
$0 = 4y^2 + 2y$
$0 = 2y(2y+1)$
$2y = 0$ or $2y + 1 = 0$
$y = 0$ or $2y = -1$
$y = 0$ or $y = -\frac{1}{2}$
The solution set is $\left\{-\frac{1}{2}, 0\right\}$.

73. $3x^2 + 7x = 0$
$x(3x+7) = 0$
$x = 0$ or $3x + 7 = 0$
$x = 0$ or $3x = -7$
$x = 0$ $x = -\frac{7}{3}$
The solution set is $\left\{-\frac{7}{3}, 0\right\}$.

75. $4x^2 = 5x$
$4x^2 - 5x = 0$
$x(4x-5) = 0$
$x = 0$ or $4x - 5 = 0$
$x = 0$ or $4x = 5$
$x = 0$ or $x = \frac{5}{4}$
The solution set is $\left\{0, \frac{5}{4}\right\}$.

77. $x - 4x^2 = 0$
$x(1-4x) = 0$
$x = 0$ or $1 - 4x = 0$
$x = 0$ or $1 = 4x$
$x = 0$ or $\frac{1}{4} = x$
The solution set is $\left\{0, \frac{1}{4}\right\}$.

79. $12a = -a^2$
$12a + a^2 = 0$
$a(12+a) = 0$
$a = 0$ or $12 + a = 0$
$a = 0$ or $a = -12$
The solution set is $\{-12, 0\}$.

81. $5bx^2 - 3ax = 0$ for x
$x(5bx - 3a) = 0$
$x = 0$ or $5bx - 3a = 0$
$x = 0$ or $5bx = 3a$
$x = 0$ or $x = \frac{3a}{5b}$
The solution set is $\left\{0, \frac{3a}{5b}\right\}$.

83. $2by^2 = -3ay$ for y
$2by^2 + 3ay = 0$
$y(2by + 3a) = 0$
$y = 0$ or $2by + 3a = 0$
$y = 0$ or $2by = -3a$
$y = 0$ or $y = -\frac{3a}{2b}$
The solution set is $\left\{-\frac{3a}{2b}, 0\right\}$.

85. $y^2 - ay + 2by - 2ab = 0$ for y
$y(y-a) + 2b(y-a) = 0$
$(y-a)(y+2b) = 0$
$y - a = 0$ or $y + 2b = 0$
$y = a$ or $y = -2b$
The solution set is $\{-2b, a\}$.

87. Let $n =$ number.
$n^2 = 7n$
$n^2 - 7n = 0$
$n(n-7) = 0$
$n = 0$ or $n - 7 = 0$
$n = 0$ or $n = 7$
The numbers are 0 and 7.

89. $A = \pi r^2$; $C = 2\pi r$
$\pi r^2 = 3(2\pi r)$
$\pi r^2 = 6\pi r$
$\pi r^2 - 6\pi r = 0$
$\pi r(r-6) = 0$
$\pi r = 0$ or $r - 6 = 0$
$r = 0$ or $r = 6$
Discard the root $r = 0$, therefore the radius should be 6 units.

91. Let $x =$ the length of the radius.
Area of circle $= \pi x^2$
Perimeter of square $= 4x$

$\pi x^2 = 4x$
$\pi x^2 - 4x = 0$
$x(\pi x - 4) = 0$
$x = 0$ or $\pi x - 4 = 0$
$x = 0$ or $\pi x = 4$
$x = 0$ or $x = \frac{4}{\pi}$
Discard the root $x = 0$, therefore the side of the square should be $\frac{4}{\pi}$ units.

93. Area of square $= x^2$
Area of rectangular lot $= x(50)$
Area of square lot $=$ Area of rectangular lot
$x^2 = 2[x(50)]$
$x^2 = 100x$
$x^2 - 100x = 0$
$x(x - 100) = 0$
$x = 0$ or $x - 100 = 0$
$x = 0$ or $x = 100$
The square is 100 feet by 100 feet and the rectangular lot is 50 feet by 100 feet.

95. Volume of sphere $= \frac{4}{3}\pi r^3$
Surface area of sphere $= 4\pi r^2$

$\frac{4}{3}\pi r^3 = 2(4\pi r^2)$
$\frac{4}{3}\pi r^3 = 8\pi r^2$
$\frac{4}{3}\pi r^3 - 8\pi r^2 = 0$
$4\pi r^2\left(\frac{1}{3}r - 2\right) = 0$
$4\pi r^2 = 0$ or $\frac{1}{3}r - 2 = 0$
$r^2 = 0$ or $\frac{1}{3}r = 2$
$r = 0$ or $r = 6$
Discard the root $r = 0$, therefore the radius should be 6 units.

101. $2x^{2a} - 3x^a$
$x^a(2x^a) + x^a(-3)$
$x^a(2x^a - 3)$

103. $y^{3m} + 5y^{2m}$
$y^{2m}(y^m) + y^{2m}(5)$
$y^{2m}(y^m + 5)$

105. $2x^{6a} - 3x^{5a} + 7x^{4a}$
$x^{4a}(2x^{2a}) + x^{4a}(-3x^a) + x^{4a}(7)$
$x^{4a}(2x^{2a} - 3x^a + 7)$

PROBLEM SET 3.5 Factoring: Difference of Two Squares and Sum or Difference of Two Cubes

1. $x^2 - 1$
$(x)^2 - (1)^2$
$(x + 1)(x - 1)$

3. $16x^2 - 25$
$(4x)^2 - (5)^2$
$(4x + 5)(4x - 5)$

5. $9x^2 - 25y^2$
$(3x)^2 - (5y)^2$
$(3x + 5y)(3x - 5y)$

7. $25x^2y^2 - 36$
$(5xy)^2 - (6)^2$
$(5xy + 6)(5xy - 6)$

9. $4x^2 - y^4$
$(2x)^2 - (y^2)^2$
$(2x + y^2)(2x - y^2)$

11. $1 - 144n^2$
$(1)^2 - (12n)^2$
$(1 + 12n)(1 - 12n)$

13. $(x + 2)^2 - y^2$
$(x + 2)^2 - (y)^2$
$(x + 2 + y)(x + 2 - y)$

15. $4x^2 - (y + 1)^2$
$(2x)^2 - (y + 1)^2$

$[2x + (y + 1)][2x - (y + 1)]$
$(2x + y + 1)(2x - y - 1)$

17. $9a^2 - (2b + 3)^2$
$(3a)^2 - (2b + 3)^2$
$[3a + (2b + 3)][3a - (2b + 3)]$
$(3a + 2b + 3)(3a - 2b - 3)$

19. $(x + 2)^2 - (x + 7)^2$
$[(x + 2) + (x + 7)][(x + 2) - (x + 7)]$
$(x + 2 + x + 7)(x + 2 - x - 7)$
$(2x + 9)(-5)$
$-5(2x + 9)$

21. $9x^2 - 36$
$9(x^2 - 4)$
$9(x + 2)(x - 2)$

23. $5x^2 + 5$
$5(x^2 + 1)$

25. $8y^2 - 32$
$8(y^2 - 4)$
$8(y + 2)(y - 2)$

27. $a^3b - 9ab$
$ab(a^2 - 9)$
$ab(a + 3)(a - 3)$

29. $16x^2 + 25$
Not factorable

31. $n^4 - 81$
$(n^2 + 9)(n^2 - 9)$
$(n^2 + 9)(n + 3)(n - 3)$

33. $3x^3 + 27x$
$3x(x^2 + 9)$

35. $4x^3y - 64xy^3$
$4xy(x^2 - 16y^2)$
$4xy(x + 4y)(x - 4y)$

37. $6x - 6x^3$
$6x(1 - x^2)$
$6x(1 + x)(1 - x)$

39. $1 - x^4y^4$
$(1 + x^2y^2)(1 - x^2y^2)$
$(1 + x^2y^2)(1 + xy)(1 - xy)$

41. $4x^2 - 64y^2$
$4(x^2 - 16y^2)$
$4(x + 4y)(x - 4y)$

43. $3x^4 - 48$
$3(x^4 - 16)$
$3(x^2 + 4)(x^2 - 4)$
$3(x^2 + 4)(x + 2)(x - 2)$

45. $a^3 - 64$
$(a)^3 - (4)^3$
$(a - 4)(a^2 + 4a + 16)$

47. $x^3 + 1$
$(x)^3 + (1)^3$
$(x + 1)(x^2 - x + 1)$

49. $27x^3 + 64y^3$
$(3x)^3 + (4y)^3$
$(3x + 4y)(9x^2 - 12xy + 16y^2)$

51. $1 - 27a^3$
$(1)^3 - (3a)^3$
$(1 - 3a)(1 + 3a + 9a^2)$

53. $x^3y^3 - 1$
$(xy)^3 - (1)^3$
$(xy - 1)(x^2y^2 + xy + 1)$

55. $x^6 - y^6$
$(x^3 + y^3)(x^3 - y^3)$
$(x + y)(x^2 - xy + y^2)(x - y)((x^2 + xy + y^2)$

57. $x^2 - 25 = 0$
$(x + 5)(x - 5) = 0$
$x + 5 = 0$ or $x - 5 = 0$
$x = -5$ or $x = 5$
The solution set is $\{-5, 5\}$.

59. $9x^2 - 49 = 0$
$(3x + 7)(3x - 7) = 0$
$3x + 7 = 0$ or $3x - 7 = 0$
$3x = -7$ or $3x = 7$

$x = -\frac{7}{3}$ or $x = \frac{7}{3}$

The solution set is $\left\{-\frac{7}{3}, \frac{7}{3}\right\}$.

61. $8x^2 - 32 = 0$
$8(x^2 - 4) = 0$
$8(x + 2)(x - 2) = 0$
$x + 2 = 0$ or $x - 2 = 0$
$x = -2$ or $x = 2$
The solution set is $\{-2, 2\}$.

63. $3x^3 = 3x$
$3x^3 - 3x = 0$
$3x(x^2 - 1) = 0$
$3x(x + 1)(x - 1) = 0$
$3x = 0$ or $x + 1 = 0$ or $x - 1 = 0$
$x = 0$ or $x = -1$ or $x = 1$
The solution set is $\{-1, 0, 1\}$.

65. $20 - 5x^2 = 0$
$5(4 - x^2) = 0$
$5(2 + x)(2 - x) = 0$
$2 + x = 0$ or $2 - x = 0$
$x = -2$ or $2 = x$
The solution set is $\{-2, 2\}$.

67. $x^4 - 81 = 0$
$(x^2 + 9)(x^2 - 9) = 0$
$(x^2 + 9)(x + 3)(x - 3) = 0$
$x^2 + 9 = 0$ or $x + 3 = 0$ or $x - 3 = 0$
$x^2 = -9$ or $x = -3$ or $x = 3$
not a real number $x = -3$ or $x = 3$
The solution set is $\{-3, 3\}$.

69. $6x^3 + 24x = 0$
$6x(x^2 + 4) = 0$
$6x = 0$ or $x^2 + 4 = 0$
$x = 0$ or $x^2 = -4$
$x = 0$ not a real number
The solution set is $\{0\}$.

71. Let x = number
$x^3 = 9x$
$x^3 - 9x = 0$
$x(x^2 - 9) = 0$
$x(x + 3)(x - 3) = 0$
$x = 0$ or $x + 3 = 0$ or $x - 3 = 0$
$x = 0$ or $x = -3$ or $x = 3$
The numbers are -3, 0, or 3.

73. Let r = radius of first circle and
$2r$ = radius of second circle.

Area of first circle $= \pi r^2$
Area of second circle $= \pi(2r)^2 = 4\pi r^2$

$\pi r^2 + 4\pi r^2 = 80\pi$
$5\pi r^2 = 80\pi$
$5\pi r^2 - 80\pi = 0$
$5\pi(r^2 - 16) = 0$
$r^2 - 16 = 0$
$(r + 4)(r - 4) = 0$
$r + 4 = 0$ or $r - 4 = 0$
$r = -4$ or $r = 4$
Discard the root $r = -4$. The radius of the first circle is 4 centimeters and the radius of the second circle is 8 centimeters.

75. Let x = width
and $2x$ = length.

$A = x(2x) = 2x^2$
$2x^2 = 50$
$2x^2 - 50 = 0$
$2(x^2 - 25) = 0$
$2(x + 5)(x - 5) = 0$
$x + 5 = 0$ or $x - 5 = 0$
$x = -5$ or $x = 5$
Discard the root $x = -5$. The width is 5 meters and the length is 10 meters.

77. Let x = radius
and $2x$ = altitude.

$SA = 2\pi r^2 + 2\pi rh$
$54\pi = 2\pi(x)^2 + 2\pi(x)(2x)$
$54\pi = 2\pi x^2 + 4\pi x^2$
$54\pi = 6\pi x^2$
$9 = x^2$
$0 = x^2 - 9$
$0 = (x + 3)(x - 3)$

$x + 3 = 0$ or $x - 3 = 0$
$x = -3$ or $x = 3$
Discard the root $x = -3$. The radius should be 3 inches and the altitude should be 6 inches.

79. Let x = radius of circle and $2x$ = side of square.

Area of square $= (2x)^2 = 4x^2$
Area of circle $= \pi x^2$

$4x^2 + \pi x^2 = 16\pi + 64$
$4x^2 + \pi x^2 - 16\pi - 64 = 0$
$x^2(4 + \pi) - 16(\pi + 4) = 0$
$x^2(\pi + 4) - 16(\pi + 4) = 0$
$(\pi + 4)(x^2 - 16) = 0$
$(\pi + 4)(x + 4)(x - 4) = 0$
$x + 4 = 0$ or $x - 4 = 0$
$x = -4$ or $x = 4$
Discard the root $x = -4$. The length of the side of the square should be 8 yards.

PROBLEM SET 3.6 Factoring Trinomials

1. $x^2 + 9x + 20$
We need two integers whose product is 20 and whose sum is 9. They are 4 and 5.
$(x + 5)(x + 4)$

3. $x^2 - 11x + 28$
We need two integers whose product is 28 and whose sum is -11. They are -4 and -7.
$(x - 4)(x - 7)$

5. $a^2 + 5a - 36$
We need two integers whose product is -36 and whose sum is 5. They are 9 and -4.
$(a + 9)(a - 4)$

7. $y^2 + 20y + 84$
We need two integers whose product is 84 and whose sum is 20. They are 6 and 14.
$(y + 6)(y + 14)$

9. $x^2 - 5x - 14$
We need two integers whose product is -14 and whose sum is -5. They are -7 and 2.
$(x - 7)(x + 2)$

11. $x^2 + 9x + 12$
We need two integers whose product is 12 and whose sum is 9. No such integers exist.
Not factorable

13. $6 + 5x - x^2$
We need two integers whose product is 6 and whose sum is 55. They are 6 and -1.
$6 + 6x - x - x^2$
$6(1 + x) - x(1 + x)$
$(1 + x)(6 - x)$

15. $x^2 + 15xy + 36y^2$
We need two integers whose product is 36 and whose sum is 15. They are 12 and 3.
$(x + 12y)(x + 3y)$

17. $a^2 - ab - 56b^2$
We need two integers whose product is -56 and whose sum is -1. They are -8 and 7.
$(a - 8b)(a + 7b)$

19. $15x^2 + 23x + 6$
$15(6) = 90$
We need two integers whose product is 90 and whose sum is 23. They are 5 and 18.
$15x^2 + 5x + 18x + 6$
$5x(3x + 1) + 6(3x + 1)$
$(3x + 1)(5x + 6)$

21. $12x^2 - x - 6$
$12(-6) = -72$
We need two integers whose product is -72 and whose sum is -1. They are -9 and 8.

$12x^2 - 9x + 8x - 6$
$3x(4x - 3) + 2(4x - 3)$
$(4x - 3)(3x + 2)$

23. $4a^2 + 3a - 27$
$4(-27) = -108$
We need two integers whose product is -108 and whose sum is 3. They are -9 and 12.
$4a^2 - 9a + 12a - 27$
$a(4a - 9) + 3(4a - 9)$
$(4a - 9)(a + 3)$

25. $3n^2 - 7n - 20$
$3(-20) = -60$
We need two integers whose product is -60 and whose sum is -7. They are -12 and 5.
$3n^2 - 12n + 5n - 20$
$3n(n - 4) + 5(n - 4)$
$(n - 4)(3n + 5)$

27. $3x^2 + 10x + 4$
$3(4) = 12$
We need two integers whose product is 12 and whose sum is 10. No such integers exist.
Not factorable

29. $10n^2 - 29n - 21$
$10(-21) = -210$
We need two integers whose product is -210 and whose sum is -29. They are -35 and 6.
$10n^2 - 35n + 6n - 21$
$5n(2n - 7) + 3(2n - 7)$
$(2n - 7)(5n + 3)$

31. $8x^2 + 26x - 45$
$8(-45) = -360$
We need two integers whose product is -360 and whose sum is 26. They are 36 and -10.
$8x^2 + 36x - 10x - 45$
$4x(2x + 9) - 5(2x + 9)$
$(2x + 9)(4x - 5)$

33. $6 - 35x - 6x^2$
$6(-6) = -36$
We need two integers whose product is -36 and whose sum is -35. They are -36 and 1.
$6 - 36x + x - 6x^2$
$6(1 - 6x) + x(1 - 6x)$
$(1 - 6x)(6 + x)$

35. $20y^2 + 31y - 9$
$20(-9) = -180$
We need two integers whose product is -180 and whose sum is 31. They are 36 and -5.
$20y^2 + 36y - 5y - 9$
$4y(5y + 9) - 1(5y + 9)$
$(5y + 9)(4y - 1)$

37. $24n^2 - 2n - 5$
$24(-5) = -120$
We need two integers whose product is -120 and whose sum is -2. They are -12 and 10.
$24n^2 - 12n + 10n - 5$
$12n(2n - 1) + 5(2n - 1)$
$(2n - 1)(12n + 5)$

39. $5n^2 + 33n + 18$
$5(18) = 90$
We need two integers whose product is 90 and whose sum is 33. They are 30 and 3.
$5n^2 + 30n + 3n + 18$
$5n(n + 6) + 3(n + 6)$
$(n + 6)(5n + 3)$

41. $x^2 + 25x + 150$
We need two integers whose product is 150 and whose sum is 25. They are 10 and 15.
$(x + 10)(x + 15)$

43. $n^2 - 36n + 320$
We need two integers whose product is 320 and whose sum is -36. They are -16 and -20.
$(n - 16)(n - 20)$

45. $t^2 + 3t - 180$
We need two integers whose product is -180 and whose sum is 3. They are 15 and -12.
$(t + 15)(t - 12)$

47. $t^4 - 5t^2 + 6$
We need two integers whose product is 6 and whose sum is -5. They are -2 and -3.
$(t^2 - 2)(t^2 - 3)$

49. $10x^4 + 3x^2 - 4$
$10(-4) = -40$
We need two integers whose product is 40 and whose sum is 3.They are 8 and -5.
$10x^4 + 8x^2 - 5x^2 - 4$
$2x^2(5x^2 + 4) - 1(5x^2 + 4)$
$(5x^2 + 4)(2x^2 - 1)$

51. $x^4 - 9x^2 + 8$
We need two integers whose product is 8 and whose sum is -9. They are -1 and -8.
$(x^2 - 1)(x^2 - 8)$
$(x + 1)(x - 1)(x^2 - 8)$

53. $18n^4 + 25n^2 - 3$
$18(-3) = -54$
We need two integers whose product is -54 and whose sum is 25. They are 27 and -2.
$18n^4 + 27n^2 - 2n^2 - 3$
$9n^2(2n^2 + 3) - 1(2n^2 + 3)$
$(2n^2 + 3)(9n^2 - 1)$
$(2n^2 + 3)(3n + 1)(3n - 1)$

55. $x^4 - 17x^2 + 16$
We need two integers whose product is 16 and whose sum is -17. They are -16 and -1.
$(x^2 - 16)(x^2 - 1)$
$(x + 4)(x - 4)(x + 1)(x - 1)$

57. $2t^2 - 8$
$2(t^2 - 4)$
$2(t + 2)(t - 2)$

59. $12x^2 + 7xy - 10y^2$
$12(-10) = -120$
We need two integers whose product is -120 and whose sum is 7. They are 15 and -8.
$12x^2 + 15xy - 8xy - 10y^2$
$3x(4x + 5y) - 2y(4x + 5y)$
$(4x + 5y)(3x - 2y)$

61. $18n^3 + 39n^2 - 15n$
$3n(6n^2 + 13n - 5)$
$6(-5) = -30$
We need two integers whose product is -30 and whose sum is 13. They are 15 and -2.
$3n(6n^2 + 15n - 2n - 5)$
$3n[3n(2n + 5) - 1(2n + 5)]$
$3n[(2n + 5)(3n - 1)]$
$3n(2n + 5)(3n - 1)$

63. $n^2 - 17n + 60$
We need two integers whose product is 60 and whose sum is -17. They are -5 and -12.
$n^2 - 5n - 12n + 60$
$n(n - 5) - 12(n - 5)$
$(n - 5)(n - 12)$

65. $36a^2 - 12a + 1$
$36(1) = 36$
We need two integers whose product is 36 and whose sum is -12. They are -6 and -6.
$36a^2 - 6a - 6a + 1$
$6a(6a - 1) - 1(6a - 1)$
$(6a - 1)(6a - 1)$
$(6a - 1)^2$

67. $6x^2 + 54$
$6(x^2 + 9)$

69. $3x^2 + x - 5$
$3(-5) = -15$
We need two integers whose product is -15 and whose sum is 1. No such integers exist.
Not factorable

71. $x^2 - (y-7)^2$
$[x + (y-7)][x - (y-7)]$
$(x + y - 7)(x - y + 7)$

73. $1 - 16x^4$
$(1 + 4x^2)(1 - 4x^2)$
$(1 + 4x^2)(1 - 2x)(1 + 2x)$

75. $4n^2 + 25n + 36$
$4(36) = 144$
We need two integers whose product is 144 and whose sum is 25. They are 16 and 9.
$4n^2 + 9n + 16n + 36$
$n(4n + 9) + 4(4n + 9)$
$(4n + 9)(n + 4)$

77. $n^3 - 49n$
$n(n^2 - 49)$
$n(n + 7)(n - 7)$

79. $x^2 - 7x - 8$
We need two integers whose product is -8 and whose sum is -7. They are -8 and 1.
$(x - 8)(x + 1)$

81. $3x^4 - 81x$
$3x(x^3 - 27)$
$3x(x - 3)(x^2 + 3x + 9)$

83. $x^4 + 6x^2 + 9$
We need two integers whose product is 9 and whose sum is 6. They are 3 and 3.
$(x^2 + 3)(x^2 + 3)$
$(x^2 + 3)^2$

85. $x^4 - 5x^2 - 36$
We need two integers whose product is -36 and whose sum is -5. They are -9 and 4.
$(x^2 - 9)(x^2 + 4)$
$(x + 3)(x - 3)(x^2 + 4)$

87. $6w^2 - 11w - 35$
$6(-35) = -210$
We need two integers whose product is -210 and whose sum is -11. They are -21 and 10.
$6w^2 - 21w + 10w - 35$
$3w(2w - 7) + 5(2w - 7)$
$(2w - 7)(3w + 5)$

89. $25n^2 + 64$
Not factorable

91. $2n^3 + 14n^2 - 20n$
$2n(n^2 + 7n - 10)$
We need two integers whose product is -10 and whose sum is 7. No such integers exist.
$2n(n^2 + 7n - 10)$

93. $2xy + 6x + y + 3$
$2x(y + 3) + 1(y + 3)$
$(y + 3)(2x + 1)$

99. $x^{2a} + 10x^a + 21$
$(x^a + 3)(x^a + 7)$

101. $4x^{2a} + 20x^a + 25$
$(2x^a + 5)^2$

103. $20x^{2n} + 21x^n - 5$
$20x^{2n} + 25x^n - 4x^n - 5$
$5x^n(4x^n + 5) - 1(4x^n + 5)$
$(4x^n + 5)(5x^n - 1)$

105. $(x + 1)^2 - 8(x + 1) + 15$
Let $y = (x + 1)$.
$y^2 - 8y + 15$
$(y - 3)(y - 5)$
$(x + 1 - 3)(x + 1 - 5)$
$(x - 2)(x - 4)$

107. $(3x - 2)^2 - 5(3x - 2) - 36$
Let $y = 3x - 2$.
$y^2 - 5y - 36$
$(y - 9)(y + 4)$
$(3x - 2 - 9)(3x - 2 + 4)$
$(3x - 11)(3x + 2)$

109. $15(x + 2)^2 - 13(x + 2) + 2$
Let $y = x + 2$.
$15y^2 - 13y + 2$
$(5y - 1)(3y - 2)$
$[5(x + 2) - 1][3(x + 2) - 2]$
$[5x + 10 - 1][3x + 6 - 2]$
$(5x + 9)(3x + 4)$

Problem Set 3.7

PROBLEM SET 3.7 Equations and Problem Solving

1. $x^2 + 4x + 3 = 0$
$(x + 3)(x + 1) = 0$
$x + 3 = 0$ or $x + 1 = 0$
$x = -3$ or $x = -1$
The solution set is $\{-3, -1\}$.

3. $x^2 + 18x + 72 = 0$
$(x + 12)(x + 6) = 0$
$x + 12 = 0$ or $x + 6 = 0$
$x = -12$ or $x = -6$
The solution set is $\{-12, -6\}$.

5. $n^2 - 13n + 36 = 0$
$(n - 9)(n - 4) = 0$
$n - 9 = 0$ or $n - 4 = 0$
$n = 9$ or $n = 4$
The solution set is $\{4, 9\}$.

7. $x^2 + 4x - 12 = 0$
$(x + 6)(x - 2) = 0$
$x + 6 = 0$ or $x - 2 = 0$
$x = -6$ or $x = 2$
The solution set is $\{-6, 2\}$.

9. $w^2 - 4w = 5$
$w^2 - 4w - 5 = 0$
$(w - 5)(w + 1) = 0$
$w - 5 = 0$ or $w + 1 = 0$
$w = 5$ or $w = -1$
The solution set is $\{-1, 5\}$.

11. $n^2 + 25n + 156 = 0$
$(n + 12)(n + 13) = 0$
$n + 12 = 0$ or $n + 13 = 0$
$n = -12$ or $n = -13$
The solution set is $\{-13, -12\}$.

13. $3t^2 + 14t - 5 = 0$
$(3t - 1)(t + 5) = 0$
$3t - 1 = 0$ or $t + 5 = 0$
$3t = 1$ or $t = -5$
$t = \frac{1}{3}$ or $t = -5$
The solution set is $\left\{-5, \frac{1}{3}\right\}$.

15. $6x^2 + 25x + 14 = 0$
$(2x + 7)(3x + 2) = 0$
$2x + 7 = 0$ or $3x + 2 = 0$
$2x = -7$ or $3x = -2$
$x = -\frac{7}{2}$ or $x = -\frac{2}{3}$
The solution set is $\left\{-\frac{7}{2}, -\frac{2}{3}\right\}$.

17. $3t(t - 4) = 0$
$3t = 0$ or $t - 4 = 0$
$t = 0$ or $t = 4$
The solution set is $\{0, 4\}$.

19. $-6n^2 + 13n - 2 = 0$
$-1(-6n^2 + 13n - 2) = -1(0)$
$6n^2 - 13n + 2 = 0$
$(6n - 1)(n - 2) = 0$
$6n - 1 = 0$ or $n - 2 = 0$
$6n = 1$ or $n = 2$
$n = \frac{1}{6}$ or $n = 2$
The solution set is $\left\{\frac{1}{6}, 2\right\}$.

21. $2n^3 = 72n$
$2n^3 - 72n = 0$
$2n(n^2 - 36) = 0$
$2n(n + 6)(n - 6) = 0$
$2n = 0$ or $n + 6 = 0$ or $n - 6 = 0$
$n = 0$ or $n = -6$ or $n = 6$
The solution set is $\{-6, 0, 6\}$.

23. $(x - 5)(x + 3) = 9$
$x^2 - 2x - 15 = 9$
$x^2 - 2x - 24 = 0$
$(x - 6)(x + 4) = 0$
$x - 6 = 0$ or $x + 4 = 0$
$x = 6$ or $x = -4$
The solution set is $\{-4, 6\}$.

25. $16 - x^2 = 0$
$(4 - x)(4 + x) = 0$
$4 - x = 0$ or $4 + x = 0$
$4 = x$ or $x = -4$
The solution set is $\{-4, 4\}$.

27. $n^2 + 7n - 44 = 0$
$(n + 11)(n - 4) = 0$
$n + 11 = 0$ or $n - 4 = 0$
$n = -11$ or $n = 4$
The solution set is $\{-11, 4\}$.

29. $3x^2 = 75$
$3x^2 - 75 = 0$
$3(x^2 - 25) = 0$
$3(x + 5)(x - 5) = 0$
$x + 5 = 0$ or $x - 5 = 0$
$x = -5$ or $x = 5$
The solution set is $\{-5, 5\}$.

31. $15x^2 + 34x + 15 = 0$
$(5x + 3)(3x + 5) = 0$
$5x + 3 = 0$ or $3x + 5 = 0$
$5x = -3$ or $3x = -5$
$x = -\frac{3}{5}$ or $x = -\frac{5}{3}$
The solution set is $\left\{-\frac{5}{3}, -\frac{3}{5}\right\}$.

33. $8n^2 - 47n - 6 = 0$
$(8n + 1)(n - 6) = 0$
$8n + 1 = 0$ or $n - 6 = 0$
$8n = -1$ or $n = 6$
$n = -\frac{1}{8}$ or $n = 6$
The solution set is $\left\{-\frac{1}{8}, 6\right\}$.

35. $28n^2 - 47n + 15 = 0$
$(7n - 3)(4n - 5) = 0$
$7n - 3 = 0$ or $4n - 5 = 0$
$7n = 3$ or $4n = 5$
$n = \frac{3}{7}$ or $n = \frac{5}{4}$
The solution set is $\left\{\frac{3}{7}, \frac{5}{4}\right\}$.

37. $35n^2 - 18n - 8 = 0$
$(7n + 2)(5n - 4) = 0$
$7n + 2 = 0$ or $5n - 4 = 0$
$7n = -2$ or $5n = 4$
$n = -\frac{2}{7}$ or $n = \frac{4}{5}$
The solution set is $\left\{-\frac{2}{7}, \frac{4}{5}\right\}$.

39. $-3x^2 - 19x + 14 = 0$
$-1(-3x^2 - 19x + 14) = -1(0)$
$3x^2 + 19x - 14 = 0$
$(3x - 2)(x + 7) = 0$
$3x - 2 = 0$ or $x + 7 = 0$
$3x = 2$ or $x = -7$
$x = \frac{2}{3}$ or $x = -7$
The solution set is $\left\{-7, \frac{2}{3}\right\}$.

41. $n(n + 2) = 360$
$n^2 + 2n = 360$
$n^2 + 2n - 360 = 0$
$(n + 20)(n - 18) = 0$
$n + 20 = 0$ or $n - 18 = 0$
$n = -20$ or $n = 18$
The solution set is $\{-20, 18\}$.

43. $9x^4 - 37x^2 + 4 = 0$
$(9x^2 - 1)(x^2 - 4) = 0$
$(3x + 1)(3x - 1)(x + 2)(x - 2) = 0$
$3x+1=0$ or $3x - 1=0$ or $x+2=0$ or $x - 2=0$
$3x = -1$ or $3x = 1$ or $x = -2$ or $x = 2$
$x = -\frac{1}{3}$ or $x = \frac{1}{3}$ or $x = -2$ or $x = 2$
The solution set is $\left\{-2, -\frac{1}{3}, \frac{1}{3}, 2\right\}$.

45. $3x^2 - 46x - 32 = 0$
$(3x + 2)(x - 16) = 0$
$3x + 2 = 0$ or $x - 16 = 0$
$3x = -2$ or $x = 16$
$x = -\frac{2}{3}$ or $x = 16$
The solution set is $\left\{-\frac{2}{3}, 16\right\}$.

47. $2x^2 + x - 3 = 0$
$(2x + 3)(x - 1) = 0$
$2x + 3 = 0$ or $x - 1 = 0$
$2x = -3$ or $x = 1$
$x = -\frac{3}{2}$ or $x = 1$
The solution set is $\left\{-\frac{3}{2}, 1\right\}$.

49. $12x^3 + 46x^2 + 40x = 0$
$2x(6x^2 + 23x + 20) = 0$
$2x(2x + 5)(3x + 4) = 0$
$2x = 0$ or $2x + 5 = 0$ or $3x + 4 = 0$
$x = 0$ or $2x = -5$ or $3x = -4$
$x = -\frac{5}{2}$ or $x = -\frac{4}{3}$
The solution set is $\left\{-\frac{5}{2}, -\frac{4}{3}, 0\right\}$.

51. $(3x - 1)^2 - 16 = 0$
$[(3x - 1) - 4][(3x - 1) + 4] = 0$
$(3x - 5)(3x + 3) = 0$
$3x - 5 = 0$ or $3x + 3 = 0$
$3x = 5$ or $3x = -3$
$x = \frac{5}{3}$ or $x = -1$
The solution set is $\left\{-1, \frac{5}{3}\right\}$.

53. $4a(a + 1) = 3$
$4a^2 + 4a = 3$
$4a^2 + 4a - 3 = 0$
$(2a + 3)(2a - 1) = 0$
$2a + 3 = 0$ or $2a - 1 = 0$
$2a = -3$ or $2a = 1$
$a = -\frac{3}{2}$ or $a = \frac{1}{2}$
The solution set is $\left\{-\frac{3}{2}, \frac{1}{2}\right\}$.

55. Let $x = 1^{st}$ consecutive integer
$x + 1 = 2^{nd}$ consecutive integer
$x(x + 1) = 72$
$x^2 + x = 72$
$x^2 + x - 72 = 0$
$(x + 9)(x - 8) = 0$
$x + 9 = 0$ or $x - 8 = 0$
$x = -9$ or $x = 8$
The integers are -9 and -8 or 8 and 9.

57. Let x = first integer
$2x + 1$ = second integer

$x(2x + 1) = 105$
$2x^2 + x = 105$
$2x^2 + x - 105 = 0$
$(2x + 15)(x - 7) = 0$
$2x + 15 = 0$ or $x - 7 = 0$
$2x = -15$ or $x = 7$
$x = -\frac{15}{2}$ or $x = 7$
Not an integer or $x = 7$

first integer = 7
second integer = 2(7) + 1 = 15
The integers are 7 and 15.

59. Let x = width
$P = 2l + 2w$
$32 = 2l + 2x$
$32 - 2x = 2l$
$\frac{1}{2}(32 - 2x) = \frac{1}{2}(2l)$
$16 - x = l$
So x = width and $16 - x$ = length.

$A = lw$
$60 = (16 - x)(x)$
$60 = 16x - x^2$
$x^2 - 16x + 60 = 0$
$(x - 10)(x - 6) = 0$
$x - 10 = 0$ or $x - 6 = 0$
$x = 10$ or $x = 6$
width = 10 or width = 6
length = 16 − 10 or length = 16 − 6
length = 6 or length = 10
So the rectangle is 6 inches by 10 inches.

61. Let $x = 1^{st}$ integer
$x + 1 = 2^{nd}$ integer

$x^2 + (x + 1)^2 = 85$
$x^2 + x^2 + 2x + 1 = 85$
$2x^2 + 2x - 84 = 0$
$2(x^2 + x - 42) = 0$
$2(x + 7)(x - 6) = 0$
$x + 7 = 0$ or $x - 6 = 0$
$x = -7$ or $x = 6$
The integers are -7 and -6 or 6 and 7.

63. Let x = side of square
$x + 2$ = width of rectangle
$x + 4$ = length of rectangle

Area of square $= x^2$
Area of rectangle $= (x + 2)(x + 4)$
$= x^2 + 6x + 8$

$x^2 + x^2 + 6x + 8 = 64$
$2x^2 + 6x + 8 = 64$
$2x^2 + 6x - 56 = 0$
$2(x^2 + 3x - 28) = 0$
$2(x + 7)(x - 4) = 0$
$x + 7 = 0$ or $x - 4 = 0$
$x = -7$ or $x = 4$
Discard the root $x = -7$.
The length of the side of the square is 4 centimeters. The dimensions of the rectangle are 6 centimeters by 8 centimeters.

65. Let $x = 1^{\text{st}}$ leg
$x + 1 = 2^{\text{nd}}$ leg
$x + 2 =$ hypotenuse

Use the Pythagorean Theorem.
$x^2 + (x + 1)^2 = (x + 2)^2$
$x^2 + x^2 + 2x + 1 = x^2 + 4x + 4$
$x^2 - 2x - 3 = 0$
$(x - 3)(x + 1) = 0$

$x - 3 = 0$ or $x + 1 = 0$
$x = 3$ or $x = -1$
Discard the root $x = -1$.
The sides of the triangle are 3 units, 4 units, and 5 units.

67. Let $x = 1^{\text{st}}$ leg
$x + 3 = 2^{\text{nd}}$ leg
$15 =$ hypotenuse

Use the Pythagorean Theorem.
$x^2 + (x + 3)^2 = 15^2$
$x^2 + x^2 + 6x + 9 = 225$
$2x^2 + 6x - 216 = 0$
$2(x^2 + 3x - 108) = 0$
$2(x + 12)(x - 9) = 0$
$x + 12 = 0$ or $x - 9 = 0$
$x = -12$ or $x = 9$
Discard the root $x = -12$.
The lengths of the legs of the triangle are 9 inches and 12 inches.

69. Let $x =$ altitude
$3x + 2 =$ side (base)
$\frac{1}{2}x(3x + 2) = 28$
$2\left[\frac{1}{2}x(3x + 2)\right] = 2(28)$
$x(3x + 2) = 56$
$3x^2 + 2x = 56$
$3x^2 + 2x - 56 = 0$
$(3x + 14)(x - 4) = 0$
$3x + 14 = 0$ or $x - 4 = 0$
$3x = -14$ or $x = 4$
$x = -\frac{14}{3}$ or $x = 4$
Discard the root $x = -\frac{14}{3}$.
The altitude is 4 inches and the side is 14 inches.

CHAPTER 3 Review Problem Set

1. $(3x - 2) + (4x - 6) + (-2x + 5)$
$3x - 2 + 4x - 6 - 2x + 5$
$3x + 4x - 2x - 2 - 6 + 5$
$5x - 3$

2. $(8x^2 + 9x - 3) - (5x^2 - 3x - 1)$
$8x^2 + 9x - 3 - 5x^2 + 3x + 1$
$3x^2 + 12x - 2$

3. $(6x^2 - 2x - 1) + (4x^2 + 2x + 5)$
$-(-2x^2 + x - 1)$

$6x^2 - 2x - 1 + 4x^2 + 2x + 5$
$+ 2x^2 - x + 1$

$6x^2 + 4x^2 + 2x^2 - 2x + 2x - x$
$- 1 + 5 + 1$

$12x^2 - x + 5$

4. $(-5x^2y^3)(4x^3y^4)$
$-20x^{2+3}y^{3+4}$
$-20x^5y^7$

5. $(-2a^2)(3ab^2)(a^2b^3)$
$-6a^{2+1+2}b^{2+3}$
$-6a^5b^5$

6. $5a^2(3a^2-2a-1)$
$15a^4-10a^3-5a^2$

7. $(4x-3y)(6x+5y)$
$24x^2+20xy-18xy-15y^2$
$24x^2+2xy-15y^2$

8. $(x+4)(3x^2-5x-1)$
$x(3x^2-5x-1)+4(3x^2-5x-1)$
$3x^3-5x^2-x+12x^2-20x-4$
$3x^3+7x^2-21x-4$

9. $(4x^2y^3)^4$
$4^4x^{2(4)}y^{3(4)}$
$256x^8y^{12}$

10. $(3x-2y)^2$
$(3x)^2+2(3x)(-2y)+(-2y)^2$
$9x^2-12xy+4y^2$

11. $(-2x^2y^3z)^3$
$(-2)^3x^{2(3)}y^{3(3)}z^3$
$-8x^6y^9z^3$

12. $\dfrac{-39x^3y^4}{3xy^3}$
$-13x^{3-1}y^{4-3}$
$-13x^2y$

13. $[3x-(2x-3y+1)]-[2y-(x-1)]$
$[3x-2x+3y-1]-[2y-x+1]$
$[x+3y-1]-2y+x-1$
$x+3y-1-2y+x-1$
$2x+y-2$

14. $(x^2-2x-5)(x^2+3x-7)$
$x^2(x^2+3x-7)-2x(x^2+3x-7)-5(x^2+3x-7)$
$x^4+3x^3-7x^2-2x^3-6x^2+14x-5x^2-15x+35$
$x^4+x^3-18x^2-x+35$

15. $(7-3x)(3+5x)$
$21+35x-9x-15x^2$
$21+26x-15x^2$

16. $-(3ab)(2a^2b^3)^2$
$-(3ab)(2^2a^4b^6)$
$-(3ab)(4a^4b^6)$
$-12a^5b^7$

17. $\left(\frac{1}{2}ab\right)(8a^3b^2)(-2a^3)$
$-8a^{1+3+3}b^{1+2}$
$-8a^7b^3$

18. $(7x-9)(x+4)$
$7x^2+28x-9x-36$
$7x^2+19x-36$

19. $(3x+2)(2x^2-5x+1)$
$3x(2x^2-5x+1)+2(2x^2-5x+1)$
$6x^3-15x^2+3x+4x^2-10x+2$
$6x^3-11x^2-7x+2$

20. $(3x^{n+1})(2x^{3n-1})$
$6x^{n+1+3n-1}$
$6x^{4n}$

21. $(2x+5y)^2$
$(2x)^2+2(2x)(5y)+(5y)^2$
$4x^2+20xy+25y^2$

22. $(x-2)^3$
$(x-2)(x-2)(x-2)$
$(x-2)(x^2-4x+4)$
$x(x^2-4x+4)-2(x^2-4x+4)$
$x^3-4x^2+4x-2x^2+8x-8$
$x^3-6x^2+12x-8$

23. $(2x+5)^3$
$(2x+5)^2(2x+5)$
$[(2x)^2+2(2x)(5)+(5)^2](2x+5)$
$(4x^2+20x+25)(2x+5)$
$(2x+5)(4x^2+20x+25)$
$2x(4x^2+20x+25)+5(4x^2+20x+25)$
$8x^3+40x^2+50x+20x^2+100x+125$
$8x^3+60x^2+150x+125$

24. $x^2+3x-28$
$(x+7)(x-4)$

25. $2t^2 - 18$
$2(t^2 - 9)$
$2(t + 3)(t - 3)$

26. $4n^2 + 9$
Not factorable

27. $12n^2 - 7n + 1$
$12(1) = 12$
We need two integers whose product is 12 and whose sum is -7. They are -3 and -4.
$12n^2 - 3n - 4n + 1$
$3n(4n - 1) - 1(4n - 1)$
$(4n - 1)(3n - 1)$

28. $x^6 - x^2$
$x^2(x^4 - 1)$
$x^2(x^2 + 1)(x^2 - 1)$
$x^2(x^2 + 1)(x + 1)(x - 1)$

29. $x^3 - 6x^2 - 72x$
$x(x^2 - 6x - 72)$
$x(x - 12)(x + 6)$

30. $6a^3b + 4a^2b^2 - 2a^2bc$
$2a^2b(3a + 2b - c)$

31. $x^2 - (y - 1)^2$
$[x + (y - 1)][x - (y - 1)]$
$(x + y - 1)(x - y + 1)$

32. $8x^2 + 12$
$4(2x^2 + 3)$

33. $12x^2 + x - 35$
$12(-35) = -420$
We need two integers whose product is -420 and whose sum is 1. They are 21 and -20.
$12x^2 + 21x - 20x - 35$
$3x(4x + 7) - 5(4x + 7)$
$(4x + 7)(3x - 5)$

34. $16n^2 - 40n + 25$
$(4n - 5)^2$

35. $4n^2 - 8n$
$4n(n - 2)$

36. $3w^3 + 18w^2 - 24w$
$3w(w^2 + 6w - 8)$

37. $20x^2 + 3xy - 2y^2$
$20(-2) = -40$
We need two integers whose product is -40 and whose sum is 3. They are 8 and -5.
$20x^2 + 8xy - 5xy - 2y^2$
$4x(5x + 2y) - y(5x + 2y)$
$(5x + 2y)(4x - y)$

38. $16a^2 - 64a$
$16a(a - 4)$

39. $3x^3 - 15x^2 - 18x$
$3x(x^2 - 5x - 6)$
$3x(x - 6)(x + 1)$

40. $n^2 - 8n - 128$
$(n + 8)(n - 16)$

41. $t^4 - 22t^2 - 75$
$(t^2 - 25)(t^2 + 3)$
$(t + 5)(t - 5)(t^2 + 3)$

42. $35x^2 - 11x - 6$
We need two integers whose product is -210 and whose sum is -11. They are $-21$and 10.
$35x^2 - 21x + 10x - 6$
$7x(5x - 3) + 2(5x - 3)$
$(5x - 3)(7x + 2)$

43. $15 - 14x + 3x^2$
$15(3) = 45$
We need two integers whose product is 45 and whose sum is -14. They are -9 and -5.
$15 - 9x - 5x + 3x^2$
$3(5 - 3x) - x(5 - 3x)$
$(5 - 3x)(3 - x)$

44. $64n^3 - 27$
$(4n - 3)(16n^2 + 12n + 9)$

45. $16x^3 + 250$
$2(8x^3 + 125)$
$2(2x + 5)(4x^2 - 10x + 25)$

46. $4x^2 - 36 = 0$
$4(x^2 - 9) = 0$
$4(x + 3)(x - 3) = 0$
$x + 3 = 0$ or $x - 3 = 0$
$x = -3$ or $x = 3$
The solution set is $\{-3, 3\}$.

47. $x^2 + 5x - 6 = 0$
$(x + 6)(x - 1) = 0$
$x + 6 = 0$ or $x - 1 = 0$
$x = -6$ or $x = 1$
The solution set is $\{-6, 1\}$.

48. $49n^2 - 28n + 4 = 0$
$(7n - 2)^2 = 0$
$7n - 2 = 0$
$7n = 2$
$n = \frac{2}{7}$
The solution set is $\left\{\frac{2}{7}\right\}$.

49. $(3x - 1)(5x + 2) = 0$
$3x - 1 = 0$ or $5x + 2 = 0$
$3x = 1$ or $5x = -2$
$x = \frac{1}{3}$ or $x = -\frac{2}{5}$
The solution set is $\left\{-\frac{2}{5}, \frac{1}{3}\right\}$.

50. $(3x - 4)^2 - 25 = 0$
$[(3x - 4) + 5][(3x - 4) - 5] = 0$
$(3x + 1)(3x - 9) = 0$
$3x + 1 = 0$ or $3x - 9 = 0$
$3x = -1$ or $3x = 9$
$x = -\frac{1}{3}$ or $x = 3$
The solution set is $\left\{-\frac{1}{3}, 3\right\}$.

51. $6a^3 = 54a$
$6a^3 - 54a = 0$
$6a(a^2 - 9) = 0$
$6a(a + 3)(a - 3) = 0$
$6a = 0$ or $a + 3 = 0$ or $a - 3 = 0$
$a = 0$ or $a = -3$ or $a = 3$
The solution set is $\{-3, 0, 3\}$.

52. $x^5 = x$
$x^5 - x = 0$
$x(x^4 - 1) = 0$
$x(x^2 + 1)(x^2 - 1) = 0$
$x(x^2 + 1)(x + 1)(x - 1) = 0$
$x = 0$ or $x^2 + 1 = 0$ or $x+1 = 0$ or $x - 1 = 0$
$x = 0$ or $x^2 = -1$ or $x = -1$ or $x = 1$
not a real number
The solution set is $\{-1, 0, 1\}$.

53. $-n^2 + 2n + 63 = 0$
$-1(-n^2 + 2n + 63) = -1(0)$
$n^2 - 2n - 63 = 0$
$(n - 9)(n + 7) = 0$
$n - 9 = 0$ or $n + 7 = 0$
$n = 9$ or $n = -7$
The solution set is $\{-7, 9\}$.

54. $7n(7n + 2) = 8$
$49n^2 + 14n = 8$
$49n^2 + 14n - 8 = 0$
$(7n + 4)(7n - 2) = 0$
$7n + 4 = 0$ or $7n - 2 = 0$
$7n = -4$ or $7n = 2$
$n = -\frac{4}{7}$ or $n = \frac{2}{7}$
The solution set is $\left\{-\frac{4}{7}, \frac{2}{7}\right\}$.

55. $30w^2 - w - 20 = 0$
$30(-20) = -600$
We need two integers whose product is -600 and whose sum is -1. They are -25 and 24.
$30w^2 - 25w + 24w - 20 = 0$
$5w(6w - 5) + 4(6w - 5) = 0$
$(6w - 5)(5w + 4) = 0$
$6w - 5 = 0$ or $5w + 4 = 0$
$6w = 5$ or $5w = -4$
$w = \frac{5}{6}$ or $w = -\frac{4}{5}$
The solution set is $\left\{-\frac{4}{5}, \frac{5}{6}\right\}$.

56. $5x^4 - 19x^2 - 4 = 0$
$(5x^2 + 1)(x^2 - 4) = 0$
$(5x^2 + 1)(x + 2)(x - 2) = 0$
$5x^2 + 1 = 0$ or $x + 2 = 0$ or $x - 2 = 0$
$5x^2 = -1$ or $x = -2$ or $x = 2$
not a real number
The solution set is $\{-2, 2\}$.

57. $9n^2 - 30n + 25 = 0$
$9(25) = 225$
We need two integers whose product is 225 and whose sum is -30. They are -15 and -15.
$9n^2 - 15n - 15n + 25 = 0$
$3n(3n - 5) - 5(3n - 5) = 0$
$(3n - 5)(3n - 5) = 0$
$(3n - 5)^2 = 0$
$3n - 5 = 0$
$3n = 5$
$n = \frac{5}{3}$
The solution set is $\left\{\frac{5}{3}\right\}$.

58. $n(2n + 4) = 96$
$2n^2 + 4n = 96$
$2n^2 + 4n - 96 = 0$
$2(n^2 + 2n - 48) = 0$
$2(n + 8)(n - 6) = 0$
$n + 8 = 0$ or $n - 6 = 0$
$n = -8$ or $n = 6$
The solution set is $\{-8, 6\}$.

59. $7x^2 + 33x - 10 = 0$
$7(-10) = -70$
We need two integers whose product is -70 and whose sum is 33. They are 35 and -2.
$7x^2 + 35x - 2x - 10 = 0$
$7x(x + 5) - 2(x + 5) = 0$
$(x + 5)(7x - 2) = 0$
$x + 5 = 0$ or $7x - 2 = 0$
$x = -5$ or $7x = 2$
$x = -5$ or $x = \frac{2}{7}$
The solution set is $\left\{-5, \frac{2}{7}\right\}$.

60. $(x + 1)(x + 2) = 42$
$x^2 + 3x + 2 = 42$
$x^2 + 3x - 40 = 0$
$(x + 8)(x - 5) = 0$
$x + 8 = 0$ or $x - 5 = 0$
$x = -8$ or $x = 5$
The solution set is $\{-8, 5\}$.

61. $x^2 + 12x - x - 12 = 0$
$x(x + 12) - 1(x + 12) = 0$
$(x + 12)(x - 1) = 0$
$x + 12 = 0$ or $x - 1 = 0$
$x = -12$ or $x = 1$
The solution is $\{-12, 1\}$.

62. $2x^4 + 9x^2 + 4 = 0$
$(2x^2 + 1)(x^2 + 4) = 0$
$2x^2 + 1 = 0$ or $x^2 + 4 = 0$
$2x^2 = -1$ or $x^2 = -4$
not real numbers
The solution set is $\emptyset$.

63. $30 - 19x - 5x^2 = 0$
$30(-5) = -150$
We need two integers whose product is -150 and whose sum is -19. They are -25 and 6.
$30 - 25x + 6x - 5x^2 = 0$
$5(6 - 5x) + x(6 - 5x) = 0$
$(6 - 5x)(5 + x) = 0$
$6 - 5x = 0$ or $5 + x = 0$
$6 = 5x$ or $x = -5$
$\frac{6}{5} = x$ or $x = -5$
The solution set is $\left\{-5, \frac{6}{5}\right\}$.

64. $3t^3 - 27t^2 + 24t = 0$
$3t(t^2 - 9t + 8) = 0$
$3t(t - 8)(t - 1) = 0$
$3t = 0$ or $t - 8 = 0$ or $t - 1 = 0$
$t = 0$ or $t = 8$ or $t = 1$
The solution set is $\{0, 1, 8\}$.

65. $-4n^2 - 39n + 10 = 0$
$-1(-4n^2 - 39n + 10) = -1(0)$
$4n^2 + 39n - 10 = 0$
$4(-10) = -40$

We need two integers whose product is -40 and whose sum is 39. They are 40 and -1.

$4n^2 + 40n - n - 10 = 0$
$4n(n + 10) - 1(n + 10) = 0$
$(n + 10)(4n - 1) = 0$
$n + 10 = 0$ or $4n - 1 = 0$
$n = -10$ or $4n = 1$
$n = -10$ or $n = \frac{1}{4}$
The solution set is $\left\{-10, \frac{1}{4}\right\}$.

66. Let $x =$ 1st integer
$x + 1 =$ 2nd integer
$x + 2 =$ 3rd integer

$x(x + 2) = 9(x + 1) - 1$
$x^2 + 2x = 9x + 9 - 1$
$x^2 + 2x = 9x + 8$
$x^2 - 7x - 8 = 0$
$(x - 8)(x + 1) = 0$
$x - 8 = 0$ or $x + 1 = 0$
$x = 8$ or $x = -1$
The integers are 8, 9, and 10 or -1, 0, and 1.

67. Let $x =$ 1st integer
$2 - x =$ 2nd integer

$x(2 - x) = -48$
$2x - x^2 = -48$
$0 = x^2 - 2x - 48$
$0 = (x - 8)(x + 6)$
$x - 8 = 0$ or $x + 6 = 0$
$x = 8$ or $x = -6$
The integers are 8 and -6.

68. Let $x =$ 1st odd whole number
$x + 2 =$ 2nd odd whole number

$x(x + 2) = 195$
$x^2 + 2x = 195$
$x^2 + 2x - 195 = 0$
$(x - 13)(x + 15) = 0$
$x - 13 = 0$ or $x + 15 = 0$
$x = 13$ or $x = -15$
Discard the root $x = -15$.
The numbers are 13 and 15.

69. Let $x =$ distance of the northbound car
$x + 4 =$ distance of the eastbound car

Use the Pythagorean theorem.
$x^2 + (x + 4)^2 = 20^2$
$x^2 + x^2 + 8x + 16 = 400$
$2x^2 + 8x - 384 = 0$
$2(x^2 + 4x - 192) = 0$
$2(x + 16)(x - 12) = 0$
$x + 16 = 0$ or $x - 12 = 0$
$x = -16$ or $x = 12$
Discard the root $x = -16$.
The northbound car traveled 12 miles and the eastbound car traveled 16 miles.

70. Use $P = 2l + 2w$.
Let $x =$ width
$32 = 2l + 2x$
$32 - 2x = 2l$
$16 - x = l =$ length

Use $A = lw$.
$x(16 - x) = 48$
$16x - x^2 = 48$
$0 = x^2 - 16x + 48$
$0 = (x - 4)(x - 12)$
$x - 4 = 0$ or $x - 12 = 0$
$x = 4$ or $x = 12$

The width is 4 meters and the length is 12 meters.

71. Let $x =$ number of rows and
$2x - 2 =$ number of chairs per row.

$x(2x - 2) = 144$
$2x^2 - 2x = 144$
$2x^2 - 2x - 144 = 0$
$2(x^2 - x - 72) = 0$
$2(x - 9)(x + 8) = 0$
$x - 9 = 0$ or $x + 8 = 0$
$x = 9$ or $x = -8$
Discard the root $x = -8$.
Number of rows $= 9$
Number of chairs per row $=$
$2(9) - 2 = 16$
There are 9 rows of 16 chairs per row.

72. Let $x =$ altitude
$2x + 1 =$ side (base)

$\frac{1}{2}x(2x+1) = 39$

$2[\frac{1}{2}x(2x+1)] = 2(39)$
$x(2x+1) = 78$
$2x^2 + x = 78$
$2x^2 + x - 78 = 0$
$(2x+13)(x-6) = 0$
$2x + 13 = 0$ or $x - 6 = 0$
$2x = -13$ or $x = 6$
$x = -\frac{13}{2}$
Discard the root $x = -\frac{13}{2}$.

The length of the altitude is 6 feet and the side is 13 feet.

73. Let $x =$ width of sidewalk
Area of pool $= 20(30) = 600$
Area of sidewalk $= 336$

For the area of the sidewalk and pool
width $= 20 + 2x$
length $= 30 + 2x$

Area of sidewalk and pool $-$ Area of pool $=$ Area of sidewalk

$(20+2x)(30+2x) - 600 = 336$
$600 + 100x + 4x^2 - 600 = 336$
$4x^2 + 100\text{x} - 336 = 0$
$4(x^2 + 25x - 84) = 0$
$4(x-3)(x+28) = 0$
$x - 3 = 0$ and $x + 28 = 0$
$x = 3$ and $x = -28$
Discard the root $x = -28$. The width of the walk is 3 feet.

74. Let $x =$ side of smaller square
$x + 3 =$ side of larger square

Area of smaller square $= x^2$
Area of larger square $= (x+3)^2$

$x^2 + (x+3)^2 = 89$
$x^2 + x^2 + 6x + 9 = 89$
$2x^2 + 6x - 80 = 0$
$2(x^2 + 3x - 40) = 0$
$2(x+8)(x-5) = 0$
$(x+8)(x-5) = 0$
$x + 8 = 0$ or $x - 5 = 0$
$x = -8$ or $x = 5$
The sides of the small square are 5 centimeters and the sides of the large square are 8 centimeters.

75. Let $x =$ radius
and $3x =$ height.

$S.A. = 2\pi r^2 + 2\pi rh$
$32\pi = 2\pi x^2 + 2\pi x(3x)$
$32\pi = 2\pi x^2 + 6\pi x^2$
$\frac{1}{\pi}(32\pi) = \frac{1}{\pi}(2\pi x^2) + \frac{1}{\pi}(6\pi x^2)$
$32 = 2x^2 + 6x^2$
$32 = 8x^2$
$0 = 8x^2 - 32$
$0 = 8(x^2 - 4)$
$0 = 8(x+2)(x-2)$
$x + 2 = 0$ or $x - 2 = 0$
$x = -2$ or $x = 2$
Discard the root $x = -2$. The length of the radius is 2 inches and the altitude is 6 inches.

CHAPTER 3 Test

1. $(-3x - 1) + (9x - 2) - (4x + 8)$
$-3x - 1 + 9x - 2 - 4x - 8$
$2x - 11$

2. $(-6xy^2)(8x^3y^2)$
$-48x^4y^4$

3. $(-3x^2y^4)^3$
$(-3)^3x^{2(3)}y^{4(3)}$
$-27x^6y^{12}$

4. $(5x-7)(4x+9)$
$20x^2+45x-28x-63$
$20x^2+17x-63$

5. $(3n-2)(2n-3)$
$6n^2-9n-4n+6$
$6n^2-13n+6$

6. $(x-4y)^3$
$(x-4y)(x-4y)^2$
$(x-4y)(x^2-8xy+16y^2)$
$x(x^2-8xy+16y^2)-4y(x^2-8xy+16y^2)$
$x^3-8x^2y+16xy^2-4x^2y+32xy^2-64y^3$
$x^3-12x^2y+48xy^2-64y^3$

7. $(x+6)(2x^2-x-5)$
$x(2x^2-x-5)+6(2x^2-x-5)$
$2x^3-x^2-5x+12x^2-6x-30$
$2x^3+11x^2-11x-30$

8. $\dfrac{-70x^4y^3}{5xy^2}$
$-14x^{4-1}y^{3-2}$
$-14x^3y$

9. $6x^2+19x-20$
$6x^2+24x-5x-20$
$6x(x+4)-5(x+4)$
$(x+4)(6x-5)$

10. $12x^2-3$
$3(4x^2-1)$
$3(2x+1)(2x-1)$

11. $64+t^3$
$(4+t)(16-4t+t^2)$

12. $30x+4x^2-16x^3$
$2x(15+2x-8x^2)$
$2x(3-2x)(5+4x)$

13. $x^2-xy+4x-4y$
$x(x-y)+4(x-y)$
$(x-y)(x+4)$

14. $24n^2+55n-24$
$(3n+8)(8n-3)$

15. $x^2+8x-48=0$
$(x+12)(x-4)=0$
$x+12=0$ or $x-4=0$
$x=-12$ or $x=4$
The solution set is $\{-12, 4\}$.

16. $4n^2=n$
$4n^2-n=0$
$n(4n-1)=0$
$n=0$ or $4n-1=0$
$n=0$ $\quad 4n=1$
$n=0$ $\quad n=\frac{1}{4}$
The solution set is $\left\{0, \frac{1}{4}\right\}$.

17. $4x^2-12x+9=0$
$(2x-3)^2=0$
$2x-3=0$
$x=\frac{3}{2}$
The solution set is $\left\{\frac{3}{2}\right\}$.

18. $(n-2)(n+7)=-18$
$n^2+7n-2n-14=-18$
$n^2+5n-14=-18$
$n^2+5n+4=0$
$(n+1)(n+4)=0$
$n+1=0$ or $n+4=0$
$n=-1$ or $n=-4$
The solution set is $\{-4, -1\}$.

19. $3x^3+21x^2-54x=0$
$3x(x^2+7x-18)=0$
$3x(x+9)(x-2)=0$
$3x=0$ or $x+9=0$ or $x-2=0$
$x=0$ or $x=-9$ or $x=2$
The solution set is $\{-9, 0, 2\}$.

20. $12+13x-35x^2=0$
$(3+7x)(4-5x)=0$
$3+7x=0$ or $4-5x=0$
$7x=-3$ or $4=5x$
$x=-\frac{3}{7}$ or $\frac{4}{5}=x$
The solution set is $\left\{-\frac{3}{7}, \frac{4}{5}\right\}$.

21. $n(3n-5)=2$
$3n^2-5n=2$
$3n^2-5n-2=0$
$(3n+1)(n-2)=0$
$3n+1=0$ or $n-2=0$
$3n=-1$ or $n=2$
$n=-\frac{1}{3}$ or $n=2$
The solution set is $\left\{-\frac{1}{3}, 2\right\}$.

22. $9x^2-36=0$
$9(x^2-4)=0$
$9(x+2)(x-2)=0$
$x+2=0$ or $x-2=0$
$x=-2$ or $x=2$
The solution set is $\{-2, 2\}$.

23. Use $P=2l+2w$
Let $x=$ width
$30=2l+2x$
$30-2x=2l$
$15-x=l=$ length
Use $A=lw$
$x(15-x)=54$
$15x-x^2=54$
$0=x^2-15x+54$
$0=(x-9)(x-6)$
$x-9=0$ or $x-6=0$
$x=9$ or $x=6$
The width is 6 inches and the length is 9 inches.

24. Let $x=$ number of chairs per row
$2x+1=$ number of rows
$x(2x+1)=105$
$2x^2+x=105$
$2x^2+x-105=0$
$(2x+15)(x-7)=0$
$2x+15=0$ or $x-7=0$
$2x=-15$ or $x=7$
$x=-\frac{15}{2}$
Discard the root $x=-\frac{15}{2}$.
The number of rows is $2(7)+1=15$.

25. Let $x=$ side of square
$x+3=$ width of rectangle
$x+5=$ length of rectangle

$x^2+(x+3)(x+5)=57$
$x^2+x^2+8x+15=57$
$2x^2+8x-42=0$
$2(x^2+4x-21)=0$
$2(x+7)(x-3)=0$
$x+7=0$ or $x-3=0$
$x=-7$ or $x=3$
Discard the root $x=-7$.
The length of the rectangle is $3+5=8$ feet.

CHAPTER 3 Cumulative Review Chapters 1-3

1. $x^2-2xy+y^2$ for $x=-2$ and $y=-4$
$(-2)^2-2(-2)(-4)+(-4)^2$
$4-16+16$
4

2. $-n^2+2n-4$ for $n=-3$
$-(-3)^2+2(-3)-4$
$-9-6-4$
-19

3. $2x^2-5x+6$ for $x=3$
$2(3)^2-5(3)+6$
$2(9)-15+6$
$18-15+6$
9

4. $3(2x-1)-2(x+4)-4(2x-7)$
$6x-3-2x-8-8x+28$
$-4x+17$ for $x=-1$
$-4(-1)+17$
$4+17$
21

5. $-(2n-1)+5(2n-3)-6(3n+4)$
$-2n+1+10n-15-18n-24$
$-10n-38$ for $n=4$
$-10(4)-38$
$-40-38$
-78

6. $2(a-4)-(a-1)+(3a-6)$
$2a-8-a+1+3a-6$
$4a-13$ for $a=-5$
$4(-5)-13$
$-20-13$
-33

7. $(3x^2-4x-7)-(4x^2-7x+8)$
$3x^2-4x-7-4x^2+7x-8$
$-x^2+3x-15$ for $x=-4$
$-(-4)^2+3(-4)-15$
$-(16)-12-15$
$-16-12-15$
-43

8. $-2(3x-5y)-4(x+2y)+3(-2x-3y)$
$-6x+10y-4x-8y-6x-9y$
$-16x-7y$ for $x=2$ and $y=-3$
$-16(2)-7(-3)$
$-32+21$
-11

9. $5(-x^2-x+3)-(2x^2-x+6)$
$\quad -2(x^2+4x-6)$

$-5x^2-5x+15-2x^2+x-6$
$\quad -2x^2-8x+12$

$-9x^2-12x+21$ for $x=2$
$-9(2)^2-12(2)+21$
$-9(4)-24+21$
$-36-24+21$
$-60+21=-39$

10. $3(x^2-4xy+2y^2)-2(x^2-6xy-y^2)$
$3x^2-12xy+6y^2-2x^2+12xy+2y^2$
x^2+8y^2 for $x=-5$ and $y=-2$
$(-5)^2+8(-2)^2$
$25+8(4)$
$25+32$
57

11. $4(3x-2)-2(4x-1)-(2x+5)$
$12x-8-8x+2-2x-5$
$2x-11$

12. $(-6ab^2)(2ab)(-3b^3)$
$36a^{1+1}b^{2+1+3}$
$36a^2b^6$

13. $(5x-7)(6x+1)$
$30x^2+5x-42x-7$
$30x^2-37x-7$

14. $(-2x-3)(x+4)$
$-2x^2-8x-3x-12$
$-2x^2-11x-12$

15. $(-4a^2b^3)^3$
$(-4)^3a^{2(3)}b^{3(3)}$
$-64a^6b^9$

16. $(x+2)(5x-6)(x-2)$
$(x+2)(5x^2-16x+12)$
$x(5x^2-16x+12)+2(5x^2-16x+12)$
$5x^3-16x^2+12x+10x^2-32x+24$
$5x^3-6x^2-20x+24$

17. $(x-3)(x^2-x-4)$
$x(x^2-x-4)-3(x^2-x-4)$
$x^3-x^2-4x-3x^2+3x+12$
x^3-4x^2-x+12

18. $(x^2+x+4)(2x^2-3x-7)$

$x^2(2x^2-3x-7)+x(2x^2-3x-7)$
$\quad +4(2x^2-3x-7)$

$2x^4-3x^3-7x^2+2x^3-3x^2-7x$
$\quad +8x^2-12x-28$

$2x^4-x^3-2x^2-19x-28$

19. $7x^2-7$
$7(x^2-1)$
$7(x+1)(x-1)$

20. $4a^2-4ab+b^2$
$(2a-b)(2a-b)$
$(2a-b)^2$

21. $3x^2 - 17x - 56$
$3x^2 - 24x + 7x - 56$
$3x(x - 8) + 7(x - 8)$
$(x - 8)(3x + 7)$

22. $1 - x^3$
$(1 - x)(1 + x + x^2)$

23. $xy - 5x + 2y - 10$
$x(y - 5) + 2(y - 5)$
$(y - 5)(x + 2)$

24. $3x^2 - 24x + 48$
$3(x^2 - 8x + 16)$
$3(x - 4)(x - 4)$
$3(x - 4)^2$

25. $4n^4 - n^2 - 3$
$(4n^2 + 3)(n^2 - 1)$
$(4n^2 + 3)(n + 1)(n - 1)$

26. $32x^4 + 108x$
$4x(8x^3 + 27)$
$4x(2x + 3)(4x^2 - 6x + 9)$

27. $4x^2 + 36$
$4(x^2 + 9)$

28. $6x^2 + 5x - 4$
$6x^2 + 8x - 3x - 4$
$2x(3x + 4) - 1(3x + 4)$
$(3x + 4)(2x - 1)$

29. $9x^2 - 30x + 25$
$9x^2 - 15x - 15x + 25$
$3x(3x - 5) - 5(3x - 5)$
$(3x - 5)(3x - 5)$
$(3x - 5)^2$

30. $2x^2 + 6xy + x + 3y$
$2x(x + 3y) + 1(x + 3y)$
$(x + 3y)(2x + 1)$

31. $8a^3 + 27b^3$
$(2a)^3 + (3b)^3$
$(2a + 3b)(4a^2 - 6ab + 9b^2)$

32. $x^4 - 16$
$(x^2 + 4)(x^2 - 4)$
$(x^2 + 4)(x + 2)(x - 2)$

33. $10m^4n^2 - 2m^3n^3 - 4m^2n^4$
$2m^2n^2(5m^2 - mn - 2n^2)$

34. $5x(2y + 7z) - 12(2y + 7z)$
$(2y + 7z)(5x - 12)$

35. $3x^2 - x - 10$
$3x^2 - 6x + 5x - 10$
$3x(x - 2) + 5(x - 2)$
$(x - 2)(3x + 5)$

36. $25 - 4a^2$
$(5)^2 - (2a)^2$
$(5 - 2a)(5 + 2a)$

37. $36x^2 + 60x + 25$
$36x^2 + 30x + 30x + 25$
$6x(6x + 5) + 5(6x + 5)$
$(6x + 5)(6x + 5)$
$(6x + 5)^2$

38. $64y^3 + 1$
$(4y)^3 + (1)^3$
$(4y + 1)(16y^2 - 4y + 1)$

39. $5x - 2y = 6$ for x
$5x = 2y + 6$
$\frac{1}{5}(5x) = \frac{1}{5}(2y + 6)$
$x = \frac{2y + 6}{5}$

40. $3x + 4y = 12$ for y
$4y = 12 - 3x$
$\frac{1}{4}(4y) = \frac{1}{4}(12 - 3x)$
$y = \frac{12 - 3x}{4}$

41. $V = 2\pi rh + 2\pi r^2$ for h

$V - 2\pi r^2 = 2\pi rh$

$\frac{1}{2\pi r}(V - 2\pi r^2) = \frac{1}{2\pi r}(2\pi rh)$

$\frac{V - 2\pi r^2}{2\pi r} = h$

42. $\frac{1}{R} = \frac{1}{R_1} + \frac{1}{R_2}$ for R_1

$RR_1R_2\left(\frac{1}{R}\right) = RR_1R_2\left(\frac{1}{R_1} + \frac{1}{R_2}\right)$

$R_1R_2 = RR_1R_2\left(\frac{1}{R_1}\right) + RR_1R_2\left(\frac{1}{R_2}\right)$

$R_1R_2 = RR_2 + RR_1$

$R_1R_2 - RR_1 = RR_2$

$R_1(R_2 - R) = RR_2$

$R_1 = \frac{RR_2}{R_2 - R}$

43. $A = P + Prt$ for r, if $A = 4997$

$P = 3800$

$t = 3$ years

$4997 = 3800 + 3800(r)(3)$

$4997 = 3800 + 11400r$

$1197 = 11400r$

$\frac{1197}{11400} = r$

$0.105 = r$

The rate is 10.5%.

44. $C = \frac{5}{9}(F - 32)$

$C = \frac{5}{9}(5 - 32)$

$C = \frac{5}{9}(-27)$

$C = -15$

The temperature is $-15°C$.

45. $(x - 2)(x + 5) = 8$

$x^2 + 5x - 2x - 10 = 8$

$x^2 + 3x - 10 = 8$

$x^2 + 3x - 18 = 0$

$(x + 6)(x - 3) = 0$

$x + 6 = 0$ or $x - 3 = 0$

$x = -6$ or $x = 3$

The solution set is $\{-6, 3\}$.

46. $(5n - 2)(3n + 7) = 0$

$5n - 2 = 0$ or $3n + 7 = 0$

$5n = 2$ or $3n = -7$

$n = \frac{2}{5}$ or $n = -\frac{7}{3}$

The solution set is $\left\{-\frac{7}{3}, \frac{2}{5}\right\}$.

47. $-2(n - 1) + 3(2n + 1) = -11$

$-2n + 2 + 6n + 3 = -11$

$4n + 5 = -11$

$4n = -16$

$n = -4$

The solution set is $\{-4\}$.

48. $x^2 + 7x - 18 = 0$

$(x + 9)(x - 2) = 0$

$x + 9 = 0$ or $x - 2 = 0$

$x = -9$ or $x = 2$

The solution set is $\{-9, 2\}$.

49. $8x^2 - 8 = 0$

$8(x^2 - 1) = 0$

$8(x + 1)(x - 1) = 0$

$x + 1 = 0$ or $x - 1 = 0$

$x = -1$ or $x = 1$

The solution set is $\{-1, 1\}$.

50. $\frac{3}{4}(x - 2) - \frac{2}{5}(2x - 3) = \frac{1}{5}$

$20[\frac{3}{4}(x - 2) - \frac{2}{5}(2x - 3)] = 20\left(\frac{1}{5}\right)$

$20[\frac{3}{4}(x - 2)] - 20[\frac{2}{5}(2x - 3)] = 4$

$15(x - 2) - 8(2x - 3) = 4$

$15x - 30 - 16x + 24 = 4$

$-x - 6 = 4$

$-x = 10$

$x = -10$

The solution set is $\{-10\}$.

51. $0.1(x - 0.1) - 0.4(x + 2) = -5.31$

$100[0.1(x - 0.1) - 0.4(x + 2)] = 100(-5.31)$

$100[0.1(x - 0.1)] - 100[0.4(x + 2)] = -531$

$10(x - 0.1) - 40(x + 2) = -531$

$10x - 1 - 40x - 80 = -531$
$-30x - 81 = -531$
$-30x = -450$
$x = 15$
The solution set is $\{15\}$.

52. $\frac{2x-1}{2} - \frac{5x+2}{3} = 3$
$6\left(\frac{2x-1}{2} - \frac{5x+2}{3}\right) = 6(3)$
$6\left(\frac{2x-1}{2}\right) - 6\left(\frac{5x+2}{3}\right) = 18$
$3(2x-1) - 2(5x+2) = 18$
$6x - 3 - 10x - 4 = 18$
$-4x - 7 = 18$
$-4x = 25$
$x = -\frac{25}{4}$
The solution set is $\left\{-\frac{25}{4}\right\}$.

53. $|3n - 2| = 7$
$3n - 2 = 7$ or $3n - 2 = -7$
$3n = 9$ or $3n = -5$
$n = 3$ or $n = -\frac{5}{3}$
The solution set is $\left\{-\frac{5}{3}, 3\right\}$.

54. $|2x - 1| = |x + 4|$
$2x - 1 = x + 4$ or $2x - 1 = -(x + 4)$
$2x = x + 5$ or $2x - 1 = -x - 4$
$x = 5$ or $2x = -x - 3$
$3x = -3$
$x = -1$
The solution set is $\{-1, 5\}$.

55. $0.08(x + 200) = 0.07x + 20$
$100[0.08(x + 200)] = 100[0.07x + 20]$
$8(x + 200) = 100(0.07x) + 100(20)$
$8x + 1600 = 7x + 2000$
$x + 1600 = 2000$
$x = 400$
The solution set is $\{400\}$.

56. $2x^2 - 12x - 80 = 0$
$2(x^2 - 6x - 40) = 0$
$2(x - 10)(x + 4) = 0$
$x - 10 = 0$ or $x + 4 = 0$
$x = 10$ or $x = -4$
The solution set is $\{-4, 10\}$.

57. $x^3 = 16x$
$x^3 - 16x = 0$
$x(x^2 - 16) = 0$
$x(x + 4)(x - 4) = 0$
$x = 0$ or $x + 4 = 0$ or $x - 4 = 0$
$x = 0$ or $x = -4$ or $x = 4$
The solution set is $\{-4, 0, 4\}$.

58. $x(x + 2) - 3(x + 2) = 0$
$(x + 2)(x - 3) = 0$
$x + 2 = 0$ or $x - 3 = 0$
$x = -2$ or $x = 3$
The solution set is $\{-2, 3\}$.

59. $-12n^2 + 5n + 2 = 0$
$12n^2 - 5n - 2 = 0$
$(4n + 1)(3n - 2) = 0$
$4n + 1 = 0$ or $3n - 2 = 0$
$4n = -1$ or $3n = 2$
$n = -\frac{1}{4}$ or $n = \frac{2}{3}$
The solution set is $\left\{-\frac{1}{4}, \frac{2}{3}\right\}$.

60. $3y(y + 1) = 90$
$3y^2 + 3y = 90$
$3y^2 + 3y - 90 = 0$
$3(y^2 + y - 30) = 0$
$3(y + 6)(y - 5) = 0$
$y + 6 = 0$ or $y - 5 = 0$
$y = -6$ or $y = 5$
The solution set is $\{-6, 5\}$.

61. $2x^3 + 6x^2 - 20x = 0$
$2x(x^2 + 3x - 10) = 0$
$2x(x + 5)(x - 2) = 0$
$2x = 0$ or $x + 5 = 0$ or $x - 2 = 0$
$x = 0$ or $x = -5$ or $x = 2$
The solution set is $\{-5, 0, 2\}$.

Chapter 3 Cumulative Review Problem Set

62. $(3n-1)(2n+3) = (n+4)(6n-5)$
$6n^2 + 7n - 3 = 6n^2 + 19n - 20$
$7n - 3 = 19n - 20$
$7n = 19n - 17$
$-12n = -17$
$n = \frac{-17}{-12} = \frac{17}{12}$
The solution set is $\left\{\frac{17}{12}\right\}$.

63. $-5(3n+4) < -2(7n-1)$
$-15n - 20 < -14n + 2$
$-15n < -14n + 22$
$-n < 22$
$-1(-n) > -1(22)$
$n > -22$
The solution set is $(-22, \infty)$.

64. $7(x+1) - 8(x-2) < 0$
$7x + 7 - 8x + 16 < 0$
$-x + 23 < 0$
$-x < -23$
$-1(-x) > -1(-23)$
$x > 23$
The solution set is $(23, \infty)$.

65. $|2x-1| > 7$
$2x - 1 < -7$ or $2x - 1 > 7$
$2x < -6$ or $2x > 8$
$x < -3$ or $x > 4$
The solution set is $(-\infty, -3) \cup (4, \infty)$.

66. $|3x+7| < 14$
$-14 < 3x + 7 < 14$
$-21 < 3x < 7$
$-7 < x < \frac{7}{3}$
The solution set is $\left(-7, \frac{7}{3}\right)$.

67. $0.09x + 0.1(x+200) > 77$
$100[0.09x + 0.1(x+200)] > 100(77)$
$100[0.09x] + 100[0.1(x+200)] > 7700$
$9x + 10(x+200) > 7700$
$9x + 10x + 2000 > 7700$
$19x + 2000 > 7700$
$19x > 5700$
$x > 300$
The solution set is $(300, \infty)$.

68. $\frac{2x-1}{4} - \frac{x-2}{6} \leq \frac{3}{8}$
$24\left(\frac{2x-1}{4} - \frac{x-2}{6}\right) \leq 24\left(\frac{3}{8}\right)$
$24\left(\frac{2x-1}{4}\right) - 24\left(\frac{x-2}{6}\right) \leq 9$
$6(2x-1) - 4(x-2) \leq 9$
$12x - 6 - 4x + 8 \leq 9$
$8x + 2 \leq 9$
$8x \leq 7$
$x \leq \frac{7}{8}$
The solution set is $\left(-\infty, \frac{7}{8}\right]$.

69. $-(x-1) + 2(3x-1) \geq 2(x+4) - (x-1)$
$-x + 1 + 6x - 2 \geq 2x + 8 - x + 1$
$5x - 1 \geq x + 9$
$5x \geq x + 10$
$4x \geq 10$
$x \geq \frac{10}{4}$
$x \geq \frac{5}{2}$
The solution set is $\left[\frac{5}{2}, \infty\right)$.

70. $\frac{1}{4}(x-2) + \frac{3}{7}(2x-1) < \frac{3}{14}$
$28[\frac{1}{4}(x-2) + \frac{3}{7}(2x-1)] < 28\left(\frac{3}{14}\right)$
$28[\frac{1}{4}(x-2)] + 28[\frac{3}{7}(2x-1)] < 6$
$7(x-2) + 12(2x-1) < 6$
$7x - 14 + 24x - 12 < 6$
$31x - 26 < 6$
$31x < 32$
$x < \frac{32}{31}$
The solution set is $\left(-\infty, \frac{32}{31}\right)$.

71. Let $x =$ 1st odd integer,
$x + 2 =$ 2nd odd integer and
$x + 4 =$ 3rd odd integer.

$3x - (x + 2) = x + 4 + 1$
$3x - x - 2 = x + 5$
$2x - 2 = x + 5$
$2x = x + 7$
$x = 7$
The integers are 7, 9, and 11.

72. Let $x =$ number of nickels,
$2x - 1 =$ number of dimes and
$2x + 9 =$ number of quarters.

$x + 2x - 1 + 2x + 9 = 48$
$5x + 8 = 48$
$5x = 40$
$x = 8$
number of nickels $= 8$
number of dimes $= 2(8) - 1 = 15$
number of quarters $= 2(8) + 9 = 25$

There are 8 nickels, 15 dimes, and 25 quarters.

73. Present Ages
Joey $= x$
Mother $= 46 - x$

In four years
Joey $= x + 4$
Mother $= 46 - x + 4 = 50 - x$

$x + 4 = \frac{1}{2}(50 - x) - 3$

$2(x + 4) = 2\left[\frac{1}{2}(50 - x) - 3\right]$

$2x + 8 = 2\left[\frac{1}{2}(50 - x)\right] - 2(3)$

$2x + 8 = 50 - x - 6$
$2x + 8 = 44 - x$
$2x = 36 - x$
$3x = 36$
$x = 12$
Joey's present age is 12 and his mother's present age is 34.

74. Let $x =$ one angle and
$180 - x =$ the other angle (supplementary).
$x - (180 - x) = 56$
$x - 180 + x = 56$
$2x - 180 = 56$
$2x = 236$
$x = 118$
The angles are 118° and 62°.

75. Let $x =$ money invested at 8% and
$x + 200 =$ money invested at 9%.
$(8\%)(x) + (9\%)(x + 200) = 86$
$.08x + .09(x + 200) = 86$
$100[.08x + .09(x + 200)] = 100(86)$
$100(.08x) + 100[.09(x + 200)] = 8600$
$8x + 9(x + 200) = 8600$
$8x + 9x + 1800 = 8600$
$17x + 1800 = 8600$
$17x = 6800$
$x = 400$
There should be \$400 invested at 8% and \$600 invested at 9%.

76. Let $x =$ number of pennies,
$x + 5 =$ number of nickels and
$2x =$ number of dimes.

$0.01x + 0.05(x + 5) + 0.10(2x) = 9.35$
$100[0.01x + 0.05(x + 5) + 0.10(2x)] = 100(9.35)$
$100(0.01x) + 100[0.05(x + 5)] + 100[0.10(2x)] = 935$
$x + 5(x + 5) + 10(2x) = 935$
$x + 5x + 25 + 20x = 935$
$26x + 25 = 935$
$26x = 910$
$x = 35$
number of pennies $= 35$
number of nickels $= 35 + 5 = 40$
number of dimes $= 2(35) = 70$

There are 35 pennies, 40 nickels, and 70 dimes.

77. Let x = time in hours Billie travels and

$x + \frac{5}{6}$ = time in hours Sandy travels.

$12x$ = Billie's distance

$8\left(x + \frac{5}{6}\right)$ = Sandy's distance

$$12x = 8\left(x + \frac{5}{6}\right)$$
$$12x = 8x + 8\left(\frac{5}{6}\right)$$
$$12x = 8x + \frac{20}{3}$$
$$4x = \frac{20}{3}$$
$$\frac{1}{4}(4x) = \frac{1}{4}\left(\frac{20}{3}\right)$$
$$x = \frac{5}{3} = 1\frac{2}{3} \text{ hours}$$

It takes Billie 1 hour 40 minutes to overtake Sandy.

78. Let x = amount of pure acid.

$$(100\%)(x) + (30\%)(150) = (40\%)(x + 150)$$
$$1.00x + .30(150) = .40(x + 150)$$
$$100[1.00x + .30(150)] = 100[.40(x + 150)]$$
$$100(1.00x) + 100[.30(150)] = 40(x + 150)$$
$$100x + 30(150) = 40x + 6000$$
$$100x + 4500 = 40x + 6000$$
$$100x = 40x + 1500$$
$$60x = 1500$$
$$x = 25$$

The amount of pure acid to be added is 25 milliliters.

79. Let x = the rate of profit.

selling price = cost + profit

$$30 = 18 + 30x$$
$$12 = 30x$$
$$\frac{12}{30} = x$$
$$0.40 = x$$

The rate of profit based on selling price is 40%.

80. Let x = score on fifth test

$$\frac{88 + 92 + 93 + 89 + x}{5} > 90$$
$$\frac{362 + x}{5} > 90$$
$$5\left(\frac{362 + x}{5}\right) > 5(90)$$
$$362 + x > 450$$
$$x > 88$$

The score should be better than 88.

81. Let x = side of square.

x = altitude

16 = side (base)

Area of square = x^2

Area of triangle = $\frac{1}{2}x(16) = 8x$

$$x^2 = \frac{1}{2}(8x)$$
$$x^2 = 4x$$
$$x^2 - 4x = 0$$
$$x(x - 4) = 0$$

$x = 0$ or $x - 4 = 0$

$x = 0$ or $x = 4$

Discard the root $x = 0$.

The length of the side of the square is 4 inches.

82. Let x = width

and $2x$ = length.

$$x(2x) = 98$$
$$2x^2 = 98$$
$$2x^2 - 98 = 0$$
$$2(x^2 - 49) = 0$$
$$2(x + 7)(x - 7) = 0$$

$x + 7 = 0$ or $x - 7 = 0$

$x = -7$ or $x = 7$

Discard the root $x = -7$.

width = 7

length = 2(7) = 14

The width should be 7 meters and the length should be 14 meters.

83. Let $x =$ number of rows and
$x + 4 =$ number of chairs per row.

$x(x + 4) = 96$
$x^2 + 4x = 96$
$x^2 + 4x - 96 = 0$
$(x + 12)(x - 8) = 0$
$x + 12 = 0$ or $x - 8 = 0$
$x = -12$ or $x = 8$
Discard the root $x = -12$. The number of rows is 8 and the number of chairs per row is 12.

84. Let $x =$ 1st leg
$x + 3 =$ 2nd leg
$x + 6 =$ hypotenuse

Use the Pythagorean Theorm.
$x^2 + (x + 3)^2 = (x + 6)^2$
$x^2 + x^2 + 6x + 9 = x^2 + 12x + 36$
$2x^2 + 6x + 9 = x^2 + 12x + 36$
$x^2 - 6x - 27 = 0$
$(x - 9)(x + 3) = 0$
$x - 9 = 0$ or $x + 3 = 0$
$x = 9$ or $x = -3$
Discard the root $x = -3$.
The legs are 9 and 12 feet and the hypotenuse is 15 feet.

Chapter 4 Rational Expressions

PROBLEM SET 4.1 Simplifying Rational Expressions

1. $\frac{27}{36} = \frac{9 \bullet 3}{9 \bullet 4} = \frac{3}{4}$

3. $\frac{45}{54} = \frac{9 \bullet 5}{9 \bullet 6} = \frac{5}{6}$

5. $\frac{24}{-60} = \frac{12 \bullet 2}{12 \bullet -5} = -\frac{2}{5}$

7. $\frac{-16}{-56} = \frac{-8 \bullet 2}{-8 \bullet 7} = \frac{2}{7}$

9. $\frac{12xy}{42y} = \frac{6y}{6y} \bullet \frac{2x}{7} = \frac{2x}{7}$

11. $\frac{18a^2}{45ab} = \frac{9a}{9a} \bullet \frac{2a}{5b} = \frac{2a}{5b}$

13. $\frac{-14y^3}{56xy^2} = \frac{14y^2}{14y^2} \bullet \frac{-y}{4x} = -\frac{y}{4x}$

15. $\frac{54c^2d}{-78cd^2} = \frac{6cd}{6cd} \bullet \frac{9c}{-13d} = -\frac{9c}{13d}$

17. $\frac{-40x^3y}{-24xy^4} = \frac{-8xy}{-8xy} \bullet \frac{5x^2}{3y^3} = \frac{5x^2}{3y^3}$

19. $\frac{x^2-4}{x^2+2x} = \frac{(x+2)(x-2)}{x(x+2)} = \frac{x-2}{x}$

21. $\frac{18x+12}{12x-6} = \frac{6(3x+2)}{6(2x-1)} = \frac{3x+2}{2x-1}$

23. $\frac{a^2+7a+10}{a^2-7a-18}$

$\frac{(a+5)(a+2)}{(a-9)(a+2)}$

$\frac{a+5}{a-9}$

25. $\frac{2n^2+n-21}{10n^2+33n-7}$

$\frac{(2n+7)(n-3)}{(2n+7)(5n-1)}$

$\frac{n-3}{5n-1}$

27. $\frac{5x^2+7}{10x}$

29. $\frac{6x^2+x-15}{8x^2-10x-3}$

$\frac{(3x+5)(2x-3)}{(4x+1)(2x-3)}$

$\frac{3x+5}{4x+1}$

31. $\frac{3x^2-12x}{x^3-64}$

$\frac{3x(x-4)}{(x-4)(x^2+4x+16)}$

$\frac{3x}{x^2+4x+16}$

33. $\frac{3x^2+17x-6}{9x^2-6x+1}$

$\frac{(3x-1)(x+6)}{(3x-1)(3x-1)}$

$\frac{x+6}{3x-1}$

35. $\frac{2x^3+3x^2-14x}{x^2y+7xy-18y}$

$\frac{x(2x^2+3x-14)}{y(x^2+7x-18)}$

$$\frac{x(2x+7)(x-2)}{y(x+9)(x-2)}$$

$$\frac{x(2x+7)}{y(x+9)}$$

37. $$\frac{5y^2+22y+8}{25y^2-4}$$

$$\frac{(5y+2)(y+4)}{(5y+2)(5y-2)}$$

$$\frac{y+4}{5y-2}$$

39. $$\frac{15x^3-15x^2}{5x^3+5x}$$

$$\frac{15x^2(x-1)}{5x(x^2+1)}$$

$$\frac{3x(x-1)}{x^2+1}$$

41. $$\frac{4x^2y+8xy^2-12y^3}{18x^3y-12x^2y^2-6xy^3}$$

$$\frac{4y(x^2+2xy-3y^2)}{6xy(3x^2-2xy-y^2)}$$

$$\frac{4y(x+3y)(x-y)}{6xy(3x+y)(x-y)}$$

$$\frac{2(x+3y)}{3x(3x+y)}$$

43. $$\frac{3n^2+16n-12}{7n^2+44n+12}$$

$$\frac{(3n-2)(n+6)}{(7n+2)(n+6)}$$

$$\frac{3n-2}{7n+2}$$

45. $$\frac{8+18x-5x^2}{10+31x+15x^2}$$

$$\frac{(4-x)(2+5x)}{(5+3x)(2+5x)}$$

$$\frac{4-x}{5+3x}$$

47. $$\frac{27x^4-x}{6x^3+10x^2-4x}$$

$$\frac{x(27x^3-1)}{2x(3x^2+5x-2)}$$

$$\frac{x(3x-1)(9x^2+3x+1)}{2x(3x-1)(x+2)}$$

$$\frac{9x^2+3x+1}{2(x+2)}$$

49. $$\frac{-40x^3+24x^2+16x}{20x^3+28x^2+8x}$$

$$\frac{-8x(5x^2-3x-2)}{4x(5x^2+7x+2)}$$

$$\frac{-8x(5x+2)(x-1)}{4x(5x+2)(x+1)}$$

$$\frac{-2(x-1)}{x+1}$$

51. $$\frac{xy+ay+bx+ab}{xy+ay+cx+ac}$$

$$\frac{y(x+a)+b(x+a)}{y(x+a)+c(x+a)}$$

$$\frac{(x+a)(y+b)}{(x+a)(y+c)}$$

$$\frac{y+b}{y+c}$$

53. $$\frac{ax-3x+2ay-6y}{2ax-6x+ay-3y}$$

$$\frac{x(a-3)+2y(a-3)}{2x(a-3)+y(a-3)}$$

Problem Set 4.1

$$\frac{(a-3)(x+2y)}{(a-3)(2x+y)}$$

$$\frac{x+2y}{2x+y}$$

55. $$\frac{5x^2+5x+3x+3}{5x^2+3x-30x-18}$$

$$\frac{5x(x+1)+3(x+1)}{x(5x+3)-6(5x+3)}$$

$$\frac{(x+1)(5x+3)}{(5x+3)(x-6)}$$

$$\frac{x+1}{x-6}$$

57. $$\frac{2st-30-12s+5t}{3st-6-18s+t}$$

$$\frac{2st-12s+5t-30}{3st-18s+t-6}$$

$$\frac{2s(t-6)+5(t-6)}{3s(t-6)+1(t-6)}$$

$$\frac{(t-6)(2s+5)}{(t-6)(3s+1)}$$

$$\frac{2s+5}{3s+1}$$

59. $$\frac{5x-7}{7-5x}$$

$$\frac{-1(-5x+7)}{(7-5x)}$$

$$\frac{-1(7-5x)}{(7-5x)}$$

$$-1$$

61. $$\frac{n^2-49}{7-n}$$

$$\frac{(n+7)(n-7)}{-1(-7+n)}$$

$$\frac{(n+7)(n-7)}{-1(n-7)}$$

$$\frac{n+7}{-1}$$

$$-n-7$$

63. $$\frac{2y-2xy}{x^2y-y}$$

$$\frac{2y(1-x)}{y(x^2-1)}$$

$$\frac{2y(1-x)}{y(x+1)(x-1)}$$

$$\frac{2y(-1)(-1+x)}{y(x+1)(x-1)}$$

$$\frac{-2y(x-1)}{y(x+1)(x-1)}$$

$$\frac{-2y}{y(x+1)}=\frac{-2}{x+1}$$

65. $$\frac{2x^3-8x}{4x-x^3}$$

$$\frac{2x(x^2-4)}{-x(-4+x^2)}$$

$$\frac{2x(x^2-4)}{-x(x^2-4)}$$

$$-2$$

67. $$\frac{n^2-5n-24}{40+3n-n^2}$$

$$\frac{(n-8)(n+3)}{(8-n)(5+n)}$$

$$\frac{(n-8)(n+3)}{-1(-8+n)(5+n)}$$

$$\frac{(n-8)(n+3)}{-1(n-8)(n+5)}$$

$$\frac{-(n+3)}{n+5} = -\frac{n+3}{n+5}$$

PROBLEM SET 4.2 Multiplying and Dividing Rational Expressions

1. $\frac{7}{12} \bullet \frac{6}{35}$

$\frac{7}{2 \bullet 6} \bullet \frac{6}{5 \bullet 7}$

$\frac{1}{10}$

3. $\frac{-4}{9} \bullet \frac{18}{30}$

$\frac{-4}{9} \bullet \frac{9 \bullet 2}{2 \bullet 15}$

$-\frac{4}{15}$

5. $\frac{3}{-8} \bullet \frac{-6}{12}$

$\frac{3}{-8} \bullet \frac{-6}{2 \bullet 6}$

$\frac{3}{16}$

7. $\left(-\frac{5}{7}\right) \div \frac{6}{7}$

$-\frac{5}{7} \bullet \frac{7}{6}$

$-\frac{5}{6}$

9. $-\frac{9}{5} \div \frac{27}{10}$

$-\frac{9}{5} \bullet \frac{10}{27}$

$-\frac{9}{5} \bullet \frac{5 \bullet 2}{3 \bullet 9}$

$-\frac{2}{3}$

11. $\frac{4}{9} \bullet \frac{6}{11} \div \frac{4}{15}$

$\frac{4}{9} \bullet \frac{6}{11} \bullet \frac{15}{4}$

$\frac{4}{3 \bullet 3} \bullet \frac{2 \bullet 3}{11} \bullet \frac{3 \bullet 5}{4}$

$\frac{10}{11}$

13. $\frac{6xy}{9y^4} \bullet \frac{30x^3y}{-48x}$

$\frac{6(30)x^4y^2}{9(-48)xy^4}$

$\frac{6}{9} \bullet \frac{30x^4y^2}{-48xy^4}$

$\frac{2}{3} \bullet \frac{5x^3}{-8y^2}$

$\frac{-5x^3}{12y^2}$

15. $\frac{5a^2b^2}{11ab} \bullet \frac{22a^3}{15ab^2}$

$\frac{5(22)a^5b^2}{11(15)a^2b^3}$

Problem Set 4.2

$$\frac{22}{11} \bullet \frac{5a^3}{15b}$$

$$\frac{2}{1} \bullet \frac{a^3}{3b}$$

$$\frac{2a^3}{3b}$$

17. $$\frac{5xy}{8y^2} \bullet \frac{18x^2y}{15}$$

$$\frac{5(18)x^3y^2}{8(15)y^2}$$

$$\frac{5}{15} \bullet \frac{18x^3}{8}$$

$$\frac{1}{3} \bullet \frac{9x^3}{4}$$

$$\frac{3x^3}{4}$$

19. $$\frac{5x^4}{12x^2y^3} \div \frac{9}{5xy}$$

$$\frac{5x^4}{12x^2y^3} \bullet \frac{5xy}{9}$$

$$\frac{5(5)x^5y}{12(9)x^2y^3}$$

$$\frac{25x^3}{108y^2}$$

21. $$\frac{9a^2c}{12bc^2} \div \frac{21ab}{14c^3}$$

$$\frac{9a^2c}{12bc^2} \bullet \frac{14c^3}{21ab}$$

$$\frac{9(14)a^2c^4}{12(21)ab^2c^2}$$

$$\frac{3(2)ac^2}{4(3)b^2}$$

$$\frac{ac^2}{2b^2}$$

23. $$\frac{9x^2y^3}{14x} \bullet \frac{21y}{15xy^2} \bullet \frac{10x}{12y^3}$$

$$\frac{9(21)(10)x^3y^4}{14(15)(12)x^2y^5}$$

$$\frac{21}{14} \bullet \frac{9}{12} \bullet \frac{10}{15} \bullet \frac{x}{y}$$

$$\frac{3}{2} \bullet \frac{3}{4} \bullet \frac{2}{3} \bullet \frac{x}{y}$$

$$\frac{3x}{4y}$$

25. $$\frac{3x+6}{5y} \bullet \frac{x^2+4}{x^2+10x+16}$$

$$\frac{3(x+2)}{5y} \bullet \frac{x^2+4}{(x+8)(x+2)}$$

$$\frac{3(x^2+4)}{5y(x+8)}$$

27. $$\frac{5a^2+20a}{a^3-2a^2} \bullet \frac{a^2-a-12}{a^2-16}$$

$$\frac{5a(a+4)}{a^2(a-2)} \bullet \frac{(a-4)(a+3)}{(a-4)(a+4)}$$

$$\frac{5(a+3)}{a(a-2)}$$

29. $$\frac{3n^2+15n-18}{3n^2+10n-48} \bullet \frac{6n^2-n-40}{4n^2+6n-10}$$

$$\frac{3(n^2+5n-6)}{3n^2+10n-48} \bullet \frac{6n^2-n-40}{2(2n^2+3-5)}$$

$$\frac{3(n+6)(n-1)}{(3n-8)(n+6)} \bullet \frac{(2n+5)(3n-8)}{2(2n+5)(n-1)}$$

$$\frac{3}{2}$$

31. $\frac{9y^2}{x^2+12x+36} \div \frac{12y}{x^2+6x}$

$\frac{9y^2}{x^2+12x+36} \bullet \frac{x^2+6x}{12y}$

$\frac{9y^2}{(x+6)(x+6)} \bullet \frac{x(x+6)}{12y}$

$\frac{3xy}{4(x+6)}$

33. $\frac{x^2-4xy+4y^2}{7xy^2} \div \frac{4x^2-3xy-10y^2}{20x^2y+25xy^2}$

$\frac{x^2-4xy+4y^2}{7xy^2} \bullet \frac{20x^2y+25xy^2}{4x^2-3xy-10y^2}$

$\frac{(x-2y)(x-2y)}{7xy^2} \bullet \frac{5xy(4x+5y)}{(4x+5y)(x-2y)}$

$\frac{5(x-2y)}{7y}$

35. $\frac{5-14n-3n^2}{1-2n-3n^2} \bullet \frac{9+7n-2n^2}{27-15n+2n^2}$

$\frac{(5+n)(1-3n)}{(1+n)(1-3n)} \bullet \frac{(9-2n)(1+n)}{(9-2n)(3-n)}$

$\frac{5+n}{3-n}$

37. $\frac{3x^4+2x^2-1}{3x^4+14x^2-5} \bullet \frac{x^4-2x^2-35}{x^4-17x^2+70}$

$\frac{(3x^2-1)(x^2+1)}{(3x^2-1)(x^2+5)} \bullet \frac{(x^2-7)(x^2+5)}{(x^2-7)(x^2-10)}$

$\frac{x^2+1}{x^2-10}$

39. $\frac{3x^2-20x+25}{2x^2-7x-15} \div \frac{9x^2-3x-20}{12x^2+28x+15}$

$\frac{3x^2-20x+25}{2x^2-7x-15} \bullet \frac{12x^2+28x+15}{9x^2-3x-20}$

$\frac{(3x-5)(x-5)}{(2x+3)(x-5)} \bullet \frac{(2x+3)(6x+5)}{(3x+4)(3x-5)}$

$\frac{6x+5}{3x+4}$

41. $\frac{10t^3+25t}{20t+10} \bullet \frac{2t^2-t-1}{t^5-\mathrm{t}}$

$\frac{5t(2t^2+5)}{10(2t+1)} \bullet \frac{(2t+1)(t-1)}{t(t^4-1)}$

$\frac{5t(2t^2+5)}{10(2t+1)} \bullet \frac{(2t+1)(t-1)}{t(t^2+1)(t^2-1)}$

$\frac{5t(2t^2+5)}{10(2t+1)} \bullet \frac{(2t+1)(t-1)}{t(t^2+1)(t+1)(t-1)}$

$\frac{2t^2+5}{2(t^2+1)(t+1)}$

43. $\frac{4t^2+t-5}{t^3-t^2} \bullet \frac{t^4+6t^3}{16t^2+40t+25}$

$\frac{(4t+5)(t-1)}{t^2(t-1)} \bullet \frac{t^3(t+6)}{(4t+5)(4t+5)}$

$\frac{t(t+6)}{4t+5}$

45. $\frac{nr+3n+2r+6}{nr+3n-3r-9} \bullet \frac{n^2-9}{n^3-4n}$

$\frac{n(r+3)+2(r+3)}{n(r+3)-3(r+3)} \bullet \frac{(n+3)(n-3)}{n(n^2-4)}$

$\frac{(r+3)(n+2)}{(r+3)(n-3)} \bullet \frac{(n+3)(n-3)}{n(n+2)(n-2)}$

$\frac{n+3}{n(n-2)}$

47. $\frac{x^2-x}{4y} \bullet \frac{10xy^2}{2x-2} \div \frac{3x^2+3x}{15x^2y^2}$

$\frac{x^2-x}{4y} \bullet \frac{10xy^2}{2x-2} \bullet \frac{15x^2y^2}{3x^2+3x}$

$\frac{x(x-1)}{4y} \bullet \frac{10xy^2}{2(x-1)} \bullet \frac{15x^2y^2}{3x(x+1)}$

$\frac{25x^3y^3}{4(x+1)}$

49. $\frac{a^2-4ab+4b^2}{6a^2-4ab} \bullet \frac{3a^2+5ab-2b^2}{6a^2+ab-b^2} \div \frac{a^2-4b^2}{8a+4b}$

$\frac{a^2-4ab+4b^2}{6a^2-4ab} \bullet \frac{3a^2+5ab-2b^2}{6a^2+ab-b^2} \bullet \frac{8a+4b}{a^2-4b^2}$

$\frac{(a-2b)(a-2a)}{2a(3a-2b)} \bullet \frac{(3a-b)(a+2b)}{(3a-b)(2a+b)} \bullet \frac{4(2a+b)}{(a+2b)(a-2b)}$

$\frac{2(a-2b)}{a(3a-2b)}$

PROBLEM SET 4.3 Adding and Subtracting Rational Expressions

1. $\frac{1}{4}+\frac{5}{6}$; LCD = 12

$\frac{1}{4}\left(\frac{3}{3}\right)+\frac{5}{6}\left(\frac{2}{2}\right)$

$\frac{3}{12}+\frac{10}{12}$

$\frac{13}{12}$

3. $\frac{7}{8}-\frac{3}{5}$; LCD = 40

$\frac{7}{8}\left(\frac{5}{5}\right)-\frac{3}{5}\left(\frac{8}{8}\right)$

$\frac{35}{40}-\frac{24}{40}$

$\frac{11}{40}$

5. $\frac{6}{5}+\frac{1}{-4}$; LCD = 20

$\frac{6}{5}\left(\frac{4}{4}\right)+\left(-\frac{1}{4}\right)\left(\frac{5}{5}\right)$

$\frac{24}{20}+\left(-\frac{5}{20}\right)$

$\frac{19}{20}$

7. $\frac{8}{15}+\frac{3}{25}$; LCD = 75

$\frac{8}{15}\left(\frac{5}{5}\right)+\frac{3}{25}\left(\frac{3}{3}\right)$

$\frac{40}{75}+\frac{9}{75}$

$\frac{49}{75}$

9. $\frac{1}{5}+\frac{5}{6}-\frac{7}{15}$; LCD = 30

$\frac{1}{5}\left(\frac{6}{6}\right)+\frac{5}{6}\left(\frac{5}{5}\right)-\frac{7}{15}\left(\frac{2}{2}\right)$

$\frac{6}{30}+\frac{25}{30}-\frac{14}{30}$

$\frac{31}{30}-\frac{14}{30}$

$\frac{17}{30}$

11. $\frac{1}{3}-\frac{1}{4}-\frac{3}{14}$; LCD = 84

$$\frac{1}{3}\left(\frac{28}{28}\right)-\frac{1}{4}\left(\frac{21}{21}\right)-\frac{3}{14}\left(\frac{6}{6}\right)$$

$$\frac{28}{84}-\frac{21}{84}-\frac{18}{84}$$

$$\frac{7}{84}-\frac{18}{84}$$

$$-\frac{11}{84}$$

13. $\dfrac{2x}{x-1}+\dfrac{4}{x-1}$

$$\frac{2x+4}{x-1}$$

15. $\dfrac{4a}{a+2}+\dfrac{8}{a+2}$

$$\frac{4a+8}{a+2}$$

$$\frac{4(a+2)}{a+2}$$

$$4$$

17. $\dfrac{3(y-2)}{7y}+\dfrac{4(y-1)}{7y}$

$$\frac{3(y-2)+4(y-1)}{7y}$$

$$\frac{3y-6+4y-4}{7y}$$

$$\frac{7y-10}{7y}$$

19. $\dfrac{x-1}{2}+\dfrac{x+3}{3}$; LCD $= 6$

$$\frac{(x-1)}{2}\left(\frac{3}{3}\right)+\frac{(x+3)}{3}\left(\frac{2}{2}\right)$$

$$\frac{3(x-1)}{6}+\frac{2(x+3)}{6}$$

$$\frac{3(x-1)+2(x+3)}{6}$$

$$\frac{3x-3+2x+6}{6}$$

$$\frac{5x+3}{6}$$

21. $\dfrac{2a-1}{4}+\dfrac{3a+2}{6}$; LCD $= 12$

$$\frac{(2a-1)}{4}\left(\frac{3}{3}\right)+\frac{(3a+2)}{6}\left(\frac{2}{2}\right)$$

$$\frac{3(2a-1)}{12}+\frac{2(3a+2)}{12}$$

$$\frac{3(2a-1)+2(3a+2)}{12}$$

$$\frac{6a-3+6a+4}{12}$$

$$\frac{12a+1}{12}$$

23. $\dfrac{n+2}{6}-\dfrac{n-4}{9}$; LCD $= 18$

$$\frac{(n+2)}{6}\left(\frac{3}{3}\right)-\frac{(n-4)}{9}\left(\frac{2}{2}\right)$$

$$\frac{3(n+2)}{18}-\frac{2(n-4)}{18}$$

$$\frac{3(n+2)-2(n-4)}{18}$$

$$\frac{3n+6-2n+8}{18}$$

$$\frac{n+14}{18}$$

Problem Set 4.3

25. $\frac{3x-1}{3}-\frac{5x+2}{5}$; LCD $= 15$

$\frac{(3x-1)}{3}\left(\frac{5}{5}\right)-\frac{(5x+2)}{5}\left(\frac{3}{3}\right)$

$\frac{5(3x-1)}{15}-\frac{3(5x+2)}{15}$

$\frac{5(3x-1)-3(5x+2)}{15}$

$\frac{15x-5-15x-6}{15}$

$-\frac{11}{15}$

27. $\frac{x-2}{5}-\frac{x+3}{6}+\frac{x+1}{15}$; LCD $= 30$

$\frac{(x-2)}{5}\left(\frac{6}{6}\right)-\frac{(x+3)}{6}\left(\frac{5}{5}\right)+\frac{(x+1)}{15}\left(\frac{2}{2}\right)$

$\frac{6(x-2)}{30}-\frac{5(x+3)}{30}+\frac{2(x+1)}{30}$

$\frac{6(x-2)-5(x+3)+2(x+1)}{30}$

$\frac{6x-12-5x-15+2x+2}{30}$

$\frac{3x-25}{30}$

29. $\frac{3}{8x}+\frac{7}{10x}$; LCD $= 40x$

$\frac{3}{8x}\left(\frac{5}{5}\right)+\frac{7}{10x}\left(\frac{4}{4}\right)$

$\frac{15}{40x}+\frac{28}{40x}$

$\frac{43}{40x}$

31. $\frac{5}{7x}-\frac{11}{4y}$; LCD $= 28xy$

$\frac{5}{7x}\left(\frac{4y}{4y}\right)-\frac{11}{4y}\left(\frac{7x}{7x}\right)$

$\frac{20y}{28xy}-\frac{77x}{28xy}$

$\frac{20y-77x}{28xy}$

33. $\frac{4}{3x}+\frac{5}{4y}-1$; LCD $= 12xy$

$\frac{4}{3x}\left(\frac{4y}{4y}\right)+\frac{5}{4y}\left(\frac{3x}{3x}\right)-\left(\frac{12xy}{12xy}\right)$

$\frac{16y}{12xy}+\frac{15x}{12xy}-\frac{12xy}{12xy}$

$\frac{16y+15x-12xy}{12xy}$

35. $\frac{7}{10x^2}+\frac{11}{15x}$; LCD $= 30x^2$

$\frac{7}{10x^2}\left(\frac{3}{3}\right)+\frac{11}{15x}\left(\frac{2x}{2x}\right)$

$\frac{21}{30x^2}+\frac{22x}{30x^2}$

$\frac{21+22x}{30x^2}$

37. $\frac{10}{7n}-\frac{12}{4n^2}$; LCD $= 28n^2$

$\frac{10}{7n}\left(\frac{4n}{4n}\right)-\frac{12}{4n^2}\left(\frac{7}{7}\right)$

$\frac{40n}{28n^2}-\frac{84}{28n^2}$

$$\frac{40n - 84}{28n^2}$$

$$\frac{4(10n - 21)}{28n^2}$$

$$\frac{10n - 21}{7n^2}$$

39. $\dfrac{3}{n^2} - \dfrac{2}{5n} + \dfrac{4}{3}$; LCD $= 15n^2$

$$\frac{3}{n^2}\left(\frac{15}{15}\right) - \frac{2}{5n}\left(\frac{3n}{3n}\right) + \frac{4}{3}\left(\frac{5n^2}{5n^2}\right)$$

$$\frac{45}{15n^2} - \frac{6n}{15n^2} + \frac{20n^2}{15n^2}$$

$$\frac{45 - 6n + 20n^2}{15n^2}$$

41. $\dfrac{3}{x} - \dfrac{5}{3x^2} - \dfrac{7}{6x}$; LCD $= 6x^2$

$$\frac{3}{x}\left(\frac{6x}{6x}\right) - \frac{5}{3x^2}\left(\frac{2}{2}\right) - \frac{7}{6x}\left(\frac{x}{x}\right)$$

$$\frac{18x}{6x^2} - \frac{10}{6x^2} - \frac{7x}{6x^2}$$

$$\frac{18x - 10 - 7x}{6x^2}$$

$$\frac{11x - 10}{6x^2}$$

43. $\dfrac{6}{5t^2} - \dfrac{4}{7t^3} + \dfrac{9}{5t^3}$; LCD $= 35t^3$

$$\frac{6}{5t^2}\left(\frac{7t}{7t}\right) - \frac{4}{7t^3}\left(\frac{5}{5}\right) + \frac{9}{5t^3}\left(\frac{7}{7}\right)$$

$$\frac{42t}{35t^3} - \frac{20}{35t^3} + \frac{63}{35t^3}$$

$$\frac{42t - 20 + 63}{35t^3}$$

$$\frac{42t + 43}{35t^3}$$

45. $\dfrac{5b}{24a^2} - \dfrac{11a}{32b}$; LCD $= 96a^2b$

$$\frac{5b}{24a^2}\left(\frac{4b}{4b}\right) - \frac{11a}{32b}\left(\frac{3a^2}{3a^2}\right)$$

$$\frac{20b^2}{96a^2b} - \frac{33a^3}{96a^2b}$$

$$\frac{20b^2 - 33a^3}{96a^2b}$$

47. $\dfrac{7}{9xy^3} - \dfrac{4}{3x} + \dfrac{5}{2y^2}$; LCD $= 18xy^3$

$$\frac{7}{9xy^3}\left(\frac{2}{2}\right) - \frac{4}{3x}\left(\frac{6y^3}{6y^3}\right) + \frac{5}{2y^2}\left(\frac{9xy}{9xy}\right)$$

$$\frac{14}{18xy^3} - \frac{24y^3}{18xy^3} + \frac{45xy}{18xy^3}$$

$$\frac{14 - 24y^3 + 45xy}{18xy^3}$$

49. $\dfrac{2x}{x-1} + \dfrac{3}{x}$; LCD $= x(x-1)$

$$\frac{2x}{(x-1)}\left(\frac{x}{x}\right) + \frac{3}{x}\left(\frac{x-1}{x-1}\right)$$

$$\frac{2x(x)}{x(x-1)} + \frac{3(x-1)}{x(x-1)}$$

$$\frac{2x(x) + 3(x-1)}{x(x-1)}$$

$$\frac{2x^2 + 3x - 3}{x(x-1)}$$

Problem Set 4.3

51. $\frac{a-2}{a}-\frac{3}{a+4}$; LCD $= a(a+4)$

$$\frac{(a-2)}{a}\left(\frac{a+4}{a+4}\right)-\frac{3}{(a+4)}\left(\frac{a}{a}\right)$$

$$\frac{(a-2)(a+4)}{a(a+4)}-\frac{3a}{a(a+4)}$$

$$\frac{(a-2)(a+4)-3a}{a(a+4)}$$

$$\frac{a^2+2a-8-3a}{a(a+4)}$$

$$\frac{a^2-a-8}{a(a+4)}$$

53. $\frac{-3}{4n+5}-\frac{8}{3n+5}$;

LCD $= (4n+5)(3n+5)$

$$\frac{-3}{(4n+5)}\left(\frac{3n+5}{3n+5}\right)-\frac{8}{(3n+5)}\left(\frac{4n+5}{4n+5}\right)$$

$$\frac{-3(3n+5)}{(4n+5)(3n+5)}-\frac{8(4n+5)}{(4n+5)(3n+5)}$$

$$\frac{-3(3n+5)-8(4n+5)}{(4n+5)(3n+5)}$$

$$\frac{-9n-15-32n-40}{(4n+5)(3n+5)}$$

$$\frac{-41n-55}{(4n+5)(3n+5)}$$

55. $\frac{-1}{x+4}+\frac{4}{7x-1}$; LCD $= (x+4)(7x-1)$

$$\frac{-1}{(x+4)}\left(\frac{7x-1}{7x-1}\right)+\frac{4}{(7x-1)}\left(\frac{x+4}{x+4}\right)$$

$$\frac{-1(7x-1)}{(x+4)(7x-1)}+\frac{4(x+4)}{(7x-1)(x+4)}$$

$$\frac{-1(7x-1)+4(x+4)}{(7x-1)(x+4)}$$

$$\frac{-7x+1+4x+16}{(7x-1)(x+4)}$$

$$\frac{-3x+17}{(7x-1)(x+4)}$$

57. $\frac{7}{3x-5}-\frac{5}{2x+7}$; LCD $= (3x-5)(2x+7)$

$$\frac{7}{(3x-5)}\left(\frac{2x+7}{2x+7}\right)-\frac{5}{(2x+7)}\left(\frac{3x-5}{3x-5}\right)$$

$$\frac{7(2x+7)}{(3x-5)(2x+7)}-\frac{5(3x-5)}{(2x+7)(3x-5)}$$

$$\frac{7(2x+7)-5(3x-5)}{(3x-5)(2x+7)}$$

$$\frac{14x+49-15x+25}{(3x-5)(2x+7)}$$

$$\frac{-x+74}{(3x-5)(2x+7)}$$

59. $\frac{5}{3x-2}+\frac{6}{4x+5}$; LCD $= (3x-2)(4x+5)$

$$\frac{5}{(3x-2)}\left(\frac{4x+5}{4x+5}\right)+\frac{6}{(4x+5)}\left(\frac{3x-2}{3x-2}\right)$$

$$\frac{5(4x+5)}{(3x-2)(4x+5)}+\frac{6(3x-2)}{(4x+5)(3x-2)}$$

$$\frac{5(4x+5)+6(3x-2)}{(3x-2)(4x+5)}$$

$$\frac{20x+25+18x-12}{(3x-2)(4x+5)}$$

$$\frac{38x+13}{(3x-2)(4x+5)}$$

61. $\frac{3x}{2x+5}+1$; LCD $=2x+5$

$\frac{3x}{2x+5}+1\left(\frac{2x+5}{2x+5}\right)$

$\frac{3x}{2x+5}+\frac{2x+5}{2x+5}$

$\frac{3x+2x+5}{2x+5}$

$\frac{5x+5}{2x+5}$

63. $\frac{4x}{x-5}-3$; LCD $=x-5$

$\frac{4x}{x-5}-3\left(\frac{x-5}{x-5}\right)$

$\frac{4x}{x-5}-\frac{3(x-5)}{x-5}$

$\frac{4x-3(x-5)}{x-5}$

$\frac{4x-3x+15}{x-5}$

$\frac{x+15}{x-5}$

65. $-1-\frac{3}{2x+1}$; LCD $=2x+1$

$-\left(\frac{2x+1}{2x+1}\right)-\frac{3}{2x+1}$

$\frac{-1(2x+1)}{2x+1}-\frac{3}{2x+1}$

$\frac{-1(2x+1)-3}{2x+1}$

$\frac{-2x-1-3}{2x+1}$

$\frac{-2x-4}{2x+1}$

67a. $\frac{1}{x-1}-\frac{x}{x-1}$

$\frac{1-x}{x-1}$

-1

b. $\frac{3}{2x-3}-\frac{2x}{2x-3}$

$\frac{3-2x}{2x-3}$

-1

c. $\frac{4}{x-4}-\frac{x}{x-4}+1$

$\frac{4-x}{x-4}+1$

$-1+1$

0

d. $-1+\frac{2}{x-2}-\frac{x}{x-2}$

$-1+\frac{2-x}{x-2}$

$-1+(-1)$

-2

PROBLEM SET 4.4 More on Rational Expressions and Complex Fractions

1. $\dfrac{2x}{x^2+4x}+\dfrac{5}{x}$

$\dfrac{2x}{x(x+4)}+\dfrac{5}{x}$; LCD $= x(x+4)$

$\dfrac{2x}{x(x+4)}+\dfrac{5}{x}\left(\dfrac{x+4}{x+4}\right)$

$\dfrac{2x+5(x+4)}{x(x+4)}$

$\dfrac{2x+5x+20}{x(x+4)}$

$\dfrac{7x+20}{x(x+4)}$

3. $\dfrac{4}{x^2+7x}-\dfrac{1}{x}$

$\dfrac{4}{x(x+7)}-\dfrac{1}{x}$; LCD $= x(x+7)$

$\dfrac{4}{x(x+7)}-\dfrac{1}{x}\left(\dfrac{x+7}{x+7}\right)$

$\dfrac{4-1(x+7)}{x(x+7)}$

$\dfrac{4-x-7}{x(x+7)}$

$\dfrac{-x-3}{x(x+7)}$

5. $\dfrac{x}{x^2-1}+\dfrac{5}{x+1}$

$\dfrac{x}{(x+1)(x-1)}+\dfrac{5}{x+1}$

LCD $= (x+1)(x-1)$

$\dfrac{x}{(x+1)(x-1)}+\dfrac{5}{(x+1)}\left(\dfrac{x-1}{x-1}\right)$

$\dfrac{x+5(x-1)}{(x+1)(x-1)}$

$\dfrac{x+5x-5}{(x+1)(x-1)}$

$\dfrac{6x-5}{(x+1)(x-1)}$

7. $\dfrac{6a+4}{a^2-1}-\dfrac{5}{a-1}$

$\dfrac{6a+4}{(a+1)(a-1)}-\dfrac{5}{a-1}$

LCD $= (a+1)(a-1)$

$\dfrac{6a+4}{(a+1)(a-1)}-\dfrac{5}{(a-1)}\left(\dfrac{a+1}{a+1}\right)$

$\dfrac{6a+4-5(a+1)}{(a+1)(a-1)}$

$\dfrac{6a+4-5a-5}{(a+1)(a-1)}$

$\dfrac{a-1}{(a+1)(a-1)}$

$\dfrac{1}{a+1}$

9. $\dfrac{2n}{n^2-25}-\dfrac{3}{4n+20}$

$\dfrac{2n}{(n+5)(n-5)}-\dfrac{3}{4(n+5)}$

LCD $= 4(n+5)(n-5)$

$\dfrac{2n}{(n+5)(n-5)}\left(\dfrac{4}{4}\right)-\dfrac{3}{4(n+5)}\left(\dfrac{n-5}{n-5}\right)$

$\dfrac{2n(4)-3(n-5)}{4(n+5)(n-5)}$

$$\frac{8n-3n+15}{4(n+5)(n-5)}$$

$$\frac{5n+15}{4(n+5)(n-5)}$$

11. $$\frac{5}{x}-\frac{5x-30}{x^2+6x}+\frac{x}{x+6}$$

$$\frac{5}{x}-\frac{5x-30}{x(x+6)}+\frac{x}{x+6}$$

$\text{LCD} = x(x+6)$

$$\frac{5}{x}\left(\frac{x+6}{x+6}\right)-\frac{5x-30}{x(x+6)}+\frac{x}{x+6}\left(\frac{x}{x}\right)$$

$$\frac{5(x+6)-(5x-30)+x^2}{x(x+6)}$$

$$\frac{5x+30-5x+30+x^2}{x(x+6)}$$

$$\frac{x^2+60}{x(x+6)}$$

13. $$\frac{3}{x^2+9x+14}+\frac{5}{2x^2+15x+7}$$

$$\frac{3}{(x+2)(x+7)}+\frac{5}{(2x+1)(x+7)}$$

$\text{LCD} = (x+2)(x+7)(2x+1)$

$$\frac{3}{(x+2)(x+7)}\left(\frac{2x+1}{2x+1}\right)+\frac{5}{(2x+1)(x+7)}\left(\frac{x+2}{x+2}\right)$$

$$\frac{3(2x+1)+5(x+2)}{(x+2)(2x+1)(x+7)}$$

$$\frac{6x+3+5x+10}{(x+2)(2x+1)(x+7)}$$

$$\frac{11x+13}{(x+2)(2x+1)(x+7)}$$

15. $$\frac{1}{a^2-3a-10}-\frac{4}{a^2+4a-45}$$

$$\frac{1}{(a-5)(a+2)}-\frac{4}{(a+9)(a-5)}$$

$\text{LCD} = (a-5)(a+2)(a+9)$

$$\frac{1}{(a-5)(a+2)}\left(\frac{a+9}{a+9}\right)-\frac{4}{(a+9)(a-5)}\left(\frac{a+2}{a+2}\right)$$

$$\frac{1(a+9)-4(a+2)}{(a-5)(a+2)(a+9)}$$

$$\frac{a+9-4a-8}{(a-5)(a+2)(a+9)}$$

$$\frac{-3a+1}{(a-5)(a+2)(a+9)}$$

17. $$\frac{3a}{8a^2-2a-3}+\frac{1}{4a^2+13a-12}$$

$$\frac{3a}{(2a+1)(4a-3)}+\frac{1}{(4a-3)(a+4)}$$

$\text{LCD} = (2a+1)(4a-3)(a+4)$

$$\frac{3a}{(2a+1)(4a-3)}\left(\frac{a+4}{a+4}\right)+\frac{1}{(4a-3)(a+4)}\left(\frac{2a+1}{2a+1}\right)$$

$$\frac{3a(a+4)+1(2a+1)}{(2a+1)(4a-3)(a+4)}$$

$$\frac{3a^2+12a+2a+1}{(2a+1)(4a-3)(a+4)}$$

$$\frac{3a^2+14a+1}{(2a+1)(4a-3)(a+4)}$$

19. $$\frac{5}{x^2+3}-\frac{2}{x^2+4x-21}$$

$$\frac{5}{x^2+3}-\frac{2}{(x+7)(x-3)}$$

$$\text{LCD} = (x^2+3)(x+7)(x-3)$$

$$\frac{5}{(x^2+3)}\left(\frac{x+7}{x+7}\right)\left(\frac{x-3}{x-3}\right) - \frac{2}{(x+7)(x-3)}\left(\frac{x^2+3}{x^2+3}\right)$$

$$\frac{5(x+7)(x-3)-2(x^2+3)}{(x^2+3)(x+7)(x-3)}$$

$$\frac{5(x^2+4x-21)-2x^2-6}{(x^2+3)(x+7)(x-3)}$$

$$\frac{5x^2+20x-105-2x^2-6}{(x^2+3)(x+7)(x-3)}$$

$$\frac{3x^2+20x-111}{(x^2+3)(x+7)(x-3)}$$

21. $$\frac{3x}{x^2-6x+9} - \frac{2}{x-3}$$

$$\frac{3x}{(x-3)(x-3)} - \frac{2}{(x-3)}$$

$$\text{LCD} = (x-3)(x-3) = (x-3)^2$$

$$\frac{3x}{(x-3)(x-3)} - \frac{2}{(x-3)}\left(\frac{x-3}{x-3}\right)$$

$$\frac{3x-2(x-3)}{(x-3)(x-3)}$$

$$\frac{3x-2x+6}{(x-3)^2}$$

$$\frac{x+6}{(x-3)^2}$$

23. $$\frac{5}{x^2-1} + \frac{9}{x^2+2x+1}$$

$$\frac{5}{(x+1)(x-1)} + \frac{9}{(x+1)(x+1)}$$

$$\text{LCD} = (x-1)(x+1)^2$$

$$\frac{5}{(x+1)(x-1)}\left(\frac{x+1}{x+1}\right) + \frac{9}{(x+1)^2}\left(\frac{x-1}{x-1}\right)$$

$$\frac{5(x+1)+9(x-1)}{(x+1)(x+1)(x-1)}$$

$$\frac{5x+5+9x-9}{(x+1)^2(x-1)}$$

$$\frac{14x-4}{(x+1)^2(x-1)}$$

25. $$\frac{2}{y^2+6y-16} - \frac{4}{y+8} - \frac{3}{y-2}$$

$$\frac{2}{(y+8)(y-2)} - \frac{4}{y+8} - \frac{3}{y-2}$$

$$\text{LCD} = (y+8)(y-2)$$

$$\frac{2}{(y+8)(y-2)} - \frac{4}{(y+8)}\left(\frac{y-2}{y-2}\right) - \frac{3}{(y-2)}\left(\frac{y+8}{y+8}\right)$$

$$\frac{2-4(y-2)-3(y+8)}{(y+8)(y-2)}$$

$$\frac{2-4y+8-3y-24}{(y+8)(y-2)}$$

$$\frac{-7y-14}{(y+8)(y-2)}$$

27. $$x - \frac{x^2}{x-2} + \frac{3}{x^2-4}$$

$$x - \frac{x^2}{x-2} + \frac{3}{(x+2)(x-2)}$$

$$\text{LCD} = (x+2)(x-2)$$

$$x\left(\frac{x+2}{x+2}\right)\left(\frac{x-2}{x-2}\right) - \frac{x^2}{(x-2)}\left(\frac{x+2}{x+2}\right) + \frac{3}{(x+2)(x-2)}$$

$$\frac{x(x+2)(x-2)-x^2(x+2)+3}{(x+2)(x-2)}$$

$$\frac{x(x^2-4)-x^3-2x^2+3}{(x+2)(x-2)}$$

$$\frac{x^3 - 4x - x^3 - 2x^2 + 3}{(x+2)(x-2)}$$

$$\frac{-2x^2 - 4x + 3}{(x+2)(x-2)}$$

29. $\dfrac{x+3}{x+10} + \dfrac{4x-3}{x^2+8x-20} + \dfrac{x-1}{x-2}$

$$\frac{x+3}{x+10} + \frac{4x-3}{(x+10)(x-2)} + \frac{x-1}{x-2};\ \text{LCD} = (x+10)(x-2)$$

$$\frac{(x+3)}{(x+10)}\left(\frac{x-2}{x-2}\right) + \frac{4x-3}{(x+10)(x-2)} + \frac{(x-1)}{(x-2)}\left(\frac{x+10}{x+10}\right)$$

$$\frac{(x+3)(x-2) + 4x - 3 + (x-1)(x+10)}{(x+10)(x-2)}$$

$$\frac{x^2 + x - 6 + 4x - 3 + x^2 + 9x - 10}{(x+10)(x-2)}$$

$$\frac{2x^2 + 14x - 19}{(x+10)(x-2)}$$

31. $\dfrac{n}{n-6} + \dfrac{n+3}{n+8} + \dfrac{12n+26}{n^2+2n-48}$

$$\frac{n}{n-6} + \frac{n+3}{n+8} + \frac{12n+26}{(n+8)(n-6)};\ \text{LCD} = (n-6)(n+8)$$

$$\frac{n}{(n-6)}\left(\frac{n+8}{n+8}\right) + \frac{(n+3)}{(n+8)}\left(\frac{n-6}{n-6}\right) + \frac{12n+26}{(n+8)(n-6)}$$

$$\frac{n(n+8) + (n+3)(n-6) + 12n + 26}{(n-6)(n+8)}$$

$$\frac{n^2 + 8n + n^2 - 3n - 18 + 12n + 26}{(n-6)(n+8)}$$

$$\frac{2n^2 + 17n + 8}{(n-6)(n+8)}$$

$$\frac{(2n+1)(n+8)}{(n-6)(n+8)} = \frac{2n+1}{n-6}$$

Problem Set 4.4

33. $\dfrac{4x-3}{2x^2+x-1}-\dfrac{2x+7}{3x^2+x-2}-\dfrac{3}{3x-2}$

$\dfrac{4x-3}{(2x-1)(x+1)}-\dfrac{2x+7}{(3x-2)(x+1)}-\dfrac{3}{3x-2}$; LCD $=(2x-1)(x+1)(3x-2)$

$$\frac{(4x-3)}{(2x-1)(x+1)}\left(\frac{3x-2}{3x-2}\right)-\frac{(2x+7)}{(3x-2)(x+1)}\left(\frac{2x-1}{2x-1}\right)-\frac{3}{(3x-2)}\left(\frac{2x-1}{2x-1}\right)\left(\frac{x+1}{x+1}\right)$$

$$\frac{(4x-3)(3x-2)-(2x+7)(2x-1)-3(2x-1)(x+1)}{(2x-1)(3x-2)(x+1)}$$

$$\frac{12x^2-17x+6-(4x^2+12x-7)-3(2x^2+x-1)}{(2x-1)(3x-2)(x+1)}$$

$$\frac{12x^2-17x+6-4x^2-12x+7-6x^2-3x+3}{(2x-1)(3x-2)(x+1)}$$

$$\frac{2x^2-32x+16}{(2x-1)(3x-2)(x+1)}$$

35. $\dfrac{n}{n^2+1}+\dfrac{n^2+3n}{n^4-1}-\dfrac{1}{n-1}$

$\dfrac{n}{n^2+1}+\dfrac{n^2+3n}{(n^2+1)(n+1)(n-1)}-\dfrac{1}{n-1}$; LCD $=(n^2+1)(n+1)(n-1)$

$$\frac{n}{(n^2+1)}\left(\frac{n+1}{n+1}\right)\left(\frac{n-1}{n-1}\right)+\frac{n^2+3n}{(n^2+1)(n+1)(n-1)}-\frac{1}{(n-1)}\left(\frac{n^2+1}{n^2+1}\right)\left(\frac{n+1}{n+1}\right)$$

$$\frac{n(n+1)(n-1)+n^2+3n-(n^2+1)(n+1)}{(n^2+1)(n+1)(n-1)}$$

$$\frac{n(n^2-1)+n^2+3n-(n^3+n^2+n+1)}{(n^2+1)(n+1)(n-1)}$$

$$\frac{n^3-n+n^2+3n-n^3-n^2-n-1}{(n^2+1)(n+1)(n-1)}$$

$$\frac{n-1}{(n^2+1)(n+1)(n-1)}=\frac{1}{(n^2+1)(n+1)}$$

37. $\dfrac{15x^2-10}{5x^2-7x+2}-\dfrac{3x+4}{x-1}-\dfrac{2}{5x-2}$

$\dfrac{15x^2-10}{(5x-2)(x-1)}-\dfrac{3x+4}{x-1}-\dfrac{2}{5x-2}$; LCD $=(5x-2)(x-1)$

$$\frac{15x^2-10}{(5x-2)(x-1)}-\frac{(3x+4)}{(x-1)}\left(\frac{5x-2}{5x-2}\right)-\frac{2}{(5x-2)}\left(\frac{x-1}{x-1}\right)$$

$$\frac{15x^2-10-(3x+4)(5x-2)-2(x-1)}{(5x-2)(x-1)}$$

$$\frac{15x^2-10-(15x^2+14x-8)-2x+2}{(5x-2)(x-1)}$$

$$\frac{15x^2-10-15x^2-14x+8-2x+2}{(5x-2)(x-1)}$$

$$\frac{-16x}{(5x-2)(x-1)}$$

39. $\dfrac{t+3}{3t-1}+\dfrac{8t^2+8t+2}{3t^2-7t+2}-\dfrac{2t+3}{t-2}$

$\dfrac{t+3}{3t-1}+\dfrac{8t^2+8t+2}{(3t-1)(t-2)}-\dfrac{2t+3}{t-2}$; LCD $=(3t-1)(t-2)$

$$\frac{(t+3)}{(3t-1)}\left(\frac{t-2}{t-2}\right)+\frac{8t^2+8t+2}{(3t-1)(t-2)}-\frac{(2t+3)}{(t-2)}\left(\frac{3t-1}{3t-1}\right)$$

$$\frac{(t+3)(t-2)+8t^2+8t+2-(2t+3)(3t-1)}{(3t-1)(t-2)}$$

$$\frac{t^2+t-6+8t^2+8t+2-(6t^2+7t-3)}{(3t-1)(t-2)}$$

$$\frac{t^2+t-6+8t^2+8t+2-6t^2-7t+3}{(3t-1)(t-2)}$$

$$\frac{3t^2+2t-1}{(3t-1)(t-2)}=\frac{(3t-1)(t+1)}{(3t-1)(t-2)}=\frac{t+1}{t-2}$$

Problem Set 4.4

41. $\dfrac{\dfrac{1}{2}-\dfrac{1}{4}}{\dfrac{5}{8}+\dfrac{3}{4}}$

$\dfrac{8}{8}\bullet\dfrac{\left(\dfrac{1}{2}-\dfrac{1}{4}\right)}{\left(\dfrac{5}{8}+\dfrac{3}{4}\right)}$

$\dfrac{8\left(\dfrac{1}{2}\right)-8\left(\dfrac{1}{4}\right)}{8\left(\dfrac{5}{8}\right)+8\left(\dfrac{3}{4}\right)}$

$\dfrac{4-2}{5+6}=\dfrac{2}{11}$

43. $\dfrac{\dfrac{3}{28}-\dfrac{5}{14}}{\dfrac{5}{7}+\dfrac{1}{4}}$

$\dfrac{28}{28}\bullet\dfrac{\left(\dfrac{3}{28}-\dfrac{5}{14}\right)}{\left(\dfrac{5}{7}+\dfrac{1}{4}\right)}$

$\dfrac{28\left(\dfrac{3}{28}\right)-28\left(\dfrac{5}{14}\right)}{28\left(\dfrac{5}{7}\right)+28\left(\dfrac{1}{4}\right)}$

$\dfrac{3-10}{20+7}=\dfrac{-7}{27}$

45. $\dfrac{\dfrac{5}{6y}}{\dfrac{10}{3xy}}$

$\dfrac{5}{6y}\bullet\dfrac{3xy}{10}$

$\dfrac{15xy}{60y}$

$\dfrac{x}{4}$

47. $\dfrac{\dfrac{3}{x}-\dfrac{2}{y}}{\dfrac{4}{y}-\dfrac{7}{xy}}$

$\dfrac{xy}{xy}\bullet\dfrac{\left(\dfrac{3}{x}-\dfrac{2}{y}\right)}{\left(\dfrac{4}{y}-\dfrac{7}{xy}\right)}$

$\dfrac{xy\left(\dfrac{3}{x}\right)-xy\left(\dfrac{2}{y}\right)}{xy\left(\dfrac{4}{y}\right)-xy\left(\dfrac{7}{xy}\right)}$

$\dfrac{3y-2x}{4x-7}$

49. $\dfrac{\dfrac{6}{a}-\dfrac{5}{b^2}}{\dfrac{12}{a^2}+\dfrac{2}{b}}$

$\dfrac{a^2b^2}{a^2b^2}\bullet\dfrac{\left(\dfrac{6}{a}-\dfrac{5}{b^2}\right)}{\left(\dfrac{12}{a^2}+\dfrac{2}{b}\right)}$

$\dfrac{a^2b^2\left(\dfrac{6}{a}\right)-a^2b^2\left(\dfrac{5}{b^2}\right)}{a^2b^2\left(\dfrac{12}{a^2}\right)+a^2b^2\left(\dfrac{2}{b}\right)}$

$\dfrac{6ab^2-5a^2}{12b^2+2a^2b}$

51. $\dfrac{\dfrac{2}{x}-3}{\dfrac{3}{y}+4}$

$$\frac{xy}{xy} \bullet \frac{\left(\frac{2}{x} - 3\right)}{\left(\frac{3}{y} + 4\right)}$$

$$\frac{xy\left(\frac{2}{x}\right) - xy(3)}{xy\left(\frac{3}{y}\right) + xy(4)}$$

$$\frac{2y - 3xy}{3x + 4xy}$$

53. $$\frac{3 + \frac{2}{n+4}}{5 - \frac{1}{n+4}}$$

$$\frac{(n+4)}{(n+4)} \bullet \frac{\left(3 + \frac{2}{n+4}\right)}{\left(5 - \frac{1}{n+4}\right)}$$

$$\frac{(n+4)(3) + (n+4)\left(\frac{2}{n+4}\right)}{(n+4)(5) - (n+4)\left(\frac{1}{n+4}\right)}$$

$$\frac{3n + 12 + 2}{5n + 20 - 1}$$

$$\frac{3n + 14}{5n + 19}$$

55. $$\frac{5 - \frac{2}{n-3}}{4 - \frac{1}{n-3}}$$

$$\frac{(n-3)}{(n-3)} \bullet \frac{\left(5 - \frac{2}{n-3}\right)}{\left(4 - \frac{1}{n-3}\right)}$$

$$\frac{(n-3)(5) - (n-3)\left(\frac{2}{n-3}\right)}{(n-3)(4) - (n-3)\left(\frac{1}{n-3}\right)}$$

$$\frac{5n - 15 - 2}{4n - 12 - 1} = \frac{5n - 17}{4n - 13}$$

57. $$\frac{\frac{-1}{y-2} + \frac{5}{x}}{\frac{3}{x} - \frac{4}{xy - 2x}}$$

$$\frac{-\frac{1}{y-2} + \frac{5}{x}}{\frac{3}{x} - \frac{4}{x(y-2)}}$$

$$\frac{x(y-2)}{x(y-2)} \bullet \frac{\left(-\frac{1}{y-2} + \frac{5}{x}\right)}{\left[\frac{3}{x} - \frac{4}{x(y-2)}\right]}$$

$$\frac{x(y-2)\left(\frac{-1}{y-2}\right) + x(y-2)\left(\frac{5}{x}\right)}{x(y-2)\left(\frac{3}{x}\right) - x(y-2)\left[\frac{4}{x(y-2)}\right]}$$

$$\frac{-x + 5(y-2)}{3(y-2) - 4}$$

$$\frac{-x + 5y - 10}{3y - 6 - 4}$$

$$\frac{-x + 5y - 10}{3y - 10}$$

Problem Set 4.4

59. $$\frac{\frac{2}{x-3}-\frac{3}{x+3}}{\frac{5}{x^2-9}-\frac{2}{x-3}}$$

$$\frac{\frac{2}{x-3}-\frac{3}{x+3}}{\frac{5}{(x+3)(x-3)}-\frac{2}{x-3}}$$

$$\frac{(x+3)(x-3)}{(x+3)(x-3)}\bullet\frac{\left(\frac{2}{x-3}-\frac{3}{x+3}\right)}{\left[\frac{5}{(x+3)(x-3)}-\frac{2}{x-3}\right]}$$

$$\frac{(x+3)(x-3)\left(\frac{2}{x-3}\right)-(x+3)(x-3)\left(\frac{3}{x+3}\right)}{(x+3)(x-3)\left[\frac{5}{(x+3)(x-3)}\right]-(x+3)(x-3)\left(\frac{2}{x-3}\right)}$$

$$\frac{2(x+3)-3(x-3)}{5-2(x+3)}$$

$$\frac{2x+6-3x+9}{5-2x-6}$$

$$\frac{-x+15}{-2x-1}$$

61. $$\frac{3a}{2-\frac{1}{a}}-1$$

$$\frac{a}{a}\bullet\frac{(3a)}{\left(2-\frac{1}{a}\right)}-1$$

$$\frac{a(3a)}{a(2)-a\left(\frac{1}{a}\right)}-1$$

$$\frac{3a^2}{2a-1}-1$$

$$\frac{3a^2}{2a-1}-1\left(\frac{2a-1}{2a-1}\right)$$

$$\frac{3a^2}{2a-1}-\frac{2a-1}{2a-1}$$

$$\frac{3a^2-(2a-1)}{2a-1}$$

$$\frac{3a^2-2a+1}{2a-1}$$

63. $2-\dfrac{x}{3-\dfrac{2}{x}}$

$2-\dfrac{x}{x}\bullet\dfrac{(x)}{\left(3-\dfrac{2}{x}\right)}$

$2-\dfrac{x(x)}{x(3)-x\left(\dfrac{2}{x}\right)}$

$2-\dfrac{x^2}{3x-2}$

$2\left(\dfrac{3x-2}{3x-2}\right)-\dfrac{x^2}{3x-2}$

$\dfrac{6x-4}{3x-2}-\dfrac{x^2}{3x-2}$

$\dfrac{6x-4-x^2}{3x-2}$

$\dfrac{-x^2+6x-4}{3x-2}$

PROBLEM SET 4.5 Dividing Polynomials

1. $\dfrac{9x^4+18x^3}{3x}$

$\dfrac{9x^4}{3x}+\dfrac{18x^3}{3x}$

$3x^3+6x^2$

3. $\dfrac{-24x^6+36x^8}{4x^2}$

$\dfrac{-24x^6}{4x^2}+\dfrac{36x^8}{4x^2}$

$-6x^4+9x^6$

5. $\dfrac{15a^3-25a^2-40a}{5a}$

$\dfrac{15a^3}{5a}-\dfrac{25a^2}{5a}-\dfrac{40a}{5a}$

$3a^2-5a-8$

7. $\dfrac{13x^3-17x^2+28x}{-x}$

$\dfrac{13x^3}{-x}+\dfrac{-17x^2}{-x}+\dfrac{28x}{-x}$

$-13x^2+17x-28$

9. $\dfrac{-18x^2y^2+24x^3y^2-48x^2y^3}{6xy}$

$\dfrac{-18x^2y^2}{6xy}+\dfrac{24x^3y^2}{6xy}-\dfrac{48x^2y^3}{6xy}$

$-3xy+4x^2y-8xy^2$

11.

$$\begin{array}{r|l} & x-13 \\ x+6 & x^2-7x-78 \\ & \underline{x^2+6x} \\ & -13x-78 \\ & \underline{-13x-78} \\ & 0 \end{array}$$

$x-13$

13.

$$\begin{array}{r|l} & x+20 \\ x-8 & x^2+12x-160 \\ & \underline{x^2-\ \ 8x} \\ & 20x-160 \\ & \underline{20x-160} \\ & 0 \end{array}$$

$x+20$

Problem Set 4.5

15.

$$\begin{array}{r|r}
 & 2x+1 \\ \cline{2-2}
x-1 & 2x^2 - x - 4 \\
 & 2x^2 - 2x \\ \cline{2-2}
 & x - 4 \\
 & x - 1 \\ \cline{2-2}
 & -3
\end{array}$$

$$2x + 1 - \frac{3}{x-1}$$

17.

$$\begin{array}{r|r}
 & 5x-1 \\ \cline{2-2}
3x+5 & 15x^2 + 22x - 5 \\
 & 15x^2 + 25x \\ \cline{2-2}
 & -3x - 5 \\
 & -3x - 5 \\ \cline{2-2}
 & 0
\end{array}$$

$5x - 1$

19.

$$\begin{array}{r|r}
 & 3x^2 - 2x - 7 \\ \cline{2-2}
x+3 & 3x^3 + 7x^2 - 13x - 21 \\
 & 3x^3 + 9x^2 \\ \cline{2-2}
 & -2x^2 - 13x \\
 & -2x^2 - 6x \\ \cline{2-2}
 & -7x - 21 \\
 & -7x - 21 \\ \cline{2-2}
 & 0
\end{array}$$

$3x^2 - 2x - 7$

21.

$$\begin{array}{r|r}
 & x^2 + 5x - 6 \\ \cline{2-2}
2x-1 & 2x^3 + 9x^2 - 17x + 6 \\
 & 2x^3 - x^2 \\ \cline{2-2}
 & 10x^2 - 17x \\
 & 10x^2 - 5x \\ \cline{2-2}
 & -12x + 6 \\
 & -12x + 6 \\ \cline{2-2}
 & 0
\end{array}$$

$x^2 + 5x - 6$

23.

$$\begin{array}{r|r}
 & 4x^2 + 7x + 12 \\ \cline{2-2}
x-2 & 4x^3 - x^2 - 2x + 6 \\
 & 4x^3 - 8x^2 \\ \cline{2-2}
 & 7x^2 - 2x \\
 & 7x^2 - 14x \\ \cline{2-2}
 & 12x + 6 \\
 & 12x - 24 \\ \cline{2-2}
 & 30
\end{array}$$

$$4x^2 + 7x + 12 + \frac{30}{x-2}$$

25.

$$\begin{array}{r|r}
 & x^3 - 4x^2 - 5x + 3 \\ \cline{2-2}
x-6 & x^4 - 10x^3 + 19x^2 + 33x - 18 \\
 & x^4 - 6x^3 \\ \cline{2-2}
 & -4x^3 + 19x^2 \\
 & -4x^3 + 24x^2 \\ \cline{2-2}
 & -5x^2 + 33x \\
 & -5x^2 + 30x \\ \cline{2-2}
 & 3x - 18 \\
 & 3x - 18 \\ \cline{2-2}
 & 0
\end{array}$$

$x^3 - 4x^2 - 5x + 3$

27.

$$\begin{array}{r|r}
 & x^2 + 5x + 25 \\ \cline{2-2}
x-5 & x^3 + 0x^2 + 0x - 125 \\
 & x^3 - 5x^2 \\ \cline{2-2}
 & 5x^2 + 0x \\
 & 5x^2 - 25x \\ \cline{2-2}
 & 25x - 125 \\
 & 25x - 125 \\ \cline{2-2}
 & 0
\end{array}$$

$x^2 + 5x + 25$

29.

$$\begin{array}{r} x^2 - x + 1 \\ x+1 \,\overline{)\, x^3 + 0x^2 + 0x + 64} \\ \underline{x^3 + x^2 } \\ -x^2 + 0x \\ \underline{-x^2 - x } \\ x + 64 \\ \underline{x + 1} \\ 63 \end{array}$$

$$x^2 - x + 1 + \frac{63}{x+1}$$

31.

$$\begin{array}{r} 2x^2 - 4x + 7 \\ x+2 \,\overline{)\, 2x^3 + 0x^2 - x - 6} \\ \underline{2x^3 + 4x^2 } \\ -4x^2 - x \\ \underline{-4x^2 - 8x } \\ 7x - 6 \\ \underline{7x + 14} \\ -20 \end{array}$$

$$2x^2 - 4x + 7 - \frac{20}{x+2}$$

33.

$$\begin{array}{r} 4a - 4b \\ a-b \,\overline{)\, 4a^2 - 8ab + 4b^2} \\ \underline{4a^2 - 4ab } \\ -4ab + 4b^2 \\ \underline{-4ab + 4b^2} \\ 0 \end{array}$$

$$4a - 4b$$

35.

$$\begin{array}{r} 4x + 7 \\ x^2 - 3x \,\overline{)\, 4x^3 - 5x^2 + 2x - 6} \\ \underline{4x^3 - 12x^2 } \\ 7x^2 + 2x \\ \underline{7x^2 - 21x } \\ 23x - 6 \end{array}$$

$$4x + 7 + \frac{23x - 6}{x^2 - 3x}$$

37.

$$\begin{array}{r} 8y - 9 \\ y^2 + y \,\overline{)\, 8y^3 - y^2 - y + 5} \\ \underline{8y^3 + 8y^2 } \\ -9y^2 - y \\ \underline{-9y^2 - 9y } \\ 8y + 5 \end{array}$$

$$8y - 9 + \frac{8y + 5}{y^2 + y}$$

39.

$$\begin{array}{r} 2x - 1 \\ x^2 + x - 1 \,\overline{)\, 2x^3 + x^2 - 3x + 1} \\ \underline{2x^3 + 2x^2 - 2x } \\ -x^2 - x + 1 \\ \underline{-x^2 - x + 1} \\ 0 \end{array}$$

$$2x - 1$$

41.

$$\begin{array}{r} x - 3 \\ 4x^2 - x + 5 \,\overline{)\, 4x^3 - 13x^2 + 8x - 15} \\ \underline{4x^3 - x^2 + 5x } \\ -12x^2 + 3x - 15 \\ \underline{-12x^2 + 3x - 15} \\ 0 \end{array}$$

$$x - 3$$

43.

$$\begin{array}{r} 5a - 8 \\ a^2 + 3a - 4 \,\overline{)\, 5a^3 + 7a^2 - 2a - 9} \\ \underline{5a^3 + 15a^2 - 20a } \\ -8a^2 + 18a - 9 \\ \underline{-8a^2 - 24a + 32} \\ 42a - 41 \end{array}$$

$$5a - 8 + \frac{42a - 41}{a^2 + 3a - 4}$$

45.

$$\begin{array}{r|l} & 2n^2+3n-4 \\ \hline n^2+1 & 2n^4+3n^3-2n^2+3n-4 \\ & 2n^4 \qquad\quad +2n^2 \\ \hline & \qquad 3n^3-4n^2+3n \\ & \qquad 3n^3 \qquad\; +3n \\ \hline & \qquad\qquad -4n^2 \qquad -4 \\ & \qquad\qquad -4n^2 \qquad -4 \\ \hline & \qquad\qquad\qquad\qquad 0 \end{array}$$

$2n^2+3n-4$

47.

$$\begin{array}{r|l} & x^4+x^3+x^2+x+1 \\ \hline x-1 & x^5+0x^4+0x^3+0x^2+0x-1 \\ & x^5-x^4 \\ \hline & \quad x^4+0x^3 \\ & \quad x^4-\;x^3 \\ \hline & \qquad x^3+0x^2 \\ & \qquad x^3-\;x^2 \\ \hline & \qquad\quad x^2+0x \\ & \qquad\quad x^2-\;x \\ \hline & \qquad\qquad x-1 \\ & \qquad\qquad x-1 \\ \hline & \qquad\qquad\quad 0 \end{array}$$

$x^4+x^3+x^2+x+1$

49.

$$\begin{array}{r|l} & x^3-x^2+x-1 \\ \hline x+1 & x^4+0x^3+0x^2+0x-1 \\ & x^4+\;x^3 \\ \hline & \quad -x^3+0x^2 \\ & \quad -x^3-\;x^2 \\ \hline & \qquad x^2+0x \\ & \qquad x^2+\;x \\ \hline & \qquad\quad -x-1 \\ & \qquad\quad -x-1 \\ \hline & \qquad\qquad 0 \end{array}$$

x^3-x^2+x-1

51.

$$\begin{array}{r|l} & 3x^2+x+1 \\ \hline x^2-1 & 3x^4+x^3-2x^2-x+6 \\ & 3x^4 \qquad -3x^2 \\ \hline & \qquad x^3+\;x^2-x \\ & \qquad x^3 \qquad\; -x \\ \hline & \qquad\qquad x^2+\quad 6 \\ & \qquad\qquad x^2-\quad 1 \\ \hline & \qquad\qquad\qquad 7 \end{array}$$

$3x^2+x+1+\dfrac{7}{x^2-1}$

53. $(x^2-8x+12)\div(x-2)$

$$\begin{array}{r|rrr} 2 & 1 & -8 & 12 \\ & & 2 & -12 \\ \hline & 1 & -6 & 0 \end{array}$$

$x-6$

55. $(x^2+2x-10)\div(x-4)$

$$\begin{array}{r|rrr} 4 & 1 & 2 & -10 \\ & & 4 & 24 \\ \hline & 1 & 6 & 14 \end{array}$$

$x+6+\dfrac{14}{x-4}$

57. $(x^3-2x^2-x+2)\div(x-2)$

$$\begin{array}{r|rrrr} 2 & 1 & -2 & -1 & 2 \\ & & 2 & 0 & -2 \\ \hline & 1 & 0 & -1 & 0 \end{array}$$

x^2-1

59. $(x^3-7x-6)\div(x+2)$

$$\begin{array}{r|rrrr} -2 & 1 & 0 & -7 & -6 \\ & & -2 & 4 & 6 \\ \hline & 1 & -2 & -3 & 0 \end{array}$$

x^2-2x-3

61. $(2x^3 - 5x^2 - 4x + 6) \div (x - 2)$

2	2	−5	−4	6
		4	−2	−12
	2	−1	−6	−6

$2x^2 - x - 6 + \frac{-6}{x-2}$

63. $(x^4 + 4x^3 - 7x - 1) \div (x - 3)$

3	1	4	0	−7	−1
		3	21	63	168
	1	7	21	56	167

$x^3 + 7x^2 + 21x + 56 + \frac{167}{x-3}$

PROBLEM SET 4.6 Fractional Equations

1. $\frac{x+1}{4} + \frac{x-2}{6} = \frac{3}{4}$

$$12\left(\frac{x+1}{4} + \frac{x-2}{6}\right) = 12\left(\frac{3}{4}\right)$$

$$12\left(\frac{x+1}{4}\right) + 12\left(\frac{x-2}{6}\right) = 9$$

$3(x+1) + 2(x-2) = 9$
$3x + 3 + 2x - 4 = 9$
$5x - 1 = 9$
$5x = 10$
$x = 2$
The solution set is $\{2\}$.

3. $\frac{x+3}{2} - \frac{x-4}{7} = 1$

$$14\left(\frac{x+3}{2} - \frac{x-4}{7}\right) = 14(1)$$

$$14\left(\frac{x+3}{2}\right) - 14\left(\frac{x-4}{7}\right) = 14$$

$7(x+3) - 2(x-4) = 14$
$7x + 21 - 2x + 8 = 14$
$5x + 29 = 14$
$5x = -15$
$x = -3$
The solution set is $\{-3\}$.

5. $\frac{5}{n} + \frac{1}{3} = \frac{7}{n}; n \neq 0$

$$3n\left(\frac{5}{n} + \frac{1}{3}\right) = 3n\left(\frac{7}{n}\right)$$

$$3n\left(\frac{5}{n}\right) + 3n\left(\frac{1}{3}\right) = 21$$

$15 + n = 21$
$n = 6$
The solution set is $\{6\}$.

7. $\frac{7}{2x} + \frac{3}{5} = \frac{2}{3x}; x \neq 0$

$$30x\left(\frac{7}{2x} + \frac{3}{5}\right) = 30x\left(\frac{2}{3x}\right)$$

$$30x\left(\frac{7}{2x}\right) + 30x\left(\frac{3}{5}\right) = 20$$

$15(7) + 6x(3) = 20$
$105 + 18x = 20$
$18x = -85$
$x = -\frac{85}{18}$
The solution set is $\left\{-\frac{85}{18}\right\}$.

9. $\frac{3}{4x} + \frac{5}{6} = \frac{4}{3x}; x \neq 0$

$$12x\left(\frac{3}{4x} + \frac{5}{6}\right) = 12x\left(\frac{4}{3x}\right)$$

$$12x\left(\frac{3}{4x}\right) + 12x\left(\frac{5}{6}\right) = 16$$

$3(3) + 2x(5) = 16$
$9 + 10x = 16$
$10x = 7$

$x = \frac{7}{10}$

The solution set is $\left\{\frac{7}{10}\right\}$.

11. $\frac{47-n}{n} = 8 + \frac{2}{n}; n \neq 0$

$n\left(\frac{47-n}{n}\right) = n\left(8 + \frac{2}{n}\right)$

$47 - n = n(8) + n\left(\frac{2}{n}\right)$

$47 - n = 8n + 2$

$47 = 9n + 2$

$45 = 9n$

$5 = n$

The solution set is $\{5\}$.

13. $\frac{n}{65-n} = 8 + \frac{2}{65-n}; n \neq 65$

$(65-n)\left(\frac{n}{65-n}\right) = (65-n)\left(8 + \frac{2}{65-n}\right)$

$n = (65-n)(8) + (65-n)\left(\frac{2}{65-n}\right)$

$n = 520 - 8n + 2$

$9n = 522$

$n = 58$

The solution set is $\{58\}$.

15. $n + \frac{1}{n} = \frac{17}{4}; n \neq 0$

$4n\left(n + \frac{1}{n}\right) = 4n\left(\frac{17}{4}\right)$

$4n(n) + 4n\left(\frac{1}{n}\right) = 17n$

$4n^2 + 4 = 17n$

$4n^2 - 17n + 4 = 0$

$(4n-1)(n-4) = 0$

$4n - 1 = 0$ or $n - 4 = 0$

$4n = 1$ or $n = 4$

$n = \frac{1}{4}$ or $n = 4$

The solution set is $\left\{\frac{1}{4}, 4\right\}$.

17. $n - \frac{2}{n} = \frac{23}{5}; n \neq 0$

$5n\left(n - \frac{2}{n}\right) = 5n\left(\frac{23}{5}\right)$

$5n(n) - 5n\left(\frac{2}{n}\right) = 23n$

$5n^2 - 10 = 23n$

$5n^2 - 23n - 10 = 0$

$(5n+2)(n-5) = 0$

$5n + 2 = 0$ or $n - 5 = 0$

$5n = -2$ or $n = 5$

$n = -\frac{2}{5}$ or $n = 5$

The solution set is $\left\{-\frac{2}{5}, 5\right\}$.

19. $\frac{5}{7x-3} = \frac{3}{4x-5}; x \neq \frac{3}{7}, x \neq \frac{5}{4}$

$5(4x-5) = 3(7x-3)$

$20x - 25 = 21x - 9$

$20x = 21x + 16$

$-x = 16$

$x = -16$

The solution set is $\{-16\}$.

21. $\frac{-2}{x-5} = \frac{1}{x+9}; x \neq 5, x \neq -9$

$-2(x+9) = 1(x-5)$

$-2x - 18 = x - 5$

$-2x = x + 13$

$-3x = 13$

$x = -\frac{13}{3}$

The solution set is $\left\{-\frac{13}{3}\right\}$.

23. $\frac{x}{x+1} - 2 = \frac{3}{x-3}; x \neq -1, x \neq 3$

$(x+1)(x-3)\left(\frac{x}{x+1} - 2\right) = (x+1)(x-3)\left(\frac{3}{x-3}\right)$

$(x+1)(x-3)\left(\frac{x}{x+1}\right) - 2(x+1)(x-3) = 3(x+1)$

$x(x-3) - 2(x^2 - 2x - 3) = 3x + 3$

$x^2 - 3x - 2x^2 + 4x + 6 = 3x + 3$
$-x^2 + x + 6 = 3x + 3$
$-x^2 - 2x + 3 = 0$
$x^2 + 2x - 3 = 0$
$(x + 3)(x - 1) = 0$
$x + 3 = 0$ or $x - 1 = 0$
$x = -3$ or $x = 1$
The solution set is $\{-3, 1\}$.

25. $\dfrac{a}{a+5} - 2 = \dfrac{3a}{a+5}; a \neq -5$

$(a+5)\left(\dfrac{a}{a+5} - 2\right) = (a+5)\left(\dfrac{3a}{a+5}\right)$

$(a+5)\left(\dfrac{a}{a+5}\right) - 2(a+5) = 3a$

$a - 2a - 10 = 3a$
$-a - 10 = 3a$
$-10 = 4a$

$-\dfrac{10}{4} = a$

$-\dfrac{5}{2} = a$

The solution set is $\left\{-\dfrac{5}{2}\right\}$.

27. $\dfrac{5}{x+6} = \dfrac{6}{x-3}; x \neq -6, x \neq 3$

$5(x-3) = 6(x+6)$
$5x - 15 = 6x + 36$
$5x = 6x + 51$
$-x = 51$
$x = -51$
The solution set is $\{-51\}$.

29. $\dfrac{3x-7}{10} = \dfrac{2}{x}; x \neq 0$

$x(3x-7) = 10(2)$
$3x^2 - 7x = 20$
$3x^2 - 7x - 20 = 0$
$(3x+5)(x-4) = 0$
$3x + 5 = 0$ or $x - 4 = 0$
$3x = -5$ or $x = 4$

$x = -\dfrac{5}{3}$ or $x = 4$

The solution set is $\left\{-\dfrac{5}{3}, 4\right\}$.

31. $\dfrac{x}{x-6} - 3 = \dfrac{6}{x-6}; x \neq 6$

$(x-6)\left(\dfrac{x}{x-6} - 3\right) = (x-6)\left(\dfrac{6}{x-6}\right)$

$(x-6)\left(\dfrac{x}{x-6}\right) - 3(x-6) = 6$

$x - 3(x-6) = 6$
$x - 3x + 18 = 6$
$-2x + 18 = 6$
$-2x = -12$
$x = 6$
The solution set is $\emptyset$.

Problem Set 4.6

33. $\frac{3s}{s+2} + 1 = \frac{35}{2(3s+1)}; s \neq -\frac{1}{3}, s \neq -2$

$$2(s+2)(3s+1)\left[\frac{3s}{s+2} + 1\right] = 2(s+2)(3s+1)\left[\frac{35}{2(3s+1)}\right]$$

$$2(s+2)(3s+1)\left(\frac{3s}{s+2}\right) + 2(s+2)(3s+1)(1) = 35(s+2)$$

$6s(3s+1) + 2(3s^2+7s+2) = 35s+70$
$18s^2+6s+6s^2+14s+4 = 35s+70$
$24s^2+20s+4 = 35s+70$
$24s^2-15s-66 = 0$
$3(8s^2-5s-22) = 0$
$3(8s+11)(s-2) = 0$
$8s+11=0$ or $s-2=0$
$8s = -11$ or $s=2$
$s = -\frac{11}{8}$ or $s=2$
The solution set is $\left\{-\frac{11}{8}, 2\right\}$.

35. $2 - \frac{3x}{x-4} = \frac{14}{x+7}; x \neq 4, x \neq -7$

$$(x-4)(x+7)\left(2 - \frac{3x}{x-4}\right) = (x-4)(x+7)\left(\frac{14}{x+7}\right)$$

$$(x-4)(x+7)(2) - (x-4)(x+7)\left(\frac{3x}{x-4}\right) = 14(x-4)$$

$2(x^2+3x-28) - 3x(x+7) = 14x-56$
$2x^2+6x-56-3x^2-21x = 14x-56$
$-x^2-15x-56 = 14x-56$
$0 = x^2+29x$
$0 = x(x+29)$
$x=0$ or $x+29=0$
$x=0$ or $x=-29$
The solution set is $\{-29, 0\}$.

37. $\frac{n+6}{27} = \frac{1}{n}; n \neq 0$

$n(n+6) = 1(27)$
$n^2+6n = 27$
$n^2+6n-27 = 0$
$(n+9)(n-3) = 0$

$n + 9 = 0$ or $n - 3 = 0$
$n = -9$ or $n = 3$
The solution set is $\{-9, 3\}$.

39. $\frac{3n}{n-1} - \frac{1}{3} = \frac{-40}{3n-18}$

$\frac{3n}{n-1} - \frac{1}{3} = \frac{-40}{3(n-6)}; n \neq 1, n \neq 6$

$3(n-1)(n-6)\left(\frac{3n}{n-1} - \frac{1}{3}\right) = 3(n-1)(n-6)\left[\frac{-40}{3(n-6)}\right]$

$3(n-1)(n-6)\left(\frac{3n}{n-1}\right) - 3(n-1)(n-6)\left(\frac{1}{3}\right) = -40(n-1)$

$9n(n-6) - (n-1)(n-6) = -40n + 40$
$9n^2 - 54n - (n^2 - 7n + 6) = -40n + 40$
$9n^2 - 54n - n^2 + 7n - 6 = -40n + 40$
$8n^2 - 47n - 6 = -40n + 40$
$8n^2 - 7n - 46 = 0$
$(8n - 23)(n + 2) = 0$
$8n - 23 = 0$ or $n + 2 = 0$
$8n = 23$ or $n = -2$

$n = \frac{23}{8}$ or $n = -2$

The solution set is $\left\{-2, \frac{23}{8}\right\}$.

41. $\frac{-3}{4x+5} = \frac{2}{5x-7}; x \neq -\frac{5}{4}, x \neq \frac{7}{5}$

$-3(5x - 7) = 2(4x + 5)$
$-15x + 21 = 8x + 10$
$-15x = 8x - 11$
$-23x = -11$
$x = \frac{-11}{-23} = \frac{11}{23}$

The solution set is $\left\{\frac{11}{23}\right\}$.

43. $\frac{2x}{x-2} + \frac{15}{x^2 - 7x + 10} = \frac{3}{x-5}$

$\frac{2x}{x-2} + \frac{15}{(x-5)(x-2)} = \frac{3}{x-5}; x \neq 2, x \neq 5$

$(x-5)(x-2)\left[\frac{2x}{x-2} + \frac{15}{(x-5)(x-2)}\right] = (x-5)(x-2)\left(\frac{3}{x-5}\right)$

$$(x-5)(x-2)\left(\frac{2x}{x-2}\right)+(x-5)(x-2)\left[\frac{15}{(x-5)(x-2)}\right]=3(x-2)$$

$2x(x-5)+15=3x-6$
$2x^2-10x+15=3x-6$
$2x^2-13x+21=0$
$(2x-7)(x-3)=0$
$2x-7=0$ or $x-3=0$
$2x=7$ or $x=3$
$x=\frac{7}{2}$ or $x=3$
The solution set is $\left\{3, \frac{7}{2}\right\}$.

45. Let d = first amount and
$1750-d$ = second amount.

$\frac{d}{1750-d}=\frac{3}{4}; d \neq 1750$
$4d=3(1750-d)$
$4d=5250-3d$
$7d=5250$
$d=750$
The amounts are \$750 and \$1000.

47. Let one angle = x and
other angle $=180-x-60=120-x$.

$\frac{x}{120-x}=\frac{2}{3}; x \neq 120$
$3x=2(120-x)$
$3x=240-2x$
$5x=240$
$x=48$
The angles are 48° and 72°.

49. Let n = number.

$n+\frac{1}{n}=\frac{53}{14}; n \neq 0$

$14n\left(n+\frac{1}{n}\right)=14n\left(\frac{53}{14}\right)$

$14n(n)+14n\left(\frac{1}{n}\right)=53n$

49. Continued

$14n^2+14=53n$
$14n^2-53n+14=0$
$(2n-7)(7n-2)=0$
$2n-7=0$ or $7n-2=0$
$2n=7$ or $7n=2$
$n=\frac{7}{2}$ or $n=\frac{2}{7}$
The numbers are $\frac{7}{2}$ and $\frac{2}{7}$.

51. $\frac{150{,}000}{2500}=\frac{210{,}000}{x}; x \neq 0$
$150000x=210000(2500)$
$150000x=525000000$
$x=3500$
The taxes are \$3500.

53. Let x = Laura's sales and
$120.75-x$ = Tammy's sales.

$\frac{120.75-x}{x}=\frac{4}{3}; x \neq 0$
$4x=3(120.75-x)$
$4x=362.25-3x$
$7x=362.25$
$x=51.75$
Laura's sales are \$51.75 and
Tammy's sales are \$69.00.

55. Let x = smaller number and $90 - x$ = larger number.

$$\frac{90-x}{x} = 10 + \frac{2}{x};\ x \neq 0$$

$$x\left(\frac{90-x}{x}\right) = x\left(10 + \frac{2}{x}\right)$$

$$90 - x = x(10) + x\left(\frac{2}{x}\right)$$

$90 - x = 10x + 2$
$-x = 10x - 88$
$-11x = -88$
$x = 8$
The numbers are 8 and 82.

57. Let x = first piece and $20 - x$ = second piece.

$$\frac{x}{20-x} = \frac{7}{3};\ x \neq 20$$

$3x = 7(20 - x)$
$3x = 140 - 7x$
$10x = 140$
$x = 14$
The lengths are 14 feet and 6 feet.

59. Let x = female voters and $1150 - x$ = male voters.

$$\frac{x}{1150-x} = \frac{3}{2};\ x \neq 1150$$

$2x = 3(1150 - x)$
$2x = 3450 - 3x$
$5x = 3450$
$x = 690$
They were 690 female voters and 460 male voters.

PROBLEM SET 4.7 More Fractional Equations and Applications

1. $$\frac{x}{4x-4} + \frac{5}{x^2-1} = \frac{1}{4}$$

$$\frac{x}{4(x-1)} + \frac{5}{(x+1)(x-1)} = \frac{1}{4};\ x \neq 1, x \neq -1$$

$$4(x-1)(x+1)\left[\frac{x}{4(x-1)} + \frac{5}{(x+1)(x-1)}\right] = 4(x+1)(x-1)\left(\frac{1}{4}\right)$$

$$4(x-1)(x+1)\left[\frac{x}{4(x-1)}\right] + 4(x-1)(x+1)\left[\frac{5}{(x+1)(x-1)}\right] = (x+1)(x-1)$$

$x(x+1) + 4(5) = x^2 - 1$
$x^2 + x + 20 = x^2 - 1$
$x = -21$
The solution set is $\{-21\}$.

3. $$3 + \frac{6}{t-3} = \frac{6}{t^2-3t}$$

$$3 + \frac{6}{t-3} = \frac{6}{t(t-3)};\ t \neq 0, t \neq 3$$

Problem Set 4.7

$$t(t-3)\left(3+\frac{6}{t-3}\right)=t(t-3)\left[\frac{6}{t(t-3)}\right]$$

$3t(t-3)+6t=6$
$3t^2-9t+6t=6$
$3t^2-3t=6$
$3t^2-3t-6=0$
$3(t^2-t-2)=0$
$3(t-2)(t+1)=0$
$t-2=0 \quad \text{or} \quad t+1=0$
$t=2 \quad \text{or} \quad t=-1$
The solution set is $\{-1, 2\}$.

5. $$\frac{3}{n-5}+\frac{4}{n+7}=\frac{2n+11}{n^2+2n-35}$$

$$\frac{3}{n-5}+\frac{4}{n+7}=\frac{2n+11}{(n+7)(n-5)};\ n\neq -7, n\neq 5$$

$$(n+7)(n-5)\left(\frac{3}{n-5}+\frac{4}{n+7}\right)=(n+7)(n-5)\left[\frac{2n+11}{(n+7)(n-5)}\right]$$

$$(n+7)(n-5)\left(\frac{3}{n-5}\right)+(n+7)(n-5)\left(\frac{4}{n+7}\right)=2n+11$$

$3(n+7)+4(n-5)=2n+11$
$3n+21+4n-20=2n+11$
$7n+1=2n+11$
$7n=2n+10$
$5n=10$
$n=2$
The solution set is $\{2\}$.

7. $$\frac{5x}{2x+6}-\frac{4}{x^2-9}=\frac{5}{2}$$

$$\frac{5x}{2(x+3)}-\frac{4}{(x+3)(x-3)}=\frac{5}{2};\ x\neq 3, x\neq -3$$

$$2(x+3)(x-3)\left[\frac{5x}{2(x+3)}-\frac{4}{(x+3)(x-3)}\right]=2(x+3)(x-3)\left(\frac{5}{2}\right)$$

$$2(x+3)(x-3)\left[\frac{5x}{2(x+3)}\right]-2(x+3)(x-3)\left[\frac{4}{(x+3)(x-3)}\right]=5(x+3)(x-3)$$

$5x(x-3)-8=5(x^2-9)$
$5x^2-15x-8=5x^2-45$
$-15x-8=-45$
$-15x=-37$

$x = \frac{37}{15}$

The solution set is $\left\{\frac{37}{15}\right\}$.

9. $1 + \frac{1}{n-1} = \frac{1}{n^2 - n}$

$1 + \frac{1}{n-1} = \frac{1}{n(n-1)}; n \neq 0, n \neq 1$

$n(n-1)\left(1 + \frac{1}{n-1}\right) = n(n-1)\left[\frac{1}{n(n-1)}\right]$

$n(n-1) + n(n-1)\left(\frac{1}{n-1}\right) = 1$

$n^2 - n + n = 1$

$n^2 = 1$

$n^2 - 1 = 0$

$(n+1)(n-1) = 0$

$n + 1 = 0$ or $n - 1 = 0$

$n = -1$ or $n = 1$

Discard the root $n = 1$.

The solution set is $\{-1\}$.

11. $\frac{2}{n-2} - \frac{n}{n+5} = \frac{10n+15}{n^2+3n-10}$

$\frac{2}{n-2} - \frac{n}{n+5} = \frac{10n+15}{(n+5)(n-2)}; n \neq -5, n \neq 2$

$(n+5)(n-2)\left(\frac{2}{n-2} - \frac{n}{n+5}\right) = (n+5)(n-2)\left[\frac{10n+15}{(n+5)(n-2)}\right]$

$(n+5)(n-2)\left(\frac{2}{n-2}\right) - (n+5)(n-2)\left(\frac{n}{n+5}\right) = 10n + 15$

$2(n+5) - n(n-2) = 10n + 15$

$2n + 10 - n^2 + 2n = 10n + 15$

$-n^2 + 4n + 10 = 10n + 15$

$0 = n^2 + 6n + 5$

$0 = (n+5)(n+1)$

$n + 5 = 0$ or $n + 1 = 0$

$n = -5$ or $n = -1$

Discard the root $n = -5$.

The solution set is $\{-1\}$.

Problem Set 4.7

13. $$\frac{2}{2x-3}-\frac{2}{10x^2-13x-3}=\frac{x}{5x+1}$$

$$\frac{2}{2x-3}-\frac{2}{(5x+1)(2x-3)}=\frac{x}{5x+1};\ x\neq\frac{3}{2},\ x\neq-\frac{1}{5}$$

$$(2x-3)(5x+1)\left[\frac{2}{2x-3}-\frac{2}{(5x+1)(2x-3)}\right]=(2x-3)(5x+1)\left(\frac{x}{5x+1}\right)$$

$$(2x-3)(5x+1)\left(\frac{2}{2x-3}\right)-(2x+3)(5x+1)\left[\frac{2}{(5x+1)(2x-3)}\right]=x(2x-3)$$

$2(5x+1)-2=2x^2-3x$
$10x+2-2=2x^2-3x$
$0=2x^2-13x$
$0=x(2x-13)$
$x=0$ or $2x-13=0$
$x=0$ or $2x=13$
$x=0$ or $x=\frac{13}{2}$

The solution set is $\left\{0, \frac{13}{2}\right\}$.

15. $$\frac{2x}{x+3}-\frac{3}{x-6}=\frac{29}{x^2-3x-18}$$

$$\frac{2x}{x+3}-\frac{3}{x-6}=\frac{29}{(x-6)(x+3)};\ x\neq 6,\ x\neq-3$$

$$(x-6)(x+3)\left(\frac{2x}{x+3}-\frac{3}{x-6}\right)=(x-6)(x+3)\left[\frac{29}{(x-6)(x+3)}\right]$$

$$(x-6)(x+3)\left(\frac{2x}{x+3}\right)-(x-6)(x+3)\left(\frac{3}{x-6}\right)=29$$

$2x(x-6)-3(x+3)=29$
$2x^2-12x-3x-9=29$
$2x^2-15x-38=0$
$(2x-19)(x+2)=0$
$2x-19=0$ or $x+2=0$
$2x=19$ or $x=-2$
$x=\frac{19}{2}$ or $x=-2$

The solution set is $\left\{-2, \frac{19}{2}\right\}$.

17. $$\frac{a}{a-5}+\frac{2}{a-6}=\frac{2}{a^2-11a+30}$$

$$\frac{a}{a-5}+\frac{2}{a-6}=\frac{2}{(a-6)(a-5)};\ a\neq 6,\ a\neq 5$$

$$(a-6)(a-5)\left(\frac{a}{a-5}+\frac{2}{a-6}\right)=(a-6)(a-5)\left[\frac{2}{(a-6)(a-5)}\right]$$

$$(a-6)(a-5)\left(\frac{a}{a-5}\right)+(a-6)(a-5)\left(\frac{2}{a-6}\right)=2$$

$a(a-6)+2(a-5)=2$
$a^2-6a+2a-10=2$
$a^2-4a-12=0$
$(a-6)(a+2)=0$
$a-6=0$ or $a+2=0$
$a=6$ or $a=-2$
Discard the root $a=6$.
The solution set is $\{-2\}$.

19. $$\frac{-1}{2x-5}+\frac{2x-4}{4x^2-25}=\frac{5}{6x+15}$$

$$\frac{-1}{2x-5}+\frac{2x-4}{(2x+5)(2x-5)}=\frac{5}{3(2x+5)};\ x\neq\frac{5}{2},\ x\neq-\frac{5}{2}$$

$$3(2x+5)(2x-5)\left[\frac{-1}{2x-5}+\frac{2x-4}{(2x+5)(2x-5)}\right]=3(2x+5)(2x-5)\left[\frac{5}{3(2x+5)}\right]$$

$$3(2x+5)(2x-5)\left(\frac{-1}{2x-5}\right)+3(2x+5)(2x-5)\left[\frac{2x-4}{(2x+5)(2x-5)}\right]=5(2x-5)$$

$-3(2x+5)+3(2x-4)=10x-25$
$-6x-15+6x-12=10x-25$
$-27=10x-25$
$-2=10x$
$-\frac{2}{10}=x$
$-\frac{1}{5}=x$
The solution set is $\left\{-\frac{1}{5}\right\}$.

21. $$\frac{7y+2}{12y^2+11y-15}-\frac{1}{3y+5}=\frac{2}{4y-3}$$

$$\frac{7y+2}{(3y+5)(4y-3)}-\frac{1}{3y+5}=\frac{2}{4y-3};\ y\neq-\frac{5}{3},\ y\neq\frac{3}{4}$$
$$(3y+5)(4y-3)\left[\frac{7y+2}{(3y+5)(4y-3)}-\frac{1}{3y+5}\right]=(3y+5)(4y-3)\left(\frac{2}{4y-3}\right)$$

$$(3y+5)(4y-3)\left[\frac{7y+2}{(3y+5)(4y-3)}\right]-(3y+5)(4y-3)\left(\frac{1}{3y+5}\right)=2(3y+5)$$

$7y+2-(4y-3)=2(3y+5)$
$7y+2-4y+3=6y+10$
$3y+5=6y+10$
$3y=6y+5$
$-3y=5$
$y=-\frac{5}{3}$

Discard the root $y=-\frac{5}{3}$.

The solution set is $\emptyset$.

23. $$\frac{2n}{6n^2+7n-3}-\frac{n-3}{3n^2+11n-4}=\frac{5}{2n^2+11n+12}$$

$$\frac{2n}{(3n-1)(2n+3)}-\frac{n-3}{(3n-1)(n+4)}=\frac{5}{(2n+3)(n+4)};\ n\neq\frac{1}{3},\ n\neq-\frac{3}{2},\ n\neq-4$$

$$(3n-1)(2n+3)(n+4)\left[\frac{2n}{(3n-1)(2n+3)}-\frac{n-3}{(3n-1)(n+4)}\right]=(3n-1)(2n+3)(n+4)\left[\frac{5}{(2n+3)(n+4)}\right]$$

$$(3n\text{-}1)(2n\text{+}3)(n\text{+}4)\left[\frac{2n}{(3n\text{-}1)(2n\text{+}3)}\right]-(3n\text{-}1)(2n\text{+}3)(n\text{+}4)\left[\frac{n-3}{(3n-1)(n\text{+}4)}\right]=(3n\text{-}1)(2n\text{+}3)(n\text{+}4)\left[\frac{5}{(2n\text{+}3)(n\text{+}4)}\right]$$

$2n(n+4)-(2n+3)(n-3)=5(3n-1)$
$2n^2+8n-(2n^2-3n-9)=15n-5$
$2n^2+8n-2n^2+3n+9=15n-5$
$11n+9=15n-5$
$11n=15n-14$
$-4n=-14$
$n=\frac{-14}{-4}=\frac{7}{2}$

The solution set is $\left\{\frac{7}{2}\right\}$.

25. $$\frac{1}{2x^2-x-1}+\frac{3}{2x^2+x}=\frac{2}{x^2-1}$$

$$\frac{1}{(2x+1)(x-1)}+\frac{3}{x(2x+1)}=\frac{2}{(x+1)(x-1)};\ x\neq-\frac{1}{2},\ x\neq1,\ x\neq-1,\ x\neq0$$

$$x(2x+1)(x+1)(x-1)\left[\frac{1}{(2x+1)(x-1)}+\frac{3}{x(2x+1)}\right]=x(2x+1)(x-1)(x+1)\left[\frac{2}{(x+1)(x-1)}\right]$$

$$x(2x+1)(x+1)(x-1)\left[\frac{1}{(2x+1)(x-1)}\right]+x(2x+1)(x+1)(x-1)\left[\frac{3}{x(2x+1)}\right]=2x(2x+1)$$

$x(x+1)+3(x+1)(x-1)=2x(2x+1)$
$x^2+x+3(x^2-1)=4x^2+2x$

$x^2 + x + 3x^2 - 3 = 4x^2 + 2x$
$4x^2 + x - 3 = 4x^2 + 2x$
$x - 3 = 2x$
$-3 = x$
The solution set is $\{-3\}$.

27. $\dfrac{x+1}{x^3 - 9x} - \dfrac{1}{2x^2 + x - 21} = \dfrac{1}{2x^2 + 13x + 21}$

$\dfrac{x+1}{x(x+3)(x-3)} - \dfrac{1}{(2x+7)(x-3)} = \dfrac{1}{(2x+7)(x+3)}; x \neq 0, x \neq 3, x \neq -3, x \neq -\dfrac{7}{2}$

$x(2x+7)(x+3)(x-3)\left[\dfrac{x+1}{x(x+3)(x-3)} - \dfrac{1}{(2x+7)(x-3)}\right] = x(2x+7)(x+3)(x-3)\left[\dfrac{1}{(2x+7)(x+3)}\right]$

$x(2x+7)(x+3)(x-3)\left[\dfrac{x+1}{x(x+3)(x-3)}\right] - x(2x+7)(x+3)(x-3)\left[\dfrac{1}{(2x+7)(x-3)}\right] = x(x-3)$

$(2x+7)(x+1) - x(x+3) = x(x-3)$
$2x^2 + 9x + 7 - x^2 - 3x = x^2 - 3x$
$x^2 + 6x + 7 = x^2 - 3x$
$6x + 7 = -3x$
$7 = -9x$
$-\dfrac{7}{9} = x$
The solution set is $\left\{-\dfrac{7}{9}\right\}$.

29. $\dfrac{4t}{4t^2 - t - 3} + \dfrac{2 - 3t}{3t^2 - t - 2} = \dfrac{1}{12t^2 + 17t + 6}$

$\dfrac{4t}{(4t+3)(t-1)} + \dfrac{2-3t}{(3t+2)(t-1)} = \dfrac{1}{(4t+3)(3t+2)}; t \neq -\dfrac{3}{4}, t \neq 1, t \neq -\dfrac{2}{3}$

$(4t+3)(t-1)(3t+2)\left[\dfrac{4t}{(4t+3)(t-1)} + \dfrac{2-3t}{(3t+2)(t-1)}\right] = (4t+3)(t-1)(3t+2)\left[\dfrac{1}{(4t+3)(3t+2)}\right]$

$(4t+3)(t-1)(3t+2)\left[\dfrac{4t}{(4t+3)(t-1)}\right] + (4t+3)(t-1)(3t+2)\left[\dfrac{2-3t}{(3t+2)(t-1)}\right] = (4t+3)(t-1)(3t+2)\left[\dfrac{1}{(4t3)(3t+2)}\right]$

$4t(3t+2) + (4t+3)(2-3t) = t - 1$
$12t^2 + 8t - 12t^2 - t + 6 = t - 1$
$7t + 6 = t - 1$
$7t = t - 7$
$6t = -7$
$t = -\dfrac{7}{6}$ The solution set is $\left\{-\dfrac{7}{6}\right\}$.

31. $y = \frac{5}{6}x + \frac{2}{9}$ for x

$18(y) = 18\left(\frac{5}{6}x + \frac{2}{9}\right)$

$18y = 18\left(\frac{5}{6}x\right) + 18\left(\frac{2}{9}\right)$

$18y = 15x + 4$
$18y - 4 = 15x$

$\frac{1}{15}(18y - 4) = \frac{1}{15}(15x)$

$\frac{18y - 4}{15} = x$

33. $\frac{-2}{x-4} = \frac{5}{y-1}$ for y

$-2(y-1) = 5(x-4)$
$-2y + 2 = 5x - 20$
$-2y = 5x - 22$

$-\frac{1}{2}(-2y) = -\frac{1}{2}(5x - 22)$

$y = \frac{-5x + 22}{2}$

35. $I = \frac{100M}{C}$ for M

$\frac{CI}{100} = M$

$M = \frac{CI}{100}$

37. $\frac{R}{S} = \frac{T}{S+T}$ for R

$S\left(\frac{R}{S}\right) = S\left(\frac{T}{S+T}\right)$

$R = \frac{ST}{S+T}$

39. $\frac{y-1}{x-3} = \frac{b-1}{a-3}$ for y

$(y-1)(a-3) = (x-3)(b-1)$
$y(a-3) - 1(a-3) = x(b-1) - 3(b-1)$
$y(a-3) - a + 3 = bx - x - 3b + 3$
$y(a-3) = bx - x - 3b + 3 + a - 3$
$y(a-3) = bx - x - 3b + a$

$y = \frac{bx - x - 3b + a}{a-3}$

41. $\frac{x}{a} + \frac{y}{b} = 1$ for y

$ab\left(\frac{x}{a} + \frac{y}{b}\right) = ab(1)$

$ab\left(\frac{x}{a}\right) + ab\left(\frac{y}{b}\right) = ab$

$bx + ay = ab$
$ay = ab - bx$

$y = \frac{ab - bx}{a}$

43. $\frac{y-1}{x+6} = \frac{-2}{3}$ for y

$3(y-1) = -2(x+6)$
$3y - 3 = -2x - 12$
$3y = -2x - 9$

$y = \frac{-2x - 9}{3}$

45.

	Rate	Time	Distance
Kent	$x+4$	$\frac{270}{x+4}$	270
Dave	x	$\frac{250}{x}$	250

$\frac{270}{x+4} = \frac{250}{x}$

$270x = 250(x+4)$
$270x = 250x + 1000$
$20x = 1000$
$x = 50$

Dave's rate is 50 mph and Kent's rate is 54 mph.

47.

	Time in Minutes	Rate
Inlet	10	$\frac{1}{10}$
Outlet	12	$\frac{1}{12}$
Together	x	$\frac{1}{x}$

$$\frac{1}{10}-\frac{1}{12}=\frac{1}{x}$$

$$60x\left(\frac{1}{10}-\frac{1}{12}\right)=60x\left(\frac{1}{x}\right)$$

$$60x\left(\frac{1}{10}\right)-60x\left(\frac{1}{12}\right)=60$$

$6x-5x=60$

$x=60$

The tank will overflow in 60 minutes.

49.

	Rate	Time	Words
Connie	$x+20$	$\frac{600}{x+20}$	600
Katie	x	$\frac{600}{x}$	600

$$\frac{600}{x+20}+5=\frac{600}{x}$$

$$x(x+20)\left(\frac{600}{x+20}+5\right)=x(x+20)\left(\frac{600}{x}\right)$$

$$x(x+20)\left(\frac{600}{x+20}\right)+5x(x+20)=600(x+20)$$

$600x+5x^2+100x=600x+12000$

$5x^2+700x=600x+12000$

$5x^2+100x-12000=0$

$5(x^2+20x-2400)=0$

$5(x+60)(x-40)=0$

$x+60=0$ or $x-40$

$x=-60$ or $x=40$

Discard the root $x=-60$.
Katie's rate is 40 words per minute and Connie's rate is 60 words per minute.

51.

	Rate	Time	Distance
Plane A	x	$\frac{1400}{x}$	1400
Plane B	$x+50$	$\frac{2000}{x+50}$	2000

$$\frac{1400}{x}+1=\frac{2000}{x+50}$$

$$x(x+50)\left(\frac{1400}{x}+1\right)=x(x+50)\left(\frac{2000}{x+50}\right)$$

$$x(x+50)\left(\frac{1400}{x}\right)+x(x+50)=2000x$$

$1400(x+50)+x^2+50x=2000x$

$1400x+70000+x^2+50x=2000x$

$x^2+1450x+70000=2000x$

$x^2-550x+70000=0$

$(x-350)(x-200)=0$

$x-350=0$ or $x-200=0$

$x=350$ or $x=200$

Case 1: Plane B travels at 400 mph for 5 hours and Plane A travels at 350 mph for 4 hours.

Case 2: Plane B travels at 250mph for 8 hours and Plane A travels at 200 mph for 7 hours.

53.

	Time in Minutes	Rate
Amy	$2x$	$\frac{1}{2x}$
Nancy	x	$\frac{1}{x}$
Together	40	$\frac{1}{40}$

$$\frac{1}{2x}+\frac{1}{x}=\frac{1}{40}$$

$$40x\left(\frac{1}{2x}+\frac{1}{x}\right)=40x\left(\frac{1}{40}\right)$$

$$40x\left(\frac{1}{2x}\right)+40x\left(\frac{1}{x}\right)=x$$

$20 + 40 = x$
$60 = x$
It would take Nancy 60 minutes and Amy would take 120 minutes.

55.

	Rate	Time	Money
Anticipated	$\frac{12}{x}$	x	12
Actual	$\frac{12}{x+1}$	$x+1$	12

$$\frac{12}{x} - 1 = \frac{12}{x+1}$$

$$x(x+1)\left(\frac{12}{x} - 1\right) = x(x+1)\left(\frac{12}{x+1}\right)$$

$$x(x+1)\left(\frac{12}{x}\right) - x(x+1) = 12x$$

$12(x+1) - x^2 - x = 12x$
$12x + 12 - x^2 - x = 12x$
$-x^2 + 11x + 12 = 12x$
$0 = x^2 + x - 12$
$(x+4)(x-3) = 0$
$x + 4 = 0$ or $x - 3 = 0$
$x = -4$ or $x = 3$
Discard the root $x = -4$.
The anticipated time was 3 hours.

57.

	Rate	Time	Distance
Out	$x+4$	$\frac{24}{x+4}$	24
Back	x	$\frac{12}{x}$	12

$$\frac{24}{x+4} - \frac{1}{2} = \frac{12}{x}$$

$$2x(x+4)\left(\frac{24}{x+4} - \frac{1}{2}\right) = 2x(x+4)\left(\frac{12}{x}\right)$$

$$2x(x+4)\left(\frac{24}{x+4}\right) - 2x(x+4)\left(\frac{1}{2}\right) = 24(x+4)$$

$2x(24) - x(x+4) = 24x + 96$
$48x - x^2 - 4x = 24x + 96$
$-x^2 + 44x = 24x + 96$
$0 = x^2 - 20x + 96$
$0 = (x-12)(x-8)$
$x - 12 = 0$ or $x - 8 = 0$
$x = 12$ or $x = 8$

Case 1: 16 mph on the way out and 12 mph back
Case 2: 12 mph on the way out and 8 mph back

CHAPTER 4 Review Problem Set

1. $\frac{26x^2y^3}{39x^4y^2} = \frac{2 \bullet 13x^2y^3}{3 \bullet 13x^4y^2} = \frac{2y}{3x^2}$

2. $\frac{a^2 - 9}{a^2 + 3a}$

$\frac{(a+3)(a-3)}{a(a+3)}$

$\frac{a-3}{a}$

3. $\frac{n^2 - 3n - 10}{n^2 + n - 2}$

$\frac{(n-5)(n+2)}{(n-1)(n+2)}$

$\frac{n-5}{n-1}$

4. $\frac{x^4 - 1}{x^3 - x}$

$\frac{(x^2+1)(x+1)(x-1)}{x(x+1)(x-1)}$

$\frac{x^2+1}{x}$

5. $\dfrac{8x^3 - 2x^2 - 3x}{12x^2 - 9x}$

$\dfrac{x(8x^2 - 2x - 3)}{3x(4x - 3)}$

$\dfrac{x(4x - 3)(2x + 1)}{3x(4x - 3)}$

$\dfrac{2x + 1}{3}$

6. $\dfrac{x^4 - 7x^2 - 30}{2x^4 + 7x^2 + 3}$

$\dfrac{(x^2 - 10)(x^2 + 3)}{(2x^2 + 1)(x^2 + 3)}$

$\dfrac{x^2 - 10}{2x^2 + 1}$

7. $\dfrac{\frac{5}{8} - \frac{1}{2}}{\frac{1}{6} + \frac{3}{4}} = \dfrac{24}{24} \bullet \dfrac{\left(\frac{5}{8} - \frac{1}{2}\right)}{\left(\frac{1}{6} + \frac{3}{4}\right)}$

$\dfrac{24\left(\frac{5}{8}\right) - 24\left(\frac{1}{2}\right)}{24\left(\frac{1}{6}\right) + 24\left(\frac{3}{4}\right)} = \dfrac{15 - 12}{4 + 18} = \dfrac{3}{22}$

8. $\dfrac{\frac{3}{2x} + \frac{5}{3y}}{\frac{4}{x} - \frac{3}{4y}}$

$\dfrac{12xy}{12xy} \bullet \dfrac{\left(\frac{3}{2x} + \frac{5}{3y}\right)}{\left(\frac{4}{x} - \frac{3}{4y}\right)}$

$\dfrac{12xy\left(\frac{3}{2x}\right) + 12xy\left(\frac{5}{3y}\right)}{12xy\left(\frac{4}{x}\right) - 12xy\left(\frac{3}{4y}\right)}$

$\dfrac{18y + 20x}{48y - 9x}$

9. $\dfrac{\frac{3}{x-2} - \frac{4}{x^2-4}}{\frac{2}{x+2} + \frac{1}{x-2}}$

$\dfrac{\frac{3}{x-2} - \frac{4}{(x+2)(x-2)}}{\frac{2}{x+2} + \frac{1}{x-2}}$

$\dfrac{(x+2)(x-2)}{(x+2)(x-2)} \bullet \dfrac{\left[\frac{3}{x-2} - \frac{4}{(x+2)(x-2)}\right]}{\left(\frac{2}{x+2} + \frac{1}{x-2}\right)}$

$\dfrac{(x+2)(x-2)\left(\frac{3}{x-2}\right) - (x+2)(x-2)\left[\frac{4}{(x+2)(x-2)}\right]}{(x+2)(x-2)\left(\frac{2}{x+2}\right) + (x+2)(x-2)\left(\frac{1}{x-2}\right)}$

$\dfrac{3(x+2) - 4}{2(x-2) + 1(x+2)}$

$\dfrac{3x + 6 - 4}{2x - 4 + x + 2}$

$\dfrac{3x + 2}{3x - 2}$

10. $1 - \dfrac{1}{2 - \frac{1}{x}}$

$1 - \dfrac{x}{x} \bullet \dfrac{(1)}{\left(2 - \frac{1}{x}\right)}$

$1 - \dfrac{x}{2(x) - x\left(\frac{1}{x}\right)}$

$1 - \dfrac{x}{2x - 1}$

Chapter 4 Review Problem Set

$$\frac{2x-1}{2x-1} - \frac{x}{2x-1}$$

$$\frac{2x-1-x}{2x-1}$$

$$\frac{x-1}{2x-1}$$

11. $$\frac{6xy^2}{7y^3} \div \frac{15x^2y}{5x^2}$$

$$\frac{6xy^2}{7y^3} \bullet \frac{5x^2}{15x^2y}$$

$$\frac{6(5)x^3y^2}{7(15)x^2y^4}$$

$$\frac{6}{15} \bullet \frac{5x}{7y^2}$$

$$\frac{2}{5} \bullet \frac{5x}{7y^2}$$

$$\frac{2x}{7y^2}$$

12. $$\frac{9ab}{3a+6} \bullet \frac{a^2-4a-12}{a^2-6a}$$

$$\frac{9ab}{3(a+2)} \bullet \frac{(a-6)(a+2)}{a(a-6)}$$

$3b$

13. $$\frac{n^2+10n+25}{n^2-n} \bullet \frac{5n^3-3n^2}{5n^2+22n-15}$$

$$\frac{(n+5)(n+5)}{n(n-1)} \bullet \frac{n^2(5n-3)}{(5n-3)(n+5)}$$

$$\frac{n(n+5)}{n-1}$$

14. $$\frac{x^2-2xy-3y^2}{x^2+9y^2} \div \frac{2x^2+xy-y^2}{2x^2-xy}$$

$$\frac{x^2-2xy-3y^2}{x^2+9y^2} \bullet \frac{2x^2-xy}{2x^2+xy-y^2}$$

$$\frac{(x-3y)(x+y)}{x^2+9y^2} \bullet \frac{x(2x-y)}{(2x-y)(x+y)}$$

$$\frac{x(x-3y)}{x^2+9y^2}$$

15. $$\frac{2x+1}{5} + \frac{3x-2}{4}; \text{LCD} = 20$$

$$\frac{(2x+1)}{5} \bullet \frac{4}{4} + \frac{(3x-2)}{4} \bullet \frac{5}{5}$$

$$\frac{4(2x+1)}{20} + \frac{5(3x-2)}{20}$$

$$\frac{4(2x+1)+5(3x-2)}{20}$$

$$\frac{8x+4+15x-10}{20}$$

$$\frac{23x-6}{20}$$

16. $$\frac{3}{2n} + \frac{5}{3n} - \frac{1}{9}; \text{LCD} = 18n$$

$$\frac{3}{2n} \bullet \frac{9}{9} + \frac{5}{3n} \bullet \frac{6}{6} - \frac{1}{9} \bullet \frac{2n}{2n}$$

$$\frac{27}{18n} + \frac{30}{18n} - \frac{2n}{18n}$$

$$\frac{57-2n}{18n}$$

17. $\dfrac{3x}{x+7} - \dfrac{2}{x}$; LCD $= x(x+7)$

$$\frac{3x}{(x+7)} \bullet \frac{x}{x} - \frac{2}{x} \bullet \frac{(x+7)}{(x+7)}$$

$$\frac{3x^2}{x(x+7)} - \frac{2(x+7)}{x(x+7)}$$

$$\frac{3x^2 - 2(x+7)}{x(x+7)}$$

$$\frac{3x^2 - 2x - 14}{x(x+7)}$$

18. $\dfrac{10}{x^2 - 5x} + \dfrac{2}{x}$

$\dfrac{10}{x(x-5)} + \dfrac{2}{x}$; LCD $= x(x-5)$

$$\frac{10}{x(x-5)} + \frac{2}{x} \bullet \frac{(x-5)}{(x-5)}$$

$$\frac{10 + 2x - 10}{x(x-5)}$$

$$\frac{2x}{x(x-5)}$$

$$\frac{2}{x-5}$$

19. $\dfrac{3}{n^2 - 5n - 36} + \dfrac{2}{n^2 + 3n - 4}$

$$\frac{3}{(n-9)(n+4)} + \frac{2}{(n+4)(n-1)}$$

LCD $= (n-9)(n+4)(n-1)$

$$\frac{3}{(n-9)(n+4)} \bullet \frac{(n-1)}{(n-1)} + \frac{2}{(n+4)(n-1)} \bullet \frac{(n-9)}{(n-9)}$$

$$\frac{3(n-1)}{(n-9)(n+4)(n-1)} + \frac{2(n-9)}{(n-9)(n+4)(n-1)}$$

$$\frac{3(n-1) + 2(n-9)}{(n-9)(n+4)(n-1)}$$

$$\frac{3n - 3 + 2n - 18}{(n-9)(n+4)(n-1)}$$

$$\frac{5n - 21}{(n-9)(n+4)(n-1)}$$

20. $\dfrac{3}{2y+3} + \dfrac{5y-2}{2y^2 - 9y - 18} - \dfrac{1}{y-6}$

$$\frac{3}{2y+3} + \frac{5y-2}{(2y+3)(y-6)} - \frac{1}{y-6}$$

LCD $= (2y+3)(y-6)$

$$\frac{3}{(2y+3)} \bullet \frac{(y-6)}{(y-6)} + \frac{5y-2}{(2y+3)(y-6)} - \frac{1}{(y-6)} \bullet \frac{(2y+3)}{(2y+3)}$$

$$\frac{3(y-6) + 5y - 2 - 1(2y+3)}{(2y+3)(y-6)}$$

$$\frac{3y - 18 + 5y - 2 - 2y - 3}{(2y+3)(y-6)}$$

$$\frac{6y - 23}{(2y+3)(y-6)}$$

21.

$$\begin{array}{r|l} & \ 6x - 1 \\ \hline 3x+2 & 18x^2 + 9x - 2 \\ & 18x^2 + 12x \\ \hline & - 3x - 2 \\ & - 3x - 2 \\ \hline & 0 \end{array}$$

$6x - 1$

22.

$$\begin{array}{r|l} & 3x^2 - 7x + 22 \\ \hline x+4 & 3x^3 + 5x^2 - 6x - 2 \\ & 3x^3 + 12x^2 \\ \hline & -7x^2 - 6x \\ & -7x^2 - 28x \\ \hline & 22x - 2 \\ & 22x + 88 \\ \hline & -90 \end{array}$$

$$3x^2 - 7x + 22 - \frac{90}{x+4}$$

23. $\frac{4x+5}{3} + \frac{2x-1}{5} = 2$

$$15\left(\frac{4x+5}{3} + \frac{2x-1}{5}\right) = 15(2)$$

$$15\left(\frac{4x+5}{3}\right) + 15\left(\frac{2x-1}{5}\right) = 30$$

$5(4x+5) + 3(2x-1) = 30$
$20x + 25 + 6x - 3 = 30$
$26x + 22 = 30$
$26x = 8$
$x = \frac{8}{26} = \frac{4}{13}$
The solution set is $\left\{\frac{4}{13}\right\}$.

24. $\frac{3}{4x} + \frac{4}{5} = \frac{9}{10x}$; $x \neq 0$

$$20x\left(\frac{3}{4x} + \frac{4}{5}\right) = 20x\left(\frac{9}{10x}\right)$$

$$20x\left(\frac{3}{4x}\right) + 20x\left(\frac{4}{5}\right) = 18$$

$15 + 16x = 18$
$16x = 3$
$x = \frac{3}{16}$
The solution set is $\left\{\frac{3}{16}\right\}$.

25. $\frac{a}{a-2} - \frac{3}{2} = \frac{2}{a-2}$; $a \neq 2$

$$2(a-2)\left(\frac{a}{a-2} - \frac{3}{2}\right) = 2(a-2)\left(\frac{2}{a-2}\right)$$

$$2(a-2)\left(\frac{a}{a-2}\right) - 2(a-2)\left(\frac{3}{2}\right) = 4$$

$2a - 3(a-2) = 4$
$2a - 3a + 6 = 4$
$-a + 6 = 4$
$-a = -2$
$a = 2$
Discard the root $a = 2$.
The solution set is $\emptyset$.

26. $\frac{4}{5y-3} = \frac{2}{3y+7}$;$y \neq \frac{3}{5}$, $y \neq -\frac{7}{3}$

$4(3y+7) = 2(5y-3)$
$12y + 28 = 10y - 6$
$2y = -34$
$y = -17$
The solution set is $\{-17\}$.

27. $n + \frac{1}{n} = \frac{53}{14}$; $n \neq 0$

$$14n\left(n + \frac{1}{n}\right) = 14n\left(\frac{53}{14}\right)$$

$$14n(n) + 14n\left(\frac{1}{n}\right) = 53n$$

$14n^2 + 14 = 53n$
$14n^2 - 53n + 14 = 0$
$(2n-7)(7n-2) = 0$
$2n - 7 = 0$ or $7n - 2 = 0$
$2n = 7$ or $7n = 2$
$n = \frac{7}{2}$ or $n = \frac{2}{7}$
The solution set is $\left\{\frac{2}{7}, \frac{7}{2}\right\}$.

28. $\frac{1}{2x-7}+\frac{x-5}{4x^2-49}=\frac{4}{6x-21}$

$\frac{1}{2x-7}+\frac{x-5}{(2x+7)(2x-7)}=\frac{4}{3(2x-7)}$; $x \neq \frac{7}{2}, x \neq -\frac{7}{2}$

$3(2x+7)(2x-7)[\frac{1}{2x-7}+\frac{x-5}{(2x+7)(2x-7)}]=3(2x+7)(2x-7)[\frac{4}{3(2x-7)}]$

$3(2x+7)+3(x-5)=4(2x+7)$
$6x+21+3x-15=8x+28$
$9x+6=8x+28$
$x=22$
The solution set is $\{22\}$.

29. $\frac{x}{2x+1}-1=\frac{-4}{7(x-2)}; x \neq -\frac{1}{2}, x \neq 2$

$7(2x+1)(x-2)\left(\frac{x}{2x+1}-1\right)=7(2x+1)(x-2)\left[\frac{-4}{7(x-2)}\right]$

$7(2x+1)(x-2)\left(\frac{x}{2x+1}\right)-1(7)(2x+1)(x-2)=-4(2x+1)$

$7x(x-2)-7(2x^2-3x-2)=-8x-4$
$7x^2-14x-14x^2+21x+14=-8x-4$
$-7x^2+7x+14=-8x-4$
$0=7x^2-15x-18$
$0=(7x+6)(x-3)$
$7x+6=0$ or $x-3=0$
$7x=-6$ or $x=3$
$x=-\frac{6}{7}$ or $x=3$
The solution set is $\left\{-\frac{6}{7}, 3\right\}$.

30. $\frac{2x}{-5}=\frac{3}{4x-13}$; $x \neq \frac{13}{4}$

$2x(4x-13)=-5(3)$
$8x^2-26x=-15$
$8x^2-26x+15=0$
$(4x-3)(2x-5)=0$
$4x-3=0$ or $2x-5=0$
$4x=3$ or $2x=5$
$x=\frac{3}{4}$ or $x=\frac{5}{2}$
The solution set is $\left\{\frac{3}{4}, \frac{5}{2}\right\}$.

31. $\dfrac{2n}{2n^2+11n-21}-\dfrac{n}{n^2+5n-14}=\dfrac{3}{n^2+5n-14}$

$\dfrac{2n}{(2n-3)(n+7)}-\dfrac{n}{(n+7)(n-2)}=\dfrac{3}{(n+7)(n-2)}; n\neq\dfrac{3}{2}, n\neq -7, n\neq 2$

$(2n-3)(n+7)(n-2)\left[\dfrac{2n}{(2n-3)(n+7)}-\dfrac{n}{(n+7)(n-2)}\right]=(2n-3)(n+7)(n-2)\left[\dfrac{3}{(n+7)(n-2)}\right]$

$(2n-3)(n+7)(n-2)\left[\dfrac{2n}{(2n-3)(n+7)}\right]-(2n-3)(n+7)(n-2)\left[\dfrac{n}{(n+7)(n-2)}\right]=3(2n-3)$

$2n(n-2)-n(2n-3)=3(2n-3)$
$2n^2-4n-2n^2+3n=6n-9$
$-n=6n-9$
$-7n=-9$
$n=\dfrac{-9}{-7}=\dfrac{9}{7}$
The solution set is $\left\{\dfrac{9}{7}\right\}$.

32. $\dfrac{2}{t^2-t-6}+\dfrac{t+1}{t^2+t-12}=\dfrac{t}{t^2+6t+8}$

$\dfrac{2}{(t-3)(t+2)}+\dfrac{t+1}{(t+4)(t-3)}=\dfrac{t}{(t+4)(t+2)}\quad ;t\neq 3, t\neq -2, t\neq -4$

$(t-3)(t+2)(t+4)[\dfrac{2}{(t-3)(t+2)}+\dfrac{t+1}{(t+4)(t-3)}]=(t-3)(t+2)(t+4)[\dfrac{\text{t}}{(t+4)(t+2)}]$

$2(t+4)+(t+2)(t+1)=t(t-3)$
$2t+8+t^2+3t+2=t^2-3t$
$t^2+5t+10=t^2-3t$
$5t+10=-3t$
$10=-8t$
$-\dfrac{10}{8}=t$
$-\dfrac{5}{4}=t$
The solution set is $\left\{-\dfrac{5}{4}\right\}$.

33. $\dfrac{y-6}{x+1}=\dfrac{3}{4}\quad$ for y

$4(y-6)=3(x+1)$
$4y-24=3x+3$
$4y=3x+27\qquad y=\dfrac{3x+27}{4}$

34. $\dfrac{x}{a} - \dfrac{y}{b} = 1$ for y

$ab\left(\dfrac{x}{a} - \dfrac{y}{b}\right) = ab(1)$

$bx - ay = ab$

$-ay = -bx + ab$

$y = \dfrac{-bx + ab}{-a}$

$y = \dfrac{bx - ab}{a}$

35. Let $x =$ one part and $1400 - x =$ other part.

$\dfrac{x}{1400 - x} = \dfrac{3}{5}$

$5x = 3(1400 - x)$

$5x = 4200 - 3x$

$8x = 4200$

$x = 525$

One part is \$525 and the other part is \$875.

36.

	Time in Minutes	Rate
Dan	x	$\frac{1}{x}$
Julio	$x - 10$	$\frac{1}{x-10}$
Together	12	$\frac{1}{12}$

$\dfrac{1}{x} + \dfrac{1}{x - 10} = \dfrac{1}{12}$

$12x(x-10)\left(\dfrac{1}{x} + \dfrac{1}{x - 10}\right) = 12x(x-10)\left(\dfrac{1}{12}\right)$

$12x(x-10)\left(\dfrac{1}{x}\right) + 12x(x-10)\left(\dfrac{1}{x - 10}\right) = x(x-10)$

$12(x - 10) + 12x = x^2 - 10x$

$12x - 120 + 12x = x^2 - 10x$

$24x - 120 = x^2 - 10x$

$0 = x^2 - 34x + 120$

$0 = (x - 30)(x - 4)$

$x - 30 = 0$ or $x - 4 = 0$

$x = 30$ or $x = 4$

Discard the root $x = 4$ since $x - 10 = 4 - 10 = -6$ and time cannot be negative.
Dan's time is 30 minutes and Julio's time is 20 minutes.

37.

	Rate	Time	Distance
Car A	x	$\frac{250}{x}$	250
Car B	$x + 5$	$\frac{440}{x+5}$	440

$\dfrac{250}{x} + 3 = \dfrac{440}{x + 5}$

$x(x+5)\left(\dfrac{250}{x} + 3\right) = x(x+5)\left(\dfrac{440}{x+5}\right)$

$x(x+5)\left(\dfrac{250}{x}\right) + 3x(x+5) = 440x$

$250(x+5) + 3x^2 + 15x = 440x$

$250x + 1250 + 3x^2 + 15x = 440x$

$3x^2 - 175x + 1250 = 0$

$(3x - 25)(x - 50) = 0$

$3x - 25 = 0$ or $x - 50 = 0$

$3x = 25$ or $x = 50$

$x = \dfrac{25}{3}$ or $x = 50$

Case 1: Car A would travel at $\dfrac{25}{3} = 8\dfrac{1}{3}$ mph and Car B would travel at $13\dfrac{1}{3}$ mph.

Case 2: Car A would travel at 50 mph and Car B would travel at 55 mph.

38.

	Time in Hours	Rate
Mark	20	$\frac{1}{20}$
Phil	30	$\frac{1}{30}$

In 5 hours Mark would finish $5\left(\dfrac{1}{20}\right) = \dfrac{5}{20} = \dfrac{1}{4}$ of the job.

Then $\dfrac{3}{4}$ of the job is remaining.

$\frac{1}{20}t+\frac{1}{30}t=\frac{3}{4}$

$60\left(\frac{1}{20}t\right)+60\left(\frac{1}{30}t\right)=60\left(\frac{3}{4}\right)$

$3t+2t=45$

$5t=45$

$t=9$

It would take 9 hours.

39.

	Rate	Time	Pay
Anticipated	$\frac{640}{x}$	x	640
Actual	$\frac{640}{x+20}$	$x+20$	640

$\frac{640}{x}-1.60=\frac{640}{x+20}$

$10x(x+20)\left(\frac{640}{x}-1.6\right)=10x(x+20)\left(\frac{640}{x+20}\right)$

$10x(x+20)\left(\frac{640}{x}\right)-10x(x+20)(1.6)=10x(640)$

$6400(x+20)-16x(x+20)=6400x$

$6400x+128000-16x^2-320x=6400x$

$-16x^2+6080x+128000=6400x$

$-16x^2-320x+128000=0$

$16x^2+320x-128000=0$

$16(x^2+20x-8000)=0$

$16(x-80)(x+100)=0$

$x-80=0$ or $x+100=0$

$x=80$ or $x=-100$

Discard the root $x=-100$.

The anticipated time would take 80 hours.

40.

	Rate	Time	Distance
1st part	x	$\frac{40}{x}$	40
2nd part	$x-3$	$\frac{26}{x-3}$	26

$\frac{40}{x}+\frac{26}{x-3}=4\frac{1}{2}$

$2x(x-3)\left(\frac{40}{x}+\frac{26}{x-3}\right)=2x(x-3)\left(\frac{9}{2}\right)$

$80(x-3)+52x=9x(x-3)$

$80x-240+52x=9x^2-27x$

$132x-240=9x^2-27x$

$0=9x^2-159x+240$

$0=3(3x^2-53x+80)$

$0=3(3x-5)(x-16)$

$3x-5=0$ or $x-16=0$

$3x=5$ or $x=16$

$x=\frac{5}{3}$

Discard the root $x=\frac{5}{3}$ since

$x-3=\frac{5}{3}-3=-\frac{4}{3}$ and rate cannot be negative.

He travels 13 mph for the last part.

CHAPTER 4 Test

1. $\frac{39x^2y^3}{72x^3y}=\frac{3(13)x^2y^3}{3(24)x^3y}=\frac{13y^2}{24x}$

2. $\frac{3x^2+17x-6}{x^3-36x}$

$\frac{(3x-1)(x+6)}{x(x+6)(x-6)}$

$\frac{3x-1}{x(x-6)}$

3. $\frac{6n^2-5n-6}{3n^2+14n+8}$

$\frac{(2n-3)(3n+2)}{(3n+2)(n+4)}$

$\frac{2n-3}{n+4}$

4. $\dfrac{2x - 2x^2}{x^2 - 1}$

$\dfrac{2x(1 - x)}{(x + 1)(x - 1)}$

$\dfrac{-2x(x - 1)}{(x + 1)(x - 1)}$

$\dfrac{-2x}{x + 1}$

5. $\dfrac{5x^2y}{8x} \bullet \dfrac{12y^2}{20xy} = \dfrac{5(12)x^2y^3}{8(20)x^2y} = \dfrac{3y^2}{8}$

6. $\dfrac{5a + 5b}{20a + 10b} \bullet \dfrac{a^2 - ab}{2a^2 + 2ab}$

$\dfrac{5(a + \mathrm{b})}{10(2a + b)} \bullet \dfrac{a(a - b)}{2a(a + b)}$

$\dfrac{a - b}{4(2a + b)}$

7. $\dfrac{3x^2 + 10x - 8}{5x^2 + 19x - 4} \div \dfrac{3x^2 - 23x + 14}{x^2 - 3x - 28}$

$\dfrac{3x^2 + 10x - 8}{5x^2 + 19x - 4} \bullet \dfrac{x^2 - 3x - 28}{3x^2 - 23x + 14}$

$\dfrac{(3x - 2)(x + 4)}{(5x - 1)(x + 4)} \bullet \dfrac{(x - 7)(x + 4)}{(3x - 2)(x - 7)}$

$\dfrac{x + 4}{5x - 1}$

8. $\dfrac{3x - 1}{4} + \dfrac{2x + 5}{6}$; LCD = 12

$\dfrac{(3x - 1)}{4} \bullet \dfrac{3}{3} + \dfrac{(2x + 5)}{6} \bullet \dfrac{2}{2}$

$\dfrac{3(3x - 1) + 2(2x + 5)}{12}$

$\dfrac{9x - 3 + 4x + 10}{12}$

$\dfrac{13x + 7}{12}$

9. $\dfrac{5x - 6}{3} - \dfrac{x - 12}{6}$; LCD = 6

$\dfrac{(5x - 6)}{3} \bullet \dfrac{2}{2} - \dfrac{x - 12}{6}$

$\dfrac{2(5x - 6)}{6} - \dfrac{x - 12}{6}$

$\dfrac{2(5x - 6) - (x - 12)}{6}$

$\dfrac{10x - 12 - x + 12}{6}$

$\dfrac{9x}{6} = \dfrac{3x}{2}$

10. $\dfrac{3}{5n} + \dfrac{2}{3} - \dfrac{7}{3n}$; LCD = $15n$

$\dfrac{3}{5n} \bullet \dfrac{3}{3} + \dfrac{2}{3} \bullet \dfrac{5n}{5n} - \dfrac{7}{3n} \bullet \dfrac{5}{5}$

$\dfrac{9}{15n} + \dfrac{10n}{15n} - \dfrac{35}{15n}$

$\dfrac{10n - 26}{15n}$

11. $\dfrac{3x}{x - 6} + \dfrac{2}{x}$; LCD = $x(x - 6)$

$\dfrac{3x}{(x - 6)} \bullet \dfrac{x}{x} + \dfrac{2}{x} \bullet \dfrac{(x - 6)}{(x - 6)}$

$\dfrac{3x^2}{x(x - 6)} + \dfrac{2(x - 6)}{x(x - 6)}$

$\dfrac{3x^2 + 2(x - 6)}{x(x - 6)}$

$\dfrac{3x^2 + 2x - 12}{x(x - 6)}$

12. $\dfrac{9}{x^2 - x} - \dfrac{2}{x}$

$\dfrac{9}{x(x-1)} - \dfrac{2}{x}$; LCD $= x(x-1)$

$\dfrac{9}{x(x-1)} - \dfrac{2}{x} \bullet \dfrac{(x-1)}{(x-1)}$

$\dfrac{9 - 2(x-1)}{x(x-1)}$

$\dfrac{9 - 2x + 2}{x(x-1)}$

$\dfrac{-2x + 11}{x(x-1)}$

13. $\dfrac{3}{2n^2 + n + 10} + \dfrac{5}{n^2 + 5n - 14}$

$\dfrac{3}{(2n+5)(n-2)} + \dfrac{5}{(n+7)(n-2)}$

LCD $= (2n+5)(n-2)(n+7)$

$\dfrac{3}{(2n+5)(n-2)} \bullet \dfrac{(n+7)}{(n+7)} + \dfrac{5}{(n+7)(n-2)} \bullet \dfrac{(2n+5)}{(2n+5)}$

$\dfrac{3(n+7)}{(2n+5)(n-2)(n+7)} + \dfrac{5(2n+5)}{(n+7)(n+2)(2n+5)}$

$\dfrac{3(n+7) + 5(2n+5)}{(2n+5)(n-2)(n+7)}$

$\dfrac{3n + 21 + 10n + 25}{(2n+5)(n-2)(n+7)}$

$\dfrac{13n + 46}{(2n+5)(n-2)(n+7)}$

14.

$$\begin{array}{r|l} & \phantom{3x^3 + {}} 3x^2 - 2x - 1 \\ \hline x + 4 & 3x^3 + 10x^2 - 9x - 4 \\ & 3x^3 + 12x^2 \\ \hline & -2x^2 - 9x \\ & -2x^2 - 8x \\ \hline & -x - 4 \\ & -x - 4 \\ \hline & 0 \end{array}$$

$3x^2 - 2x - 1$

15. $\dfrac{\dfrac{3}{2x} - \dfrac{1}{6}}{\dfrac{2}{3x} + \dfrac{3}{4}}$

$\dfrac{12x}{12x} \bullet \dfrac{\left(\dfrac{3}{2x} - \dfrac{1}{6}\right)}{\left(\dfrac{2}{3x} + \dfrac{3}{4}\right)}$

$\dfrac{12x\left(\dfrac{3}{2x}\right) - 12x\left(\dfrac{1}{6}\right)}{12x\left(\dfrac{2}{3x}\right) + 12x\left(\dfrac{3}{4}\right)}$

$\dfrac{18 - 2x}{8 + 9x}$

16. $\dfrac{x+2}{y-4} = \dfrac{3}{4}$ for y

$3(y-4) = 4(x+2)$

$3y - 12 = 4x + 8$

$3y = 4x + 20$

$y = \dfrac{4x + 20}{3}$

17. $\dfrac{x-1}{2} - \dfrac{x+2}{5} = -\dfrac{3}{5}$

$10\left(\dfrac{x-1}{2} - \dfrac{x+2}{5}\right) = 10\left(-\dfrac{3}{5}\right)$

$10\left(\dfrac{x-1}{2}\right) - 10\left(\dfrac{x+2}{5}\right) = -6$

$5(x-1) - 2(x+2) = -6$

$5x - 5 - 2x - 4 = -6$

$3x - 9 = -6$
$3x = 3$
$x = 1$
The solution set is $\{1\}$.

18. $\frac{5}{4x} + \frac{3}{2} = \frac{7}{5x}$; $x \neq 0$

$20x\left(\frac{5}{4x} + \frac{3}{2}\right) = 20x\left(\frac{7}{5x}\right)$

$20x\left(\frac{5}{4x}\right) + 20x\left(\frac{3}{2}\right) = 28$

$25 + 30x = 28$
$30x = 3$
$x = \frac{3}{30} = \frac{1}{10}$
The solution set is $\left\{\frac{1}{10}\right\}$.

19. $\frac{-3}{4n-1} = \frac{-2}{3n+11}$; $n \neq \frac{1}{4}, n \neq -\frac{11}{3}$
$-3(3n+11) = -2(4n-1)$
$-9n - 33 = -8n + 2$
$-9n = -8n + 35$
$-n = 35$
$n = -35$
The solution set is $\{-35\}$.

20. $n - \frac{5}{n} = 4$; $n \neq 0$

$n\left(n - \frac{5}{n}\right) = n(4)$

$n(n) - n\left(\frac{5}{n}\right) = 4n$

$n^2 - 5 = 4n$
$n^2 - 4n - 5 = 0$
$(n-5)(n+1) = 0$
$n - 5 = 0$ or $n + 1 = 0$
$n = 5$ or $n = -1$
The solution set is $\{-1, 5\}$.

21. $\frac{6}{x-4} - \frac{4}{x+3} = \frac{8}{x-4}$; $x \neq 4, x \neq -3$

$(x\text{-}4)(x\text{+}3)\left(\frac{6}{x\text{-}4} - \frac{4}{x\text{+}3}\right) = (x\text{-}4)(x\text{+}3)\left(\frac{8}{x\text{-}4}\right)$

$(x\text{-}4)(x\text{+}3)\left(\frac{6}{x\text{-}4}\right)\text{-}(x\text{-}4)(x\text{+}3)\left(\frac{4}{x\text{+}3}\right) = 8(x+3)$

$6(x+3) - 4(x-4) = 8x + 24$
$6x + 18 - 4x + 16 = 8x + 24$
$2x + 34 = 8x + 24$
$2x = 8x - 10$
$-6x = -10$

$x = \frac{-10}{-6} = \frac{5}{3}$
The solution set is $\left\{\frac{5}{3}\right\}$.

22. $\frac{1}{3x-1} + \frac{x-2}{9x^2-1} = \frac{7}{6x-2}$

$\frac{1}{3x-1} + \frac{x-2}{(3x+1)(3x-1)} = \frac{7}{2(3x-1)}$; $x \neq \frac{1}{3}, x \neq -\frac{1}{3}$

$2(3x-1)(3x+1)\left[\frac{1}{3x-1} + \frac{x-2}{(3x+1)(3x-1)}\right] = 2(3x+1)(3x-1)\left[\frac{7}{2(3x-1)}\right]$

$2(3x+1) + 2(x-2) = 7(3x+1)$
$6x + 2 + 2x - 4 = 21x + 7$

$8x - 2 = 21x + 7$
$8x = 21x + 9$
$-13x = 9$
$x = -\frac{9}{13}$
The solution set is $\left\{-\frac{9}{13}\right\}$.

23. Let x = numerator and
$3x - 9$ = denominator.

$\frac{x}{3x-9} = \frac{3}{8}; x \neq 3$

$8x = 3(3x - 9)$
$8x = 9x - 27$
$-x = -27$
$x = 27$
The number is $\frac{27}{72}$.

24.

	Time in Minutes	Rate
Jodie	$3x$	$\frac{1}{3x}$
Jannie	x	$\frac{1}{x}$
Together	15	$\frac{1}{15}$

$\frac{1}{3x} + \frac{1}{x} = \frac{1}{15}$

$15x\left(\frac{1}{3x} + \frac{1}{x}\right) = 15x\left(\frac{1}{15}\right)$

$5 + 15 = x$
$20 = x$
It would take Jodie 60 minutes.

25.

	Rate	Time	Distance
Rene	$x + 3$	$\frac{60}{x+3}$	60
Sue	x	$\frac{60}{x}$	60

$\frac{60}{x+3} + 1 = \frac{60}{x}$

$x(x+3)\left(\frac{60}{x+3} + 1\right) = x(x+3)\left(\frac{60}{x}\right)$

$x(x+3)\left(\frac{60}{x+3}\right) + 1x(x+3) = 60(x+3)$

$60x + x^2 + 3x = 60x + 180$
$x^2 + 63x = 60x + 180$
$x^2 + 3x - 180 = 0$
$(x + 15)(x - 12) = 0$
$x + 15 = 0$ or $x - 12 = 0$
$x = -15$ or $x = 12$
Discard the root $x = -15$.
Rene's rate is $12 + 3 = 15$ mph.

Chapter 5 Exponents and Radicals

PROBLEM SET 5.1 Using Integers as Exponents

1. $3^{-3} = \frac{1}{3^3} = \frac{1}{27}$

3. $-10^{-2} = \frac{-1}{10^2} = -\frac{1}{100}$

5. $\frac{1}{3^{-4}} = 3^4 = 81$

7. $-\left(\frac{1}{3}\right)^{-3} = -\left(\frac{3}{1}\right)^3 =$

$-\frac{3^3}{1^3} = -\frac{27}{1} = -27$

9. $\left(-\frac{1}{2}\right)^{-3} = \left(-\frac{2}{1}\right)^3 =$

$\frac{(2)^3}{(-1)^3} = \frac{8}{-1} = -8$

11. $\left(-\frac{3}{4}\right)^0 = 1$

13. $\frac{1}{\left(\frac{3}{7}\right)^{-2}} = \left(\frac{3}{7}\right)^2 = \frac{3^2}{7^2} = \frac{9}{49}$

15. $2^7 \bullet 2^{-3} = 2^{7-3} = 2^4 = 16$

17. $10^{-5} \bullet 10^2 = 10^{-5+2} = 10^{-3}$

$= \frac{1}{10^3} = \frac{1}{1000}$

19. $10^{-1} \bullet 10^{-2} = 10^{-1-2} = 10^{-3}$

$= \frac{1}{10^3} = \frac{1}{1000}$

21. $(3^{-1})^{-3} = 3^{-1(-3)} = 3^3 = 27$

23. $(5^3)^{-1} = 5^{3(-1)} = 5^{-3} = \frac{1}{5^3} = \frac{1}{125}$

25. $(2^3 \bullet 3^{-2})^{-1}$

$2^{3(-1)} \bullet 3^{-2(-1)}$

$2^{-3} \bullet 3^2$

$\frac{3^2}{2^3} = \frac{9}{8}$

27. $(4^2 \bullet 5^{-1})^2$

$4^{2(2)} \bullet 5^{-1(2)}$

$4^4 \bullet 5^{-2}$

$\frac{4^4}{5^2} = \frac{256}{25}$

29. $\left(\frac{2^{-1}}{5^{-2}}\right)^{-1}$

$\frac{2^{-1(-1)}}{5^{-2(-1)}}$

$\frac{2}{5^2} = \frac{2}{25}$

31. $\left(\frac{2^{-1}}{3^{-2}}\right)^2$

$\frac{2^{-1(2)}}{3^{-2(2)}}$

$\frac{2^{-2}}{3^{-4}}$

$\frac{3^4}{2^2}$

$\frac{81}{4}$

33. $\frac{3^3}{3^{-1}}$

$3^{3-(-1)}$

$3^4 = 81$

35. $\dfrac{10^{-2}}{10^{2}}$

$10^{-2-2} = 10^{-4}$

$\dfrac{1}{10^4} = \dfrac{1}{10{,}000}$

37. $2^{-2} + 3^{-2}$

$\dfrac{1}{2^2} + \dfrac{1}{3^2}$

$\dfrac{1}{4} + \dfrac{1}{9}$; LCD = 36

$\dfrac{1}{4} \bullet \dfrac{9}{9} + \dfrac{1}{9} \bullet \dfrac{4}{4}$

$\dfrac{9}{36} + \dfrac{4}{36}$

$\dfrac{13}{36}$

39. $\left(\dfrac{1}{3}\right)^{-1} - \left(\dfrac{2}{5}\right)^{-1}$

$\left(\dfrac{3}{1}\right)^{1} - \left(\dfrac{5}{2}\right)^{1}$

$\dfrac{3}{1} - \dfrac{5}{2}$; LCD = 2

$\dfrac{3}{1} \bullet \dfrac{2}{2} - \dfrac{5}{2}$

$\dfrac{6}{2} - \dfrac{5}{2} = \dfrac{1}{2}$

41. $(2^{-3} + 3^{-2})^{-1}$

$\left(\dfrac{1}{2^3} + \dfrac{1}{3^2}\right)^{-1}$

$\left(\dfrac{1}{8} + \dfrac{1}{9}\right)^{-1}$

$\left(\dfrac{9}{72} + \dfrac{8}{72}\right)^{-1}$

$\left(\dfrac{17}{72}\right)^{-1}$

$\left(\dfrac{72}{17}\right)^{1} = \dfrac{72}{17}$

43. $x^2 \bullet x^{-8}$

x^{2-8}

$x^{-6} = \dfrac{1}{x^6}$

45. $a^3 \bullet a^{-5} \bullet a^{-1}$

a^{3-5-1}

$a^{-3} = \dfrac{1}{a^3}$

47. $(a^{-4})^2$

$a^{-4(2)}$

$a^{-8} = \dfrac{1}{a^8}$

49. $(x^2y^{-6})^{-1}$

$x^{2(-1)}y^{-6(-1)}$

$x^{-2}y^6 = \dfrac{y^6}{x^2}$

51. $(ab^3c^{-2})^{-4}$

$a^{1(-4)}b^{3(-4)}c^{-2(-4)}$

$a^{-4}b^{-12}c^{8}$

$\dfrac{c^8}{a^4b^{12}}$

53. $(2x^3y^{-4})^{-3}$

$2^{-3}x^{3(-3)}y^{-4(-3)}$

$2^{-3}x^{-9}y^{12}$

$\frac{y^{12}}{2^3x^9}$

$\frac{y^{12}}{8x^9}$

55. $\left(\frac{x^{-1}}{y^{-4}}\right)^{-3}$

$\frac{x^{-1(-3)}}{y^{-4(-3)}}$

$\frac{x^3}{y^{12}}$

57. $\left(\frac{3a^{-2}}{2b^{-1}}\right)^{-2}$

$\frac{3^{-2}a^{-2(-2)}}{2^{-2}b^{-1(-2)}}$

$\frac{3^{-2}a^4}{2^{-2}b^2}$

$\frac{2^2a^4}{3^2b^2}$

$\frac{4a^4}{9b^2}$

59. $\frac{x^{-6}}{x^{-4}}$

$x^{-6-(-4)}$

$x^{-2} = \frac{1}{x^2}$

61. $\frac{a^3b^{-2}}{a^{-2}b^{-4}}$

$a^{3-(-2)}b^{-2-(-4)}$

a^5b^2

63. $(2xy^{-1})(3x^{-2}y^4)$

$6x^{1-2}y^{-1+4}$

$6x^{-1}y^3$

$\frac{6y^3}{x}$

65. $(-7a^2b^{-5})(-a^{-2}b^7)$

$7a^{2-2}b^{-5+7}$

$7a^0b^2$

$7(1)b^2$

$7b^2$

67. $\frac{28x^{-2}y^{-3}}{4x^{-3}y^{-1}}$

$7x^{-2-(-3)}y^{-3-(-1)}$

$7x^1y^{-2}$

$\frac{7x}{y^2}$

69. $\frac{-72a^2b^{-4}}{6a^3b^{-7}}$

$-12a^{2-3}b^{-4-(-7)}$

$-12a^{-1}b^3$

$\frac{-12b^3}{a}$

71. $\left(\frac{35x^{-1}y^{-2}}{7x^4y^3}\right)^{-1}$

$(5x^{-1-4}y^{-2-3})^{-1}$

$(5x^{-5}y^{-5})^{-1}$

$5^{-1}x^{-5(-1)}y^{-5(-1)}$

$5^{-1}x^5y^5$

$\frac{x^5y^5}{5}$

73. $\left(\frac{-36a^{-1}b^{-6}}{4a^{-1}b^4}\right)^{-2}$

$[-9a^{-1-(-1)}b^{-6-4}]^{-2}$

$(-9a^0b^{-10})^{-2}$

$(-9b^{-10})^{-2}$

$(-9)^{-2}b^{-10(-2)}$

$(-9)^{-2}b^{20}$

$\frac{b^{20}}{(-9)^2}$

$\frac{b^{20}}{81}$

75. $x^{-2}+x^{-3}$

$\frac{1}{x^2}+\frac{1}{x^3}$; LCD $= x^3$

$\frac{1}{x^2}\bullet\frac{x}{x}+\frac{1}{x^3}$

$\frac{x}{x^3}+\frac{1}{x^3}$

$\frac{x+1}{x^3}$

77. $x^{-3}-y^{-1}$

$\frac{1}{x^3}-\frac{1}{y}$; LCD $= x^3y$

$\frac{1}{x^3}\bullet\frac{y}{y}-\frac{1}{y}\bullet\frac{x^3}{x^3}$

$\frac{y}{x^3y}-\frac{x^3}{x^3y}$

$\frac{y-x^3}{x^3y}$

79. $3a^{-2}+4b^{-1}$

$\frac{3}{a^2}+\frac{4}{b}$; LCD $= a^2b$

$\frac{3}{a^2}\bullet\frac{b}{b}+\frac{4}{b}\bullet\frac{a^2}{a^2}$

$\frac{3b}{a^2b}+\frac{4a^2}{a^2b}$

$\frac{3b+4a^2}{a^2b}$

81. $x^{-1}y^{-2}-xy^{-1}$

$\frac{1}{xy^2}-\frac{x}{y}$; LCD $= xy^2$

$\frac{1}{xy^2}-\frac{x}{y}\bullet\frac{xy}{xy}$

$\frac{1}{xy^2}-\frac{x^2y}{xy^2}$

$\frac{1-x^2y}{xy^2}$

83. $2x^{-1}-3x^{-2}$

$\frac{2}{x}-\frac{3}{x^2}$; LCD $= x^2$

$\frac{2}{x}\bullet\frac{x}{x}-\frac{3}{x^2}$

$\frac{2x}{x^2}-\frac{3}{x^2}$

$\frac{2x-3}{x^2}$

PROBLEM SET 5.2 Roots and Radicals

1. $\sqrt{64} = 8$

3. $-\sqrt{100} = -(10) = -10$

5. $\sqrt[3]{27} = 3$

7. $\sqrt[3]{-64} = -4$

9. $\sqrt[4]{81} = 3$

11. $\sqrt{\frac{16}{25}} = \frac{4}{5}$

13. $-\sqrt{\frac{36}{49}} = -\left(\frac{6}{7}\right) = -\frac{6}{7}$

15. $\sqrt{\frac{9}{36}} = \frac{3}{6} = \frac{1}{2}$

17. $\sqrt[3]{\frac{27}{64}} = \frac{3}{4}$

19. $\sqrt[3]{8^3} = 8$

21. $\sqrt{27} = \sqrt{9}\sqrt{3} = 3\sqrt{3}$

23. $\sqrt{32} = \sqrt{16}\sqrt{2} = 4\sqrt{2}$

25. $\sqrt{80} = \sqrt{16}\sqrt{5} = 4\sqrt{5}$

27. $\sqrt{160} = \sqrt{16}\sqrt{10} = 4\sqrt{10}$

29. $4\sqrt{18} = 4\sqrt{9}\sqrt{2} = 4(3)\sqrt{2} = 12\sqrt{2}$

31. $-6\sqrt{20} = -6\sqrt{4}\sqrt{5} = -6(2)\sqrt{5} = -12\sqrt{5}$

33. $\frac{2}{5}\sqrt{75} = \frac{2}{5}\sqrt{25}\sqrt{3} = \frac{2}{5}(5)\sqrt{3} = 2\sqrt{3}$

35. $\frac{3}{2}\sqrt{24} = \frac{3}{2}\sqrt{4}\sqrt{6} = \frac{3}{2}(2)\sqrt{6} = 3\sqrt{6}$

37. $-\frac{5}{6}\sqrt{28} = -\frac{5}{6}\sqrt{4}\sqrt{7} = -\frac{5}{6}(2)\sqrt{7} = -\frac{5}{3}\sqrt{7}$

39. $\sqrt{\frac{19}{4}} = \frac{\sqrt{19}}{\sqrt{4}} = \frac{\sqrt{19}}{2}$

41. $\sqrt{\frac{27}{16}} = \frac{\sqrt{27}}{\sqrt{16}} = \frac{\sqrt{9}\sqrt{3}}{4} = \frac{3\sqrt{3}}{4}$

43. $\sqrt{\frac{75}{81}} = \frac{\sqrt{75}}{\sqrt{81}} = \frac{\sqrt{25}\sqrt{3}}{9} = \frac{5\sqrt{3}}{9}$

45. $\sqrt{\frac{2}{7}} = \frac{\sqrt{2}}{\sqrt{7}} \bullet \frac{\sqrt{7}}{\sqrt{7}} = \frac{\sqrt{14}}{\sqrt{49}} = \frac{\sqrt{14}}{7}$

47. $\sqrt{\frac{2}{3}} = \frac{\sqrt{2}}{\sqrt{3}} \bullet \frac{\sqrt{3}}{\sqrt{3}} = \frac{\sqrt{6}}{\sqrt{9}} = \frac{\sqrt{6}}{3}$

49. $\frac{\sqrt{5}}{\sqrt{12}} = \frac{\sqrt{5}}{\sqrt{4}\sqrt{3}} = \frac{\sqrt{5}}{2\sqrt{3}} = \frac{\sqrt{5}}{2\sqrt{3}} \bullet \frac{\sqrt{3}}{\sqrt{3}} = \frac{\sqrt{15}}{2\sqrt{9}} = \frac{\sqrt{15}}{6}$

51. $\frac{\sqrt{11}}{\sqrt{24}} = \frac{\sqrt{11}}{\sqrt{4}\sqrt{6}} = \frac{\sqrt{11}}{2\sqrt{6}} = \frac{\sqrt{11}}{2\sqrt{6}} \bullet \frac{\sqrt{6}}{\sqrt{6}} = \frac{\sqrt{66}}{2\sqrt{36}} = \frac{\sqrt{66}}{2(6)} = \frac{\sqrt{66}}{12}$

53. $\frac{\sqrt{18}}{\sqrt{27}} = \sqrt{\frac{18}{27}} = \sqrt{\frac{2}{3}} = \frac{\sqrt{2}}{\sqrt{3}} = \frac{\sqrt{2}}{\sqrt{3}} \bullet \frac{\sqrt{3}}{\sqrt{3}} = \frac{\sqrt{6}}{\sqrt{9}} = \frac{\sqrt{6}}{3}$

55. $\frac{\sqrt{35}}{\sqrt{7}} = \sqrt{\frac{35}{7}} = \sqrt{5}$

57. $\frac{2\sqrt{3}}{\sqrt{7}} = \frac{2\sqrt{3}}{\sqrt{7}} \bullet \frac{\sqrt{7}}{\sqrt{7}} = \frac{2\sqrt{21}}{\sqrt{49}} = \frac{2\sqrt{21}}{7}$

59. $\frac{-4\sqrt{12}}{\sqrt{5}} = \frac{-4\sqrt{4}\sqrt{3}}{\sqrt{5}} = \frac{-4(2)\sqrt{3}}{\sqrt{5}} =$

$\frac{-8\sqrt{3}}{\sqrt{5}} = \frac{-8\sqrt{3}}{\sqrt{5}} \bullet \frac{\sqrt{5}}{\sqrt{5}} =$

$\frac{-8\sqrt{15}}{\sqrt{25}} = \frac{-8\sqrt{15}}{5}$

61. $\frac{3\sqrt{2}}{4\sqrt{3}} = \frac{3\sqrt{2}}{4\sqrt{3}} \bullet \frac{\sqrt{3}}{\sqrt{3}} = \frac{3\sqrt{6}}{4(3)} = \frac{\sqrt{6}}{4}$

63. $\frac{-8\sqrt{18}}{10\sqrt{50}} = \frac{-4}{5}\sqrt{\frac{18}{50}} = \frac{-4}{5}\sqrt{\frac{9}{25}} =$

$-\frac{4}{5}\sqrt{\frac{9}{25}} = -\frac{4}{5}\left(\frac{3}{5}\right) = -\frac{12}{25}$

65. $\sqrt[3]{16} = \sqrt[3]{8}\sqrt[3]{2} = 2\sqrt[3]{2}$

67. $2\sqrt[3]{81} = 2\sqrt[3]{27}\sqrt[3]{3} = 2(3)\sqrt[3]{3} = 6\sqrt[3]{3}$

69. $\frac{2}{\sqrt[3]{9}} = \frac{2}{\sqrt[3]{9}} \bullet \frac{\sqrt[3]{3}}{\sqrt[3]{3}} = \frac{2\sqrt[3]{3}}{\sqrt[3]{27}} = \frac{2\sqrt[3]{3}}{3}$

71. $\frac{\sqrt[3]{27}}{\sqrt[3]{4}} = \frac{3}{\sqrt[3]{4}} = \frac{3}{\sqrt[3]{4}} \bullet \frac{\sqrt[3]{2}}{\sqrt[3]{2}} =$

$\frac{3\sqrt[3]{2}}{\sqrt[3]{8}} = \frac{3\sqrt[3]{2}}{2}$

73. $\frac{\sqrt[3]{6}}{\sqrt[3]{4}} = \sqrt[3]{\frac{6}{4}} = \sqrt[3]{\frac{3}{2}} = \frac{\sqrt[3]{3}}{\sqrt[3]{2}} \bullet \frac{\sqrt[3]{4}}{\sqrt[3]{4}} =$

$\frac{\sqrt[3]{12}}{\sqrt[3]{8}} = \frac{\sqrt[3]{12}}{2}$

75. $S = \sqrt{30Df}$

$S = \sqrt{30(150)(0.4)} = \sqrt{1800} = 42$ mph

$S = \sqrt{30(200)(0.4)} = \sqrt{2400} = 49$ mph

$S = \sqrt{30(350)(0.4)} = \sqrt{4200} = 65$ mph

77. $K = \sqrt{s(s-a)(s-b)(s-c)}$

$s = \frac{a+b+c}{2} = \frac{14+16+18}{2} = \frac{48}{2} = 24$

$K = \sqrt{24(24-14)(24-16)(24-18)}$

$K = \sqrt{24(10)(8)(6)}$

$K = \sqrt{11520} = 107$ square centimeters

79. $K = \sqrt{s(s-a)(s-b)(s-c)}$

$s = \frac{a+b+c}{2} = \frac{18+18+18}{2} = \frac{54}{2} = 27$

$K = \sqrt{27(27-18)(27-18)(27-18)}$

$K = \sqrt{27(9)(9)(9)}$

$K = \sqrt{19683} = 140$ square inches.

Further Investigations

85a) $\sqrt{2} = 1.414$
b) $\sqrt{75} = 8.660$
c) $\sqrt{156} = 12.490$
d) $\sqrt{691} = 26.287$
e) $\sqrt{3249} = 57.000$
f) $\sqrt{45123} = 212.422$
g) $\sqrt{0.14} = 0.374$
h) $\sqrt{0.023} = 0.152$
i) $\sqrt{0.8649} = 0.930$

PROBLEM SET 5.3 Combining Radicals and Simplifying Radicals That Contain Variables

1. $5\sqrt{18} - 2\sqrt{2}$
$5\sqrt{9}\sqrt{2} - 2\sqrt{2}$
$5(3)\sqrt{2} - 2\sqrt{2}$
$15\sqrt{2} - 2\sqrt{2}$
$13\sqrt{2}$

3. $7\sqrt{12} + 10\sqrt{48}$
$7\sqrt{4}\sqrt{3} + 10\sqrt{16}\sqrt{3}$
$7(2)\sqrt{3} + 10(4)\sqrt{3}$
$14\sqrt{3} + 40\sqrt{3}$
$54\sqrt{3}$

5. $-2\sqrt{50} - 5\sqrt{32}$
$-2\sqrt{25}\sqrt{2} - 5\sqrt{16}\sqrt{2}$
$-2(5)\sqrt{2} - 5(4)\sqrt{2}$
$-10\sqrt{2} - 20\sqrt{2}$
$-30\sqrt{2}$

7. $3\sqrt{20} - \sqrt{5} - 2\sqrt{45}$
$3\sqrt{4}\sqrt{5} - \sqrt{5} - 2\sqrt{9}\sqrt{5}$
$3(2)\sqrt{5} - \sqrt{5} - 2(3)\sqrt{5}$
$6\sqrt{5} - \sqrt{5} - 6\sqrt{5}$
$-\sqrt{5}$

9. $-9\sqrt{24} + 3\sqrt{54} - 12\sqrt{6}$
$-9\sqrt{4}\sqrt{6} + 3\sqrt{9}\sqrt{6} - 12\sqrt{6}$
$-9(2)\sqrt{6} + 3(3)\sqrt{6} - 12\sqrt{6}$
$-18\sqrt{6} + 9\sqrt{6} - 12\sqrt{6}$
$-21\sqrt{6}$

11. $\frac{3}{4}\sqrt{7} - \frac{2}{3}\sqrt{28}$

$\frac{3}{4}\sqrt{7} - \frac{2}{3}\sqrt{4}\sqrt{7}$

$\frac{3}{4}\sqrt{7} - \frac{2}{3}(2)\sqrt{7}$

$\frac{3}{4}\sqrt{7} - \frac{4}{3}\sqrt{7}$

$\frac{9}{12}\sqrt{7} - \frac{16}{12}\sqrt{7}$

$-\frac{7\sqrt{7}}{12}$

13. $\frac{3}{5}\sqrt{40} + \frac{5}{6}\sqrt{90}$

$\frac{3}{5}\sqrt{4}\sqrt{10} + \frac{5}{6}\sqrt{9}\sqrt{10}$

$\frac{3}{5}(2)\sqrt{10} + \frac{5}{6}(3)\sqrt{10}$

$\frac{6}{5}\sqrt{10} + \frac{5}{2}\sqrt{10}$

$\frac{12}{10}\sqrt{10} + \frac{25}{10}\sqrt{10}$

$\frac{37\sqrt{10}}{10}$

15. $\frac{3\sqrt{18}}{5} - \frac{5\sqrt{72}}{6} + \frac{3\sqrt{98}}{4}$

$\frac{3\sqrt{9}\sqrt{2}}{5} - \frac{5\sqrt{36}\sqrt{2}}{6} + \frac{3\sqrt{49}\sqrt{2}}{4}$

$\frac{3(3)\sqrt{2}}{5} - \frac{5(6)\sqrt{2}}{6} + \frac{3(7)\sqrt{2}}{4}$

$\frac{9\sqrt{2}}{5} - 5\sqrt{2} + \frac{21\sqrt{2}}{4}$

$\frac{(9\sqrt{2})}{5} \bullet \frac{4}{4} - (5\sqrt{2}) \bullet \frac{20}{20} + \frac{(21\sqrt{2})}{4} \bullet \frac{5}{5}$

$\frac{36\sqrt{2}}{20} - \frac{100\sqrt{2}}{20} + \frac{105\sqrt{2}}{20}$

$\frac{41\sqrt{2}}{20}$

17. $5\sqrt[3]{3} + 2\sqrt[3]{24} - 6\sqrt[3]{81}$

$5\sqrt[3]{3} + 2\sqrt[3]{8}\sqrt[3]{3} - 6\sqrt[3]{27}\sqrt[3]{3}$

Problem Set 5.3

$5\sqrt[3]{3} + 2(2)\sqrt[3]{3} - 6(3)\sqrt[3]{3}$

$5\sqrt[3]{3} + 4\sqrt[3]{3} - 18\sqrt[3]{3}$

$-9\sqrt[3]{3}$

19. $-\sqrt[3]{16} + 7\sqrt[3]{54} - 9\sqrt[3]{2}$

$-\sqrt[3]{8}\sqrt[3]{2} + 7\sqrt[3]{27}\sqrt[3]{2} - 9\sqrt[3]{2}$

$-2\sqrt[3]{2} + 7(3)\sqrt[3]{2} - 9\sqrt[3]{2}$

$-2\sqrt[3]{2} + 21\sqrt[3]{2} - 9\sqrt[3]{2}$

$10\sqrt[3]{2}$

21. $\sqrt{32x}$

$\sqrt{16}\sqrt{2x}$

$4\sqrt{2x}$

23. $\sqrt{75x^2}$

$\sqrt{25x^2}\sqrt{3}$

$5x\sqrt{3}$

25. $\sqrt{20x^2y}$

$\sqrt{4x^2}\sqrt{5y}$

$2x\sqrt{5y}$

27. $\sqrt{64x^3y^7}$

$\sqrt{64x^2y^6}\sqrt{xy}$

$8xy^3\sqrt{xy}$

29. $\sqrt{54a^4b^3}$

$\sqrt{9a^4b^2}\sqrt{6b}$

$3a^2b\sqrt{6b}$

31. $\sqrt{63x^6y^8}$

$\sqrt{9x^6y^8}\sqrt{7}$

$3x^3y^4\sqrt{7}$

33. $2\sqrt{40a^3}$

$2\sqrt{4a^2}\sqrt{10a}$

$2(2a)\sqrt{10a}$

$4a\sqrt{10a}$

35. $\frac{2}{3}\sqrt{96xy^3}$

$\frac{2}{3}\sqrt{16y^2}\sqrt{6xy}$

$\frac{2}{3}(4y)\sqrt{6xy}$

$\frac{8y}{3}\sqrt{6xy}$

37. $\sqrt{\frac{2x}{5y}} = \frac{\sqrt{2x}}{\sqrt{5y}} \bullet \frac{\sqrt{5y}}{\sqrt{5y}} = \frac{\sqrt{10xy}}{\sqrt{25y^2}} =$

$\frac{\sqrt{10xy}}{5y}$

39. $\sqrt{\frac{5}{12x^4}} = \frac{\sqrt{5}}{\sqrt{4x^4}\sqrt{3}} =$

$\frac{\sqrt{5}}{2x^2\sqrt{3}} \bullet \frac{\sqrt{3}}{\sqrt{3}} = \frac{\sqrt{15}}{2x^2\sqrt{9}} =$

$\frac{\sqrt{15}}{2x^2(3)} = \frac{\sqrt{15}}{6x^2}$

41. $\frac{5}{\sqrt{18y}} = \frac{5}{\sqrt{9}\sqrt{2y}} =$

$\frac{5}{3\sqrt{2y}} \bullet \frac{\sqrt{2y}}{\sqrt{2y}} = \frac{5\sqrt{2y}}{3\sqrt{4y^2}} =$

$\frac{5\sqrt{2y}}{3(2y)} = \frac{5\sqrt{2y}}{6y}$

43. $\frac{\sqrt{7x}}{\sqrt{8y^5}} = \frac{\sqrt{7x}}{\sqrt{4y^4}\sqrt{2y}} =$

$\frac{\sqrt{7x}}{2y^2\sqrt{2y}} \bullet \frac{\sqrt{2y}}{\sqrt{2y}} = \frac{\sqrt{14xy}}{2y^2\sqrt{4y^2}} =$

$\frac{\sqrt{14xy}}{(2y^2)(2y)} = \frac{\sqrt{14xy}}{4y^3}$

45. $\frac{\sqrt{18y^3}}{\sqrt{16x}} = \frac{\sqrt{9y^2}\sqrt{2y}}{\sqrt{16}\sqrt{x}} =$

$\frac{3y\sqrt{2y}}{4\sqrt{x}} \bullet \frac{\sqrt{x}}{\sqrt{x}} = \frac{3y\sqrt{2xy}}{4\sqrt{x^2}} = \frac{3y\sqrt{2xy}}{4x}$

47. $\frac{\sqrt{24a^2b^3}}{\sqrt{7ab^6}} = \frac{\sqrt{4a^2b^2}\sqrt{6b}}{\sqrt{b^6}\sqrt{7a}} =$

$\frac{2a\sqrt{6b}}{b^2\sqrt{7a}} \bullet \frac{\sqrt{7a}}{\sqrt{7a}} = \frac{2a\sqrt{42ab}}{b^2\sqrt{49a^2}} =$

$\frac{2a\sqrt{42ab}}{b^2(7a)} = \frac{2a\sqrt{42ab}}{7ab^2} = \frac{2\sqrt{42ab}}{7b^2}$

49. $\sqrt[3]{24y} = \sqrt[3]{8}\sqrt[3]{3y} = 2\sqrt[3]{3y}$

51. $\sqrt[3]{16x^4} = \sqrt[3]{8x^3}\sqrt[3]{2x} = 2x\sqrt[3]{2x}$

53. $\sqrt[3]{56x^6y^8} = \sqrt[3]{8x^6y^6}\sqrt[3]{7y^2} = 2x^2y^2\sqrt[3]{7y^2}$

55. $\sqrt[3]{\frac{7}{9x^2}} = \frac{\sqrt[3]{7}}{\sqrt[3]{9x^2}} \bullet \frac{\sqrt[3]{3x}}{\sqrt[3]{3x}} =$

$\frac{\sqrt[3]{21x}}{\sqrt[3]{27x^3}} = \frac{\sqrt[3]{21x}}{3x}$

57. $\frac{\sqrt[3]{3y}}{\sqrt[3]{16x^4}} = \frac{\sqrt[3]{3y}}{\sqrt[3]{8x^3}\sqrt[3]{2x}} =$

$\frac{\sqrt[3]{3y}}{2x\sqrt[3]{2x}} \bullet \frac{\sqrt[3]{4x^2}}{\sqrt[3]{4x^2}} = \frac{\sqrt[3]{12x^2y}}{2x\sqrt[3]{8x^3}} =$

$\frac{\sqrt[3]{12x^2y}}{2x(2x)} = \frac{\sqrt[3]{12x^2y}}{4x^2}$

59. $\frac{\sqrt[3]{12xy}}{\sqrt[3]{3x^2y^5}} = \sqrt[3]{\frac{12xy}{3x^2y^5}} = \sqrt[3]{\frac{4}{xy^4}} = \frac{\sqrt[3]{4}}{\sqrt[3]{xy^4}} =$

$\frac{\sqrt[3]{4}}{\sqrt[3]{y^3}\sqrt[3]{xy}} = \frac{\sqrt[3]{4}}{y\sqrt[3]{xy}} \bullet \frac{\sqrt[3]{x^2y^2}}{\sqrt[3]{x^2y^2}} =$

$\frac{\sqrt[3]{4x^2y^2}}{y\sqrt[3]{x^3y^3}} = \frac{\sqrt[3]{4x^2y^2}}{y(xy)} = \frac{\sqrt[3]{4x^2y^2}}{xy^2}$

61. $\sqrt{8x+12y}$

$\sqrt{4(2x+3y)}$

$\sqrt{4}\sqrt{2x+3y}$

$2\sqrt{2x+3y}$

63. $\sqrt{16x+48y}$

$\sqrt{16(x+3y)}$

$\sqrt{16}\sqrt{x+3y}$

$4\sqrt{x+3y}$

65. $-3\sqrt{4x}+5\sqrt{9x}+6\sqrt{16x}$

$-3\sqrt{4}\sqrt{x}+5\sqrt{9}\sqrt{x}+6\sqrt{16}\sqrt{x}$

$-3(2)\sqrt{x}+5(3)\sqrt{x}+6(4)\sqrt{x}$

$-6\sqrt{x}+15\sqrt{x}+24\sqrt{x}$

$33\sqrt{x}$

67. $2\sqrt{18x}-3\sqrt{8x}-6\sqrt{50x}$

$2\sqrt{9}\sqrt{2x}-3\sqrt{4}\sqrt{2x}-6\sqrt{25}\sqrt{2x}$

$2(3)\sqrt{2x}-3(2)\sqrt{2x}-6(5)\sqrt{2x}$

$6\sqrt{2x}-6\sqrt{2x}-30\sqrt{2x}$

$-30\sqrt{2x}$

69. $5\sqrt{27n} - \sqrt{12n} - 6\sqrt{3n}$
$5\sqrt{9}\sqrt{3n} - \sqrt{4}\sqrt{3n} - 6\sqrt{3n}$
$5(3)\sqrt{3n} - 2\sqrt{3n} - 6\sqrt{3n}$
$15\sqrt{3n} - 2\sqrt{3n} - 6\sqrt{3n}$
$7\sqrt{3n}$

71. $7\sqrt{4ab} - \sqrt{16ab} - 10\sqrt{25ab}$
$7\sqrt{4}\sqrt{ab} - \sqrt{16}\sqrt{ab} - 10\sqrt{25}\sqrt{ab}$
$7(2)\sqrt{ab} - 4\sqrt{ab} - 10(5)\sqrt{ab}$
$14\sqrt{ab} - 4\sqrt{ab} - 50\sqrt{ab}$
$-40\sqrt{ab}$

73. $-3\sqrt{2x^3} + 4\sqrt{8x^3} - 3\sqrt{32x^3}$
$-3\sqrt{x^2}\sqrt{2x} + 4\sqrt{4x^2}\sqrt{2x} - 3\sqrt{16x^2}\sqrt{2x}$
$-3(x)\sqrt{2x} + 4(2x)\sqrt{2x} - 3(4x)\sqrt{2x}$
$-3x\sqrt{2x} + 8x\sqrt{2x} - 12x\sqrt{2x}$
$-7x\sqrt{2x}$

79. **(a)** $\sqrt{125x^2} = \sqrt{25x^2}\sqrt{5} = 5|x|\sqrt{5}$

(b) $\sqrt{16x^4} = 4x^2$

(c) $\sqrt{8b^3} = \sqrt{4b^2}\sqrt{2b} = 2b\sqrt{2b}$

(d) $\sqrt{3y^5} = \sqrt{y^4}\sqrt{3y} = y^2\sqrt{3y}$

(e) $\sqrt{288x^6} = \sqrt{144x^6}\sqrt{2} = 12\,|x^3|\,\sqrt{2}$

(f) $\sqrt{28m^8} = \sqrt{4m^8}\sqrt{7} = 2m^4\sqrt{7}$

(g) $\sqrt{128c^{10}} = \sqrt{64c^{10}}\sqrt{2} = 8\,|c^5|\,\sqrt{2}$

(h) $\sqrt{18d^7} = \sqrt{9d^6}\sqrt{2d} = 3d^3\sqrt{2d}$

(i) $\sqrt{49x^2} = 7|x|$

(j) $\sqrt{80n^{20}} = \sqrt{16n^{20}}\sqrt{5} = 4n^{10}\sqrt{5}$

(k) $\sqrt{81h^3} = \sqrt{81h^2}\sqrt{h} = 9h\sqrt{h}$

PROBLEM SET 5.4 Products and Quotients Involving Radicals

1. $\sqrt{6}\sqrt{12} =$
$\sqrt{72} = \sqrt{36}\sqrt{2} = 6\sqrt{2}$

3. $(3\sqrt{3})(2\sqrt{6}) =$
$6\sqrt{18} =$
$6\sqrt{9}\sqrt{2} = 6(3)\sqrt{2} = 18\sqrt{2}$

5. $(4\sqrt{2})(-6\sqrt{5}) =$
$-24\sqrt{10}$

7. $(-3\sqrt{3})(-4\sqrt{8}) =$
$12\sqrt{24} =$
$12\sqrt{4}\sqrt{6} = 12(2)\sqrt{6} = 24\sqrt{6}$

9. $(5\sqrt{6})(4\sqrt{6}) =$
$20\sqrt{36} =$
$20(6) = 120$

11. $\left(2\sqrt[3]{4}\right)\left(6\sqrt[3]{2}\right) =$
$12\sqrt[3]{8} =$
$12(2) = 24$

13. $\left(4\sqrt[3]{6}\right)\left(7\sqrt[3]{4}\right) =$
$28\sqrt[3]{24} =$
$28\sqrt[3]{8}\sqrt[3]{3} = 28(2)\sqrt[3]{3} = 56\sqrt[3]{3}$

15. $\sqrt{2}(\sqrt{3} + \sqrt{5}) =$
$\sqrt{6} + \sqrt{10}$

17. $3\sqrt{5}(2\sqrt{2} - \sqrt{7}) =$
$6\sqrt{10} - 3\sqrt{35}$

19. $2\sqrt{6}(3\sqrt{8} - 5\sqrt{12})$
$6\sqrt{48} - 10\sqrt{72}$
$6\sqrt{16}\sqrt{3} - 10\sqrt{36}\sqrt{2}$

$6(4)\sqrt{3} - 10(6)\sqrt{2}$
$24\sqrt{3} - 60\sqrt{2}$

21. $-4\sqrt{5}(2\sqrt{5} + 4\sqrt{12})$
$-8\sqrt{25} - 16\sqrt{60}$
$-8(5) - 16\sqrt{4}\sqrt{15}$
$-8(5) - 16(2)\sqrt{15}$
$-40 - 32\sqrt{15}$

23. $3\sqrt{x}\left(5\sqrt{2} + \sqrt{y}\right)$
$15\sqrt{2x} + 3\sqrt{xy}$

25. $\sqrt{xy}\left(5\sqrt{xy} - 6\sqrt{x}\right)$
$5\sqrt{x^2y^2} - 6\sqrt{x^2y}$
$5xy - 6\sqrt{x^2}\sqrt{y}$
$5xy - 6x\sqrt{y}$

27. $\sqrt{5y}\left(\sqrt{8x} + \sqrt{12y^2}\right)$
$\sqrt{40xy} + \sqrt{60y^3}$
$\sqrt{4}\sqrt{10xy} + \sqrt{4y^2}\sqrt{15y}$
$2\sqrt{10xy} + 2y\sqrt{15y}$

29. $5\sqrt{3}\left(2\sqrt{8} - 3\sqrt{18}\right)$
$10\sqrt{24} - 15\sqrt{54}$
$10\sqrt{4}\sqrt{6} - 15\sqrt{9}\sqrt{6}$
$10(2)\sqrt{6} - 15(3)\sqrt{6}$
$20\sqrt{6} - 45\sqrt{6}$
$-25\sqrt{6}$

31. $(\sqrt{3} + 4)(\sqrt{3} - 7)$
$\sqrt{9} - 7\sqrt{3} + 4\sqrt{3} - 28$
$3 - 3\sqrt{3} - 28$
$-25 - 3\sqrt{3}$

33. $(\sqrt{5} - 6)(\sqrt{5} - 3)$
$\sqrt{25} - 3\sqrt{5} - 6\sqrt{5} + 18$
$5 - 9\sqrt{5} + 18$
$23 - 9\sqrt{5}$

35. $(3\sqrt{5} - 2\sqrt{3})(2\sqrt{7} + \sqrt{2})$
$6\sqrt{35} + 3\sqrt{10} - 4\sqrt{21} - 2\sqrt{6}$

37. $(2\sqrt{6} + 3\sqrt{5})(\sqrt{8} - 3\sqrt{12})$
$2\sqrt{48} - 6\sqrt{72} + 3\sqrt{40} - 9\sqrt{60}$
$2\sqrt{16}\sqrt{3} - 6\sqrt{36}\sqrt{2} + 3\sqrt{4}\sqrt{10} - 9\sqrt{4}\sqrt{15}$
$2(4)\sqrt{3} - 6(6)\sqrt{2} + 3(2)\sqrt{10} - 9(2)\sqrt{15}$
$8\sqrt{3} - 36\sqrt{2} + 6\sqrt{10} - 18\sqrt{15}$

39. $(2\sqrt{6} + 5\sqrt{5})(3\sqrt{6} - \sqrt{5})$
$6\sqrt{36} - 2\sqrt{30} + 15\sqrt{30} - 5\sqrt{25}$
$6(6) - 2\sqrt{30} + 15\sqrt{30} - 5(5)$
$36 + 13\sqrt{30} - 25$
$11 + 13\sqrt{30}$

41. $(3\sqrt{2} - 5\sqrt{3})(6\sqrt{2} - 7\sqrt{3})$
$18\sqrt{4} - 21\sqrt{6} - 30\sqrt{6} + 35\sqrt{9}$
$36 - 51\sqrt{6} + 105$
$141 - 51\sqrt{6}$

43. $(\sqrt{6} + 4)(\sqrt{6} - 4)$
$\sqrt{36} - 4\sqrt{6} + 4\sqrt{6} - 16$
$6 - 16$
-10

45. $(\sqrt{2} + \sqrt{10})(\sqrt{2} - \sqrt{10})$
$\sqrt{4} - \sqrt{20} + \sqrt{20} - \sqrt{100}$
$2 - 10$
-8

47. $(\sqrt{2x} + \sqrt{3y})(\sqrt{2x} - \sqrt{3y})$
$\sqrt{4x^2} - \sqrt{6xy} + \sqrt{6xy} - \sqrt{9y^2}$
$2x - 3y$

49. $2\sqrt[3]{3}\left(5\sqrt[3]{4} + \sqrt[3]{6}\right)$
$10\sqrt[3]{12} + 2\sqrt[3]{18}$

51. $3\sqrt[3]{4}\left(2\sqrt[3]{2} - 6\sqrt[3]{4}\right)$
$6\sqrt[3]{8} - 18\sqrt[3]{16}$
$6(2) - 18\sqrt[3]{8}\sqrt[3]{2}$
$6(2) - 18(2)\sqrt[3]{2}$
$12 - 36\sqrt[3]{2}$

Problem Set 5.4

53. $\dfrac{2}{\sqrt{7}+1} =$

$\dfrac{2}{(\sqrt{7}+1)} \bullet \dfrac{(\sqrt{7}-1)}{(\sqrt{7}-1)} =$

$\dfrac{2(\sqrt{7}-1)}{\sqrt{49}-1} = \dfrac{2(\sqrt{7}-1)}{7-1} =$

$\dfrac{2(\sqrt{7}-1)}{6} = \dfrac{\sqrt{7}-1}{3}$

55. $\dfrac{3}{\sqrt{2}-5} = \dfrac{3}{(\sqrt{2}-5)} \bullet \dfrac{(\sqrt{2}+5)}{(\sqrt{2}+5)} =$

$\dfrac{3(\sqrt{2}+5)}{\sqrt{4}-25} = \dfrac{3(\sqrt{2}+5)}{2-25} =$

$\dfrac{3(\sqrt{2}+5)}{-23} = \dfrac{3\sqrt{2}+15}{-23} = \dfrac{-3\sqrt{2}-15}{23}$

57. $\dfrac{1}{\sqrt{2}+\sqrt{7}} = \dfrac{1}{(\sqrt{2}+\sqrt{7})} \bullet \dfrac{(\sqrt{2}-\sqrt{7})}{(\sqrt{2}-\sqrt{7})} =$

$\dfrac{\sqrt{2}-\sqrt{7}}{\sqrt{4}-\sqrt{49}} = \dfrac{\sqrt{2}-\sqrt{7}}{2-7} = \dfrac{\sqrt{2}-\sqrt{7}}{-5} =$

$\dfrac{-\sqrt{2}+\sqrt{7}}{5} = \dfrac{\sqrt{7}-\sqrt{2}}{5}$

59. $\dfrac{\sqrt{2}}{\sqrt{10}-\sqrt{3}} = \dfrac{\sqrt{2}}{(\sqrt{10}-\sqrt{3})} \bullet \dfrac{(\sqrt{10}+\sqrt{3})}{(\sqrt{10}+\sqrt{3})} =$

$\dfrac{\sqrt{2}(\sqrt{10}+\sqrt{3})}{\sqrt{100}-\sqrt{9}} = \dfrac{\sqrt{20}+\sqrt{6}}{10-3} =$

$\dfrac{\sqrt{4}\sqrt{5}+\sqrt{6}}{7} = \dfrac{2\sqrt{5}+\sqrt{6}}{7}$

61. $\dfrac{\sqrt{3}}{2\sqrt{5}+4} = \dfrac{\sqrt{3}}{(2\sqrt{5}+4)} \bullet \dfrac{(2\sqrt{5}-4)}{(2\sqrt{5}-4)} =$

$\dfrac{\sqrt{3}(2\sqrt{5}-4)}{4\sqrt{25}-16} = \dfrac{2\sqrt{15}-4\sqrt{3}}{20-16} =$

$\dfrac{2\sqrt{15}-4\sqrt{3}}{4} = \dfrac{2(\sqrt{15}-2\sqrt{3})}{4} =$

$\dfrac{\sqrt{15}-2\sqrt{3}}{2}$

63. $\dfrac{6}{3\sqrt{7}-2\sqrt{6}} =$

$\dfrac{6}{(3\sqrt{7}-2\sqrt{6})} \bullet \dfrac{(3\sqrt{7}+2\sqrt{6})}{(3\sqrt{7}+2\sqrt{6})} =$

$\dfrac{6(3\sqrt{7}+2\sqrt{6})}{9\sqrt{49}-4\sqrt{36}} = \dfrac{18\sqrt{7}+12\sqrt{6}}{9(7)-4(6)} =$

$\dfrac{18\sqrt{7}+12\sqrt{6}}{63-24} = \dfrac{18\sqrt{7}+12\sqrt{6}}{39} =$

$\dfrac{3(6\sqrt{7}+4\sqrt{6})}{39} = \dfrac{6\sqrt{7}+4\sqrt{6}}{13}$

65. $\dfrac{\sqrt{6}}{3\sqrt{2}+2\sqrt{3}} =$

$\dfrac{\sqrt{6}}{(3\sqrt{2}+2\sqrt{3})} \bullet \dfrac{(3\sqrt{2}-2\sqrt{3})}{(3\sqrt{2}-2\sqrt{3})} =$

$\dfrac{\sqrt{6}(3\sqrt{2}-2\sqrt{3})}{9\sqrt{4}-4\sqrt{9}} = \dfrac{3\sqrt{12}-2\sqrt{18}}{9(2)-4(3)} =$

$\dfrac{3\sqrt{4}\sqrt{3}-2\sqrt{9}\sqrt{2}}{18-12} = \dfrac{6\sqrt{3}-6\sqrt{2}}{6} =$

$\dfrac{6(\sqrt{3}-\sqrt{2})}{6} = \sqrt{3}-\sqrt{2}$

67. $\dfrac{2}{\sqrt{x}+4} =$

$\dfrac{2}{\sqrt{x}+4} \bullet \dfrac{(\sqrt{x}-4)}{(\sqrt{x}-4)} =$

$\dfrac{2(\sqrt{x}-4)}{\sqrt{x^2}-16} = \dfrac{2\sqrt{x}-8}{x-16}$

69. $\dfrac{\sqrt{x}}{\sqrt{x}-5} =$

$\dfrac{\sqrt{x}}{(\sqrt{x}-5)} \bullet \dfrac{(\sqrt{x}+5)}{(\sqrt{x}+5)} =$

$\dfrac{\sqrt{x}(\sqrt{x}+5)}{\sqrt{x^2}-25} = \dfrac{\sqrt{x^2}+5\sqrt{x}}{x-25} =$

$\dfrac{x+5\sqrt{x}}{x-25}$

71. $\dfrac{\sqrt{x}-2}{\sqrt{x}+6} =$

$\dfrac{(\sqrt{x}-2)}{(\sqrt{x}+6)} \bullet \dfrac{(\sqrt{x}-6)}{(\sqrt{x}-6)} =$

$\dfrac{\sqrt{x^2}-6\sqrt{x}-2\sqrt{x}+12}{\sqrt{x^2}-36} =$

$\dfrac{x-8\sqrt{x}+12}{x-36}$

73. $\dfrac{\sqrt{x}}{\sqrt{x}+2\sqrt{y}} = =$

$\dfrac{\sqrt{x}}{(\sqrt{x}+2\sqrt{y})} \bullet \dfrac{(\sqrt{x}-2\sqrt{y})}{(\sqrt{x}-2\sqrt{y})} =$

$\dfrac{\sqrt{x^2}-2\sqrt{xy}}{\sqrt{x^2}-4\sqrt{y^2}} = \dfrac{x-2\sqrt{xy}}{x-4y}$

75. $\dfrac{3\sqrt{y}}{2\sqrt{x}-3\sqrt{y}} =$

$\dfrac{3\sqrt{y}}{(2\sqrt{x}-3\sqrt{y})} \bullet \dfrac{(2\sqrt{x}+3\sqrt{y})}{(2\sqrt{x}+3\sqrt{y})} =$

$\dfrac{6\sqrt{xy}+9\sqrt{y^2}}{4\sqrt{x^2}-9\sqrt{y^2}} = \dfrac{6\sqrt{xy}+9y}{4x-9y}$

PROBLEM SET 5.5 Equations Involving Radicals

1. $\sqrt{5x} = 10$
$(\sqrt{5x})^2 = 10^2$
$5x = 100$
$x = 20$
Check:
$\sqrt{5(20)} \stackrel{?}{=} 10$
$\sqrt{100} \stackrel{?}{=} 10$
$10 = 10$
The solution set is $\{20\}$.

3. $\sqrt{2x}+4 = 0$
$\sqrt{2x} = -4$
$(\sqrt{2x})^2 = (-4)^2$
$2x = 16$
$x = 8$
Check:
$\sqrt{2(8)}+4 \stackrel{?}{=} 0$
$\sqrt{16}+4 \stackrel{?}{=} 0$
$4+4 \stackrel{?}{=} 0$
$8 \neq 0$
The solution set is $\emptyset$.

5. $2\sqrt{n} = 5$
$\sqrt{n} = \dfrac{5}{2}$
$(\sqrt{n})^2 = \left(\dfrac{5}{2}\right)^2$
$n = \dfrac{25}{4}$
Check:
$2\sqrt{\dfrac{25}{4}} \stackrel{?}{=} 5$
$2\left(\dfrac{5}{2}\right) \stackrel{?}{=} 5$
$5 = 5$
The solution set is $\left\{\dfrac{25}{4}\right\}$.

7. $3\sqrt{n}-2 = 0$
$3\sqrt{n} = 2$

$\sqrt{n} = \frac{2}{3}$
$(\sqrt{n})^2 = \left(\frac{2}{3}\right)^2$
$n = \frac{4}{9}$
Check:
$3\sqrt{\frac{4}{9}} - 2 \stackrel{?}{=} 0$
$3\left(\frac{2}{3}\right) - 2 \stackrel{?}{=} 0$
$2 - 2 \stackrel{?}{=} 0$
$0 = 0$
The solution set is $\left\{\frac{4}{9}\right\}$.

9. $\sqrt{3y+1} = 4$
$(\sqrt{3y+1})^2 = 4^2$
$3y + 1 = 16$
$3y = 15$
$y = 5$
Check:
$\sqrt{3(5)+1} \stackrel{?}{=} 4$
$\sqrt{16} \stackrel{?}{=} 4$
$4 = 4$
The solution set is $\{5\}$.

11. $\sqrt{4y-3} - 6 = 0$
$\sqrt{4y-3} = 6$
$(\sqrt{4y-3})^2 = 6^2$
$4y - 3 = 36$
$4y = 39$
$y = \frac{39}{4}$
Check:
$\sqrt{4\left(\frac{39}{4}\right) - 3} - 6 \stackrel{?}{=} 0$
$\sqrt{39-3} - 6 \stackrel{?}{=} 0$
$\sqrt{36} - 6 \stackrel{?}{=} 0$
$6 - 6 \stackrel{?}{=} 0$
$0 = 0$
The solution set is $\left\{\frac{39}{4}\right\}$.

13. $\sqrt{3x-1} + 1 = 4$
$\sqrt{3x-1} = 3$
$(\sqrt{3x-1})^2 = (3)^2$
$3x - 1 = 9$
$3x = 10$
$x = \frac{10}{3}$
Check:
$\sqrt{3(\frac{10}{3}) - 1} + 1 \stackrel{?}{=} 40$
$\sqrt{10-1} + 1 \stackrel{?}{=} 4$
$\sqrt{9} + 1 \stackrel{?}{=} 4$
$3 + 1 \stackrel{?}{=} 4$
$4 = 4$
The solution set is $\left\{\frac{10}{3}\right\}$.

15. $\sqrt{2n+3} - 2 = -1$
$\sqrt{2n+3} = 1$
$(\sqrt{2n+3})^2 = (1)^2$
$2n + 3 = 1$
$2n = -2$
$n = -1$
Check:
$\sqrt{2(-1)+3} - 2 \stackrel{?}{=} -1$
$\sqrt{-2+3} - 2 \stackrel{?}{=} -1$
$\sqrt{1} - 2 \stackrel{?}{=} -1$
$1 - 2 \stackrel{?}{=} -1$
$-1 = -1$
The solution set is $\{-1\}$.

17. $\sqrt{2x-5} = -1$
$(\sqrt{2x-5})^2 = (-1)^2$
$2x - 5 = 1$
$2x = 6$
$x = 3$
Check:
$\sqrt{2(3)-5} \stackrel{?}{=} -1$
$\sqrt{1} \stackrel{?}{=} -1$
$1 \neq -1$
The solution set is $\emptyset$.

19. $\sqrt{5x+2}=\sqrt{6x+1}$

$(\sqrt{5x+2})^2=(\sqrt{6x+1})^2$

$5x+2=6x+1$

$2=x+1$

$1=x$

Check:

$\sqrt{5(1)+2}\stackrel{?}{=}\sqrt{6(1)+1}$

$\sqrt{7}=\sqrt{7}$

The solution set is $\{1\}$.

21. $\sqrt{3x+1}=\sqrt{7x-5}$

$(\sqrt{3x+1})^2=(\sqrt{7x-5})^2$

$3x+1=7x-5$

$3x=7x-6$

$-4x=-6$

$x=\frac{-6}{-4}=\frac{3}{2}$

Check:

$\sqrt{3\left(\frac{3}{2}\right)+1}\stackrel{?}{=}\sqrt{7\left(\frac{3}{2}\right)-5}$

$\sqrt{\frac{9}{2}+1}\stackrel{?}{=}\sqrt{\frac{21}{2}-5}$

$\sqrt{\frac{11}{2}}=\sqrt{\frac{11}{2}}$

The solution set is $\left\{\frac{3}{2}\right\}$.

23. $\sqrt{3x-2}-\sqrt{x+4}=0$

$\sqrt{3x-2}=\sqrt{x+4}$

$(\sqrt{3x-2})^2=(\sqrt{x+4})^2$

$3x-2=x+4$

$3x=x+6$

$2x=6$

$x=3$

Check:

$\sqrt{3(3)-2}-\sqrt{3+4}\stackrel{?}{=}0$

$\sqrt{7}-\sqrt{7}\stackrel{?}{=}0$

$0=0$

The solution set is $\{3\}$.

25. $5\sqrt{t-1}=6$

$\sqrt{t-1}=\frac{6}{5}$

$(\sqrt{t-1})^2=\left(\frac{6}{5}\right)^2$

$t-1=\frac{36}{25}$

$t=\frac{36}{25}+1$

$t=\frac{61}{25}$

Check:

$5\sqrt{\frac{61}{25}-1}\stackrel{?}{=}6$

$5\sqrt{\frac{36}{25}}\stackrel{?}{=}6$

$5\left(\frac{6}{5}\right)\stackrel{?}{=}6$

$6=6$

The solution set is $\left\{\frac{61}{25}\right\}$.

27. $\sqrt{x^2+7}=4$

$(\sqrt{x^2+7})^2=4^2$

$x^2+7=16$

$x^2-9=0$

$(x+3)(x-3)=0$

$x+3=0$ or $x-3=0$

$x=-3$ or $x=3$

Checking $x=-3$:

$\sqrt{(-3)^2+7}\stackrel{?}{=}4$

$\sqrt{16}\stackrel{?}{=}4$

$4=4$

Checking $x=3$:

$\sqrt{3^2+7}\stackrel{?}{=}4$

$\sqrt{16}\stackrel{?}{=}4$

$4=4$

The solution set is $\{-3, 3\}$.

29. $\sqrt{x^2+13x+37}=1$

$(\sqrt{x^2+13x+37})^2=1^2$

$x^2+13x+37=1$

$x^2 + 13x + 36 = 0$
$(x + 4)(x + 9) = 0$
$x + 4 = 0$ or $x + 9 = 0$
$x = -4$ or $x = -9$
Checking $x = -4$:
$\sqrt{(-4)^2 + 13(-4) + 37} \stackrel{?}{=} 1$
$\sqrt{16 - 52 + 37} \stackrel{?}{=} 1$
$\sqrt{1} \stackrel{?}{=} 1$
$1 = 1$
Checking $x = -9$:
$\sqrt{(-9)^2 + 13(-9) + 37} \stackrel{?}{=} 1$
$\sqrt{81 - 117 + 37} \stackrel{?}{=} 1$
$\sqrt{1} \stackrel{?}{=} 1$
$1 = 1$
The solution set is $\{-9, -4\}$.

31. $\sqrt{x^2 - x + 1} = x + 1$
$(\sqrt{x^2 - x + 1})^2 = (x + 1)^2$
$x^2 - x + 1 = x^2 + 2x + 1$
$-x + 1 = 2x + 1$
$-x = 2x$
$-3x = 0$
$x = 0$
Check:
$\sqrt{0^2 - 0 + 1} \stackrel{?}{=} 0 + 1$
$\sqrt{1} \stackrel{?}{=} 1$
$1 = 1$
The solution set is $\{0\}$.

33. $\sqrt{x^2 + 3x + 7} = x + 2$
$(\sqrt{x^2 + 3x + 7})^2 = (x + 2)^2$
$x^2 + 3x + 7 = x^2 + 4x + 4$
$3x + 7 = 4x + 4$
$3x + 3 = 4x$
$3 = x$
Check:
$\sqrt{3^2 + 3(3) + 7} \stackrel{?}{=} 3 + 2$
$\sqrt{9 + 9 + 7} \stackrel{?}{=} 5$
$\sqrt{25} \stackrel{?}{=} 5$
$5 = 5$
The solution set is $\{3\}$.

35. $\sqrt{-4x + 17} = x - 3$
$(\sqrt{-4x + 17})^2 = (x - 3)^2$
$-4x + 17 = x^2 - 6x + 9$
$0 = x^2 - 2x - 8$
$0 = (x - 4)(x + 2)$
$x - 4 = 0$ or $x + 2 = 0$
$x = 4$ or $x = -2$
Checking $x = 4$:
$\sqrt{-4(4) + 17} \stackrel{?}{=} 4 - 3$
$\sqrt{1} \stackrel{?}{=} 1$
$1 = 1$
Checking $x = -2$:
$\sqrt{-4(-2) + 17} \stackrel{?}{=} -2 - 3$
$\sqrt{25} \stackrel{?}{=} -5$
$5 \neq -5$
The solution set is $\{4\}$.

37. $\sqrt{n + 4} = n + 4$
$(\sqrt{n + 4})^2 = (n + 4)^2$
$n + 4 = n^2 + 8n + 16$
$0 = n^2 + 7n + 12$
$0 = (n + 4)(n + 3)$
$n + 4 = 0$ or $n + 3 = 0$
$n = -4$ or $n = -3$
Checking $n = -4$:
$\sqrt{0} \stackrel{?}{=} 0$
$0 = 0$
Checking $n = -3$:
$\sqrt{-3 + 4} \stackrel{?}{=} -3 + 4$
$\sqrt{1} \stackrel{?}{=} 1$
$1 = 1$
The solution set is $\{-4, -3\}$.

39. $\sqrt{3y} = y - 6$
$(\sqrt{3y})^2 = (y - 6)^2$
$3y = y^2 - 12y + 36$
$0 = y^2 - 15y + 36$
$0 = (y - 12)(y - 3)$
$y - 12 = 0$ or $y - 3 = 0$
$y = 12$ or $y = 3$

Checking $y = 12$:
$\sqrt{3(12)} \stackrel{?}{=} 12 - 6$
$\sqrt{36} \stackrel{?}{=} 6$
$6 = 6$
Checking $y = 3$:
$\sqrt{3(3)} \stackrel{?}{=} 3 - 6$
$\sqrt{9} \stackrel{?}{=} -3$
$3 \neq -3$
The solution set is $\{12\}$.

41. $4\sqrt{x} + 5 = x$
$4\sqrt{x} = x - 5$
$(4\sqrt{x})^2 = (x - 5)^2$
$16x = x^2 - 10x + 25$
$0 = x^2 - 26x + 25$
$0 = (x - 25)(x - 1)$
$x - 25 = 0$ or $x - 1 = 0$
$x = 25$ or $x = 1$
Checking $x = 25$:
$4\sqrt{25} + 5 \stackrel{?}{=} 25$
$4(5) + 5 \stackrel{?}{=} 25$
$25 = 25$
Checking $x = 1$:
$4\sqrt{1} + 5 \stackrel{?}{=} 1$
$4 + 5 \stackrel{?}{=} 1$
$9 \neq 1$
The solution set is $\{25\}$.

43. $\sqrt[3]{x - 2} = 3$
$\left(\sqrt[3]{x - 2}\right)^3 = 3^3$
$x - 2 = 27$
$x = 29$
Check:
$\sqrt[3]{29 - 2} \stackrel{?}{=} 3$
$\sqrt[3]{27} \stackrel{?}{=} 3$
$3 = 3$
The solution set is $\{29\}$.

45. $\sqrt[3]{2x + 3} = -3$
$\left(\sqrt[3]{2x + 3}\right)^3 = (-3)^3$
$2x + 3 = -27$
$2x = -30$
$x = -15$
Check:
$\sqrt[3]{2(-15) + 3} \stackrel{?}{=} -3$
$\sqrt[3]{-27} \stackrel{?}{=} -3$
$-3 = -3$
The solution set is $\{-15\}$.

47. $\sqrt[3]{2x + 5} = \sqrt[3]{4 - x}$
$\left(\sqrt[3]{2x + 5}\right)^3 = \left(\sqrt[3]{4 - x}\right)^3$
$2x + 5 = 4 - x$
$2x = -1 - x$
$3x = -1$
$x = -\frac{1}{3}$
Check:
$\sqrt[3]{2\left(-\frac{1}{3}\right) + 5} \stackrel{?}{=} \sqrt[3]{4 - \left(-\frac{1}{3}\right)}$
$\sqrt[3]{-\frac{2}{3} + 5} \stackrel{?}{=} \sqrt[3]{4 + \frac{1}{3}}$
$\sqrt[3]{4\frac{1}{3}} = \sqrt[3]{4\frac{1}{3}}$
The solution set is $\left\{-\frac{1}{3}\right\}$.

49. $\sqrt{x + 19} - \sqrt{x + 28} = -1$
$\sqrt{x + 19} = \sqrt{x + 28} - 1$
$(\sqrt{x + 19})^2 = (\sqrt{x + 28} - 1)^2$
$x + 19 = (\sqrt{x + 28})^2 + 2(-1)\sqrt{x + 28} + 1$
$x + 19 = x + 28 - 2\sqrt{x + 28} + 1$
$x + 19 = x + 29 - 2\sqrt{x + 28}$
$-10 = -2\sqrt{x + 28}$
$5 = \sqrt{x + 28}$
$(5)^2 = (\sqrt{x + 28})^2$
$25 = x + 28$
$-3 = x$
Check:
$\sqrt{-3 + 19} - \sqrt{-3 + 28} \stackrel{?}{=} -1$

$\sqrt{16} - \sqrt{25} \stackrel{?}{=} -1$
$4 - 5 \stackrel{?}{=} -1$
$-1 = -1$
The solution set is $\{-3\}$.

51. $\sqrt{3x+1} + \sqrt{2x+4} = 3$
$\sqrt{3x+1} = 3 - \sqrt{2x+4}$
$(\sqrt{3x+1})^2 = (3 - \sqrt{2x+4})^2$
$3x+1 = 9 + 2(3)(-\sqrt{2x+4}) + (-\sqrt{2x+4})^2$
$3x+1 = 9 - 6\sqrt{2x+4} + 2x + 4$
$3x+1 = 2x + 13 - 6\sqrt{2x+4}$
$x - 12 = -6\sqrt{2x+4}$
$(x-12)^2 = (-6\sqrt{2x+4})^2$
$x^2 - 24x + 144 = 36(2x+4)$
$x^2 - 24x + 144 = 72x + 144$
$x^2 - 96x = 0$
$x(x-96) = 0$
$x = 0$ or $x - 96 = 0$
$x = 0$ or $x = 96$
Checking $x = 0$:
$\sqrt{3(0)+1} + \sqrt{2(0)+4} \stackrel{?}{=} 3$
$\sqrt{1} + \sqrt{4} \stackrel{?}{=} 3$
$1 + 2 \stackrel{?}{=} 3$
$3 = 3$
Checking $x = 96$:
$\sqrt{3(96)+1} + \sqrt{2(96)+4} \stackrel{?}{=} 3$
$\sqrt{289} + \sqrt{196} \stackrel{?}{=} 3$
$17 + 14 \stackrel{?}{=} 3$
$31 \neq 3$
The solution set is $\{0\}$.

53. $\sqrt{n-4} + \sqrt{n+4} = 2\sqrt{n-1}$
$(\sqrt{n-4} + \sqrt{n+4})^2 = (2\sqrt{n-1})^2$
$(\sqrt{n\text{-}4})^2 + 2\sqrt{n\text{-}4}\sqrt{n\text{+}4} + (\sqrt{n\text{+}4})^2 = 4(n\text{-}1)$
$n - 4 + 2\sqrt{(n-4)(n+4)} + n + 4 = 4n - 4$
$2n + 2\sqrt{n^2-16} = 4n - 4$
$2\sqrt{n^2-16} = 2n - 4$
$2\sqrt{n^2-16} = 2(n-2)$
$\sqrt{n^2-16} = n - 2$
$(\sqrt{n^2-16})^2 = (n-2)^2$
$n^2 - 16 = n^2 - 4n + 4$
$-16 = -4n + 4$
$-20 = -4n$
$5 = n$
Check:
$\sqrt{5-4} + \sqrt{5+4} \stackrel{?}{=} 2\sqrt{5-1}$
$\sqrt{1} + \sqrt{9} \stackrel{?}{=} 2\sqrt{4}$
$1 + 3 \stackrel{?}{=} 2(2)$
$4 = 4$
The solution set is $\{5\}$.

55. $\sqrt{t+3} - \sqrt{t-2} = \sqrt{7-t}$
$(\sqrt{t+3} - \sqrt{t-2})^2 = (\sqrt{7-t})^2$
$(\sqrt{t\text{+}3})^2 \text{-} 2\sqrt{t\text{+}3}\sqrt{t\text{-}2} + (\text{-}\sqrt{t\text{-}2})^2 = 7 - t$
$t + 3 - 2\sqrt{(t+3)(t-2)} + t - 2 = 7 - t$
$2t + 1 - 2\sqrt{t^2+t-6} = 7 - t$
$-2\sqrt{t^2+t-6} = -3t + 6$
$(-2\sqrt{t^2+t-6})^2 = (-3t+6)^2$
$4(t^2+t-6) = 9t^2 - 36t + 36$
$4t^2 + 4t - 24 = 9t^2 - 36t + 36$
$0 = 5t^2 - 40t + 60$
$0 = 5(t^2 - 8t + 12)$
$0 = 5(t-6)(t-2)$
$t - 6 = 0$ or $t - 2 = 0$
$t = 6$ or $t = 2$
Checking $t = 6$:
$\sqrt{6+3} - \sqrt{6-2} \stackrel{?}{=} \sqrt{7-6}$
$\sqrt{9} - \sqrt{4} \stackrel{?}{=} \sqrt{1}$
$3 - 2 \stackrel{?}{=} 1$
$1 = 1$
Checking $t = 2$:
$\sqrt{2+3} - \sqrt{2-2} \stackrel{?}{=} \sqrt{7-2}$
$\sqrt{5} - \sqrt{0} \stackrel{?}{=} \sqrt{5}$
$\sqrt{5} = \sqrt{5}$
The solution set is $\{2, 6\}$.

57. $\sqrt{30Df} = S$
$\sqrt{30D(0.95)} = S$
$\sqrt{28.5D} = S$
$(\sqrt{28.5D})^2 = S^2$
$28.5D = S^2$
$D = \dfrac{S^2}{28.5}$
$D = \dfrac{40^2}{28.5} = 56$ feet
$D = \dfrac{55^2}{28.5} = 106$ feet
$D = \dfrac{65^2}{28.5} = 148$ feet

59. $L = \dfrac{8T^2}{\pi^2}$
$L = \dfrac{8(2)^2}{\pi^2} = 3.2$ feet
$L = \dfrac{8(2.5)^2}{\pi^2} = 5.1$ feet
$L = \dfrac{8(3)^2}{\pi^2} = 7.3$ feet

PROBLEM SET 5.6 Merging Exponents and Roots

1. $81^{\frac{1}{2}} = \sqrt{81} = 9$

3. $27^{\frac{1}{3}} = \sqrt[3]{27} = 3$

5. $(-8)^{\frac{1}{3}} = \sqrt[3]{-8} = -2$

7. $-25^{\frac{1}{2}} = -\sqrt{25} = -5$

9. $36^{-\frac{1}{2}} = \dfrac{1}{36^{\frac{1}{2}}} = \dfrac{1}{\sqrt{36}} = \dfrac{1}{6}$

11. $\left(\dfrac{1}{27}\right)^{-\frac{1}{3}} = \left(\dfrac{27}{1}\right)^{\frac{1}{3}} = 27^{\frac{1}{3}} = \sqrt[3]{27} = 3$

13. $4^{\frac{3}{2}} = \left(\sqrt{4}\right)^3 = 2^3 = 8$

15. $27^{\frac{4}{3}} = \left(\sqrt[3]{27}\right)^4 = 3^4 = 81$

17. $(-1)^{\frac{7}{3}} = \left(\sqrt[3]{-1}\right)^7 = (-1)^7 = -1$

19. $-4^{\frac{5}{2}} = -(\sqrt{4})^5 = -(2)^5 = -32$

21. $\left(\dfrac{27}{8}\right)^{\frac{4}{3}} = \left(\sqrt[3]{\dfrac{27}{8}}\right)^4 = \left(\dfrac{3}{2}\right)^4 = \dfrac{3^4}{2^4} = \dfrac{81}{16}$

23. $\left(\dfrac{1}{8}\right)^{-\frac{2}{3}} = \left(\dfrac{8}{1}\right)^{\frac{2}{3}} = 8^{\frac{2}{3}} = \left(\sqrt[3]{8}\right)^2$
$= (2)^2 = 4$

25. $64^{-\frac{7}{6}} = \dfrac{1}{64^{\frac{7}{6}}} = \dfrac{1}{\left(\sqrt[6]{64}\right)^7} = \dfrac{1}{2^7} = \dfrac{1}{128}$

27. $-25^{\frac{3}{2}} = -(\sqrt{25})^3 = -(5)^3 = -125$

29. $125^{\frac{4}{3}} = \left(\sqrt[3]{125}\right)^4 = 5^4 = 625$

31. $x^{\frac{4}{3}} = \sqrt[3]{x^4}$

33. $3x^{\frac{1}{2}} = 3\sqrt{x}$

35. $(2y)^{\frac{1}{3}} = \sqrt[3]{2y}$

37. $(2x - 3y)^{\frac{1}{2}} = \sqrt{2x - 3y}$

39. $(2a - 3b)^{\frac{2}{3}} = \sqrt[3]{(2a - 3b)^2}$

41. $x^{\frac{2}{3}}y^{\frac{1}{3}} = \sqrt[3]{x^2y}$

43. $-3x^{\frac{1}{5}}y^{\frac{2}{5}} = -3\sqrt[5]{xy^2}$

45. $\sqrt{5y} = (5y)^{\frac{1}{2}} = 5^{\frac{1}{2}}y^{\frac{1}{2}}$

Problem Set 5.6

47. $3\sqrt{y} = 3y^{\frac{1}{2}}$

49. $\sqrt[3]{xy^2} = (xy^2)^{\frac{1}{3}} = x^{\frac{1}{3}}y^{\frac{2}{3}}$

51. $\sqrt[4]{a^2b^3} = (a^2b^3)^{\frac{1}{4}} = a^{\frac{2}{4}}b^{\frac{3}{4}} = a^{\frac{1}{2}}b^{\frac{3}{4}}$

53. $\sqrt[5]{(2x-y)^3} = (2x-y)^{\frac{3}{5}}$

55. $5x\sqrt{y} = 5xy^{\frac{1}{2}}$

57. $-\sqrt[3]{x+y} = -(x+y)^{\frac{1}{3}}$

59. $(2x^{\frac{2}{5}})(6x^{\frac{1}{4}})$

$12x^{\frac{2}{5}+\frac{1}{4}}$

$12x^{\frac{8}{20}+\frac{5}{20}}$

$12x^{\frac{13}{20}}$

61. $(y^{\frac{2}{3}})(y^{-\frac{1}{4}}) =$

$y^{\frac{2}{3}-\frac{1}{4}} = y^{\frac{8}{12}-\frac{3}{12}} = y^{\frac{5}{12}}$

63. $(x^{\frac{2}{5}})(4x^{-\frac{1}{2}}) =$

$4x^{\frac{2}{5}-\frac{1}{2}} = 4x^{\frac{4}{10}-\frac{5}{10}} = 4x^{-\frac{1}{10}} = \dfrac{4}{x^{\frac{1}{10}}}$

65. $(4x^{\frac{1}{2}}y)^2$

$4^2x^{\frac{1}{2}(2)}y^2$

$16xy^2$

67. $(8x^6y^3)^{\frac{1}{3}}$

$8^{\frac{1}{3}}x^{6(\frac{1}{3})}y^{3(\frac{1}{3})}$

$2x^2y$

69. $\dfrac{24x^{\frac{3}{5}}}{6x^{\frac{1}{3}}} =$

$4x^{\frac{3}{5}-\frac{1}{3}} = 4x^{\frac{9}{15}-\frac{5}{15}} = 4x^{\frac{4}{15}}$

71. $\dfrac{48b^{\frac{1}{3}}}{12b^{\frac{3}{4}}} =$

$4b^{\frac{1}{3}-\frac{3}{4}} = 4b^{\frac{4}{12}-\frac{9}{12}} = 4b^{-\frac{5}{12}} = \dfrac{4}{b^{\frac{5}{12}}}$

73. $\left(\dfrac{6x^{\frac{2}{5}}}{7y^{\frac{2}{3}}}\right)^2 = \dfrac{6^2x^{\frac{2}{5}(2)}}{7^2y^{\frac{2}{3}(2)}} = \dfrac{36x^{\frac{4}{5}}}{49y^{\frac{4}{3}}}$

75. $\left(\dfrac{x^2}{y^3}\right)^{-\frac{1}{2}} = \dfrac{x^{2(-\frac{1}{2})}}{y^{3(-\frac{1}{2})}} = \dfrac{x^{-1}}{y^{-\frac{3}{2}}} = \dfrac{y^{\frac{3}{2}}}{x}$

77. $\left(\dfrac{18x^{\frac{1}{3}}}{9x^{\frac{1}{4}}}\right)^2 = (2x^{\frac{1}{3}-\frac{1}{4}})^2 =$

$(2x^{\frac{4}{12}-\frac{3}{12}})^2 = (2x^{\frac{1}{12}})^2 =$

$2^2x^{\frac{1}{12}(2)} = 4x^{\frac{2}{12}} = 4x^{\frac{1}{6}}$

79. $\left(\dfrac{60a^{\frac{1}{5}}}{15a^{\frac{3}{4}}}\right)^2 = (4a^{\frac{1}{5}-\frac{3}{4}})^2 = (4a^{\frac{4}{20}-\frac{15}{20}})^2$

$(4a^{-\frac{11}{20}})^2 = 4^2a^{-\frac{11}{20}(2)} = 16a^{-\frac{11}{10}} = \dfrac{16}{a^{\frac{11}{10}}}$

81. $\sqrt[3]{3}\sqrt{3} =$
$3^{\frac{1}{3}} \bullet 3^{\frac{1}{2}} = 3^{\frac{1}{3}+\frac{1}{2}} = 3^{\frac{2}{6}+\frac{3}{6}} = 3^{\frac{5}{6}} =$
$\sqrt[6]{3^5} = \sqrt[6]{243}$

83. $\sqrt[4]{6}\sqrt{6}$
$6^{\frac{1}{4}} \bullet 6^{\frac{1}{2}} = 6^{\frac{1}{4}+\frac{1}{2}} = 6^{\frac{1}{4}+\frac{2}{4}} = 6^{\frac{3}{4}} =$
$\sqrt[4]{6^3} = \sqrt[4]{216}$

85. $\dfrac{\sqrt[3]{3}}{\sqrt[4]{3}} = \dfrac{3^{\frac{1}{3}}}{3^{\frac{1}{4}}} = 3^{\frac{1}{3}-\frac{1}{4}} = 3^{\frac{4}{12}-\frac{3}{12}} =$
$3^{\frac{1}{12}} = \sqrt[12]{3}$

87. $\dfrac{\sqrt[3]{8}}{\sqrt[4]{4}} = \dfrac{\sqrt[3]{2^3}}{\sqrt[4]{2^2}} = \dfrac{2^{\frac{3}{3}}}{2^{\frac{2}{4}}} = \dfrac{2^1}{2^{\frac{1}{2}}} =$

$2^{1-\frac{1}{2}} = 2^{\frac{1}{2}} = \sqrt{2}$

89. $\frac{\sqrt[4]{27}}{\sqrt{3}} = \frac{\sqrt[4]{3^3}}{\sqrt{3}} = \frac{3^{\frac{3}{4}}}{3^{\frac{1}{2}}} =$

$3^{\frac{3}{4}-\frac{1}{2}} = 3^{\frac{3}{4}-\frac{2}{4}} = 3^{\frac{1}{4}} = \sqrt[4]{3}$

Further Investigations

93a) $\sqrt[3]{1728} = 12$

b) $\sqrt[3]{5832} = 18$

c) $\sqrt[4]{2401} = 7$

d) $\sqrt[4]{65,536} = 16$

e) $\sqrt[5]{161,051} = 11$

f) $\sqrt[5]{6,436,343} = 23$

95a) $16^{\frac{5}{2}} = 1024$

b) $25^{\frac{7}{2}} = 78125$

c) $16^{\frac{9}{4}} = 512$

d) $27^{\frac{5}{3}} = 243$

e) $343^{\frac{2}{3}} = 49$

f) $512^{\frac{4}{3}} = 4096$

PROBLEM SET 5.7 Scientific Notation

1. $89 = (8.9)(10)^1$

3. $4290 = (4.29)(10)^3$

5. $6,120,000 = (6.12)(10)^6$

7. $40,000,000 = (4)(10)^7$

9. $376.4 = (3.764)(10)^2$

11. $0.347 = (3.47)(10)^{-1}$

13. $0.0214 = (2.14)(10)^{-2}$

15. $0.00005 = (5)(10)^{-5}$

17. $0.00000000194 = (1.94)(10)^{-9}$

19. $(2.3)(10)^1 = 23$

21. $(4.19)(10)^3 = 4190$

23. $(5)(10)^8 = 500,000,000$

25. $(3.14)(10)^{10} = 31,400,000,000$

27. $(4.3)(10)^{-1} = 0.43$

29. $(9.14)(10)^{-4} = 0.000914$

31. $(5.123)(10)^{-8} = 0.00000005123$

33. $(0.0037)(0.00002)$

$(3.7)(10)^{-3}(2)(10)^{-5}$

$(7.4)(10)^{-8}$

0.000000074

35. $(0.00007)(11,000)$

$(7)(10)^{-5}(1.1)(10)^4$

$(7.7)(10)^{-1}$

0.77

37. $\frac{360,000,000}{0.0012} = \frac{(3.6)(10)^8}{(1.2)(10)^{-3}} =$

$(3)(10)^{11} = 300,000,000,000$

39. $\frac{0.000064}{16,000} = \frac{(6.4)(10)^{-5}}{(1.6)(10)^4} =$

$(4)(10)^{-9} = 0.000000004$

41. $\frac{(60,000)(0.006)}{(0.0009)(400)} = \frac{(6)(10)^4(6)(10)^{-3}}{(9)(10)^{-4}(4)(10)^2} =$

$\frac{(36)(10)^1}{(36)(10)^{-2}} = (1)(10)^3 = 1000$

43. $\frac{(0.0045)(60000)}{(1800)(0.00015)} =$

$\frac{(4.5)(10)^{-3}(6)(10)^4}{(1.8)(10)^3(1.5)(10)^{-4}} =$

$\frac{(27)(10)^1}{(2.7)(10)^{-1}} = \frac{(2.7)(10)^1(10)^1}{(2.7)(10)^{-1}} =$

$(1)(10)^3 = 1000$

Problem Set 5.7

45. $\sqrt{9{,}000{,}000} = \sqrt{(9)(10)^6} =$

$[(9)(10)^6]^{\frac{1}{2}} = 9^{\frac{1}{2}}(10)^{6(\frac{1}{2})} =$

$(3)(10)^3 = 3000$

47. $\sqrt[3]{8000} = \sqrt[3]{(8)(10)^3} = [(8)(10)^3]^{\frac{1}{3}} =$

$8^{\frac{1}{3}}(10)^{3(\frac{1}{3})} = 2(10) = 20$

49. $(90{,}000)^{\frac{3}{2}} = [(9)(10)^4]^{\frac{3}{2}} =$

$(9)^{\frac{3}{2}}(10)^{4(\frac{3}{2})} = (27)(10)^6 = 27{,}000{,}000$

51. $602,000,000,000,000,000,000,000 =$
$(6.02)(10^{23})$

53. Let $x =$ the number of times his laptop computer is faster than his first computer.

$(1.33)(10^9) = (1.6)(10^6)\,x$

$\dfrac{(1.33)(10^9)}{(1.6)(10^6)} = x$

$(0.83125)(10^3) = x$
$831.25 = x$
$x \approx 831$
The laptop is approximately 831 times faster than the first computer.

55. Let $x =$ debt per person.

$x = \dfrac{5,700,000,000,000}{275,000,000}$

$x = \dfrac{(5.7)(10^{12})}{(2.75)(10^8)}$

$x = (2.07)(10^4)$

The average debt is $(2.07)(10^4)$ dollars.

57. Let $x =$ the mass of a carbon atom.
One amu $= \frac{1}{12}$(the mass of a carbon atom)

$(1.66)(10^{-27}) = \dfrac{1}{12}x$

$12(1.66)(10^{-27}) = x$

$(1.2)(10^1)(1.66)(10^{-27}) = x$

$(1.992)(10^{-26}) = x$

The mass of a carbon atom is $(1.992)(10^{-26})$ kilograms.

59. Let $x =$ the number of times that the weight of a proton is more than the weight of an electron.

$(1.67)(10^{-27}) = (9.11)(10^{-31})\,x$

$\dfrac{(1.67)(10^{-27})}{(9.11)(10^{-31})} = x$

$(0.1833150)(10^4) = x$

$(0.1833150)(10^4) = x$
$1833.15 = x$
$x \approx 1833$
The weight of a proton is approximately 1833 times the weight of an electron.

Further Investigations

63(a) $\sqrt{49{,}000{,}000} = \sqrt{(49)(10)^6}$
$[(49)(10)^6]^{\frac{1}{2}} = (49)^{\frac{1}{2}}(10)^{6(\frac{1}{2})}$
$(7)(10)^3 = 7{,}000$

(b) $\sqrt{0.0025} = \sqrt{(25)(10)^{-4}} =$
$[(25)(10)^{-4}]^{\frac{1}{2}} = (25)^{\frac{1}{2}}(10)^{-4(\frac{1}{2})} =$
$(5)(10)^{-2} = 0.05$

(c) $\sqrt{14400} = \sqrt{(144)(10)^2} =$
$[(144)(10)^2]^{\frac{1}{2}} = (144)^{\frac{1}{2}}(10)^{2(\frac{1}{2})} =$
$(12)(10)^1 = 120$

(d) $\sqrt{0.000121} = \sqrt{(121)(10)^{-6}} =$
$[(121)(10)^{-6}]^{\frac{1}{2}} = (121)^{\frac{1}{2}}(10)^{-6(\frac{1}{2})} =$
$(11)(10)^{-3} = 0.011$

(e) $\sqrt[3]{27000} = \sqrt[3]{(27)(10)^3}$
$[(27)(10)^3]^{\frac{1}{3}} = (27)^{\frac{1}{3}}(10)^{3(\frac{1}{3})} =$
$(3)(10)^1 = 30$

(f) $\sqrt[3]{0.00064} = \sqrt[3]{(64)(10)^{-6}}$
$[(64)(10)^{-6}]^{\frac{1}{3}} = (64)^{\frac{1}{3}}(10)^{-6(\frac{1}{3})} =$
$(4)(10)^{-2} = 0.04$

65 **(a)** $(4576)^4 = (4.385)(10)^{14}$
(b) $(719)^{10} = (3.692)(10)^{28}$
(c) $(28)^{12} = (2.322)(10)^{17}$
(d) $(8619)^6 = (4.100)(10)^{23}$
(e) $(314)^5 = (3.052)(10)^{12}$
(f) $(145{,}723)^2 = (2.124)(10)^{10}$

CHAPTER 5 Review Problem Set

1. $4^{-3} = \frac{1}{4^3} = \frac{1}{64}$

2. $\left(\frac{2}{3}\right)^{-2} = \left(\frac{3}{2}\right)^2 = \frac{3^2}{2^2} = \frac{9}{4}$

3. $(3^2 \bullet 3^{-3})^{-1} =$
$(3^{-1})^{-1} = 3^{-1(-1)} = 3^1 = 3$

4. $\sqrt[3]{-8} = -2$

5. $\sqrt[4]{\frac{16}{81}} = \frac{2}{3}$

6. $4^{\frac{5}{2}} = (\sqrt{4})^5 = 2^5 = 32$

7. $(-1)^{\frac{2}{3}} = \left(\sqrt[3]{-1}\right)^2 = (-1)^2 = 1$

8. $\left(\frac{8}{27}\right)^{\frac{2}{3}} = \left(\sqrt[3]{\frac{8}{27}}\right)^2 = \left(\frac{2}{3}\right)^2 = \frac{4}{9}$

9. $-16^{\frac{3}{2}} = -(\sqrt{16})^3 = -(4)^3 = -64$

10. $\frac{2^3}{2^{-2}} = 2^{3-(-2)} = 2^5 = 32$

11. $(4^{-2} \bullet 4^2)^{-1} = (4^0)^{-1} = (1)^{-1} = \frac{1}{1} = 1$

12. $\left(\frac{3^{-1}}{3^2}\right)^{-1} = (3^{-1-2})^{-1} = (3^{-3})^{-1} =$
$3^3 = 27$

13. $\sqrt{54} = \sqrt{9}\sqrt{6} = 3\sqrt{6}$

14. $\sqrt{48x^3y}$
$\sqrt{16x^2}\sqrt{3xy}$
$4x\sqrt{3xy}$

15. $\frac{4\sqrt{3}}{\sqrt{6}} = 4\sqrt{\frac{3}{6}} = 4\sqrt{\frac{1}{2}} = (4)\frac{\sqrt{1}}{\sqrt{2}} =$

$\frac{4}{\sqrt{2}} \bullet \frac{\sqrt{2}}{\sqrt{2}} = \frac{4\sqrt{2}}{\sqrt{4}} = \frac{4\sqrt{2}}{2} = 2\sqrt{2}$

16. $\sqrt{\frac{5}{12x^3}} = \frac{\sqrt{5}}{\sqrt{12x^3}} = \frac{\sqrt{5}}{\sqrt{4x^2}\sqrt{3x}} =$

$\frac{\sqrt{5}}{2x\sqrt{3x}} \bullet \frac{\sqrt{3x}}{\sqrt{3x}} = \frac{\sqrt{15x}}{2x\sqrt{9x^2}} =$

$\frac{\sqrt{15x}}{2x(3x)} = \frac{\sqrt{15x}}{6x^2}$

17. $\sqrt[3]{56} = \sqrt[3]{8}\sqrt[3]{7} = 2\sqrt[3]{7}$

18. $\frac{\sqrt[3]{2}}{\sqrt[3]{9}} = \frac{\sqrt[3]{2}}{\sqrt[3]{9}} \bullet \frac{\sqrt[3]{3}}{\sqrt[3]{3}} = \frac{\sqrt[3]{6}}{\sqrt[3]{27}} = \frac{\sqrt[3]{6}}{3}$

19. $\sqrt{\frac{9}{5}} = \frac{\sqrt{9}}{\sqrt{5}} = \frac{3}{\sqrt{5}} = \frac{3}{\sqrt{5}} \bullet \frac{\sqrt{5}}{\sqrt{5}} =$
$\frac{3\sqrt{5}}{\sqrt{25}} = \frac{3\sqrt{5}}{5}$

20. $\sqrt{\frac{3x^3}{7}} = \frac{\sqrt{3x^3}}{\sqrt{7}} = \frac{\sqrt{x^2}\sqrt{3x}}{\sqrt{7}} = \frac{x\sqrt{3x}}{\sqrt{7}} =$

$\frac{x\sqrt{3x}}{\sqrt{7}} \bullet \frac{\sqrt{7}}{\sqrt{7}} = \frac{x\sqrt{21x}}{\sqrt{49}} = \frac{x\sqrt{21x}}{7}$

21. $\sqrt[3]{108x^4y^8}$

$\sqrt[3]{27x^3y^6}\sqrt[3]{4xy^2}$

$3xy^2\sqrt[3]{4xy^2}$

22. $\frac{3}{4}\sqrt{150}$

$\frac{3}{4}\sqrt{25}\sqrt{6}$

$\frac{3}{4}(5)\sqrt{6}$

$\frac{15\sqrt{6}}{4}$

23. $\frac{2}{3}\sqrt{45xy^3}$

$\frac{2}{3}\sqrt{9y^2}\sqrt{5xy}$

$\frac{2}{3}(3y)\sqrt{5xy}$

$2y\sqrt{5xy}$

24. $\frac{\sqrt{8x^2}}{\sqrt{2x}}$

$\sqrt{\frac{8x^2}{2x}}$

$\sqrt{4x}$

$\sqrt{4}\sqrt{x}$

$2\sqrt{x}$

25. $(3\sqrt{8})(4\sqrt{5})$

$12\sqrt{40}$

$12\sqrt{4}\sqrt{10}$

$12(2)\sqrt{10}$

$24\sqrt{10}$

26. $(5\sqrt[3]{2})(6\sqrt[3]{4})$

$30\sqrt[3]{8}$

$30(2)$

60

27. $3\sqrt{2}(4\sqrt{6}-2\sqrt{7})$

$12\sqrt{12}-6\sqrt{14}$

$12\sqrt{4}\sqrt{3}-6\sqrt{14}$

$12(2)\sqrt{3}-6\sqrt{14}$

$24\sqrt{3}-6\sqrt{14}$

28. $(\sqrt{x}+3)(\sqrt{x}-5)$

$\sqrt{x^2}-5\sqrt{x}+3\sqrt{x}-15$

$x-2\sqrt{x}-15$

29. $(2\sqrt{5}-\sqrt{3})(2\sqrt{5}+\sqrt{3})$

$4\sqrt{25}+2\sqrt{15}-2\sqrt{15}-\sqrt{9}$

$4(5)-3$

$20-3$

17

30. $(3\sqrt{2}+\sqrt{6})(5\sqrt{2}-3\sqrt{6})$

$15\sqrt{4}-9\sqrt{12}+5\sqrt{12}-3\sqrt{36}$

$15(2)-4\sqrt{12}-3(6)$

$30-4\sqrt{12}-18$

$12-8\sqrt{3}$

31. $(2\sqrt{a}+\sqrt{b})(3\sqrt{a}-4\sqrt{b})$

$6\sqrt{a^2}-8\sqrt{ab}+3\sqrt{ab}-4\sqrt{b^2}$

$6a-5\sqrt{ab}-4b$

32. $(4\sqrt{8}-\sqrt{2})(\sqrt{8}+3\sqrt{2})$

$4\sqrt{64}+12\sqrt{16}-\sqrt{16}-3\sqrt{4}$

$4(8)+11\sqrt{16}-3(2)$

$32+44-6$

70

33. $\dfrac{4}{\sqrt{7}-1}=\dfrac{4}{(\sqrt{7}-1)}\bullet\dfrac{(\sqrt{7}+1)}{(\sqrt{7}+1)}=$

$\dfrac{4(\sqrt{7}+1)}{\sqrt{49}-1}=\dfrac{4(\sqrt{7}+1)}{7-1}=$

$\dfrac{4(\sqrt{7}+1)}{6}=\dfrac{2(\sqrt{7}+1)}{3}$

34. $\dfrac{\sqrt{3}}{\sqrt{8}+\sqrt{5}}=$

$\dfrac{\sqrt{3}}{\sqrt{8}+\sqrt{5}}\bullet\dfrac{(\sqrt{8}-\sqrt{5})}{(\sqrt{8}-\sqrt{5})}=$

$\dfrac{\sqrt{24}-\sqrt{15}}{\sqrt{64}-\sqrt{25}}=\dfrac{\sqrt{24}-\sqrt{15}}{8-5}=$

$\dfrac{2\sqrt{6}-\sqrt{15}}{3}$

35. $\dfrac{3}{2\sqrt{3}+3\sqrt{5}}=$

$\dfrac{3}{(2\sqrt{3}+3\sqrt{5})}\bullet\dfrac{(2\sqrt{3}-3\sqrt{5})}{(2\sqrt{3}-3\sqrt{5})}=$

$\dfrac{3(2\sqrt{3}-3\sqrt{5})}{4\sqrt{9}-9\sqrt{25}}=\dfrac{3(2\sqrt{3}-3\sqrt{5})}{4(3)-9(5)}=$

$\dfrac{3(2\sqrt{3}-3\sqrt{5})}{12-45}=\dfrac{3(2\sqrt{3}-3\sqrt{5})}{-33}=$

$\dfrac{2\sqrt{3}-3\sqrt{5}}{-11}=\dfrac{-2\sqrt{3}+3\sqrt{5}}{11}=$

$\dfrac{3\sqrt{5}-2\sqrt{3}}{11}$

36. $\dfrac{3\sqrt{2}}{2\sqrt{6}-\sqrt{10}}=$

$\dfrac{3\sqrt{2}}{(2\sqrt{6}-\sqrt{10})}\bullet\dfrac{(2\sqrt{6}+\sqrt{10})}{(2\sqrt{6}+\sqrt{10})}=$

$\dfrac{6\sqrt{12}+3\sqrt{20}}{4\sqrt{36}-\sqrt{100}}=\dfrac{6\sqrt{4}\sqrt{3}+3\sqrt{4}\sqrt{5}}{4(6)-10}=$

$\dfrac{12\sqrt{3}+6\sqrt{5}}{14}=\dfrac{2(6\sqrt{3}+3\sqrt{5})}{14}=$

$\dfrac{6\sqrt{3}+3\sqrt{5}}{7}$

37. $(x^{-3}y^{4})^{-2}=x^{-3(-2)}y^{4(-2)}=$

$x^{6}y^{-8}=\dfrac{x^{6}}{y^{8}}$

38. $\left(\dfrac{2a^{-1}}{3b^{4}}\right)^{-3}=\dfrac{2^{-3}a^{3}}{3^{-3}b^{-12}}=$

$\dfrac{3^{3}a^{3}b^{12}}{2^{3}}=\dfrac{27a^{3}b^{12}}{8}$

39. $(4x^{\frac{1}{2}})(5x^{\frac{1}{5}})$

$20x^{\frac{1}{2}+\frac{1}{5}}$

$20x^{\frac{7}{10}}$

40. $\dfrac{42a^{\frac{3}{4}}}{6a^{\frac{1}{3}}}=7a^{\frac{3}{4}-\frac{1}{3}}=7a^{\frac{5}{12}}$

41. $\left(\dfrac{x^{3}}{y^{4}}\right)^{-\frac{1}{3}}=\left(\dfrac{y^{4}}{x^{3}}\right)^{\frac{1}{3}}=\dfrac{y^{\frac{4}{3}}}{x}$

42. $\left(\dfrac{6x^{-2}}{2x^{4}}\right)^{-2}=(3x^{-6})^{-2}=3^{-2}x^{12}=$

$\dfrac{x^{12}}{3^{2}}=\dfrac{x^{12}}{9}$

43. $3\sqrt{45}-2\sqrt{20}-\sqrt{80}$

$3\sqrt{9}\sqrt{5}-2\sqrt{4}\sqrt{5}-\sqrt{16}\sqrt{5}$

$3(3)\sqrt{5}-2(2)\sqrt{5}-4\sqrt{5}$

$9\sqrt{5}-4\sqrt{5}-4\sqrt{5}$

$\sqrt{5}$

44. $4\sqrt[3]{24}+3\sqrt[3]{3}-2\sqrt[3]{81}$

$4\sqrt[3]{8}\sqrt[3]{3}+3\sqrt[3]{3}-2\sqrt[3]{27}\sqrt[3]{3}$

$4(2)\sqrt[3]{3}+3\sqrt[3]{3}-2(3)\sqrt[3]{3}$

$8\sqrt[3]{3}+3\sqrt[3]{3}-6\sqrt[3]{3}$

$5\sqrt[3]{3}$

45. $3\sqrt{24}-\frac{2}{5}\sqrt{54}+\frac{1}{4}\sqrt{96}$

$3\sqrt{4}\sqrt{6}-\frac{2}{5}\sqrt{9}\sqrt{6}+\frac{1}{4}\sqrt{16}\sqrt{6}$

$3(2)\sqrt{6}-\frac{2}{5}(3)\sqrt{6}+\frac{1}{4}(4)\sqrt{6}$

$6\sqrt{6}-\frac{6}{5}\sqrt{6}+\sqrt{6}$

$7\sqrt{6}-\frac{6}{5}\sqrt{6}$

$\frac{35\sqrt{6}}{5}-\frac{6\sqrt{6}}{5}$

$\frac{29\sqrt{6}}{5}$

46. $-2\sqrt{12x}+3\sqrt{27x}-5\sqrt{48x}$

$-2\sqrt{4}\sqrt{3x}+3\sqrt{9}\sqrt{3x}-5\sqrt{16}\sqrt{3x}$

$-2(2)\sqrt{3x}+3(3)\sqrt{3x}-5(4)\sqrt{3x}$

$-4\sqrt{3x}+9\sqrt{3x}-20\sqrt{3x}$

$-15\sqrt{3x}$

47. $x^{-2}+y^{-1}$

$\frac{1}{x^2}+\frac{1}{y}$; LCD $=x^2y$

$\frac{1}{x^2}\bullet\frac{y}{y}+\frac{1}{y}\bullet\frac{x^2}{x^2}$

$\frac{y}{x^2y}+\frac{x^2}{x^2y}$

$\frac{y+x^2}{x^2y}$

48. $a^{-2}-2a^{-1}b^{-1}$

$\frac{1}{a^2}-\frac{2}{ab}$; LCD $=a^2b$

$\frac{1}{a^2}\bullet\frac{b}{b}-\frac{2}{ab}\bullet\frac{a}{a}$

$\frac{b-2a}{a^2b}$

49. $\sqrt{7x-3}=4$

$(\sqrt{7x-3})^2=4^2$

$7x-3=16$

$7x=19$

$x=\frac{19}{7}$

Check :

$\sqrt{7\left(\frac{19}{7}\right)-3}\stackrel{?}{=}4$

$\sqrt{19-3}\stackrel{?}{=}4$

$\sqrt{16}\stackrel{?}{=}4$

$4=4$

The solution set is $\left\{\frac{19}{7}\right\}$.

50. $\sqrt{2y+1}=\sqrt{5y-11}$

$(\sqrt{2y+1})^2=(\sqrt{5y-11})^2$

$2y+1=5y-11$

$-3y=-12$

$y=4$

Check :

$\sqrt{2(4)+1} \stackrel{?}{=} \sqrt{5(4)-11}$

$\sqrt{9} = \sqrt{9}$

The solution set is $\{4\}$.

51. $\sqrt{2x} = x - 4$

$(\sqrt{2x})^2 = (x-4)^2$

$2x = x^2 - 8x + 16$

$0 = x^2 - 10x + 16$

$0 = (x-8)(x-2)$

$x - 8 = 0$ or $x - 2 = 0$

$x = 8$ or $x = 2$

Checking $x = 8$:

$\sqrt{2(8)} \stackrel{?}{=} 8 - 4$

$\sqrt{16} \stackrel{?}{=} 4$

$4 = 4$

Checking $x = 2$:

$\sqrt{2(2)} \stackrel{?}{=} 2 - 4$

$\sqrt{4} \stackrel{?}{=} -2$

$2 \neq -2$

The solution set is $\{8\}$.

52. $\sqrt{n^2 - 4n - 4} = n$

$(\sqrt{n^2 - 4n - 4})^2 = n^2$

$n^2 - 4n - 4 = n^2$

$-4n - 4 = 0$

$-4n = 4$

$n = -1$

Check:

$\sqrt{(-1)^2 - 4(-1) - 4} \stackrel{?}{=} -1$

$\sqrt{1 + 4 - 4} \stackrel{?}{=} -1$

$\sqrt{1} \stackrel{?}{=} -1$

$1 \neq -1$

The solution set is $\emptyset$.

53. $\sqrt[3]{2x - 1} = 3$

$\left(\sqrt[3]{2x-1}\right)^3 = 3^3$

$2x - 1 = 27$

$2x = 28$

$x = 14$

Checking $x = 14$:

$\sqrt[3]{2(14) - 1} \stackrel{?}{=} 3$

$\sqrt[3]{27} \stackrel{?}{=} 3$

$3 = 3$

The solution set is $\{14\}$.

54. $\sqrt{t^2 + 9t - 1} = 3$

$(\sqrt{t^2 + 9t - 1})^2 = 3^2$

$t^2 + 9t - 1 = 9$

$t^2 + 9t - 10 = 0$

$(t + 10)(t - 1) = 0$

$t + 10 = 0$ or $t - 1 = 0$

$t = -10$ or $t = 1$

Checking $t = -10$:

$\sqrt{(-10)^2 + 9(-10) - 1} \stackrel{?}{=} 3$

$\sqrt{100 - 90 - 1} \stackrel{?}{=} 3$

$\sqrt{9} \stackrel{?}{=} 3$

$3 = 3$

Checking $t = 1$:

$\sqrt{(1)^2 + 9(1) - 1} \stackrel{?}{=} 3$

$\sqrt{1 + 9 - 1} \stackrel{?}{=} 3$

$\sqrt{9} \stackrel{?}{=} 3$

$3 = 3$

The solution set is $\{-10, 1\}$.

55. $\sqrt{x^2 + 3x - 6} = x$

$(\sqrt{x^2 + 3x - 6})^2 = x^2$

$x^2 + 3x - 6 = x^2$

$3x - 6 = 0$

$3x = 6$

$x = 2$

Check :

$\sqrt{2^2 + 3(2) - 6} \stackrel{?}{=} 2$

$\sqrt{4 + 6 - 6} \stackrel{?}{=} 2$

$\sqrt{4} \stackrel{?}{=} 2$

$2 = 2$

The solution set is $\{2\}$.

56. $\sqrt{x+1} - \sqrt{2x} = -1$

$\sqrt{x+1} = \sqrt{2x} - 1$

$(\sqrt{x+1})^2 = (\sqrt{2x} - 1)^2$

$x + 1 = (\sqrt{2x})^2 - 2\sqrt{2x} + 1$

$x + 1 = 2x - 2\sqrt{2x} + 1$

$-x = -2\sqrt{2x}$

$(-x)^2 = (-2\sqrt{2x})^2$

$x^2 = 4(2x)$

$x^2 = 8x$

$x^2 - 8x = 0$

$x(x - 8) = 0$

$x = 0$ or $x - 8 = 0$

$x = 0$ or $x = 8$

Checking $x = 0$:

$\sqrt{0+1} - \sqrt{2(0)} \stackrel{?}{=} -1$

$\sqrt{1} - \sqrt{0} \stackrel{?}{=} -1$

$1 \neq -1$

Checking $x = 8$:

$\sqrt{8+1} - \sqrt{2(8)} \stackrel{?}{=} -1$

$\sqrt{9} - \sqrt{16} \stackrel{?}{=} -1$

$3 - 4 \stackrel{?}{=} 1$

$-1 = -1$

The solution set is $\{8\}$.

57. $(0.00002)(0.0003)$

$(2)(10)^{-5}(3)(10)^{-4}$

$(6)(10)^{-9}$

0.000000006

58. $(120{,}000)(300{,}000)$

$(1.2)(10)^5(3)(10)^5$

$(3.6)(10)^{10}$

$36{,}000{,}000{,}000$

59. $(0.000015)(400{,}000)$

$(1.5)(10)^{-5}(4)(10)^5$

$(6.0)(10)^0$

6

60. $\dfrac{0.000045}{0.0003} = \dfrac{(4.5)(10)^{-5}}{(3)(10)^{-4}} =$

$(1.5)(10)^{-1} = 0.15$

61. $\dfrac{(0.00042)(0.0004)}{0.006} =$

$\dfrac{(4.2)(10)^{-4}(4)(10)^{-4}}{(6)(10)^{-3}} =$

$\dfrac{(16.8)(10)^{-8}}{(6)(10)^{-3}} = (2.8)(10)^{-5} =$

0.000028

62. $\sqrt{0.000004} = \sqrt{(4)(10)^{-6}} =$

$[(4)(10)^{-6}]^{\frac{1}{2}} = (4)^{\frac{1}{2}}(10)^{-6(\frac{1}{2})} =$

$(2)(10)^{-3} = 0.002$

63. $\sqrt[3]{0.000000008} = \sqrt[3]{(8)(10)^{-9}} =$

$[(8)(10)^{-9}]^{\frac{1}{3}} = (8)^{\frac{1}{3}}(10)^{-9(\frac{1}{3})} =$

$(2)(10)^{-3} = 0.002$

64. $(4000000)^{\frac{3}{2}} =$

$[(4)(10)^6]^{\frac{3}{2}} = (4)^{\frac{3}{2}}(10)^{6(\frac{3}{2})} =$

$(8)(10)^9 = 8{,}000{,}000{,}000$

CHAPTER 5 Test

1. $(4)^{-\frac{5}{2}} = \frac{1}{4^{\frac{5}{2}}} = \frac{1}{(4^{\frac{1}{2}})^5} = \frac{1}{(2)^5} = \frac{1}{32}$

2. $-16^{\frac{5}{4}} = -(16^{\frac{1}{4}})^5 = -(2^5) = -32$

3. $\left(\frac{2}{3}\right)^{-4} = \left(\frac{3}{2}\right)^4 = \frac{3^4}{2^4} = \frac{81}{16}$

4. $\left(\frac{2^{-1}}{2^{-2}}\right)^{-2} = (2^1)^{-2} = \frac{1}{2^2} = \frac{1}{4}$

5. $\sqrt{63} = \sqrt{9}\sqrt{7} = 3\sqrt{7}$

6. $\sqrt[3]{108}$
$\sqrt[3]{27}\sqrt[3]{4}$
$3\sqrt[3]{4}$

7. $\sqrt{52x^4y^3} = \sqrt{4x^4y^2}\sqrt{13y} = 2x^2y\sqrt{13y}$

8. $\frac{5\sqrt{18}}{3\sqrt{12}} = \frac{5\sqrt{9}\sqrt{2}}{3\sqrt{4}\sqrt{3}} = \frac{15\sqrt{2}}{6\sqrt{3}} =$

$\frac{5\sqrt{2}}{2\sqrt{3}} \bullet \frac{\sqrt{3}}{\sqrt{3}} = \frac{5\sqrt{6}}{2\sqrt{9}} = \frac{5\sqrt{6}}{6}$

9. $\sqrt{\frac{7}{24x^3}} = \frac{\sqrt{7}}{\sqrt{24x^3}} = \frac{\sqrt{7}}{\sqrt{4x^2}\sqrt{6x}} =$

$\frac{\sqrt{7}}{2x\sqrt{6x}} = \frac{\sqrt{7}}{2x\sqrt{6x}} \bullet \frac{\sqrt{6x}}{\sqrt{6x}} =$

$\frac{\sqrt{42x}}{2x\sqrt{36x^2}} = \frac{\sqrt{42x}}{2x(6x)} = \frac{\sqrt{42x}}{12x^2}$

10. $(4\sqrt{6})(3\sqrt{12})$
$12\sqrt{72}$
$12\sqrt{36}\sqrt{2}$
$12(6)\sqrt{2}$
$72\sqrt{2}$

11. $(3\sqrt{2} + \sqrt{3})(\sqrt{2} - 2\sqrt{3})$
$3\sqrt{4} - 6\sqrt{6} + \sqrt{6} - 2\sqrt{9}$
$3(2) - 5\sqrt{6} - 2(3)$
$6 - 5\sqrt{6} - 6$
$-5\sqrt{6}$

12. $2\sqrt{50} - 4\sqrt{18} - 9\sqrt{32}$
$2\sqrt{25}\sqrt{2} - 4\sqrt{9}\sqrt{2} - 9\sqrt{16}\sqrt{2}$
$2(5)\sqrt{2} - 4(3)\sqrt{2} - 9(4)\sqrt{2}$
$10\sqrt{2} - 12\sqrt{2} - 36\sqrt{2}$
$-38\sqrt{2}$

13. $\frac{3\sqrt{2}}{4\sqrt{3} - \sqrt{8}} = \frac{3\sqrt{2}}{4\sqrt{3} - \sqrt{4}\sqrt{2}} =$

$\frac{3\sqrt{2}}{4\sqrt{3} - 2\sqrt{2}} =$

$\frac{3\sqrt{2}}{(4\sqrt{3} - 2\sqrt{2})} \bullet \frac{(4\sqrt{3} + 2\sqrt{2})}{(4\sqrt{3} + 2\sqrt{2})} =$

$\frac{12\sqrt{6} + 6\sqrt{4}}{16\sqrt{9} - 4\sqrt{4}} = \frac{12\sqrt{6} + 6(2)}{16(3) - 4(2)} =$

$\frac{12\sqrt{6} + 12}{48 - 8} = \frac{12\sqrt{6} + 12}{40} =$

$\frac{4(3\sqrt{6} + 3)}{4(10)} = \frac{3\sqrt{6} + 3}{10}$

14. $\left(\frac{2x^{-1}}{3y}\right)^{-2} = \frac{2^{-2}x^2}{3^{-2}y^{-2}} = \frac{3^2x^2y^2}{2^2} = \frac{9x^2y^2}{4}$

15. $\frac{-84a^{\frac{1}{2}}}{7a^{\frac{4}{5}}} = -12a^{\frac{1}{2} - \frac{4}{5}} =$

$-12a^{-\frac{3}{10}} = -\frac{12}{a^{\frac{3}{10}}}$

16. $x^{-1}+y^{-3}$

$\frac{1}{x}+\frac{1}{y^3}$;LCD $= xy^3$

$\frac{1}{x}\bullet\frac{y^3}{y^3}+\frac{1}{y^3}\bullet\frac{x}{x}$

$\frac{y^3}{xy^3}+\frac{x}{xy^3}=\frac{y^3+x}{xy^3}$

17. $(3x^{-\frac{1}{2}})(-4x^{\frac{3}{4}}) = -12x^{-\frac{1}{2}+\frac{3}{4}} =$

$-12x^{\frac{1}{4}}$

18. $(3\sqrt{5}-2\sqrt{3})(3\sqrt{5}+2\sqrt{3})$

$9\sqrt{25}+6\sqrt{15}-6\sqrt{15}-4\sqrt{9}$

$9(5)-4(3)$

$45-12=33$

19. $\frac{(0.00004)(300)}{0.00002}=\frac{(4)(10)^{-5}(3)(10)^2}{(2)(10)^{-5}}=$

$\frac{(12)(10)^{-3}}{(2)(10)^{-5}}=(6)(10)^2=600$

20. $\sqrt{0.000009}=\sqrt{(9)(10)^{-6}}=$

$[(9)(10)^{-6}]^{\frac{1}{2}}=(9)^{\frac{1}{2}}(10)^{-6(\frac{1}{2})}=$

$(3)(10)^{-3}=0.003$

21. $\sqrt{3x+1}=3$

$(\sqrt{3x+1})^2=3^2$

$3x+1=9$

$3x=8$

$x=\frac{8}{3}$

Check:

$\sqrt{3\left(\frac{8}{3}\right)+1}\stackrel{?}{=}3$

$\sqrt{8+1}\stackrel{?}{=}3$

$\sqrt{9}\stackrel{?}{=}3$

$3=3$

The solution set is $\left\{\frac{8}{3}\right\}$.

22. $\sqrt[3]{3x+2}=2$

$(\sqrt[3]{3x+2})^3=(2)^3$

$3x+2=8$

$3x=6$

$x=2$

Check:

$\sqrt[3]{3(2)+2}\stackrel{?}{=}2$

$\sqrt[3]{8}\stackrel{?}{=}2$

$2=2$

The solution set is $\{2\}$.

23. $\sqrt{x}=x-2$

$(\sqrt{x})^2=(x-2)^2$

$x=x^2-4x+4$

$0=x^2-5x+4$

$0=(x-4)(x-1)$

$x-4=0$ or $x-1=0$

$x=4$ or $x=1$

Checking $x=4$:

$\sqrt{4}\stackrel{?}{=}4-2$

$2=2$

Checking $x=1$:

$\sqrt{1}\stackrel{?}{=}1-2$

$1\neq-1$

The solution set is $\{4\}$.

24. $\sqrt{5x-2}=\sqrt{3x+8}$

$(\sqrt{5x-2})^2=(\sqrt{3x+8})^2$

$5x-2=3x+8$

$2x=10$

$x=5$

Check:

$\sqrt{5(5)-2}\stackrel{?}{=}\sqrt{3(5)+8}$

$\sqrt{23}=\sqrt{23}$

The solution set is $\{5\}$.

25. $\sqrt{x^2 - 10x + 28} = 2$

$(\sqrt{x^2 - 10x + 28})^2 = 2^2$

$x^2 - 10x + 28 = 4$

$x^2 - 10x + 24 = 0$

$(x - 6)(x - 4) = 0$

$x - 6 = 0$ or $x - 4 = 0$

$x = 6$ or $x = 4$

Checking $x = 6$:

$\sqrt{6^2 - 10(6) + 28} \stackrel{?}{=} 2$

$\sqrt{36 - 60 + 28} \stackrel{?}{=} 2$

$\sqrt{4} \stackrel{?}{=} 2$

$2 = 2$

Checking $x = 4$:

$\sqrt{4^2 - 10(4) + 28} \stackrel{?}{=} 2$

$\sqrt{16 - 40 + 28} \stackrel{?}{=} 2$

$\sqrt{4} \stackrel{?}{=} 2$

$2 = 2$

The solution set is $\{4, 6\}$.

Chapter 6 Quadratic Equations and Inequalities

PROBLEM SET 6.1 Complex Numbers

1. False
3. True
5. True
7. True

9. $(6+3i)+(4+5i)$
$(6+4)+(3+5)i$
$10+8i$

11. $(-8+4i)+(2+6i)$
$(-8+2)+(4+6)i$
$-6+10i$

13. $(3+2i)-(5+7i)$
$3+2i-5-7i$
$(3-5)+(2-7)i$
$-2-5i$

15. $(-7+3i)-(5-2i)$
$-7+3i-5+2i$
$(-7-5)+(3+2)i$
$-12+5i$

17. $(-3-10i)+(2-13i)$
$(-3+2)+(-10-13)i$
$-1-23i$

19. $(4-8i)-(8-3i)$
$4-8i-8+3i$
$(4-8)+(-8+3)i$
$-4-5i$

21. $(-1-i)-(-2-4i)$
$-1-i+2+4i$
$(-1+2)+(-1+4)i$
$1+3i$

23. $\left(\frac{3}{2}+\frac{1}{3}i\right)+\left(\frac{1}{6}-\frac{3}{4}i\right)$

$\left(\frac{3}{2}+\frac{1}{6}\right)+\left(\frac{1}{3}-\frac{3}{4}\right)i$

$\frac{5}{3}-\frac{5}{12}i$

25. $\left(-\frac{5}{9}+\frac{3}{5}i\right)-\left(\frac{4}{3}-\frac{1}{6}i\right)$

$-\frac{5}{9}+\frac{3}{5}i-\frac{4}{3}+\frac{1}{6}i$

$\left(-\frac{5}{9}-\frac{4}{3}\right)+\left(\frac{3}{5}+\frac{1}{6}\right)i$

$-\frac{17}{9}+\frac{23}{30}i$

27. $\sqrt{-81}=\sqrt{-1}\sqrt{81}=i\sqrt{81}=9i$

29. $\sqrt{-14}=\sqrt{-1}\sqrt{14}=i\sqrt{14}$

31. $\sqrt{-\frac{16}{25}}=\sqrt{-1}\sqrt{\frac{16}{25}}=i\left(\frac{4}{5}\right)=\frac{4}{5}i$

33. $\sqrt{-18}=\sqrt{-1}\sqrt{18}=i\sqrt{9}\sqrt{2}=$
$i(3)\sqrt{2}=3i\sqrt{2}$

35. $\sqrt{-75}=\sqrt{-1}\sqrt{75}=i\sqrt{25}\sqrt{3}=$
$i(5)\sqrt{3}=5i\sqrt{3}$

37. $3\sqrt{-28}=3\sqrt{-1}\sqrt{28}=3i\sqrt{4}\sqrt{7}=$
$3i(2)\sqrt{7}=6i\sqrt{7}$

39. $-2\sqrt{-80}=-2\sqrt{-1}\sqrt{80}=$
$-2i\sqrt{16}\sqrt{5}=-2i(4)\sqrt{5}=-8i\sqrt{5}$

41. $12\sqrt{-90}=12\sqrt{-1}\sqrt{90}=$
$12i\sqrt{9}\sqrt{10}=12i(3)\sqrt{10}=36i\sqrt{10}$

43. $\sqrt{-4}\sqrt{-16}=(i\sqrt{4})(i\sqrt{16})=$
$(2i)(4i)=8i^2=8(-1)=-8$

45. $\sqrt{-3}\sqrt{-5}=(i\sqrt{3})(i\sqrt{5})=$
$i^2\sqrt{15}=-\sqrt{15}$

47. $\sqrt{-9}\sqrt{-6} = (i\sqrt{9})(i\sqrt{6}) =$
$(3i)(i\sqrt{6}) = 3i^2\sqrt{6} =$
$3(-1)\sqrt{6} = -3\sqrt{6}$

49. $\sqrt{-15}\sqrt{-5} = (i\sqrt{15})(i\sqrt{5}) =$
$i^2\sqrt{75} = -1\sqrt{25}\sqrt{3} =$
$-1(5)\sqrt{3} = -5\sqrt{3}$

51. $\sqrt{-2}\sqrt{-27} = (i\sqrt{2})(i\sqrt{27}) =$
$i^2\sqrt{54} = -1\sqrt{9}\sqrt{6} =$
$-1(3)\sqrt{6} = -3\sqrt{6}$

53. $\sqrt{6}\sqrt{-8} = (\sqrt{6})(i\sqrt{8}) = i\sqrt{48}$
$i\sqrt{16}\sqrt{3} = 4i\sqrt{3}$

55. $\dfrac{\sqrt{-25}}{\sqrt{-4}} = \dfrac{i\sqrt{25}}{i\sqrt{4}} = \dfrac{5i}{2i} = \dfrac{5}{2}$

57. $\dfrac{\sqrt{-56}}{\sqrt{-7}} = \dfrac{i\sqrt{56}}{i\sqrt{7}} = \sqrt{\dfrac{56}{7}} =$

$\sqrt{8} = \sqrt{4}\sqrt{2} = 2\sqrt{2}$

59. $\dfrac{\sqrt{-24}}{\sqrt{6}} = \dfrac{i\sqrt{24}}{\sqrt{6}} = i\sqrt{\dfrac{24}{6}} = i\sqrt{4} = 2i$

61. $(5i)(4i) = 20i^2 = 20(-1) = -20 + 0i$

63. $(7i)(-6i) = -42i^2 =$
$-42(-1) = 42 + 0i$

65. $(3i)(2-5i) = 6i - 15i^2 = 6i - 15(-1) =$
$6i + 15 = 15 + 6i$

67. $(-6i)(-2-7i) = 12i + 42i^2 =$
$12i + 42(-1) = 12i - 42 = -42 + 12i$

69. $(3+2i)(5+4i) = 15 + 12i + 10i + 8i^2 =$
$15 + 22i + 8(-1) = 15 + 22i - 8 =$
$7 + 22i$

71. $(6-2i)(7-i) = 42 - 6i - 14i + 2i^2 =$
$42 - 20i + 2(-1) = 42 - 20i - 2 =$
$40 - 20i$

73. $(-3-2i)(5+6i) =$
$-15 - 18i - 10i - 12i^2 =$
$-15 - 28i - 12(-1) =$
$-15 - 28i + 12 = -3 - 28i$

75. $(9+6i)(-1-i) = -9 - 9i - 6i - 6i^2 =$
$-9 - 15i - 6(-1) = -9 - 15i + 6 =$
$-3 - 15i$

77. $(4+5i)^2$
$(4)^2 + 2(4)(5i) + (5i)^2$
$16 + 40i + 25i^2$
$16 + 40i + 25(-1)$
$16 + 40i - 25$
$-9 + 40i$

79. $(-2-4i)^2$
$(-2)^2 + 2(-2)(-4i) + (-4i)^2$
$4 + 16i + 16i^2$
$4 + 16i + 16(-1)$
$4 + 16i - 16$
$-12 + 16i$

81. $(6+7i)(6-7i)$
$36 - 42i + 42i - 49i^2$
$36 - 49(-1) + 0i$
$36 + 49 + 0i$
$85 + 0i$

83. $(-1+2i)(-1-2i)$
$1 + 2i - 2i - 4i^2$
$1 - 4(-1) + 0i$
$1 + 4 + 0i$
$5 + 0i$

85. $\dfrac{3i}{2+4i} = \dfrac{3i}{(2+4i)} \bullet \dfrac{(2-4i)}{(2-4i)} =$

$\dfrac{6i - 12i^2}{4 - 16i^2} = \dfrac{6i - 12(-1)}{4 - 16(-1)} =$

$\dfrac{6i + 12}{4 + 16} = \dfrac{12 + 6i}{20} =$

$\dfrac{12}{20} + \dfrac{6}{20}i = \dfrac{3}{5} + \dfrac{3}{10}i$

87. $\frac{-2i}{3-5i}=\frac{-2i}{(3-5i)}\bullet\frac{(3+5i)}{(3+5i)}=$

$\frac{-6i-10i^2}{9-25i^2}=\frac{-6i-10(-1)}{9-25(-1)}=$

$\frac{-6i+10}{9+25}=\frac{10-6i}{34}=$

$\frac{10}{34}-\frac{6}{34}i=\frac{5}{17}-\frac{3}{17}i$

89. $\frac{-2+6i}{3i}=\frac{(-2+6i)}{3i}\bullet\frac{(-i)}{(-i)}=$

$\frac{2i-6i^2}{-3i^2}=\frac{2i-6(-1)}{-3(-1)}=\frac{2i+6}{3}=$

$=\frac{6+2i}{3}=\frac{6}{3}+\frac{2}{3}i=2+\frac{2}{3}i$

91. $\frac{2}{7i}=\frac{2}{7i}\bullet\frac{(-i)}{(-i)}=\frac{-2i}{-7i^2}=$

$\frac{-2i}{-7(-1)}=0-\frac{2}{7}i$

93. $\frac{2+6i}{1+7i}=\frac{(2+6i)}{(1+7i)}\bullet\frac{(1-7i)}{(1-7i)}=$

$\frac{2-14i+6i-42i^2}{1-49i^2}=\frac{2-8i-42(-1)}{1-49(-1)}=$

$\frac{2-8i+42}{1+49}=\frac{44-8i}{50}=\frac{44}{50}-\frac{8}{50}i=$

$\frac{22}{25}-\frac{4}{25}i$

95. $\frac{3+6i}{4-5i}=\frac{(3+6i)}{(4-5i)}\bullet\frac{(4+5i)}{(4+5i)}=$

$\frac{12+15i+24i+30i^2}{16-25i^2}=$

$\frac{12+39i+30(-1)}{16+25}=\frac{12+39i-30}{41}=$

$\frac{-18+39i}{41}=-\frac{18}{41}+\frac{39}{41}i$

97. $\frac{-2+7i}{-1+i}=\frac{(-2+7i)}{(-1+i)}\bullet\frac{(-1-i)}{(-1-i)}=$

$\frac{2+2i-7i-7i^2}{1-i^2}=\frac{2-5i-7(-1)}{1-(-1)}=$

$\frac{2-5i+7}{1+1}=\frac{9-5i}{2}=\frac{9}{2}-\frac{5}{2}i$

99. $\frac{-1-3i}{-2-10i}=\frac{(-1-3i)}{(-2-10i)}\bullet\frac{(-2+10i)}{(-2+10i)}=$

$\frac{2-10i+6i-30i^2}{4-100i^2}=\frac{2-4i-30(-1)}{4-100(-1)}=$

$\frac{2-4i+30}{4+100}=\frac{32-4i}{104}=$

$\frac{32}{104}-\frac{4}{104}i=\frac{4}{13}-\frac{1}{26}i$

101a. $\frac{-4-\sqrt{-12}}{2}=\frac{-4-i\sqrt{12}}{2}=$

$\frac{-4-2i\sqrt{3}}{2}=\frac{2(-2-i\sqrt{3})}{2}=$

$-2-i\sqrt{3}$

b. $\frac{6+\sqrt{-24}}{4}=\frac{6+i\sqrt{24}}{4}=$

$\frac{6+2i\sqrt{6}}{4}=\frac{2(3+i\sqrt{6})}{4}=$

$\frac{3+i\sqrt{6}}{2}$

c. $\frac{-1-\sqrt{-18}}{2}=\frac{-1-i\sqrt{18}}{2}=$

$\frac{-1-3i\sqrt{2}}{2}$

d. $\frac{-6+\sqrt{-27}}{3} = \frac{-6+i\sqrt{27}}{3} =$

$\frac{-6+3i\sqrt{3}}{3} = \frac{3(-2+i\sqrt{3})}{3} =$

$-2+i\sqrt{3}$

e. $\frac{10+\sqrt{-45}}{4} = \frac{10+i\sqrt{45}}{4} =$

$\frac{10+3i\sqrt{5}}{4}$

f. $\frac{4-\sqrt{-48}}{2} = \frac{4-i\sqrt{48}}{2} =$

$\frac{4-4i\sqrt{3}}{2} = \frac{4(1-i\sqrt{3})}{2} =$

$2(1-i\sqrt{3}) = 2-2i\sqrt{3}$

PROBLEM SET 6.2 Quadratic Equations

1. $x^2 - 9x = 0$
$x(x-9) = 0$
$x = 0$ or $x - 9 = 0$
$x = 0$ or $x = 9$
The solution set is $\{0, 9\}$.

3. $x^2 = -3x$
$x^2 + 3x = 0$
$x(x+3) = 0$
$x = 0$ or $x + 3 = 0$
$x = 0$ or $x = -3$
The solution set is $\{-3, 0\}$.

5. $3y^2 + 12y = 0$
$3y(y+4) = 0$
$3y = 0$ or $y + 4 = 0$
$y = 0$ or $y = -4$
The solution set is $\{-4, 0\}$.

7. $5n^2 - 9n = 0$
$n(5n-9) = 0$
$n = 0$ or $5n - 9 = 0$
$n = 0$ or $5n = 9$
$n = 0$ or $n = \frac{9}{5}$
The solution set is $\left\{0, \frac{9}{5}\right\}$.

9. $x^2 + x - 30 = 0$
$(x+6)(x-5) = 0$
$x + 6 = 0$ or $x - 5 = 0$
$x = -6$ or $x = 5$
The solution set is $\{-6, 5\}$.

11. $x^2 - 19x + 84 = 0$
$(x-12)(x-7) = 0$
$x - 12 = 0$ or $x - 7 = 0$
$x = 12$ or $x = 7$
The solution set is $\{7, 12\}$.

13. $2x^2 + 19x + 24 = 0$
$(2x+3)(x+8) = 0$
$2x + 3 = 0$ or $x + 8 = 0$
$2x = -3$ or $x = -8$
$x = -\frac{3}{2}$ or $x = -8$
The solution set is $\left\{-8, -\frac{3}{2}\right\}$.

15. $15x^2 + 29x - 14 = 0$
$(3x+7)(5x-2) = 0$
$3x + 7 = 0$ or $5x - 2 = 0$
$3x = -7$ or $5x = 2$
$x = -\frac{7}{3}$ or $x = \frac{2}{5}$
The solution set is $\left\{-\frac{7}{3}, \frac{2}{5}\right\}$.

17. $25x^2 - 30x + 9 = 0$
$(5x-3)(5x-3) = 0$
$(5x-3)^2 = 0$
$5x - 3 = 0$
$5x = 3$
$x = \frac{3}{5}$
The solution set is $\left\{\frac{3}{5}\right\}$.

19. $6x^2 - 5x - 21 = 0$
$(2x + 3)(3x - 7) = 0$
$2x + 3 = 0$ or $3x - 7 = 0$
$2x = -3$ or $3x = 7$
$x = -\frac{3}{2}$ or $x = \frac{7}{3}$
The solution set is $\left\{-\frac{3}{2}, \frac{7}{3}\right\}$.

21. $3\sqrt{x} = x + 2$
$(3\sqrt{x})^2 = (x + 2)^2$
$9x = x^2 + 4x + 4$
$0 = x^2 - 5x + 4$
$0 = (x - 4)(x - 1)$
$x - 4 = 0$ or $x - 1 = 0$
$x = 4$ or $x = 1$
Checking $x = 4$:
$3\sqrt{4} \stackrel{?}{=} 4 + 2$
$3(2) \stackrel{?}{=} 6$
$6 = 6$
Checking $x = 1$:
$3\sqrt{1} \stackrel{?}{=} 1 + 2$
$3(1) \stackrel{?}{=} 3$
$3 = 3$
The solution set is $\{1, 4\}$.

23. $\sqrt{2x} = x - 4$
$(\sqrt{2x})^2 = (x - 4)^2$
$2x = x^2 - 8x + 16$
$0 = x^2 - 10x + 16$
$0 = (x - 8)(x - 2)$
$x - 8 = 0$ or $x - 2 = 0$
$x = 8$ or $x = 2$
Checking $x = 8$:
$\sqrt{2(8)} \stackrel{?}{=} 8 - 4$
$\sqrt{16} \stackrel{?}{=} 4$
$4 = 4$
Checking $x = 2$:
$\sqrt{2(2)} \stackrel{?}{=} 2 - 4$
$\sqrt{4} \stackrel{?}{=} -2$
$2 \neq -2$
The solution set is $\{8\}$.

25. $\sqrt{3x} + 6 = x$
$\sqrt{3x} = x - 6$
$(\sqrt{3x})^2 = (x - 6)^2$
$3x = x^2 - 12x + 36$
$0 = x^2 - 15x + 36$
$0 = (x - 12)(x - 3)$
$x - 12 = 0$ or $x - 3 = 0$
$x = 12$ or $x = 3$
Checking $x = 12$:
$\sqrt{3(12)} + 6 \stackrel{?}{=} 12$
$\sqrt{36} + 6 \stackrel{?}{=} 12$
$12 = 12$
Checking $x = 3$:
$\sqrt{3(3)} + 6 \stackrel{?}{=} 3$
$\sqrt{9} + 6 \stackrel{?}{=} 3$
$3 + 6 \stackrel{?}{=} 3$
$9 \neq 3$
The solution set is $\{12\}$.

27. $x^2 - 5kx = 0$
$x(x - 5k) = 0$
$x = 0$ or $x - 5k = 0$
$x = 0$ or $x = 5k$
The solution set is $\{0, 5k\}$.

29. $x^2 = 16k^2x$
$x^2 - 16k^2x = 0$
$x(x - 16k^2) = 0$
$x = 0$ or $x - 16k^2 = 0$
$x = 0$ or $x = 16k^2$
The solution set is $\{0, 16k^2\}$.

31. $x^2 - 12kx + 35k^2 = 0$
$(x - 5k)(x - 7k) = 0$
$x - 5k = 0$ or $x - 7k = 0$
$x = 5k$ or $x = 7k$
The solution set is $\{5k, 7k\}$.

33. $2x^2 + 5kx - 3k^2 = 0$
$(2x - k)(x + 3k) = 0$
$2x - k = 0$ or $x + 3k = 0$
$2x = k$ or $x = -3k$
$x = \frac{k}{2}$ or $x = -3k$
The solution set is $\left\{-3k, \frac{k}{2}\right\}$.

35. $x^2 = 1$
$x = \pm\sqrt{1} = \pm 1$
The solution set is $\{-1, 1\}$.

37. $x^2 = -36$
$x = \pm\sqrt{-36}$
$x = \pm i\sqrt{36}$
$x = \pm 6i$
The solution set is $\{-6i, 6i\}$.

39. $x^2 = 14$
$x = \pm\sqrt{14}$
The solution set is $\{-\sqrt{14}, \sqrt{14}\}$.

41. $n^2 - 28 = 0$
$n^2 = 28$
$n = \pm\sqrt{28}$
$n = \pm 2\sqrt{7}$
The solution set is $\{-2\sqrt{7}, 2\sqrt{7}\}$.

43. $3t^2 = 54$
$t^2 = 18$
$t = \pm\sqrt{18}$
$t = \pm 3\sqrt{2}$
The solution set is $\{-3\sqrt{2}, 3\sqrt{2}\}$.

45. $2t^2 = 7$

$$t^2 = \frac{7}{2}$$

$$t = \pm\sqrt{\frac{7}{2}}$$

$$t = \pm\frac{\sqrt{7}}{\sqrt{2}} \bullet \frac{\sqrt{2}}{\sqrt{2}}$$

$$t = \pm\frac{\sqrt{14}}{2}$$

The solution set is $\left\{-\frac{\sqrt{14}}{2}, \frac{\sqrt{14}}{2}\right\}$.

47. $15y^2 = 20$

$$y^2 = \frac{20}{15}$$

$$y^2 = \frac{4}{3}$$

$$y = \pm\sqrt{\frac{4}{3}}$$

$$y = \pm\frac{2}{\sqrt{3}}$$

$$y = \pm\frac{2}{\sqrt{3}} \bullet \frac{\sqrt{3}}{\sqrt{3}}$$

$$y = \pm\frac{2\sqrt{3}}{3}$$

The solution set is $\left\{-\frac{2\sqrt{3}}{3}, \frac{2\sqrt{3}}{3}\right\}$.

49. $10x^2 + 48 = 0$

$$10x^2 = -48$$

$$x^2 = -\frac{48}{10}$$

$$x^2 = -\frac{24}{5}$$

$$x = \pm\sqrt{-\frac{24}{5}}$$

$$x = \pm\frac{i\sqrt{24}}{\sqrt{5}}$$

$$x = \pm\frac{i\sqrt{4}\sqrt{6}}{\sqrt{5}}$$

$$x = \pm\frac{2i\sqrt{6}}{\sqrt{5}} \bullet \frac{\sqrt{5}}{\sqrt{5}}$$

$$x = \pm\frac{2i\sqrt{30}}{5}$$

The solution set is $\left\{-\frac{2i\sqrt{30}}{5}, \frac{2i\sqrt{30}}{5}\right\}$.

51. $24x^2 = 36$

$$x^2 = \frac{36}{24}$$

$$x^2 = \frac{3}{2}$$

$$x = \pm\sqrt{\frac{3}{2}}$$

$$x = \pm\frac{\sqrt{3}}{\sqrt{2}} \bullet \frac{\sqrt{2}}{\sqrt{2}}$$

$$x = \pm\frac{\sqrt{6}}{2}$$

The solution set is $\left\{-\frac{\sqrt{6}}{2}, \frac{\sqrt{6}}{2}\right\}$.

53. $(x-2)^2 = 9$
$x-2 = \pm\sqrt{9}$
$x-2 = \pm 3$
$x = 2 \pm 3$
$x = 2+3$ or $x = 2-3$
$x = 5$ or $x = -1$
The solution set is $\{-1, 5\}$.

55. $(x+3)^2 = 25$
$x+3 = \pm\sqrt{25}$
$x+3 = \pm 5$
$x = -3 \pm 5$
$x = -3+5$ or $x = -3-5$
$x = 2$ or $x = -8$
The solution set is $\{-8, 2\}$.

57. $(x+6)^2 = -4$
$x+6 = \pm\sqrt{-4}$
$x+6 = \pm i\sqrt{4}$
$x+6 = \pm 2i$
$x = -6 \pm 2i$
The solution set is $\{-6-2i, -6+2i\}$.

59. $(2x-3)^2 = 1$
$2x-3 = \pm\sqrt{1}$
$2x-3 = \pm 1$
$2x = 3 \pm 1$
$x = \dfrac{3 \pm 1}{2}$
$x = \dfrac{3+1}{2}$ or $x = \dfrac{3-1}{2}$
$x = 2$ or $x = 1$
The solution set is $\{1, 2\}$.

61. $(n-4)^2 = 5$
$n-4 = \pm\sqrt{5}$
$n = 4 \pm \sqrt{5}$
The solution set is $\{4-\sqrt{5}, 4+\sqrt{5}\}$.

63. $(t+5)^2 = 12$
$t+5 = \pm\sqrt{12}$
$t+5 = \pm\sqrt{4}\sqrt{3}$
$t+5 = \pm 2\sqrt{3}$
$t = -5 \pm 2\sqrt{3}$
The solution set is
$\{-5-2\sqrt{3}, -5+2\sqrt{3}\}$.

65. $(3y-2)^2 = -27$
$3y-2 = \pm\sqrt{-27}$
$3y-2 = \pm i\sqrt{27}$
$3y-2 = \pm i\sqrt{9}\sqrt{3}$
$3y-2 = \pm 3i\sqrt{3}$
$3y = 2 \pm 3i\sqrt{3}$
$y = \dfrac{2 \pm 3i\sqrt{3}}{3}$
The solution set is $\left\{\dfrac{2 \pm 3i\sqrt{3}}{3}\right\}$.

67. $3(x+7)^2 + 4 = 79$
$3(x+7)^2 = 75$
$(x+7)^2 = 25$
$x+7 = \pm\sqrt{25}$
$x+7 = \pm 5$
$x = -7 \pm 5$
$x = -7-5$ or $x = -7+5$
$x = -12$ or $x = -2$
The solution set is $\{-12, -2\}$.

69. $2(5x-2)^2 + 5 = 25$
$2(5x-2)^2 = 20$
$(5x-2)^2 = 10$
$5x-2 = \pm\sqrt{10}$
$5x = 2 \pm \sqrt{10}$
$x = \dfrac{2 \pm \sqrt{10}}{5}$
The solution set is $\left\{\dfrac{2 \pm \sqrt{10}}{5}\right\}$.

71. $4^2 + 6^2 = c^2$
$16 + 36 = c^2$
$52 = c^2$
$\sqrt{52} = c$
$\sqrt{4}\sqrt{13} = c$
$2\sqrt{13} = c$
$c = 2\sqrt{13}$ centimeters

73. $a^2 + 8^2 = 12^2$
$a^2 + 64 = 144$
$a^2 = 80$
$a = \sqrt{80}$
$a = \sqrt{16}\sqrt{5}$
$a = 4\sqrt{5}$ inches

75. $15^2 + b^2 = 17^2$
$225 + b^2 = 289$
$b^2 = 64$
$b = \sqrt{64} = 8$
$b = 8$ yards

77. $6^2 + 6^2 = c^2$
$36 + 36 = c^2$
$72 = c^2$
$c = \sqrt{72} = \sqrt{36}\sqrt{2} = 6\sqrt{2}$
$c = 6\sqrt{2}$ inches

79. $a^2 + a^2 = 8^2$
$2a^2 = 64$
$a^2 = 32$
$a = \sqrt{32}$
$a = \sqrt{16}\sqrt{2}$
$a = b = 4\sqrt{2}$ meters

81. $a = 3$ then $c = 2a = 2(3) = 6$
$3^2 + b^2 = 6^2$
$9 + b^2 = 36$
$b^2 = 27$
$b = \sqrt{27}$
$b = \sqrt{9}\sqrt{3} = 3\sqrt{3}$ inches
$c = 6$ inches and $b = 3\sqrt{3}$ inches.

83. $c = 14$ then $a = \frac{1}{2}c = \frac{1}{2}(14) = 7$
$7^2 + b^2 = 14^2$
$49 + b^2 = 196$
$b^2 = 147$
$b = \sqrt{147} = \sqrt{49}\sqrt{3} = 7\sqrt{3}$
$a = 7$ centimeters and $b = 7\sqrt{3}$ centimeters

85. $b = 10 \quad a = \frac{1}{2}c$

$$\left(\frac{1}{2}c\right)^2 + 10^2 = c^2$$

$$\frac{1}{4}c^2 + 100 = c^2$$

$$4\left(\frac{1}{4}c^2 + 100\right) = 4(c^2)$$

$$c^2 + 400 = 4c^2$$

$$400 = 3c^2$$

$$\frac{400}{3} = c^2$$

$$\sqrt{\frac{400}{3}} = c$$

$$c = \frac{\sqrt{400}}{\sqrt{3}} = \frac{20}{\sqrt{3}} \bullet \frac{\sqrt{3}}{\sqrt{3}} = \frac{20\sqrt{3}}{3}$$

$$a = \frac{1}{2}\left(\frac{20\sqrt{3}}{3}\right) = \frac{10\sqrt{3}}{3}$$

$c = \frac{20\sqrt{3}}{3}$ feet and $a = \frac{10\sqrt{3}}{3}$ feet.

87. $a^2 + 16^2 = 24^2$
$a^2 + 256 = 576$
$a^2 = 320$
$a = \sqrt{320} \approx 17.9$ feet
The ladder is 17.9 feet from the foundation of the house.

89. $16^2 + 34^2 = c^2$
$256 + 1156 = c^2$
$1412 = c^2$
$c = \sqrt{1412} \approx 38$ meters
The diagonal is 38 meters.

91. Let $x =$ length of a side.
$x^2 + x^2 = 75^2$
$2x^2 = 5625$
$x^2 = 2812.5$
$x = \sqrt{2812.5} \approx 53$ meters
The length of a side is 53 meters.

95. The diagonal of the base:
$8^2 + 6^2 = c^2$
$64 + 36 = c^2$
$100 = c^2$
$\sqrt{100} = c$
$10 = c$
The right triangle to find the diagonal has sides $a = 10$ and $b = 4$.

$c^2 = 10^2 + 4^2$
$c^2 = 100 + 16$
$c^2 = 116$
$c = \sqrt{116}$
$c \approx 10.8$ centimeters

97. $s^2 + s^2 = h^2$
$2s^2 = h^2$
$\sqrt{2s^2} = h$
$\sqrt{s^2}\sqrt{2} = h$
$s\sqrt{2} = h$

PROBLEM SET 6.3 Completing the Square

1a. $x^2 - 4x - 60 = 0$
$(x - 10)(x + 6) = 0$
$x - 10 = 0$ or $x - 6 = 0$
$x = 10$ or $x = -6$

b. $x^2 - 4x - 60 = 0$
$x^2 - 4x = 60$
$x^2 - 4x + 4 = 60 + 4$
$(x - 2)^2 = 64$
$x - 2 = \pm\sqrt{64}$
$x - 2 = \pm 8$
$x = 2 \pm 8$
$x = 2 + 8$ or $x = 2 - 8$
$x = 10$ or $x = -6$
The solution set is $\{-6, 10\}$.

3a. $x^2 - 14x = -40$
$x^2 - 14x + 40 = 0$
$(x - 10)(x - 4) = 0$
$x - 10 = 0$ or $x - 4 = 0$
$x = 10$ or $x = 4$

b. $x^2 - 14x = -40$
$x^2 - 14x + 49 = -40 + 49$
$(x - 7)^2 = 9$
$x - 7 = \pm\sqrt{9}$
$x - 7 = \pm 3$
$x = 7 \pm 3$
$x = 7 + 3$ or $x = 7 - 3$
$x = 10$ or $x = 4$
The solution set is $\{4, 10\}$.

5a. $x^2 - 5x - 50 = 0$
$(x - 10)(x + 5) = 0$
$x - 10 = 0$ or $x + 5 = 0$
$x = 10$ or $x = -5$

b. $x^2 - 5x - 50 = 0$
$x^2 - 5x = 50$
$x^2 - 5x + \frac{25}{4} = 50 + \frac{25}{4}$
$\left(x - \frac{5}{2}\right)^2 = \frac{225}{4}$
$x - \frac{5}{2} = \pm\sqrt{\frac{225}{4}}$
$x - \frac{5}{2} = \pm\frac{15}{2}$
$x = \frac{5}{2} \pm \frac{15}{2}$
$x = \frac{5}{2} + \frac{15}{2}$ or $x = \frac{5}{2} - \frac{15}{2}$
$x = \frac{20}{2}$ or $x = \frac{-10}{2}$
$x = 10$ or $x = -5$
The solution set is $\{-5, 10\}$.

7a. $x(x + 7) = 8$
$x^2 + 7x = 8$
$x^2 + 7x - 8 = 0$
$(x + 8)(x - 1) = 0$
$x + 8 = 0$ or $x - 1 = 0$
$x = -8$ or $x = 1$

b. $x(x + 7) = 8$
$x^2 + 7x = 8$
$x^2 + 7x + \frac{49}{4} = 8 + \frac{49}{4}$
$\left(x + \frac{7}{2}\right)^2 = \frac{81}{4}$
$x + \frac{7}{2} = \pm\sqrt{\frac{81}{4}}$
$x + \frac{7}{2} = \pm\frac{9}{2}$
$x = -\frac{7}{2} + \frac{9}{2}$ or $x = -\frac{7}{2} - \frac{9}{2}$

$x = \frac{2}{2} = 1$ or $x = -\frac{16}{2} = -8$

The solution set is $\{-8, 1\}$.

9a. $2n^2 - n - 15 = 0$

$(2n + 5)(n - 3) = 0$

$2n + 5 = 0$ or $n - 3 = 0$

$2n = -5$ or $n = 3$

$n = -\frac{5}{2}$ or $n = 3$

b. $2n^2 - n - 15 = 0$

$n^2 - \frac{1}{2}n - \frac{15}{2} = 0$

$n^2 - \frac{1}{2}n = \frac{15}{2}$

$n^2 - \frac{1}{2}n + \frac{1}{16} = \frac{15}{2} + \frac{1}{16}$

$\left(n - \frac{1}{4}\right)^2 = \frac{121}{16}$

$n - \frac{1}{4} = \pm\sqrt{\frac{121}{16}}$

$n - \frac{1}{4} = \pm\frac{11}{4}$

$n = \frac{1}{4} \pm \frac{11}{4}$

$n = \frac{1}{4} + \frac{11}{4}$ or $n = \frac{1}{4} - \frac{11}{4}$

$n = \frac{12}{4}$ or $n = -\frac{10}{4}$

$n = 3$ or $n = -\frac{5}{2}$

The solution set is $\left\{-\frac{5}{2}, 3\right\}$.

11a. $3n^2 + 7n - 6 = 0$

$(3n - 2)(n + 3) = 0$

$3n - 2 = 0$ or $n + 3 = 0$

$3n = 2$ or $n = -3$

$n = \frac{2}{3}$ or $n = -3$

b. $3n^2 + 7n - 6 = 0$

$n^2 + \frac{7}{3}n - 2 = 0$

$n^2 + \frac{7}{3}n = 2$

$n^2 + \frac{7}{3}n + \frac{49}{36} = 2 + \frac{49}{36}$

$\left(n + \frac{7}{6}\right)^2 = \frac{121}{36}$

$n + \frac{7}{6} = \pm\sqrt{\frac{121}{36}}$

$n + \frac{7}{6} = \pm\frac{11}{6}$

$n = -\frac{7}{6} \pm \frac{11}{6}$

$n = -\frac{7}{6} + \frac{11}{6}$ or $n = -\frac{7}{6} - \frac{11}{6}$

$n = \frac{4}{6} = \frac{2}{3}$ or $n = -\frac{18}{6} = -3$

The solution set is $\left\{-3, \frac{2}{3}\right\}$.

13a. $n(n + 6) = 160$

$n^2 + 6n = 160$

$n^2 + 6n - 160 = 0$

$(n + 16)(n - 10) = 0$

$n + 16 = 0$ or $n - 10 = 0$

$n = -16$ or $n = 10$

b. $n(n + 6) = 160$

$n^2 + 6n = 160$

$n^2 + 6n + 9 = 160 + 9$

$(n + 3)^2 = 169$

$n + 3 = \pm\sqrt{169}$

$n + 3 = \pm 13$

$n = -3 + 13$ or $n = -3 - 13$

$n = 10$ or $n = -16$

The solution set is $\{-16, 10\}$.

15. $x^2 + 4x - 2 = 0$

$x^2 + 4x = 2$

$x^2 + 4x + 4 = 2 + 4$

$(x + 2)^2 = 6$

$x + 2 = \pm\sqrt{6}$

$x = -2 \pm \sqrt{6}$

The solution set is $\{-2 \pm \sqrt{6}\}$.

17. $x^2 + 6x - 3 = 0$
$x^2 + 6x = 3$
$x^2 + 6x + 9 = 3 + 9$
$(x + 3)^2 = 12$
$x + 3 = \pm\sqrt{12}$
$x + 3 = \pm 2\sqrt{3}$
$x = -3 \pm 2\sqrt{3}$
The solution set is $\{-3 \pm 2\sqrt{3}\}$.

19. $y^2 - 10y = 1$
$y^2 - 10y + 25 = 1 + 25$
$(y - 5)^2 = 26$
$y - 5 = \pm\sqrt{26}$
$y = 5 \pm \sqrt{26}$
The solution set is $\{5 \pm \sqrt{26}\}$.

21. $n^2 - 8n + 17 = 0$
$n^2 - 8n = -17$
$n^2 - 8n + 16 = -17 + 16$
$(n - 4)^2 = -1$
$n - 4 = \pm\sqrt{-1}$
$n - 4 = \pm i$
$n = 4 \pm i$
The solution set is $\{4 \pm i\}$.

23. $n(n + 12) = -9$
$n^2 + 12n = -9$
$n^2 + 12n + 36 = -9 + 36$
$(n + 6)^2 = 27$
$n + 6 = \pm\sqrt{27}$
$n + 6 = \pm 3\sqrt{3}$
$n = -6 \pm 3\sqrt{3}$
The solution set is $\{-6 \pm 3\sqrt{3}\}$.

25. $n^2 + 2n + 6 = 0$
$n^2 + 2n = -6$
$n^2 + 2n + 1 = -6 + 1$
$(n + 1)^2 = -5$
$n + 1 = \pm\sqrt{-5}$
$n + 1 = \pm i\sqrt{5}$
$n = -1 \pm i\sqrt{5}$
The solution set is $\{-1 \pm i\sqrt{5}\}$.

27. $x^2 + 3x - 2 = 0$
$x^2 + 3x = 2$
$x^2 + 3x + \frac{9}{4} = 2 + \frac{9}{4}$
$\left(x + \frac{3}{2}\right)^2 = \frac{17}{4}$
$x + \frac{3}{2} = \pm\sqrt{\frac{17}{4}}$
$x + \frac{3}{2} = \pm\frac{\sqrt{17}}{2}$
$x = -\frac{3}{2} \pm \frac{\sqrt{17}}{2}$
$x = \frac{-3 \pm \sqrt{17}}{2}$
The solution set is $\left\{\frac{-3 \pm \sqrt{17}}{2}\right\}$.

29. $x^2 + 5x + 1 = 0$
$x^2 + 5x = -1$
$x^2 + 5x + \frac{25}{4} = -1 + \frac{25}{4}$
$\left(x + \frac{5}{2}\right)^2 = \frac{21}{4}$
$x + \frac{5}{2} = \pm\sqrt{\frac{21}{4}}$
$x + \frac{5}{2} = \pm\frac{\sqrt{21}}{2}$
$x = -\frac{5}{2} \pm \frac{\sqrt{21}}{2}$
$x = \frac{-5 \pm \sqrt{21}}{2}$
The solution set is $\left\{\frac{-5 \pm \sqrt{21}}{2}\right\}$.

31. $y^2 - 7y + 3 = 0$
$y^2 - 7y = -3$
$y^2 - 7y + \frac{49}{4} = -3 + \frac{49}{4}$
$\left(y + \frac{7}{2}\right)^2 = \frac{37}{4}$
$y + \frac{7}{2} = \pm\sqrt{\frac{37}{4}}$
$y + \frac{7}{2} = \pm\frac{\sqrt{37}}{2}$
$y = -\frac{7}{2} \pm \frac{\sqrt{37}}{2}$

$$y = \frac{-7 \pm \sqrt{37}}{2}$$

The solution set is $\left\{\frac{-7 \pm \sqrt{37}}{2}\right\}$.

33. $2x^2 + 4x - 3 = 0$

$$x^2 + 2x - \frac{3}{2} = 0$$

$$x^2 + 2x = \frac{3}{2}$$

$$x^2 + 2x + 1 = \frac{3}{2} + 1$$

$$(x + 1)^2 = \frac{5}{2}$$

$$x + 1 = \pm\sqrt{\frac{5}{2}}$$

$$x + 1 = \pm\frac{\sqrt{5}}{\sqrt{2}} \bullet \frac{\sqrt{2}}{\sqrt{2}}$$

$$x + 1 = \pm\frac{\sqrt{10}}{2}$$

$$x = -1 \pm \frac{\sqrt{10}}{2}$$

$$x = -\frac{2}{2} \pm \frac{\sqrt{10}}{2}$$

$$x = \frac{-2 \pm \sqrt{10}}{2}$$

The solution set is $\left\{\frac{-2 \pm \sqrt{10}}{2}\right\}$.

35. $3n^2 - 6n + 5 = 0$

$$n^2 - 2n + \frac{5}{3} = 0$$

$$n^2 - 2n = -\frac{5}{3}$$

$$n^2 - 2n + 1 = -\frac{5}{3} + 1$$

$$(n - 1)^2 = -\frac{2}{3}$$

$$n - 1 = \pm\sqrt{-\frac{2}{3}}$$

$$n - 1 = \pm\frac{i\sqrt{2}}{\sqrt{3}}$$

$$n - 1 = \pm\frac{i\sqrt{2}}{\sqrt{3}} \bullet \frac{\sqrt{3}}{\sqrt{3}}$$

$$n - 1 = \pm\frac{i\sqrt{6}}{3}$$

$$n = 1 \pm \frac{i\sqrt{6}}{3}$$

$$n = \frac{3}{3} \pm \frac{i\sqrt{6}}{3}$$

$$n = \frac{3 \pm i\sqrt{6}}{3}$$

The solution set is $\left\{\frac{3 \pm i\sqrt{6}}{3}\right\}$.

37. $3x^2 + 5x - 1 = 0$

$$x^2 + \frac{5}{3}x - \frac{1}{3} = 0$$

$$x^2 + \frac{5}{3}x = \frac{1}{3}$$

$$x^2 + \frac{5}{3}x + \frac{25}{36} = \frac{1}{3} + \frac{25}{36}$$

$$\left(x + \frac{5}{6}\right)^2 = \frac{37}{36}$$

$$x + \frac{5}{6} = \pm\sqrt{\frac{37}{36}}$$

$$x + \frac{5}{6} = \pm\frac{\sqrt{37}}{6}$$

$$x = -\frac{5}{6} \pm \frac{\sqrt{37}}{6}$$

$$x = \frac{-5 \pm \sqrt{37}}{6}$$

The solution set is $\left\{\frac{-5 \pm \sqrt{37}}{6}\right\}$.

39. $x^2 + 8x - 48 = 0$

$(x + 12)(x - 4) = 0$

$x + 12 = 0$ or $x - 4 = 0$

$x = -12$ or $x = 4$

The solution set is $\{-12, 4\}$.

41. $2n^2 - 8n = -3$

$n^2 - 4n = -\frac{3}{2}$

$n^2 - 4n + 4 = -\frac{3}{2} + 4$

$(n-2)^2 = \frac{5}{2}$

$n - 2 = \pm\sqrt{\frac{5}{2}}$

$n - 2 = \pm\frac{\sqrt{5}}{\sqrt{2}} \bullet \frac{\sqrt{2}}{\sqrt{2}}$

$n - 2 = \pm\frac{\sqrt{10}}{2}$

$n = 2 \pm \frac{\sqrt{10}}{2}$

$n = \frac{4}{2} \pm \frac{\sqrt{10}}{2}$

$n = \frac{4 \pm \sqrt{10}}{2}$

The solution set is $\left\{\frac{4 \pm \sqrt{10}}{2}\right\}$.

43. $(3x - 1)(2x + 9) = 0$

$3x - 1 = 0$ or $2x + 9 = 0$

$3x = 1$ or $2x = -9$

$x = \frac{1}{3}$ or $x = -\frac{9}{2}$

The solution set is $\left\{-\frac{9}{2}, \frac{1}{3}\right\}$.

45. $(x + 2)(x - 7) = 10$

$x^2 - 5x - 14 = 10$

$x^2 - 5x - 24 = 0$

$(x - 8)(x + 3) = 0$

$x - 8 = 0$ or $x + 3 = 0$

$x = 8$ or $x = -3$

The solution set is $\{-3, 8\}$.

47. $(x - 3)^2 = 12$

$x - 3 = \pm\sqrt{12}$

$x - 3 = \pm 2\sqrt{3}$

$x = 3 \pm 2\sqrt{3}$

The solution set is $\{3 \pm 2\sqrt{3}\}$.

49. $3n^2 - 6n + 4 = 0$

$n^2 - 2n + \frac{4}{3} = 0$

$n^2 - 2n = -\frac{4}{3}$

$n^2 - 2n + 1 = -\frac{4}{3} + 1$

$(n - 1)^2 = -\frac{1}{3}$

$n - 1 = \pm\sqrt{-\frac{1}{3}}$

$n - 1 = \pm\frac{i}{\sqrt{3}}$

$n - 1 = \pm\frac{i}{\sqrt{3}} \bullet \frac{\sqrt{3}}{\sqrt{3}}$

$n - 1 = \pm\frac{i\sqrt{3}}{3}$

$n = 1 \pm \frac{i\sqrt{3}}{3}$

$n = \frac{3}{3} \pm \frac{i\sqrt{3}}{3}$

$n = \frac{3 \pm i\sqrt{3}}{3}$

The solution set is $\left\{\frac{3 \pm i\sqrt{3}}{3}\right\}$.

51. $n(n + 8) = 240$

$n^2 + 8n = 240$

$n^2 + 8n - 240 = 0$

$(n + 20)(n - 12) = 0$

$n + 20 = 0$ or $n - 12 = 0$

$n = -20$ or $n = 12$

The solution set is $\{-20, 12\}$.

53. $3x^2 + 5x = -2$

$3x^2 + 5x + 2 = 0$

$(3x + 2)(x + 1) = 0$

$3x + 2 = 0$ or $x + 1 = 0$

$3x = -2$ or $x = -1$

$x = -\frac{2}{3}$ or $x = -1$

The solution set is $\left\{-1, -\frac{2}{3}\right\}$.

55. $4x^2 - 8x + 3 = 0$
$(2x - 1)(2x - 3) = 0$
$2x - 1 = 0$ or $2x - 3 = 0$
$2x = 1$ or $2x = 3$
$x = \frac{1}{2}$ or $x = \frac{3}{2}$
The solution set is $\left\{\frac{1}{2}, \frac{3}{2}\right\}$.

57. $x^2 + 12x = 4$
$x^2 + 12x + 36 = 4 + 36$
$(x + 6)^2 = 40$
$x + 6 = \pm\sqrt{40}$
$x + 6 = \pm 2\sqrt{10}$
$x = -6 \pm 2\sqrt{10}$
The solution set is $\{-6 \pm 2\sqrt{10}\}$.

59. $4(2x + 1)^2 - 1 = 11$
$4(2x + 1)^2 = 12$
$(2x + 1)^2 = 3$
$2x + 1 = \pm\sqrt{3}$
$2x = -1 \pm \sqrt{3}$
$x = \frac{-1 \pm \sqrt{3}}{2}$
The solution set is $\left\{\frac{-1 \pm \sqrt{3}}{2}\right\}$.

61. $ax^2 + bx + c = 0$

$$x^2 + \frac{b}{a}x + \frac{c}{a} = 0$$

$$x^2 + \frac{b}{a}x = -\frac{c}{a}$$

$$x^2 + \frac{b}{a}x + \frac{b^2}{4a^2} = -\frac{c}{a} + \frac{b^2}{4a^2}$$

$$\left(x + \frac{b}{2a}\right)^2 = \frac{b^2}{4a^2} - \frac{c}{a}$$

$$\left(x + \frac{b}{2a}\right)^2 = \frac{b^2}{4a^2} - \frac{c}{a} \bullet \frac{4a}{4a}$$

$$\left(x + \frac{b}{2a}\right)^2 = \frac{b^2}{4a^2} - \frac{4ac}{4a^2}$$

$$\left(x + \frac{b}{2a}\right)^2 = \frac{b^2 - 4ac}{4a^2}$$

$$x + \frac{b}{2a} = \pm\sqrt{\frac{b^2 - 4ac}{4a^2}}$$

$$x + \frac{b}{2a} = \pm\frac{\sqrt{b^2 - 4ac}}{2a}$$

$$x = -\frac{b}{2a} \pm \frac{\sqrt{b^2 - 4ac}}{2a}$$

$$x = \frac{-b \pm \sqrt{b^2 - 4ac}}{2a}$$

The solution set is
$\left\{\frac{-b \pm \sqrt{b^2 - 4ac}}{2a}\right\}$.

65. $\frac{x^2}{a^2} + \frac{y^2}{b^2} = 1$ for x

$$\frac{x^2}{a^2} = 1 - \frac{y^2}{b^2}$$

$$\frac{x^2}{a^2} = \frac{b^2 - y^2}{b^2}$$

$$x^2 = \frac{a^2(b^2 - y^2)}{b^2}$$

$$x = \sqrt{\frac{a^2(b^2 - y^2)}{b^2}}$$

$$x = \frac{a\sqrt{b^2 - y^2}}{b}$$

67. $A = \pi r^2$ for r

$$\frac{A}{\pi} = r^2$$

$$\sqrt{\frac{A}{\pi}} = r$$

$r = \frac{\sqrt{A}}{\sqrt{\pi}} \bullet \frac{\sqrt{\pi}}{\sqrt{\pi}}$

$r = \frac{\sqrt{A\pi}}{\pi}$

69. $x^2 - 5ax + 6a^2 = 0$
$(x - 2a)(x - 3a) = 0$
$x - 2a = 0$ or $x - 3a = 0$
$x = 2a$ or $x = 3a$
The solution set is $\{2a, 3a\}$.

71. $6x^2 + ax - 2a^2 = 0$
$(3x + 2a)(2x - a) = 0$
$3x + 2a = 0$ or $2x - a = 0$
$3x = -2a$ or $2x = a$
$x = -\frac{2a}{3}$ or $x = \frac{a}{2}$
The solution set is $\left\{-\frac{2a}{3}, \frac{a}{2}\right\}$.

73. $9x^2 - 12bx + 4b^2 = 0$
$(3x - 2b)(3x - 2b) = 0$
$(3x - 2b)^2 = 0$
$3x - 2b = 0$
$3x = 2b$
$x = \frac{2b}{3}$
The solution set is $\left\{\frac{2b}{3}\right\}$.

PROBLEM SET 6.4 Quadratic Formula

1. $x^2 + 4x - 21 = 0$
$b^2 - 4ac$
$4^2 - 4(1)(-21) = 16 + 84 = 100$
Since $b^2 - 4ac > 0$, there will be two real solutions.
$(x + 7)(x - 3) = 0$
$x + 7 = 0$ or $x - 3 = 0$
$x = -7$ or $x = 3$
The solution set is $\{-7, 3\}$.

3. $9x^2 - 6x + 1 = 0$
$b^2 - 4ac$
$(-6)^2 - 4(9)(1) = 36 - 36 = 0$
Since $b^2 - 4ac = 0$, there will be one real solution.
$(3x - 1)(3x - 1) = 0$
$(3x - 1)^2 = 0$
$3x - 1 = 0$
$3x = 1$
$x = \frac{1}{3}$
The solution set is $\left\{\frac{1}{3}\right\}$.

5. $x^2 - 7x + 13 = 0$
$b^2 - 4ac$
$(-7)^2 - 4(1)(13) = 49 - 52 = -3$
Since $b^2 - 4ac < 0$, there will be two complex solutions.

$x = \frac{-(-7) \pm \sqrt{(-7)^2 - 4(1)(13)}}{2(1)}$

$x = \frac{7 \pm \sqrt{49 - 52}}{2}$

$x = \frac{7 \pm \sqrt{-3}}{2}$

$x = \frac{7 \pm i\sqrt{3}}{2}$

The solution set is $\left\{\frac{7 \pm i\sqrt{3}}{2}\right\}$.

7. $15x^2 + 17x - 4 = 0$
$b^2 - 4ac$
$(17)^2 - 4(15)(-4) = 289 + 240 = 529$
Since $b^2 - 4ac > 0$, there will be two real solutions.
$(3x + 4)(5x - 1) = 0$
$3x + 4 = 0$ or $5x - 1 = 0$
$3x = -4$ or $5x = 1$
$x = -\frac{4}{3}$ or $x = \frac{1}{5}$
The solution set is $\left\{-\frac{4}{3}, \frac{1}{5}\right\}$.

9. $3x^2 + 4x = 2$
$3x^2 + 4x - 2 = 0$
$b^2 - 4ac$
$(4)^2 - 4(3)(-2) = 16 + 24 = 40$

Since $b^2 - 4ac > 0$, there will be two real solutions.

$$x = \frac{-(4) \pm \sqrt{40}}{2(3)}$$
$$x = \frac{-4 \pm 2\sqrt{10}}{6}$$
$$x = \frac{2(-2 \pm \sqrt{10})}{6}$$
$$x = \frac{-2 \pm \sqrt{10}}{3}$$
The solution set is $\left\{\frac{-2 \pm \sqrt{10}}{3}\right\}$.

11. $x^2 + 2x - 1 = 0$

$$x = \frac{-(2) \pm \sqrt{(2)^2 - 4(1)(-1)}}{2(1)}$$
$$x = \frac{-2 \pm \sqrt{4+4}}{2}$$
$$x = \frac{-2 \pm \sqrt{8}}{2}$$
$$x = \frac{-2 \pm 2\sqrt{2}}{2}$$
$$x = \frac{2(-1 \pm \sqrt{2})}{2}$$
$$x = -1 \pm \sqrt{2}$$
The solution set is $\{-1 \pm \sqrt{2}\}$.

13. $n^2 + 5n - 3 = 0$

$$n = \frac{-(5) \pm \sqrt{(5)^2 - 4(1)(-3)}}{2(1)}$$
$$n = \frac{-5 \pm \sqrt{25+12}}{2}$$
$$n = \frac{-5 \pm \sqrt{37}}{2}$$
The solution set is $\left\{\frac{-5 \pm \sqrt{37}}{2}\right\}$.

15. $a^2 - 8a = 4$
$a^2 - 8a - 4 = 0$

$$a = \frac{-(-8) \pm \sqrt{(-8)^2 - 4(1)(-4)}}{2(1)}$$
$$a = \frac{8 \pm \sqrt{64+16}}{2}$$
$$a = \frac{8 \pm \sqrt{80}}{2}$$
$$a = \frac{8 \pm 4\sqrt{5}}{2}$$
$$a = \frac{2(4 \pm 2\sqrt{5})}{2}$$
$$a = 4 \pm 2\sqrt{5}$$
The solution set is $\{4 \pm 2\sqrt{5}\}$.

17. $n^2 + 5n + 8 = 0$

$$n = \frac{-(5) \pm \sqrt{(5)^2 - 4(1)(8)}}{2(1)}$$
$$n = \frac{-5 \pm \sqrt{25-32}}{2}$$
$$n = \frac{-5 \pm \sqrt{-7}}{2}$$
$$n = \frac{-5 \pm i\sqrt{7}}{2}$$
The solution set is $\left\{\frac{-5 \pm i\sqrt{7}}{2}\right\}$.

19. $x^2 - 18x + 80 = 0$

$$x = \frac{-(-18) \pm \sqrt{(-18)^2 - 4(1)(80)}}{2(1)}$$
$$x = \frac{18 \pm \sqrt{324-320}}{2}$$
$$x = \frac{18 \pm \sqrt{4}}{2}$$
$$x = \frac{18 \pm 2}{2}$$

$x = \dfrac{18+2}{2}$ or $x = \dfrac{18-2}{2}$

$x = 10$ or $x = 8$

The solution set is $\{8, 10\}$.

21. $-y^2 = -9y + 5$

$0 = y^2 - 9y + 5$

$$y = \frac{-(-9) \pm \sqrt{(-9)^2 - 4(1)(5)}}{2(1)}$$

$$y = \frac{9 \pm \sqrt{81 - 20}}{2}$$

$$y = \frac{9 \pm \sqrt{61}}{2}$$

The solution set is $\left\{\dfrac{9 \pm \sqrt{61}}{2}\right\}$.

23. $2x^2 + x - 4 = 0$

$$x = \frac{-(1) \pm \sqrt{(1)^2 - 4(2)(-4)}}{2(2)}$$

$$x = \frac{-1 \pm \sqrt{1 + 32}}{4}$$

$$x = \frac{-1 \pm \sqrt{33}}{4}$$

The solution set is $\left\{\dfrac{-1 \pm \sqrt{33}}{4}\right\}$.

25. $4x^2 + 2x + 1 = 0$

$$x = \frac{-2 \pm \sqrt{2^2 - 4(4)(1)}}{2(4)}$$

$$x = \frac{-2 \pm \sqrt{4 - 16}}{8}$$

$$x = \frac{-2 \pm \sqrt{-12}}{8}$$

$$x = \frac{-2 \pm i\sqrt{12}}{8}$$

$$x = \frac{-2 \pm 2i\sqrt{3}}{8}$$

$$x = \frac{2(-1 \pm i\sqrt{3})}{8}$$

$$x = \frac{-1 \pm i\sqrt{3}}{4}$$

The solution set is $\left\{\dfrac{-1 \pm i\sqrt{3}}{4}\right\}$.

27. $3a^2 - 8a + 2 = 0$

$$a = \frac{-(-8) \pm \sqrt{(-8)^2 - 4(3)(2)}}{2(3)}$$

$$a = \frac{8 \pm \sqrt{64 - 24}}{6}$$

$$a = \frac{8 \pm \sqrt{40}}{6}$$

$$a = \frac{8 \pm 2\sqrt{10}}{6}$$

$$a = \frac{2(4 \pm \sqrt{10})}{6}$$

$$a = \frac{4 \pm \sqrt{10}}{3}$$

The solution set is $\left\{\dfrac{4 \pm \sqrt{10}}{3}\right\}$.

29. $-2n^2 + 3n + 5 = 0$

$$n = \frac{-(3) \pm \sqrt{(3)^2 - 4(-2)(5)}}{2(-2)}$$

$$n = \frac{-3 \pm \sqrt{9 + 40}}{-4}$$

$$n = \frac{-3 \pm \sqrt{49}}{-4}$$

$$n = \frac{-3 \pm 7}{-4}$$

$n = \dfrac{-3 + 7}{-4}$ or $n = \dfrac{-3 - 7}{-4}$

$n = -1 \quad \text{or} \quad n = \frac{-10}{-4} = \frac{5}{2}$

The solution set is $\left\{-1, \frac{5}{2}\right\}$.

31. $3x^2 + 19x + 20 = 0$

$x = \frac{-19 \pm \sqrt{19^2 - 4(3)(20)}}{2(3)}$

$x = \frac{-19 \pm \sqrt{361 - 240}}{6}$

$x = \frac{-19 \pm \sqrt{121}}{6}$

$x = \frac{-19 \pm 11}{6}$

$x = \frac{-19 + 11}{6} \quad \text{or} \quad x = \frac{-19 - 11}{6}$

$x = \frac{-8}{6} = -\frac{4}{3} \quad \text{or} \quad x = \frac{-30}{6} = -5$

The solution set is $\left\{-5, -\frac{4}{3}\right\}$.

33. $36n^2 - 60n + 25 = 0$

$n = \frac{-(-60) \pm \sqrt{(-60)^2 - 4(36)(25)}}{2(36)}$

$n = \frac{60 \pm \sqrt{3600 - 3600}}{72}$

$n = \frac{60}{72} = \frac{5}{6}$

The solution set is $\left\{\frac{5}{6}\right\}$.

35. $4x^2 - 2x = 3$

$4x^2 - 2x - 3 = 0$

$x = \frac{-(-2) \pm \sqrt{(-2)^2 - 4(4)(-3)}}{2(4)}$

$x = \frac{2 \pm \sqrt{4 + 48}}{8}$

$x = \frac{2 \pm \sqrt{52}}{8}$

$x = \frac{2 \pm 2\sqrt{13}}{8}$

$x = \frac{2(1 \pm \sqrt{13})}{8}$

$x = \frac{1 \pm \sqrt{13}}{4}$

The solution set is $\left\{\frac{1 \pm \sqrt{13}}{4}\right\}$.

37. $5x^2 - 13x = 0$

$x = \frac{-(-13) \pm \sqrt{(-13)^2 - 4(5)(0)}}{2(5)}$

$x = \frac{13 \pm \sqrt{169}}{10}$

$x = \frac{13 \pm 13}{10}$

$x = \frac{13 + 13}{10} \quad \text{or} \quad x = \frac{13 - 13}{10}$

$x = \frac{26}{10} = \frac{13}{5} \quad \text{or} \quad x = 0$

The solution set is $\left\{0, \frac{13}{5}\right\}$.

39. $3x^2 = 5$

$3x^2 - 5 = 0$

$x = \frac{-0 \pm \sqrt{0^2 - 4(3)(-5)}}{2(3)}$

$x = \frac{\pm\sqrt{60}}{6} = \frac{\pm 2\sqrt{15}}{6} = \pm\frac{\sqrt{15}}{3}$

The solution set is $\left\{\pm\frac{\sqrt{15}}{3}\right\}$.

41. $6t^2 + t - 3 = 0$

$t = \frac{-1 \pm \sqrt{1^2 - 4(6)(-3)}}{2(6)}$

$t = \frac{-1 \pm \sqrt{1 + 72}}{12}$

$t = \frac{-1 \pm \sqrt{73}}{12}$

The solution set is $\left\{\frac{-1 \pm \sqrt{73}}{12}\right\}$.

43. $n^2 + 32n + 252 = 0$

$$n = \frac{-32 \pm \sqrt{32^2 - 4(1)(252)}}{2(1)}$$

$$n = \frac{-32 \pm \sqrt{1024 - 1008}}{2}$$

$$n = \frac{-32 \pm \sqrt{16}}{2}$$

$$n = \frac{-32 \pm 4}{2}$$

$$n = \frac{-32 + 4}{2} \quad \text{or} \quad n = \frac{-32 - 4}{2}$$

$$n = -14 \quad \text{or} \quad n = -18$$

The solution set is $\{-18, -14\}$.

45. $12x^2 - 73x + 110 = 0$

$$x = \frac{-(-73) \pm \sqrt{(-73)^2 - 4(12)(110)}}{2(12)}$$

$$x = \frac{73 \pm \sqrt{5329 - 5280}}{24}$$

$$x = \frac{73 \pm \sqrt{49}}{24}$$

$$x = \frac{73 \pm 7}{24}$$

$$x = \frac{73 + 7}{24} \quad \text{or} \quad x = \frac{73 - 7}{24}$$

$$x = \frac{80}{24} \quad \text{or} \quad x = \frac{66}{24}$$

$$x = \frac{10}{3} \quad \text{or} \quad x = \frac{11}{4}$$

The solution set is $\left\{\frac{11}{4}, \frac{10}{3}\right\}$.

47. $-2x^2 + 4x - 3 = 0$

$$x = \frac{-4 \pm \sqrt{4^2 - 4(-2)(-3)}}{2(-2)}$$

$$x = \frac{-4 \pm \sqrt{16 - 24}}{-4}$$

$$x = \frac{-4 \pm \sqrt{-8}}{-4}$$

$$x = \frac{-4 \pm i\sqrt{8}}{-4}$$

$$x = \frac{-4 \pm 2i\sqrt{2}}{-4}$$

$$x = \frac{-2(2 \pm i\sqrt{2})}{-4}$$

$$x = \frac{2 \pm i\sqrt{2}}{2}$$

The solution set is $\left\{\frac{2 \pm i\sqrt{2}}{2}\right\}$.

49. $-6x^2 + 2x + 1 = 0$

$$x = \frac{-2 \pm \sqrt{2^2 - 4(-6)(1)}}{2(-6)}$$

$$x = \frac{-2 \pm \sqrt{4 + 24}}{-12}$$

$$x = \frac{-2 \pm \sqrt{28}}{-12}$$

$$x = \frac{-2 \pm 2\sqrt{7}}{-12}$$

$$x = \frac{-2(1 \pm \sqrt{7})}{-12}$$

$$x = \frac{1 \pm \sqrt{7}}{6}$$

The solution set is $\left\{\frac{1 \pm \sqrt{7}}{6}\right\}$.

55. $x^2 - 16x - 24 = 0$

$$x = \frac{-(-16) \pm \sqrt{(-16)^2 - 4(1)(-24)}}{2(1)}$$

$$x = \frac{16 \pm \sqrt{256 + 96}}{2}$$

$$x = \frac{16 \pm \sqrt{352}}{2}$$

$$x = \frac{16 + \sqrt{352}}{2} \quad \text{or} \quad x = \frac{16 - \sqrt{352}}{2}$$

$x = 17.381$ or $x = -1.381$

The solution set is $\{-1.381, 17.381\}$.

57. $x^2 + 10x - 46 = 0$

$$x = \frac{-10 \pm \sqrt{(10)^2 - 4(1)(-46)}}{2(1)}$$

$$x = \frac{-10 \pm \sqrt{100 + 184}}{2}$$

$$x = \frac{-10 \pm \sqrt{284}}{2}$$

$$x = \frac{-10 + \sqrt{284}}{2} \quad \text{or} \quad x = \frac{-10 - \sqrt{284}}{2}$$

$x = 3.426$ or $x = -13.426$

The solution set is $\{-13.426, 3.426\}$.

59. $x^2 + 9x + 3 = 0$

$$x = \frac{-9 \pm \sqrt{(9)^2 - 4(1)(3)}}{2(1)}$$

$$x = \frac{-9 \pm \sqrt{81 - 12}}{2}$$

$$x = \frac{-9 \pm \sqrt{69}}{2}$$

$$x = \frac{-9 + \sqrt{69}}{2} \quad \text{or} \quad x = \frac{-9 - \sqrt{69}}{2}$$

$x = -0.347$ or $x = -8.653$

The solution set is $\{-8.653, -.347\}$.

61. $5x^2 - 9x + 1 = 0$

$$x = \frac{-(-9) \pm \sqrt{(-9)^2 - 4(5)(1)}}{2(5)}$$

$$x = \frac{9 \pm \sqrt{81 - 20}}{10}$$

$$x = \frac{9 \pm \sqrt{61}}{10}$$

$$x = \frac{9 + \sqrt{61}}{10} \quad \text{or} \quad x = \frac{9 - \sqrt{61}}{10}$$

$x = 1.681$ or $x = 0.119$

The solution set is $\{0.119, 1.681\}$.

63. $3x^2 - 12x - 10 = 0$

$$x = \frac{-(-12) \pm \sqrt{(-12)^2 - 4(3)(-10)}}{2(3)}$$

$$x = \frac{12 \pm \sqrt{144 + 120}}{6}$$

$$x = \frac{12 \pm \sqrt{264}}{6}$$

$$x = \frac{12 + \sqrt{264}}{6} \quad \text{or} \quad x = \frac{12 - \sqrt{264}}{6}$$

$x = 4.708$ or $x = -0.708$

The solution set is $\{-0.708, 4.708\}$.

65. $4x^2 - kx + 1 = 0$

$b^2 - 4ac = 0$

$(-k)^2 - 4(4)(1) = 0$

$k^2 - 16 = 0$

$(k + 4)(k - 4) = 0$

$k + 4 = 0$ or $k - 4 = 0$

$k = -4$ or $k = 4$

PROBLEM SET 6.5 More Quadratic Equations and Applications

1. $x^2 - 4x - 6 = 0$
$x^2 - 4x = 6$
$x^2 - 4x + 4 = 6 + 4$
$(x-2)^2 = 10$
$x - 2 = \pm\sqrt{10}$
$x = 2 \pm \sqrt{10}$
The solution set is $\{2 \pm \sqrt{10}\}$.

3. $3x^2 + 23x - 36 = 0$
$(3x-4)(x+9) = 0$
$3x - 4 = 0$ or $x + 9 = 0$
$3x = 4$ or $x = -9$
$x = \frac{4}{3}$ or $x = -9$
The solution set is $\left\{-9, \frac{4}{3}\right\}$.

5. $x^2 - 18x = 9$
$x^2 - 18x + 81 = 9 + 81$
$(x-9)^2 = 90$
$x - 9 = \pm\sqrt{90}$
$x - 9 = \pm 3\sqrt{10}$
$x = 9 \pm 3\sqrt{10}$
The solution set is $\{9 \pm 3\sqrt{10}\}$.

7. $2x^2 - 3x + 4 = 0$

$$x = \frac{-(-3) \pm \sqrt{(-3)^2 - 4(2)(4)}}{2(2)}$$

$$x = \frac{3 \pm \sqrt{9-32}}{4}$$

$$x = \frac{3 \pm \sqrt{-23}}{4}$$

$$x = \frac{3 \pm i\sqrt{23}}{4}$$

The solution set is $\left\{\frac{3 \pm i\sqrt{23}}{4}\right\}$.

9. $135 + 24n + n^2 = 0$
$n^2 + 24n + 135 = 0$
$(n+15)(n+9) = 0$
$n + 15 = 0$ or $n + 9 = 0$
$n = -15$ or $n = -9$
The solution set is $\{-15, -9\}$.

11. $(x-2)(x+9) = -10$
$x^2 + 7x - 18 = -10$
$x^2 + 7x - 8 = 0$
$(x+8)(x-1) = 0$
$x + 8 = 0$ or $x - 1 = 0$
$x = -8$ or $x = 1$
The solution set is $\{-8, 1\}$.

13. $2x^2 - 4x + 7 = 0$

$$x = \frac{-(-4) \pm \sqrt{(-4)^2 - 4(2)(7)}}{2(2)}$$

$$x = \frac{4 \pm \sqrt{16-56}}{4}$$

$$x = \frac{4 \pm \sqrt{-40}}{4}$$

$$x = \frac{4 \pm 2i\sqrt{10}}{4}$$

$$x = \frac{2(2 \pm i\sqrt{10})}{4}$$

$$x = \frac{2 \pm i\sqrt{10}}{2}$$

The solution set is $\left\{\frac{2 \pm i\sqrt{10}}{2}\right\}$.

15. $x^2 - 18x + 15 = 0$
$x^2 - 18x = -15$
$x^2 - 18x + 81 = -15 + 81$
$(x-9)^2 = 66$
$x - 9 = \pm\sqrt{66}$
$x = 9 \pm \sqrt{66}$
The solution set is $\{9 \pm \sqrt{66}\}$.

17. $20y^2 + 17y - 10 = 0$
$(4y+5)(5y-2) = 0$
$4y + 5 = 0$ or $5y - 2 = 0$
$4y = -5$ or $5y = 2$

$y = -\frac{5}{4}$ or $y = \frac{2}{5}$

The solution set is $\left\{-\frac{5}{4}, \frac{2}{5}\right\}$.

19. $4t^2 + 4t - 1 = 0$

$t = \frac{-4 \pm \sqrt{4^2 - 4(4)(-1)}}{2(4)}$

$t = \frac{-4 \pm \sqrt{16 + 16}}{8}$

$t = \frac{-4 \pm \sqrt{32}}{8}$

$t = \frac{-4 \pm 4\sqrt{2}}{8}$

$t = \frac{4(-1 \pm \sqrt{2})}{8}$

$t = \frac{-1 \pm \sqrt{2}}{2}$

The solution set is $\left\{\frac{-1 \pm \sqrt{2}}{2}\right\}$.

21. $n + \frac{3}{n} = \frac{19}{4}; n \neq 0$

$4n\left(n + \frac{3}{n}\right) = 4n\left(\frac{19}{4}\right)$

$4n^2 + 12 = 19n$

$4n^2 - 19n + 12 = 0$

$(4n - 3)(n - 4) = 0$

$4n - 3 = 0$ or $n - 4 = 0$

$4n = 3$ or $n = 4$

$n = \frac{3}{4}$

The solution set is $\left\{\frac{3}{4}, 4\right\}$.

23. $\frac{3}{x} + \frac{7}{x - 1} = 1; x \neq 0, x \neq 1$

$x(x - 1)\left(\frac{3}{x} + \frac{7}{x - 1}\right) = x(x - 1)(1)$

$x(x - 1)\left(\frac{3}{x}\right) + x(x - 1)\left(\frac{7}{x - 1}\right) = x^2 - x$

$3(x - 1) + 7x = x^2 - x$

$3x - 3 + 7x = x^2 - x$

$10x - 3 = x^2 - x$

$0 = x^2 - 11x + 3$

$x = \frac{-(-11) \pm \sqrt{(-11)^2 - 4(1)(3)}}{2(1)}$

$x = \frac{11 \pm \sqrt{121 - 12}}{2}$

$x = \frac{11 \pm \sqrt{109}}{2}$

The solution set is $\left\{\frac{11 \pm \sqrt{109}}{2}\right\}$.

25. $\frac{12}{x - 3} + \frac{8}{x} = 14; x \neq 3, x \neq 0$

$x(x - 3)\left(\frac{12}{x - 3} + \frac{8}{x}\right) = x(x - 3)(14)$

$12x + 8(x - 3) = 14x^2 - 42x$

$12x + 8x - 24 = 14x^2 - 42x$

$20x - 24 = 14x^2 - 42x$

$0 = 14x^2 - 62x + 24$

$0 = 2(7x^2 - 31x + 12)$

$0 = 2(7x - 3)(x - 4)$

$7x - 3 = 0$ or $x - 4 = 0$

$7x = 3$ or $x = 4$

$x = \frac{3}{7}$ or $x = 4$

The solution set is $\left\{\frac{3}{7}, 4\right\}$.

27. $\frac{3}{x - 1} - \frac{2}{x} = \frac{5}{2}; x \neq 0, x \neq 1$

$2x(x - 1)\left(\frac{3}{x - 1} - \frac{2}{x}\right) = 2x(x - 1)\left(\frac{5}{2}\right)$

$2x(x-1)\left(\frac{3}{x - 1}\right) - 2x(x-1)\left(\frac{2}{x}\right) = 5x(x-1)$

$6x - 4(x - 1) = 5x^2 - 5x$

$6x - 4x + 4 = 5x^2 - 5x$

$0 = 5x^2 - 7x - 4$

$x = \frac{-(-7) \pm \sqrt{(-7)^2 - 4(5)(-4)}}{2(5)}$

Problem Set 6.5

$$x = \frac{7 \pm \sqrt{49 + 80}}{10}$$

$$x = \frac{7 \pm \sqrt{129}}{10}$$

The solution set is $\left\{\frac{7 \pm \sqrt{129}}{10}\right\}$.

29. $\frac{6}{x} + \frac{40}{x+5} = 7; x \neq 0, x \neq -5$

$$x(x+5)\left(\frac{6}{x} + \frac{40}{x+5}\right) = x(x+5)(7)$$

$6(x+5) + 40x = 7x(x+5)$
$6x + 30 + 40x = 7x^2 + 35x$
$0 = 7x^2 - 11x - 30$
$0 = (7x + 10)(x - 3)$
$7x + 10 = 0$ or $x - 3 = 0$
$7x = -10$ or $x = 3$
$x = -\frac{10}{7}$ or $x = 3$
The solution set is $\left\{-\frac{10}{7}, 3\right\}$.

31. $\frac{5}{n-3} - \frac{3}{n+3} = 1; n \neq 3, n \neq -3$

$$(n\text{-}3)(n\text{+}3)\left(\frac{5}{n\text{-}3} - \frac{3}{n\text{+}3}\right) = (n\text{-}3)(n\text{+}3)(1)$$

$5(n+3) - 3(n-3) = n^2 - 9$
$5n + 15 - 3n + 9 = n^2 - 9$
$2n + 24 = n^2 - 9$
$0 = n^2 - 2n - 33$
$n^2 - 2n - 33 = 0$
$n^2 - 2n = 33$
$n^2 - 2n + 1 = 33 + 1$
$(n-1)^2 = 34$
$n - 1 = \pm\sqrt{34}$
$n = 1 \pm \sqrt{34}$
The solution set is $\{1 \pm \sqrt{34}\}$.

33. $x^4 - 18x^2 + 72 = 0$
$(x^2 - 12)(x^2 - 6) = 0$
$x^2 - 12 = 0$ or $x^2 - 6 = 0$
$x^2 = 12$ or $x^2 = 6$
$x = \pm\sqrt{12}$ or $x = \pm\sqrt{6}$
$x = \pm 2\sqrt{3}$ or $x = \pm\sqrt{6}$
The solution set is $\{\pm\sqrt{6}, \pm 2\sqrt{3}\}$.

35. $3x^4 - 35x^2 + 72 = 0$
$(3x^2 - 8)(x^2 - 9) = 0$
$3x^2 - 8 = 0$ or $x^2 - 9 = 0$
$3x^2 = 8$ or $x^2 = 9$
$x^2 = \frac{8}{3}$ or $x = \pm\sqrt{9}$
$x = \pm\sqrt{\frac{8}{3}}$ or $x = \pm 3$
$x = \pm\frac{2\sqrt{2}}{\sqrt{3}}$ or $x = \pm 3$
$x = \pm\frac{2\sqrt{2}}{\sqrt{3}} \bullet \frac{\sqrt{3}}{\sqrt{3}}$ or $x = \pm 3$
$x = \pm\frac{2\sqrt{6}}{3}$ or $x = \pm 3$
The solution set is $\left\{\pm\frac{2\sqrt{6}}{3}, \pm 3\right\}$.

37. $3x^4 + 17x^2 + 20 = 0$
$(3x^2 + 5)(x^2 + 4) = 0$
$3x^2 + 5 = 0$ or $x^2 + 4 = 0$
$3x^2 = -5$ or $x^2 = -4$
$x^2 = -\frac{5}{3}$ or $x = \pm\sqrt{-4}$
$x = \pm\sqrt{-\frac{5}{3}}$ or $x = \pm 2i$
$x = \pm\frac{i\sqrt{5}}{\sqrt{3}} \bullet \frac{\sqrt{3}}{\sqrt{3}}$ or $x = \pm 2i$
$x = \pm\frac{i\sqrt{15}}{3}$ or $x = \pm 2i$
The solution set is $\left\{\pm\frac{i\sqrt{15}}{3}, \pm 2i\right\}$.

39. $6x^2 - 29x^2 + 28 = 0$
$(3x^2 - 4)(2x^2 - 7) = 0$
$3x^2 - 4 = 0$ or $2x^2 - 7 = 0$
$3x^2 = 4$ or $2x^2 = 7$

$x^2 = \frac{4}{3}$ or $x^2 = \frac{7}{2}$

$x = \pm\sqrt{\frac{4}{3}}$ or $x = \pm\sqrt{\frac{7}{2}}$

$x = \pm\frac{2}{\sqrt{3}}$ or $x = \pm\frac{\sqrt{7}}{\sqrt{2}} \bullet \frac{\sqrt{2}}{\sqrt{2}}$

$x = \pm\frac{2}{\sqrt{3}} \bullet \frac{\sqrt{3}}{\sqrt{3}}$ or $x = \pm\frac{\sqrt{14}}{2}$

$x = \pm\frac{2\sqrt{3}}{3}$ or $x = \pm\frac{\sqrt{14}}{2}$

The solution set is $\left\{\pm\frac{2\sqrt{3}}{3}, \pm\frac{\sqrt{14}}{2}\right\}$.

41. Let $x = 1^{\text{st}}$ whole number and
$x + 1 = 2^{\text{nd}}$ whole number.

$x^2 + (x + 1)^2 = 145$
$x^2 + x^2 + 2x + 1 = 145$
$2x^2 + 2x - 144 = 0$
$2(x^2 + x - 72) = 0$
$2(x + 9)(x - 8) = 0$
$x + 9 = 0$ or $x - 8 = 0$
$x = -9$ or $x = 8$
Discard the root $x = -9$.
The numbers are 8 and 9.

43. Let $x = 1^{\text{st}}$ positive integer and
$x - 3 = 2^{\text{nd}}$ positive integer.

$x(x - 3) = 108$
$x^2 - 3x = 108$
$x^2 - 3x - 108 = 0$
$(x - 12)(x + 9) = 0$
$x - 12 = 0$ or $x + 9 = 0$
$x = 12$ or $x = -9$
Discard the root $x = -9$.
The numbers are 12 and 9.

45. Let $x = 1^{\text{st}}$ number and
$10 - x = 2^{\text{nd}}$ number.

$x(10 - x) = 22$
$10x - x^2 = 22$
$x^2 - 10x = -22$
$x^2 - 10x + 25 = -22 + 25$
$(x - 5)^2 = 3$
$x - 5 = \pm\sqrt{3}$
$x = 5 \pm \sqrt{3}$
$x_1 = 5 + \sqrt{3}$ or $x_1 = 5 - \sqrt{3}$
$x_2 = 10 - (5 + \sqrt{3})$ $\quad x_2 = 10 - (5 - \sqrt{3})$
$x_2 = 10 - 5 - \sqrt{3}$ $\quad x_2 = 10 - 5 + \sqrt{3}$
$x_2 = 5 - \sqrt{3}$ $\quad x_2 = 5 + \sqrt{3}$
The numbers are $5 + \sqrt{3}$ and $5 - \sqrt{3}$.

47.

Numbers	Reciprocal
x	$\frac{1}{x}$
$9 - x$	$\frac{1}{9 - x}$

$\frac{1}{x} + \frac{1}{9 - x} = \frac{1}{2}$

$2x(9 - x)\left(\frac{1}{x} + \frac{1}{9 - x}\right) = 2x(9 - x)\left(\frac{1}{2}\right)$

$2(9 - x) + 2x = x(9 - x)$
$18 - 2x + 2x = 9x - x^2$
$18 = 9x - x^2$
$x^2 - 9x + 18 = 0$
$(x - 6)(x - 3) = 0$
$x - 6 = 0$ or $x - 3 = 0$
$x = 6$ or $x = 3$
The numbers are 6 and 3.

49. Let $x =$ one leg and
$21 - x =$ other leg.

$x^2 + (21 - x)^2 = 15^2$
$x^2 + 441 - 42x + x^2 = 225$
$2x^2 - 42x + 216 = 0$
$2(x^2 - 21x + 108) = 0$
$2(x - 9)(x - 12) = 0$
$x - 9 = 0$ or $x - 12 = 0$
$x = 9$ or $x = 12$
The legs are 9 inches and 12 inches.

51. Let $x =$ width of walk
Area of plot and sidewalk
$= (12 + 2x)(20 + 2x)$
Area of plot $= 12(20) = 240$

Area of Plot and Sidewalk $-$ Area of Plot $=$
Area of sidewalk

$(12+2x)(20+2x)-240=68$
$240+24x+40x+4x^2-240=68$
$4x^2+64x-68=0$
$4(x^2+16x-17)=0$
$4(x+17)(x-1)=0$
$x+17=0$ or $x-1=0$
$x=-17$ or $x=1$
Discard the root $x=-17$.
The width of the walk is 1 meter.

53. Let $x=$ width
$P=2l+2w$
$44=2l+2x$
$44-2x=2l$
$22-x=$ length

Area = length · width
$x(22-x)=112$
$22x-x^2=112$
$0=x^2-22x+112$
$0=(x-8)(x-14)$
$x-8=0$ or $x-14=0$
$x=8$ or $x=14$
The dimensions of the rectangle are 8 inches by 14 inches.

55.

	Rate	Time	Distance
Charlotte	$x+5$	$\frac{250}{x+5}$	250
Lorraine	x	$\frac{180}{x}$	180

$$\frac{250}{x+5}=\frac{180}{x}+1$$

$$x(x+5)\left(\frac{250}{x+5}\right)=x(x+5)\left(\frac{180}{x}+1\right)$$

$250x=180(x+5)+1x(x+5)$
$250x=180x+900+x^2+5x$
$250x=x^2+185x+900$
$0=x^2-65x+900$
$0=(x-45)(x-20)$
$x-45=0$ or $x-20=0$
$x=45$ or $x=20$

Case 1: Lorraine drives at 20 mph and Charlotte drives at 25 mph.
Case 2: Lorraine drives at 45 mph and Charlotte drives at 50 mph.

57.

	Rate	Time	Distance
1st Part	x	$\frac{330}{x}$	330
2nd Part	$x+5$	$\frac{240}{x+5}$	240

$$\frac{330}{x}+\frac{240}{x+5}=10$$

$$x(x+5)\left(\frac{330}{x}+\frac{240}{x+5}\right)=10x(x+5)$$

$330(x+5)+240x=10x^2+50x$
$330x+1650+240x=10x^2+50x$
$570x+1650=10x^2+50x$
$0=10x^2-520x-1650$
$0=10(x^2-52x-165)$
$0=10(x-55)(x+3)$
$x-55=0$ or $x+3=0$
$x=55$ or $x=-3$
Discard the root $x=-3$.
He drove at 55 mph for the first part.

59.

	Time for job	Rate
Terry	$x+2$	$\frac{1}{x+2}$
Tom	x	$\frac{1}{x}$

$$\left(\frac{1}{x+2}+\frac{1}{x}\right)3+\left(\frac{1}{x+2}\right)1=1$$

$$\frac{3}{x+2}+\frac{3}{x}+\frac{1}{x+2}=1$$

$$\frac{4}{x+2}+\frac{3}{x}=1$$

$$x(x+2)\left(\frac{4}{x+2}+\frac{3}{x}\right)=x(x+2)(1)$$

$4x+3(x+2)=x^2+2x$
$4x+3x+6=x^2+2x$

$0 = x^2 - 5x - 6$
$0 = (x-6)(x+1)$
$x - 6 = 0 \quad \text{or} \quad x + 1 = 0$
$x = 6 \quad \text{or} \quad x = -1$
Discard the root $x = -1$.
It takes Tom 6 hours and Terry 8 hours.

61.

	rate pay/hour	time in hours	Pay
Anticipated	$\frac{24}{x}$	x	24
Actual	$\frac{24}{x+1}$	$x+1$	24

$$\frac{24}{x} - 4 = \frac{24}{x+1}$$

$$x(x+1)\left(\frac{24}{x} - 4\right) = x(x+1)\left(\frac{24}{x+1}\right)$$

$24(x+1) - 4x(x+1) = 24x$
$24x + 24 - 4x^2 - 4x = 24x$
$-4x^2 + 20x + 24 = 24x$
$-4x^2 - 4x + 24 = 0$
$-4(x^2 + x - 6) = 0$
$-4(x+3)(x-2) = 0$
$x + 3 = 0 \quad \text{or} \quad x - 2 = 0$
$x = -3 \quad \text{or} \quad x = 2$
Discard the root $x = -3$.
The time he anticipated it would take was 2 hours.

63.

	number of persons	price	price/ person
Group	x	80	$\frac{80}{x}$
Group Less 2	$x-2$	80	$\frac{80}{x-2}$

$$\frac{80}{x-2} = \frac{80}{x} + 2$$

$$x(x-2)\left(\frac{80}{x-2}\right) = x(x-2)\left(\frac{80}{x} + 2\right)$$

$x(80) = 80(x-2) + 2x(x-2)$
$80x = 80x - 160 + 2x^2 - 4x$
$0 = 2x^2 - 4x - 160$
$0 = x^2 - 2x - 80$
$0 = (x-10)(x+8)$
$x - 10 = 0 \quad \text{or} \quad x + 8 = 0$
$x = 10 \quad \text{or} \quad x = -8$
Discard the root $x = -8$.
There were $10 - 2 = 8$ people that actually contributed.

65.

	price/ share	number of shares	price
Purchase	$\frac{720}{x}$	x	720
Sell	$\frac{800}{x-20}$	$x-20$	800

$$\frac{720}{x} + 8 = \frac{800}{x-20}$$

$$x(x-20)\left(\frac{720}{x} + 8\right) = x(x-20)\left(\frac{800}{x-20}\right)$$

$720(x-20) + 8x(x-20) = 800x$
$720x - 14400 + 8x^2 - 160x = 800x$
$8x^2 + 560x - 14400 = 800x$
$8x^2 - 240x - 14400 = 0$
$8(x^2 - 30x - 1800) = 0$
$8(x-60)(x+30) = 0$
$x - 60 = 0 \quad \text{or} \quad x + 30 = 0$
$x = 60 \quad \text{or} \quad x = -30$
Discard the root $x = -30$.
He sold 40 shares at \$20 per share.

67. $S = \frac{n(n+1)}{2}$

$$1275 = \frac{n(n+1)}{2}$$

$$2(1275) = 2\left[\frac{n(n+1)}{2}\right]$$

$2550 = n(n+1)$
$2550 = n^2 + n$
$0 = n^2 + n - 2550$
$0 = (n+51)(n-50)$
$n + 51 = 0 \quad \text{or} \quad n - 50 = 0$
$n = -51 \quad \text{or} \quad n = 50$
Discard the root $n = -51$.
There needs to be 50 consecutive numbers.

69. $A = P(1+r)^t$
$594.05 = 500(1+r)^2$
$1.1881 = (1+r)^2$
$(1+r)^2 = 1.1881$
$1 + r = \pm\sqrt{1.1881}$
$r = -1 \pm \sqrt{1.1881}$
$r = -1 + \sqrt{1.1881}$ or $r = -1 - \sqrt{1.1881}$
$r = 0.09$ or $r = -2.09$
Discard the root $r = -2.09$.
The rate is 9%.

75. $x - 9\sqrt{x} + 18 = 0$
Let $y = \sqrt{x}$, then $y^2 = x$.
$y^2 - 9y + 18 = 0$
$(y-6)(y-3) = 0$
$y - 6 = 0$ or $y - 3 = 0$
$y = 6$ or $y = 3$
Substituting $\sqrt{x}$ for y:
$\sqrt{x} = 6$ or $\sqrt{x} = 3$
$(\sqrt{x})^2 = 6^2$ or $(\sqrt{x})^2 = 3^2$
$x = 36$ or $x = 9$
Checking $x = 36$:
$36 - 9\sqrt{36} + 18 \stackrel{?}{=} 0$
$36 - 54 + 18 \stackrel{?}{=} 0$
$0 = 0$
Checking $x = 9$:
$9 - 9\sqrt{9} + 18 \stackrel{?}{=} 0$
$9 - 27 + 18 \stackrel{?}{=} 0$
$0 = 0$
The solution set is $\{9, 36\}$.

77. $x + \sqrt{x} - 2 = 0$
Let $y = \sqrt{x}$, then $y^2 = x$.
$y^2 + y - 2 = 0$
$(y+2)(y-1) = 0$
$y + 2 = 0$ or $y - 1 = 0$
$y = -2$ or $y = 1$
Substituting $\sqrt{x}$ for y:
$\sqrt{x} = -2$ or $\sqrt{x} = 1$
$(\sqrt{x})^2 = (-2)^2$ or $(\sqrt{x})^2 = 1^2$
$x = 4$ or $x = 1$
Checking $x = 4$:
$4 + \sqrt{4} - 2 \stackrel{?}{=} 0$
$4 + 2 - 2 \stackrel{?}{=} 0$
$4 \neq 0$
Checking $x = 1$:
$1 + \sqrt{1} - 2 \stackrel{?}{=} 0$
$1 + 1 - 2 \stackrel{?}{=} 0$
$0 = 0$
The solution set is $\{1\}$.

79. $6x^{\frac{2}{3}} - 5x^{\frac{1}{3}} - 6 = 0$
Let $y = x^{\frac{1}{3}}$, then $y^2 = x^{\frac{2}{3}}$.
$6y^2 - 5y - 6 = 0$
$(3y+2)(2y-3) = 0$
$3y + 2 = 0$ or $2y - 3 = 0$
$3y = -2$ or $2y = 3$
$y = -\frac{2}{3}$ or $y = \frac{3}{2}$
Substituting $x^{\frac{1}{3}}$ for y:
$x^{\frac{1}{3}} = -\frac{2}{3}$ or $x^{\frac{1}{3}} = \frac{3}{2}$
$(x^{\frac{1}{3}})^3 = \left(-\frac{2}{3}\right)^3$ or $(x^{\frac{1}{3}})^3 = \left(\frac{3}{2}\right)^3$
$x = -\frac{8}{27}$ or $x = \frac{27}{8}$
The solution set is $\left\{-\frac{8}{27}, \frac{27}{8}\right\}$.

81. $12x^{-2} - 17x^{-1} - 5 = 0$
Let $y = x^{-1}$, then $y^2 = x^{-2}$.
$12y^2 - 17y - 5 = 0$
$(4y+1)(3y-5) = 0$
$4y + 1 = 0$ or $3y - 5 = 0$
$4y = -1$ or $3y = 5$
$y = -\frac{1}{4}$ or $y = \frac{5}{3}$

Substitute x^{-1} for y:
$x^{-1} = -\frac{1}{4}$ or $x^{-1} = \frac{5}{3}$
$x = -4$ or $x = \frac{3}{5}$
The solution set is $\left\{-4, \frac{3}{5}\right\}$.

PROBLEM SET 6.6 Quadratic and Other Nonlinear Inequalities

1. $(x+2)(x-1) > 0$
$(x+2)(x-1) = 0$
$x+2=0$ or $x-1=0$
$x=-2$ or $x=1$

Test Point	-3	-2 0	1 2
$x+2$:	negative	positive	positive
$x-1$:	negative	negative	positive
product:	positive	negative	positive

The solution set is $(-\infty, -2) \cup (1, \infty)$.

3. $(x+1)(x+4) < 0$
$(x+1)(x+4) = 0$
$x+1=0$ or $x+4=0$
$x=-1$ or $x=-4$

Test Point	-5	-4 -2	-1 0
$x+1$:	negative	negative	positive
$x+4$:	negative	positive	positive
product:	positive	negative	positive

The solution set is $(-4, -1)$.

5. $(2x-1)(3x+7) \geq 0$
$(2x-1)(3x+7) = 0$
$2x-1=0$ or $3x+7=0$
$2x=1$ or $3x=-7$
$x=\frac{1}{2}$ or $x=-\frac{7}{3}$

Test Point	-3	$-\frac{7}{3}$ 0	$\frac{1}{2}$ 1
$2x-1$:	negative	negative	positive
$3x+7$:	negative	positive	positive
product:	positive	negative	positive

The solution set is $\left(-\infty, -\frac{7}{3}\right] \cup \left[\frac{1}{2}, \infty\right)$.

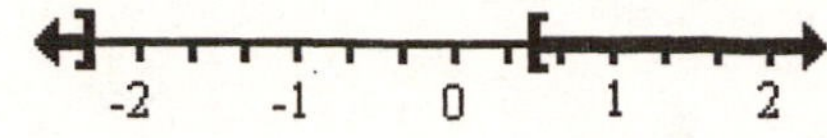

7. $(x+2)(4x-3) \leq 0$
$(x+2)(4x-3) = 0$
$x+2=0$ or $4x-3=0$
$x=-2$ or $4x=3$
$x=-2$ or $x=\frac{3}{4}$

Test Point	-3	-2 0	$\frac{3}{4}$ 1
$x+2$:	negative	positive	positive
$4x-3$:	negative	negative	positive
product:	positive	negative	positive

The solution set is $\left[-2, \frac{3}{4}\right]$.

9. $(x+1)(x-1)(x-3) > 0$
$(x+1)(x-1)(x-3) = 0$
$x+1=0$ or $x-1=0$ or $x-3=0$
$x=-1$ or $x=1$ or $x=3$

Test Point	-2	-1 0	1 2	3 4
$x+1$:	neg	pos	pos	pos
$x-1$:	neg	neg	pos	pos
$x-3$:	neg	neg	neg	pos
product:	neg	pos	neg	pos

The solution set is $(-1, 1) \cup (3, \infty)$.

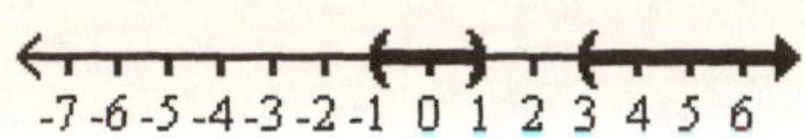

11. $x(x+2)(x-4) \le 0$
$x(x+2)(x-4) = 0$
$x = 0$ or $x + 2 = 0$ or $x - 4 = 0$
$x = 0$ or $x = -2$ or $x = 4$

	-2	0	4	
Test Point	-3	-1	2	5
x:	neg	neg	pos	pos
$x+2$:	neg	pos	pos	pos
$x-4$:	neg	neg	neg	pos
product:	neg	pos	neg	pos

The solution set is $(-\infty, -2] \cup [0, 4]$.

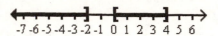

13. $\dfrac{x+1}{x-2} > 0; x \neq 2$

$x + 1 = 0$ or $x - 2 = 0$
$x = -1$ or $x = 2$

	-1	2	
Test Point	-2	0	3
$x+1$:	negative	positive	positive
$x-2$:	negative	negative	positive
quotient:	positive	negative	positive

The solution set is $(-\infty, -1) \cup (2, \infty)$.

15. $\dfrac{x-3}{x+2} < 0; x \neq -2$

$x - 3 = 0$ or $x + 2 = 0$
$x = 3$ or $x = -2$

	-2	3	
Test Point	-3	0	4
$x-3$:	negative	negative	positive
$x+2$:	negative	positive	positive
quotient:	positive	negative	positive

The solution set is $(-2, 3)$.

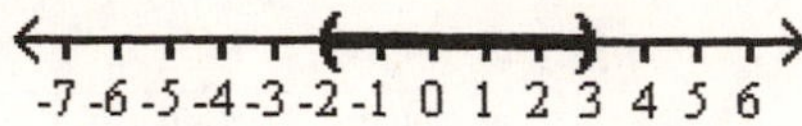

17. $\dfrac{2x-1}{x} \ge 0; x \neq 0$

$2x - 1 = 0$ or $x = 0$
$2x = 1$ or $x = 0$
$x = \dfrac{1}{2}$ or $x = 0$

	0	$\frac{1}{2}$	
Test Point	-1	$\frac{1}{4}$	1
$2x-1$:	negative	negative	positive
x:	negative	positive	positive
quotient:	positive	negative	positive

The solution set is $(-\infty, 0) \cup \left[\frac{1}{2}, \infty\right)$.

19. $\dfrac{-x+2}{x-1} \le 0; x \neq 1$

$-x + 2 = 0$ or $x - 1 = 0$
$2 = x$ or $x = 1$

	1	2	
Test Point	0	$\frac{3}{2}$	3
$-x+2$:	positive	positive	negative
$x-1$:	negative	positive	positive
quotient:	negative	positive	negative

The solution set is $(-\infty, 1) \cup [2, \infty)$.

-7 -6 -5 -4 -3 -2 -1 0 1 2 3 4 5 6

21. $x^2 + 2x - 35 < 0$
$(x + 7)(x - 5) < 0$
$(x + 7)(x - 5) = 0$
$x + 7 = 0$ or $x - 5 = 0$
$x = -7$ or $x = 5$

	-7		5
Test Point	-8	0	6
$x + 7$:	negative	positive	positive
$x - 5$:	negative	negative	positive
product:	positive	negative	positive

The solution set is $(-7, 5)$.

23. $x^2 - 11x + 28 > 0$
$(x - 7)(x - 4) > 0$
$(x - 7)(x - 4) = 0$
$x - 7 = 0$ or $x - 4 = 0$
$x = 7$ or $x = 4$

	4		7
Test Point	0	6	8
$x - 7$:	negative	negative	positive
$x - 4$:	negative	positive	positive
product:	positive	negative	positive

The solution set is $(-\infty, 4) \cup (7, \infty)$.

25. $3x^2 + 13x - 10 \leq 0$
$(3x - 2)(x + 5) \leq 0$
$(3x - 2)(x + 5) = 0$
$3x - 2 = 0$ or $x + 5 = 0$
$3x = 2$ or $x = -5$
$x = \frac{2}{3}$ or $x = -5$

	-5		$\frac{2}{3}$
Test Point	-6	0	1
$3x - 2$:	negative	negative	positive
$x + 5$:	negative	positive	positive
product:	positive	negative	positive

The solution set is $\left[-5, \frac{2}{3}\right]$.

27. $8x^2 + 22x + 5 \geq 0$
$(2x + 5)(4x + 1) \geq 0$
$(2x + 5)(4x + 1) = 0$
$2x + 5 = 0$ or $4x + 1 = 0$
$2x = -5$ or $4x = -1$
$x = -\frac{5}{2}$ or $x = -\frac{1}{4}$

	$-\frac{5}{2}$		$-\frac{1}{4}$
Test Point	-3	-1	0
$2x + 5$:	negative	positive	positive
$4x + 1$:	negative	negative	positive
product:	positive	negative	positive

The solution set is $\left(-\infty, -\frac{5}{2}\right] \cup \left[-\frac{1}{4}, \infty\right)$.

29. $x(5x - 36) > 32$
$5x^2 - 36x > 32$
$5x^2 - 36x - 32 > 0$
$(5x + 4)(x - 8) > 0$
$(5x + 4)(x - 8) = 0$
$5x + 4 = 0$ or $x - 8 = 0$
$5x = -4$ or $x = 8$
$x = -\frac{4}{5}$ or $x = 8$

	$-\frac{4}{5}$		8
Test Point	-1	0	10
$5x + 4$:	negative	positive	positive
$x - 8$:	negative	negative	positive
product:	positive	negative	positive

The solution set is $\left(-\infty, -\frac{4}{5}\right) \cup (8, \infty)$.

31. $x^2 - 14x + 49 \geq 0$
$(x - 7)(x - 7) \geq 0$
$(x - 7)(x - 7) = 0$
$(x - 7)^2 = 0$
$x - 7 = 0$
$x = 7$

		7
Test Point	5	9
$x-7$:	negative	positive
$x-7$:	negative	positive
product:	positive	positive

The solution set is $(-\infty, \infty)$.

33. $4x^2+20x+25 \le 0$
$(2x+5)(2x+5) \le 0$
$(2x+5)(2x+5) = 0$
$(2x+5)^2 = 0$
$2x+5=0$
$2x = -5$
$x = -\frac{5}{2}$

		$-\frac{5}{2}$
Test Point	-3	0
$2x+5$:	negative	positive
$2x+5$:	negative	positive
product:	positive	positive

The solution set is $\left\{-\frac{5}{2}\right\}$.

35. $(x+1)(x-3)^2 > 0$
$(x+1)(x-3)(x-3) > 0$
$(x+1)(x-3)^2 = 0$
$x+1=0$ or $x-3=0$
$x=-1$ or $x=3$

		-1	3
Test Point	-2	0	4
$x+1$:	negative	positive	positive
$x-3$:	negative	negative	positive
$x-3$:	negative	negative	positive
product:	negative	positive	positive

The solution set is $(-1, 3) \cup (3, \infty)$.

37. $4-x^2<0$
$(2-x)(2+x)<0$
$(2-x)(2+x)=0$
$2-x=0$ $\quad$ $2+x=0$
$2=x$ $\quad$ $x=-2$

		-2	2
Test Point	-3	0	3
$2-x$:	positive	positive	negative
$2+x$:	negative	positive	positive
product:	negative	positive	negative

The solution set is $(-\infty, -2) \cup (2, \infty)$.

39. $4(x^2-36)<0$
$4(x-6)(x+6)<0$
$x-6=0$ $\quad$ $x+6=0$
$x=6$ $\quad$ $x=-6$

		-6	6
Test Point	-7	0	7
$x-6$:	negative	negative	positive
$x+6$:	negative	positive	positive
product:	positive	negative	positive

The solution set is $(-6, 6)$.

41. $5x^2+20>0$
$5(x^2+4)>0$
Since $5>0$ and for
any value of x, $x^2+4>0$,
then the solution set is $(-\infty, \infty)$.

43. $x^2-2x \ge 0$
$x(x-2) \ge 0$
$x=0$ $\quad$ $x-2=0$
$x=0$ $\quad$ $x=2$

		0	2
Test Point	-1	1	3
x:	negative	positive	positive
$x-2$:	negative	negative	positive
product:	positive	negative	positive

The solution set is $(-\infty, 0] \cup [2, \infty)$.

45. $3x^3+12x^2>0$
$3x^2(x+4)>0$
$3x^2=0$ $\quad$ $x+4=0$
$x=0$ $\quad$ $x=-4$

Test Point	-5	-2	2
	(boundaries: -4, 0)		
$3x^2$:	positive	positive	positive
$x+4$:	negative	positive	positive
product:	negative	positive	positive

The solution set is $(-4, 0) \cup (0, \infty)$.

47. $\dfrac{2x}{x+3} > 4$

$\dfrac{2x}{x+3} - 4 > 0$

$\dfrac{2x}{x+3} - \dfrac{4(x+3)}{x+3} > 0$

$\dfrac{2x - 4x - 12}{x+3} > 0$

$\dfrac{-2x-12}{x+3} > 0; x \neq -3$

$-2x - 12 = 0$ or $x + 3 = 0$
$-2x = 12$ or $x = -3$
$x = -6$ or $x = -3$

Test Point	-7	-5	0
	(boundaries: -6, -3)		
$-2x-12$:	positive	negative	negative
$x+3$:	negative	negative	positive
quotient:	negative	positive	negative

The solution set is $(-6, -3)$.

49. $\dfrac{x-1}{x-5} \leq 2$

$\dfrac{x-1}{x-5} - 2 \leq 0$

$\dfrac{x-1}{x-5} - \dfrac{2(x-5)}{x-5} \leq 0$

$\dfrac{x-1-2(x-5)}{x-5} \leq 0$

$\dfrac{x-1-2x+10}{x-5} \leq 0$

$\dfrac{-x+9}{x-5} \leq 0; x \neq 5$

$-x + 9 = 0$ or $x - 5 = 0$
$9 = x$ or $x = 5$

Test Point	4	6	10
	(boundaries: 5, 9)		
$-x+9$:	positive	positive	negative
$x-5$:	negative	positive	positive
quotient:	negative	positive	negative

The solution set is $(-\infty, 5) \cup [9, \infty)$.

51. $\dfrac{x+2}{x-3} > -2$

$\dfrac{x+2}{x-3} + 2 > 0$

$\dfrac{x+2}{x-3} + \dfrac{2(x-3)}{x-3} > 0$

$\dfrac{x+2+2(x-3)}{x-3} > 0$

$\dfrac{x+2+2x-6}{x-3} > 0$

$\dfrac{3x-4}{x-3} > 0; x \neq 3$

$3x - 4 = 0$ or $x - 3 = 0$
$3x = 4$ or $x = 3$
$x = \dfrac{4}{3}$ or $x = 3$

Test Point	0	2	4
	(boundaries: $\frac{4}{3}$, 3)		
$3x-4$:	negative	positive	positive
$x-3$:	negative	negative	positive
quotient:	positive	negative	positive

The solution set is $\left(-\infty, \frac{4}{3}\right) \cup (3, \infty)$.

53. $\dfrac{3x+2}{x+4} \leq 2$

$\dfrac{3x+2}{x+4} - 2 \leq 0$

$\dfrac{3x+2}{x+4} - \dfrac{2(x+4)}{x+4} \leq 0$

$\dfrac{3x+2-2(x+4)}{x+4} \leq 0$

$\dfrac{3x+2-2x-8}{x+4} \leq 0$

$\dfrac{x-6}{x+4} \leq 0; x \neq -4$

$x-6=0$ or $x+4=0$

$x=6$ or $x=-4$

	-4		6
Test Point	-5	0	7
$x-6$:	negative	negative	positive
$x+4$:	negative	positive	positive
quotient:	positive	negative	positive

The solution set is $(-4, 6]$.

55. $\dfrac{x+1}{x-2} < 1$

$\dfrac{x+1}{x-2} - 1 < 0$

$\dfrac{x+1}{x-2} - \dfrac{1(x-2)}{x-2} < 0$

$\dfrac{x+1-1(x-2)}{x-2} < 0$

$\dfrac{x+1-x+2}{x-2} < 0$

$\dfrac{3}{x-2} < 0; x \neq 2$

$x-2=0$

$x=2$

	2	
Test Point	1	3
3:	positive	positive
$x-2$:	negative	positive
quotient:	negative	positive

The solution set is $(-\infty, 2)$.

CHAPTER 6 Review Problem Set

1. $(-7+3i)+(9-5i)$
$(-7+9)+(3-5)i$
$2-2i$

2. $(4-10i)-(7-i)$
$4-10i-7+9i$
$-3-i$

3. $5i(3-6i)$
$15i-30i^2$
$15i-30(-1)$
$15i+30$
$30+15i$

4. $(5-7i)(6+8i)$
$30+40i-42i-56i$
$30-2i+56$
$86-2i$

5. $(-2-3i)(4-8i)$
$-8+16i-12i+24i^2$
$-8+4i-24$
$-32+4i$

6. $(4-3i)(4+3i)$
$16+12i-12i-9i^2$
$16+9+0i$
$25+0i$

7. $\dfrac{4+3i}{6-2i} = \dfrac{(4+3i)}{(6-2i)} \bullet \dfrac{(6+2i)}{(6+2i)}$

$\dfrac{24+8i+18i+6i^2}{36-4i^2} = \dfrac{24+26i-6}{36+4}$

$\dfrac{18+26i}{40} = \dfrac{18}{40} + \dfrac{26}{40}i = \dfrac{9}{20} + \dfrac{13}{20}i$

8. $\dfrac{-1-i}{-2+5i}$

$\dfrac{(-1-i)}{(-2+5i)} \bullet \dfrac{(-2-5i)}{(-2-5i)} = \dfrac{2+7i+5i^2}{4-25i^2}$

$\dfrac{2+7i-5}{4+25} = \dfrac{-3+7i}{29} = -\dfrac{3}{29} + \dfrac{7}{29}i$

9. $4x^2 - 20x + 25 = 0$
$b^2 - 4ac$
$(-20)^2 - 4(4)(25)$
$400 - 400 = 0$
Since $b^2 - 4ac = 0$, there will be one real solution with multiplicity of two.

10. $5x^2 - 7x + 31 = 0$
$b^2 - 4ac$
$(-7)^2 - 4(5)(31)$
$49 - 620 = -571$
Since $b^2 - 4ac < 0$, there will be two nonreal complex solutions.

11. $7x^2 - 2x - 14 = 0$
$b^2 - 4ac$
$(-2)^2 - 4(7)(-14)$
$4 + 392 = 396$
Since $b^2 - 4ac > 0$, there will be two unequal real solutions.

12. $5x^2 - 2x = 4$
$5x^2 - 2x - 4 = 0$
$b^2 - 4ac$
$(-2)^2 - 4(5)(-4)$
$4 + 80 = 84$
Since $b^2 - 4ac > 0$, there will be two unequal real solutions.

13. $x^2 - 17x = 0$
$x(x - 17) = 0$
$x = 0$ or $x - 17 = 0$
$x = 0$ or $x = 17$
The solution set is $\{0, 17\}$.

14. $(x-2)^2 = 36$
$x - 2 = \pm 6$
$x = 2 + 6$ or $x = 2 - 6$
$x = 8$ or $x = -4$
The solution set is $\{-4, 8\}$.

15. $(2x-1)^2 = -64$
$2x - 1 = \pm\sqrt{-64}$
$2x - 1 = \pm 8i$
$2x = 1 \pm 8i$
$x = \dfrac{1 \pm 8i}{2}$
The solution set is $\left\{\dfrac{1 \pm 8i}{2}\right\}$.

16. $x^2 - 4x - 21 = 0$
$(x - 7)(x + 3) = 0$
$x - 7 = 0$ or $x + 3 = 0$
$x = 7$ or $x = -3$
The solution set is $\{-3, 7\}$.

17. $x^2 + 2x - 9 = 0$
$x^2 + 2x = 9$
$x^2 + 2x + 1 = 9 + 1$
$(x+1)^2 = 10$
$x + 1 = \pm\sqrt{10}$
$x = -1 \pm \sqrt{10}$
The solution set is $\{-1 \pm \sqrt{10}\}$.

18. $x^2 - 6x = -34$
$x^2 - 6x + 9 = -34 + 9$
$(x-3)^2 = -25$
$x - 3 = \pm\sqrt{-25}$
$x - 3 = \pm 5i$
$x = 3 \pm 5i$
The solution set is $\{3 \pm 5i\}$.

19. $4\sqrt{x} = x - 5$
$(4\sqrt{x})^2 = (x-5)^2$
$16x = x^2 - 10x + 25$
$0 = x^2 - 26x + 25$
$0 = (x - 25)(x - 1)$
$x - 25 = 0$ or $x - 1 = 0$
$x = 25$ or $x = 1$
Checking $x = 25$:
$4\sqrt{25} \stackrel{?}{=} 25 - 5$
$4(5) \stackrel{?}{=} 20$
$20 = 20$
Checking $x = 1$:
$4\sqrt{1} \stackrel{?}{=} 1 - 5$
$4(1) \stackrel{?}{=} -4$

$4 \neq -4$
The solution set is $\{25\}$.

20. $3n^2 + 10n - 8 = 0$
$(3n - 2)(n + 4) = 0$
$3n - 2 = 0$ or $n + 4 = 0$
$3n = 2$ or $n = -4$
$n = \frac{2}{3}$
The solution set is $\left\{-4, \frac{2}{3}\right\}$.

21. $n^2 - 10n = 200$
$n^2 - 10n - 200 = 0$
$(n - 20)(n + 10) = 0$
$n - 20 = 0$ or $n + 10 = 0$
$n = 20$ or $n = -10$
The solution set is $\{-10, 20\}$.

22. $3a^2 + a - 5 = 0$

$$a = \frac{-1 \pm \sqrt{1^2 - 4(3)(-5)}}{2(3)}$$
$$a = \frac{-1 \pm \sqrt{1 + 60}}{6}$$
$$a = \frac{-1 \pm \sqrt{61}}{6}$$

The solution set is $\left\{\frac{-1 \pm \sqrt{61}}{6}\right\}$.

23. $x^2 - x + 3 = 0$

$$x = \frac{-1(-1) \pm \sqrt{(-1)^2 - 4(1)(3)}}{2(1)}$$
$$x = \frac{1 \pm \sqrt{1 - 12}}{2}$$
$$x = \frac{1 \pm \sqrt{-11}}{2}$$
$$x = \frac{1 \pm i\sqrt{11}}{2}$$

The solution set is $\left\{\frac{-1 \pm i\sqrt{11}}{2}\right\}$.

24. $2x^2 - 5x + 6 = 0$

$$x = \frac{-(-5) \pm \sqrt{(-5)^2 - 4(2)(6)}}{2(2)}$$
$$x = \frac{5 \pm \sqrt{25 - 48}}{4}$$
$$x = \frac{5 \pm \sqrt{-23}}{4}$$
$$x = \frac{5 \pm i\sqrt{23}}{4}$$

The solution set is $\left\{\frac{5 \pm i\sqrt{23}}{4}\right\}$.

25. $2a^2 + 4a - 5 = 0$

$$a = \frac{-4 \pm \sqrt{4^2 - 4(2)(-5)}}{2(2)}$$
$$a = \frac{-4 \pm \sqrt{16 + 40}}{4}$$
$$a = \frac{-4 \pm \sqrt{56}}{4}$$
$$a = \frac{-4 \pm 2\sqrt{14}}{4}$$
$$a = \frac{2(-2 \pm \sqrt{14})}{4}$$
$$a = \frac{-2 \pm \sqrt{14}}{2}$$

The solution set is $\left\{\frac{-2 \pm \sqrt{14}}{2}\right\}$.

26. $t(t + 5) = 36$
$t^2 + 5t = 36$
$t^2 + 5t - 36 = 0$
$(t + 9)(t - 4) = 0$
$t + 9 = 0$ or $t - 4 = 0$
$t = -9$ or $t = 4$
The solution set is $\{-9, 4\}$.

27. $x^2 + 4x + 9 = 0$
$x^2 + 4x = -9$
$x^2 + 4x + 4 = -9 + 4$
$(x + 2)^2 = -5$
$x + 2 = \pm\sqrt{-5}$
$x + 2 = \pm i\sqrt{5}$
$x = -2 \pm i\sqrt{5}$
The solution set is $\{-2 \pm i\sqrt{5}\}$.

28. $(x - 4)(x - 2) = 80$
$x^2 - 6x + 8 = 80$

$x^2 - 6x - 72 = 0$
$(x - 12)(x + 6) = 0$
$x - 12 = 0 \quad$ or $\quad x + 6 = 0$
$x = 12 \quad$ or $\quad x = -6$
The solution set is $\{-6, 12\}$.

29. $\frac{3}{x} + \frac{2}{x+3} = 1;\ x \neq 0,\ x \neq -3$

$x(x+3)\left(\frac{3}{x} + \frac{2}{x+3}\right) = x(x+3)(1)$
$3(x+3) + 2x = x(x+3)$
$3x + 9 + 2x = x^2 + 3x$
$0 = x^2 - 2x - 9$
$x^2 - 2x = 9$
$x^2 - 2x + 1 = 9 + 1$
$(x-1)^2 = 10$
$x - 1 = \pm\sqrt{10}$
$x = 1 \pm \sqrt{10}$
The solution set is $\{1 \pm \sqrt{10}\}$.

30. $2x^4 - 23x^2 + 56 = 0$
$(2x^2 - 7)(x^2 - 8) = 0$
$2x^2 - 7 = 0 \quad$ or $\quad x^2 - 8 = 0$
$2x^2 = 7 \quad$ or $\quad x^2 = 8$
$x^2 = \frac{7}{2} \quad$ or $\quad x = \pm\sqrt{8}$
$x = \pm\sqrt{\frac{7}{2}} \quad$ or $\quad x = \pm 2\sqrt{2}$

$x = \pm\frac{\sqrt{7}}{\sqrt{2}} \bullet \frac{\sqrt{2}}{\sqrt{2}}$

$x = \pm\frac{\sqrt{14}}{2}$
The solution set is $\left\{\pm\frac{\sqrt{14}}{2}, \pm 2\sqrt{2}\right\}$.

31. $\frac{3}{n-2} = \frac{n+5}{4};\ n \neq 2$

$3(4) = (n-2)(n+5)$

$12 = n^2 + 3n - 10$

$0 = n^2 + 3n - 22$

$n = \frac{-3 \pm \sqrt{3^2 - 4(1)(-22)}}{2(1)}$

$n = \frac{-3 \pm \sqrt{9 + 88}}{2}$

$n = \frac{-3 \pm \sqrt{97}}{2}$

The solution set is $\left\{\frac{-3 \pm \sqrt{97}}{2}\right\}$.

32. $x^2 + 3x - 10 > 0$
$x^2 + 3x - 10 = 0$
$(x + 5)(x - 2) = 0$
$x + 5 = 0 \quad$ or $\quad x - 2 = 0$
$x = -5 \quad$ or $\quad x = 2$

	-5	2	
Test Point	-6	0	3
$x + 5$:	negative	positive	positive
$x - 2$:	negative	negative	positive
product:	positive	negative	positive

The solution set is $(-\infty, -5) \cup (2, \infty)$.

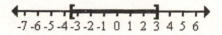

-7 -6 -5 -4 -3 -2 -1 0 1 2 3 4 5 6

33. $2x^2 + x - 21 \leq 0$
$(2x + 7)(x - 3) \leq 0$
$(2x + 7)(x - 3) = 0$
$2x + 7 = 0 \quad$ or $\quad x - 3 = 0$
$2x = -7 \quad$ or $\quad x = 3$
$x = -\frac{7}{2} \quad$ or $\quad x = 3$

	$-\frac{7}{2}$	3	
Test Point	-4	0	4
$2x + 7$:	negative	positive	positive
$x - 3$:	negative	negative	positive
product:	positive	negative	positive

The solution set is $\left[-\frac{7}{2}, 3\right]$.

-7 -6 -5 -4 -3 -2 -1 0 1 2 3 4 5 6

34. $\frac{x-4}{x+6} \geq 0; x \neq -6$

$x - 4 = 0 \qquad x + 6 = 0$

$x = 4 \qquad x = -6$

	-6		4
Test Point	-7	0	5
$x-4$:	negative	negative	positive
$x+6$:	negative	positive	positive
quotient:	positive	negative	positive

The solution set is $(-\infty, -6) \cup [4, \infty)$.

-7 -6 -5 -4 -3 -2 -1 0 1 2 3 4 5 6

35. $\frac{2x-1}{x+1} > 4$

$\frac{2x-1}{x+1} - 4 > 0$

$\frac{2x-1}{x+1} - \frac{4(x+1)}{x+1} > 0$

$\frac{2x-1-4(x+1)}{x+1} > 0$

$\frac{2x-1-4x-4}{x+1} > 0$

$\frac{-2x-5}{x+1} > 0, x \neq -1$

$-2x - 5 = 0 \quad$ or $\quad x + 1 = 0$

$-2x = 5 \quad$ or $\quad x = -1$

$x = -\frac{5}{2} \quad$ or $\quad x = -1$

	$-\frac{5}{2}$		-1
Test Point	-3	-2	0
$-2x-5$:	positive	negative	negative
$x+1$:	negative	negative	positive
quotient:	negative	positive	negative

The solution set is $\left(-\frac{5}{2}, -1\right)$.

-7 -6 -5 -4 -3 -2 -1 0 1 2 3 4 5 6

36. Let $x =$ one number and
$6 - x =$ other number.

$x(6-x) = 2$

$6x - x^2 = 2$

$0 = x^2 - 6x + 2$

$x^2 - 6x = -2$

$x^2 - 6x + 9 = -2 + 9$

$(x-3)^2 = 7$

$x - 3 = \pm\sqrt{7}$

$x = 3 \pm \sqrt{7}$

The numbers are $3 + \sqrt{7}$ and $3 - \sqrt{7}$.

37.

	price/ share	number of shares	Cost
Purchase	$\frac{250}{x}$	x	250
Sell	$\frac{300}{x-5}$	$x-5$	300

$\frac{250}{x} + 5 = \frac{300}{x-5}$

$x(x-5)\left(\frac{250}{x} + 5\right) = x(x-5)\left(\frac{300}{x-5}\right)$

$250(x-5) + 5x(x-5) = 300x$

$250x - 1250 + 5x^2 - 25x = 300x$

$5x^2 + 225x - 1250 = 300x$

$5x^2 - 75x - 1250 = 0$

$5(x^2 - 15x - 250) = 0$

$5(x-25)(x+10) = 0$

$x - 25 = 0 \quad$ or $\quad x + 10 = 0$

$x = 25 \quad$ or $\quad x = -10$

Discard the root $x = -10$.

She sold 20 shares at \$15 per share.

38.

	Rate	Time	Distance
Andre	x	$\frac{270}{x}$	270
Sandy	$x+7$	$\frac{260}{x+7}$	260

$\frac{270}{x} = \frac{260}{x+7} + 1$

$x(x+7)\left(\frac{270}{x}\right) = x(x+7)\left(\frac{260}{x+7} + 1\right)$

$270(x+7) = 260x + x(x+7)$
$270x + 1890 = 260x + x^2 + 7x$
$0 = x^2 - 3x - 1890$
$0 = (x-45)(x+42)$
$x - 45 = 0$ or $x + 42 = 0$
$x = 45$ or $x = -42$
Discard the root $x = -42$.
Andre traveled 45 mph and
Sandy traveled 52 mph.

39. Let s = length of a side.
Area of square = s^2
Perimeter of square = $4s$

$s^2 = 2(4s)$
$s^2 = 8s$
$s^2 - 8s = 0$
$s(s-8) = 0$
$s = 0$ or $s - 8 = 0$
$s = 0$ or $s = 8$
Discard the root $s = 0$.
The length of the side
of the square is 8 units.

40. Let x = 1st even whole number and
$x + 2$ = 2nd even whole number.

$x^2 + (x+2)^2 = 164$
$x^2 + x^2 + 4x + 4 = 164$
$2x^2 + 4x - 160 = 0$
$2(x^2 + 2x - 80) = 0$
$2(x+10)(x-8) = 0$
$x + 10 = 0$ or $x - 8 = 0$
$x = -10$ or $x = 8$
Discard the root $x = -10$.
The whole numbers are 8 and 10.

41. Let x = width and
$19 - x$ = length.

$x(19 - x) = 84$
$19x - x^2 = 84$
$0 = x^2 - 19x + 84$
$0 = (x-12)(x-7)$
$x - 12 = 0$ or $x - 7 = 0$
$x = 12$ or $x = 7$
The dimensions of the rectangle
are 7 inches by 12 inches.

42.

	Time in Hours	Rate
Billy	$x+2$	$\frac{1}{x+2}$
Janet	x	$\frac{1}{x}$

$\left(\frac{1}{x+2}\right)(2) + \frac{1}{x}(2) + \left(\frac{1}{x+2}\right)1 = 1$

$\frac{2}{x+2} + \frac{2}{x} + \frac{1}{x+2} = 1$

$\frac{3}{x+2} + \frac{2}{x} = 1$

$x(x+2)\left(\frac{3}{x+2} + \frac{2}{x}\right) = x(x+2)(1)$

$3x + 2(x+2) = x(x+2)$
$3x + 2x + 4 = x^2 + 2x$
$0 = x^2 - 3x + 4$
$0 = (x-4)(x+1)$
$x - 4 = 0$ or $x + 1 = 0$
$x = 4$ or $x = -1$
Discard the root $x = -1$.
It would take Janet 4 hours and
it would take Billy 6 hours.

43. Let x = width of strip.
Area of the 40×60 lot = 2400 sq. m.
width of new lot = $x + 40$
length of new lot = $x + 60$

$(x+40)(x+60) = 2400 + 1100$
$x^2 + 100x + 2400 = 3500$
$x^2 + 100x - 1100 = 0$
$(x-10)(x+110) = 0$
$x - 10 = 0$ or $x + 110 = 0$
$x = 10$ or $x = -110$
Discard the root $x = -110$.
The width of the strip should be 10 meters.

CHAPTER 6 Test

1. $(3-4i)(5+6i)$
$15+18i-20i-24i^2$
$15-2i+24$
$39-2i$

2. $\dfrac{2-3i}{3+4i}=\dfrac{(2-3i)}{(3+4i)}\bullet\dfrac{(3-4i)}{(3-4i)}=$

$\dfrac{6-17i+12i^2}{9-16i^2}=\dfrac{6-17i-12}{9+16}=$

$\dfrac{-6-17i}{25}=-\dfrac{6}{25}-\dfrac{17}{25}i$

3. $x^2=7x$
$x^2-7x=0$
$x(x-7)=0$
$x=0$ or $x-7=0$
$x=0$ or $x=7$
The solution set is $\{0, 7\}$.

4. $(x-3)^2=16$
$x-3=\pm\sqrt{16}$
$x-3=\pm 4$
$x=3\pm 4$
$x=3+4$ or $x=3-4$
$x=7$ or $x=-1$
The solution set is $\{-1, 7\}$.

5. $x^2+3x-18=0$
$(x+6)(x-3)=0$
$x+6=0$ or $x-3=0$
$x=-6$ or $x=3$
The solution set is $\{-6, 3\}$.

6. $x^2-2x-1=0$
$x^2-2x=1$
$x^2-2x+1=1+1$
$(x-1)^2=2$
$x-1=\pm\sqrt{2}$
$x=1\pm\sqrt{2}$
The solution set is $\{1\pm\sqrt{2}\}$.

7. $5x^2-2x+1=0$

$x=\dfrac{-(-2)\pm\sqrt{(-2)^2-4(5)(1)}}{2(5)}$

$x=\dfrac{2\pm\sqrt{4-20}}{10}$

$x=\dfrac{2\pm\sqrt{-16}}{10}$

$x=\dfrac{2\pm 4i}{10}$

$x=\dfrac{2(1\pm 2i)}{10}$

$x=\dfrac{1\pm 2i}{5}$

The solution set is $\left\{\dfrac{1\pm 2i}{5}\right\}$.

8. $x^2+30x=-224$
$x^2+30x+224=0$
$(x+16)(x+14)=0$
$x+16=0$ or $x+14=0$
$x=-16$ or $x=-14$
The solution set is $\{-16, -14\}$.

9. $(3x-1)^2+36=0$
$(3x-1)^2=-36$
$3x-1=\pm\sqrt{-36}$
$3x-1=\pm 6i$
$3x=1\pm 6i$
$x=\dfrac{1\pm 6i}{3}$
The solution set is $\left\{\dfrac{1\pm 6i}{3}\right\}$.

10. $(5x-6)(4x+7)=0$
$5x-6=0$ or $4x+7=0$
$5x=6$ or $4x=-7$
$x=\dfrac{6}{5}$ or $x=-\dfrac{7}{4}$
The solution set is $\left\{-\dfrac{7}{4}, \dfrac{6}{5}\right\}$.

11. $(2x+1)(3x-2)=55$
$6x^2-x-2=55$
$6x^2-x-57=0$
$(6x-19)(x+3)=0$
$6x-19=0$ or $x+3=0$
$6x=19$ or $x=-3$

$x = \frac{19}{6}$ or $x = -3$

The solution set is $\left\{-3, \frac{19}{6}\right\}$.

12. $n(3n - 2) = 40$

$3n^2 - 2n = 40$

$3n^2 - 2n - 40 = 0$

$(3n + 10)(n - 4) = 0$

$3n + 10 = 0$ or $n - 4 = 0$

$3n = -10$ or $n = 4$

$n = -\frac{10}{3}$

The solution set is $\left\{-\frac{10}{3}, 4\right\}$.

13. $x^4 + 12x - 64 = 0$

$(x^2 + 16)(x^2 - 4) = 0$

$x^2 + 16 = 0$ or $x^2 - 4 = 0$

$x^2 = -16$ or $x^2 = 4$

$x = \pm\sqrt{-16}$ or $x = \pm\sqrt{4}$

$x = \pm 4i$ or $x = \pm 2$

The solution set is $\{\pm 2, \pm 4i\}$.

14. $\frac{3}{x} + \frac{2}{x+1} = 4; x \neq 0, x \neq -1$

$x(x+1)\left(\frac{3}{x} + \frac{2}{x+1}\right) = x(x+1)(4)$

$3(x+1) + 2x = 4x(x+1)$

$3x + 3 + 2x = 4x^2 + 4x$

$0 = 4x^2 - x - 3$

$0 = (4x + 3)(x - 1)$

$4x + 3 = 0$ or $x - 1 = 0$

$4x = -3$ or $x = 1$

$x = -\frac{3}{4}$

The solution set is $\left\{-\frac{3}{4}, 1\right\}$.

15. $3x^2 - 2x - 3 = 0$

$x = \frac{-(-2) \pm \sqrt{(-2)^2 - 4(3)(-3)}}{2(3)}$

$x = \frac{2 \pm \sqrt{4 + 36}}{6}$

$x = \frac{2 \pm \sqrt{40}}{6}$

$x = \frac{2 \pm 2\sqrt{10}}{6}$

$x = \frac{2(1 \pm \sqrt{10})}{6}$

$x = \frac{1 \pm \sqrt{10}}{3}$

The solution set is $\left\{\frac{1 \pm \sqrt{10}}{3}\right\}$.

16. $4x^2 + 20x + 25 = 0$

$b^2 - 4ac$

$(20)^2 - 4(4)(25)$

$400 - 400 = 0$

Because $b^2 - 4ac = 0$, there will be two equal real solutions.

17. $4x^2 - 3x = -5$

$4x^2 - 3x + 5 = 0$

$b^2 - 4ac$

$(-3)^2 - 4(4)(5)$

$9 - 80 = -71$

Because $b^2 - 4ac < 0$, there will be two nonreal complex solutions.

18. $x^2 - 3x - 54 \leq 0$

$(x - 9)(x + 6) \leq 0$

$(x - 9)(x + 6) = 0$

$x - 9 = 0$ or $x + 6 = 0$

$x = 9$ or $x = -6$

	(−6)	(9)	
Test Point	-8	0	10
$x - 9$:	negative	negative	positive
$x + 6$:	negative	positive	positive
product:	positive	negative	positive

The solution set is $[-6, 9]$.

19. $\frac{3x - 1}{x + 2} > 0; x \neq -2$

$3x - 1 = 0$ or $x + 2 = 0$

$3x = 1$ or $x = -2$

$x = \frac{1}{3}$ or $x = -2$

	-2		$\frac{1}{3}$
Test Point	-3	0	1
$3x-1$:	negative	negative	positive
$x+2$:	negative	positive	positive
quotient:	positive	negative	positive

The solution set is $(-\infty, -2) \cup \left(\frac{1}{3}, \infty\right)$.

20. $\frac{x-2}{x+6} \geq 3$

$\frac{x-2}{x+6} - 3 \geq 0$

$\frac{x-2}{x+6} - \frac{3(x+6)}{x+6} \geq 0$

$\frac{x-2-3(x+6)}{x+6} \geq 0$

$\frac{x-2-3x-18}{x+6} \geq 0$

$\frac{-2x-20}{x+6} \geq 0; x \neq -6$

$-2x-20=0$ $\quad$ $x+6=0$
$-2x=20$ $\quad$ $x=-6$
$x=-10$

	-10		-6
Test Point	-11	-8	0
$-2x-20$:	positive	negative	negative
$x+6$:	negative	negative	positive
quotient:	negative	positive	negative

The solution set is $[-10, 6)$.

21. $c = 24$
Let a be the side opposite the $30°$ angle, then $a = \frac{1}{2}(24) = 12$.

$12^2 + b^2 = 24^2$
$144 + b^2 = 576$
$b^2 = 432$
$b = \sqrt{432}$
$b = 20.8$ feet

The ladder reaches 20.8 feet up the building.

22. $16^2 + 24^2 = c^2$
$256 + 576 = c^2$
$832 = c^2$
$\sqrt{832} = c$
$29 \approx c$
The diagonal is 29 meters.

23.

	price/ share	number of shares	Cost
Purchase	$\frac{3000}{x}$	x	3000
Sell	$\frac{3000}{x-50}$	$x-50$	3000

$\frac{3000}{x} + 5 = \frac{3000}{x-50}$

$x(x-50)\left(\frac{3000}{x} + 5\right) = x(x-50)\left(\frac{3000}{x-50}\right)$

$3000(x-50) + 5x(x-50) = 3000x$
$3000x - 150{,}000 + 5x^2 - 250x = 3000x$
$5x^2 + 2750x - 150{,}000 = 3000x$
$5x^2 - 250x - 150{,}000 = 0$
$5(x^2 - 50x - 30{,}000) = 0$
$5(x-200)(x+150) = 0$
$x - 200 = 0$ or $x + 150 = 0$
$x = 200$ or $x = -150$
Discard the root $x = -150$.
Dana sold 150 shares.

24. Let x = width and $20.5 - x$ = length.

$x(20.5 - x) = 91$
$20.5x - x^2 = 91$
$0 = x^2 - 20.5x + 91$
$2(0) = 2(x^2 - 20.5x + 91)$
$0 = 2x^2 - 41x + 182$
$0 = (2x-13)(x-14)$
$2x - 13 = 0$ or $x - 14 = 0$
$2x = 13$ or $x = 14$
$x = \frac{13}{2} = 6\frac{1}{2}$
The shortest side is $6\frac{1}{2}$ inches.

25. Let x = one number and
$6 - x$ = other number.

$x(6 - x) = 4$
$6x - x^2 = 4$
$-x^2 + 6x = 4$
$x^2 - 6x = -4$

$x^2 - 6x + 9 = -4 + 9$
$(x - 3)^2 = 5$
$x - 3 = \pm\sqrt{5}$
$x = 3 \pm \sqrt{5}$
The larger number is $3 + \sqrt{5}$.

CHAPTER 6 Cumulative Review Chapters 1-6

1. $\dfrac{4a^2b^3}{12a^3b} = \dfrac{b^2}{3a}$

$\dfrac{(-8)^2}{3(5)} = \dfrac{64}{15}$

2. $\dfrac{\frac{1}{x} + \frac{1}{y}}{\frac{1}{x} - \frac{1}{y}}$

For $x = 4$ and $y = 7$,

$\dfrac{\frac{1}{4} + \frac{1}{7}}{\frac{1}{4} - \frac{1}{7}} = \dfrac{28}{28} \bullet \dfrac{\left(\frac{1}{4} + \frac{1}{7}\right)}{\left(\frac{1}{4} - \frac{1}{7}\right)}$

$\dfrac{28\left(\frac{1}{4}\right) + 28\left(\frac{1}{7}\right)}{28\left(\frac{1}{4}\right) - 28\left(\frac{1}{7}\right)} = \dfrac{7+4}{7-4} = \dfrac{11}{3}$

3. $\dfrac{3}{n} + \dfrac{5}{2n} - \dfrac{4}{3n}$

$\dfrac{3}{25} + \dfrac{5}{2(25)} - \dfrac{4}{3(25)}$, for $n = 25$

$\dfrac{3}{25} + \dfrac{5}{50} - \dfrac{4}{75}$; LCD = 150

$\dfrac{3}{25} \bullet \dfrac{6}{6} + \dfrac{5}{50} \bullet \dfrac{3}{3} - \dfrac{4}{75} \bullet \dfrac{2}{2}$

$\dfrac{18}{150} + \dfrac{15}{150} - \dfrac{8}{150}$

$\dfrac{25}{150} = \dfrac{1}{6}$

4. $\dfrac{4}{x-1} - \dfrac{2}{x+2}$

$\dfrac{4}{\frac{1}{2} - 1} - \dfrac{2}{\frac{1}{2} + 2}$, for $x = \dfrac{1}{2}$

$\dfrac{4}{-\frac{1}{2}} - \dfrac{2}{\frac{5}{2}}$

$4\left(\dfrac{-2}{1}\right) - 2\left(\dfrac{2}{5}\right)$

$-8 - \dfrac{4}{5}$

$-8\dfrac{4}{5} = -\dfrac{44}{5}$

5. $2\sqrt{2x + y} - 5\sqrt{3x - y}$
For $x = 5$ and $y = 6$,
$2\sqrt{2(5) + 6} - 5\sqrt{3(5) - 6}$
$2\sqrt{16} - 5\sqrt{9}$
$2(4) - 5(3)$
$8 - 15$
-7

6. $(3a^2b)(-2ab)(4ab^3)$
$-24a^4b^5$

7. $(x + 3)(2x^2 - x - 4)$
$2x^3 - x^2 - 4x + 6x^2 - 3x - 12$
$2x^3 + 5x^2 - 7x - 12$

8. $\dfrac{6xy^2}{14y} \bullet \dfrac{7x^2y}{8x}$

$\dfrac{6(7)x^3y^3}{14(8)xy} = \dfrac{3x^2y^2}{8}$

9. $\dfrac{a^2+6a-40}{a^2-4a} \div \dfrac{2a^2+19a-10}{a^3+a^2}$

$\dfrac{a^2+6a-40}{a^2-4a} \bullet \dfrac{a^3+a^2}{2a^2+19a-10}$

$\dfrac{(a+10)(a-4)}{a(a-4)} \bullet \dfrac{a^2(a+1)}{(2a-1)(a+10)}$

$\dfrac{a(a+1)}{(2a-1)}$

10. $\dfrac{3x+4}{6} - \dfrac{5x-1}{9}$; LCD $= 18$

$\dfrac{(3x+4)}{6} \bullet \dfrac{3}{3} - \dfrac{(5x-1)}{9} \bullet \dfrac{2}{2}$

$\dfrac{3(3x+4)-2(5x-1)}{18}$

$\dfrac{9x+12-10x+2}{18} = \dfrac{-x+14}{18}$

11. $\dfrac{4}{x^2+3x} + \dfrac{5}{x}$

$\dfrac{4}{x(x+3)} + \dfrac{5}{x}$; LCD $= x(x+3)$

$\dfrac{4}{x(x+3)} + \dfrac{5}{x} \bullet \dfrac{(x+3)}{(x+3)}$

$\dfrac{4+5(x+3)}{x(x+3)} = \dfrac{4+5x+15}{x(x+3)}$

$\dfrac{5x+19}{x(x+3)}$

12. $\dfrac{3n^2+n}{n^2+10n+16} \bullet \dfrac{2n^2-8}{3n^3-5n^2-2n}$

$\dfrac{n(3n+1)}{(n+8)(n+2)} \bullet \dfrac{2(n+2)(n-2)}{n(3n+1)(n-2)}$

$\dfrac{2}{n+8}$

13. $\dfrac{3}{5x^2+3x-2} - \dfrac{2}{5x^2-22x+8}$

$\dfrac{3}{(5x-2)(x+1)} - \dfrac{2}{(5x-2)(x-4)}$

LCD $= (5x-2)(x+1)(x-4)$

$\dfrac{3}{(5x-2)(x+1)} \bullet \dfrac{(x-4)}{(x-4)} - \dfrac{2}{(5x-2)(x-4)} \bullet \dfrac{(x+1)}{(x+1)}$

$\dfrac{3(x-4)-2(x+1)}{(5x-2)(x+1)(x-4)}$

$\dfrac{3x-12-2x-2}{(5x-2)(x+1)(x-4)}$

$\dfrac{x-14}{(5x-2)(x+1)(x-4)}$

14.

$$\begin{array}{r|l}
 & y^2 - 5y + 6 \\ \hline
y-2 & y^3 - 7y^2 + 16y - 12 \\
 & y^3 - 2y^2 \\ \hline
 & -5y^2 + 16y \\
 & -5y^2 + 10y \\ \hline
 & 6y - 12 \\
 & 6y - 12 \\ \hline
 & 0
\end{array}$$

$y^2 - 5y + 6$

15.

$$\begin{array}{r|l}
 & x^2 - 3x - 2 \\ \hline
4x-5 & 4x^3 - 17x^2 + 7x + 10 \\
 & 4x^3 - 5x^2 \\ \hline
 & -12x^2 + 7x \\
 & -12x^2 + 15x \\ \hline
 & -8x + 10 \\
 & -8x + 10 \\ \hline
 & 0
\end{array}$$

$x^2 - 3x - 2$

16. $(3\sqrt{2}+2\sqrt{5})(5\sqrt{2}-\sqrt{5})$
$15\sqrt{4}-3\sqrt{10}+10\sqrt{10}-2\sqrt{25}$
$15(2)+7\sqrt{10}-2(5)$
$30+7\sqrt{10}-10$
$20+7\sqrt{10}$

17. $(\sqrt{x}-3\sqrt{y})(2\sqrt{x}+4\sqrt{y})$
$2\sqrt{x^2}+4\sqrt{xy}-6\sqrt{xy}-12\sqrt{y^2}$
$2x-2\sqrt{xy}-12y$

18. $-\sqrt{\frac{9}{64}}=-\frac{3}{8}$

19. $\sqrt[3]{-\frac{8}{27}}=-\frac{2}{3}$

20. $\sqrt[3]{0.008}=\sqrt[3]{(8)(10)^{-3}}=$
$[(8)(10)^{-3}]^{\frac{1}{3}}=(8)^{\frac{1}{3}}(10)^{-3(\frac{1}{3})}=$
$(2)(10)^{-1}=0.2$

21. $32^{-\frac{1}{5}}=\frac{1}{32^{\frac{1}{5}}}=\frac{1}{2}$

22. $3^0+3^{-1}+3^{-2}$

$1+\frac{1}{3}+\frac{1}{9}$; LCD=9

$\frac{9}{9}+\frac{3}{9}+\frac{1}{9}$

$\frac{13}{9}$

23. $-9^{\frac{3}{2}}=-(9^{\frac{1}{2}})^3=-(3)^3=-27$

24. $\left(\frac{3}{4}\right)^{-2}=\frac{3^{-2}}{4^{-2}}=\frac{4^2}{3^2}=\frac{16}{9}$

25. $\frac{1}{\left(\frac{2}{3}\right)^{-3}}=\left(\frac{2}{3}\right)^3=\frac{2^3}{3^3}=\frac{8}{27}$

26. $3x^4+81x$
$3x(x^3+27)$
$3x(x+3)(x^2-3x+9)$

27. $6x^2+19x-20$
$(6x-5)(x+4)$

28. $12+13x-14x^2$
$(4+7x)(3-2x)$

29. $9x^4+68x^2-32$
$(9x^2-4)(x^2+8)$
$(3x+2)(3x-2)(x^2+8)$

30. $2ax-ay-2bx+by$
$a(2x-y)-b(2x-y)$
$(2x-y)(a-b)$

31. $27x^3-8y^3$
$(3x-2y)(9x^2+6xy+4y^2)$

32. $3(x-2)-2(3x+5)=4(x-1)$
$3x-6-6x-10=4x-4$
$-3x-16=4x-4$
$-7x=12$
$x=-\frac{12}{7}$
The solution set is $\left\{-\frac{12}{7}\right\}$.

33. $0.06n+0.08(n+50)=25$
$100[0.06n+0.08(n+50)]=100(25)$
$100(0.06n)+100[0.08(n+50)]=2500$
$6n+8(n+50)=2500$
$6n+8n+400=2500$
$14n=2100$
$n=150$
The solution set is $\{150\}$.

34. $4\sqrt{x}+5=x$
$4\sqrt{x}=x-5$
$(4\sqrt{x})^2=(x-5)^2$
$16x=x^2-10x+25$
$0=x^2-26x+25$
$0=(x-25)(x-1)$
$x-25=0$ or $x-1=0$
$x=25$ or $x=1$

Checking $x = 25$:
$4\sqrt{25} + 5 \stackrel{?}{=} 25$
$4(5) + 5 \stackrel{?}{=} 25$
$25 = 25$
Checking $x = 1$:
$4\sqrt{1} + 5 \stackrel{?}{=} 1$
$4(1) + 5 \stackrel{?}{=} 1$
$9 \neq 1$
The solution set is $\{25\}$.

35. $\sqrt[3]{n^2 - 1} = -1$
$\left(\sqrt[3]{n^2 - 1}\right)^3 = (-1)^3$
$n^2 - 1 = -1$
$n^2 = 0$
$n = 0$
Check:
$\sqrt[3]{0^2 - 1} \stackrel{?}{=} -1$
$\sqrt[3]{-1} \stackrel{?}{=} -1$
$-1 = -1$
The solution set is $\{0\}$.

36. $6x^2 - 24 = 0$
$6(x^2 - 4) = 0$
$6(x + 2)(x - 2) = 0$
$x + 2 = 0$ or $x - 2 = 0$
$x = -2$ or $x = 2$
The solution set is $\{-2, 2\}$.

37. $a^2 + 14a + 49 = 0$
$(a + 7)(a + 7) = 0$
$(a + 7)^2 = 0$
$a + 7 = 0$
$a = -7$
The solution set is $\{-7\}$.

38. $3n^2 + 14n - 24 = 0$
$(3n - 4)(n + 6) = 0$
$3n - 4 = 0$ or $n + 6 = 0$
$3n = 4$ or $n = -6$
$n = \frac{4}{3}$
The solution set is $\left\{-6, \frac{4}{3}\right\}$.

39. $\frac{2}{5x - 2} = \frac{4}{6x + 1}; x \neq \frac{2}{5}, x \neq -\frac{1}{6}$
$2(6x + 1) = 4(5x - 2)$
$12x + 2 = 20x - 8$
$-8x = -10$
$x = \frac{-10}{-8} = \frac{5}{4}$
The solution set is $\left\{\frac{5}{4}\right\}$.

40. $\sqrt{2x - 1} - \sqrt{x + 2} = 0$
$\sqrt{2x - 1} = \sqrt{x + 2}$
$(\sqrt{2x - 1})^2 = (\sqrt{x + 2})^2$
$2x - 1 = x + 2$
$x = 3$
Check:
$\sqrt{2(3) - 1} - \sqrt{3 + 2} \stackrel{?}{=} 0$
$\sqrt{5} - \sqrt{5} \stackrel{?}{=} 0$
$0 = 0$
The solution set is $\{3\}$.

41. $5x - 4 = \sqrt{5x - 4}$
$(5x - 4)^2 = (\sqrt{5x - 4})^2$
$25x^2 - 40x + 16 = 5x - 4$
$25x^2 - 45x + 20 = 0$
$5(5x^2 - 9x + 4) = 0$
$5(5x - 4)(x - 1) = 0$
$5x - 4 = 0$ or $x - 1 = 0$
$5x = 4$ or $x = 1$
$x = \frac{4}{5}$ or $x = 1$
Checking $x = \frac{4}{5}$:
$5\left(\frac{4}{5}\right) - 4 \stackrel{?}{=} \sqrt{5\left(\frac{4}{5}\right) - 4}$
$4 - 4 \stackrel{?}{=} \sqrt{4 - 4}$
$0 \stackrel{?}{=} \sqrt{0}$
$0 = 0$
Checking $x = 1$:
$5(1) - 4 \stackrel{?}{=} \sqrt{5(1) - 4}$
$5 - 4 \stackrel{?}{=} \sqrt{1}$
$1 = 1$

The solution set is $\left\{\frac{4}{5}, 1\right\}$.

42. $|3x - 1| = 11$

$3x - 1 = 11$ or $3x - 1 = -11$

$3x = 12$ or $3x = -10$

$x = 4$ or $x = -\frac{10}{3}$

The solution set is $\left\{-\frac{10}{3}, 4\right\}$.

43. $(3x - 2)(4x - 1) = 0$

$3x - 2 = 0$ or $4x - 1 = 0$

$3x = 2$ or $4x = 1$

$x = \frac{2}{3}$ or $x = \frac{1}{4}$

The solution set is $\left\{\frac{1}{4}, \frac{2}{3}\right\}$.

44. $(2x + 1)(x - 2) = 7$

$2x^2 - 4x + x - 2 = 7$

$2x^2 - 3x - 9 = 0$

$(2x + 3)(x - 3) = 0$

$2x + 3 = 0$ or $x - 3 = 0$

$2x = -3$ or $x = 3$

$x = -\frac{3}{2}$

The solution set is $\left\{-\frac{3}{2}, 3\right\}$.

45. $\frac{5}{6x} - \frac{2}{3} = \frac{7}{10x}; x \neq 0$

$30x\left(\frac{5}{6x} - \frac{2}{3}\right) = 30x\left(\frac{7}{10x}\right)$

$30x\left(\frac{5}{6x}\right) - 30x\left(\frac{2}{3}\right) = 21$

$25 - 20x = 21$

$-20x = -4$

$x = \frac{-4}{-20} = \frac{1}{5}$

The solution set is $\left\{\frac{1}{5}\right\}$.

46. $\frac{3}{y+4} + \frac{2y-1}{y^2-16} = \frac{-2}{y-4}$

$\frac{3}{y+4} + \frac{2y-1}{(y+4)(y-4)} = \frac{-2}{y-4}$

LCD $= (y + 4)(y - 4); y \neq -4, y \neq 4$

$(y+4)(y-4)[\frac{3}{y+4} + \frac{2y-1}{(y+4)(y-4)}] = (y+4)(y-4)\left(\frac{-2}{y-4}\right)$

$(y+4)(y-4)\left(\frac{3}{y+4}\right) + (y+4)(y-4)[\frac{2y-1}{(y+4)(y-4)}] = (y+4)(y-4)\left(\frac{-2}{y-4}\right)$

$3(y - 4) + 2y - 1 = -2y - 8$

$3y - 12 + 2y - 1 = -2y - 8$

$5y - 13 = -2y - 8$

$7y = 5$

$y = \frac{5}{7}$

The solution set is $\left\{\frac{5}{7}\right\}$.

47. $6x^4 - 23x^2 - 4 = 0$
$(6x^2 + 1)(x^2 - 4) = 0$
$(6x^2 + 1)(x + 2)(x - 2) = 0$
$6x^2 + 1 = 0$ or $x + 2 = 0$ or $x - 2 = 0$
$6x^2 = -1$ or $x = -2$ or $x = 2$
$x^2 = -\frac{1}{6}$
not a real number
The solution set is $\{-2, 2\}$.

48. $3n^3 + 3n = 0$
$3n(n^2 + 1) = 0$
$3n = 0$ or $n^2 + 1 = 0$
$n = 0$ or $n^2 = -1$
not a real number
The solution set is $\{0\}$.

49. $n^2 - 13n - 114 = 0$
$(n + 6)(n - 19) = 0$
$n + 6 = 0$ or $n - 19 = 0$
$n = -6$ or $n = 19$
The solution set is $\{-6, 19\}$.

50. $12x^2 + x - 6 = 0$
$(4x + 3)(3x - 2) = 0$
$4x + 3 = 0$ or $3x - 2 = 0$
$4x = -3$ or $3x = 2$
$x = -\frac{3}{4}$ or $x = \frac{2}{3}$
The solution set is $\left\{-\frac{3}{4}, \frac{2}{3}\right\}$.

51. $x^2 - 2x + 26 = 0$

$$x = \frac{-(-2) \pm \sqrt{(-2)^2 - 4(1)(26)}}{2(1)}$$

$$x = \frac{2 \pm \sqrt{4 - 104}}{2}$$

$$x = \frac{2 \pm \sqrt{-100}}{2}$$

$$x = \frac{2 \pm 10i}{2} = \frac{2(1 \pm 5i)}{2}$$

$x = 1 \pm 5i$
The solution set is $\{1 - 5i, 1 + 5i\}$.

52. $(x + 2)(x - 6) = -15$
$x^2 - 4x - 12 = -15$
$x^2 - 4x + 3 = 0$
$(x - 3)(x - 1) = 0$
$x - 3 = 0$ or $x - 1 = 0$
$x = 3$ or $x = 1$
The solution set is $\{1, 3\}$.

53. $(3x - 1)(x + 4) = 0$
$3x - 1 = 0$ or $x + 4 = 0$
$3x = 1$ or $x = -4$
$x = \frac{1}{3}$ or $x = -4$
The solution set is $\left\{-4, \frac{1}{3}\right\}$.

54. $x^2 + 4x + 20 = 0$

$$x = \frac{-4 \pm \sqrt{4^2 - 4(1)(20)}}{2(1)}$$

$$x = \frac{-4 \pm \sqrt{16 - 80}}{2}$$

$$x = \frac{-4 \pm \sqrt{-64}}{2}$$

$$x = \frac{-4 \pm 8i}{2} = \frac{2(-2 \pm 4i)}{2}$$

$x = -2 \pm 4i$
The solution set is $\{-2 - 4i, -2 + 4i\}$.

55. $2x^2 - x - 4 = 0$

$$x = \frac{-(-1) \pm \sqrt{(-1)^2 - 4(2)(-4)}}{2(2)}$$

$$x = \frac{1 \pm \sqrt{1 + 32}}{4}$$

$$x = \frac{1 \pm \sqrt{33}}{4}$$

The solution set is $\left\{\frac{1 - \sqrt{33}}{4}, \frac{1 + \sqrt{33}}{4}\right\}$.

56. $6 - 2x \geq 10$
$-2x \geq 4$
$-\frac{1}{2}(-2x) \leq -\frac{1}{2}(4)$
$x \leq -2$
The solution set is $(-\infty, -2]$.

57. $4(2x - 1) < 3(x + 5)$
$8x - 4 < 3x + 15$
$8x < 3x + 19$
$5x < 19$
$x < \frac{19}{5}$
The solution set is $\left(-\infty, \frac{19}{5}\right)$.

58. $\frac{n+1}{4} + \frac{n-2}{12} > \frac{1}{6}$

$12\left(\frac{n+1}{4} + \frac{n-2}{12}\right) > 12\left(\frac{1}{6}\right)$

$12\left(\frac{n+1}{4}\right) + 12\left(\frac{n-2}{12}\right) > 2$

$3(n + 1) + n - 2 > 2$
$3n + 3 + n - 2 > 2$
$4n + 1 > 2$
$4n > 1$
$n > \frac{1}{4}$
The solution set is $\left(\frac{1}{4}, \infty\right)$.

59. $|2x - 1| < 5$
$-5 < 2x - 1 < 5$
$-4 < 2x < 6$
$-2 < x < 3$
The solution set is $(-2, 3)$.

60. $|3x + 2| > 11$
$3x + 2 < -11$ or $3x + 2 > 11$
$3x < -13$ or $3x > 9$
$x < -\frac{13}{3}$ or $x > 3$
The solution set is $\left(-\infty, -\frac{13}{3}\right) \cup (3, \infty)$.

61. $\frac{1}{2}(3x - 1) - \frac{2}{3}(x + 4) \leq \frac{3}{4}(x - 1)$

$12\left[\frac{1}{2}(3x - 1) - \frac{2}{3}(x + 4)\right] \leq 12\left[\frac{3}{4}(x - 1)\right]$

$12\left[\frac{1}{2}(3x - 1)\right] - 12\left[\frac{2}{3}(x + 4)\right] \leq 9(x - 1)$

$6(3x - 1) - 8(x + 4) \leq 9x - 9$
$18x - 6 - 8x - 32 \leq 9x - 9$
$10x - 38 \leq 9x - 9$
$10x \leq 9x + 29$
$x \leq 29$
The solution set is $(-\infty, 29]$.

62. $x^2 - 2x - 8 \leq 0$
$x^2 - 2x - 8 = 0$
$(x - 4)(x + 2) = 0$
$x - 4 = 0$ or $x + 2 = 0$
$x = 4$ or $x = -2$

	-2		4
Test Point	-3	0	5
$x - 4$:	negative	negative	positive
$x + 2$:	negative	positive	positive
product:	positive	negative	positive

The solution set is $[-2, 4]$.

63. $3x^2 + 14x - 5 > 0$
$3x^2 + 14x - 5 = 0$
$(3x - 1)(x + 5) = 0$
$3x - 1 = 0$ or $x + 5 = 0$
$x = \frac{1}{3}$ or $x = -5$

	-5		$\frac{1}{3}$
Test Point	-6	0	1
$3x - 1$:	negative	negative	positive
$x + 5$:	negative	positive	positive
product:	positive	negative	positive

The solution set is $(-\infty, -5) \cup \left(\frac{1}{3}, \infty\right)$.

64. $\frac{x+2}{x-7} \geq 0;\ x \neq 7$
$x + 2 = 0$ or $x - 7 = 0$
$x = -2$ or $x = 7$

Test Point	-3	0	8
	(-2)		(7)
$x+2$:	negative	positive	positive
$x-7$:	negative	negative	positive
quotient:	positive	negative	positive

The solution set is $(-\infty, -2] \cup (7, \infty)$.

65. $\dfrac{2x-1}{x+3} < 1$

$\dfrac{2x-1}{x+3} - 1 < 0$

$\dfrac{2x-1}{x+3} - \dfrac{x+3}{x+3} < 0$

$\dfrac{2x-1-(x+3)}{x+3} < 0$

$\dfrac{2x-1-x-3}{x+3} < 0$

$\dfrac{x-4}{x+3} < 0; x \neq -3$

$x-4=0$ or $x+3=0$
$x=4$ or $x=-3$

Test Point	-4	0	5
	(-3)		(4)
$x-4$:	negative	negative	positive
$x+3$:	negative	positive	positive
quotient:	positive	negative	positive

The solution set is $(-3, 4)$.

66. Let $x =$ liters of 60% solution.
$(60\%)(x) + (10\%)(14) = (25\%)(x+14)$
$0.60x + 0.10(14) = 0.25(x+14)$
$100[0.60x + 0.10(14)] = 100[0.25(x+14)]$
$60x + 10(14) = 25(x+14)$
$60x + 140 = 25x + 350$
$35x = 210$
$x = 6$
There needs to be 6 liters of 60% acid solution added.

67. Let $x =$ one part and
$2250 - x =$ other part.
$\dfrac{x}{2250-x} = \dfrac{2}{3}$
$3x = 2(2250 - x)$
$3x = 4500 - 2x$
$5x = 4500$
$x = 900$
It is divided \$900 and \$1350.

68. Dimensions of picture :
width $= x$
length $= 2x - 7$
Area $= x(2x-7) = 2x^2 - 7x$
Dimensions of picture and border
width $= x + 2$
length $= 2x - 5$
Area $= (x+2)(2x-5) = 2x^2 - x - 10$

Area of Border $=$
$2x^2 - x - 10 - (2x^2 - 7x) =$
$2x^2 - x - 10 - 2x^2 + 7x = 6x - 10$
$6x - 10 = 62$
$6x = 72$
$x = 12$
The picture is 12 inches by 17 inches.

69.

	Time in hours	Rate
Lolita	x	$\frac{1}{x}$
Doug	10	$\frac{1}{10}$
Together	$3\frac{20}{60} = \frac{10}{3}$	$\frac{1}{\frac{10}{3}} = \frac{3}{10}$

$\dfrac{1}{x} + \dfrac{1}{10} = \dfrac{3}{10}$

$10x\left(\dfrac{1}{x} + \dfrac{1}{10}\right) = 10x\left(\dfrac{3}{10}\right)$

$10x\left(\dfrac{1}{x}\right) + 10x\left(\dfrac{1}{10}\right) = 3x$

$10 + x = 3x$
$10 = 2x$
$5 = x$
It would take Lolita 5 hours.

70.

	Price/ golfball	Number of golfballs	Cost
1st	$\frac{14}{x}$	x	14
Re-Order	$\frac{14}{x+1}$	$x+1$	14

$\frac{14}{x} - 0.25 = \frac{14}{x+1}; x \neq -1, x \neq 0$

$\frac{14}{x} - \frac{1}{4} = \frac{14}{x+1}$

$4x(x+1)\left(\frac{14}{x} - \frac{1}{4}\right) = 4x(x+1)\left(\frac{14}{x+1}\right)$

$4(x+1)(14) - x(x+1) = 4x(14)$
$56x + 56 - x^2 - x = 56x$
$0 = x^2 + x - 56$
$0 = (x+8)(x-7)$
$x + 8 = 0$ or $x - 7 = 0$
$x = -8$ or $x = 7$
Discard the root $x = -8$.
She bought 7 golf balls.

71.

	Rate	Time	Distance
First Jogger	$\frac{1}{8}$ mpm	x	$\frac{1}{8}x$
Second Jogger	$\frac{1}{6}$ mpm	x	$\frac{1}{6}x$

$\frac{1}{2} + \frac{1}{8}x = \frac{1}{6}x$

$24\left(\frac{1}{2} + \frac{1}{8}x\right) = 24\left(\frac{1}{6}x\right)$

$24\left(\frac{1}{2}\right) + 24\left(\frac{1}{8}x\right) = 24\left(\frac{1}{6}x\right)$

$12 + 3x = 4x$
$12 = x$
It would take 12 minutes.

72. $P = 100 \quad t = 2 \quad A = 114.49$
$A = P(1+r)^t$
$114.49 = 100(1+r)^2$
$1.1449 = (1+r)^2$
$1.1449 = 1 + 2r + r^2$
$0 = -0.1449 + 2r + r^2$
$r^2 + 2r - 0.1449 = 0$

$r = \frac{-2 \pm \sqrt{2^2 - 4(1)(-0.1449)}}{2(1)}$

$r = \frac{-2 \pm \sqrt{4 + 0.5796}}{2}$

$r = \frac{-2 \pm \sqrt{4.5796}}{2}$

$r = \frac{-2 - \sqrt{4.5796}}{2}$ or $r = \frac{-2 + \sqrt{4.5796}}{2}$
$r = -2.07$ or $r = 0.07$
Discard the negative solution.
$r = 0.07$
The rate of interest is 7%.

73. Let $x =$ the number of rows,
then $2x - 1 =$ the number of chairs per row.

$x(2x-1) = 120$
$2x^2 - x = 120$
$2x^2 - x - 120 = 0$
$(2x+15)(x-8) = 0$
$2x + 15 = 0$ or $x - 8 = 0$
$2x = -15$ or $x = 8$
$x = -\frac{15}{2}$
Discard the negative solution.
There will be 8 rows with
$2x - 1 = 2(8) - 1 = 15$ chairs per row.

74. Let $x =$ number of shares purchased. Then $\frac{2800}{x}$ is the price per share when purchased and $\frac{2800}{x} + 6$ is the price per share a month later. Bjorn sold all but 60 shares so the number of shares sold can be expressed as $x - 60$. Then the price per share times the number of shares sold equals $2800.

$$\left(\frac{2800}{x} + 6\right)(x - 60) = 2800$$

$$2800 - \frac{168000}{x} + 6x - 360 = 2800$$

$$-\frac{168000}{x} + 6x - 360 = 0$$

$$x\left(\frac{-168000}{x} + 6x - 360\right) = x(0)$$

$-168000 + 6x^2 - 360x = 0$
$6x^2 - 360x - 168000 = 0$
$6(x^2 - 60x - 28000) = 0$
$x^2 - 60x - 28000 = 0$
$(x - 200)(x + 140) = 0$
$x - 200 = 0$ or $x + 140 = 0$
$x = 200$ or $x = -140$
Discard the negative solution.
Bjorn sold $x - 60 = 200 - 60 = 140$ shares.

Chapter 7 Linear Equations and Inequalities in Two Variables

PROBLEM SET 7.1 Rectangular Coordinate System and Linear Equations

1. $x + 2y = 4$

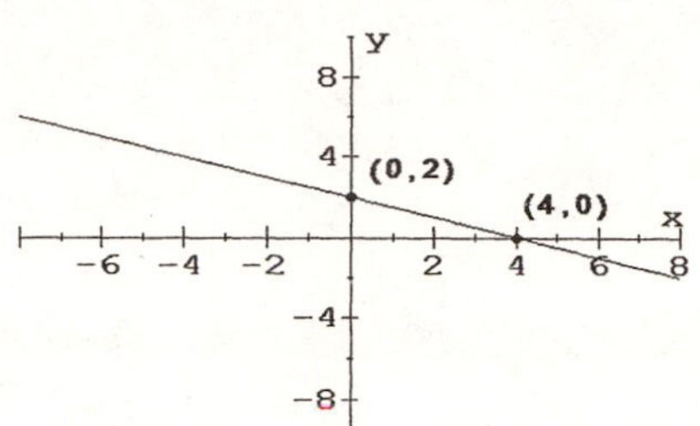

3. $2x - y = 2$

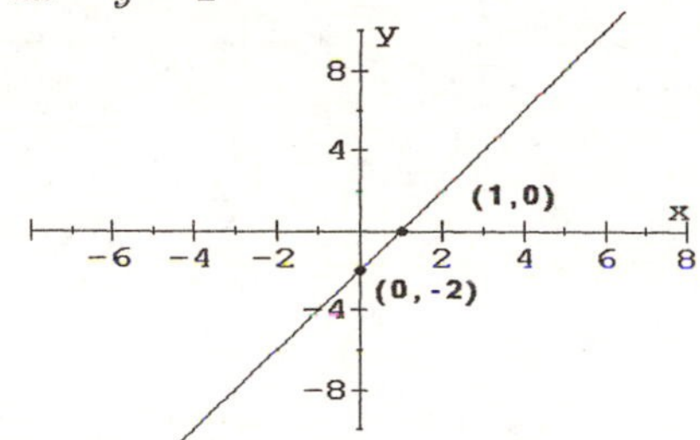

5. $3x + 2y = 6$

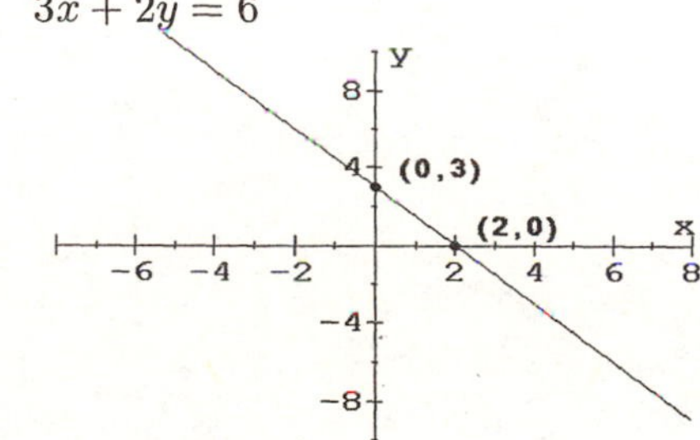

7. $5x - 4y = 20$

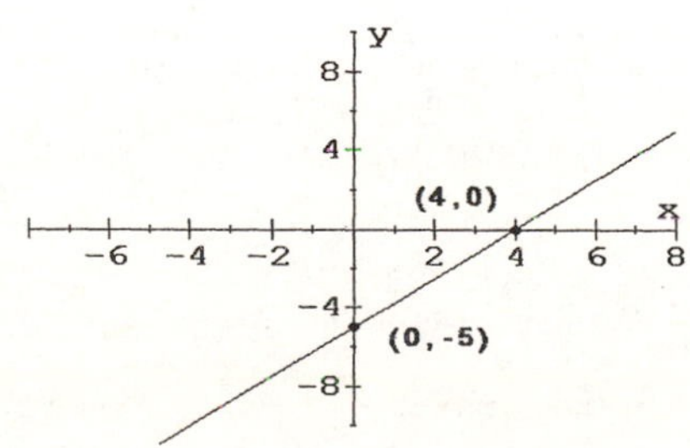

9. $x + 4y = -6$

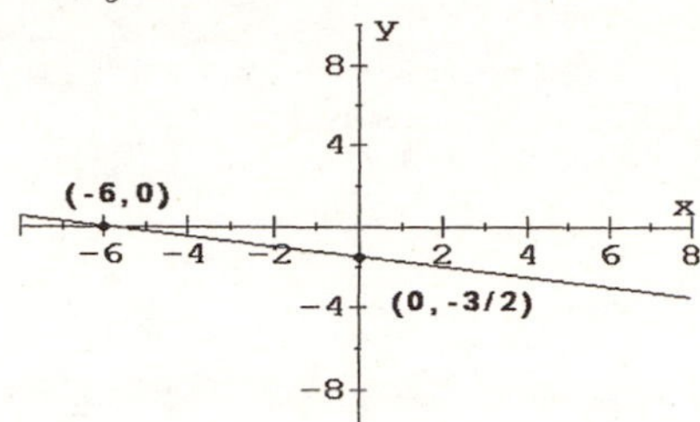

11. $-x - 2y = 3$

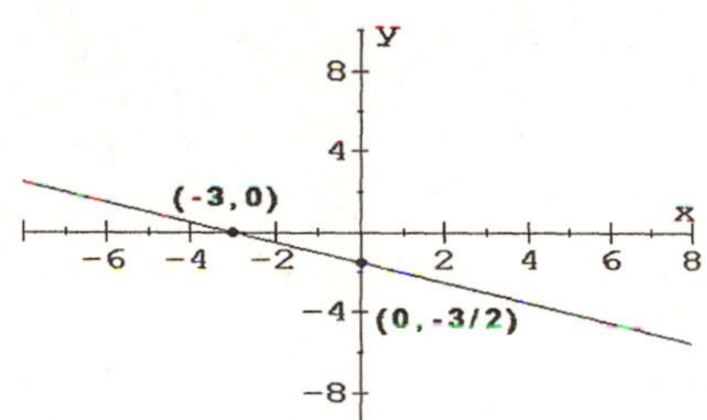

13. $y = x + 3$

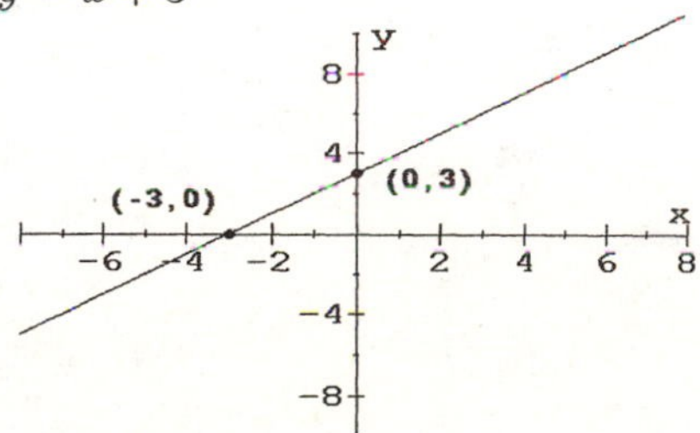

15. $y = -2x - 1$

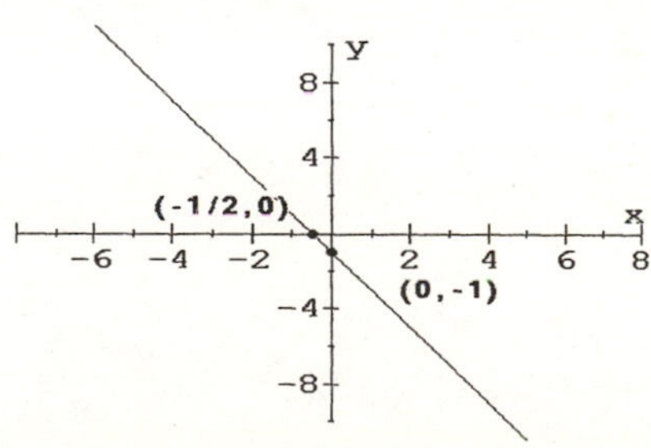

Problem Set 7.1

17. $y = \frac{1}{2}x + \frac{2}{3}$

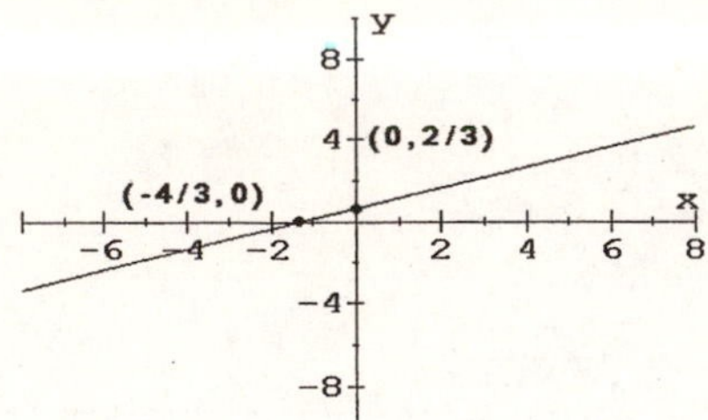

25. $y = -\frac{1}{4}x + \frac{1}{6}$

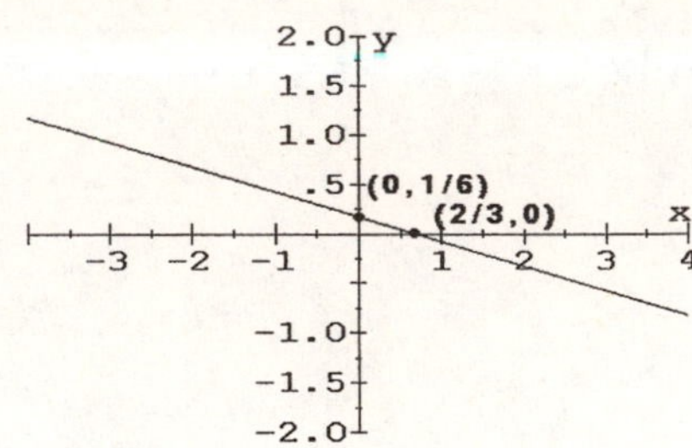

19. $y = -x$

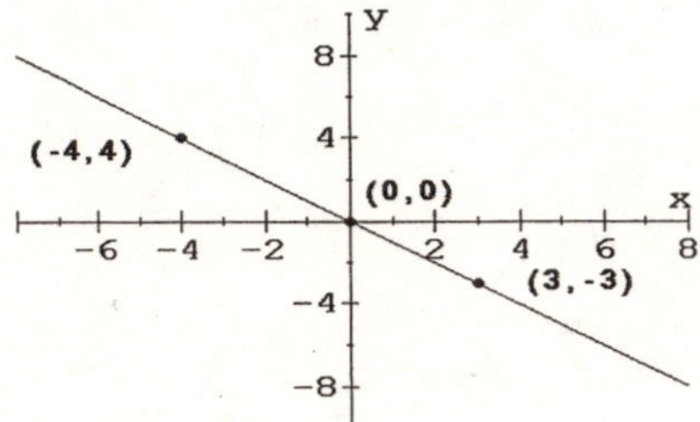

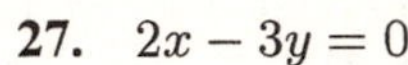

27. $2x - 3y = 0$

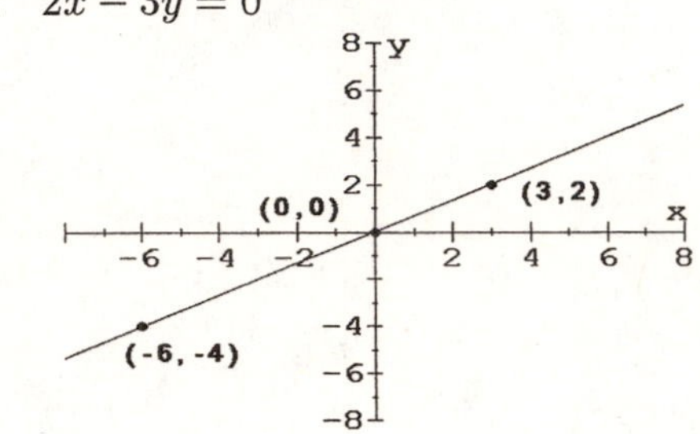

21. $y = 3x$

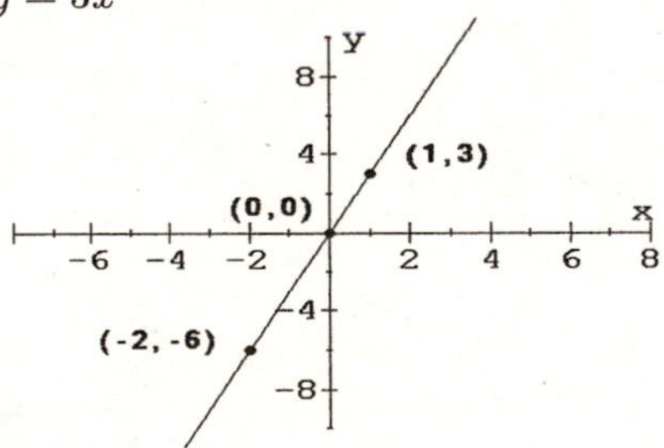

29. $x = 0$

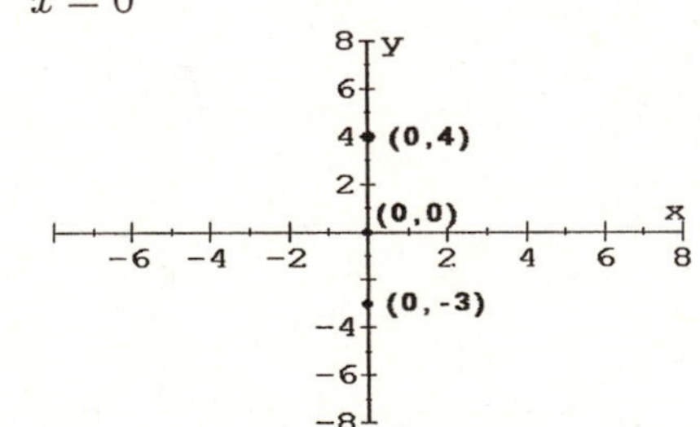

23. $x = 2y - 1$

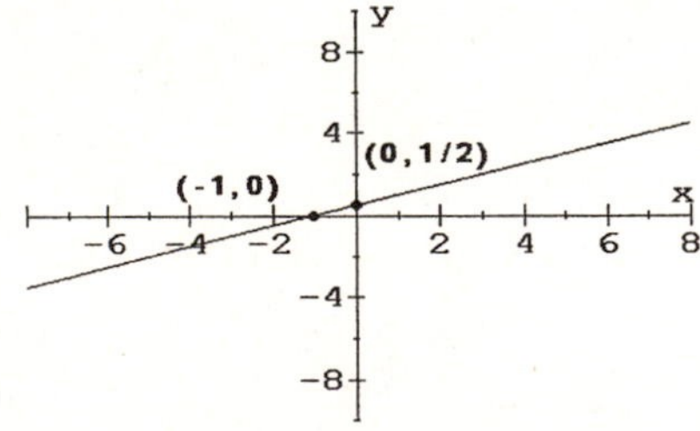

31. $y = 2$

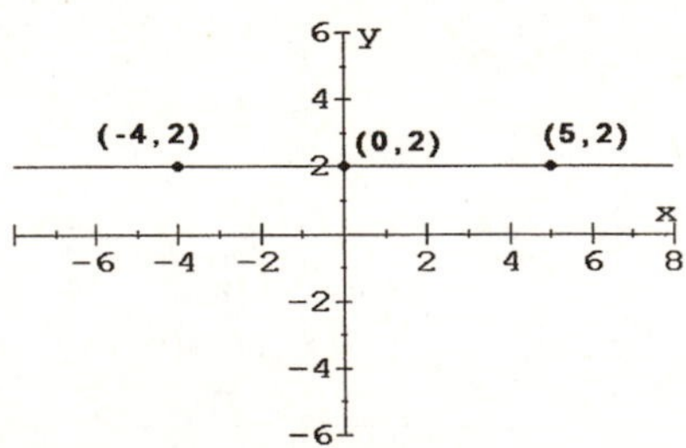

33. $-3y = -x + 3$

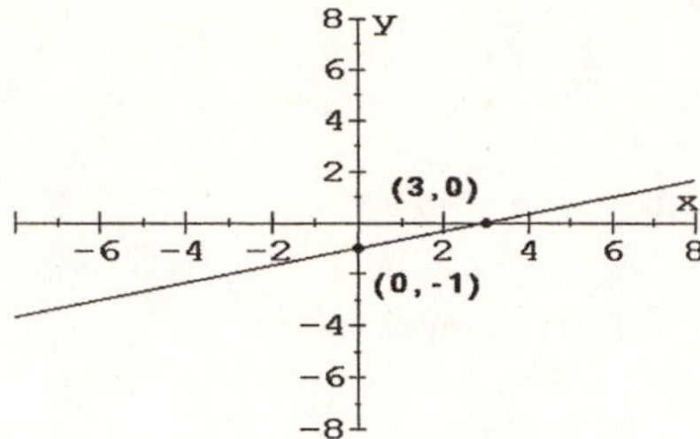

35. $c = 3t + 5$

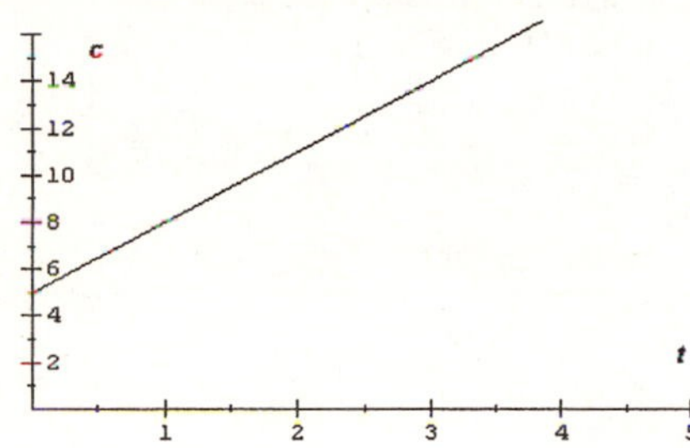

37. $C = 0.30p$

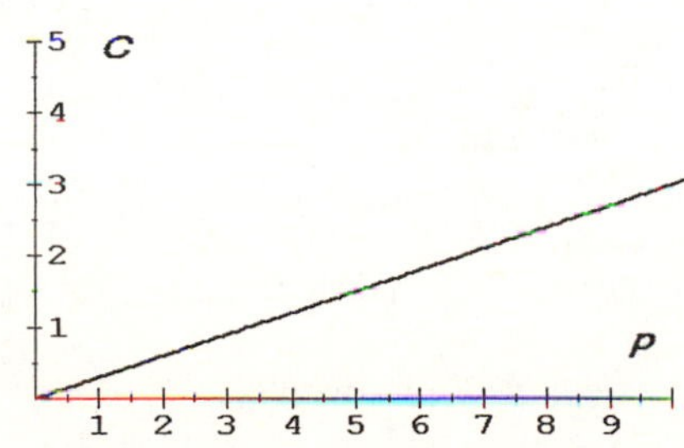

39. **(a)** $c = 0.25m + 10$

$m = 5 \quad c = 0.25(5) + 10 = 11.25$

$m = 10 \quad c = 0.25(10) + 10 = 12.50$

$m = 15 \quad c = 0.25(15) + 10 = 13.75$

$m = 20 \quad c = 0.25(20) + 10 = 15.00$

$m = 30 \quad c = 0.25(30) + 10 = 17.50$

$m = 60 \quad c = 0.25(60) + 10 = 25.00$

(b) Graph $c = 0.25m + 10$

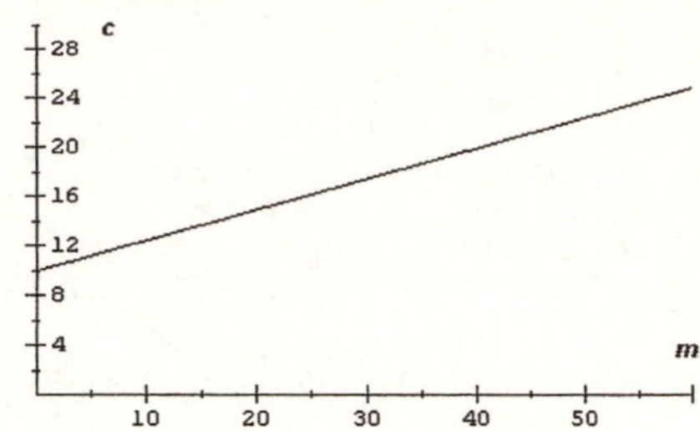

(c) approximate values for $m = 25, 40, 45$

For $m = 25$, $c \approx$ \$15.00

For $m = 40$, $c \approx$ \$18.00

For $m = 45$, $c \approx$ \$20.00

(d) $m = 25$, $c = 0.25(25) + 10 = 16.25$

$m = 40$, $c = 0.25(40) + 10 = 20.00$

$m = 45$, $c = 0.25(45) + 10 = 21.25$

Further Investigations

45. $\left|x - y\right| = 4$

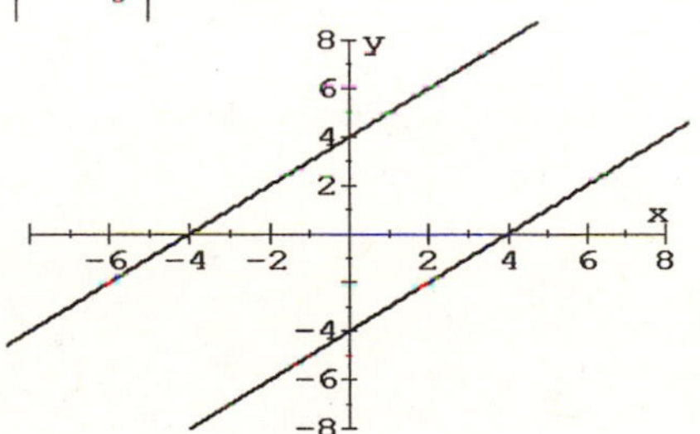

47. $\left|3x + 2y\right| = 6$

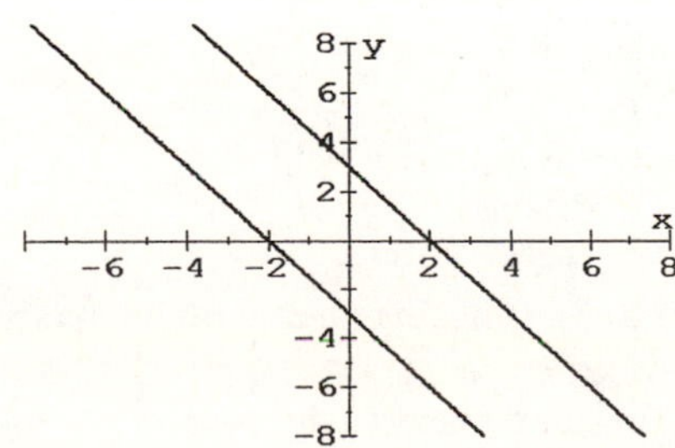

PROBLEM SET 7.2 Graphing Nonlinear Equations

1.

given point (-3, 1)
reflection across y-axis (3, 1)
reflection across x-axis (-3, -1)
reflection through origin (3, -1)

3.

reflection through origin (-7, 2)
reflection across x-axis (7, 2)
reflection across y-axis (-7, -2)
given point (7, -2)

5.

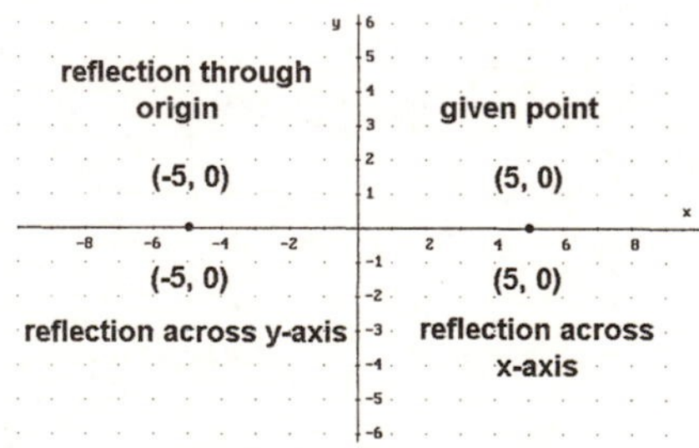

7. $-3x + 2y^2 = -4$

To test symmetry for:
x-axis replace y with $-y$
$-3x + 2(-y)^2 = -4$
$-3x + 2y^2 = -4$
Yes, it is symmetric to the x-axis.

y-axis replace x with $-x$
$-3(-x) + 2y^2 = -4$
$3x + 2y^2 = -4$
No, it is not symmetric to the y-axis.

origin replace x with $-x$
and y with $-y$.
$-3(-x) + 2(-y)^2 = -4$
$3x + 2y^2 = -4$
No, it is not symmetric to the origin.

9. $y = 4x^2 + 13$

To test symmetry for:
x-axis replace y with $-y$
$-y = 4x^2 + 13$
No, it is not symmetric to the x-axis.

y-axis replace x with $-x$
$y = 4(-x)^2 + 13$
$y = 4x^2 + 13$
Yes, it is symmetric to the y-axis.

origin replace x with $-x$
and y with $-y$
$-y = 4(-x)^2 + 13$
$-y = 4x^2 + 13$
No, it is not symmetric to the origin.

11. $2x^2y^2 = 5$

To test symmetry for:
x-axis replace y with $-y$
$2x^2(-y)^2 = 5$
$2x^2y^2 = 5$
Yes, it is symmetric to the x-axis.

y-axis replace x with $-x$
$2(-x)^2y^2 = 5$
$2x^2y^2 = 5$
Yes, it is symmetric to the y-axis.

origin replace x with $-x$
and y with $-y$
$2(-x)^2(-y)^2 = 5$
$2x^2y^2 = 5$
Yes, it is symmetric to the origin.

13. $x^2 - 2x - y^2 = 4$
To test symmetry for:
x-axis replace y with $-y$
$x^2 - 2x - (-y)^2 = 4$
$x^2 - 2x - y^2 = 4$
Yes, it is symmetric to the x-axis.

y-axis replace x with $-x$
$(-x)^2 - 2(-x) - y^2 = 4$
$x^2 + 2x - y^2 = 4$
No, it is not symmetric to the y-axis.

origin replace x with $-x$
and y with $-y$
$(-x^2) - 2(-x) - (-y)^2 = 4$
$x^2 + 2x - y^2 = 4$
No, it is not symmetric to the origin.

15. $y = 2x^2 - 7x - 3$

To test symmetry for:
x-axis replace y with $-y$
$-y = 2x^2 - 7x - 3$
No, it is not symmetric to the x-axis.

y-axis replace x with $-x$
$y = 2(-x)^2 - 7(-x) - 3$
$y = 2x^2 + 7x - 3$
No, it is not symmetric to the y-axis.

origin replace x with $-x$
and y with $-y$
$-y = 2(-x)^2 - 7(-x) - 3$
$-y = 2x^2 + 7x - 3$
No, it is not symmetric to the origin.

17. $y = 2x$

To test symmetry for:
x-axis replace y with $-y$
$-y = 2x$
No, it is not symmetric to the x-axis.

y-axis replace x with $-x$
$y = 2(-x)$
$y = -2x$
No, it is not symmetric to the y-axis.

origin replace x with $-x$
and y with $-y$
$-y = 2(-x)$
$-y = -2x$
$-1(-y) = -1(-2x)$
$y = 2x$
Yes, it is symmetric to the origin.

19. $y = x^4 - x^2 + 2$

To test symmetry for:
x-axis replace y with $-y$
$-y = x^4 - x^2 + 2$
No, it is not symmetric to the x-axis.

y-axis replace x with $-x$
$y = (-x)^4 - (-x)^2 + 2$
$y = x^4 - x^2 + 2$
Yes, it is symmetric to the y-axis.

origin replace x with $-x$
and y with $-y$
$-y = (-x)^4 - (-x)^2 + 2$
$-y = x^4 - x^2 + 2$
No, it is not symmetric to the origin.

21. $x^2 - y^2 = -6$

To test symmetry for:
x-axis replace y with $-y$
$x^2 - (-y)^2 = -6$
$x^2 - y^2 = -6$
Yes, it is symmetric to the x-axis.

y-axis replace x with $-x$
$(-x)^2 - y^2 = -6$
$x^2 - y^2 = -6$
Yes, it is symmetric to the y-axis.

origin replace x with $-x$
and y with $-y$
$(-x)^2 - (-y)^2 = -6$
$x^2 - y^2 = -6$
Yes, it is symmetric to the origin.

23. $x = -y^2 + 9$

To test symmetry for:
x-axis replace y with $-y$
$x = -(-y^2) + 9$
$x = -y^2 + 9$
Yes, it is symmetric to the x-axis.

y-axis replace x with $-x$
$-x = -y^2 + 9$

$x = y^2 - 9$
No, it is not symmetric to the y-axis.

origin replace x with $-x$
and y with $-y$
$-x = -(-y)^2 + 9$
$x = y^2 + 9$
No, it is not symmetric to the origin.

25. $2x^2 + 3y^2 + 8y + 2 = 0$

To test symmetry for:
x-axis replace y with $-y$
$2x^2 + 3(-y)^2 + 8(-y) + 2 = 0$
$2x^2 + 3y^2 - 8y + 2 = 0$
No, it is not symmetric to the x-axis.

y-axis replace x with $-x$
$2x^2 + 3y^2 + 8y + 2 = 0$
$2(-x)^2 + 3y^2 + 8y + 2 = 0$
$2x^2 + 3y^2 + 8y + 2 = 0$
Yes, it is symmetric to the y-axis.

origin replace x with $-x$
and y with $-y$
$2(-x)^2 + 3(-y)^2 + 8(-y) + 2 = 0$
$2x^2 + 3y^2 - 8y + 2 = 0$
No, it is not symmetric to the origin.

27. $y = x - 4$

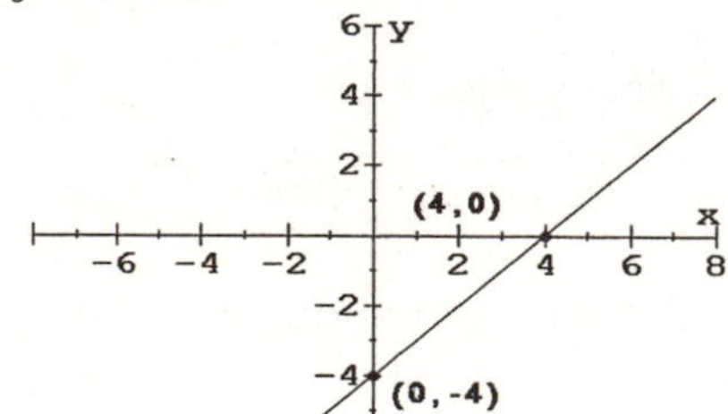

29. $y = 2x + 4$

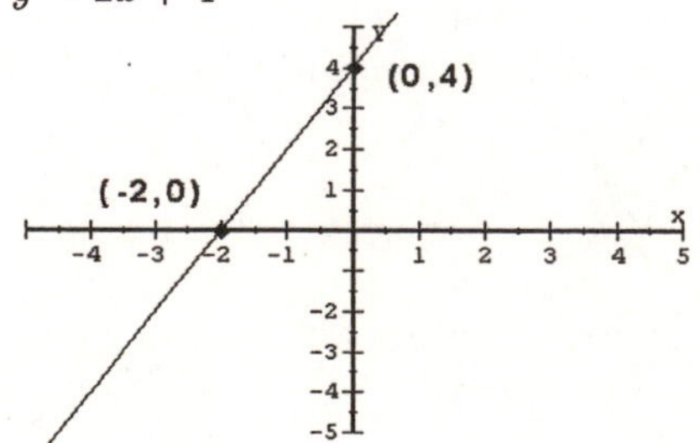

31. $y = -3x - 1$

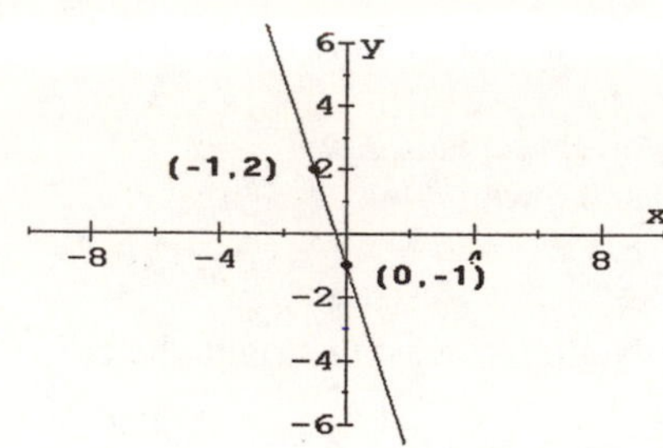

33. $y = -\frac{1}{3}x + 2$

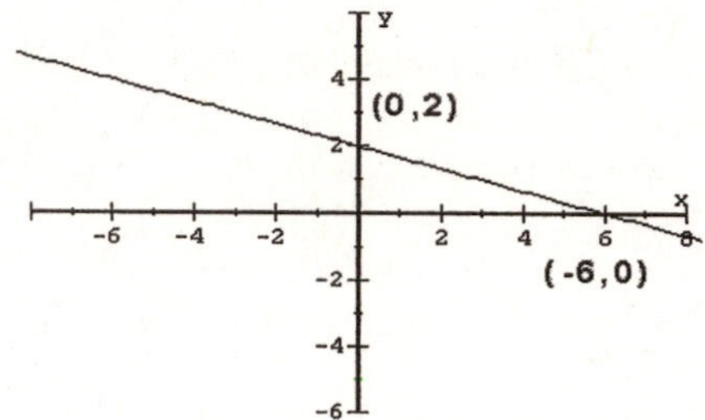

35. $y = \frac{1}{2}x$

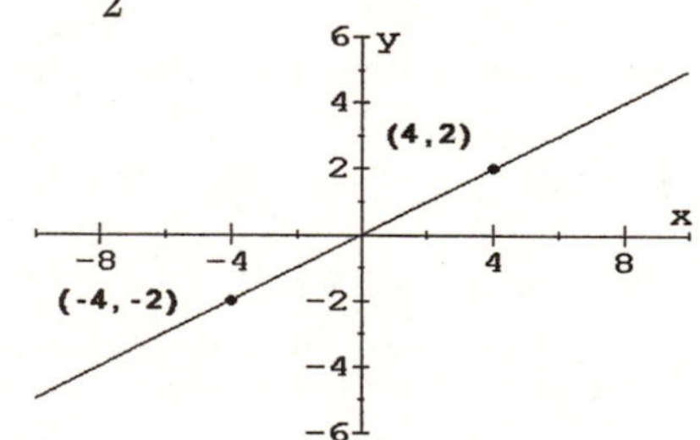

37. $2x - y = 4$

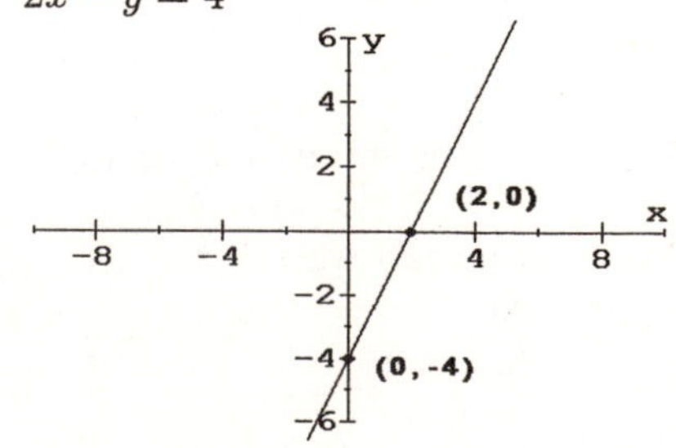

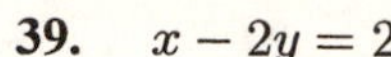

39. $x - 2y = 2$

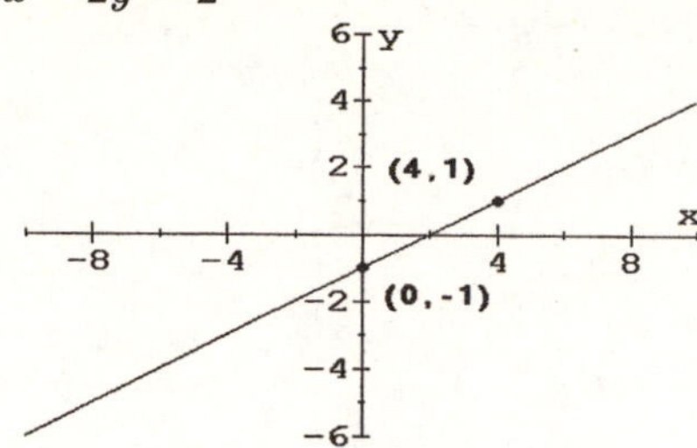

41. $y = x^2 + 2$

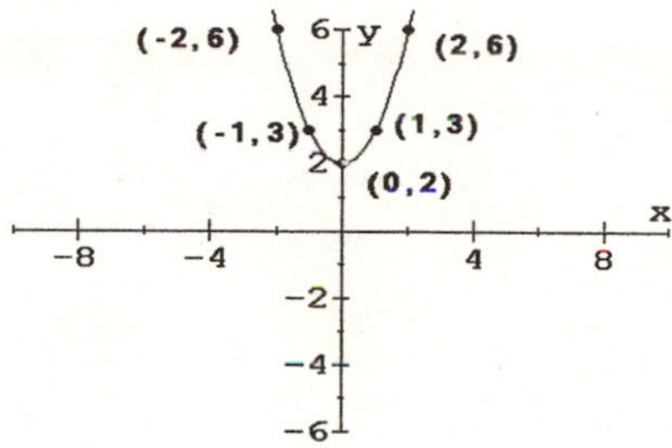

43. $y = x^3$

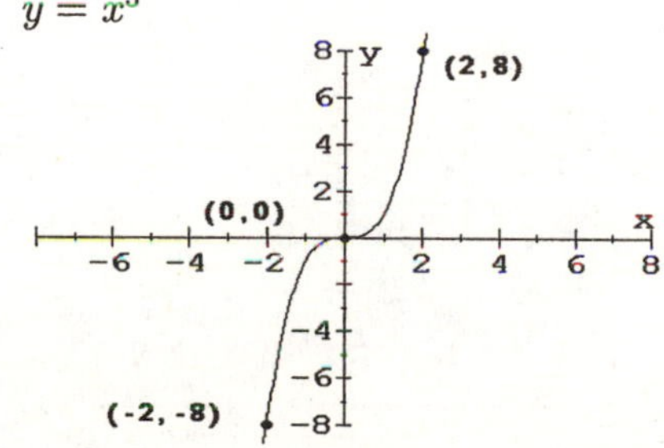

45. $y = \dfrac{-1}{x^2}$

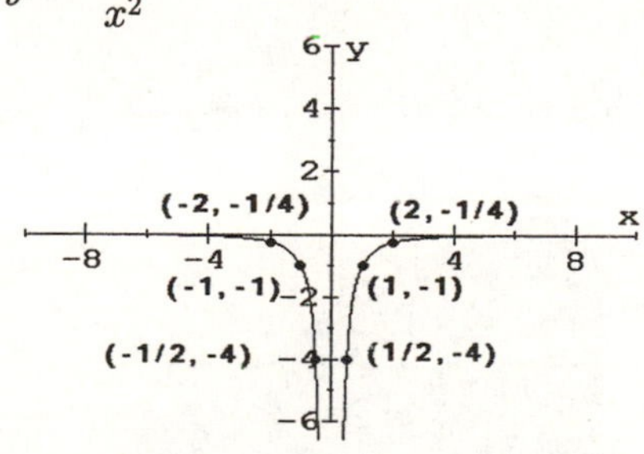

47. $y = -3x^2$

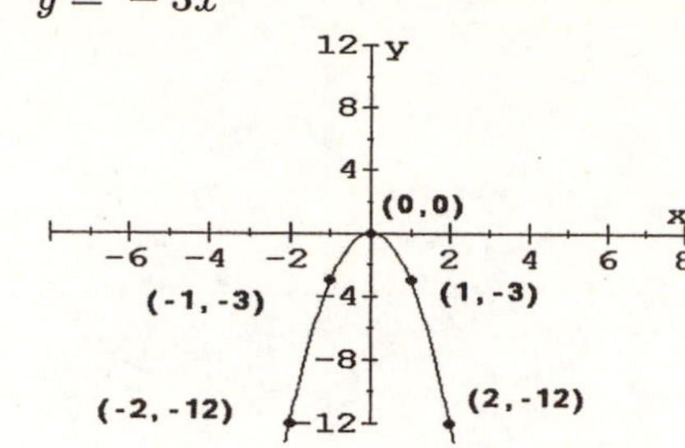

49. $xy = 2$

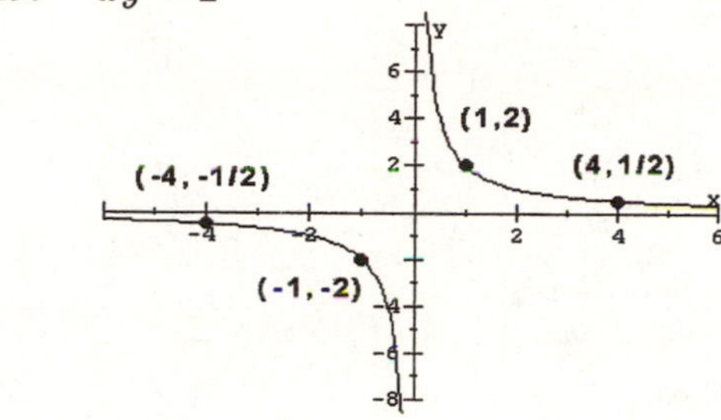

51. $xy^2 = -4$

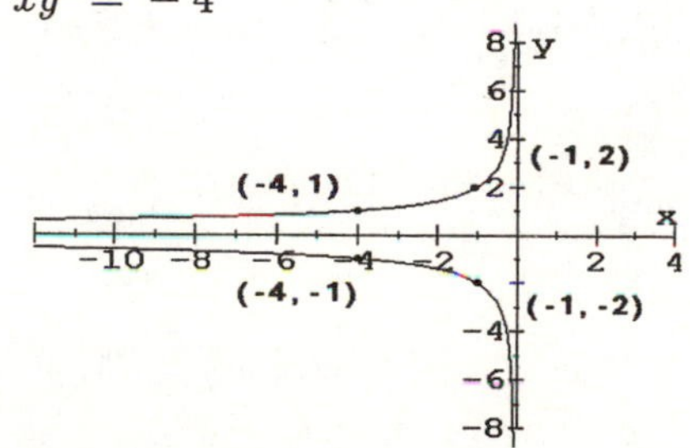

53. $y^2 = x^3$

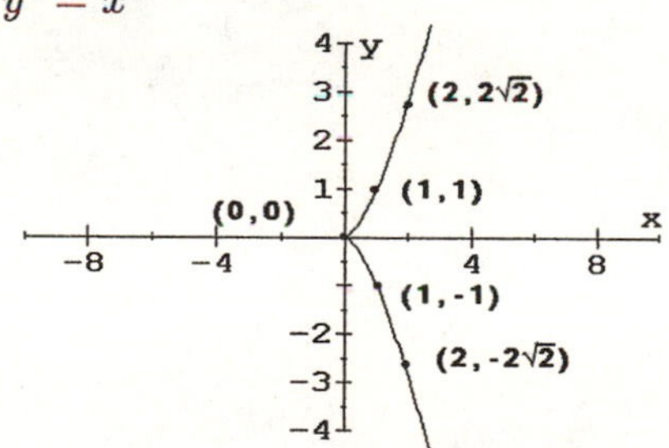

55. $y = \dfrac{4}{x^2+1}$

59. $x = -y^3 + 2$

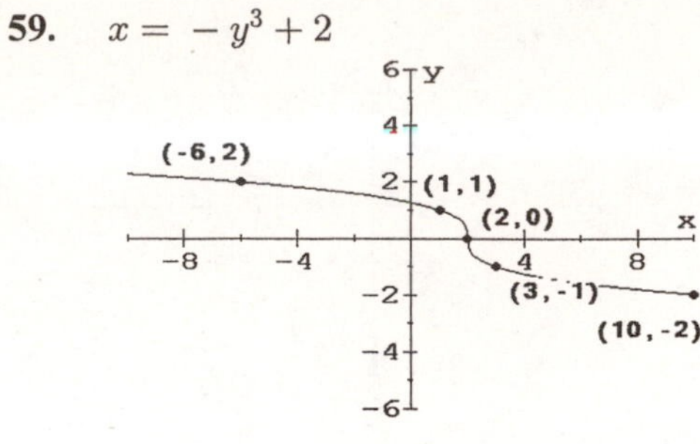

57. $y = x^4$

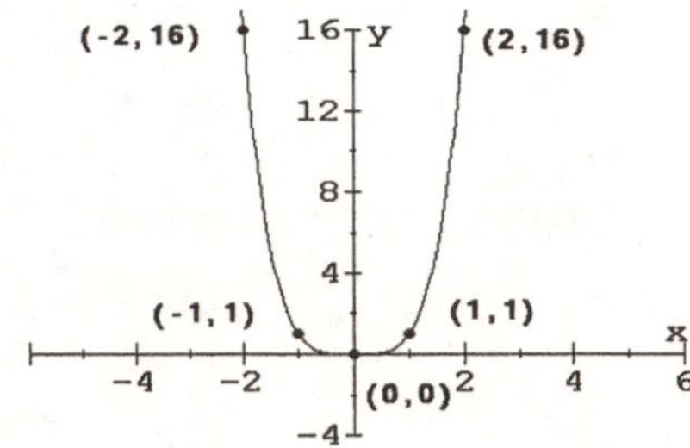

PROBLEM SET 7.3 Linear Inequalities in Two Variables

1. $x - y > 2$

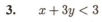

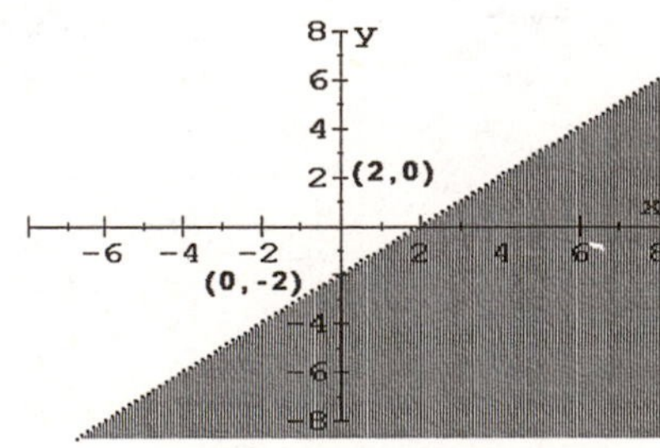

5. $2x + 5y \geq 10$

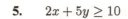

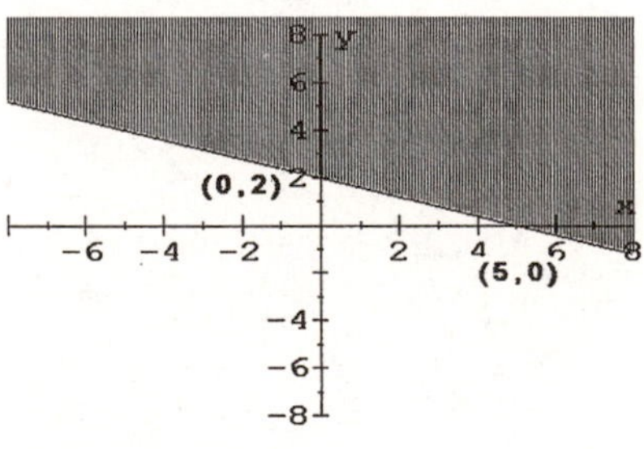

3. $x + 3y < 3$

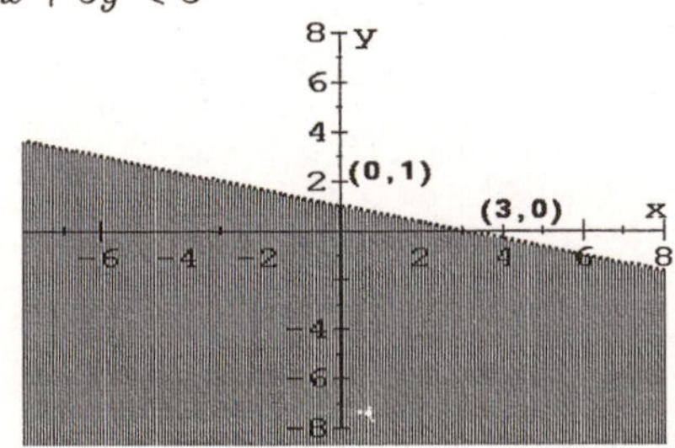

7. $y \leq -x + 2$

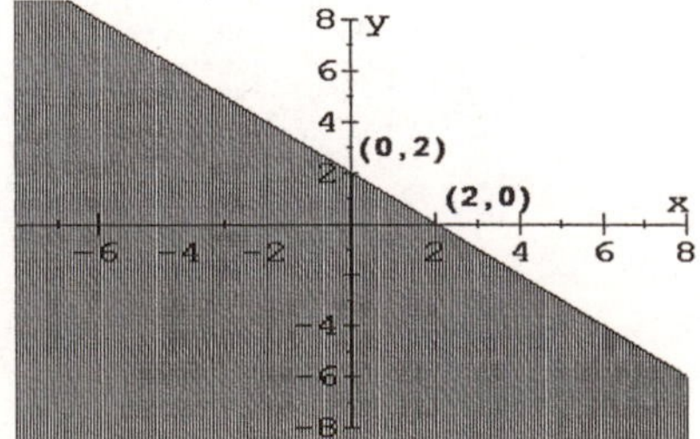

9. $y > -x$

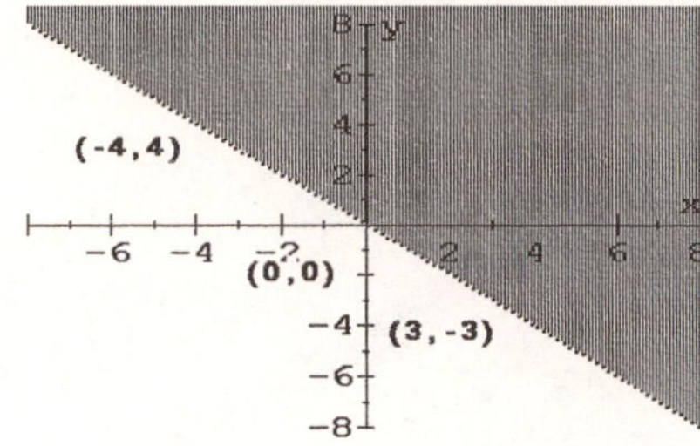

11. $2x - y \geq 0$

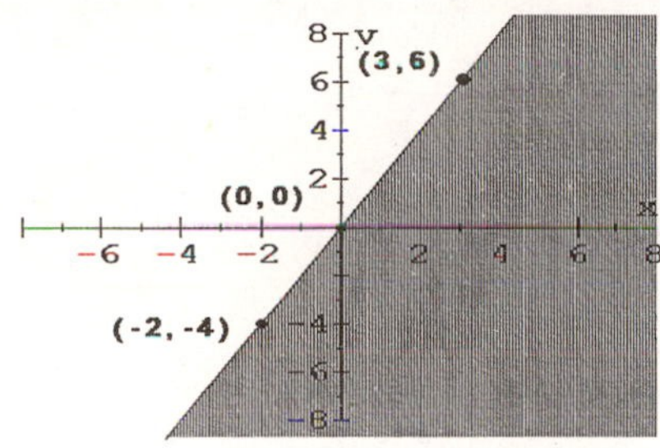

13. $-x + 4y - 4 \leq 0$

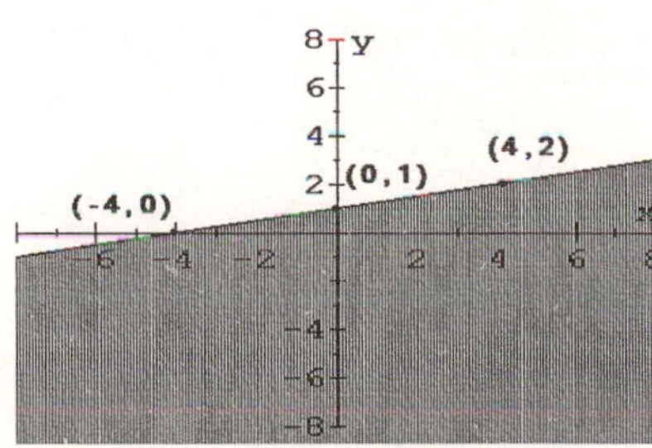

15. $y > -\frac{3}{2}x - 3$

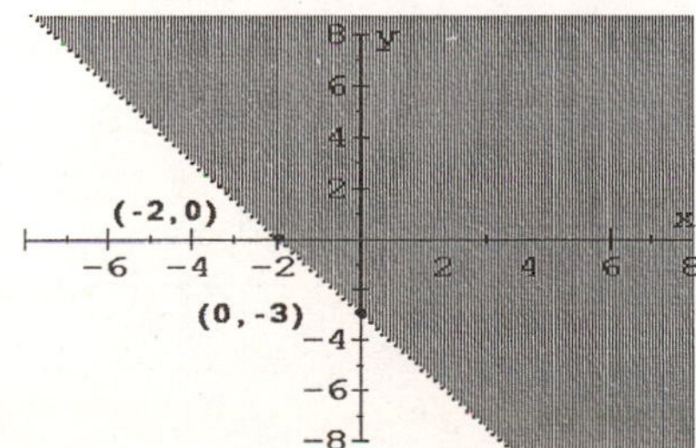

17. $y < -\frac{1}{2}x + 2$

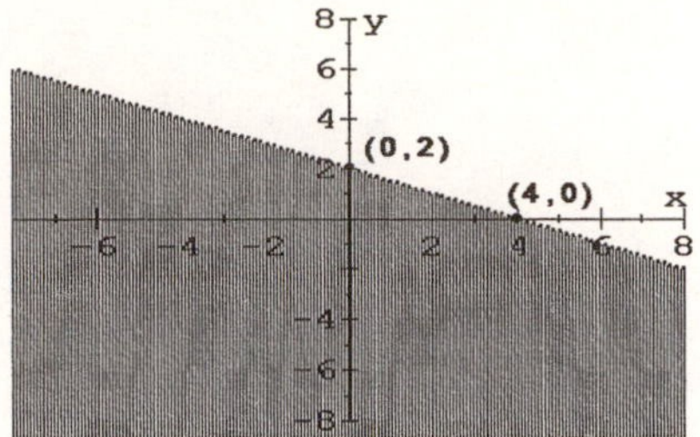

19. $x \leq 3$

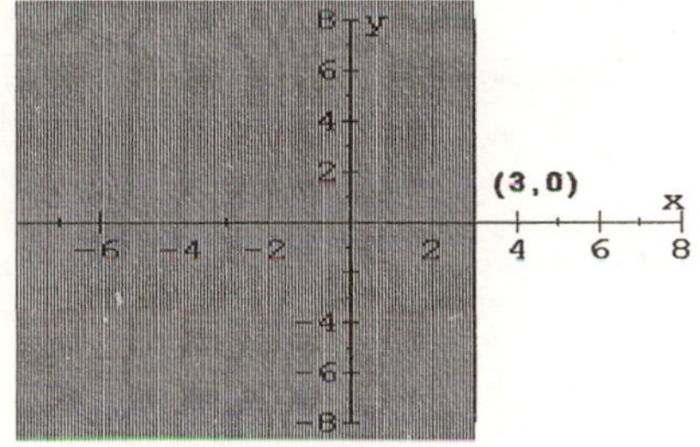

21. $x > 1$ and $y < 3$

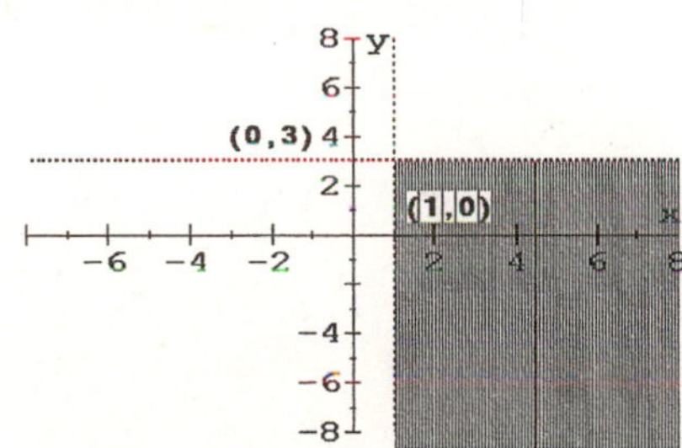

23. $x \leq -1$ and $y < 1$

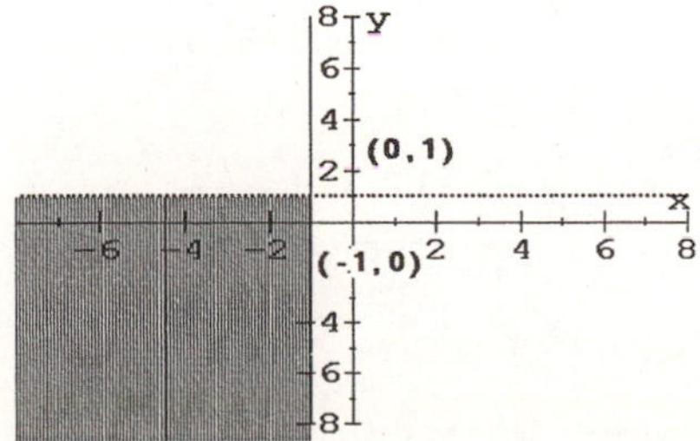

Further Investigations

27. $\left| x \right| < 2$

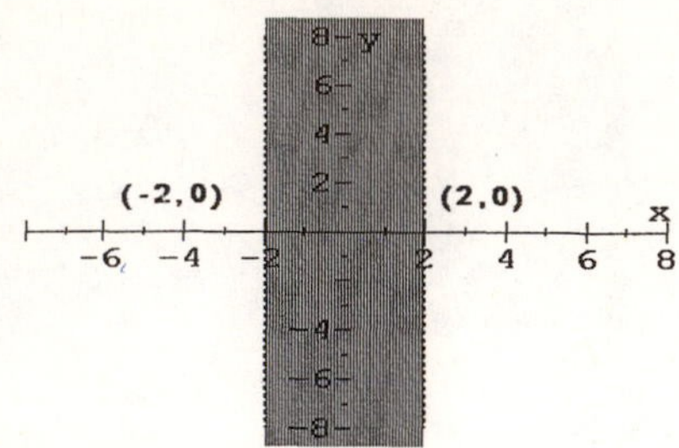

29. $\left| x + y \right| < 1$

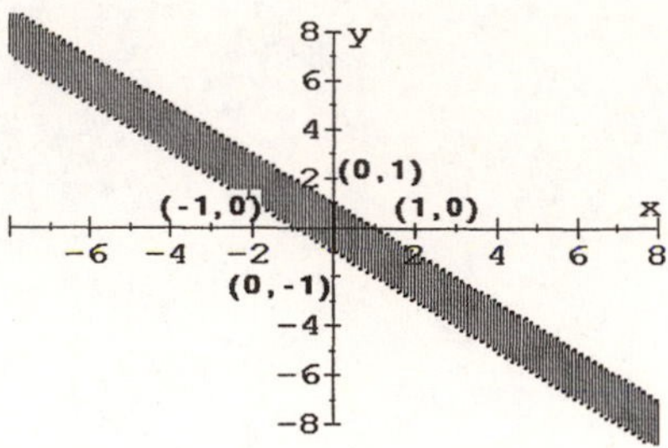

PROBLEM SET 7.4 Distance and Slope

1. $(-2, -1), (7, 11)$
$d = \sqrt{[7-(-2)]^2 + [11-(-1)]^2}$
$d = \sqrt{9^2 + 12^2}$
$d = \sqrt{81 + 144} = \sqrt{225} = 15$

3. $(1, -1), (3, -4)$
$d = \sqrt{(3-1)^2 + [-4-(-1)]^2}$
$d = \sqrt{2^2 + (-3)^2}$
$d = \sqrt{4+9} = \sqrt{13}$

5. $(6, -4), (9, -7)$
$d = \sqrt{(9-6)^2 + [-7-(-4)]^2}$
$d = \sqrt{3^2 + (-3)^2}$
$d = \sqrt{9+9} = \sqrt{18}$
$d = \sqrt{18} = \sqrt{9}\sqrt{2} = 3\sqrt{2}$

7. $(-3, 3), (0, -3)$
$d = \sqrt{[0-(-3)]^2 + (-3-3)^2}$
$d = \sqrt{(3)^2 + (-6)^2}$
$d = \sqrt{9+36} = \sqrt{45}$
$d = \sqrt{45} = \sqrt{9}\sqrt{5} = 3\sqrt{5}$

9. $(1, -6), (-5, -6)$
$d = \sqrt{(-5-1)^2 + [-6-(-6)]^2}$
$d = \sqrt{(-6)^2 + (0)^2}$
$d = \sqrt{36} = 6$

11. $(1, 7), (4, -2)$
$d = \sqrt{(4-1)^2 + [-2-7]^2}$
$d = \sqrt{3^2 + (-9)^2}$
$d = \sqrt{9+81} = \sqrt{90}$
$d = \sqrt{90} = \sqrt{9}\sqrt{10} = 3\sqrt{10}$

13. $(-3, 1), (5, 7)$, and $(8, 3)$

Distance from $(-3, 1)$ to $(5, 7)$.
$d = \sqrt{[5-(-3)]^2 + (7-1)^2}$
$d = \sqrt{8^2 + 6^2}$
$d = \sqrt{64+36} = \sqrt{100} = 10$

Distance from $(-3, 1)$ to $(8, 3)$.
$d = \sqrt{[8-(-3)]^2 + (3-1)^2}$
$d = \sqrt{11^2 + 2^2}$
$d = \sqrt{121+4} = \sqrt{125}$
$d = \sqrt{125} = \sqrt{25}\sqrt{5} = 5\sqrt{5}$

Distance from (5, 7) to (8, 3).

$d = \sqrt{(8-5)^2 + (3-7)^2}$

$d = \sqrt{3^2 + (-4)^2}$

$d = \sqrt{9+16} = \sqrt{25} = 5$

The potential legs are 10 and 5 and the hypotenuse is $5\sqrt{5}$.

$(10)^2 + (5)^2 \stackrel{?}{=} \left(5\sqrt{5}\right)^2$

$100 + 25 \stackrel{?}{=} (25)(5)$

$125 = 125$

Yes, it is a right triangle.

15. Distance from (3, 6) to (7, 12).

$d = \sqrt{(7-3)^2 + (12-6)^2}$

$d = \sqrt{4^2 + 6^2}$

$d = \sqrt{16+36} = \sqrt{52}$

$d = \sqrt{52} = \sqrt{4}\sqrt{13} = 2\sqrt{13}$

Distance from (7, 12) to (11, 18).

$d = \sqrt{(11-7)^2 + (18-12)^2}$

$d = \sqrt{4^2 + 6^2}$

$d = \sqrt{16+36} = \sqrt{52}$

$d = \sqrt{52} = \sqrt{4}\sqrt{13} = 2\sqrt{13}$

Distance from (11, 18) to (15, 24).

$d = \sqrt{(15-11)^2 + (24-18)^2}$

$d = \sqrt{4^2 + 6^2}$

$d = \sqrt{16+36} = \sqrt{52}$

$d = \sqrt{52} = \sqrt{4}\sqrt{13} = 2\sqrt{13}$

The distances are all equal, so the points do divide the line segment into three segments of equal length.

17. (1, 2), (4, 6)

$$m = \frac{6-2}{4-1} = \frac{4}{3}$$

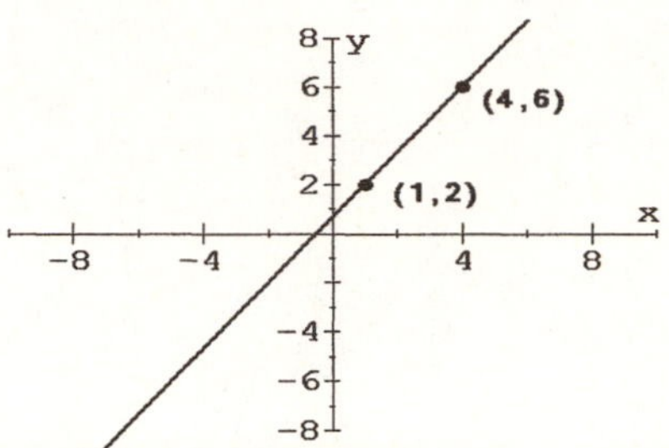

19. $(-4, 5), (-1, -2)$

$$m = \frac{-2-5}{-1-(-4)} = \frac{-7}{3}$$

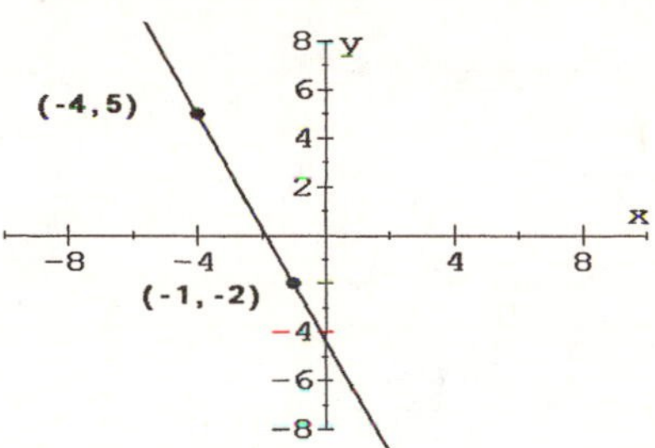

21. $(2, 6), (6, -2)$

$$m = \frac{-2-6}{6-2} = -\frac{8}{4} = -2$$

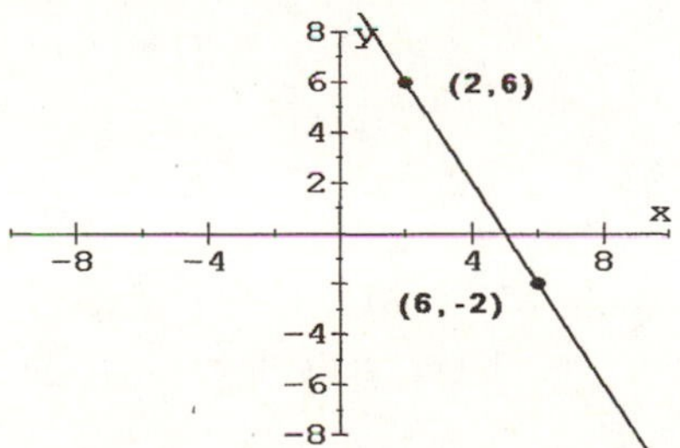

23. $(-6, 1), (-1, 4)$

$$m = \frac{4-1}{-1-(-6)} = \frac{3}{5}$$

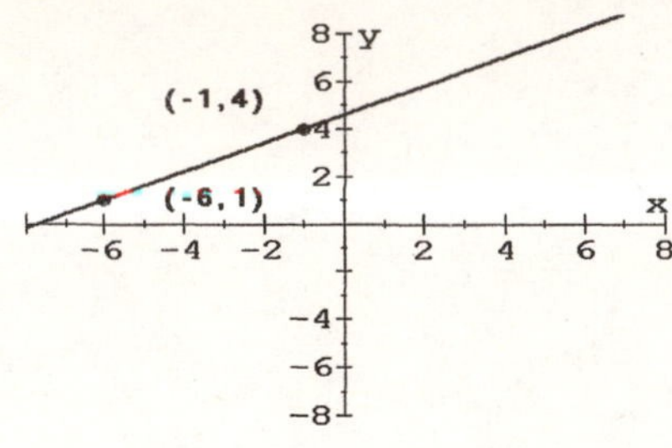

25. $(-2,-4),(2,-4)$

$$m=\frac{-4-(-4)}{2-(-2)}=\frac{0}{4}=0$$

27. $(0,-2),(4,0)$

$$m=\frac{0-(-2)}{4-0}=\frac{2}{4}=\frac{1}{2}$$

29. $(-2,4),(x,6)\ m=\frac{2}{9}$

$$\frac{6-4}{x-(-2)}=\frac{2}{9}$$

$$\frac{2}{x+2}=\frac{2}{9}$$

$2(x+2)=2(9)$

$2x+4=18$
$2x=14$
$x=7$

31. $(x,4),(2,-5)\ m=-\frac{9}{4}$

$$\frac{-5-4}{2-x}=-\frac{9}{4}$$

$$\frac{-9}{2-x}=\frac{-9}{4}$$

$-9(4)=-9(2-x)$
$-36=-18+9x$
$-18=9x$
$-2=x$

33. $(2,5)\ m=\frac{1}{2}$

Add 1 to the y-coordinate and add 2 to the x-coordinate.

$(2+2,5+1)=(4,6)$

$(4+2,6+1)=(6,7)$

$(6+2,7+1)=(8,8)$

35. $(-3,4)\ m=3=\frac{3}{1}$

Add 3 to the y-coordinate and add 1 to the x-coordinate.

$(-3+1,4+3)=(-2,7)$

$(-2+1,7+3)=(-1,10)$

$(-1+1,10+3)=(0,13)$

37. $(5,-2)\ m=\frac{-2}{3}$

Add -2 to the y-coordinate and add 3 to the x-coordinate.

$(5+3,-2-2)=(8,-4)$

$(8+3,-4-2)=(11,-6)$

$(11+3,-6-2)=(14,-8)$

39. $(-2, -4)$ $m = -2 = \frac{-2}{1}$

Add -2 to the y-coordinate and add 1 to the x-coordinate.

$(-2+1, -4-2) = (-1, -6)$

$(-1+1, -6-2) = (0, -8)$

$(0+1, -8-2) = (1, -10)$

41. $(3, 1)$ $m = \frac{2}{3}$

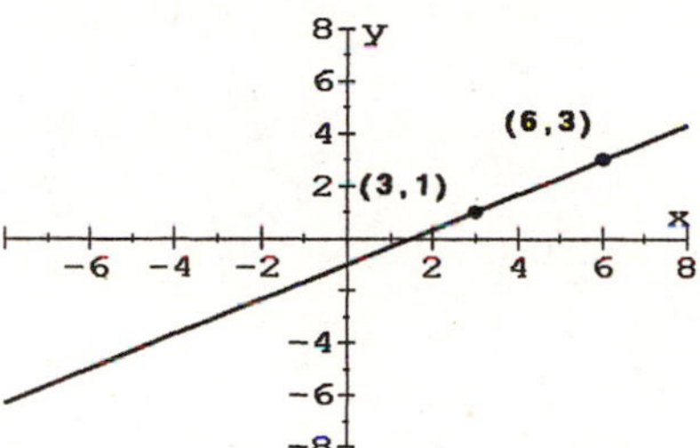

43. $(-2, 3)$ $m = -1$

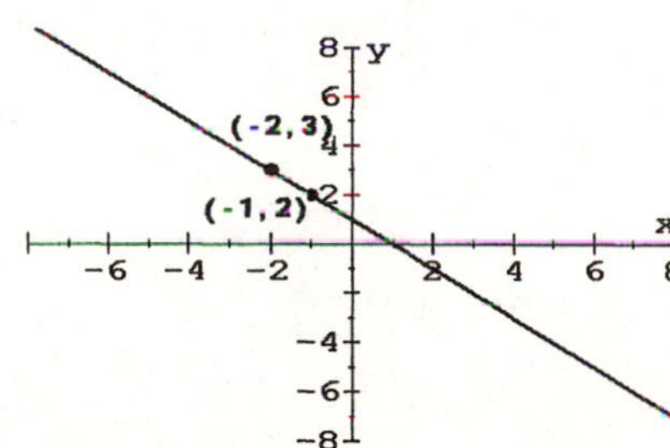

45. $(0, 5)$ $m = \frac{-1}{4}$

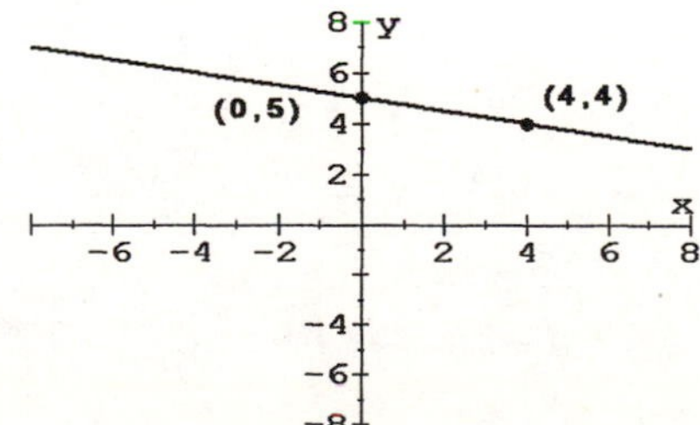

47. $(2, -2)$ $m = \frac{3}{2}$

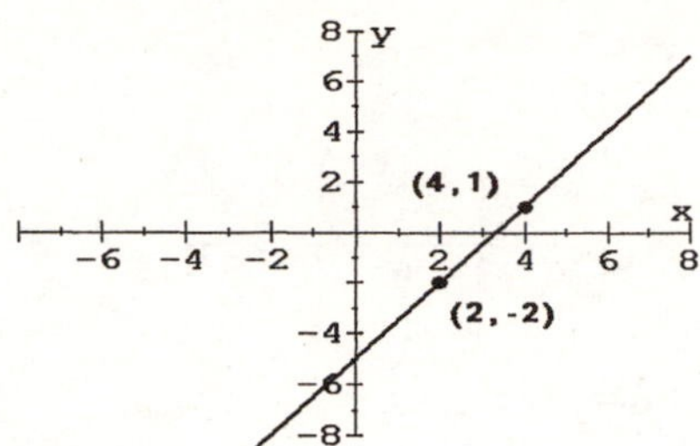

49. $2x + 3y = 6$
Let $x = 0$.
$2(0) + 3y = 6$
$3y = 6$
$y = 2$
$(0, 2)$
Let $y = 0$.
$2x + 3(0) = 6$
$2x = 6$
$x = 3$
$(3, 0)$

$$m = \frac{0-2}{3-0} = -\frac{2}{3}$$

51. $x - 2y = 4$
Let $x = 0$.
$(0) - 2y = 4$
$-2y = 4$
$y = -2$
$(0, -2)$
Let $y = 0$.
$x - 2(0) = 4$
$x = 4$
$(4, 0)$

$$m = \frac{0-(-2)}{4-0} = \frac{2}{4} = \frac{1}{2}$$

53. $4x - 7y = 12$
Let $x = 0$.
$4(0) - 7y = 12$
$-7y = 12$

$$y = \frac{12}{-7} = -\frac{12}{7}$$

$$\left(0, -\frac{12}{7}\right)$$

Let $y = 0$.
$4x - 7(0) = 12$
$4x = 12$
$x = 3$
$(3, 0)$

$$m = \frac{0 - \left(-\frac{12}{7}\right)}{3 - 0} = \frac{\frac{12}{7}}{3}$$

$$m = \frac{12}{7} \bullet \frac{1}{3} = \frac{4}{7}$$

55. $y = 4$
$(2, 4)$
$(3, 4)$

$$m = \frac{4 - 4}{3 - 2} = \frac{0}{1} = 0$$

57. $y = -5x$
Let $x = 1$.
$y = -5(1) = -5$
$(1, -5)$
Let $x = 2$.
$y = -5(2) = -10$
$(2, -10)$

$$m = \frac{-10 - (-5)}{2 - 1} = \frac{-5}{1} = -5$$

59. $\frac{2}{100} = \frac{y}{5280}$

$100y = 2(5280)$

$100y = 10560$

$y = 105.6$ feet

61. $\frac{215 \text{ feet}}{2640 \text{ feet}} = 0.081 = 8.1\%$

63. $\frac{2}{3} = \frac{y}{28}$

$3y = 2(28)$

$3y = 56$

$y \approx 19$ centimeters

Further Investigations

69a. $2 + \frac{1}{3}(5 - 2) =$

$2 + \frac{1}{3}(3) = 3$

$3 + \frac{1}{3}(9 - 3) =$

$3 + \frac{1}{3}(6) = 5$

$(3, 5)$

b. $1 + \frac{2}{3}(7 - 1) =$

$1 + \frac{2}{3}(6) = 5$

$4 + \frac{2}{3}(13 - 4) =$

$4 + \frac{2}{3}(9) = 10$

$(5, 10)$

c. $-2 + \frac{2}{5}[8 - (-2)] =$

$-2 + \frac{2}{5}(10) = 2$

$1 + \frac{2}{5}(11 - 1) =$

$1 + \frac{2}{5}(10) = 5$

$(2, 5)$

d. $2 + \frac{3}{5}(-3 - 2) =$

$2 + \frac{3}{5}(-5) = -1$

$-3 + \frac{3}{5}[8 - (-3)] =$

$-3+\frac{3}{5}(11)=\frac{18}{5}$

$\left(-1, \frac{18}{5}\right)$

e. $-1+\frac{5}{8}[4-(-1)]=$

$-1+\frac{5}{8}(5)=\frac{17}{8}$

$-2+\frac{5}{8}[-10-(-2)]=$

$-2+\frac{5}{8}(-8)=-7$

$\left(\frac{17}{8}, -7\right)$

f. $-2+\frac{7}{8}[-1-(-2)]=$

$-2+\frac{7}{8}(1)=-\frac{9}{8}$

$3+\frac{7}{8}(-9-3)=$

$3+\frac{7}{8}(-12)=-\frac{15}{2}$

$\left(-\frac{9}{8}, -\frac{15}{2}\right)$

PROBLEM SET 7.5 Determining the Equation of a Line

1. $m=\frac{1}{2}, (3, 5)$

$y-5=\frac{1}{2}(x-3)$

$2(y-5)=2\left[\frac{1}{2}(x-3)\right]$

$2y-10=x-3$

$-10=x-2y-3$

$-7=x-2y$

$x-2y=-7$

3. $m=3, (-2, 4)$

$y-4=3[x-(-2)]$

$y-4=3(x+2)$

$y-4=3x+6$

$-4=3x-y+6$

$-10=3x-y$

$3x-y=-10$

5. $m=-\frac{3}{4}, (-1, -3)$

$y-(-3)=-\frac{3}{4}[x-(-1)]$

$y+3=-\frac{3}{4}(x+1)$

$4(y+3)=4\left[-\frac{3}{4}(x+1)\right]$

$4y+12=-3(x+1)$

$4y+12=-3x-3$

$3x+4y+12=-3$

$3x+4y=-15$

7. $m=\frac{5}{4}, (4, -2)$

$y-(-2)=\frac{5}{4}(x-4)$

$y+2=\frac{5}{4}(x-4)$

$4(y+2)=4\left[\frac{5}{4}(x-4)\right]$

$4y+8=5(x-4)$

$4y+8=5x-20$

$8=5x-4y-20$

$28=5x-4y$

$5x-4y=28$

9. $(2, 1), (6, 5)$

$m=\frac{5-1}{6-2}=\frac{4}{4}=1$

$y-1=1(x-2)$

$$y - 1 = x - 2$$
$$-1 = x - y - 2$$
$$1 = x - y$$
$$x - y = 1$$

11. $(-2, -3), (2, 7)$

$$m = \frac{7 - (-3)}{2 - (-2)} = \frac{10}{4} = \frac{5}{2}$$
$$y - 7 = \frac{5}{2}(x - 2)$$
$$2(y - 7) = 2\left[\frac{5}{2}(x - 2)\right]$$
$$2y - 14 = 5(x - 2)$$
$$2y - 14 = 5x - 10$$
$$-14 = 5x - 2y - 10$$
$$-4 = 5x - 2y$$
$$5x - 2y = -4$$

13. $(-3, 2), (4, 1)$

$$m = \frac{1 - 2}{4 - (-3)} = \frac{-1}{7}$$
$$y - 1 = -\frac{1}{7}(x - 4)$$
$$7(y - 1) = 7\left[-\frac{1}{7}(x - 4)\right]$$
$$7y - 7 = -1(x - 4)$$
$$7y - 7 = -x + 4$$
$$x + 7y - 7 = 4$$
$$x + 7y = 11$$

15. $(-1, -4), (3, -6)$

$$m = \frac{-6 - (-4)}{3 - (-1)} = \frac{-2}{4} = -\frac{1}{2}$$
$$y - (-6) = -\frac{1}{2}(x - 3)$$
$$y + 6 = -\frac{1}{2}(x - 3)$$
$$2(y + 6) = 2\left[-\frac{1}{2}(x - 3)\right]$$
$$2y + 12 = -1(x - 3)$$
$$2y + 12 = -x + 3$$
$$x + 2y + 12 = 3$$
$$x + 2y = -9$$

17. $(0, 0), (5, 7)$

$$m = \frac{7 - 0}{5 - 0} = \frac{7}{5}$$
$$y - 0 = \frac{7}{5}(x - 0)$$
$$y = \frac{7}{5}x$$
$$5(y) = 5\left(\frac{7}{5}x\right)$$
$$5y = 7x$$
$$0 = 7x - 5y$$
$$7x - 5y = 0$$

19. $m = \frac{3}{7}, b = 4$

$$y = \frac{3}{7}x + 4$$

21. $m = 2, b = -3$

$$y = 2x - 3$$

23. $m = -\frac{2}{5}, b = 1$

$$y = -\frac{2}{5}x + 1$$

25. $m = 0, b = -4$

$$y = 0x - 4$$

27. x-intercept of 2 and y-intercept of -4.

$(2, 0), (0, -4)$

$$m = \frac{-4 - 0}{0 - 2} = \frac{-4}{-2} = 2$$
$$y = mx + b$$
$$y = 2x - 4$$
$$0 = 2x - y - 4$$
$$4 = 2x - y$$
$$2x - y = 4$$

29. x-intercept of -3 and slope of $-\frac{5}{8}$

$(-3, 0), m = -\frac{5}{8}$

$$y - 0 = -\frac{5}{8}[x - (-3)]$$

$$y = -\frac{5}{8}(x+3)$$
$$8(y) = 8\left[-\frac{5}{8}(x+3)\right]$$
$$8y = -5(x+3)$$
$$8y = -5x - 15$$
$$5x + 8y = -15$$

31. (2, − 4); parallel to y-axis
$$x = 2$$
$$x + 0y = 2$$

33. (5, 6); perpendicular to y-axis
(5, 6), $m = 0$
$$y - 6 = 0(x-5)$$
$$y - 6 = 0$$
$$y = 6$$
$$0x + y = 6$$

35. (1, 3); parallel to the line with equation $x + 5y = 9$
Parallel lines have the same slope.
$$x + 5y = 9$$
$$\frac{1}{5}(5y) = \frac{1}{5}(-x+9)$$
$$y = -\frac{1}{5}x - \frac{9}{5}$$
$$m = -\frac{1}{5}$$
$$y - 3 = -\frac{1}{5}(x-1)$$
$$5(y-3) = 5\left[-\frac{1}{5}(x-1)\right]$$
$$5y - 15 = -1(x-1)$$
$$5y - 15 = -x + 1$$
$$x + 5y - 15 = 1$$
$$x + 5y = 16$$

37. Contains the origin and is parallel to the line with equation $4x - 7y = 3$
Parallel lines have the same slope.
$$4x - 7y = 3$$
$$-7y = -4x + 3$$
$$-\frac{1}{7}(-7y) = -\frac{1}{7}(-4x+3)$$
$$y = \frac{4}{7}x - \frac{3}{7}$$
$$m = \frac{4}{7}$$
We have the point (0,
$$y - 0 = \frac{4}{7}(x-0)$$
$$y = \frac{4}{7}x$$
$$7(y) = 7\left(\frac{4}{7}x\right)$$
$$7y = 4x$$
$$0 = 4x - 7y$$
$$4x - 7y = 0$$

39. (− 1, 3); perpendicular to the line with equation $2x - y = 4$
Perpendicular lines have slopes that are negative reciprocals.
$$2x - y = 4$$
$$-y = -2x + 4$$
$$-(-y) = -(-2x+4)$$
$$y = 2x + 4$$
$$m = 2$$
The negative reciprocal is $-\frac{1}{2}$.
$$y - 3 = -\frac{1}{2}[x - (-1)]$$
$$y - 3 = -\frac{1}{2}(x+1)$$
$$2(y-3) = 2\left[-\frac{1}{2}(x+1)\right]$$
$$2y - 6 = -1(x+1)$$
$$2y - 6 = -x - 1$$
$$x + 2y - 6 = -1$$
$$x + 2y = 5$$

41. Contains the origin (0, 0) and is perpendicular to the line with equation $-2x + 3y = 8$
Perpendicular lines have slopes that are negative reciprocals.
$$-2x + 3y = 8$$
$$3y = 2x + 8$$
$$\frac{1}{3}(3y) = \frac{1}{3}(2x+8)$$
$$y = \frac{2}{3}x + \frac{8}{3}$$

Pro

$m = \frac{2}{3}$

The negative reciprocal is $-\frac{3}{2}$.

$y - 0 = -\frac{3}{2}(x - 0)$

$y = -\frac{3}{2}x$

$2(y) = 2\left(-\frac{3}{2}x\right)$

$2y = -3x$

$3x + 2y = 0$

43. $3x + y = 7$

$y = -3x + 7$

$m = -3$ and $b = 7$

45. $3x + 2y = 9$

$2y = -3x + 9$

$\frac{1}{2}(2y) = \frac{1}{2}(-3x + 9)$

$y = -\frac{3}{2}x + \frac{9}{2}$

$m = -\frac{3}{2}$ and $b = \frac{9}{2}$

47. $x = 5y + 12$

$x - 12 = 5y$

$\frac{1}{5}(x - 12) = \frac{1}{5}(5y)$

$y = \frac{1}{5}x - \frac{12}{5}$

$m = \frac{1}{5}$ and $b = -\frac{12}{5}$

49. $y = \frac{2}{3}x - 4$

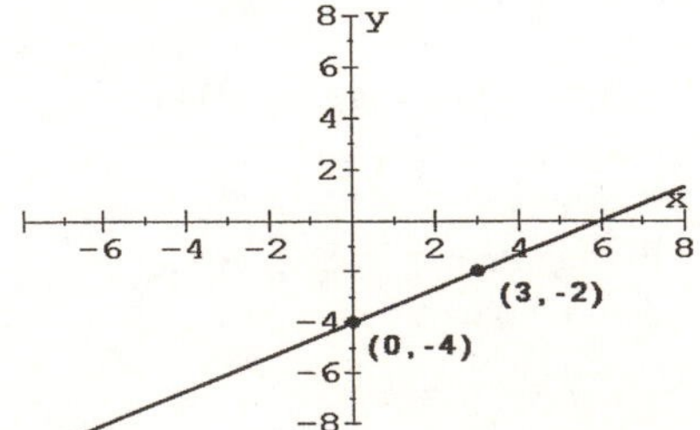

51. $y = 2x + 1$

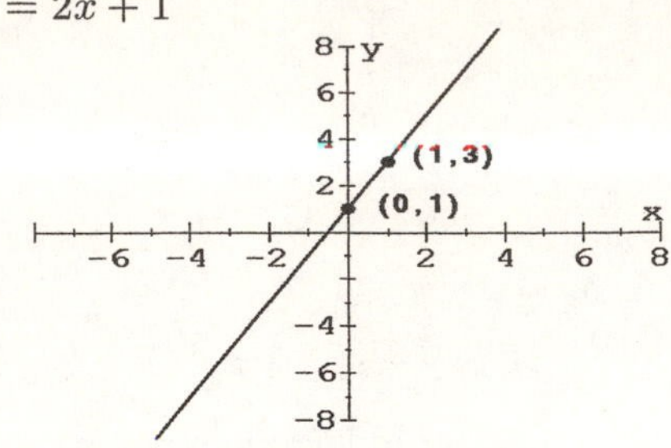

53. $y = -\frac{3}{2}x + 4$

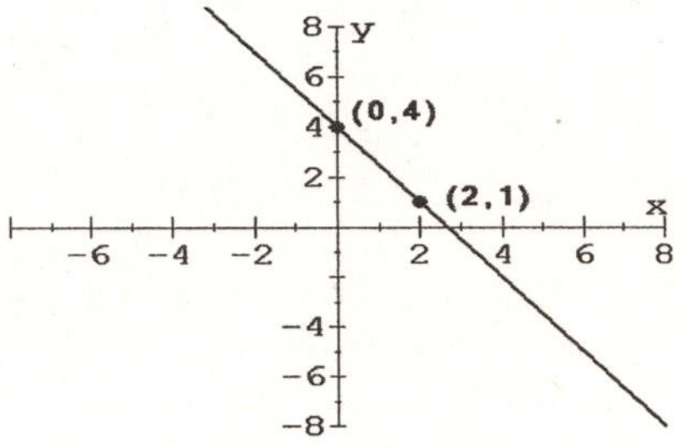

55. $y = -x + 2$

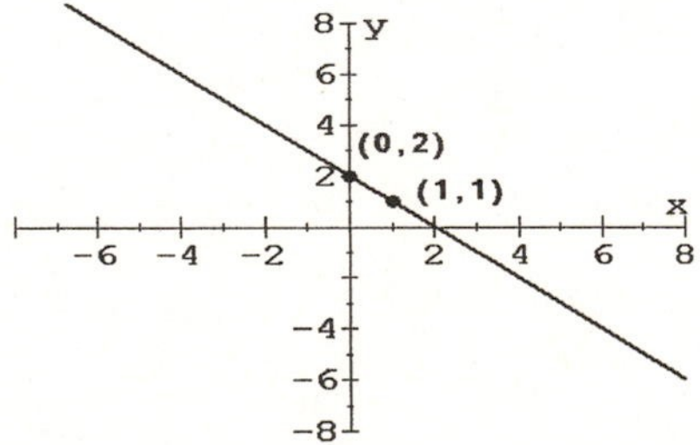

57. $y = -\frac{2}{5}x - 1$

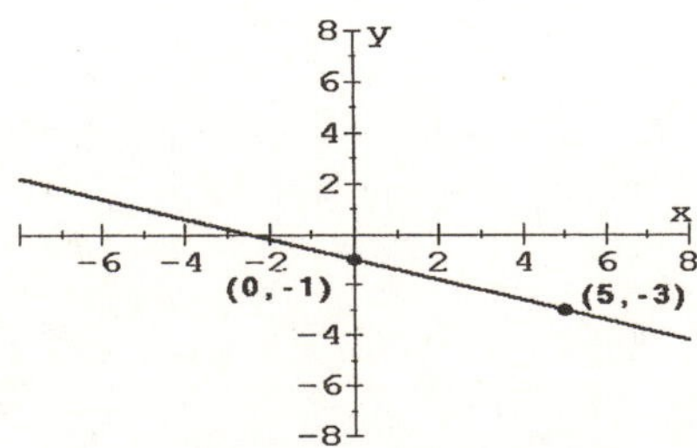

59. $x + 2y = 5$

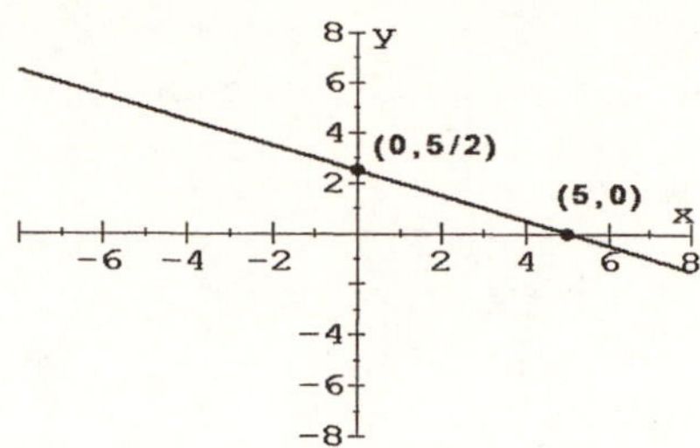

61. $-y = -4x + 7$

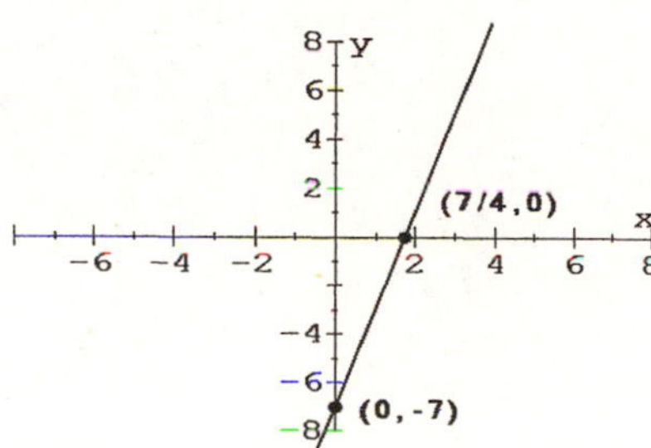

63. $7y = -2x$

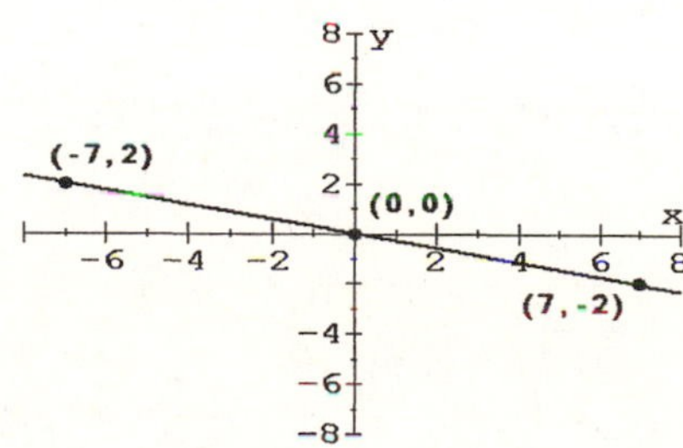

65. $x = 2$

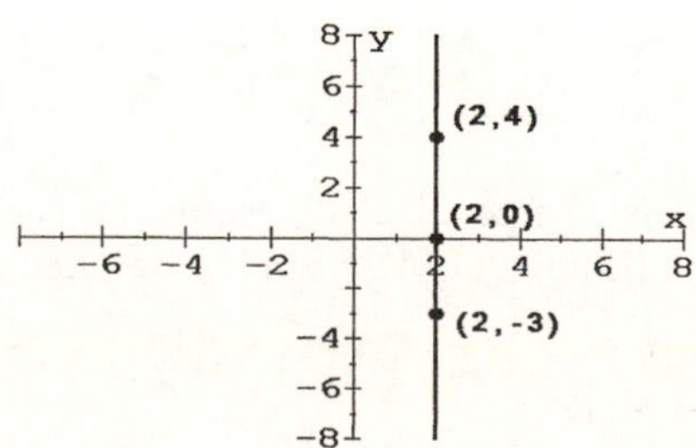

67. Let x = square footage and let y = pounds of fertilizer.
Ordered pairs are: $(5000, 7)$; $(10000, 12)$

$$m = \frac{12-7}{10000-5000} = \frac{5}{5000} = \frac{1}{1000}$$

$$y - 7 = \frac{1}{1000}\left(x - 5000\right)$$

$$1000(y-7) = 1000\left[\frac{1}{1000}\left(x-5000\right)\right]$$

$$1000y - 7000 = x - 5000$$

$$1000y = x + 2000$$

Therefore, the equation is $y = \frac{1}{1000}x + 2$.

69. Let y = temperature °F and x = temperature °C.
Ordered pairs are: $(10, 50)$, $(-5, 23)$

$$m = \frac{23-50}{-5-10} = \frac{-27}{-15} = \frac{9}{5}$$

$$y - 50 = \frac{9}{5}(x - 10)$$

$$5(y-50) = 5\left[\frac{9}{5}(x-10)\right]$$

$$5y - 250 = 9(x - 10)$$

$$5y - 250 = 9x - 90$$

$$5y = 9x + 160$$

Therefore, the equation is $y = \frac{9}{5}x + 32$.

75. $Ax + By = C$

$$By = -Ax + C$$

$$y = -\frac{A}{B}x + \frac{C}{B}$$

$$m = -\frac{A}{B}$$

$$A'x + B'y = C'$$

$$B'y = -A'x + C'$$

$$y = -\frac{A'}{B'}x + \frac{C'}{B'}$$

$$m = -\frac{A'}{B'}$$

Problem Set 7.5

a. If $\frac{A}{A'} = \frac{B}{B'}$, then $AB' = A'B$.

$AB' = A'B$

$\left(\frac{1}{BB'}\right)AB' = \left(\frac{1}{BB'}\right)A'B$

$\frac{A}{B} = \frac{A'}{B'}$

If $\frac{A}{B} = \frac{A'}{B'}$, then $-\frac{A}{B} = -\frac{A'}{B'}$ and the lines have the same slope.

If $\frac{B}{B'} \neq \frac{C}{C'}$, then $BC' \neq B'C$.

$BC' \neq B'C$

$\left(\frac{1}{BB'}\right)BC' \neq \left(\frac{1}{BB'}\right)B'C$

$\frac{C'}{B'} \neq \frac{C}{B}$

If $\frac{C'}{B'} \neq \frac{C}{B}$, then the lines have different y-intercepts.

If two lines have equal slopes and different y-intercepts, the lines are parallel.

b. If $AA' = -BB'$, then

$\left(\frac{1}{AB'}\right)AA' = \left(\frac{1}{AB'}\right) - BB'$

$\frac{A'}{B'} = -\frac{B}{A}$.

If two lines have slopes that are negative reciprocals, then the lines are perpendicular.

77a. $(-1,2)$ and $(3,0)$

$M = \left(\frac{-1+3}{2}, \frac{2+0}{2}\right) =$

$\left(\frac{2}{2}, \frac{2}{2}\right) = (1,1)$

$m = \frac{0-2}{3-(-1)} = \frac{-2}{4} = -\frac{1}{2}$

The perpendicular bisector will pass through the point (1,1) and have a slope of $m = 2$.

$y - 1 = 2(x-1)$
$y - 1 = 2x - 2$
$2x - y = 1$

b. $(6, -10)$ and $(-4, 2)$

$M = \left(\frac{6+(-4)}{2}, \frac{-10+2}{2}\right) =$

$\left(\frac{2}{2}, \frac{-8}{2}\right) = (1, -4)$

$m = \frac{2-(-10)}{-4-6} = \frac{12}{-10} = -\frac{6}{5}$

The perpendicular bisector will pass through the point (1, – 4) and have a slope of $m = \frac{5}{6}$.

$y - (-4) = \frac{5}{6}(x-1)$

$6(y+4) = 6\left[\frac{5}{6}(x-1)\right]$

$6y + 24 = 5(x-1)$
$6y + 24 = 5x - 5$
$5x - 6y = 29$

c. $(-7, -3)$ and $(5, 9)$

$M = \left(\frac{-7+5}{2}, \frac{-3+9}{2}\right) =$

$\left(\frac{-2}{2}, \frac{6}{2}\right) = (-1, 3)$

$m = \frac{9-(-3)}{5-(-7)} = \frac{12}{12} = 1$

The perpendicular bisector will pass through the point (– 1,3) and have a slope of $m = -1$.

$y - 3 = -1[(x-(-1))]$
$y - 3 = -1(x+1)$
$y - 3 = -x - 1$
$x + y = 2$

d. $(0,4)$ and $(12,-4)$

$$M=\left(\frac{0+12}{2},\frac{4+(-4)}{2}\right)=$$

$$\left(\frac{12}{2},\frac{0}{2}\right)=(6,0)$$

$$m=\frac{-4-4}{12-0}=\frac{-8}{12}=-\frac{2}{3}$$

The perpendicular bisector will pass through the point (6, 0) and have a slope of $m=\frac{3}{2}$.

$$y-0=\frac{3}{2}(x-6)$$

$$2(y-0)=2\left[\frac{3}{2}\left(x-6\right)\right]$$

$$2y=3(x-6)$$
$$2y=3x-18$$
$$3x-2y=18$$

CHAPTER 7 Review Problem Set

1a. $(3,4),(-2,-2)$

$$m=\frac{-2-4}{-2-3}=\frac{-6}{-5}=\frac{6}{5}$$

b. $(-2,3),(4,-1)$

$$m=\frac{-1-3}{4-(-2)}=\frac{-4}{6}=-\frac{2}{3}$$

2. $(-4,3)$ and $(12,y); m=\frac{1}{8}$

$$\frac{y-3}{12-(-4)}=\frac{1}{8}$$

$$8(y-3)=1(12+4)$$
$$8y-24=16$$
$$8y=40$$
$$y=5$$

3. $(x,5)$ and $(3,-1); m=-\frac{3}{2}$

$$\frac{-1-5}{3-x}=-\frac{3}{2}$$

$$\frac{-6}{3-x}=\frac{-3}{2}$$

$$-6(2)=-3(3-x)$$
$$-12=-9+3x$$
$$-3=3x$$
$$-1=x$$

4a. $4x+y=7$
$$y=-4x+7$$
$$m=-4$$

b. $2x-7y=3$
$$-7y=-2x+3$$
$$-\frac{1}{7}(-7y)=-\frac{1}{7}(-2x+3)$$
$$y=\frac{2}{7}x-\frac{3}{7};\quad m=\frac{2}{7}$$

5. $(2,3),(5,-1)$, and $(-4,-5)$

Distance from (2, 3) to (5, – 1)

$$d=\sqrt{(5-2)^2+(-1-3)^2}$$
$$d=\sqrt{3^2+(-4)^2}$$
$$d=\sqrt{9+16}=\sqrt{25}=5$$

Distance from (2, 3) to (– 4, – 5)

$$d=\sqrt{(-5-3)^2+(-4-2)^2}$$
$$d=\sqrt{(-8)^2+(-6)^2}$$
$$d=\sqrt{64+36}=\sqrt{100}=10$$

Distance from (5, – 1) to (– 4, – 5)

$$d=\sqrt{(-4-5)^2+[-5-(-1)]^2}$$
$$d=\sqrt{(-9)^2+(-4)^2}$$
$$d=\sqrt{81+16}=\sqrt{97}$$

The lengths of the sides are 5, 10, and $\sqrt{97}$ units.

Chapter 7 Review Problem Set

6a. Distance from $(-1, 4)$ to $(1, -2)$.

$$d = \sqrt{[(1-(-1)]^2 + (-2-4)^2}$$
$$d = \sqrt{(2)^2 + (-6)^2}$$
$$d = \sqrt{4+36} = \sqrt{40} = 2\sqrt{10}$$

b. Distance from $(5, 0)$ to $(2, 7)$.

$$d = \sqrt{(7-0)^2 + (2-5)^2}$$
$$d = \sqrt{(7)^2 + (-3)^2}$$
$$d = \sqrt{49+9} = \sqrt{58}$$

7. $M = \left(\frac{3+(-1)}{2}, \frac{2+10}{2}\right)$

$$M = \left(\frac{2}{2}, \frac{12}{2}\right) = (1, 6)$$

8. $(-1, 2)$ and $(3, -5)$

$$m = \frac{-5-2}{3-(-1)} = \frac{-7}{4}$$
$$y-(-5) = -\frac{7}{4}(x-3)$$
$$y+5 = -\frac{7}{4}(x-3)$$
$$4(y+5) = 4\left[-\frac{7}{4}(x-3)\right]$$
$$4y+20 = -7(x-3)$$
$$4y+20 = -7x+21$$
$$7x+4y+20 = 21$$
$$7x+4y = 1$$

9. $m = -\frac{3}{7}; b = 4$

$$y = -\frac{3}{7}x+4$$
$$7(y) = 7\left(-\frac{3}{7}x+4\right)$$
$$7y = -3x+28$$
$$3x+7y = 28$$

10. $(-1, -6); m = \frac{2}{3}$

$$y-(-6) = \frac{2}{3}\left[x-(-1)\right]$$
$$y+6 = \frac{2}{3}(x+1)$$
$$3(y+6) = 3\left[\frac{2}{3}(x+1)\right]$$
$$3y+18 = 2(x+1)$$
$$3y+18 = 2x+2$$
$$18 = 2x-3y+2$$
$$16 = 2x-3y$$
$$2x-3y = 16$$

11. $(2, 5)$; parallel to the line with equation $x-2y=4$
Parallel lines have the same slope.

$$x-2y = 4$$
$$-2y = -x+4$$
$$-\frac{1}{2}(-2y) = -\frac{1}{2}(-x+4)$$
$$y = \frac{1}{2}x-2$$
$$m = \frac{1}{2}$$
$$y-5 = \frac{1}{2}(x-2)$$
$$2(y-5) = 2\left[\frac{1}{2}(x-2)\right]$$
$$2y-10 = x-2$$
$$-10 = x-2y-2$$
$$-8 = x-2y$$
$$x-2y = -8$$

12. $(-2, -6)$; perpendicular to the line with equation $3x+2y=12$
Perpendicular lines have slopes that are negative reciprocals.

$$3x+2y = 12$$
$$2y = -3x+12$$
$$\frac{1}{2}(2y) = \frac{1}{2}(-3x+12)$$

$y = -\frac{3}{2}x + 6$

$m = -\frac{3}{2}$

The negative reciprocal is $\frac{2}{3}$.

$y - (-6) = \frac{2}{3}\left[x - (-2)\right]$

$y + 6 = \frac{2}{3}(x + 2)$

$3(y + 6) = 3\left[\frac{2}{3}(x + 2)\right]$

$3y + 18 = 2(x + 2)$
$3y + 18 = 2x + 4$
$18 = 2x - 3y + 4$
$14 = 2x - 3y$
$2x - 3y = 14$

13. (0, 4) and (2, 6)

$m = \frac{6 - 4}{2 - 0} = \frac{2}{2} = 1$

$y - 4 = 1(x - 0)$
$y - 4 = x$
$x - y = -4$

14. $(3, -5)$; $m = -1$
$y - (-5) = -1(x - 3)$
$y + 5 = -x + 3$
$x + y = -2$

15. $(-8, 3)$; parallel to the line with equation $4x + y = 7$
Parallel lines have the same slope.

$4x + y = 7$
$y = -4x + 7$
$m = -4$
$y - 3 = -4[x - (-8)]$
$y - 3 = -4(x + 8)$
$y - 3 = -4x - 32$
$4x + y = -29$

16. $2x - y = 6$

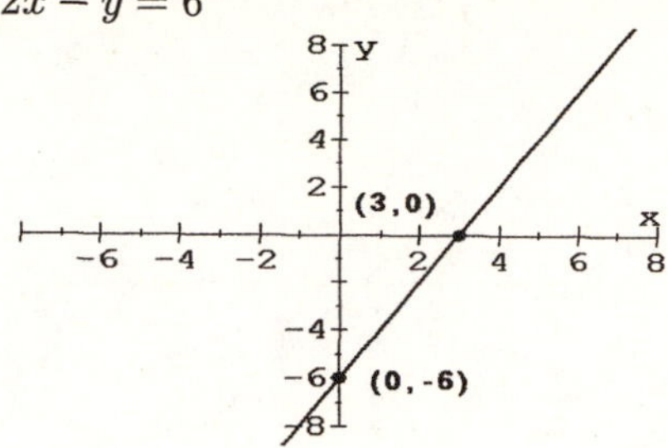

17. $y = 2x - 5$

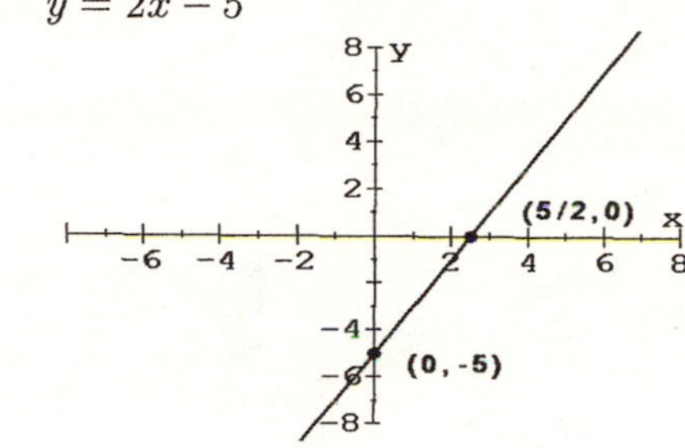

18. $y = -2x - 1$

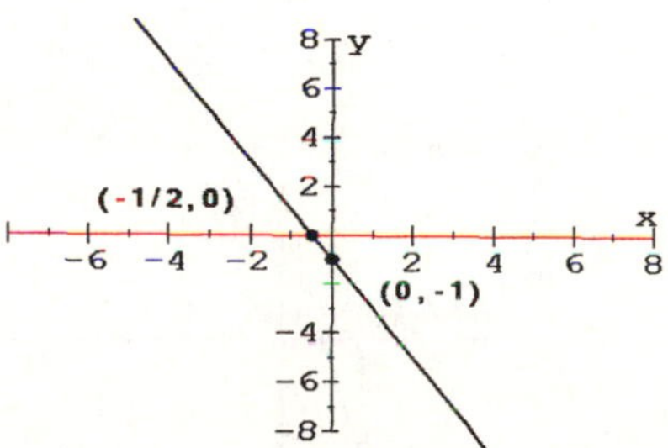

19. $y = -4x$

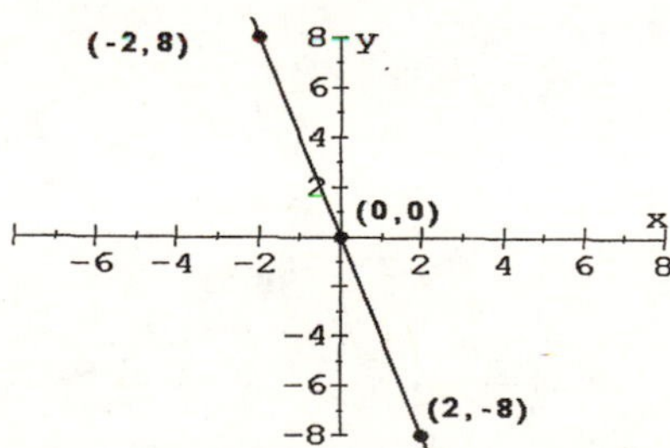

20. $-3x - 2y = 6$

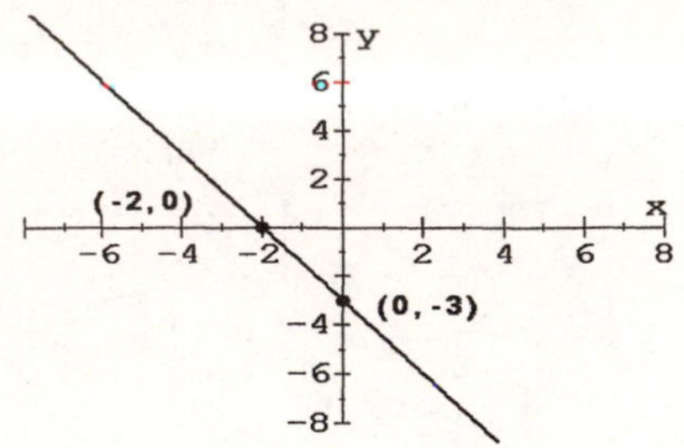

21. $x = 2y + 4$

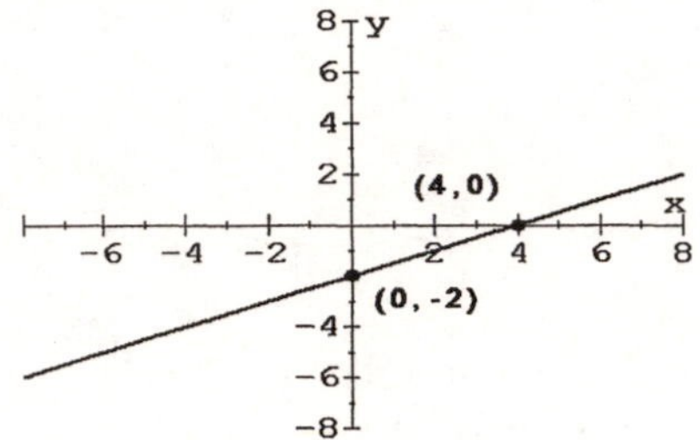

22. $5x - y = -5$

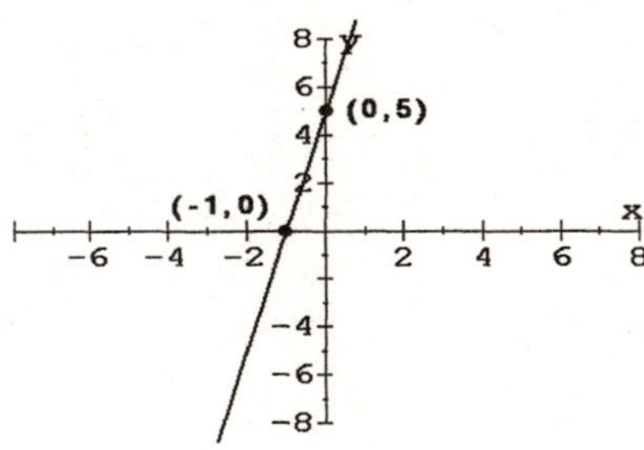

23. $y = -\frac{1}{2}x + 3$

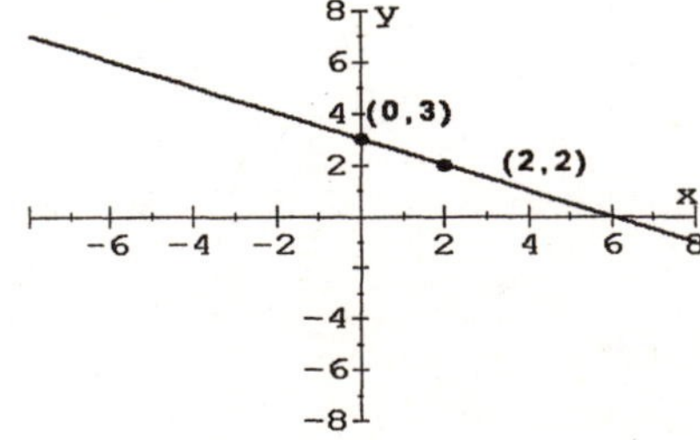

24. $y = \frac{3x - 4}{2}$

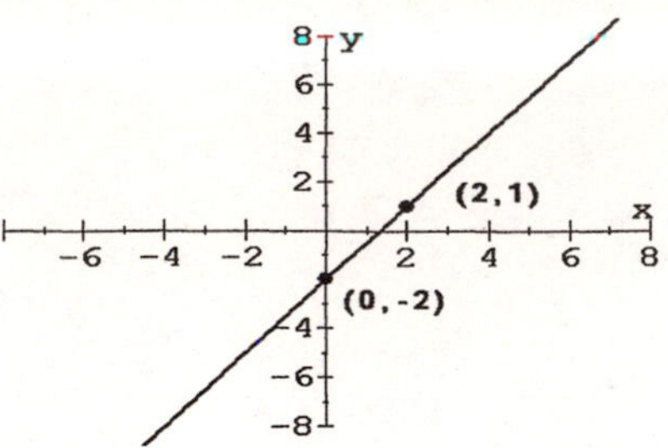

25. $y = 4$

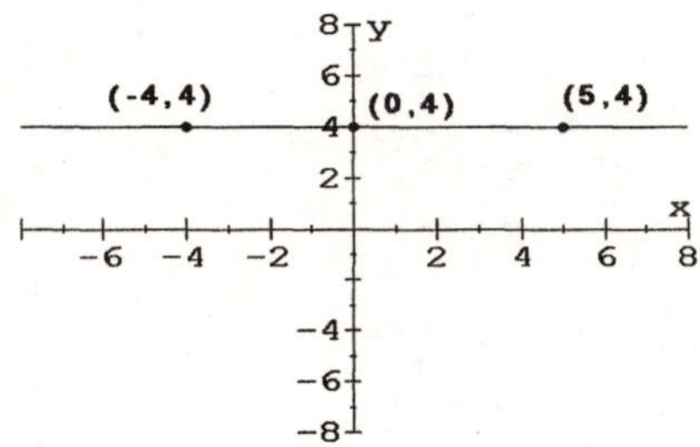

26. $2x + 3y = 0$

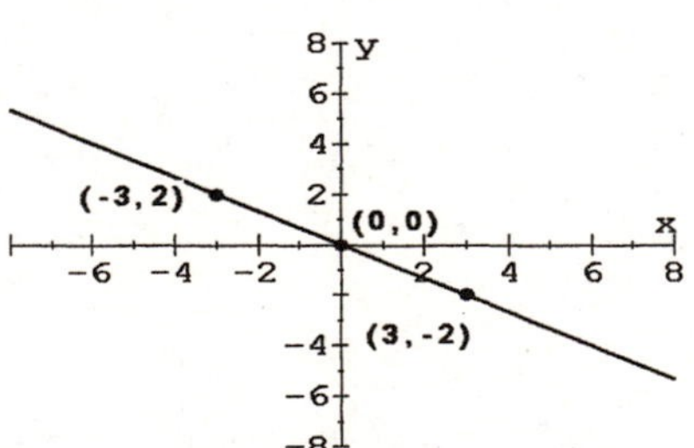

27. $y = \frac{3}{5}x - 4$

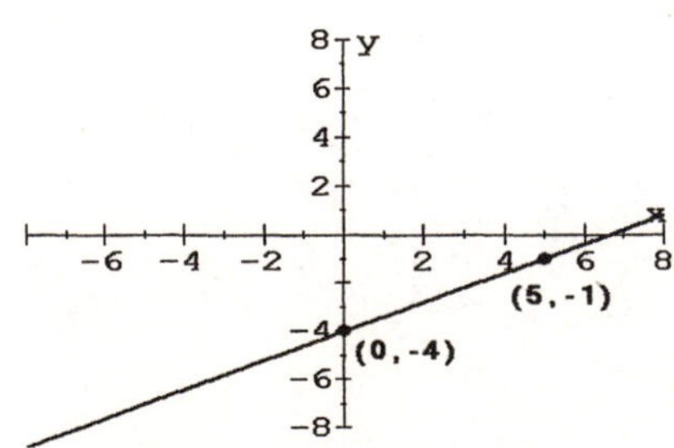

28. $x = 1$

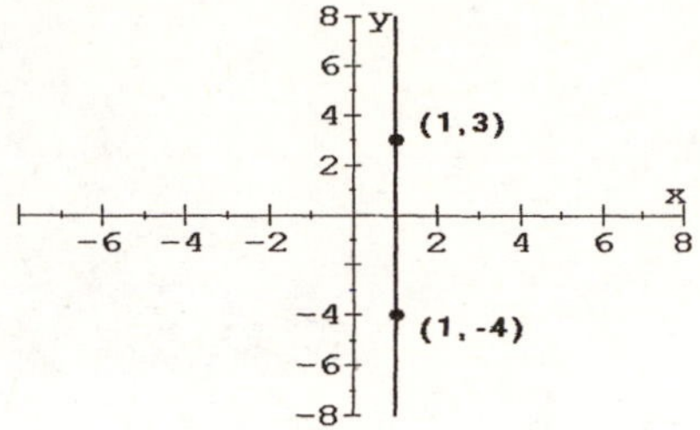

29. $x = -3$

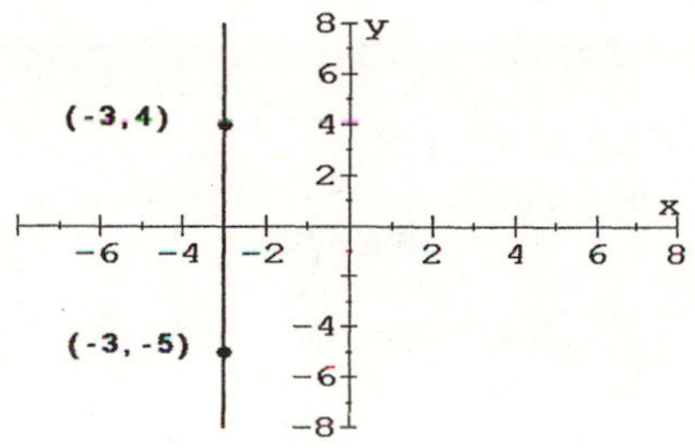

30. $y = -2$

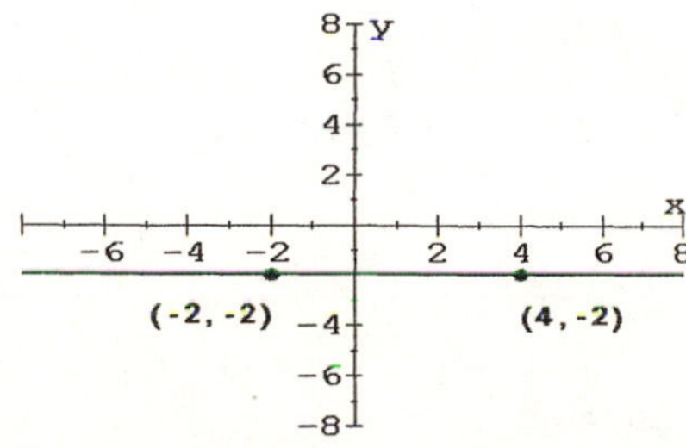

31. $2x - 3y = 3$

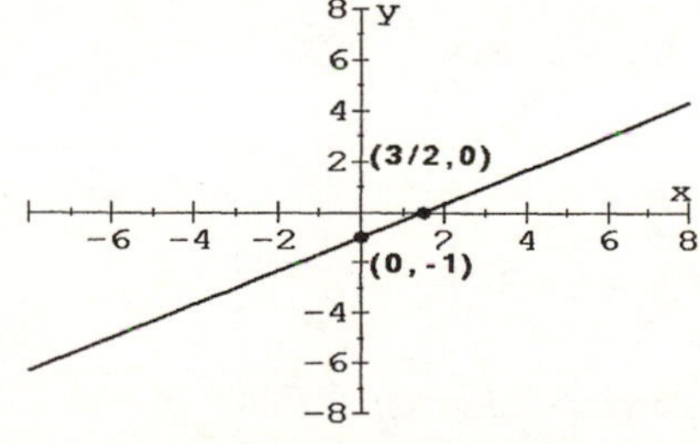

32. $y = x^3 + 2$

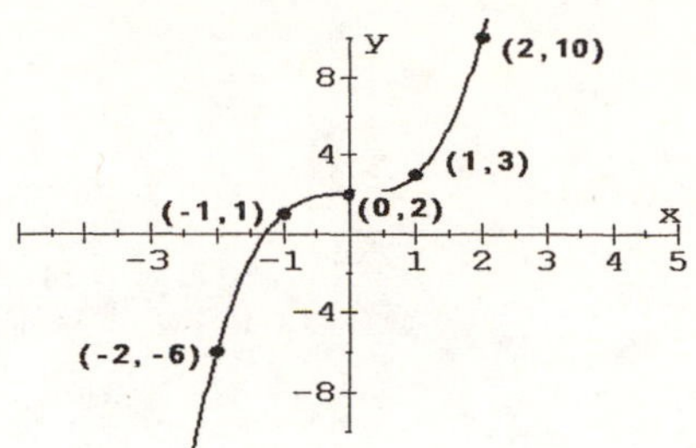

33. $y = -x^3$

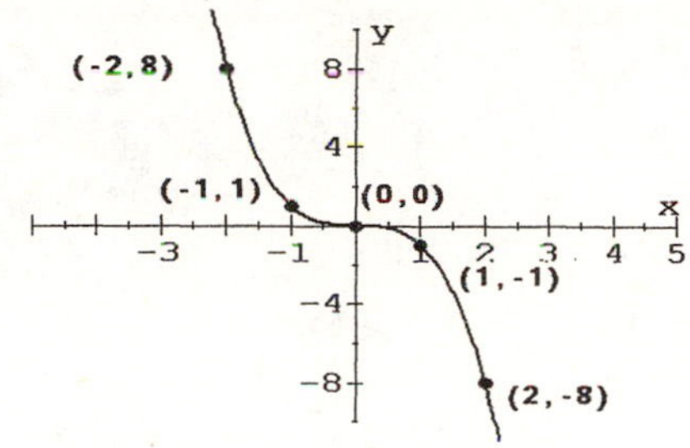

34. $y = x^2 + 3$

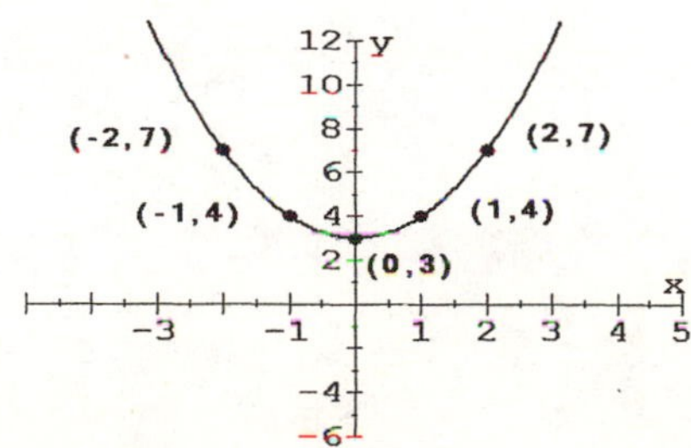

35. $y = -2x^2 - 1$

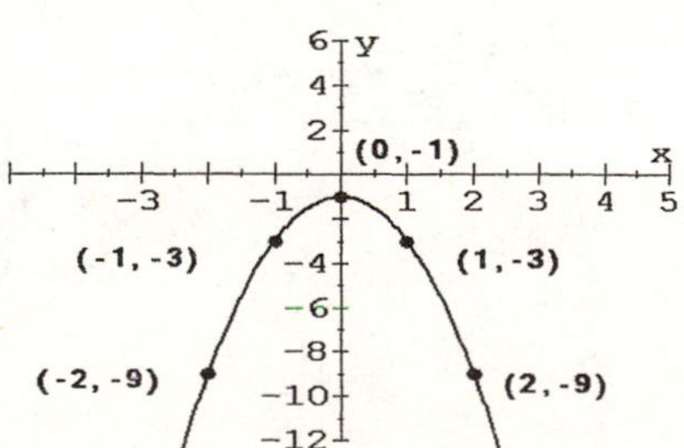

36. $-x+3y<-6$

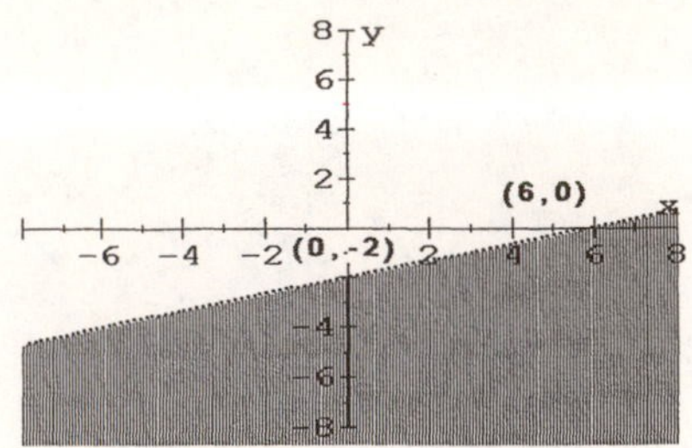

37. $x+2y\geq 4$

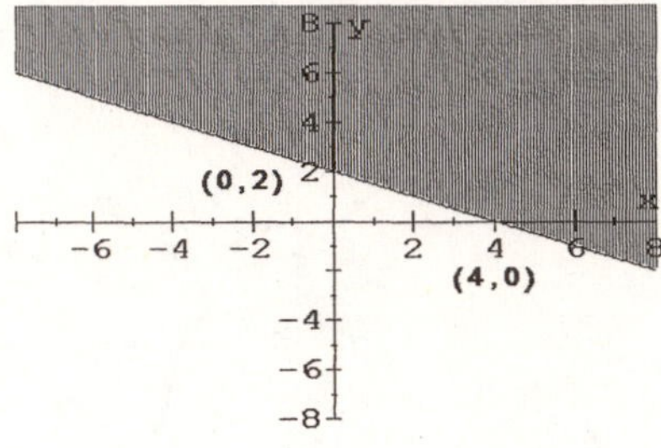

38. $2x-3y\leq 6$

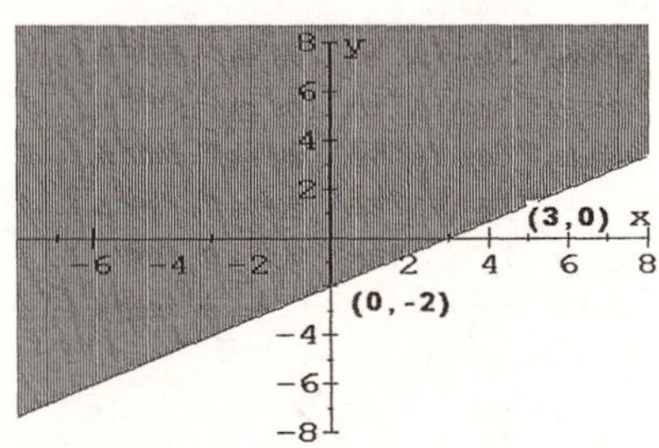

39. $y>-\frac{1}{2}x+3$

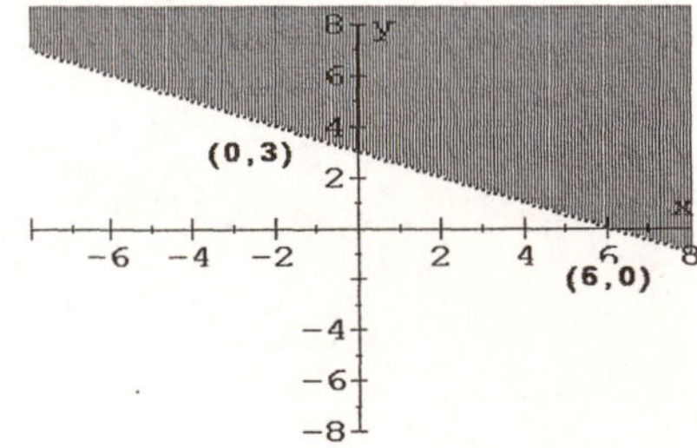

40. $y<2x-5$

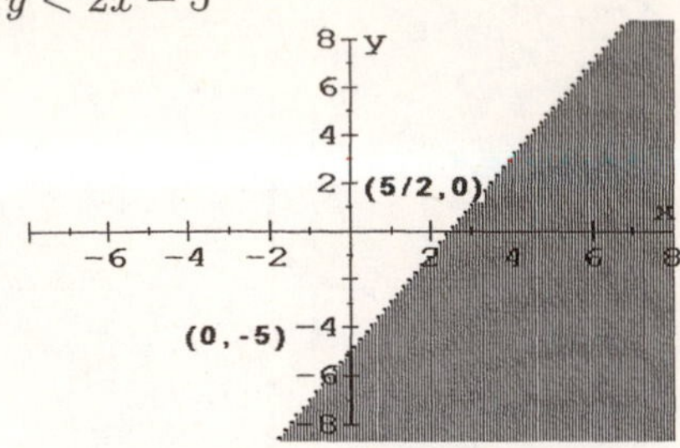

41. $y\geq \frac{2}{3}x$

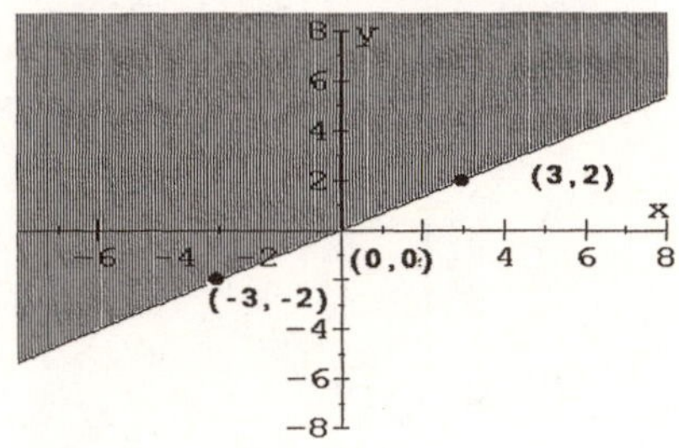

42. 1 mile = 5280 ft.
6% of 1 mile
0.06(5280) = 316.8
It will rise 316.8 feet.

43. $\frac{2}{3}=\frac{x}{12}$

$2(12)=3x$

$24=3x$

$8=x$

The rise will be 8 inches.

44. $-3x+5y=7$

$5y=3x+7$

$\frac{1}{5}(5y)=\frac{1}{5}(3x+7)$

$y=\frac{3}{5}x+\frac{7}{5}$

Slope of a perpendicular line will be $-\left(\frac{1}{\frac{3}{5}}\right)=-\frac{5}{3}$.

45. $4x + 5y = 10$
$5y = -4x + 10$
$\frac{1}{5}(5y) = \frac{1}{5}(-4x + 10)$
$y = -\frac{4}{5}x + 2$
Any line parallel would have $m = -\frac{4}{5}$.

46. Let $x =$ the value of the home
and let $y =$ taxes on the home.
Ordered pairs are:
$(200000, 2400)$; $(250000, 3150)$

$m = \frac{3150 - 2400}{250000 - 200000} = \frac{750}{50000} = \frac{3}{200}$
$y - 2400 = \frac{3}{200}(x - 200000)$
$y - 2400 = \frac{3}{200}x - 3000$
Therefore, the equation is
$y = \frac{3}{200}x - 600.$

47. Let $x =$ the number of miles
and let $y =$ cost.
Ordered pairs are: $(300, 40)$; $(1000, 180)$
$m = \frac{180 - 40}{1000 - 300} = \frac{140}{700} = \frac{1}{5}$
$y - 40 = \frac{1}{5}(x - 300)$
$y - 40 = \frac{1}{5}x - 60$
Therefore, the equation is $y = \frac{1}{5}x - 20.$

48. Let $x =$ the number of correct answers
and let $y =$ the number of points.
Ordered pairs are: $(12, 96)$; $(18, 144)$

$m = \frac{144 - 96}{18 - 12} = \frac{48}{6} = 8$
$y - 96 = 8(x - 12)$
$y - 96 = 8x - 96$
Therefore, the equation is
$y = 8x.$

49. Let $x =$ time in hours and
let $y =$ feet in cable.
Ordered pairs are: $(\frac{3}{2}, 300)$; $(4, 1050)$

$m = \frac{1050 - 300}{4 - \frac{3}{2}} = \frac{750}{\frac{5}{2}} =$

$750 \cdot \frac{2}{5} = \frac{300}{1}$
$y - 300 = 300\left(x - \frac{3}{2}\right)$
$y - 300 = 300x - 450$
Therefore, the equation is $y = 300x - 150.$

50a. $y = x^2 + 4$
To test symmetry for the
x-axis replace y with $-y$.
$-y = x^2 + 4$
$y = -x^2 - 4$
No, it is not symmetric to the x-axis.

To test symmetry for the
y-axis replace x with $-x$.
$y = (-x)^2 + 4$
$y = x^2 + 4$
Yes, it is symmetric to the y-axis.

To test symmetry for the
origin replace x with $-x$
and y with $-y$.
$-y = (-x)^2 + 4$
$-y = x^2 + 4$
$y = -x^2 - 4$
No, it is not symmetric to the origin.

b. $xy = -4$
To test symmetry for the
x-axis replace y with $-y$.
$x(-y) = -4$
$-xy = -4$
$xy = 4$
No, it is not symmetric to the x-axis.

To test symmetry for the
y-axis replace x with $-x$.
$(-x)y = -4$
$-xy = -4$
$xy = 4$
No, it is not symmetric to the y-axis.

To test symmetry for the origin replace x with $-x$ and y with $-y$.
$(-x)(-y) = -4$
$xy = -4$
Yes, it is symmetric to the origin.

c. $y = -x^3$
To test symmetry for the x-axis replace y with $-y$.
$-y = -x^3$
$y = x^3$
No, it is not symmetric to the x-axis.

To test symmetry for the y-axis replace x with $-x$.
$y = -(-x)^3$
$y = x^3$
No, it is not symmetric to the y-axis.

To test symmetry for the origin replace x with $-x$ and y with $-y$.
$-y = -(-x)^3$
$y = -x^3$
Yes, it is symmetric to the origin.

d. $x = y^4 + 2y^2$
To test symmetry for the x-axis replace y with $-y$.
$x = (-y)^4 + 2(-y)^2$
$x = y^4 + 2y^2$
Yes, it is symmetric to the x-axis.

To test symmetry for the y-axis replace x with $-x$.
$-x = y^4 + 2y^2$
No, it is not symmetric to the y-axis.

To test symmetry for the origin replace x with $-x$ and y with $-y$.
$-x = (-y)^4 + 2(-y)^2$
$-x = y^4 + 2y^2$
No, it is not symmetric to the origin.

CHAPTER 7 Test

1. $(-2, 4), (3, -2)$

$$m = \frac{-2-4}{3-(-2)} = \frac{-6}{5}$$

2. $3x - 7y = 12$

$$-7y = -3x + 12$$
$$-\frac{1}{7}(-7y) = -\frac{1}{7}(-3x + 12)$$
$$y = \frac{3}{7}x - \frac{12}{7}$$
$$m = \frac{3}{7}$$

3. $(4, 2)$ and $(-3, -1)$

$$d = \sqrt{(-3-4)^2 + (-1-2)^2}$$
$$d = \sqrt{(-7)^2 + (-3)^2}$$
$$d = \sqrt{49 + 9} = \sqrt{58}$$

4. $m = -\frac{3}{2}, (4, -5)$

$$y - (-5) = -\frac{3}{2}(x - 4)$$
$$y + 5 = -\frac{3}{2}(x - 4)$$
$$2(y + 5) = 2\left[-\frac{3}{2}(x - 4)\right]$$
$$2y + 10 = -3(x - 4)$$
$$2y + 10 = -3x + 12$$
$$3x + 2y + 10 = 12$$
$$3x + 2y = 2$$

5. $(-4, 2)$ and $(2, 1)$

$$m = \frac{1-2}{2-(-4)} = \frac{-1}{6}$$
$$y - 1 = -\frac{1}{6}(x - 2)$$

$$6(y-1)=6\left[-\frac{1}{6}(x-2)\right]$$
$$6y-6=-1(x-2)$$
$$6y-6=-x+2$$
$$\frac{1}{6}(6y)=\frac{1}{6}(-x+8)$$
$$y=-\frac{1}{6}x+\frac{4}{3}$$

6. $(-2,-4)$; parallel to the line with equation $5x+2y=7$

Parallel lines have slopes that are equal.
$$5x+2y=7$$
$$2y=-5x+7$$
$$\frac{1}{2}(2y)=\frac{1}{2}(-5x+7)$$
$$y=-\frac{5}{2}x+\frac{7}{2}$$
$$m=-\frac{5}{2}$$
$$y-(-4)=-\frac{5}{2}\left[x-(-2)\right]$$
$$y+4=-\frac{5}{2}(x+2)$$
$$2(y+4)=2\left[-\frac{5}{2}(x+2)\right]$$
$$2y+8=-5(x+2)$$
$$2y+8=-5x-10$$
$$5x+2y+8=-10$$
$$5x+2y=-18$$

7. $(4, 7)$; perpendicular to the line with equation $x-6y=9$
Perpendicular lines have slopes that are negative reciprocals.
$$x-6y=9$$
$$-6y=-x+9$$
$$-\frac{1}{6}(-6y)=-\frac{1}{6}(-x+9)$$
$$y=\frac{1}{6}x-\frac{9}{6}$$
$$m=\frac{1}{6}$$
$$y-7=-6(x-4)$$
$$y-7=-6x+24$$
$$6x+y-7=24$$
$$6x+y=31$$

8. $y=9x$

To test symmetry for x-axis replace y with $-y$.
$$(-y)=9x$$
$$y=-9x$$
No, it is not symmetric to the x-axis.

To test symmetry for y-axis replace x with $-x$.
$$y=9(-x)$$
$$y=-9x$$
No, it is not symmetric to the y-axis.

To test symmetry for origin replace x with $-x$ and y with $-y$.
$$(-y)=9(-x)$$
$$y=9x$$
Yes, it is symmetric to the origin.

9. $y^2=x^2+6$

To test symmetry for x-axis replace y with $-y$.
$$(-y)^2=x^2+6$$
$$y^2=x^2+6$$
Yes, it is symmetric to the x-axis.

To test symmetry for y-axis replace x with $-x$.
$$y^2=(-x)^2+6$$
$$y^2=x^2+6$$
Yes, it is symmetric to the y-axis.

To test symmetry for origin replace x with $-x$ and y with $-y$.
$$(-y)^2=(-x)^2+6$$
$$y^2=x^2+6$$
Yes, it is symmetric to the origin.

10. $x^2+6x+2y^2-8=0$

To test symmetry for x-axis replace y with $-y$.

$x^2 + 6x + 2(-y)^2 - 8 = 0$
$x^2 + 6x + 2y^2 - 8 = 0$
Yes, it is symmetric to the x-axis.

To test symmetry for y-axis replace x with $-x$.
$(-x)^2 + 6(-x) + 2y^2 - 8 = 0$
$x^2 - 6x + 2y^2 - 8 = 0$
No, it is not symmetric to the y-axis.

To test symmetry for origin replace x with $-x$ and y with $-y$.
$(-x)^2 + 6(-x) + 2(-y)^2 - 8 = 0$
$x^2 - 6x + 2y^2 - 8 = 0$
No, it is not symmetric to the origin.

11. $7x - 2y = 9$
$-2y = -7x + 9$
$y = \frac{7}{2}x - \frac{9}{2}$
The slope of all lines parallel is $\frac{7}{2}$.

12. $4x + 9y = -6$
$9y = -4x - 6$
$y = -\frac{4}{9}x - \frac{6}{9}$
The slope of a perpendicular line is $-\left[\frac{1}{\left(-\frac{4}{9}\right)}\right] = \frac{9}{4}$

13. $y = \frac{3}{5}x - \frac{2}{3}$
$0 = \frac{3}{5}x - \frac{2}{3}$
$\frac{2}{3} = \frac{3}{5}x$
$\frac{5}{3}\left(\frac{2}{3}\right) = \frac{5}{3}\left(\frac{3}{5}\right)x$
$\frac{10}{9} = x$
The x-intercept is $\frac{10}{9}$.

14. $\frac{3}{4}x - \frac{2}{5}y = \frac{1}{4}$
Let $x = 0$.
$\frac{3}{4}(0) - \frac{2}{5}y = \frac{1}{4}$
$-\frac{2}{5}y = \frac{1}{4}$
$-\frac{5}{2}\left(-\frac{2}{5}y\right) = -\frac{5}{2}\left(\frac{1}{4}\right)$
$y = -\frac{5}{8}$
The y-intercept is $-\frac{5}{8}$.

15. $\frac{25}{100} = \frac{120}{x}$
$25x = 12000$
$x = 480$
The horizontal change is 480 feet.

16. $\frac{200}{3000}(100\%) \approx 6.7\%$

17. $\frac{3}{4} = \frac{32}{x}$
$3x = 4(32)$
$3x = 128$
$x \approx 43$
The run would be 43 centimeters.

18. $y = -x^2 - 3$

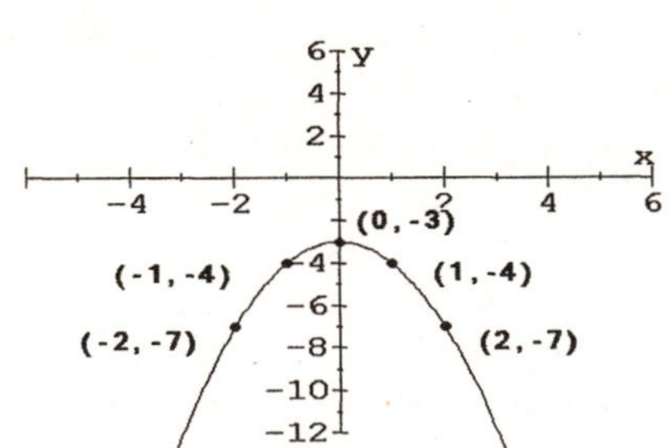

19. $y = -x - 3$

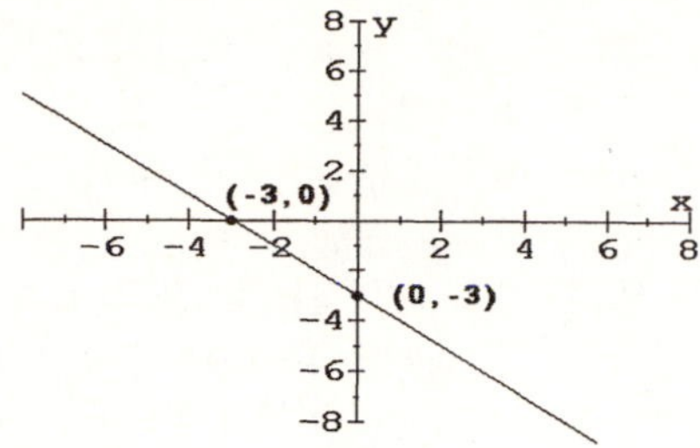

20. $-3x + y = 5$

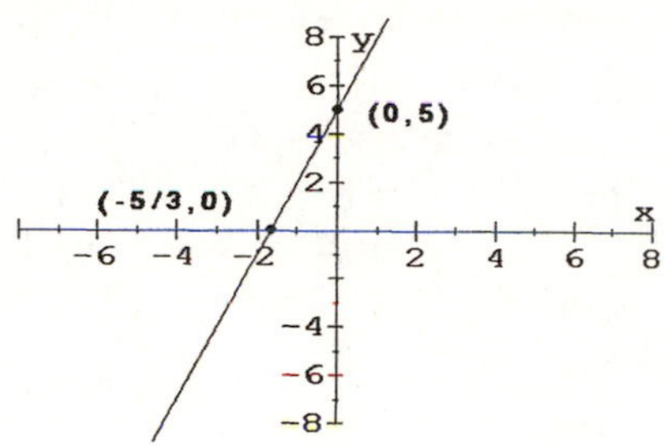

21. $3y = 2x$

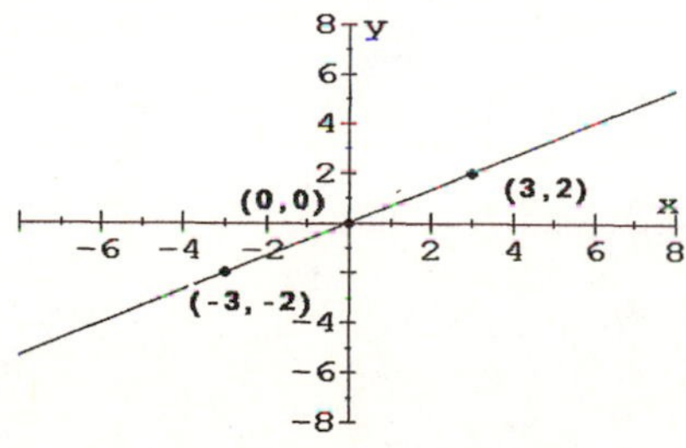

22. $\frac{1}{3}x + \frac{1}{2}y = 2$

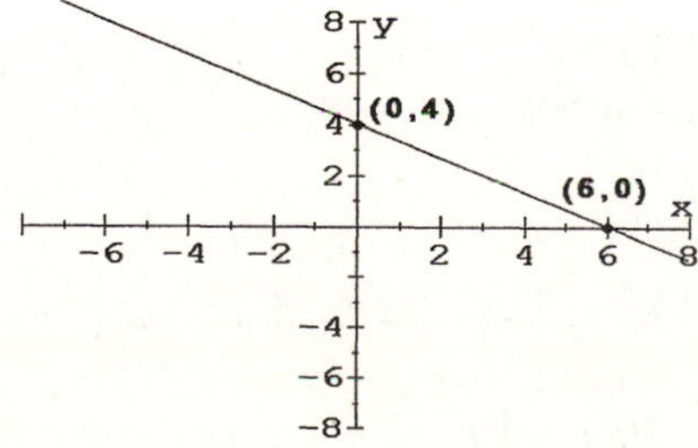

23. $y = \frac{-x - 1}{4}$

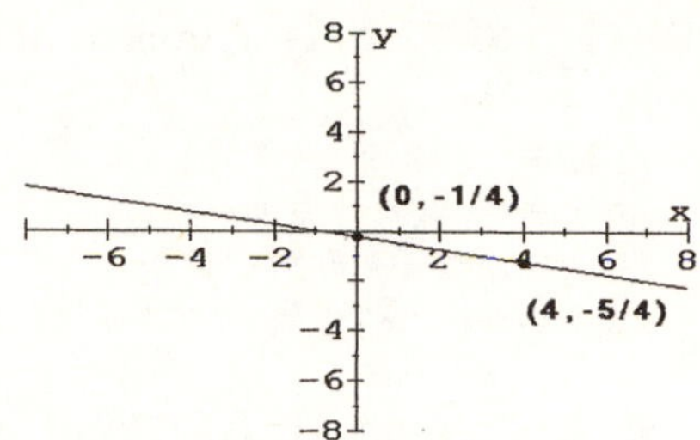

24. $2x - y < 4$

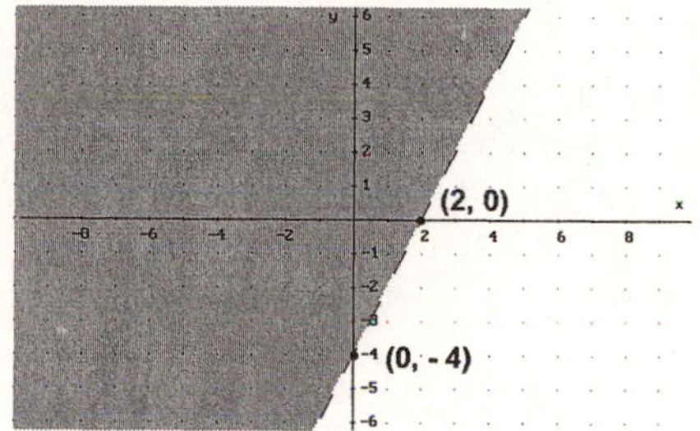

25. $3x + 2y \geq 6$

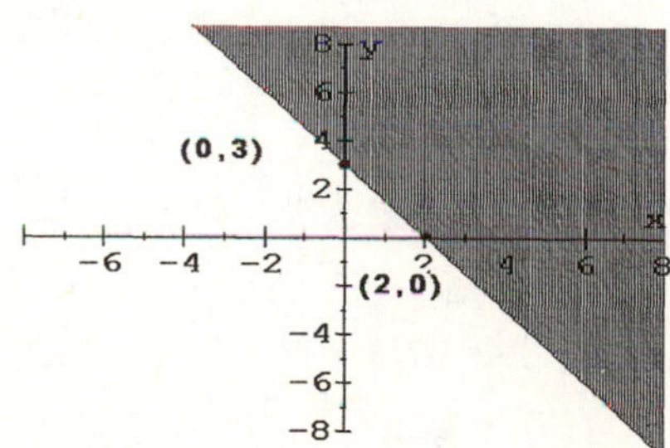

Chapter 8 Functions

PROBLEM SET 8.1 Concept of a Function

1. $f(x) = -2x + 5$
$f(3) = -2(3) + 5 = -1$
$f(5) = -2(5) + 5 = -5$
$f(-2) = -2(-2) + 5 = 9$

3. $g(x) = -2x^2 + x - 5$
$g(3) = -2(3)^2 + (3) - 5 = -20$
$g(-1) = -2(-1)^2 + (-1) - 5 = -8$
$g(2a) = -2(2a)^2 + (2a) - 5$
$= -2(4a^2) + 2a - 5$
$= -8a^2 + 2a - 5$

5. $h(x) = \frac{2}{3}x - \frac{3}{4}$

$h(3) = \frac{2}{3}(3) - \frac{3}{4} = 2 - \frac{3}{4} = \frac{5}{4}$

$h(4) = \frac{2}{3}(4) - \frac{3}{4} = \frac{8}{3} - \frac{3}{4} = \frac{23}{12}$

$h\left(-\frac{1}{2}\right) = \frac{2}{3}\left(-\frac{1}{2}\right) - \frac{3}{4} =$

$-\frac{1}{3} - \frac{3}{4} = -\frac{13}{12}$

7. $f(x) = \sqrt{2x - 1}$
$f(5) = \sqrt{2(5) - 1} = 3$
$f\left(\frac{1}{2}\right) = \sqrt{2\left(\frac{1}{2}\right) - 1} = 0$
$f(23) = \sqrt{2(23) - 1} = \sqrt{45} = 3\sqrt{5}$

9. $f(x) = -2x + 7$
$f(a) = -2a + 7$
$f(a + 2) = -2(a + 2) + 7$
$= -2a - 4 + 7$
$= -2a + 3$
$f(a + h) = -2(a + h) + 7$
$= -2a - 2h + 7$

11. $f(x) = x^2 - 4x + 10$
$f(-a) = (-a)^2 - 4(-a) + 10$
$= a^2 + 4a + 10$
$f(a - 4) = (a - 4)^2 - 4(a - 4) + 10$
$= a^2 - 8a + 16 - 4a + 16 + 10$
$= a^2 - 12a + 42$
$f(a + h) = (a + h)^2 - 4(a + h) + 10$
$= a^2 + 2ah + h^2 - 4a - 4h + 10$

13. $f(x) = -x^2 + 3x + 5$
$f(-a) = -(-a)^2 + 3(-a) + 5$
$= -a^2 - 3a + 5$
$f(a + 6) = -(a + 6)^2 + 3(a + 6) + 5$
$= -(a^2 + 12a + 36) + 3a + 18 + 5$
$= -a^2 - 12a - 36 + 3a + 23$
$= -a^2 - 9a - 13$
$f(-a + 1) = -(-a + 1)^2 + 3(-a + 1) + 5$
$= -(a^2 - 2a + 1) - 3a + 3 + 5$
$= -a^2 + 2a - 1 - 3a + 8$
$= -a^2 - a + 7$

15. $f(x) = \begin{cases} x \text{ for } x \geq 0 \\ x^2 \text{ for } x < 0 \end{cases}$

To find $f(4)$, note $x = 4$
which is ≥ 0 so use:
$f(x) = x$
$f(4) = 4$

To find $f(10)$, note $x = 10$
which is ≥ 0 so use:
$f(x) = x$
$f(10) = 10$

To find $f(-3)$, note $x = -3$
which is < 0 so use:
$f(x) = x^2$
$f(-3) = (-3)^2 = 9$

To find $f(-5)$, note $x = -5$
which is < 0 so use:
$f(x) = x^2$
$f(-5) = (-5)^2 = 25$

17. $f(x) = \begin{cases} 2x \text{ for } x \geq 0 \\ -2x \text{ for } x < 0 \end{cases}$

To find $f(3)$, note $x = 3$
which is ≥ 0 so use:
$f(x) = 2x$
$f(3) = 2(3) = 6$

To find $f(5)$, note $x = 5$
which is ≥ 0 so use:
$f(x) = 2x$
$f(5) = 2(5) = 10$

To find $f(-3)$, note $x = -3$
which is < 0 so use:
$f(x) = -2x$
$f(-3) = -2(-3) = 6$

To find $f(-5)$, note $x = -5$
which is < 0 so use:
$f(x) = -2x$
$f(-5) = -2(-5) = 10$

19. $f(x) = \begin{cases} 1 \text{ for } x > 0 \\ 0 \text{ for } -1 < x \leq 0 \\ -1 \text{ for } x \leq -1 \end{cases}$

To find $f(2)$, note $x = 2$
which is ≥ 0 so use:
$f(x) = 1$
$f(2) = 1$

To find $f(0)$, note $x = 0$
whic is $-1 < 0 \leq 0$ so use:
$f(x) = 0$
$f(0) = 0$

To find $f\left(-\frac{1}{2}\right)$, note $x = -\frac{1}{2}$
which is $-1 < -\frac{1}{2} \leq 0$ so use:
$f(x) = 0$
$f\left(-\frac{1}{2}\right) = 0$

To find $f(-4)$, note $x = -4$
which is ≤ -1 so use:
$f(x) = -1; \ f(-4) = -1$

21. $f(x) = -7x - 2$

$$\begin{aligned} f(a+h) &= -7(a+h) - 2 \\ &= -7a - 7h - 2 \end{aligned}$$

$f(a) = -7a - 2$

$$\begin{aligned} \frac{f(a+h) - f(a)}{h} &= \frac{-7a - 7h - 2 - (-7a - 2)}{h} \\ &= \frac{-7a - 7h - 2 + 7a + 2}{h} \\ &= \frac{-7h}{h} = -7 \end{aligned}$$

23. $f(x) = -x^2 + 4x - 2$

$$\begin{aligned} f(a+h) &= -(a+h)^2 + 4(a+h) - 2 \\ &= -(a^2 + 2ah + h^2) + 4a + 4h - 2 \\ &= -a^2 - 2ah - h^2 + 4a + 4h - 2 \end{aligned}$$

$$f(a) = -a^2 + 4a - 2$$

$$\frac{f(a+h) - f(a)}{h} = \frac{-a^2 - 2ah - h^2 + 4a + 4h - 2 - (-a^2 + 4a - 2)}{h}$$

$$= \frac{-a^2 - 2ah - h^2 + 4a + 4h - 2 + a^2 - 4a + 2}{h}$$

$$= \frac{-2ah - h^2 + 4h}{h}$$

$$= \frac{h(-2a - h + 4)}{h} = -2a - h + 4$$

25. $f(x) = 3x^2 - x - 4$

$$\begin{aligned} f(a+h) &= 3(a+h)^2 - (a+h) - 4 \\ &= 3(a^2 + 2ah + h^2) - (a+h) - 4 \\ &= 3a^2 + 6ah + 3h^2 - a - h - 4 \end{aligned}$$

$$f(a) = 3a^2 - a - 4$$

$$\frac{f(a+h) - f(a)}{h} = \frac{3a^2 + 6ah + 3h^2 - a - h - 4 - (3a^2 - a - 4)}{h}$$

$$= \frac{3a^2 + 6ah + 3h^2 - a - h - 4 - 3a^2 + a + 4}{h}$$

$$= \frac{6ah - h + 3h^2}{h} = 6a + 3h - 1$$

27. $f(x) = x^3 - x^2 + 2x - 1$

$$\begin{aligned} f(a+h) &= (a+h)^3 - (a+h)^2 + 2(a+h) - 1 \\ &= a^3 + 3a^2h + 3ah^2 + h^3 - (a^2 + 2ah + h^2) + 2a + 2h - 1 \\ &= a^3 + 3a^2h + 3ah^2 + h^3 - a^2 - 2ah - h^2 + 2a + 2h - 1 \end{aligned}$$

$$f(a) = a^3 - a^2 + 2a - 1$$

$$\frac{f(a+h) - f(a)}{h} = \frac{a^3+3a^2h+3ah^2+h^3 - a^2 - 2ah - h^2+2a+2h - 1 - (a^3 - a^2+2a - 1)}{h}$$

$$= \frac{a^3 + 3a^2h + 3ah^2 + h^3 - a^2 - 2ah - h^2 + 2a + 2h - 1 - a^3 + a^2 - 2a + 1}{h}$$

$$= \frac{3a^2h + 3ah^2 + h^3 - 2ah - h^2 + 2h}{h}$$

$$= \frac{h(3a^2 + 3ah + h^2 - 2a - h + 2)}{h} = 3a^2 + 3ah + h^2 - 2a - h + 2$$

29. $f(x) = \dfrac{2}{x-1}$

$$f(a+h) = \frac{2}{a+h-1}$$

$$f(a) = \frac{2}{a-1}$$

$$f(a+h) - f(a) = \frac{2}{a+h-1} - \frac{2}{a-1} = \frac{2(a-1) - 2(a+h-1)}{(a+h-1)(a-1)}$$

$$= \frac{2a - 2 - 2a - 2h + 2}{(a+h-1)(a-1)} = \frac{-2h}{(a+h-1)(a-1)}$$

$$\frac{f(a+h) - f(a)}{h} = \frac{\dfrac{-2h}{(a+h-1)(a-1)}}{h} = \frac{-2}{(a+h-1)(a-1)}$$

31. $f(x) = \dfrac{1}{x^2}$

$$f(a+h) = \frac{1}{(a+h)^2}$$

$$f(a) = \frac{1}{a^2}$$

$$\frac{f(a+h) - f(a)}{h} = \frac{\dfrac{1}{(a+h)^2} - \dfrac{1}{a^2}}{h} = \frac{\dfrac{1}{(a+h)^2} - \dfrac{1}{a^2}}{h} \cdot \frac{(a+h)^2a^2}{(a+h)^2a^2} = \frac{a^2 - (a+h)^2}{ha^2(a+h)^2}$$

$$= \frac{a^2 - (a^2 + 2ah + h^2)}{ha^2(a+h)^2} = \frac{-2ah - h^2}{ha^2(a+h)^2} = \frac{h(-2a-h)}{ha^2(a+h)^2} = \frac{-2a-h}{a^2(a+h)^2}$$

33. It is a function because no vertical line would intersect the graph in more than one point.

35. It is not a function because some vertical line would intersect the graph in more than one point.

37. It is a function because no vertical line would intersect the graph in more than one point.

39. It is a function because no vertical line would intersect the graph in more than one point.

41. $f(x) = \sqrt{3x-4}$

For domain, $3x - 4 \geq 0$

Therefore $x \geq \dfrac{4}{3}$ and $D = \left\{x \middle| x \geq \dfrac{4}{3}\right\}$.

Because $\sqrt{3x-4}$ is nonnegative, $R = \{f(x) | f(x) \geq 0\}$.

43. $f(x) = x^2 - 2$

The domain is the set of all real numbers since any real number can be substituted for x.

Note the smallest value of x^2 is 0 at $x = 0$,
hence the smallest value of $x^2 - 2$ is at $x = 0$
and it is $0^2 - 2 = -2$. Thus the range is
$f(x) \geq -2$.

$D = \{x | x \text{ is any real number}\}$
$R = \{f(x) | f(x) \geq -2\}$

45. $f(x) = |x|$
The domain is the set of all real numbers.
$D = \{x | x \text{ is any real number}\}$
Because $f(x)$ is nonnegative,
$R = \{f(x) | f(x) \geq 0\}$.

47. $f(x) = -\sqrt{x}$
Since x is under the radical $x \geq 0$. Hence
the domain is the set of all nonnegative
real numbers.
$D = \{x | x \text{ is any nonnegative real number}\}$

Since $-\sqrt{x}$ is nonpositive for all values of x,
the range is $f(x) \leq 0$.
$R = \{f(x) | f(x) \text{ is any nonpositive real number}\}$

49. $f(x) = \dfrac{-4}{x+2}$

Since the denominator cannot equal 0,
$x + 2 \neq 0$
$x \neq -2$
$D = \{x | x \neq -2\}$

51. $f(x) = \dfrac{5}{(2x-1)(x+4)}$

Since the denominator cannot equal 0,
we must discard values of x that make
$2x - 1 = 0$ or $x + 4 = 0$
$x = \frac{1}{2}$ or $x = -4$
$D = \left\{x | x \neq \frac{1}{2} \text{ and } x \neq -4\right\}$

53. $f(x) = \dfrac{1}{x^2 - 4}$

Since the denominator cannot equal 0,
we must discard values of x that make
$x^2 - 4 = 0$.
$(x+2)(x-2) = 0$
$x + 2 = 0$ or $x + 2 = 0$
$x = -2$ or $x = -2$
$D = \{x | x \neq 2 \text{ and } x \neq -2\}$

55. $f(x) = \dfrac{4x}{x^2 - x - 12}$
Since the denominator cannot equal 0,
we must discard values of x that make
$x^2 - x - 12 = 0$.
$(x-4)(x+3) = 0$
$x - 4 = 0$ or $x + 3 = 0$
$x = 4$ or $x = -3$
$D = \{x | x \neq -3 \text{ and } x \neq 4\}$

57. $f(x) = \dfrac{x}{6x^2 + 13x - 5}$

Since the denominator cannot equal 0,
we must discard values of x that make
$6x^2 + 13x - 5 = 0$.
$(3x-1)(2x+5) = 0$
$3x - 1 = 0$ or $2x + 5 = 0$
$3x = 1$ or $2x = 5$
$x = \frac{1}{3}$ or $x = -\frac{5}{2}$
$D = \left\{x | x \neq -\frac{5}{2} \text{ and } x \neq \frac{1}{3}\right\}$

59. $f(x) = \sqrt{x^2 - 16}$

The radicand $x^2 - 16$ must be nonnegative.
$x^2 - 16 \geq 0$
$(x+4)(x-4) \geq 0$
Critical values:
$x - 4 = 0$ or $x + 4 = 0$
$x = 4$ or $x = -4$

Test Point	-5	0	5
	(-4)		(4)
$x - 4$:	negative	negative	positive
$x + 4$:	negative	positive	positive
product:	positive	negative	positive

We find the product is nonnegative when
$x \geq 4$ or $x \leq -4$
The domain is $(-\infty, -4] \cup [4, \infty)$.

61. $f(x)=\sqrt{x^2+1}-4$
Since x^2+1 is always nonnegative, the domain is all real numbers.
$D=(-\infty,\infty)$.

63. $f(x)=\sqrt{x^2-3x-40}$

The radicand $x^2-3x-40$, must be nonnegative.
$x^2-3x-40\geq 0$
$(x-8)(x+5)\geq 0$
Critical values:
$x-8=0$ or $x+5=0$
$x=8$ or $x=-5$

Test Point	-6	0	9
	(-5)		(8)
$x-8$:	negative	negative	positive
$x+5$:	negative	positive	positive
product:	positive	negative	positive

We find the product is nonnegative when
$x\geq 8$ or $x\leq -5$
The domain is $(-\infty,-5]\cup[8,\infty)$.

65. $f(x)=-\sqrt{8x^2+6x-35}$
$8x^2+6x-35\geq 0$
$(4x-7)(2x+5)\geq 0$
Critical values
$4x-7=0$ or $2x+5=0$
$4x=7$ or $2x=-5$
$x=\frac{7}{4}$ or $x=-\frac{5}{2}$

To determine when $(4x-7)(2x+5)\geq 0$, test intervals $\left(-\infty,-\frac{5}{2}\right)$, $\left(-\frac{5}{2},\frac{7}{4}\right)$ and $\left(\frac{7}{4},\infty\right)$.

Critical values $-\frac{5}{2}$ and $\frac{7}{4}$:

Test Point	-3	0	3
$4x-7$:	negative	negative	positive
$2x+5$:	negative	positive	positive
product:	positive	negative	positive

The domain is $\left(-\infty,-\frac{5}{2}\right]\cup\left[\frac{7}{4},\infty\right)$.

67. $f(x)=\sqrt{11-x^2}$

The radicand $1-x^2$, must be nonnegative.
$1-x^2\geq 0$
$(1+x)(1-x)\geq 0$
Critical values:
$1-x=0$ or $1+x=0$
$-x=-1$ or $x=-1$
$x=1$

Critical values -1 and 1:

Test Point	-2	0	2
$1-x$:	positive	positive	negative
$1+x$:	negative	positive	positive
product:	negative	positive	negative

We find the product is nonnegative when
$x\geq -1$ and $x\leq 1$
The domain is $[-1,1]$.

69. $A(r)=\pi r^2$
$A(2)=\pi(2)^2=4\pi=12.57$
$A(3)=\pi(3)^2=9\pi=28.27$
$A(12)=\pi(12)^2=144\pi=452.39$
$A(17)=\pi(17)^2=289\pi=907.92$

71. $h(t)=64t-16t^2$
$h(1)=64(1)-16(1)^2=48$
$h(2)=64(2)-16(2)^2=64$
$h(3)=64(3)-16(3)^2=48$
$h(4)=64(4)-16(4)^2=0$

73. $I(r)=500r$
$I(0.11)=500(0.11)=\$55$
$I(0.12)=500(0.12)=\$60$
$I(0.135)=500(0.135)=\$67.50$
$I(0.15)=500(0.15)=\$75$

75. $A(r)=2\pi r^2+16\pi r$
$A(2)=2\pi(2)^2+16\pi(2)=125.66$
$A(4)=2\pi(4)^2+16\pi(4)=301.59$
$A(8)=2\pi(8)^2+16\pi(8)=804.25$

PROBLEM SET 8.2 Linear Functions and Applications

1. $f(x) = 2x - 4$

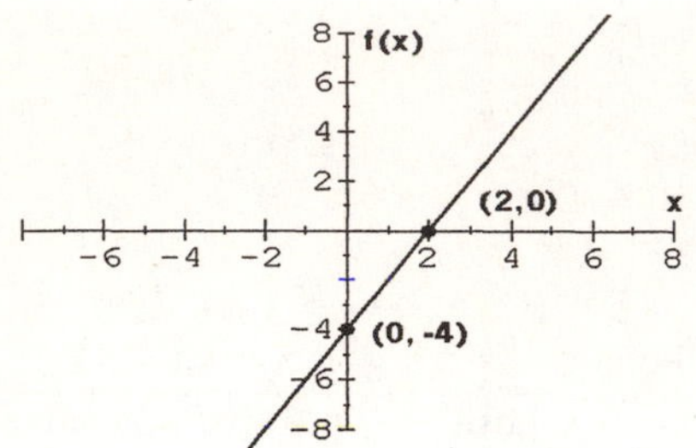

3. $f(x) = -x + 3$

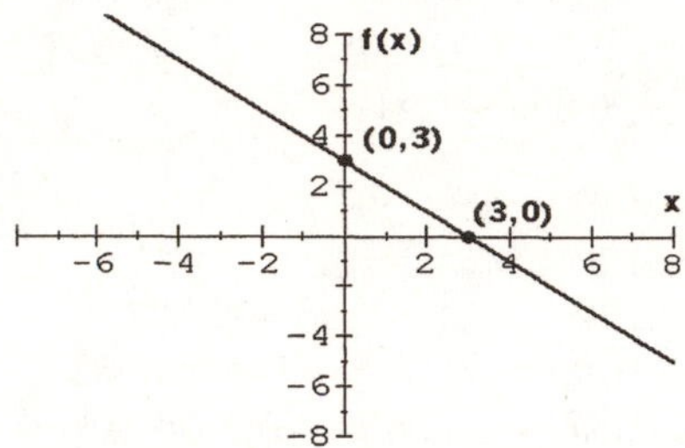

5. $f(x) = 3x + 9$

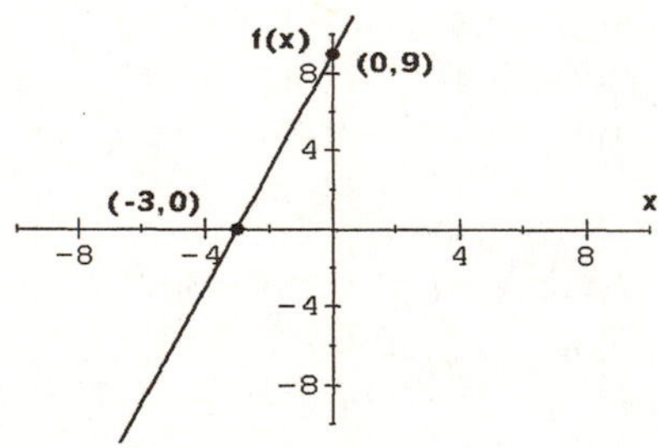

7. $f(x) = -4x - 4$

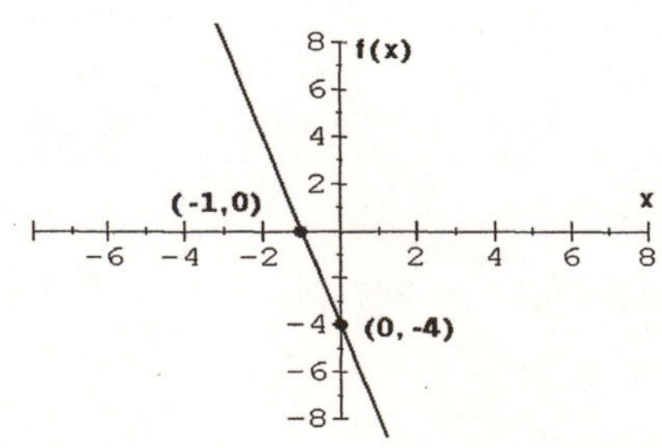

9. $f(x) = -3x$

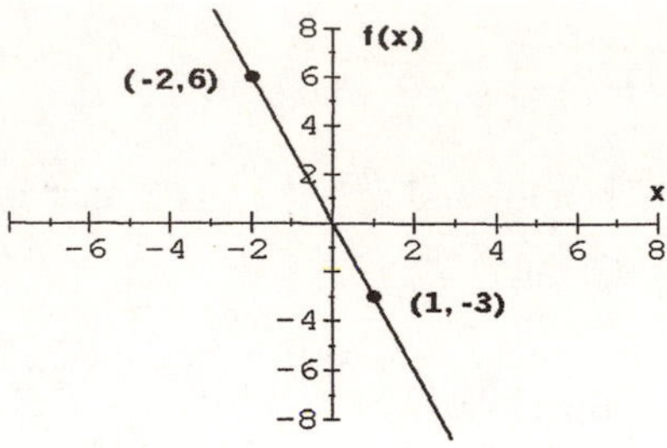

11. $f(x) = -3$

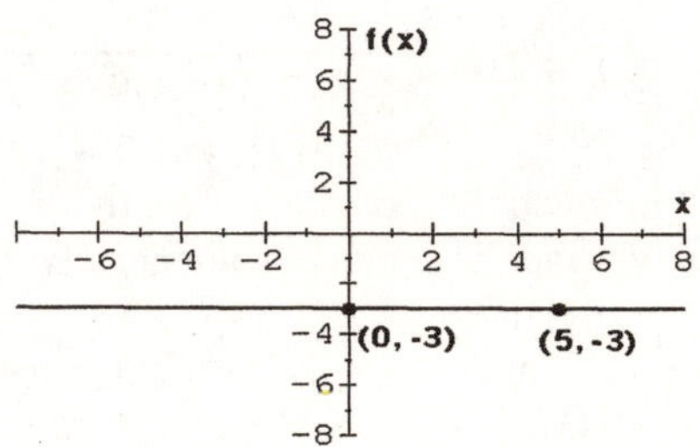

13. $f(x) = \frac{1}{2}x + 3$

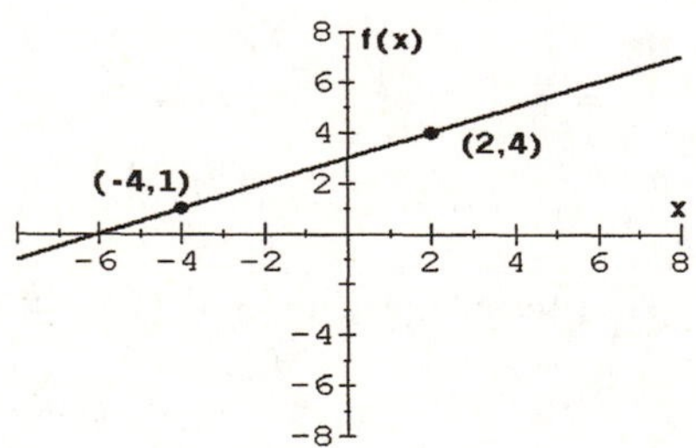

15. $f(x) = -\frac{3}{4}x - 6$

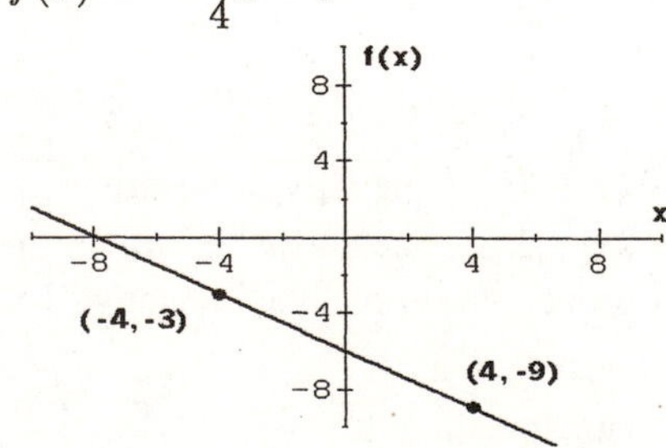

17. $m = \frac{2}{3}, (-1, 3)$

$f(x) = mx + b$

$3 = \frac{2}{3}(-1) + b$

$\frac{9}{3} + \frac{2}{3} = b$

$\frac{11}{3} = b$

$f(x) = \frac{2}{3}x + \frac{11}{3}$

19. $(-3, -1), (2, -6)$

$m = \frac{-6-(-1)}{2-(-3)} = \frac{-5}{5} = -1$

$f(x) = mx + b$
$-1 = -1(-3) + b$
$-1 = 3 + b$
$-4 = b$
$f(x) = -x - 4$

21. $g(x) = 5x - 2$, $(6, 3)$

$m = -\frac{1}{5}$, $(6, 3)$ for perpendicular line

$f(x) = mx + b$

$3 = -\frac{1}{5}(6) + b$

$\frac{15}{5} = -\frac{6}{5} + b$

$\frac{21}{5} = b$

$f(x) = -\frac{1}{5}x + \frac{21}{5}$

23a. 3 hours/night × 31 nights = 93 hours
$c(93) = 0.0045(93) = \$0.42$

b. $c(h) = 0.0045h$

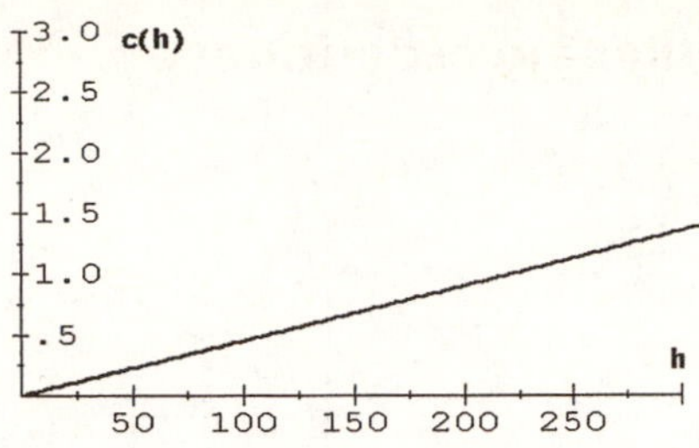

c. $c(225) \approx \$1.00$

d. $c(225) = 0.0045(225) = \$1.01$

25. $f(x) = 26,\ x \leq 200$
$g(x) = 26 + 0.15(x - 200),\ x > 200$

$f(150) = \$26$

$g(230) = 26 + 0.15(230 - 200)$
$g(230) = 26 + 0.15(30)$
$g(230) = \$30.50$

$g(360) = 26 + 0.15(360 - 200)$
$g(360) = 26 + 0.15(160)$
$g(360) = \$50.00$

$g(430) = 26 + 0.15(430 - 200)$
$g(430) = 26 + 0.15(230)$
$g(430) = \$60.50$

27. $s(c) = 1.4c$
$s(1.50) = 1.4(1.50) = \$2.10$
$s(3.25) = 1.4(3.25) = \$4.55$
$s(14.80) = 1.4(14.80) = \$20.72$
$s(21) = 1.4(21) = \$29.40$
$s(24.20) = 1.4(24.20) = \$33.88$

29. $f(p) = p - 0.2p = 0.8p$
$f(9.50) = 0.8(9.50) = \$7.60$
$f(15) = 0.8(15) = \$12.00$
$f(75) = 0.8(75) = \$60.00$
$f(12.50) = 0.8(12.50) = \$10.00$
$f(750) = 0.8(750) = \$600.00$

Further Investigations

33. $f(x) = |x|$

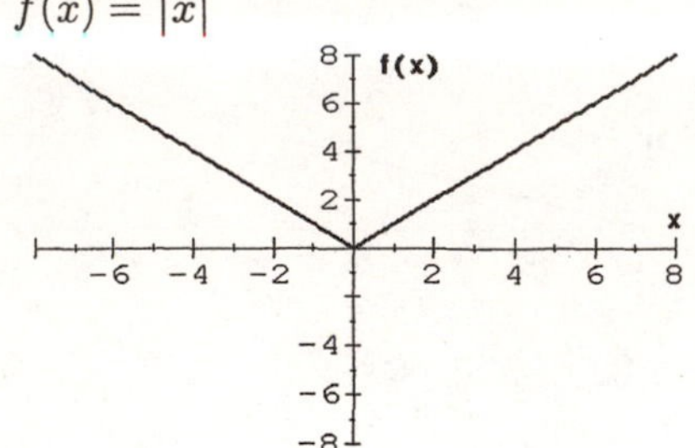

37. $f(x) = \frac{x}{|x|}$

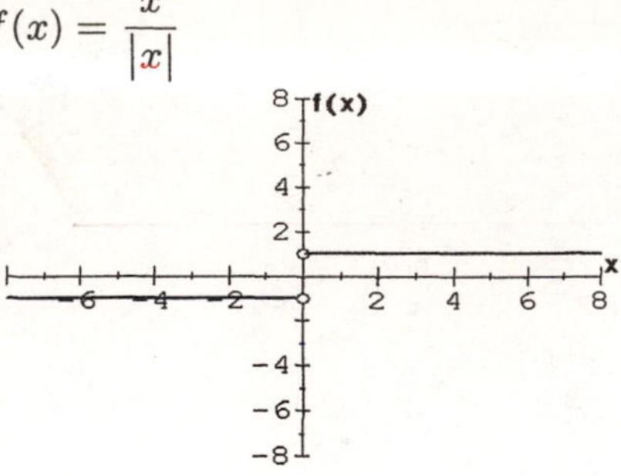

35. $f(x) = x - |x|$

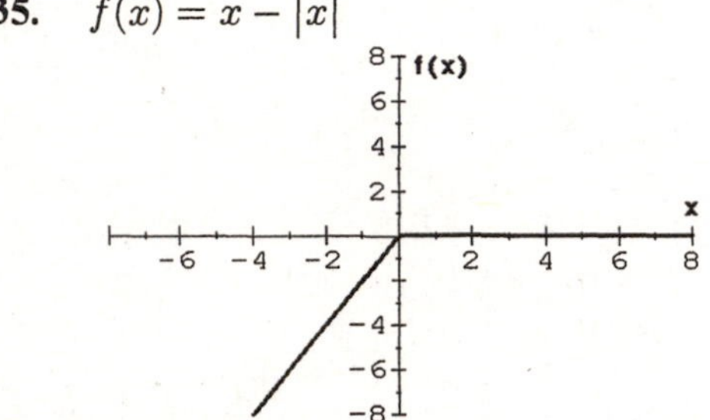

PROBLEM SET 8.3 Quadratic Functions

1. $f(x) = x^2 + 1$ the basic parabola shifted 1 unit up.

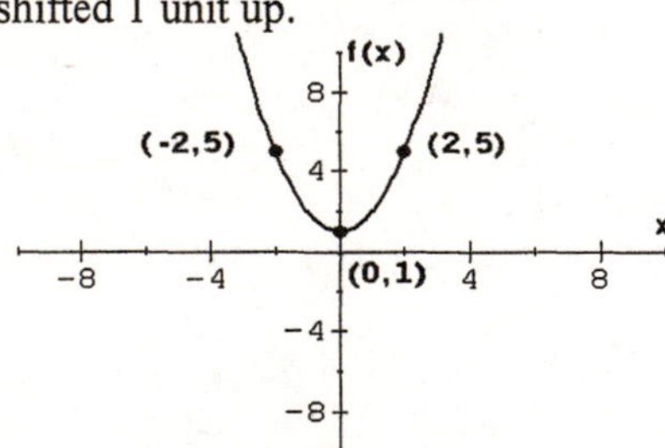

3. $f(x) = 3x^2$ the basic parabola stretched by a factor of 3.

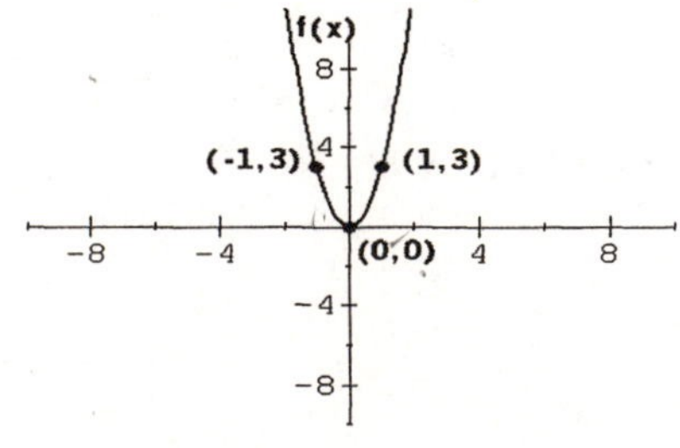

5. $f(x) = -x^2 + 2$ the basic parabola refected across the x-axis and shifted 2 units up.

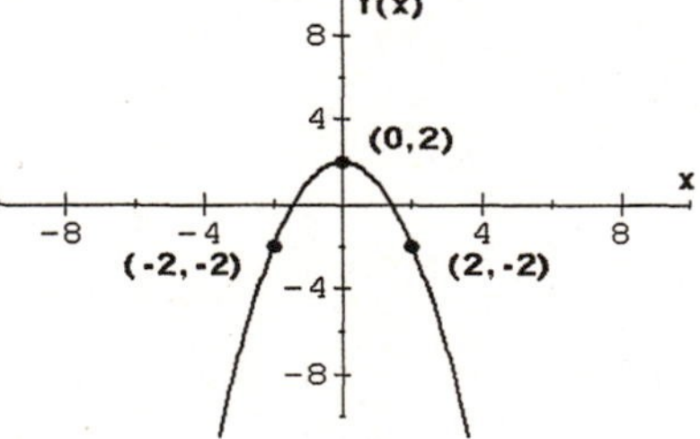

7. $f(x) = (x + 2)^2$
$f(x) = [x - (-2)]^2$ the basic parabola shifted 2 units to the left.

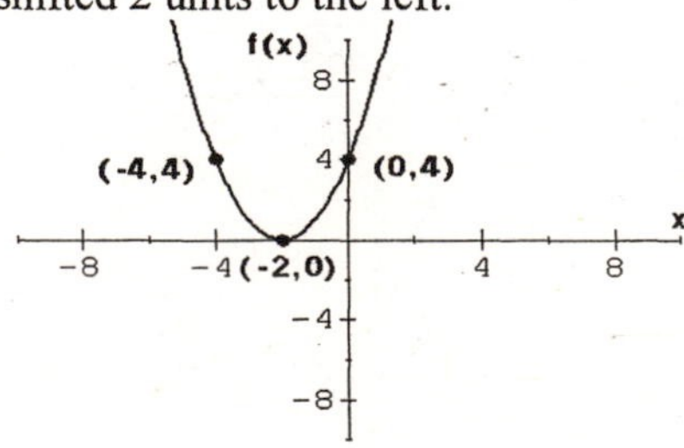

9. $f(x) = -2(x+1)^2$
$f(x) = -2[x-(-1)]^2$ the basic parabola reflected across the x-axis, stretched by a factor of 2, and shifted 1 unit to the left.

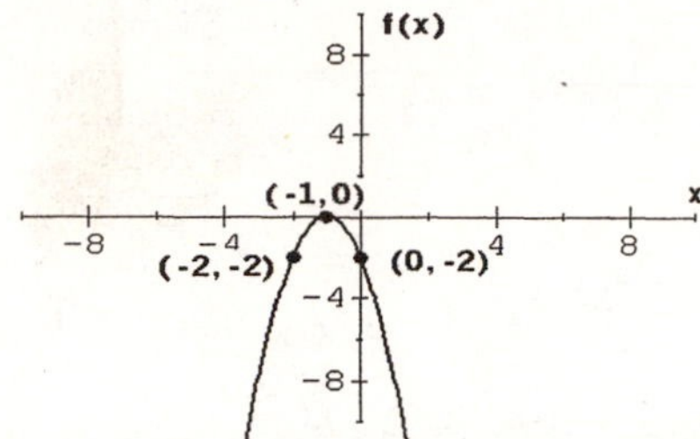

11. $f(x) = (x-1)^2 + 2$ the basic parabola shifted 1 unit to the right and 2 units up.

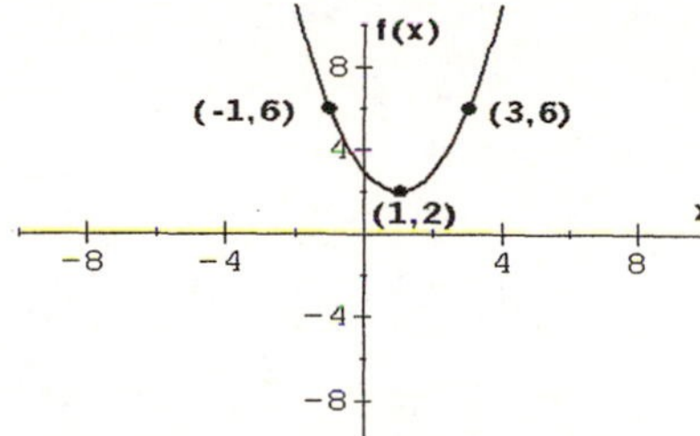

13. $f(x) = \frac{1}{2}(x-2)^2 - 3$ the basic parabola shrunk by a factor of $\frac{1}{2}$, shifted 2 units to the right, and 3 units down.

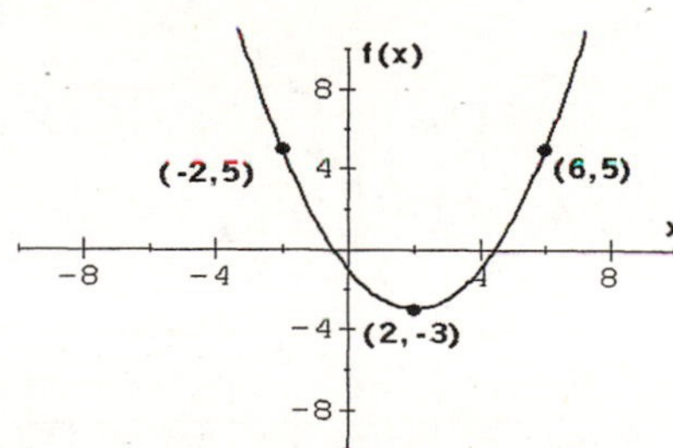

15. $f(x) = x^2 + 2x + 4$
$f(x) = x^2 + 2x + 1 + 4 - 1$
$f(x) = (x+1)^2 + 3$
This is the basic parabola shifted 1 unit to the left and 3 units up.

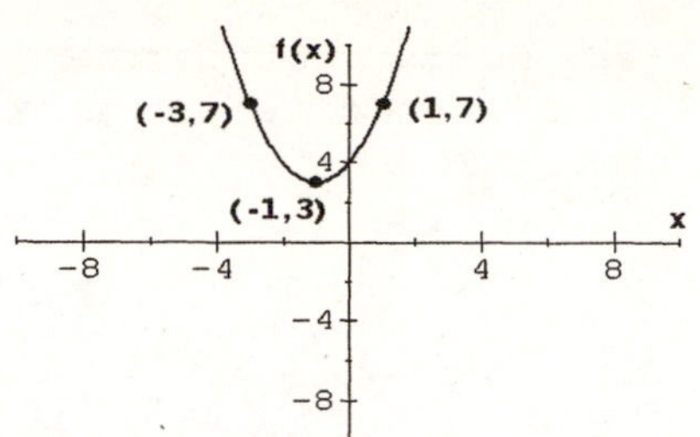

17. $f(x) = x^2 - 3x + 1$
$f(x) = x^2 - 3x + \frac{9}{4} - \frac{9}{4} + 1$
$f(x) = \left(x - \frac{3}{2}\right)^2 - \frac{5}{4}$
This is the basic parabola shifted $\frac{3}{2}$ units to the right and $\frac{5}{4}$ unit down.

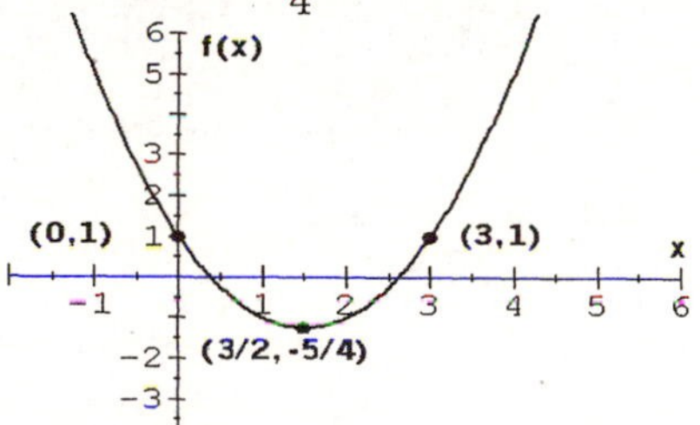

19. $f(x) = 2x^2 + 12x + 17$
$f(x) = 2(x^2 + 6x) + 17$
$f(x) = 2(x^2 + 6x + 9) - 18 + 17$
$f(x) = 2(x+3)^2 - 1$
This is the basic parabola stretched by a factor of 2, shifted 3 units to the left, and 1 unit down.

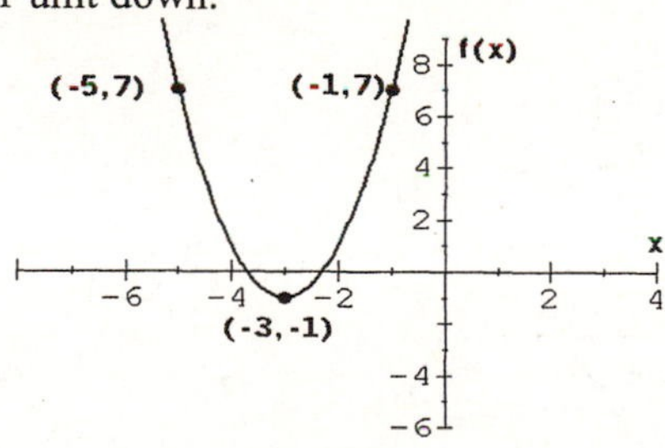

21. $f(x) = -x^2 - 2x + 1$
$f(x) = -(x^2 + 2x) + 1$
$f(x) = -(x^2 + 2x + 1) + 1 + 1$
$f(x) = -(x+1)^2 + 2$

Problem Set 8.3

This is the basic parabola reflected across the x-axis, and shifted 1 unit to the left and 2 units up.

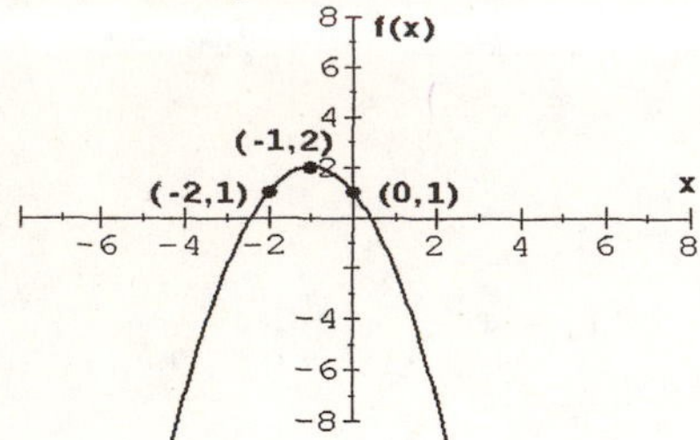

23. $f(x) = 2x^2 - 2x + 3$
$f(x) = 2(x^2 - x) + 3$
$f(x) = 2\left(x^2 - x + \frac{1}{4}\right) - \frac{1}{2} + 3$
$f(x) = 2\left(x - \frac{1}{2}\right)^2 + \frac{5}{2}$

This is the basic parabola stretched by a factor of 2, shifted $\frac{1}{2}$ unit to the right and $\frac{5}{2}$ units up.

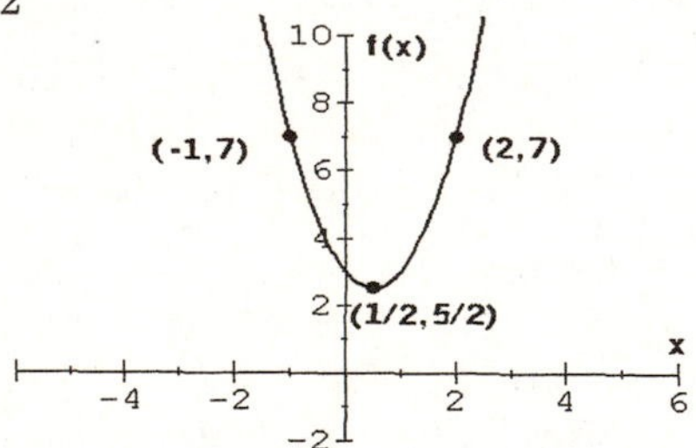

25. $f(x) = -2x^2 - 5x + 1$
$f(x) = -2\left(x^2 + \frac{5}{2}x\right) + 1$
$f(x) = -2\left(x^2 + \frac{5}{2}x + \frac{25}{16}\right) + \frac{25}{8} + 1$
$f(x) = -2\left(x + \frac{5}{4}\right)^2 + \frac{33}{8}$

This is the parabola reflected across the x-axis, stretched by a factor of 2, shifted $\frac{5}{4}$ units to the left, and $\frac{33}{8}$ units up.

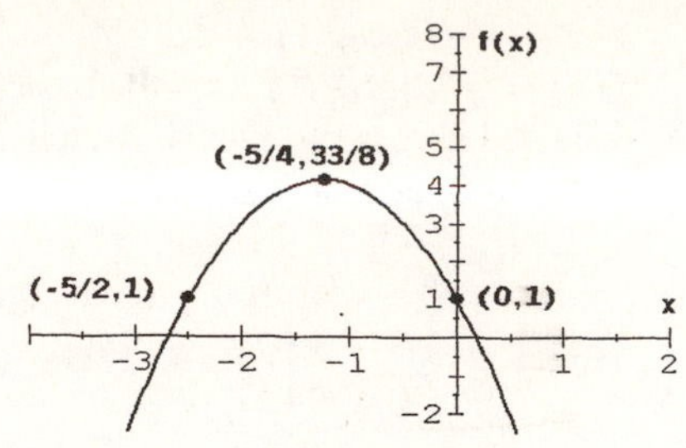

27. $f(x) = \begin{cases} x \text{ for } x \geq 0 \\ 3x \text{ for } x < 0 \end{cases}$

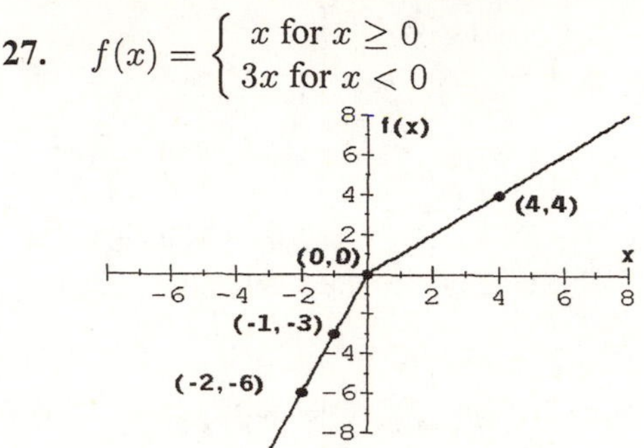

29. $f(x) = \begin{cases} 2x + 1 \text{ for } x \geq 0 \\ x^2 \text{ for } x < 0 \end{cases}$

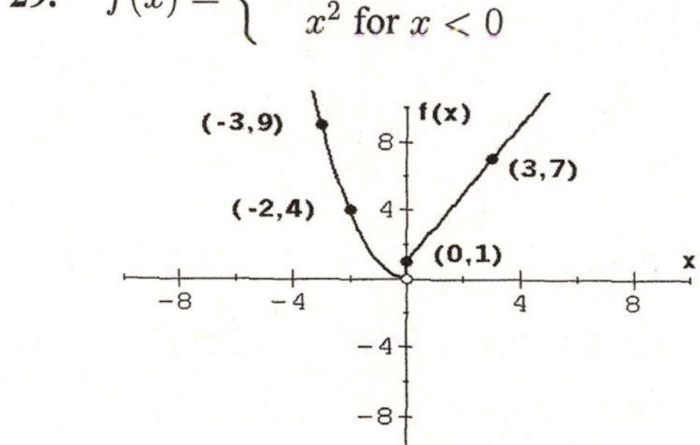

31. $f(x) = \begin{cases} 2 \text{ for } x \geq 0 \\ -1 \text{ for } x < 0 \end{cases}$

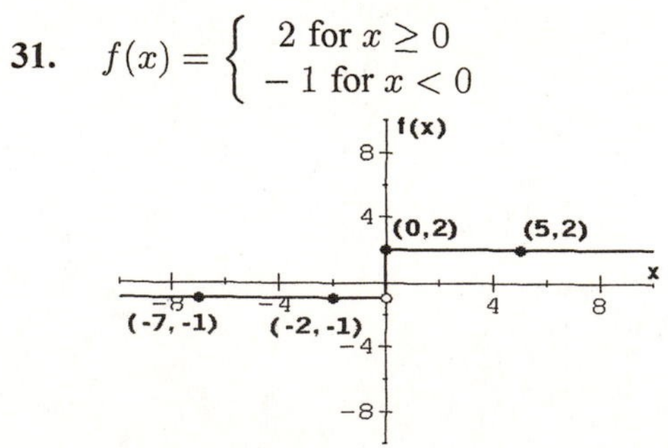

33. $f(x) = \begin{cases} 1 \text{ for } 0 \le x < 1 \\ 2 \text{ for } 1 \le x < 2 \\ 3 \text{ for } 2 \le x < 3 \\ 4 \text{ for } 3 \le x < 4 \end{cases}$

(3,4)

(0,1)

35. $f(x) = [x]$ for $-4 \le x < 4$

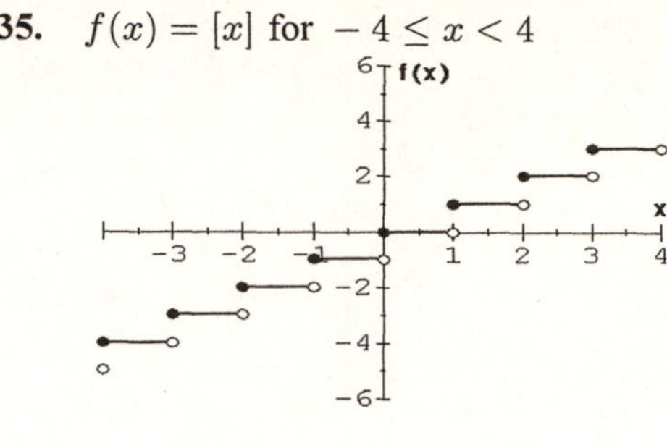

PROBLEM SET 8.4 More Quadratic Functions and Applications

1. $f(x) = x^2 - 8x + 15$

Step 1: Because $a = 1 > 0$, the parabola opens upward.

Step 2: $-\dfrac{b}{2a} = -\dfrac{-8}{2(1)} = 4$

Step 3: $f\left(-\dfrac{b}{2a}\right) = f(4)$
$= 4^2 - 8(4) + 15 = -1$
Thus the vertex is $(4, -1)$.

Step 4: Letting $x = 3$, we obtain
$f(3) = 3^2 - 8(3) + 15 = 0$
$(3, 0)$ is on the graph and so is its reflection $(5, 0)$ across the line of symmetry $x = 4$.

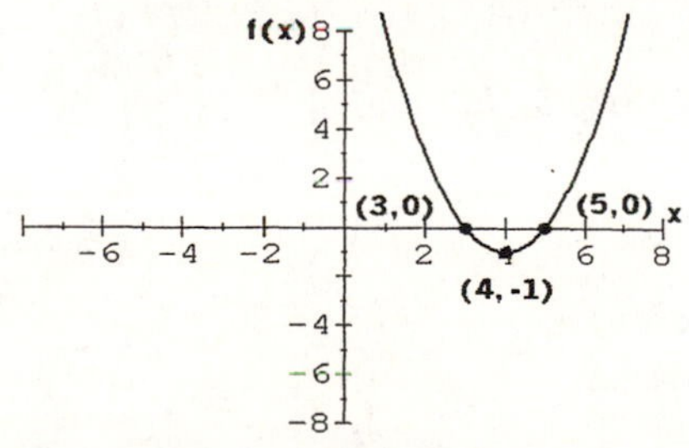

3. $f(x) = 2x^2 + 20x + 52$

Step 1: Because $a = 2 > 0$, the parabola opens upward.

Step 2: $-\dfrac{b}{2a} = -\dfrac{20}{2(2)} = -5$

Step 3: $f\left(-\dfrac{b}{2a}\right) = f(-5)$
$= 2(-5)^2 + 20(-5) + 52$
$= 50 - 100 + 52 = 2$
Thus the vertex is $(-5, 2)$.

Step 4: Letting $x = -4$, we obtain
$f(-4) = 2(-4)^2 + 20(-4) + 52$
$= 32 - 80 + 52 = 4$
$(-4, 4)$ is on the graph and so is its reflection $(-6, 4)$ across the line of symmetry $x = -5$.

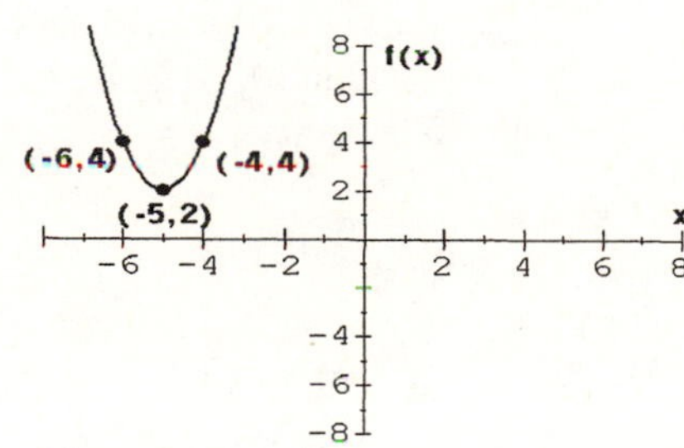

5. $f(x) = -x^2 + 4x - 7$

Step 1: Because $a = -1 < 0$, the parabola opens downward.

Step 2: $-\dfrac{b}{2a} = -\dfrac{4}{2(-1)} = 2$

Step 3: $f\left(-\dfrac{b}{2a}\right) = f(2)$
$= -2^2 + 4(2) - 7 = -3$
Thus the vertex is $(2, -3)$.

Step 4: Letting $x = 3$, we obtain
$f(3) = -3^2 + 4(3) - 7 = -4$
$(3, -4)$ is on the graph and so is its reflection $(1, -4)$ across the line of symmetry $x = 2$.

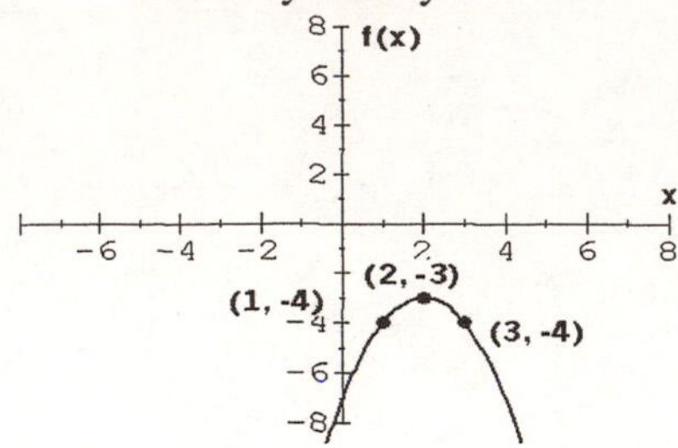

7. $f(x) = -3x^2 + 6x - 5$
Step 1: Because $a = -3 < 0$, the parabola opens downward.

Step 2: $-\frac{b}{2a} = -\frac{6}{2(-3)} = 1$

Step 3: $f\left(-\frac{b}{2a}\right) = f(1)$
$= -3(1)^2 + 6(1) - 5$
$= -3 + 6 - 5 = -2$
Thus the vertex is $(1, -2)$.

Step 4: Letting $x = 2$, we obtain
$f(2) = -3(2)^2 + 6(2) - 5$
$= -12 + 12 - 5 = -5$
$(2, -5)$ is on the graph and so is its reflection $(0, -5)$ across the line of symmetry $x = 1$.

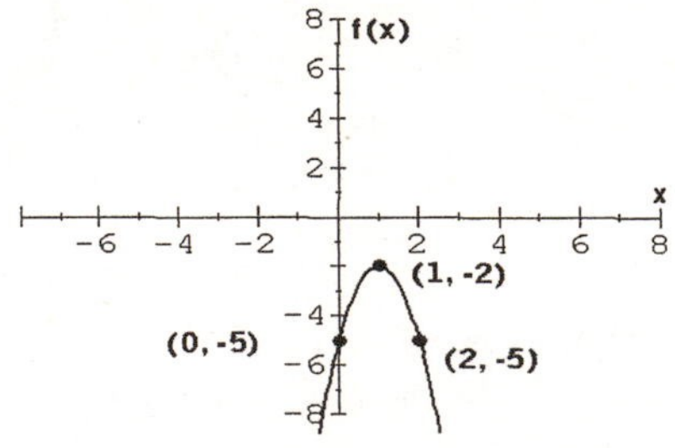

9. $f(x) = x^2 + 3x - 1$
Step 1: Because $a = 1 > 0$, the parabola opens upward.

Step 2: $-\frac{b}{2a} = -\frac{3}{2(1)} = -\frac{3}{2}$

Step 3: $f\left(-\frac{b}{2a}\right) = f\left(-\frac{3}{2}\right)$
$= \left(-\frac{3}{2}\right)^2 + 3\left(-\frac{3}{2}\right) - 1 = -\frac{13}{4}$
Thus the vertex is $\left(-\frac{3}{2}, -\frac{13}{4}\right)$.

Step 4: Letting $x = 0$, we obtain
$f(0) = 0^2 + 3(0) - 1 = -1$
$(0, -1)$ is on the graph and so is its reflection $(-3, -1)$ across the line of symmetry $x = -\frac{3}{2}$.

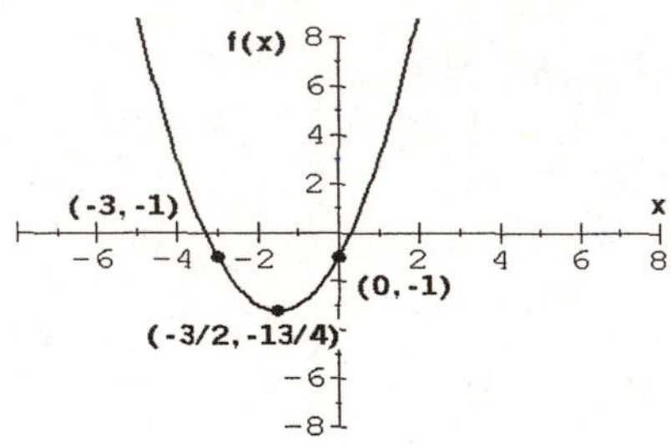

11. $f(x) = -2x^2 + 5x + 1$
Step 1: Because $a = -2 < 0$, the parabola opens downward.

Step 2: $-\frac{b}{2a} = -\frac{5}{2(-2)} = \frac{5}{4}$

Step 3: $f\left(-\frac{b}{2a}\right) = f\left(\frac{5}{4}\right)$
$= -2\left(\frac{5}{4}\right)^2 + 5\left(\frac{5}{4}\right) + 1$
$= -\frac{50}{16} + \frac{25}{4} + 1 = \frac{66}{16} = \frac{33}{8}$
Thus the vertex is $\left(\frac{5}{4}, \frac{33}{8}\right)$.

Step 4: Letting $x = 0$, we obtain
$f(0) = -2(0)^2 + 5(0) + 1 = 1$
(0, 1) is on the graph and so is its reflection $\left(\frac{5}{2}, 1\right)$ across the line of symmetry $x = \frac{5}{4}$.

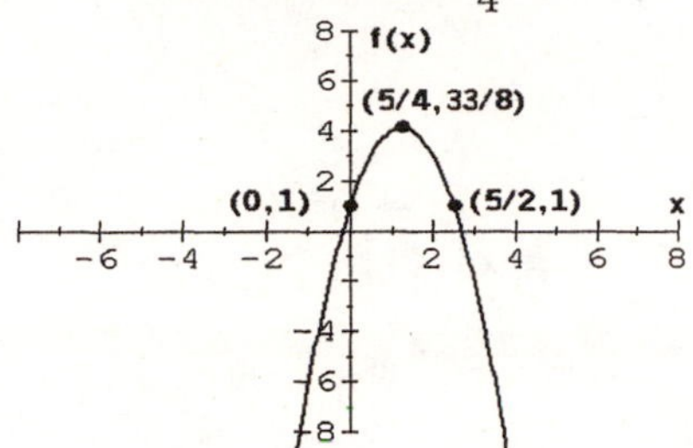

13. $f(x) - x^2 + 3$
This is the basic parabola $f(x) = x^2$ reflected across the x-axis then moved up 3 units.

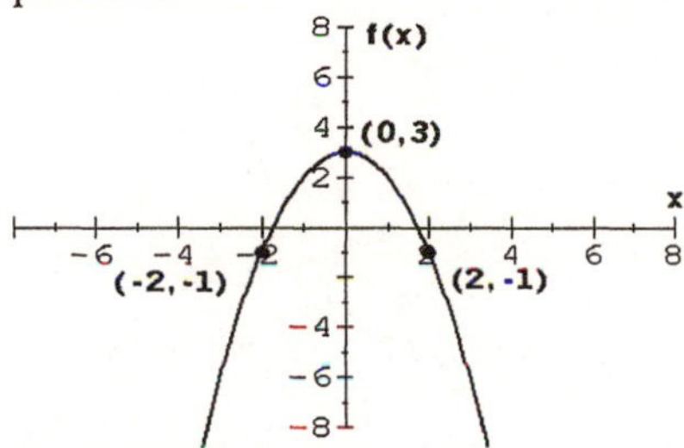

15. $f(x) = x^2 + x - 1$
Step 1: Because $a = 1 > 0$,
the parabola opens upward.

Step 2: $-\frac{b}{2a} = -\frac{1}{2(1)} = -\frac{1}{2}$

Step 3: $f\left(-\frac{b}{2a}\right) = f\left(-\frac{1}{2}\right)$
$= \left(-\frac{1}{2}\right)^2 + \left(-\frac{1}{2}\right) - 1$
$= \frac{1}{4} - \frac{1}{2} - 1 = -\frac{5}{4}$
Thus the vertex is $\left(-\frac{1}{2}, -\frac{5}{4}\right)$.

Step 4: Letting $x = 1$, we obtain
$f(1) = (1)^2 + (1) - 1 = 1$
(1, 1) is on the graph and so is its reflection $(-2, 1)$ across the line of symmetry $x = -\frac{1}{2}$.

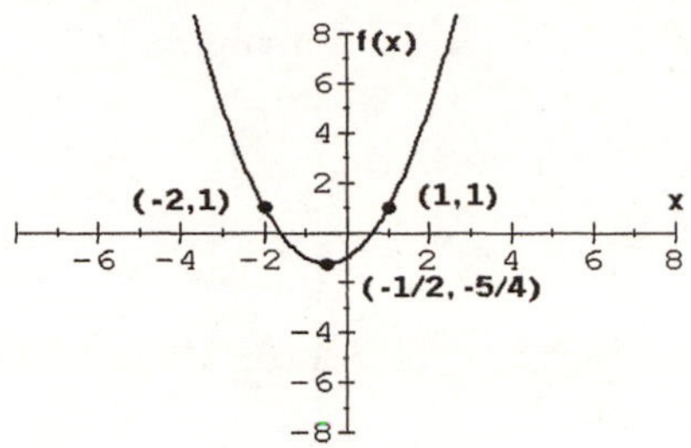

17. $f(x) = -2x^2 + 4x + 1$
Step 1: Because $a = -2 < 0$,
the parabola opens downward.

Step 2: $-\frac{b}{2a} = -\frac{4}{2(-2)} = 1$

Step 3: $f\left(-\frac{b}{2a}\right) = f(1)$
$= -2(1)^2 + 4(1) + 1 = 3$
Thus the vertex is $(1, 3)$.

Step 4: Letting $x = 0$, we obtain
$f(0) = -2(0)^2 + 4(0) + 1 = 1.$
(0, 1) is on the graph and so is its reflection (2, 1) across the line of symmetry $x = 1$.

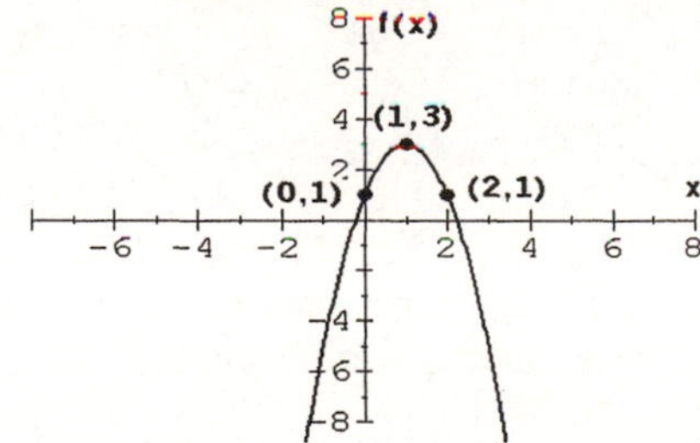

Problem Set 8.4

19. $f(x) = -\left(x + \frac{5}{2}\right)^2 + \frac{3}{2}$

This is the basic parabola reflected across the x-axis and shifted $\frac{5}{2}$ units to the left and $\frac{3}{2}$ units up.

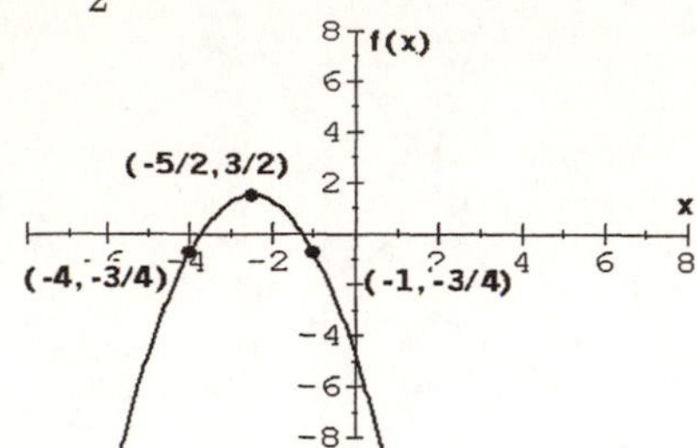

21. $f(x) = 3x^2 - 12$

x-intercepts: Let $f(x) = 0$.

$0 = 3x^2 - 12$

$0 = 3(x^2 - 4)$

$0 = 3(x - 2)(x + 2)$

$x - 2 = 0$ or $x + 2 = 0$

$x = 2$ or $x = -2$

The x-intercepts are -2 and 2.

Vertex:

$-\frac{b}{2a} = -\frac{0}{2(3)} = 0$

$f\left(-\frac{b}{2a}\right) = f(0) = 3(0)^2 - 12 = -12$

The vertex is $(0, -12)$.

23. $f(x) = 5x^2 - 10x$

x-intercepts: Let $f(x) = 0$.

$0 = 5x^2 - 10x$

$0 = 5x(x - 2)$

$5x = 0$ or $x - 2 = 0$

$x = 0$ or $x = 2$

The x-intercepts are 0 and 2.

Vertex:

$-\frac{b}{2a} = -\frac{-10}{2(5)} = 1$

$f\left(-\frac{b}{2a}\right) = f(1) = 5(1)^2 - 10 = -5$

The vertex is $(1, -5)$.

25. $f(x) = x^2 - 8x + 15$

x-intercepts: Let $f(x) = 0$.

$0 = x^2 - 8x + 15$

$0 = (x - 3)(x - 5)$

$x - 3 = 0$ or $x - 5 = 0$

$x = 3$ or $x = 5$

The x-intercepts are 3 and 5.

$$\begin{aligned} f(x) &= (x^2 - 8x) + 15 \\ &= (x^2 - 8x + 16) - 16 + 15 \\ &= (x - 4)^2 - 1 \end{aligned}$$

The vertex is $(4, -1)$.

27. $f(x) = 2x^2 - 28x + 96$

x-intercepts: Let $f(x) = 0$.

$0 = 2x^2 - 28x + 96$

$0 = 2(x^2 - 14x + 48)$

$0 = 2(x - 6)(x - 8)$

$x - 6 = 0$ or $x - 8 = 0$

$x = 6$ or $x = 8$

The x-intercepts are 6 and 8.

$$\begin{aligned} f(x) &= 2(x^2 - 14x) + 96 \\ &= 2(x^2 - 14x + 49) - 98 + 96 \\ &= 2(x - 7)^2 - 2 \end{aligned}$$

The vertex is $(7, -2)$.

29. $f(x) = -x^2 + 10x - 24$

x-intercepts: Let $f(x) = 0$.

$0 = -x^2 + 10x - 24$

$0 = x^2 - 10x + 24$

$0 = (x - 6)(x - 4)$

$x - 6 = 0$ or $x - 4 = 0$

$x = 6$ or $x = 4$

The x-intercepts are 6 and 4.

$-\frac{b}{2a} = -\frac{10}{2(-1)} = 5$

$f\left(-\frac{b}{2a}\right) = f(5)$

$= -5^2 + 10(5) - 24 = 1$

The vertex is $(5, 1)$.

31. $f(x) = x^2 - 14x + 44$
x-intercepts: Let $f(x) = 0$.
$0 = x^2 - 14x + 44$
$$x = \frac{-(-14) \pm \sqrt{(-14)^2 - 4(1)(44)}}{2(1)}$$
$$x = \frac{14 \pm \sqrt{20}}{2} = \frac{14 \pm 2\sqrt{5}}{2}$$
$x = 7 \pm \sqrt{5}$
The x-intercepts are $7 - \sqrt{5}$ and $7 + \sqrt{5}$.

$$-\frac{b}{2a} = -\frac{-14}{2(1)} = 7$$
$$f\left(-\frac{b}{2a}\right) = f(7)$$
$$= (7)^2 - 14(7) + 44 = -5$$
The vertex is $(7, -5)$.

33. $f(x) = -x^2 + 9x - 21$
x-intercepts: Let $f(x) = 0$.
$0 = -x^2 + 9x - 21$
$0 = x^2 - 9x + 21$

$$x = \frac{-(-9) \pm \sqrt{(-9)^2 - 4(1)(21)}}{2(1)}$$
$$x = \frac{9 \pm \sqrt{-3}}{2}$$
Since the solutions for x are not real numbers, there are no x-intercepts.

$$-\frac{b}{2a} = -\frac{9}{2(-1)} = \frac{9}{2}$$
$$f\left(-\frac{b}{2a}\right) = f\left(\frac{9}{2}\right)$$
$$= -\left(\frac{9}{2}\right)^2 + 9\left(\frac{9}{2}\right) - 21$$
$$= -\frac{3}{4}$$
The vertex is $\left(\frac{9}{2}, -\frac{3}{4}\right)$.

35. $f(x) = -4x^2 + 4x + 4$
x-intercepts: Let $f(x) = 0$.
$0 = -4x^2 + 4x + 4$
$0 = -4(x^2 - x - 1)$
$0 = x^2 - x - 1$
$$x = \frac{-(-1) \pm \sqrt{(-1)^2 - 4(1)(-1)}}{2(1)}$$
$$x = \frac{1 \pm \sqrt{5}}{2}$$
The x-intercepts are $\frac{1 - \sqrt{5}}{2}$ and $\frac{1 + \sqrt{5}}{2}$.
$$\begin{aligned} f(x) &= -4x^2 + 4x + 4 \\ &= -4(x^2 - x) + 4 \\ &= -4\left(x^2 - x + \frac{1}{4}\right) + 1 + 4 \\ &= -4\left(x - \frac{1}{2}\right)^2 + 5 \end{aligned}$$
The vertex is $\left(\frac{1}{2}, 5\right)$.

37. $f(x) = x^2 + 3x - 88$
$0 = x^2 + 3x - 88$
$0 = (x + 11)(x - 8)$
$x + 11 = 0$ or $x - 8 = 0$
$x = -11$ or $x = 8$
The zeros are -11 and 8.

39. $f(x) = 4x^2 - 48x + 108$
$0 = 4(x^2 - 12x + 27)$
$0 = 4(x - 9)(x - 3)$
$x - 9 = 0$ or $x - 3 = 0$
$x = 9$ or $x = 3$
The zeros are 9 and 3.

41. $f(x) = x^2 - 4x + 11$
$0 = x^2 - 4x + 11$
$$x = \frac{-(-4) \pm \sqrt{(-4)^2 - 4(1)(11)}}{2(1)}$$
$$x = \frac{4 \pm \sqrt{-28}}{2}$$
$$x = \frac{4 \pm 2i\sqrt{7}}{2}$$
$x = 2 \pm i\sqrt{7}$
The zeros are $2 - i\sqrt{7}$ and $2 + i\sqrt{7}$.

43. $p(x) = -2x^2 + 280x - 1000$

$$-\frac{b}{2a} = -\frac{280}{2(-2)} = 70$$

70 units should be sold.

45. $f(x) = 96x - 16x^2$

$$= -16x^2 + 96x$$
$$= -16(x^2 - 6x)$$
$$= -16(x^2 - 6x + 9) + 144$$
$$= -16(x - 3)^2 + 144$$

The vertex is (3, 144).
The highest point is 144 feet.

47. Let $x =$ one number and
$50 - x =$ other number.

$f(x) = x(50 - x)$
$f(x) = -x^2 + 50x$

$$-\frac{b}{2a} = -\frac{50}{2(-1)} = 25$$

The numbers are 25 and 25.

49. Let $x =$ width of playground.
Then since the length plus the width is equal to half the perimeter,
$120 - x =$ length of playground.

$f(x) = x(120 - x)$
$f(x) = -x^2 + 120x$

$$-\frac{b}{2a} = -\frac{120}{2(-1)} = 60$$

The width is 60 meters and the length is 60 meters.

53. Let x be the number of \$0.25 decreases in the monthly rate. Then $1000 + 20x$ represents the number of subscribers and $15 - 0.25x$ represents what each subscriber would pay per month.

$f(x) = (1000 + 20x)(15 - 0.25x)$
$f(x) = 15000 + 50x - 5x^2$

The function is maximized at the vertex of the parabola.

$$-\frac{b}{2a} = -\frac{50}{2(-5)} = 5$$

So 5 decreases.
New rate $= 15 - 5(0.25) =$ \$13.75.
Number of subscribers = 1000+20(5) = 1100.

PROBLEM SET 8.5 Transformations of Some Basic Curves

1. $f(x) = x^4 + 2$
This is the graph of $f(x) = x^4$ shifted 2 units up.

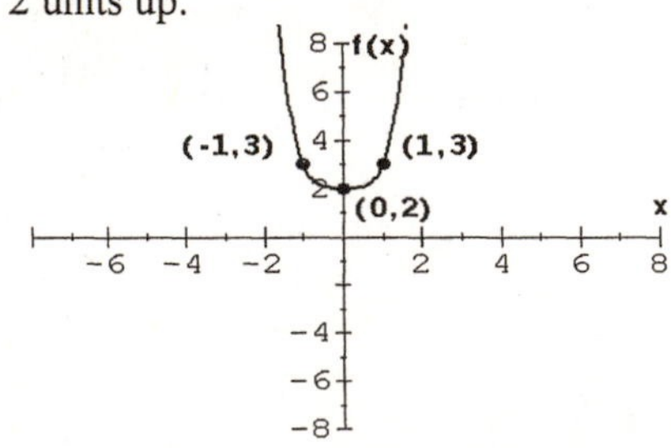

3. $f(x) = (x - 2)^4$
This is the graph of $f(x) = x^4$ shifted 2 units to the right.

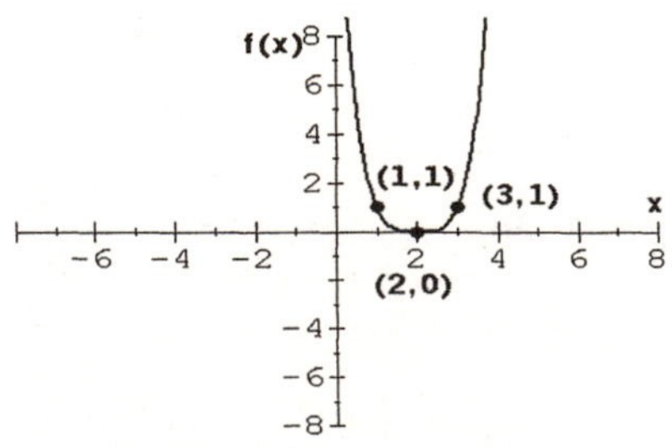

5. $f(x) = -x^3$
This is the graph of $f(x) = x^3$ reflected across the x-axis.

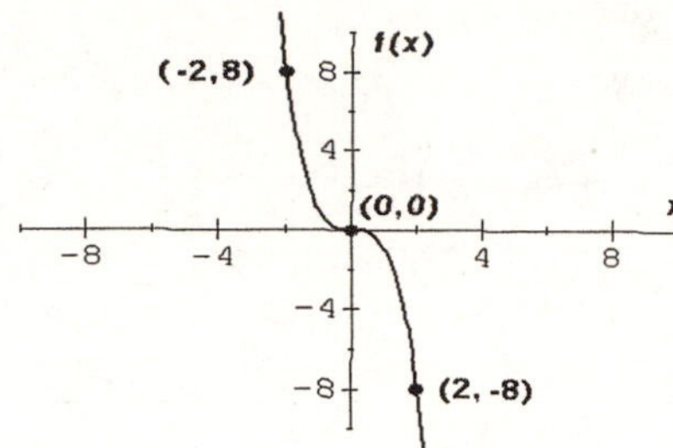

7. $f(x) = (x+2)^3$
This is the graph of $f(x) = x^3$ shifted 2 units to the left.

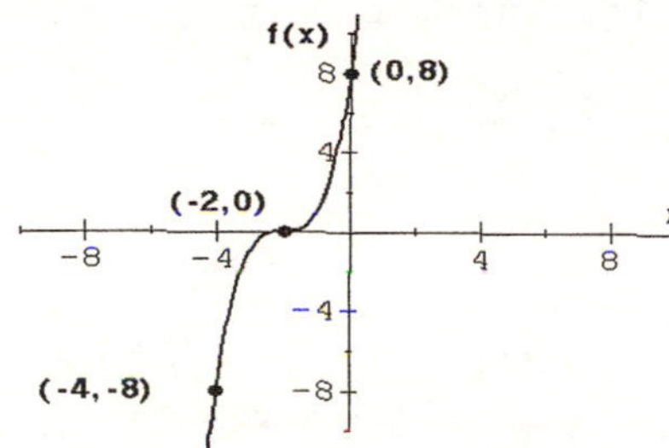

9. $f(x) = |x-1| + 2$
This is the graph of $f(x) = |x|$ shifted 1 unit to the right and 2 units up.

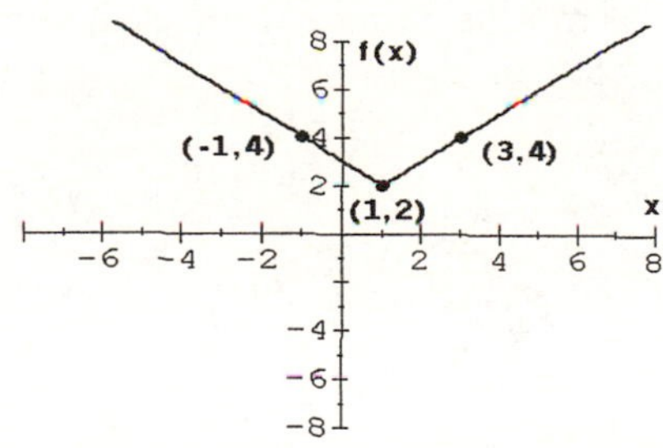

11. $f(x) = |x+1| - 3$
This is the graph of $f(x) = |x|$ shifted 1 unit to the left and 3 units down.

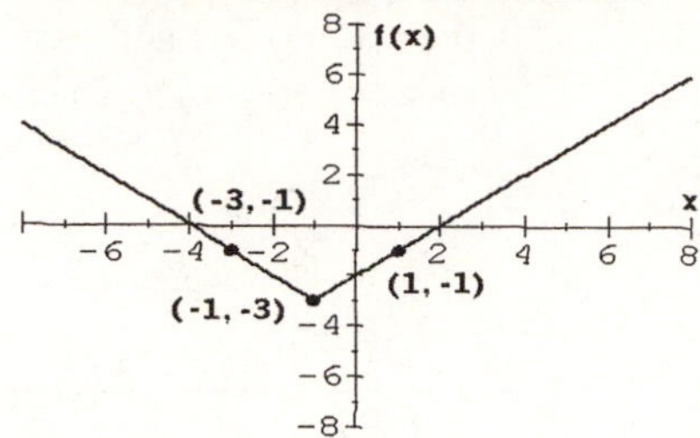

13. $f(x) = x + |x|$

If $x \geq 0$, then $|x| = x$ and
$f(x) = x + x = 2x$
If $x < 0$, then $|x| = -x$ and
$f(x) = x + (-x) = 0$

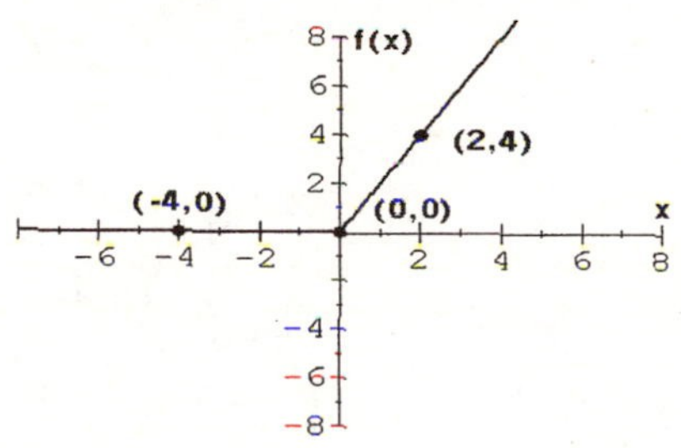

15. $f(x) = -|x-2| - 1$
This is the graph of $f(x) = |x|$ reflected across the x-axis and shifted 2 units to the right and 1 unit down.

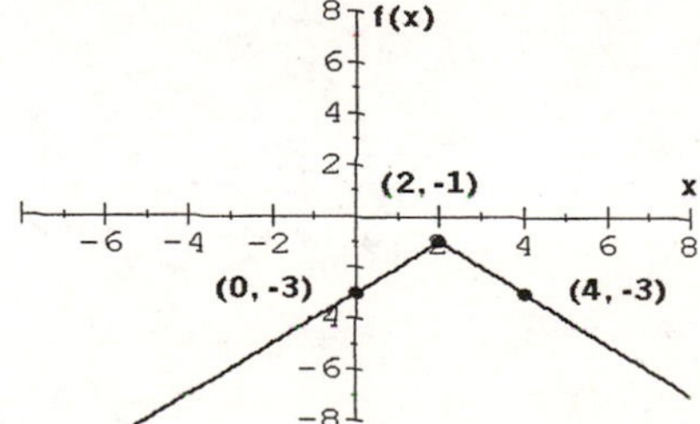

17. $f(x) = x - |x|$
If $x \geq 0$, then $|x| = x$ and
$f(x) = x - x = 0$
If $x < 0$, then $|x| = -x$ and
$f(x) = x - (-x) = 2x$

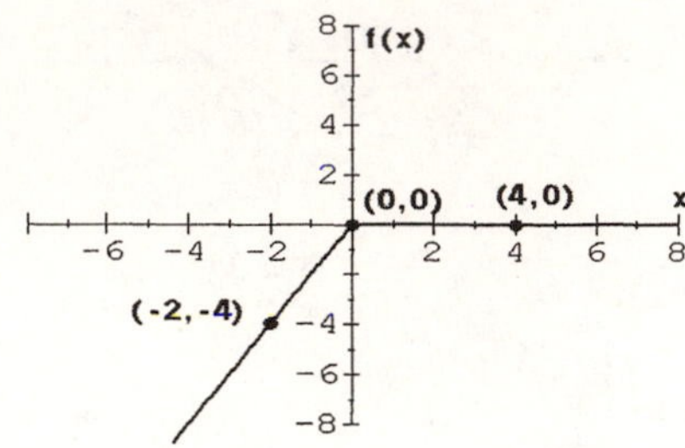

19. $f(x) = -2\sqrt{x}$
This is the graph of $f(x) = \sqrt{x}$ reflected across the x-axis and stretched by a factor of 2.

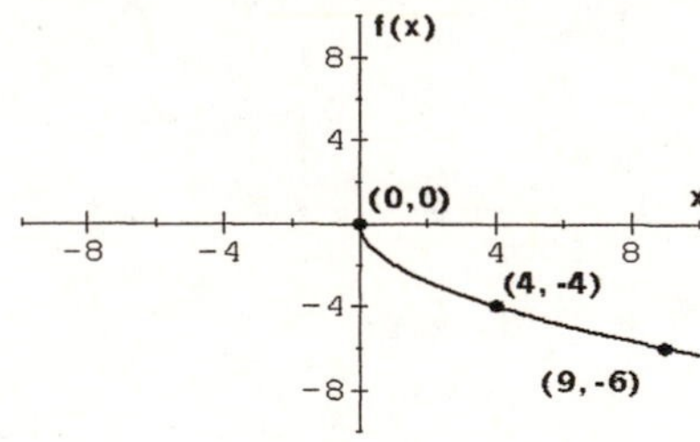

21. $f(x) = \sqrt{x+2} - 3$
This is the graph of $f(x) = \sqrt{x}$ shifted 2 units to the left and 3 units down.

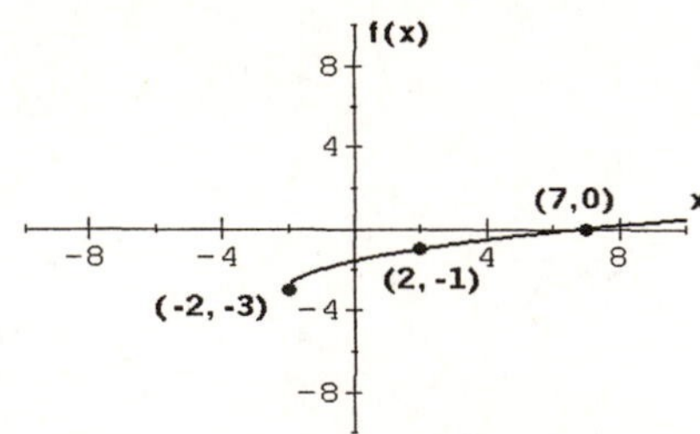

23. $f(x) = \sqrt{2-x}$
This is the graph of $f(x) = \sqrt{x}$ reflected across the y-axis and shifted 2 units to the right.

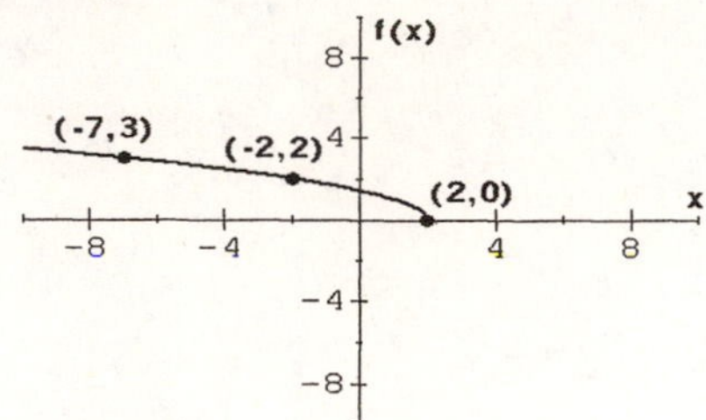

25. $f(x) = -2x^4 + 1$
This is the graph of $f(x) = x^4$ reflected across the x-axis, stretched by a factor of 2, and shifted 1 unit up.

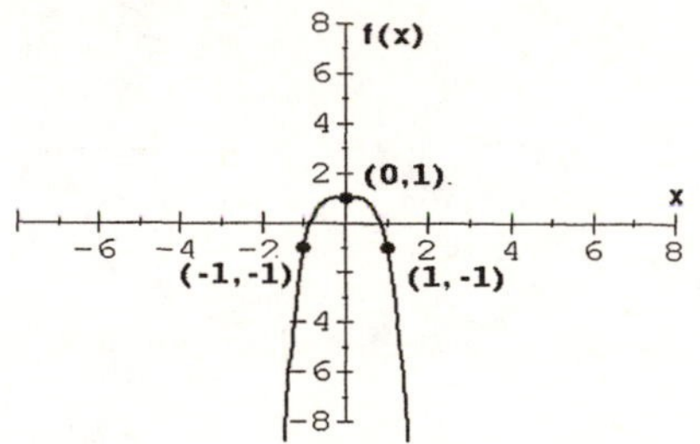

27. $f(x) = -2x^3$
This is the graph of $f(x) = x^3$ reflected across the x-axis and stretched by a factor of 2.

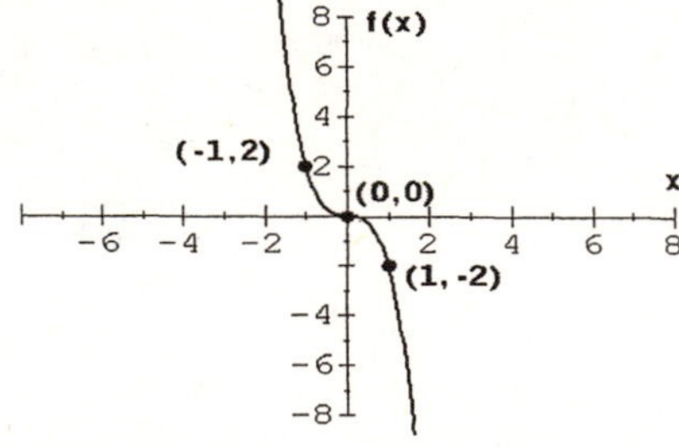

29. $f(x) = 3(x-2)^3 - 1$
This is the graph of $f(x) = x^3$ stretched by a factor of 3 and shifted 2 units to the right and 1 unit down.

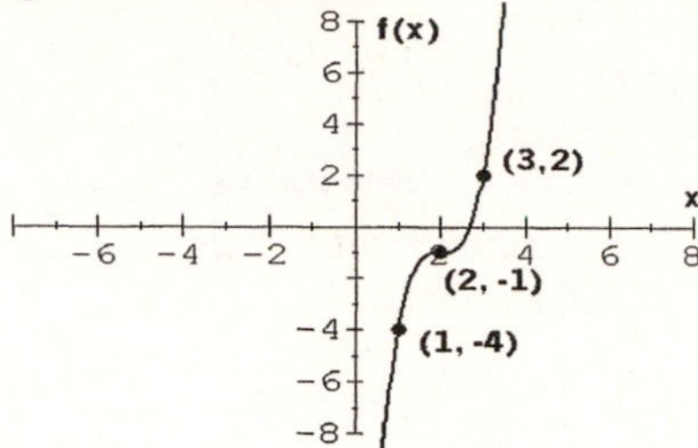

31. $y = f(x)$

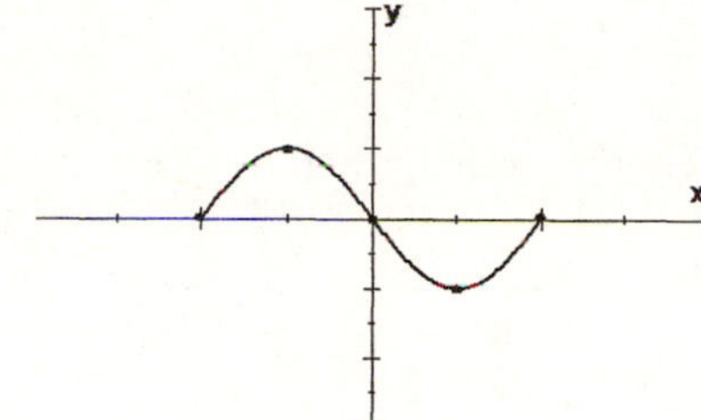

a. Shift the graph up 3 units.
$y = f(x) + 3$

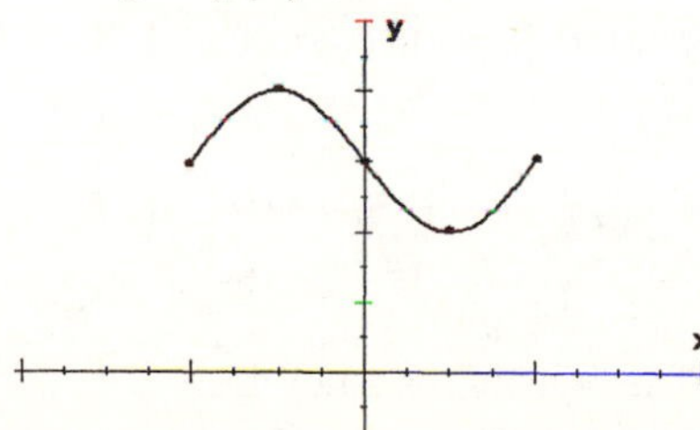

b. Shift the graph 2 units to the right.
$y = f(x-2)$

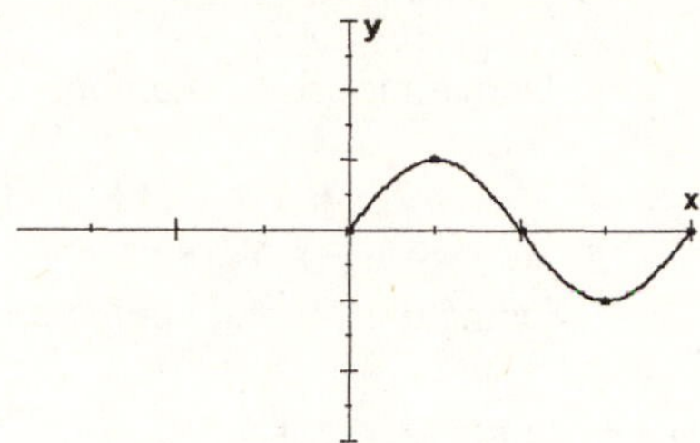

c. Reflect the graph around the x-axis.
$y = -f(x)$

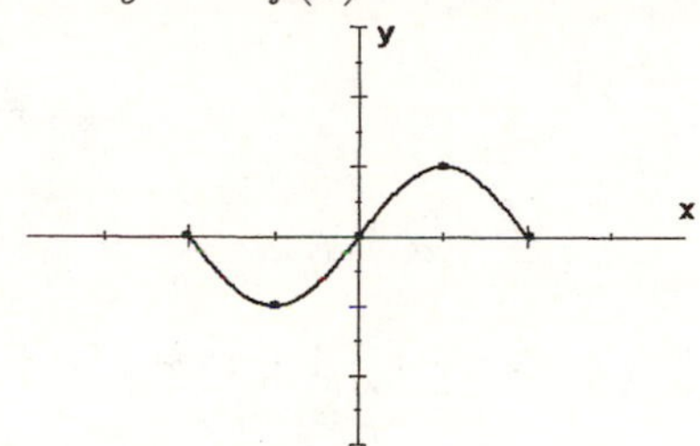

d. Shift the graph 3 units to the left and 4 units down.
$y = f(x+3) - 4$

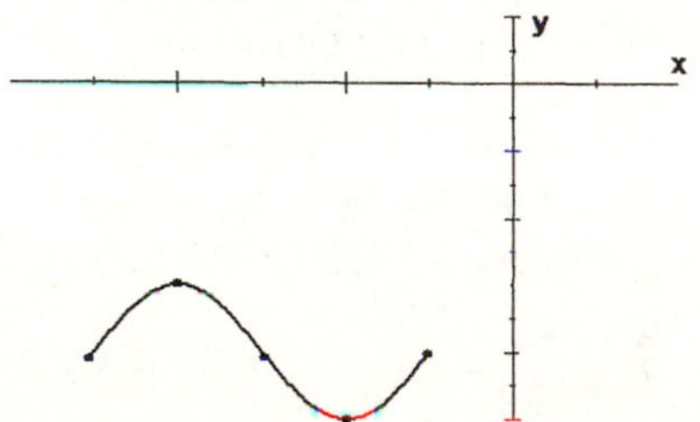

PROBLEM SET 8.6 Combining Functions

1. $f(x) = 3x - 4,\ g(x) = 5x + 2$

$f + g = 3x - 4 + 5x + 2$
$= 8x - 2$
Domain is all real numbers.

$f - g = 3x - 4 - (5x + 2)$
$= 3x - 4 - 5x - 2$
$= -2x - 6$
Domain is all real numbers.

$f \cdot g = (3x - 4)(5x + 2)$
$= 15x^2 - 14x - 8$
Domain is all real numbers.

$\frac{f}{g} = \frac{3x-4}{5x+2} \qquad D = \left\{x \middle| x \neq -\frac{5}{2}\right\}$

Problem Set 8.6

3. $f(x) = x^2 - 6x + 4,\ g(x) = -x - 1$

$$f + g = x^2 - 6x + 4 - x - 1$$
$$= x^2 - 7x + 3$$
Domain is all real numbers.

$$f - g = x^2 - 6x + 4 - (-x - 1)$$
$$= x^2 - 5x + 5$$
Domain is all real numbers.

$$f \cdot g = (x^2 - 6x + 4)(-x - 1)$$
$$= -x^3 + 5x^2 + 2x - 4$$
Domain is all real numbers.

$$\frac{f}{g} = \frac{x^2 - 6x + 4}{-x - 1} \qquad D = \{x | x \neq -1\}$$

5. $f(x) = x^2 - x - 1,\ g(x) = x^2 + 4x - 5$

$$f + g = x^2 - x - 1 + x^2 + 4x - 5$$
$$= 2x^2 + 3x - 6$$
Domain is all real numbers.

$$f - g = x^2 - x - 1 - (x^2 + 4x - 5)$$
$$= -5x + 4$$
Domain is all real numbers.

$$f \cdot g = (x^2 - x - 1)(x^2 + 4x - 5)$$
$$= x^4 + 3x^3 - 10x^2 + x + 5$$
Domain is all real numbers.

$$\frac{f}{g} = \frac{x^2 - x - 1}{x^2 + 4x - 5} = \frac{x^2 - x - 1}{(x+5)(x-1)}$$

$D = \{x | x \neq 1 \text{ and } x \neq -5\}$

7. $f(x) = \sqrt{x - 1},\ g(x) = \sqrt{x}$

$$f + g = \sqrt{x - 1} + \sqrt{x}$$
$$D = \{x | x \geq 1\}$$

$$f - g = \sqrt{x - 1} - \sqrt{x}$$
$$D = \{x | x \geq 1\}$$

$$f \cdot g = \sqrt{x - 1}\sqrt{x} = \sqrt{x^2 - x}$$
$$D = \{x | x \geq 1\}$$

$$\frac{f}{g} = \frac{\sqrt{x(x-1)}}{\sqrt{x}}$$
$$D = \{x | x \geq 1\}$$

9. $f(x) = 2x,\ g(x) = 3x - 1$
$(f \circ g)(x) = 2(3x - 1) = 6x - 2$
$(g \circ f)(x) = 3(2x) - 1 = 6x - 1$
The domain for all of these is all real numbers.

11. $f(x) = 5x - 3,\ g(x) = 2x + 1$
$(f \circ g)(x) = 5(2x + 1) - 3 = 10x + 2$
$(g \circ f)(x) = 2(5x - 3) + 1 = 10x - 5$
The domain for all of these is all real numbers.

13. $f(x) = 3x + 4,\ g(x) = x^2 + 1$
$(f \circ g)(x) = 3(x^2 + 1) + 4 = 3x^2 + 7$
$$(g \circ f)(x) = (3x + 4)^2 + 1$$
$$= 9x^2 + 24x + 16 + 1$$
$$= 9x^2 + 24x + 17$$
The domain for all of these is all real numbers.

15. $f(x) = 3x - 4,\ g(x) = x^2 + 3x - 4$
$$(f \circ g)(x) = 3(x^2 + 3x - 4) - 4$$
$$= 3x^2 + 9x - 16$$
$$(g \circ f)(x) = (3x - 4)^2 + 3(3x - 4) - 4$$
$$= 9x^2 - 24x + 16 + 9x - 12 - 4$$
$$= 9x^2 - 15x$$
The domain for all of these is all real numbers.

17. $f(x) = \frac{1}{x},\ g(x) = 2x + 7$
$$(f \circ g)(x) = \frac{1}{2x + 7}$$
$$D = \left\{x | x \neq -\frac{7}{2}\right\}$$
$$(g \circ f)(x) = 2\left(\frac{1}{x}\right) + 7$$
$$= \frac{2}{x} + \frac{7x}{x} = \frac{2 + 7x}{x}$$
$$D = \{x | x \neq 0\}$$

19. $f(x) = \sqrt{x-2},\ g(x) = 3x - 1$
$$(f \circ g)(x) = \sqrt{3x-1-2} = \sqrt{3x-3}$$
$$D = \{x|x \geq 1\}$$
$$(g \circ f)(x) = 3\sqrt{x-2} - 1$$
$$D = \{x|x \geq 2\}$$

21. $f(x) = \dfrac{1}{x-1},\ g(x) = \dfrac{2}{x}$
$$(f \circ g)(x) = \frac{1}{\frac{2}{x} - 1} = \frac{1}{\frac{2-x}{x}} = \frac{x}{2-x}$$
$$D = \{x|x \neq 0 \text{ and } x \neq 2\}$$
$$(g \circ f)(x) = \frac{2}{\frac{1}{x-1}} = 2(x-1) = 2x - 2$$
$$D = \{x|x \neq 1\}$$

23. $f(x) = 2x + 1,\ g(x) = \sqrt{x-1}$
$$(f \circ g)(x) = 2\sqrt{x-1} + 1$$
$$D = \{x|x \geq 1\}$$
$$(g \circ f)(x) = \sqrt{2x+1-1} = \sqrt{2x}$$
$$D = \{x|x \geq 0\}$$

25. $f(x) = \dfrac{1}{x-1},\ g(x) = \dfrac{x+1}{x}$
$$(f \circ g)(x) = \frac{1}{\frac{x+1}{x} - 1} = \frac{1}{\frac{x+1-x}{x}} = x$$
$$D = \{x|x \neq 0\}$$
$$(g \circ f)(x) = \frac{\frac{1}{x-1} + 1}{\frac{1}{x-1}} = \frac{\frac{1+x-1}{x-1}}{\frac{1}{x-1}}$$
$$= \frac{x}{x-1} \cdot \frac{x-1}{1} = x$$
$$D = \{x|x \neq 1\}$$

27. $f(x) = 3x - 2,\ g(x) = x^2 + 1$
$$g(-1) = (-1)^2 + 1 = 2$$
$$(f \circ g)(-1) = f(g(-1)) = f(2)$$
$$= 3(2) - 2 = 4$$
$$f(3) = 3(3) - 2 = 7$$
$$(g \circ f)(3) = g(f(3)) = g(7)$$
$$= 7^2 + 1 = 50$$

29. $f(x) = 2x - 3,\ g(x) = x^2 - 3x - 4$
$$g(-2) = (-2)^2 - 3(-2) - 4 = 6$$
$$(f \circ g)(-2) = f(g(-2)) = f(6)$$
$$= 2(6) - 3 = 9$$
$$f(1) = 2(1) - 3 = -1$$
$$(g \circ f)(1) = g(f(1)) = g(-1)$$
$$= (-1)^2 - 3(-1) - 4 = 0$$

31. $f(x) = \sqrt{x},\ g(x) = 3x - 1$
$$g(4) = 3(4) - 1 = 11$$
$$(f \circ g)(4) = f(g(4)) = f(11) = \sqrt{11}$$
$$f(4) = \sqrt{4} = 2$$
$$(g \circ f)(4) = g(f(4)) = g(2) = 3(2) - 1 = 5$$

33. $f(x) = 2x,\ g(x) = \dfrac{1}{2}x$
$$(f \circ g)(x) = 2\left(\frac{1}{2}x\right) = x$$
$$(g \circ f)(x) = \frac{1}{2}(2x) = x$$

35. $f(x) = x - 2,\ g(x) = x + 2$
$$(f \circ g)(x) = (x+2) - 2 = x$$
$$(g \circ f)(x) = (x-2) + 2 = x$$

37. $f(x) = 3x + 4,\ g(x) = \dfrac{x-4}{3}$
$$(f \circ g)(x) = 3\left(\frac{x-4}{3}\right) + 4 = x - 4 + 4 = x$$
$$(g \circ f)(x) = \frac{(3x+4) - 4}{3} = \frac{3x}{x} = x$$

43. $f(x) = x^2,\ g(x) = \sqrt{x}$
$(f \circ g)(x) = (\sqrt{x})^2 = \sqrt{x^2} = x$
$(g \circ f)(x) = \sqrt{(x^2)} = x$

PROBLEM SET 8.7 Direct and Inverse Variation

1. $y = kx^3$

3. $A = klw$

5. $V = \dfrac{k}{P}$

7. $V = khr^2$

9. $y = kx$
$72 = k(3)$
$24 = k$

11. $A = kr^2$
$154 = k(7^2)$
$154 = 49k$
$k = \dfrac{154}{49} = \dfrac{22}{7}$

13. $A = kbh$
$81 = k(9)(18)$
$81 = 162k$
$k = \dfrac{1}{2}$

15. $y = \dfrac{kxz}{w}$
$154 = \dfrac{k(6)(11)}{3}$
$154 = 22k$
$k = 7$

17. $y = \dfrac{kx^2}{w^3}$
$18 = \dfrac{k(9)^2}{(3)^3}$
$18 = \dfrac{81k}{27}$
$18 = 3k$
$k = 6$

19. $y = kx$
$5 = k(-15)$
$k = -\dfrac{1}{3}$
$y = -\dfrac{1}{3}x$
$y = -\dfrac{1}{3}(-24) = 8$

21. $V = kBh$
$96 = k(36)(8)$
$96 = 288k$
$k = \dfrac{96}{288} = \dfrac{1}{3}$
$V = \dfrac{1}{3}Bh$
$V = \dfrac{1}{3}(48)(6) = 96$

23. $t = \dfrac{k}{r}$
$3 = \dfrac{k}{50}$
$k = 150$
$t = \dfrac{150}{r}$
$t = \dfrac{150}{30} = 5$ hours

25. $P = k\sqrt{l}$
$4 = k\sqrt{12}$
$k = \dfrac{4}{\sqrt{12}} = \dfrac{4}{2\sqrt{3}} = \dfrac{2\sqrt{3}}{3}$
$P = \dfrac{2\sqrt{3}\sqrt{l}}{3}$
$P = \dfrac{2\sqrt{3}\sqrt{3}}{3} = \dfrac{2(3)}{3} = 2$ seconds

27. $d = \dfrac{km}{p}$

$32 = \dfrac{k(16)}{4}$

$32 = 4k$

$k = 8$

$d = \dfrac{8m}{p}$

$d = \dfrac{8(24)}{8} = 24$ days

29. $V = \dfrac{kT}{P}$

$48 = \dfrac{k(320)}{20}$

$48 = 16k$

$k = 3$

$V = \dfrac{3T}{P}$

$V = \dfrac{3(280)}{30} = 28$

31. $C = kwd$

$900 = k(15)(5)$

$900 = 75k$

$k = 12$

$C = 12wd$

$C = 12(20)(10) = \$2400$

Further Investigations

37. $P = k\sqrt{l}$

$2.4 = k\sqrt{9}$

$2.4 = k(3)$

$k = 0.8$

$P = 0.8\sqrt{12}$

$P = 2.8$ seconds

39. $y = \dfrac{kx}{z^2}$

$0.336 = \dfrac{k(6)}{(5)^2}$

$0.336 = \dfrac{k(6)}{25}$

$\dfrac{0.336(25)}{6} = k$

$k = 1.4$

CHAPTER 8 Review Problem Set

1. $f(x) = 3x^2 - 2x - 1$

$f(2) = 3(2)^2 - 2(2) - 1 = 7$

$f(-1) = 3(-1)^2 - 2(-1) - 1 = 4$

$f(-3) = 3(-3)^2 - 2(-3) - 1 = 32$

2a. $f(x) = -5x + 4$

$f(a+h) = -5(a+h) + 4$

$= -5a - 5h + 4$

$f(a) = -5a + 4$

$$\frac{f(a+h) - f(a)}{h} = \frac{-5a - 5h + 4 - (-5a + 4)}{h}$$

$$= \frac{-5a - 5h + 4 + 5a - 4}{h}$$

$$= \frac{-5h}{h} = -5$$

b. $f(x) = 2x^2 - x + 4$

$f(a+h) = 2(a+h)^2 - (a+h) + 4$

$= 2a^2 + 4ah + 2h^2 - a - h + 4$

$f(a) = 2a^2 - a + 4$

$$\frac{f(a+h) - f(a)}{h} = \frac{4ah + 2h^2 - h}{h}$$

$$= 4a + 2h - 1$$

c. $f(x) = -3x^2 + 2x - 5$

$f(a+h) = -3(a+h)^2 + 2(a+h) - 5$

$= -3a^2 - 6ah - 3h^2 + 2a + 2h - 5$

$f(a) = -3a^2 + 2a - 5$

$$\frac{f(a+h) - f(a)}{h} = \frac{-6ah - 3h^2 + 2h}{h}$$

$$= -6a - 3h + 2$$

3. $f(x) = x^2 + 5$
$D = \{x | x \text{ is a real number}\}$
$R = \{f(x) | f(x) \geq 5\}$

4. $f(x) = \dfrac{2}{2x^2 + 7x - 4}$

$f(x) = \dfrac{2}{(2x-1)(x+4)}$

$D = \left\{x | x \neq -4 \text{ and } x \neq \dfrac{1}{2}\right\}$

5. $f(x) = \sqrt{x^2 - 7x + 10}$
$f(x) = \sqrt{(x-5)(x-2)}$

The product of the radical must be positive or zero. The critical numbers are 2 and 5. The product needs to be analyzed in the intervals $(-\infty, 2)$, $(2, 5)$, and $(5, \infty)$.

	2		5
Test Point	0	3	6
$x - 5$:	negative	negative	positive
$x - 2$:	negative	positive	positive
product:	positive	negative	positive

Domain: $(-\infty, 2] \cup [5, \infty)$

6. $f(x) = -2x + 2$

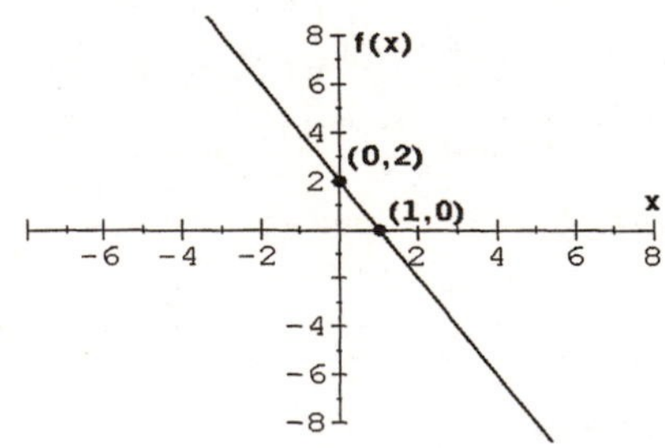

7. $f(x) = 2x^2 - 1$
This is the basic parabola $f(x) = x^2$ stretched by a factor of 2 and shifted 1 unit down.

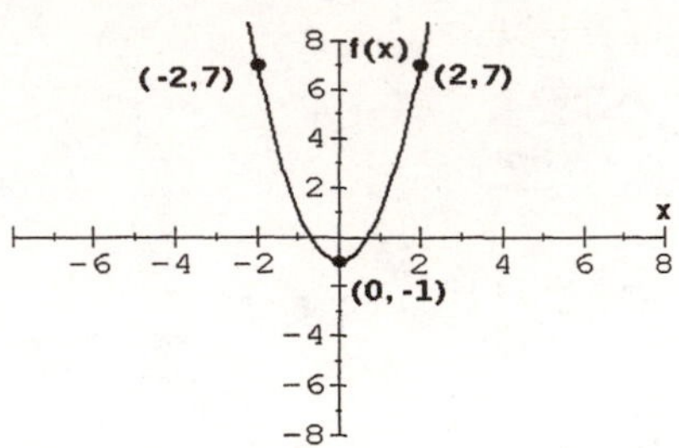

8. $f(x) = -\sqrt{x-2} + 1$
This is the basic graph $f(x) = \sqrt{x}$ reflected around the x-axis, shifted 2 units right and 1 unit up.

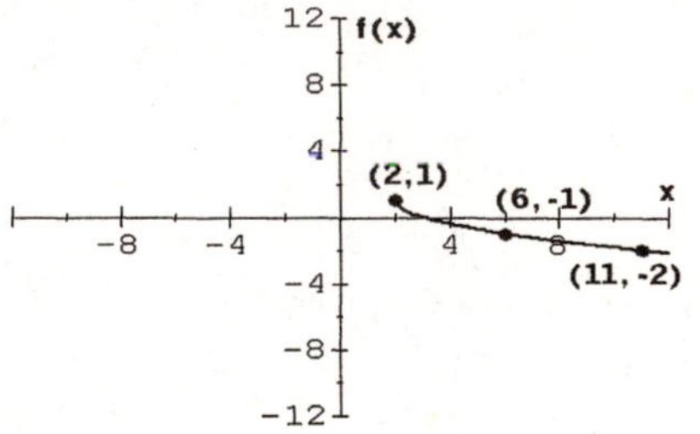

9. $f(x) = x^2 - 8x + 17$
$f(x) = x^2 - 8x + 16 - 16 + 17$
$f(x) = (x-4)^2 + 1$
The graph is the basic parabola $f(x) = x^2$ shifted 4 units to the right and 1 unit up.

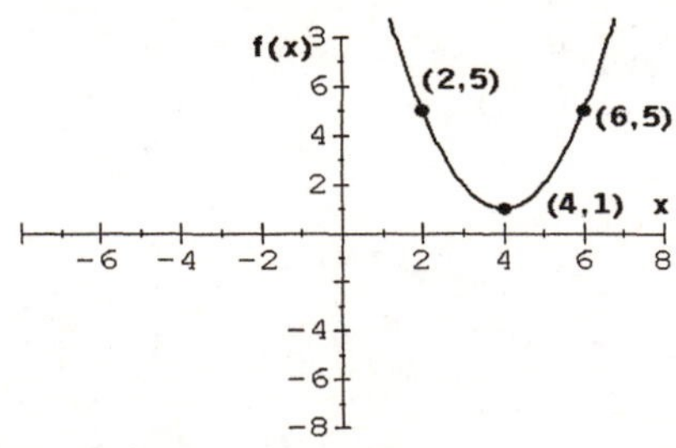

10. $f(x) = -x^3 + 2$
This is the basic graph $f(x) = x^3$ reflected around the x-axis, shifted 2 units up.

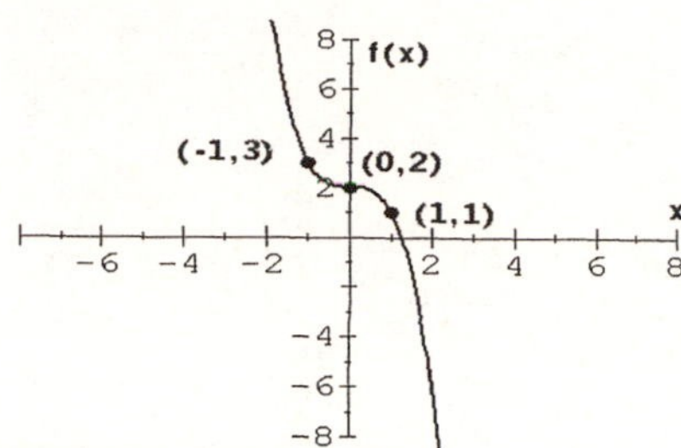

11. $f(x) = 2|x - 1| + 3$
This is the graph of $f(x) = |x|$ stretched by a factor of 2 and shifted 1 unit right and 3 units up.

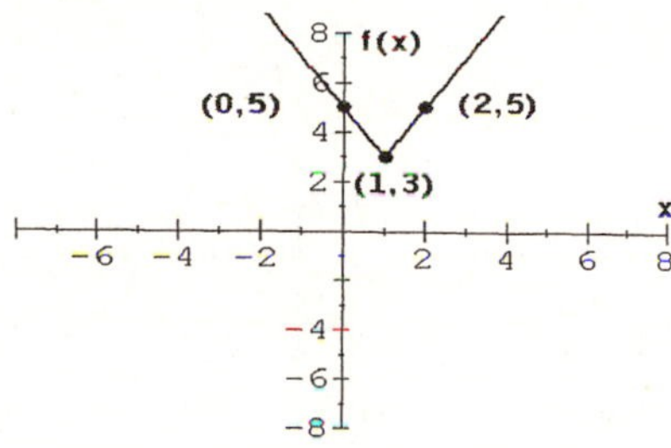

12. $f(x) = -2x^2 - 12x - 19$
$f(x) = -2(x^2 + 6x) - 19$
$f(x) = -2(x^2 + 6x + 9) + 18 - 19$
$f(x) = -2(x + 3)^2 - 1$
This is the basic graph $f(x) = x^2$ reflected around the x-axis, shifted 3 units left and 1 unit down.

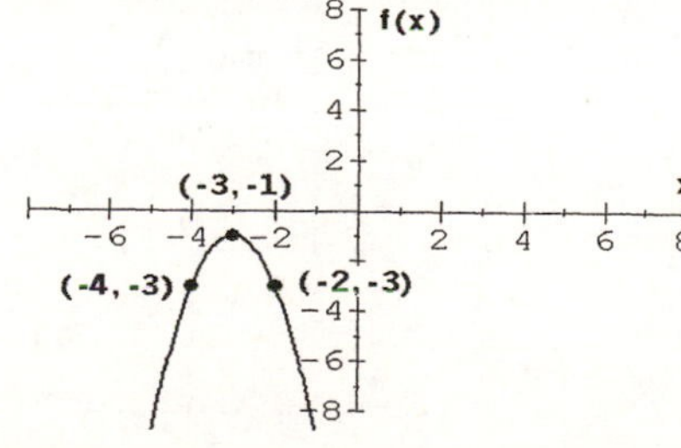

13. $f(x) = -\frac{1}{3}x + 1$
This is the graph of a straight line with y-intercept of 1 and slope $m = -\frac{1}{3}$.

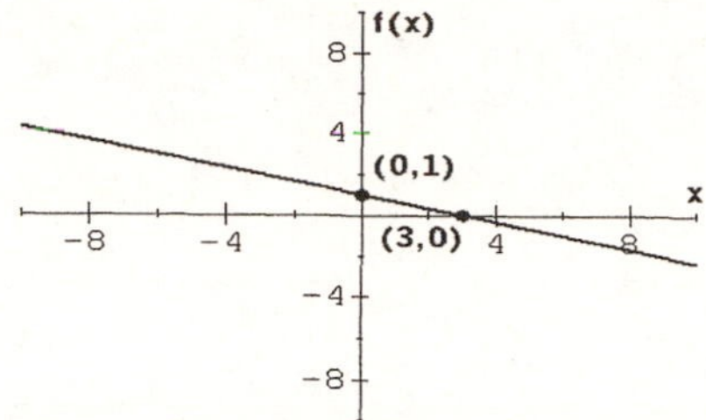

14. $f(x) = -\frac{2}{x^2}$
This is the basic graph $f(x) = \frac{1}{x^2}$ reflected around the x-axis and stretched by a factor of 2.

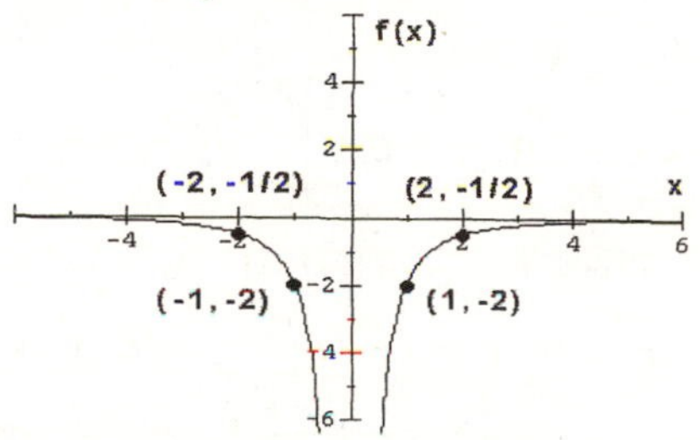

15. $f(x) = 2|x| - x$
If $x \geq 0$, then $|x| = x$ and
$f(x) = 2x - x = x$
If $x < 0$, then $|x| = -x$ and
$f(x) = 2(-x) - x = -3x$

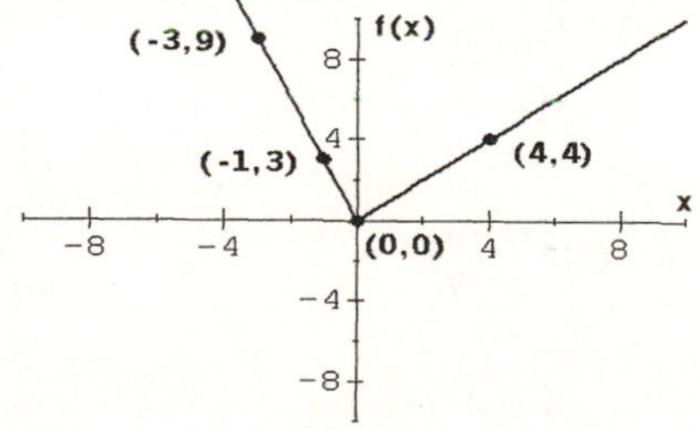

16. $f(x) = (x - 2)^2$

This is the basic graph $f(x) = x^2$ shifted 2 units right.

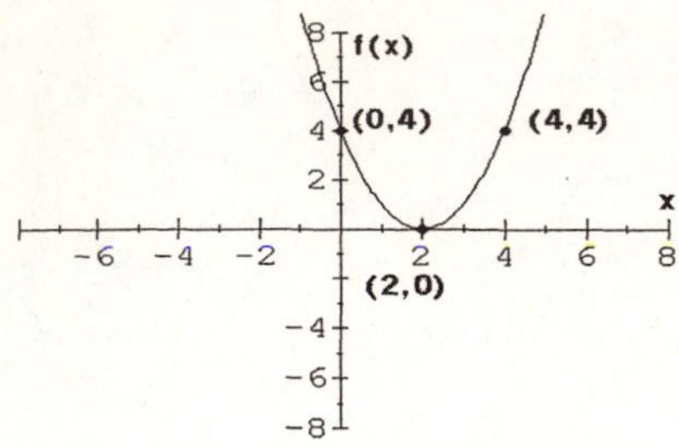

17. $f(x) = \sqrt{-x + 4}$

This is the basic graph $f(x) = \sqrt{x}$ reflected across the y-axis and shifted right 4 units.

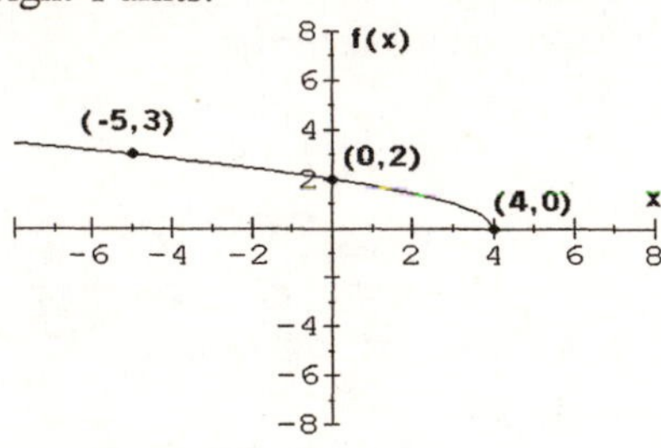

18. $f(x) = -(x + 1)^2 - 3$

This is the basic graph $f(x) = x^2$ reflected across the x-axis, shifted one unit left and 3 units down.

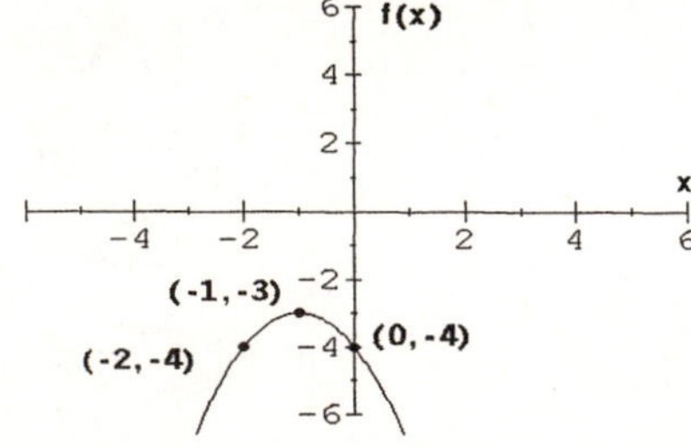

19. $f(x) = \sqrt{x + 3} - 2$

This is the basic graph $f(x) = \sqrt{x}$ shifted 3 units left and 2 units down.

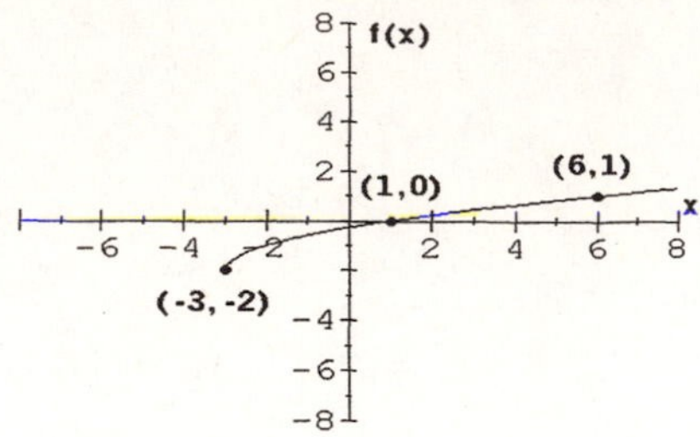

20. $f(x) = -|x| + 4$

This is the basic graph $f(x) = |x|$ reflected across the x-axis and shifted 4 units up.

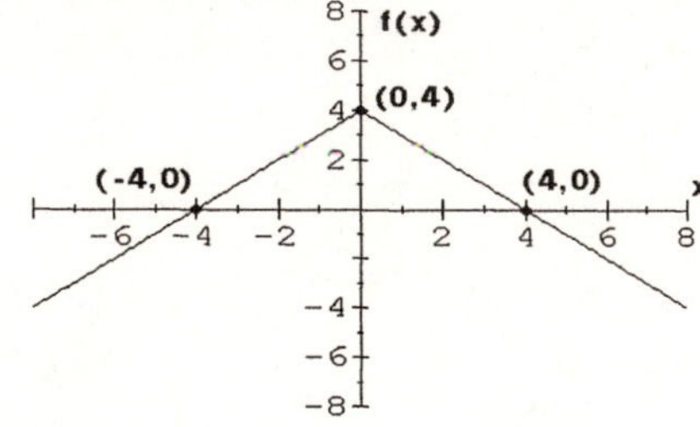

21. $f(x) = (x - 2)^3$

This is the basic graph $f(x) = x^3$ shifted 2 units right.

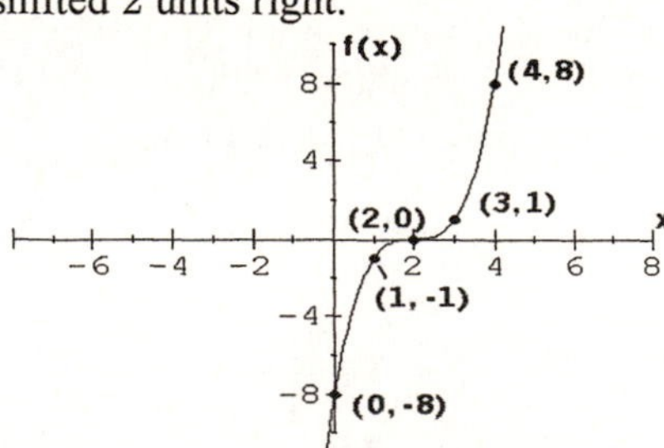

22. $f(x) = \begin{cases} x^2 - 1 & \text{for } x < 0 \\ 3x - 1 & \text{for } x \geq 0 \end{cases}$

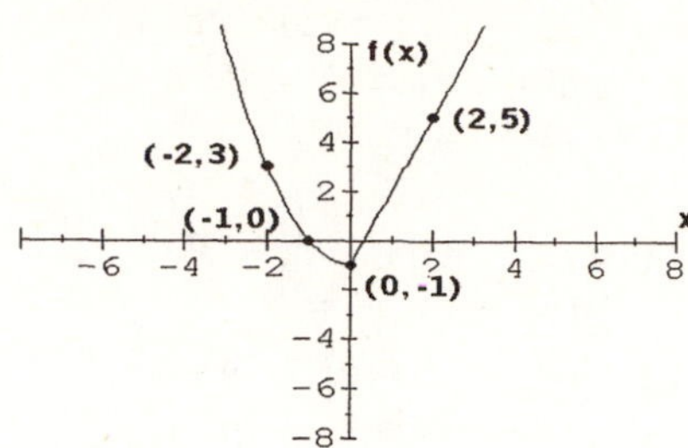

23. $f(x) = \begin{cases} 3 & \text{for } x \leq -3 \\ |x| & \text{for } -3 < x < 3 \\ 2x - 3 & \text{for } x \geq 3 \end{cases}$

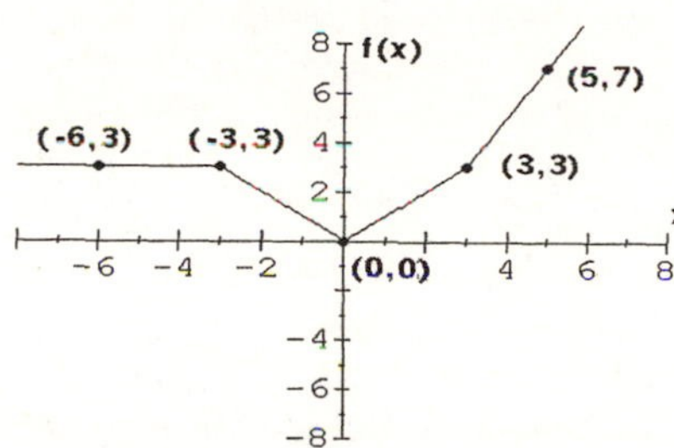

24. $(f+g)(x) = 2x + 3 + x^2 - 4x - 3$
$= x^2 - 2x$

$(f-g)(x) = 2x + 3 - (x^2 - 4x - 3)$
$= 2x + 3 - x^2 + 4x + 3$
$= -x^2 + 6x + 6$

$(f \cdot g)(x) = (2x+3)(x^2 - 4x - 3)$
$= 2x^3 - 8x^2 - 6x + 3x^2 - 12x - 9$
$= 2x^3 - 5x^2 - 18x - 9$

$\left(\frac{f}{g}\right)(x) = \frac{2x+3}{x^2 - 4x - 3}$

25. $f(x) = 3x - 9$, $g(x) = -2x + 7$

$(f \circ g)(x) = 3(-2x + 7) - 9$
$= -6x + 12$
$D = \{\text{all reals}\}$

$(g \circ f)(x) = -2(3x - 9) + 7$
$= -6x + 25$
$D = \{\text{all reals}\}$

26. $f(x) = x^2 - 5$, $g(x) = 5x - 4$

$(f \circ g)(x) = (5x - 4)^2 - 5$
$= 25x^2 - 40x + 16 - 5$
$= 25x^2 - 40x + 11$
$D = \{\text{all reals}\}$

$(g \circ f)(x) = 5(x^2 - 5) - 4$
$= 5x^2 - 25 - 4$
$= 5x^2 - 29$
$D = \{\text{all reals}\}$

27. $f(x) = \sqrt{x - 5}$, $g(x) = x + 2$

$(f \circ g)(x) = \sqrt{x + 2 - 5} = \sqrt{x - 3}$

$D = \{x | x \geq 3\}$

$(g \circ f)(x) = \sqrt{x - 5} + 2$

$D = \{x | x \geq 5\}$

28. $f(x) = \frac{1}{x}$, $g(x) = x^2 - x - 6$

$(f \circ g)(x) = \frac{1}{x^2 - x - 6} = \frac{1}{(x-3)(x+2)}$

D $= \{x | x \neq 3 \text{ and } x \neq -2\}$.

$(g \circ f)(x) = \left(\frac{1}{x}\right)^2 - \left(\frac{1}{x}\right) - 6$

$= \frac{1}{x^2} - \frac{x}{x^2} - \frac{6x^2}{x^2}$

$= \frac{1 - x - 6x^2}{x^2}$

D $= \{x | x \neq 0\}$.

29. $f(x) = x^2$, $g(x) = \sqrt{x - 1}$

$(f \circ g)(x) = (\sqrt{x - 1})^2 = x - 1$

D$=\{x | x \geq 1\}$.

$(g \circ f)(x) = \sqrt{x^2 - 1}$
$(g \circ f)(x) = \sqrt{(x-1)(x+1)}$

The product of the radical must be positive or zero. The critical numbers are 1 and -1.
The product needs to be analyzed in the intervals $(-\infty, -1)$, $(-1, 1)$, and $(1, \infty)$.

	-1		1
Test Point	-2	0	2
$x+1$:	negative	positive	positive
$x-1$:	negative	negative	positive
product:	positive	negative	positive

Domain: $(-\infty, -1] \cup [1, \infty)$.

30. $f(x) = \dfrac{1}{x-3}, \; g(x) = \dfrac{1}{x+2}$

$$(f \circ g)(x) = \frac{1}{\frac{1}{x+2} - 3}$$

$$= \frac{1}{\frac{-3x-5}{x+2}} = \frac{x+2}{-3x-5}$$

$$D = \left\{x \mid x \neq -2 \text{ and } x \neq -\frac{5}{3}\right\}$$

$$(g \circ f)(x) = \frac{1}{\frac{1}{x-3} + 2}$$

$$= \frac{1}{\frac{2x-5}{x-3}} = \frac{x-3}{2x-5}$$

$$D = \left\{x \mid x \neq 3 \text{ and } x \neq \frac{5}{2}\right\}$$

31. $f(x) = \begin{cases} x^2 - 2 \text{ for } x \geq 0 \\ -3x + 4 \text{ for } x < 0 \end{cases}$

$f(5) = 5^2 - 2 = 23$
$f(0) = 0^2 - 2 = -2$
$f(-3) = -3(-3) + 4 = 13$

32. $f(x) = -x^2 - x + 4, \; g(x) = \sqrt{x-2}$

$g(6) = \sqrt{6-2} = \sqrt{4} = 2$
$f(g(6)) = f(2) = -(2)^2 - 2 + 4 = -2$
$f(-2) = -(-2)^2 - (-2) + 4$
$= -4 + 2 + 4 = 2$
$g(f(-2)) = g(2) = \sqrt{2-2} = 0$

33. $f(x) = |x|, \; g(x) = x^2 - x - 1$

$(f \circ g)(1) = f(g(1))$
$g(1) = (1)^2 - (1) - 1 = -1$
$f(g(1)) = f(-1) = |-1| = 1$
$(f \circ g)(1) = 1$

$(g \circ f)(-3) = g(f(-3))$
$f(-3) = |-3| = 3$
$g(f(-3)) = g(3) = (3)^2 - (3) - 1 = 5$
$(g \circ f)(-3) = 5$

34. $g(x) = \dfrac{2}{3}x + 4$

$m = \dfrac{2}{3}$, $(5, -2)$ for parallel lines
$y - (-2) = \dfrac{2}{3}(x - 5)$
$y + 2 = \dfrac{2}{3}x - \dfrac{10}{3}$
$y = \dfrac{2}{3}x - \dfrac{16}{3}$
$f(x) = \dfrac{2}{3}x - \dfrac{16}{3}$

35. Since the slope of $g(x)$ is $-\dfrac{1}{2}$, the slope of a linear perpendicular function is 2.
$y - 3 = 2[x - (-6)]$
$y - 3 = 2x + 12$
$y = 2x + 15$
$f(x) = 2x + 15$

36. 4 hours/night $\times$ 30 nights $=$ 120 hours
$c(120) = 0.006(120) = 0.72$
The cost to burn the light for 30 nights would be \$0.72.

37. $f(x) = x - 0.30x = 0.70x$
$f(65) = 0.70(65) = \$45.50$
$f(48) = 0.70(48) = \$33.60$
$f(15.50) = 0.70(15.50) = \$10.85$

38. $f(x) = 3x^2 + 6x - 24$
$= 3(x^2 + 2x) - 24$
$= 3(x^2 + 2x + 1) - 3 - 24$
$= 3(x+1)^2 - 27$
The vertex is $(-1, -27)$.

$0 = 3(x+1)^2 - 27$
$27 = 3(x+1)^2$
$9 = (x+1)^2$
$\pm 3 = x + 1$
$x = -1 \pm 3$
$x = -4 \quad \text{or } x = 2$
x-intercepts are -4 and 2.

39. $f(x) = x^2 - 6x - 5$
$f(x) = x^2 - 6x + 9 - 9 - 5$
$f(x) = (x-3)^2 - 14$
The vertex is $(3, -14)$.

$0 = x^2 - 6x - 5$

$$x = \frac{-(-6) \pm \sqrt{(-6)^2 - 4(1)(-5)}}{2(1)}$$

$$x = \frac{6 \pm \sqrt{56}}{2} = \frac{6 \pm 2\sqrt{14}}{2} = 3 \pm \sqrt{14}$$

x-intercepts are $3 \pm \sqrt{14}$.

40. $f(x) = 2x^2 - 28x + 101$
$= 2(x^2 - 14x) + 101$
$= 2(x^2 - 14x + 49) - 98 + 101$
$= 2(x-7)^2 + 3$
The vertex is $(7, 3)$.

$0 = 2(x-7)^2 + 3$
$-3 = 2(x-7)^2$
Since we have a squared quantity equaling a negative number there are no real solutions. Therefore, there are no x-intercepts.

41. Let $x =$ one number and
$10 - x =$ the other number.

$f(x) = x^2 + 4(10 - x)$
$f(x) = x^2 - 4x + 40$

$$-\frac{b}{2a} = -\frac{-4}{2(1)} = 2$$

The minimum is when one number is 2 and the other number is 8.

42. Let $x =$ number of students above 100.
$f(x) = (100 + x)(496 - 4x)$
$= 49600 + 96x - 4x^2$
$= -4(x^2 - 24x) + 49600$
$= -4(x^2 - 24x + 144) + 576 + 49600$
$= -4(x - 12)^2 + 50176$
The vertex is $(12, 50176)$.
So the number of students should be 12 more than 100 or 112.

43. $y = \dfrac{kx}{w}$

$27 = \dfrac{k(18)}{6}$

$27 = 3k$
$k = 9$

44. $y = kx\sqrt{w}$
$140 = k(5)\sqrt{16}$
$140 = 20k$
$7 = k$
$y = 7x\sqrt{w}$
$y = 7(9)\sqrt{49} = 441$

45. $w = \dfrac{k}{d^2}$

$200 = \dfrac{k}{4000^2}$

$k = 200(4000)^2 = 3.2(10)^9$

$w = \dfrac{3.2(10)^9}{5000^2} = 128$ pounds

46. $h = \dfrac{kf}{p}$

$10 = \dfrac{k(20)}{3}$

$\dfrac{3}{2} = k$

$h = \dfrac{3f}{2p}$

$h = \dfrac{3(40)}{2(4)} = 15$ hours

CHAPTER 8 Test

1. $f(x) = -\frac{1}{2}x + \frac{1}{3}$

$f(-3) = -\frac{1}{2}(-3) + \frac{1}{3}$

$= \frac{3}{2} + \frac{1}{3} = \frac{11}{6}$

2. $f(-2) = -(-2)^2 - 6(-2) + 3$
$= -4 + 12 + 3$
$= 11$

3. $f(a+h) = 3(a+h)^2 + 2(a+h) - 5$

$= 3a^2 + 6ah + 3h^2 + 2a + 2h - 5$

$f(a) = 3a^2 + 2a - 5$

$f(a+h) - f(a) = 6ah + 3h^2 + 2h$

$\frac{f(a+h) - f(a)}{h} = \frac{6ah + 3h^2 + 2h}{h}$

$= 6a + 3h + 2$

4. $f(x) = \frac{-3}{2x^2 + 7x - 4}$

$f(x) = \frac{-3}{(2x-1)(x+4)}$

$D = \left\{x | x \neq \frac{1}{2} \text{ and } x \neq -4\right\}$

5. $f(x) = \sqrt{5 - 3x}$

$5 - 3x \geq 0$

$-3x \geq -5$

$x \leq \frac{5}{3}$

$D = \left\{x | x \leq \frac{5}{3}\right\}$

6. $(f+g)(x) = 3x - 1 + 2x^2 - x - 5$
$= 2x^2 + 2x - 6$

$(f-g)(x) = 3x - 1 - (2x^2 - x - 5)$
$= 3x - 1 - 2x^2 + x + 5$
$= -2x^2 + 4x + 4$

$(f \cdot g)(x) = (3x-1)(2x^2 - x - 5)$
$= 6x^3 - 3x^2 - 15x - 2x^2 + x + 5$
$= 6x^3 - 5x^2 - 14x + 5$

7. $f(x) = -3x + 4$, $g(x) = 7x + 2$

$(f \circ g)(x) = -3(7x+2) + 4$
$= -21x - 2$

8. $(g \circ f)(x) = 2(2x+5)^2 - (2x+5) + 3$
$= 2(4x^2 + 20x + 25) - 2x - 5 + 3$
$= 8x^2 + 40x + 50 - 2x - 2$
$= 8x^2 + 38x + 48$

9. $f(x) = \frac{3}{x-2}$, $g(x) = \frac{2}{x}$

$(f \circ g)(x) = \frac{3}{\frac{2}{x} - 2} = \frac{3}{\frac{2-2x}{x}} = \frac{3x}{2-2x}$

10. $f(x) = x^2 - 2x - 3$, $g(x) = |x - 3|$

$g(-2) = |-2-3| = |-5| = 5$
$f(g(-2)) = f(5) = (5)^2 - 2(5) - 3$
$= 25 - 10 - 3 = 12$

$f(1) = (1)^2 - 2(1) - 3 = -4$
$g(f(1)) = g(-4) = |-4-3| = 7$

11. $y - (-8) = -\frac{5}{6}(x-4)$
$6(y+8) = -5(x-4)$
$6y + 48 = -5x + 20$
$6y = -5x - 28$

$y = \frac{-5x - 28}{6}$

$f(x) = -\frac{5}{6}x - \frac{14}{3}$

12. $\left(\frac{f}{g}\right)(x) = \frac{\frac{3}{x}}{\frac{2}{x-1}} = \frac{3(x-1)}{2x}$

$D = \{x | x \neq 0 \text{ and } x \neq 1\}$

13. $f(x) = 2x^2 - x + 1,\ g(x) = x^2 + 3$

$(f + g)(x) = 3x^2 - x + 4$
$(f + g)(-2) = 3(-2)^2 - (-2) + 4 = 18$

$(f - g)(x) = 2x^2 - x + 1 - (x^2 + 3)$
$= x^2 - x - 2$
$(f - g)(4) = (4)^2 - (4) - 2 = 10$

$(g - f)(x) = x^2 + 3 - (2x^2 - x + 1)$
$= -x^2 + x + 2$
$(g - f)(-1) = -(-1)^2 + (-1) + 2 = 0$

14. $f(x) = x^2 + 5x - 6,\ g(x) = x - 1$

$(f \cdot g)(x) = (x^2 + 5x - 6)(x - 1)$
$= x^3 + 5x^2 - 6x - x^2 - 5x + 6$
$= x^3 + 4x^2 - 11x + 6$

$$\left(\frac{f}{g}\right)(x) = \frac{x^2 + 5x - 6}{x - 1}$$
$$= \frac{(x + 6)(x - 1)}{x - 1} = x + 6$$

15. Let $x =$ one number and
$60 - x =$ other number.

$f(x) = x^2 + 12(60 - x)$
$f(x) = x^2 - 12x + 720$
$$-\frac{b}{2a} = -\frac{-12}{2(1)} = 6$$
The minimum is when the numbers are 6 and 54.

16. $y = kxz$
$18 = k(8)(9)$
$$k = \frac{18}{72} = \frac{1}{4}$$
$$y = \frac{1}{4}xz$$
$$y = \frac{1}{4}(5)(12) = 15$$

17. $$y = \frac{k}{x}$$
$$\frac{1}{2} = \frac{k}{-8}$$
$$k = \left(\frac{1}{2}\right)(-8) = -4$$

18. $i = krt$
$140 = k(0.07)(5)$
$$k = \frac{140}{(0.07)(5)} = 400$$
$i = 400rt$
$i = 400(0.08)(3) = 96$

There would be \$96 in interest earned if the same amount was invested at 8% for 3 years.

19. $f(x) = 1x + 0.35x = 1.35x$
$f(13) = 1.35(13) = \$17.55$

20. $f(x) = 4x^2 - 16x - 48$
$0 = 4x^2 - 16x - 48$
$0 = 4(x^2 - 4x - 12)$
$0 = 4(x - 6)(x + 2)$
$x - 6 = 0$ or $x + 2 = 0$
$x = 6$ or $x = -2$
The x-intercepts are -2 and 6.

$f(x) = 4(x^2 - 4x) - 48$
$f(x) = 4(x^2 - 4x + 4) - 16 - 48$
$f(x) = 4(x - 2)^2 - 64$
The vertex is $(2, -64)$.

21. $f(x) = (x - 2)^3 - 3$
This is the graph of $f(x) = x^3$ shifted 2 units to the right and 3 units down.

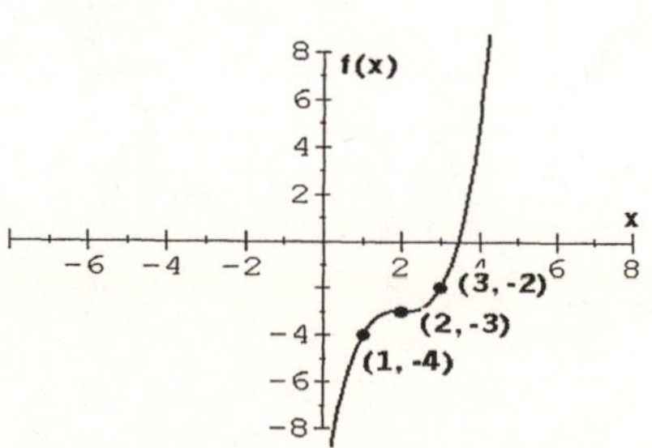

22. $f(x) = -2x^2 - 12x - 14$
$f(x) = -2(x^2 + 6x) - 14$
$f(x) = -2(x^2 + 6x + 9) + 18 - 14$
$f(x) = -2(x + 3)^2 + 4$
This is the basic graph $f(x) = x^2$ stretched by a factor of 2, reflected across the x-axis, shifted 3 units left and 4 units up.

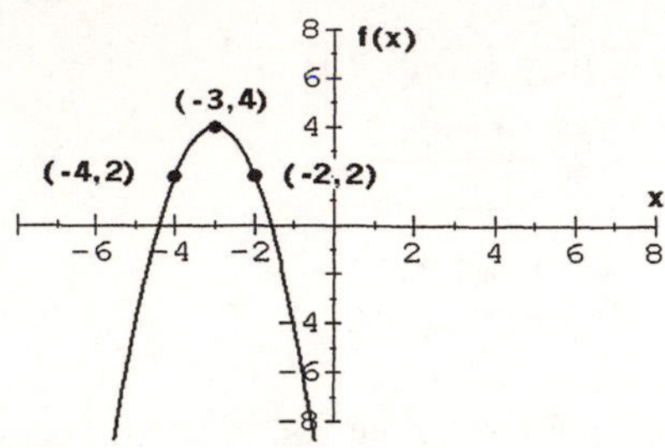

23. $f(x) = 3|x - 2| - 1$
This is the graph of $f(x) = |x|$ stretched by a factor of 3 and shifted 2 units to the right and 1 unit down.

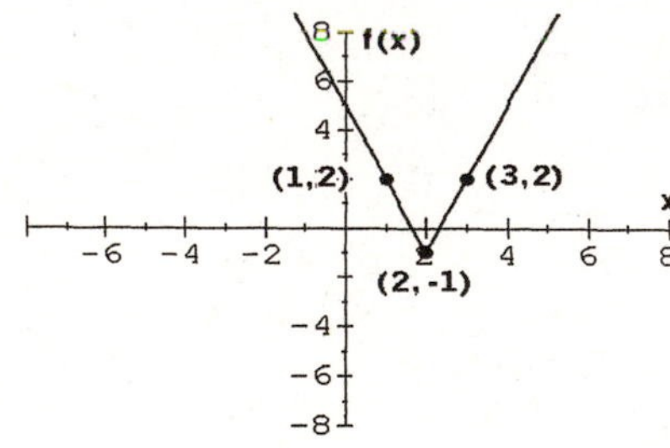

24. $f(x) = \sqrt{-x + 2}$
This is the graph of $f(x) = \sqrt{x}$, reflected across the y-axis, and shifted 2 units to the right.

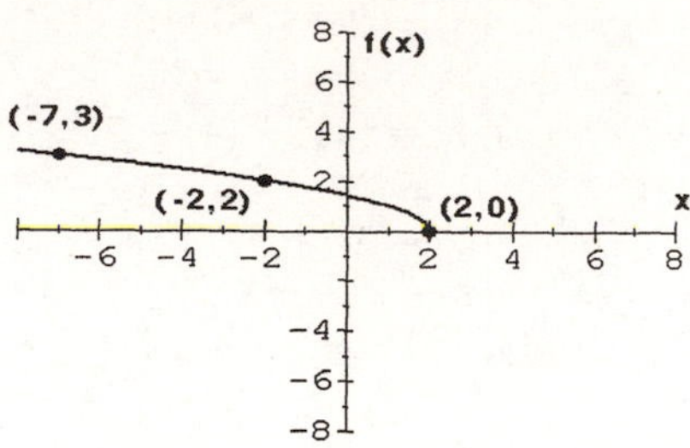

25. $f(x) = -x - 1$
This is the graph of a straight line whose y-intercept is -1 and slope is -1.

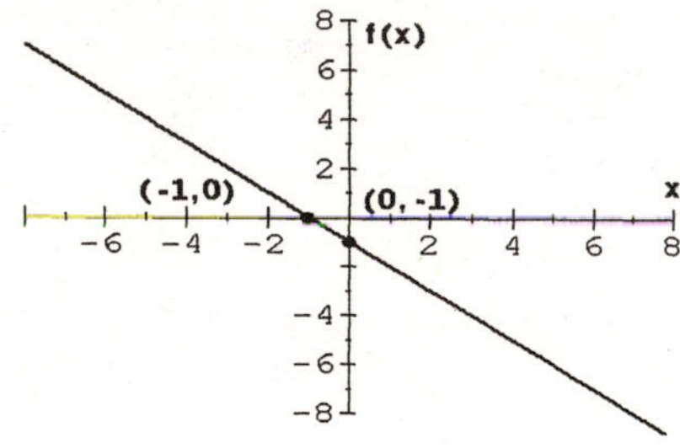

Chapter 9 Polynomial and Rational Functions

PROBLEM SET 9.1 Synthetic Division

1. $(4x^2 - 5x - 6) \div (x - 2)$

$$\begin{array}{r|rrr} 2 & 4 & -5 & -6 \\ & & 8 & 6 \\ \hline & 4 & 3 & 0 \end{array}$$

Q: $4x + 3$
R: 0

3. $(2x^2 - x - 21) \div (x + 3)$

$$\begin{array}{r|rrr} -3 & 2 & -1 & -21 \\ & & -6 & 21 \\ \hline & 2 & -7 & 0 \end{array}$$

Q: $2x - 7$
R: 0

5. $(3x^2 - 16x + 17) \div (x - 4)$

$$\begin{array}{r|rrr} 4 & 3 & -16 & 17 \\ & & 12 & -16 \\ \hline & 3 & -4 & 1 \end{array}$$

Q: $3x - 4$
R: 1

7. $(4x^2 + 19x - 32) \div (x + 6)$

$$\begin{array}{r|rrr} -6 & 4 & 19 & -32 \\ & & -24 & 30 \\ \hline & 4 & -5 & -2 \end{array}$$

Q: $4x - 5$
R: -2

9. $(x^3 + 2x^2 - 7x + 4) \div (x - 1)$

$$\begin{array}{r|rrrr} 1 & 1 & 2 & -7 & 4 \\ & & 1 & 3 & -4 \\ \hline & 1 & 3 & -4 & 0 \end{array}$$

Q: $x^2 + 3x - 4$
R: 0

11. $(3x^3 + 8x^2 - 8) \div (x + 2)$

$$\begin{array}{r|rrrr} -2 & 3 & 8 & 0 & -8 \\ & & -6 & -4 & 8 \\ \hline & 3 & 2 & -4 & 0 \end{array}$$

Q: $3x^2 + 2x - 4$
R: 0

13. $(5x^3 - 9x^2 - 3x - 2) \div (x - 2)$

$$\begin{array}{r|rrrr} 2 & 5 & -9 & -3 & -2 \\ & & 10 & 2 & -2 \\ \hline & 5 & 1 & -1 & -4 \end{array}$$

Q: $5x^2 + x - 1$
R: -4

15. $(x^3 + 6x^2 - 8x + 1) \div (x + 7)$

$$\begin{array}{r|rrrr} -7 & 1 & 6 & -8 & 1 \\ & & -7 & 7 & 7 \\ \hline & 1 & -1 & -1 & 8 \end{array}$$

Q: $x^2 - x - 1$
R: 8

17. $(-x^3 + 7x^2 - 14x + 6) \div (x - 3)$

$$\begin{array}{r|rrrr} 3 & -1 & 7 & -14 & 6 \\ & & -3 & 12 & -6 \\ \hline & -1 & 4 & -2 & 0 \end{array}$$

Q: $-x^2 + 4x - 2$
R: 0

19. $(-3x^3 + x^2 + 2x + 2) \div (x + 1)$

$$\begin{array}{r|rrrr} -1 & -3 & 1 & 2 & 2 \\ & & 3 & -4 & 2 \\ \hline & -3 & 4 & -2 & 4 \end{array}$$

Q: $-3x^2 + 4x - 2$
R: 4

Problem Set 9.1

21. $(3x^3 - 2x - 5) \div (x - 2)$

$$\begin{array}{r|rrrr} 2 & 3 & 0 & -2 & -5 \\ & & 6 & 12 & 20 \\ \hline & 3 & 6 & 10 & 15 \end{array}$$

Q: $3x^2 + 6x + 10$
R: 15

23. $(2x^4 + x^3 + 3x^2 + 2x - 2) \div (x + 1)$

$$\begin{array}{r|rrrrr} -1 & 2 & 1 & 3 & 2 & -2 \\ & & -2 & 1 & -4 & 2 \\ \hline & 2 & -1 & 4 & -2 & 0 \end{array}$$

Q: $2x^3 - x^2 + 4x - 2$
R: 0

25. $(x^4 + 4x^3 - 7x - 1) \div (x - 3)$

$$\begin{array}{r|rrrrr} 3 & 1 & 4 & 0 & -7 & -1 \\ & & 3 & 21 & 63 & 168 \\ \hline & 1 & 7 & 21 & 56 & 167 \end{array}$$

Q: $x^3 + 7x^2 + 21x + 56$
R: 167

27. $(x^4 + 5x^3 - x^2 + 25) \div (x + 5)$

$$\begin{array}{r|rrrrr} -5 & 1 & 5 & -1 & 0 & 25 \\ & & -5 & 0 & 5 & -25 \\ \hline & 1 & 0 & -1 & 5 & 0 \end{array}$$

Q: $x^3 - x + 5$
R: 0

29. $(x^4 - 16) \div (x - 2)$

$$\begin{array}{r|rrrrr} 2 & 1 & 0 & 0 & 0 & -16 \\ & & 2 & 4 & 8 & 16 \\ \hline & 1 & 2 & 4 & 8 & 0 \end{array}$$

Q: $x^3 + 2x^2 + 4x + 8$
R: 0

31. $(x^5 - 1) \div (x + 1)$

$$\begin{array}{r|rrrrrr} -1 & 1 & 0 & 0 & 0 & 0 & -1 \\ & & -1 & 1 & -1 & 1 & -1 \\ \hline & 1 & -1 & 1 & -1 & 1 & -2 \end{array}$$

Q: $x^4 - x^3 + x^2 - x + 1$
R: -2

33. $(x^5 + 1) \div (x + 1)$

$$\begin{array}{r|rrrrrr} -1 & 1 & 0 & 0 & 0 & 0 & 1 \\ & & -1 & 1 & -1 & 1 & -1 \\ \hline & 1 & -1 & 1 & -1 & 1 & 0 \end{array}$$

Q: $x^4 - x^3 + x^2 - x + 1$
R: 0

35. $(x^5 + 3x^4 - 5x^3 - 3x^2 + 3x - 4) \div (x + 4)$

$$\begin{array}{r|rrrrrr} -4 & 1 & 3 & -5 & -3 & 3 & -4 \\ & & -4 & 4 & 4 & -4 & 4 \\ \hline & 1 & -1 & -1 & 1 & -1 & 0 \end{array}$$

Q: $x^4 - x^3 - x^2 + x - 1$
R: 0

37. $(4x^5 - 6x^4 + 2x^3 + 2x^2 - 5x + 2) \div (x - 1)$

$$\begin{array}{r|rrrrrr} 1 & 4 & -6 & 2 & 2 & -5 & 2 \\ & & 4 & -2 & 0 & 2 & -3 \\ \hline & 4 & -2 & 0 & 2 & -3 & -1 \end{array}$$

Q: $4x^4 - 2x^3 + 2x - 3$
R: -1

39. $(9x^3 - 6x^2 + 3x - 4) \div \left(x - \frac{1}{3}\right)$

$$\begin{array}{r|rrrr} \frac{1}{3} & 9 & -6 & 3 & -4 \\ & & 3 & -1 & \frac{2}{3} \\ \hline & 9 & -3 & 2 & -\frac{10}{3} \end{array}$$

Q: $9x^2 - 3x + 2$
R: $-\frac{10}{3}$

41. $(3x^4 - 2x^3 + 5x^2 - x - 1) \div \left(x + \frac{1}{3}\right)$

$-\frac{1}{3}$	3	-2	5	-1	-1
		-1	1	-2	1
	3	-3	6	-3	0

Q: $3x^3 - 3x^2 + 6x - 3$
R: 0

PROBLEM SET 9.2 Remainder and Factor Theorems

1. $f(x) = x^2 + 2x - 6$

a) $f(3) = 3^2 + 2(3) - 6$
$f(3) = 9 + 6 - 6$
$f(3) = 9$

b)

3	1	2	-6
		3	15
	1	5	9

$f(3) = 9$

3. $f(x) = x^3 - 2x^2 + 3x - 1$

a) $f(-1) = (-1)^3 - 2(-1)^2 + 3(-1) - 1$
$f(-1) = -1 - 2(1) - 3 - 1$
$f(-1) = -7$

b)

-1	1	-2	3	-1
		-1	3	-6
	1	-3	6	-7

$f(-1) = -7$

5. $f(x) = 2x^4 - x^3 - 3x^2 + 4x - 1$

a) $f(2) = 2(2)^4 - (2)^3 - 3(2)^2 + 4(2) - 1$
$f(2) = 32 - 8 - 12 + 8 - 1$
$f(2) = 19$

b)

2	2	-1	-3	4	-1
		4	6	6	20
	2	3	3	10	19

$f(2) = 19$

7. $f(n) = 6n^3 - 35n^2 + 8n - 10$

a) $f(6) = 6(6)^3 - 35(6)^2 + 8(6) - 10$
$f(6) = 1296 - 1260 + 48 - 10$
$f(6) = 74$

b)

6	6	-35	8	-10
		36	6	84
	6	1	14	74

$f(6) = 74$

9. $f(n) = 2n^5 - 1$

a) $f(-2) = 2(-2)^5 - 1$
$f(-2) = -64 - 1$
$f(-2) = -65$

b)

-2	2	0	0	0	0	-1
		-4	8	-16	32	-64
	2	-4	8	-16	32	-65

$f(-2) = -65$

11. $f(x) = 6x^5 - 3x^3 + 2$

-1	6	0	-3	0	0	2
		-6	6	-3	3	-3
	6	-6	3	-3	3	-1

$f(-1) = -1$

13. $f(x) = 2x^4 - 15x^3 - 9x^2 - 2x - 3$

8	2	-15	-9	-2	-3
		16	8	-8	-80
	2	1	-1	-10	-83

$f(8) = -83$

Problem Set 9.2

15. $f(n) = 4n^7 + 3$

$f(3) = 4(3)^7 + 3$
$f(3) = 4(2187) + 3$
$f(3) = 8751$

17. $f(n) = 3n^5 + 17n^4 - 4n^3 + 10n^2 - 15n + 13$

$$\begin{array}{r|rrrrrr} -6 & 3 & 17 & -4 & 10 & -15 & 13 \\ & & -18 & 6 & -12 & 12 & 18 \\ \hline & 3 & -1 & 2 & -2 & -3 & 31 \end{array}$$

$f(-6) = 31$

19. $f(x) = -4x^4 - 6x^2 + 7$

$$\begin{array}{r|rrrrr} 4 & -4 & 0 & -6 & 0 & 7 \\ & & -16 & -64 & -280 & -1120 \\ \hline & -4 & -16 & -70 & -280 & -1113 \end{array}$$

$f(4) = -1113$

21.

$$\begin{array}{r|rrr} 2 & 5 & -17 & 14 \\ & & 10 & -14 \\ \hline & 5 & -7 & 0 \end{array}$$

$f(2) = 0$
Yes, it is a factor.

23.

$$\begin{array}{r|rrr} -3 & 6 & 13 & -14 \\ & & -18 & 15 \\ \hline & 6 & -5 & 1 \end{array}$$

$f(-3) = 1$
No, it is not a factor.

25.

$$\begin{array}{r|rrrr} 1 & 4 & -13 & 21 & -12 \\ & & 4 & -9 & 12 \\ \hline & 4 & -9 & 12 & 0 \end{array}$$

$f(1) = 0$
Yes, it is a factor.

27.

$$\begin{array}{r|rrrr} -2 & 1 & 7 & 1 & -18 \\ & & -2 & -10 & 18 \\ \hline & 1 & 5 & -9 & 0 \end{array}$$

$f(-2) = 0$
Yes, it is a factor.

29.

$$\begin{array}{r|rrrr} 3 & 3 & -5 & -17 & 17 \\ & & 9 & 12 & -15 \\ \hline & 3 & 4 & -5 & 2 \end{array}$$

$f(3) = 2$
No, it is not a factor.

31.

$$\begin{array}{r|rrrr} -2 & 1 & 0 & 0 & 8 \\ & & -2 & 4 & -8 \\ \hline & 1 & -2 & 4 & 0 \end{array}$$

$f(-2) = 0$
Yes, it is a factor.

33.

$$\begin{array}{r|rrrrr} 3 & 1 & 0 & 0 & 0 & -81 \\ & & 3 & 9 & 27 & 81 \\ \hline & 1 & 3 & 9 & 27 & 0 \end{array}$$

$f(3) = 0$
Yes, it is a factor.

35.

$$\begin{array}{r|rrrr} 2 & 1 & -6 & -13 & 42 \\ & & 2 & -8 & -42 \\ \hline & 1 & -4 & -21 & 0 \end{array}$$

$(x - 2)(x^2 - 4x - 21)$
$(x - 2)(x - 7)(x + 3)$

37.

$$\begin{array}{r|rrrr} -2 & 12 & 29 & 8 & -4 \\ & & -24 & -10 & 4 \\ \hline & 12 & 5 & -2 & 0 \end{array}$$

$(x + 2)(12x^2 + 5x - 2)$
$(x + 2)(4x - 1)(3x + 2)$

39.

$$\begin{array}{r|rrrr} -1 & 1 & -2 & -7 & -4 \\ & & -1 & 3 & 4 \\ \hline & 1 & -3 & -4 & 0 \end{array}$$

$(x + 1)(x^2 - 3x - 4)$
$(x + 1)(x - 4)(x + 1)$
$(x + 1)^2(x - 4)$

41.

$$\begin{array}{r|rrrrrr} 6 & 1 & -6 & 0 & 0 & -16 & 96 \\ & & 6 & 0 & 0 & 0 & -96 \\ \hline & 1 & 0 & 0 & 0 & -16 & 0 \end{array}$$

$(x - 6)(x^4 - 16)$
$(x - 6)(x^2 - 4)(x^2 + 4)$
$(x - 6)(x + 2)(x - 2)(x^2 + 4)$

43.

-5	9	21	-104	80
		-45	120	-80
	9	-24	16	0

$(x+5)(9x^2-24x+16)$
$(x+5)(3x-4)(3x-4)$
$(x+5)(3x-4)^2$

45.

1	k^2	0	$3k$	0	-4
		k^2	k^2	k^2+3k	k^2+3k
	k^2	k^2	k^2+3k	k^2+3k	k^2+3k-4

$k^2+3k-4=0$
$(k+4)(k-1)=0$
$k+4=0$ or $k-1=0$
$k=-4$ or $k=1$
The values are -4 or 1.

47.

-3	k	19	1	-6
		$-3k$	$9k-57$	$-27k+168$
	k	$-3k+19$	$9k-56$	$-27k+162$

$-27k+162=0$
$-27k=-162$
$k=6$
The value is 6.

49. $f(c)=3x^4+2x^2+5$
$f(c)>0$ for all values of c.

51. If $f(-1)=0$, then
$(x+1)$ is a factor.
$f(x)=x^n-1$
$f(-1)=(-1)^n-1$
$0=(-1)^n-1$
$1=(-1)^n$
This is true for all even positive integral values of n. Therefore, $(x+1)$ is a factor of x^n-1 for all even positive integral values of n.

53a. If $f(y)=0$, then
$(x-y)$ is a factor.
$f(x)=x^n-y^n$
$f(y)=y^n-y^n$
$f(y)=0$
This is true for all positive integral values of n. Therefore $(x-y)$ is a factor of x^n-y^n for all positive integral values of n.

b. If $f(-y)=0$, then
$(x+y)$ is a factor.
$f(x)=x^n-y^n$
$f(-y)=(-y)^n-y^n$
$0=(-y)^n-y^n$
$y^n=(-y)^n$
This is true for all even positive integral values on n. Therefore, $(x+y)$ is a factor of x^n-y^n for all even positive integral values of n.

c. If $f(-y)=0$, then
$(x+y)$ is a factor.
$f(x)=x^n+y^n$
$f(-y)=(-y)^n+y^n$
$(-y)^n=-y^n$
This is true for all odd positive integral values of n. Therefore $(x+y)$ is a factor of x^n+y^n for all odd positive integral values of n.

Further Investigations

57. $f(x)=x^2+4x-2$

a)

$1+i$	1	4	-2
		$1+i$	$4+6i$
	1	$5+i$	$2+6i$

$f(1+i)=2+6i$

b) $f(1+i)=(1+i)^2+4(1+i)-2$
$f(1+i)=1+2i-1+4+4i-2$
$f(1+i)=2+6i$

59. $f(x)=x^4+6x^2+8$

If $f(2i)=0$, then $(x-2i)$ is a factor of $f(x)$.

$2i$	1	0	6	0	8
		$2i$	-4	$4i$	-8
	1	$2i$	2	$4i$	0

$f(2i)=0$
Therefore, $(x-2i)$ is a factor.

61a) $f(x)=x^3+5x^2-2x+1$
$f(x)=x(x^2+5x-2)+1$
$f(x)=x[x(x+5)-2]+1$

$f(4)=4[4(4+5)-2]+1$
$f(4)=4[4(9)-2]+1$
$f(4)=4[34]+1$
$f(4)=136+1=137$

$f(-5)=-5[-5(-5+5)-2]+1$
$f(-5)=-5(-2)+1$
$f(-5)=10+1=11$

$f(7)=7[7(7+5)-2]+1$
$f(7)=7[7(12)-2]+1$
$f(7)=7[82]+1$
$f(7)=574+1=575$

b) $f(x)=2x^3-4x^2-3x+2$
$f(x)=x(2x^2-4x-3)+2$
$f(x)=x[x(2x-4)-3]+2$

$f(3)=3[3(2(3)-4)-3]+2$
$f(3)=3[3(2)-3]+2$
$f(3)=3(3)+2=11$

$f(6)=6[6(2(6)-4)-3]+2$
$f(6)=6[6(8)-3]+2$
$f(6)=6(45)+2$
$f(6)=270+2=272$

$f(-7)=-7[-7(2(-7)-4)-3]+2$
$f(-7)=-7[-7(-18)-3]+2$
$f(-7)=-7(123)+2$
$f(-7)=-861+2=-859$

c) $f(x)=-2x^3+5x^2-6x-7$
$f(x)=x(-2x^2+5x-6)-7$
$f(x)=x[x(-2x+5)-6]-7$

$f(4)=4[4(-2(4)+5)-6]-7$
$f(4)=4[4(-3)-6]-7$
$f(4)=4[-18]-7$
$f(4)=-72-7=-79$

$f(5)=5[5(-2(5)+5)-6]-7$
$f(5)=5[5(-5)-6]-7$
$f(5)=5[-31]-7=-162$

$f(-3)=-3[-3(-2(-3)+5)-6]-7$
$f(-3)=-3[-3(11)-6]-7$
$f(-3)=-3(-39)-7$
$f(-3)=117-7=110$

d) $f(x)=x^4+3x^3-2x^2+5x-1$
$f(x)=x(x^3+3x^2-2x+5)-1$
$f(x)=x[x(x^2+3x-2)+5]-1$
$f(x)=x\{x[x(x+3)-2]+5\}-1$

$f(5)=5\{5[5(5+3)-2]+5\}-1$
$f(5)=5\{5[5(8)-2]+5\}-1$
$f(5)=5\{5[38]+5\}-1$
$f(5)=5(195)-1$
$f(5)=975-1=974$

$f(6)=6\{6[6(6+3)-2]+5\}-1$
$f(6)=6\{6[52]+5\}-1$
$f(6)=6(317)-1$
$f(6)=1902-1=1901$

$f(-3)=-3\{-3[-3(-3+3)-2]+5\}-1$
$f(-3)=-3\{-3(-2)+5\}-1$
$f(-3)=-3(11)-1$
$f(-3)=-33-1=-34$

PROBLEM SET 9.3 Polynomial Equations

1. $x^3 - 2x^2 - 11x + 12 = 0$

c: $\pm 1, \pm 2, \pm 3, \pm 4, \pm 6, \pm 12$

d: ± 1

$\frac{c}{d}$: $\pm 1, \pm 2, \pm 3, \pm 4, \pm 6, \pm 12$

$$\begin{array}{r|rrrr} 1 & 1 & -2 & -11 & 12 \\ & & 1 & -1 & -12 \\ \hline & 1 & -1 & -12 & 0 \end{array}$$

$(x-1)(x^2 - x - 12) = 0$

$(x-1)(x-4)(x+3) = 0$

$x - 1 = 0$ or $x - 4 = 0$ or $x + 3 = 0$

$x = 1$ or $x = 4$ or $x = -3$

The solution set is $\{-3, 1, 4\}$.

3. $15x^3 + 14x^2 - 3x - 2 = 0$

c: $\pm 1, \pm 2$

d: $\pm 1, \pm 3, \pm 5, \pm 15$

$\frac{c}{d}$: $\pm 1, \pm 2, \pm \frac{1}{3}, \pm \frac{2}{3}, \pm \frac{1}{5}, \pm \frac{2}{5}, \pm \frac{1}{15}, \pm \frac{2}{15}$

$$\begin{array}{r|rrrr} -1 & 15 & 14 & -3 & -2 \\ & & -15 & 1 & 2 \\ \hline & 15 & -1 & -2 & 0 \end{array}$$

$(x+1)(15x^2 - x - 2) = 0$

$(x+1)(3x+1)(5x-2) = 0$

$x + 1 = 0$ or $3x + 1 = 0$ or $5x - 2 = 0$

$x = -1$ or $3x = -1$ or $5x = 2$

$x = -1$ or $x = -\frac{1}{3}$ or $x = \frac{2}{5}$

The solution set is $\left\{-1, -\frac{1}{3}, \frac{2}{5}\right\}$.

5. $8x^3 - 2x^2 - 41x - 10 = 0$

c: $\pm 1, \pm 2, \pm 5, \pm 10$

d: $\pm 1, \pm 2, \pm 4, \pm 8$

$\frac{c}{d}$: $\pm 1, \pm 2, \pm 5, \pm 10,$

$\pm \frac{1}{2}, \pm \frac{5}{2}, \pm \frac{1}{4}, \pm \frac{5}{4}, \pm \frac{1}{8}, \pm \frac{5}{8}$

$$\begin{array}{r|rrrr} -2 & 8 & -2 & -41 & -10 \\ & & -16 & 36 & 10 \\ \hline & 8 & -18 & -5 & 0 \end{array}$$

$(x+2)(8x^2 - 18x - 5) = 0$

$(x+2)(2x-5)(4x+1) = 0$

$x + 2 = 0$ or $2x - 5 = 0$ or $4x + 1 = 0$

$x = -2$ or $2x = 5$ or $4x = -1$

$x = -2$ or $x = \frac{5}{2}$ or $x = -\frac{1}{4}$

The solution set is $\left\{-2, -\frac{1}{4}, \frac{5}{2}\right\}$.

7. $x^3 - x^2 - 8x + 12 = 0$

c: $\pm 1, \pm 2, \pm 3, \pm 4, \pm 6, \pm 12$

d: ± 1

$\frac{c}{d}$: $\pm 1, \pm 2, \pm 3, \pm 4, \pm 6, \pm 12$

$$\begin{array}{r|rrrr} 2 & 1 & -1 & -8 & 12 \\ & & 2 & 2 & -12 \\ \hline & 1 & 1 & -6 & 0 \end{array}$$

$(x-2)(x^2 + x - 6) = 0$

$(x-2)(x+3)(x-2) = 0$

$x - 2 = 0$ or $x + 3 = 0$ or $x - 2 = 0$

$x = 2$ or $x = -3$ or $x = 2$

The solution set is $\{-3, 2\}$.

9. $x^3 - 4x^2 + 8 = 0$

c: $\pm 1, \pm 2, \pm 4, \pm 8$

d: ± 1

$\frac{c}{d}$: $\pm 1, \pm 2, \pm 4, \pm 8$

$$\begin{array}{r|rrrr} 2 & 1 & -4 & 0 & 8 \\ & & 2 & -4 & -8 \\ \hline & 1 & -2 & -4 & 0 \end{array}$$

$(x-2)(x^2 - 2x - 4) = 0$

$x - 2 = 0$ or $x^2 - 2x - 4 = 0$

$x = 2$ or $x = \frac{-(-2) \pm \sqrt{(-2)^2 - 4(1)(-4)}}{2(1)}$

$x = \frac{2 \pm \sqrt{4 + 16}}{2}$

$x = \frac{2 \pm \sqrt{20}}{2}$

$x = \frac{2 \pm 2\sqrt{5}}{2}$

$$x = \frac{2(1 \pm \sqrt{5})}{2}$$
$$x = 1 \pm \sqrt{5}$$
The solution set is $\{2, 1 \pm \sqrt{5}\}$.

11. $x^4 + 4x^3 - x^2 - 16x - 12 = 0$
c: $\pm 1, \pm 2, \pm 3, \pm 4, \pm 6, \pm 12$
d: ± 1
$\frac{c}{d}$: $\pm 1, \pm 2, \pm 3, \pm 4, \pm 6, \pm 12$

$$\begin{array}{r|rrrrr} -1 & 1 & 4 & -1 & -16 & -12 \\ & & -1 & -3 & 4 & 12 \\ \hline & 1 & 3 & -4 & -12 & 0 \end{array}$$

$(x + 1)(x^3 + 3x^2 - 4x - 12) = 0$
c: $\pm 1, \pm 2, \pm 3, \pm 4, \pm 6, \pm 12$
d: ± 1
$\frac{c}{d}$: $\pm 1, \pm 2, \pm 3, \pm 4, \pm 6, \pm 12$

$$\begin{array}{r|rrrr} 2 & 1 & 3 & -4 & -12 \\ & & 2 & 10 & 12 \\ \hline & 1 & 5 & 6 & 0 \end{array}$$

$(x + 1)(x - 2)(x^2 + 5x + 6) = 0$
$(x + 1)(x - 2)(x + 3)(x + 2) = 0$
$x+1 = 0$ or $x-2 = 0$ or $x+3 = 0$ or $x+2 = 0$
$x = -1$ or $x = 2$ or $x = -3$ or $x = -2$
The solution set is $\{-3, -2, -1, 2\}$.

13. $x^4 + x^3 - 3x^2 - 17x - 30 = 0$
c: $\pm 1, \pm 2, \pm 3, \pm 5, \pm 6, \pm 10, \pm 15, \pm 30$
d: ± 1
$\frac{c}{d}$: $\pm 1, \pm 2, \pm 3, \pm 5, \pm 6, \pm 10, \pm 15, \pm 30$

$$\begin{array}{r|rrrrr} -2 & 1 & 1 & -3 & -17 & -30 \\ & & -2 & 2 & 2 & 30 \\ \hline & 1 & -1 & -1 & -15 & 0 \end{array}$$

$(x + 2)(x^3 - x^2 - x - 15) = 0$
c: $\pm 1, \pm 3, \pm 5, \pm 15$
d: ± 1
$\frac{c}{d}$: $\pm 1, \pm 3, \pm 5, \pm 15$

$$\begin{array}{r|rrrr} 3 & 1 & -1 & -1 & -15 \\ & & 3 & 6 & 15 \\ \hline & 1 & 2 & 5 & 0 \end{array}$$

$(x + 2)(x - 3)(x^2 + 2x + 5) = 0$
$x + 2 = 0$ or $x - 3 = 0$
$x = -2$ or $x = 3$ or $x^2 + 2x + 5 = 0$
$$x = \frac{-2 \pm \sqrt{2^2 - 4(1)(5)}}{2(1)}$$
$$x = \frac{-2 \pm \sqrt{-16}}{2}$$
$$x = \frac{-2 \pm 4i}{2}$$
$$x = -1 \pm 2i$$
The solution set is $\{-2, 3, -1 \pm 2i\}$.

15. $x^3 - x^2 + x - 1 = 0$
c: ± 1
d: ± 1
$\frac{c}{d}$: ± 1

$$\begin{array}{r|rrrr} 1 & 1 & -1 & 1 & -1 \\ & & 1 & 0 & 1 \\ \hline & 1 & 0 & 1 & 0 \end{array}$$

$(x - 1)(x^2 + 1) = 0$
$x - 1 = 0$ or $x^2 + 1 = 0$
$x = 1$ or $x^2 = -1$
$x = 1$ or $x = \pm\sqrt{-1}$
$x = 1$ or $x = \pm i$
The solution set is $\{1, \pm i\}$.

17. $2x^4 + 3x^3 - 11x^2 - 9x + 15 = 0$
c: $\pm 1, \pm 3, \pm 5, \pm 15$
d: $\pm 1, \pm 2$
$\frac{c}{d}$: $\pm 1, \pm 3, \pm 5, \pm 15,$
$\pm \frac{1}{2}, \pm \frac{3}{2}, \pm \frac{5}{2}, \pm \frac{15}{2}$

$$\begin{array}{r|rrrrr} 1 & 2 & 3 & -11 & -9 & 15 \\ & & 2 & 5 & -6 & -15 \\ \hline & 2 & 5 & -6 & -15 & 0 \end{array}$$

$(x - 1)(2x^3 + 5x^2 - 6x - 15) = 0$
c: $\pm 1, \pm 3, \pm 5, \pm 15$
d: $\pm 1, \pm 2$
$\frac{c}{d}$: $\pm 1, \pm 3, \pm 5, \pm 15,$
$\pm \frac{1}{2}, \pm \frac{3}{2}, \pm \frac{5}{2}, \pm \frac{15}{2}$

$$\begin{array}{r|rrrr} -\frac{5}{2} & 2 & 5 & -6 & -15 \\ & & -5 & 0 & 15 \\ \hline & 2 & 0 & -6 & 0 \end{array}$$

$(x-1)\left(x+\frac{5}{2}\right)(2x^2-6)=0$

$x-1=0$ or $x+\frac{5}{2}=0$ or $2x^2-6=0$

$x=1$ or $x=-\frac{5}{2}$ or $2x^2=6$

$x=1$ or $x=-\frac{5}{2}$ or $x^2=3$

$x=1$ or $x=-\frac{5}{2}$ or $x=\pm\sqrt{3}$

The solution set is $\left\{-\frac{5}{2}, 1, \pm\sqrt{3}\right\}$.

19. $4x^4+12x^3+x^2-12x+4=0$
c: $\pm 1, \pm 2, \pm 4$
d: $\pm 1, \pm 2, \pm 4$
$\frac{c}{d}$: $\pm 1, \pm 2, \pm 4, \pm\frac{1}{2}, \pm\frac{1}{4}$

$$\begin{array}{r|rrrrr} -2 & 4 & 12 & 1 & -12 & 4 \\ & & -8 & -8 & 14 & -4 \\ \hline & 4 & 4 & -7 & 2 & 0 \end{array}$$

$(x+2)(4x^3+4x^2-7x+2)=0$
c: $\pm 1, \pm 2$
d: $\pm 1, \pm 2, \pm 4$
$\frac{c}{d}$: $\pm 1, \pm 2, \pm\frac{1}{2}, \pm\frac{1}{4}$

$$\begin{array}{r|rrrr} \frac{1}{2} & 4 & 4 & -7 & 2 \\ & & 2 & 3 & -2 \\ \hline & 4 & 6 & -4 & 0 \end{array}$$

$(x+2)\left(x-\frac{1}{2}\right)(4x^2+6x-4)=0$

$(x+2)\left(x-\frac{1}{2}\right)(2)(2x^2+3x-2)=0$

$2(x+2)\left(x-\frac{1}{2}\right)(2x-1)(x+2)=0$

$2(x+2)\left(x-\frac{1}{2}\right)(2)\left(x-\frac{1}{2}\right)(x+2)=0$

$4(x+2)^2\left(x-\frac{1}{2}\right)^2=0$

$x+2=0$ or $x-\frac{1}{2}=0$

$x=-2$ or $x=\frac{1}{2}$

The solution set is $\left\{-2, \frac{1}{2}\right\}$.

21. $x^4+3x-2=0$
c: $\pm 1, \pm 2$
d: ± 1
$\frac{c}{d}$: $\pm 1, \pm 2$

$$\begin{array}{r|rrrrr} 1 & 1 & 0 & 0 & 3 & -2 \\ & & 1 & 1 & 1 & 4 \\ \hline & 1 & 1 & 1 & 4 & 2 \end{array}$$

$$\begin{array}{r|rrrrr} -1 & 1 & 0 & 0 & 3 & -2 \\ & & -1 & 1 & -1 & -2 \\ \hline & 1 & -1 & 1 & 2 & -4 \end{array}$$

$$\begin{array}{r|rrrrr} 2 & 1 & 0 & 0 & 3 & -2 \\ & & 2 & 4 & 8 & 22 \\ \hline & 1 & 2 & 4 & 11 & 20 \end{array}$$

$$\begin{array}{r|rrrrr} -2 & 1 & 0 & 0 & 3 & -2 \\ & & -2 & 4 & -8 & 10 \\ \hline & 1 & -2 & 4 & -5 & 8 \end{array}$$

None of the possible rational roots gave a remainder of zero. Therefore, there are no rational roots.

23. $3x^4-4x^3-10x^2+3x-4=0$
c: $\pm 1, \pm 2, \pm 4$
d: $\pm 1, \pm 3$
$\frac{c}{d}$: $\pm 1, \pm 2, \pm 4, \pm\frac{1}{3}, \pm\frac{2}{3}, \pm\frac{4}{3}$

None of the possible rational roots gave a remainder of zero. Therefore, there are no rational roots.

25. $x^5+2x^4-2x^3+5x^2-2x-3=0$
c: $\pm 1, \pm 3$
d: ± 1

$\frac{c}{d}$: $\pm 1, \pm 3$

$$\begin{array}{r|rrrrrr} 1 & 1 & 2 & -2 & 5 & -2 & -3 \\ & & 1 & 3 & 1 & 6 & 4 \\ \hline & 1 & 3 & 1 & 6 & 4 & 1 \end{array}$$

$$\begin{array}{r|rrrrrr} -1 & 1 & 2 & -2 & 5 & -2 & -3 \\ & & -1 & -1 & 3 & -8 & 10 \\ \hline & 1 & 1 & -3 & 8 & -10 & 7 \end{array}$$

$$\begin{array}{r|rrrrrr} 3 & 1 & 2 & -2 & 5 & -2 & -3 \\ & & 3 & 15 & 39 & 132 & 390 \\ \hline & 1 & 5 & 13 & 44 & 130 & 387 \end{array}$$

$$\begin{array}{r|rrrrrr} -3 & 1 & 2 & -2 & 5 & -2 & -3 \\ & & -3 & 3 & -3 & -6 & 24 \\ \hline & 1 & -1 & 1 & 2 & -8 & 21 \end{array}$$

None of the possible rational roots gave a remainder of zero. Therefore, there are no rational roots.

27. $\frac{1}{10}x^3 + \frac{1}{5}x^2 - \frac{1}{2}x - \frac{3}{5} = 0$

$10\left(\frac{1}{10}x^3 + \frac{1}{5}x^2 - \frac{1}{2}x - \frac{3}{5}\right) = 10(0)$

$x^3 + 2x^2 - 5x - 6 = 0$
c: $\pm 1, \pm 2, \pm 3, \pm 6$
d: ± 1
$\frac{c}{d}$: $\pm 1, \pm 2, \pm 3, \pm 6$

$$\begin{array}{r|rrrr} -1 & 1 & 2 & -5 & -6 \\ & & -1 & -1 & 6 \\ \hline & 1 & 1 & -6 & 0 \end{array}$$

$(x + 1)(x^2 + x - 6) = 0$
$(x + 1)(x + 3)(x - 2) = 0$
$x + 1 = 0$ or $x + 3 = 0$ or $x - 2 = 0$
$x = -1$ or $x = -3$ or $x = 2$
The solution set is $\{-3, -1, 2\}$.

29. $x^3 - \frac{5}{6}x^2 - \frac{22}{3}x + \frac{5}{2} = 0$

$6\left(x^3 - \frac{5}{6}x^2 - \frac{22}{3}x + \frac{5}{2}\right) = 6(0)$

$6x^3 - 5x^2 - 44x + 15 = 0$
c: $\pm 1, \pm 3, \pm 5, \pm 15$
d: $\pm 1, \pm 2, \pm 3, \pm 6$
$\frac{c}{d}$: $\pm 1, \pm 3, \pm 5, \pm 15$

$\pm\frac{1}{2}, \pm\frac{3}{2}, \pm\frac{5}{2}, \pm\frac{15}{2}$

$\pm\frac{1}{3}, \pm\frac{5}{3}, \pm\frac{1}{6}, \pm\frac{5}{6}$

$$\begin{array}{r|rrrr} 3 & 6 & -5 & -44 & 15 \\ & & 18 & 39 & -15 \\ \hline & 6 & 13 & -5 & 0 \end{array}$$

$(x - 3)(6x^2 + 13x - 5) = 0$
$(x - 3)(2x + 5)(3x - 1) = 0$
$x - 3 = 0$ or $2x + 5 = 0$ or $3x - 1 = 0$
$x = 3$ or $2x = -5$ or $3x = 1$
$x = 3$ or $x = -\frac{5}{2}$ or $x = \frac{1}{3}$
The solution set is $\left\{-\frac{5}{2}, \frac{1}{3}, 3\right\}$.

31. $6x^2 + 7x - 20 = 0$
There is one variation in sign, so there is one positive real solution.

Replacing x with $-x$:
$6(-x)^2 + 7(-x) - 20$
$6x^2 - 7x - 20$
There is one variation in sign, so there is one negative real solution.

33. $2x^3 + x - 3 = 0$
There is one variation in sign, so there is one positive real solution.

Replacing x with $-x$:
$2(-x)^3 + (-x) - 3$
$-2x^3 - x - 3$
No variations in sign, so there are no negative real solutions.

So, there is one positive real solution and two nonreal complex solutions.

35. $3x^3 - 2x^2 + 6x + 5 = 0$
There are two variations in sign, so there are two or no positive real solutions.

Replacing x with $-x$:
$3(-x)^3 - 2(-x)^2 + 6(-x) + 5$
$-3x^3 - 2x^2 - 6x + 5$
There is one variation in sign,
so there is one negative real solution.

So, there are two cases.
I: 1 negative and 2 positive real solutions
or
II: 1 negative real solution and
2 nonreal complex solutions.

37. $x^5 - 3x^4 + 5x^3 - x^2 + 2x - 1 = 0$
There are 5 variations in sign, so there could be 5, 3, or 1 positive real solutions.

Replacing x with $-x$:
$(-x)^5 - 3(-x)^4 + 5(-x)^3 - (-x)^2 + 2(-x) - 1$
$-x^5 - 3x^4 - 5x^3 - x^2 - 2x - 1$
There are no variations in sign,
so there are no negative real solutions.

There are three cases.
I: 5 positive real solutions
or
II: 3 positive real solutions and
2 nonreal complex solutions.
or
III: 1 positive real solution and
4 nonreal complex solutions.

39. $x^5 + 32 = 0$
There are no variations in sign, so there are no positive real solutions.

Replacing x with $-x$:
$(-x)^5 + 32$
$-x^5 + 32$
There is one variation in sign,
so there is one negative real solution.

So there is one negative real solution and four nonreal complex solutions.

Further Investigations

43. $x^2 - 2 = 0$
$c: \pm 1, \pm 2$
$d: \pm 1$
$\frac{c}{d}: \pm 1, \pm 2$
None of these possible rational solutions is a solution.

Substituting $x = \sqrt{2}$ to determine if it is a solution.
$x^2 - 2 \overset{?}{=} 0$
$(\sqrt{2})^2 - 2 \overset{?}{=} 0$
$2 - 2 \overset{?}{=} 0$
$0 = 0$
$\sqrt{2}$ is a solution.

There are no rational solutions.
Therefore, $\sqrt{2}$ is not a rational number.

45. Complex solutions occur in conjugate pairs. If a polynomial equation has an odd degree, then it has an odd number of solutions. Since the complex solutions occur in pairs, at least one solution must be real.

PROBLEM SET 9.4 Graphing Polynomial Functions

1. $f(x) = -(x-3)^3$

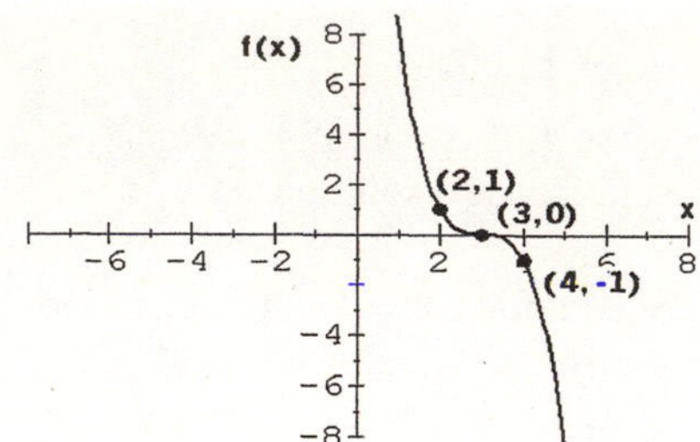

9. $f(x) = (x+1)^4 + 3$

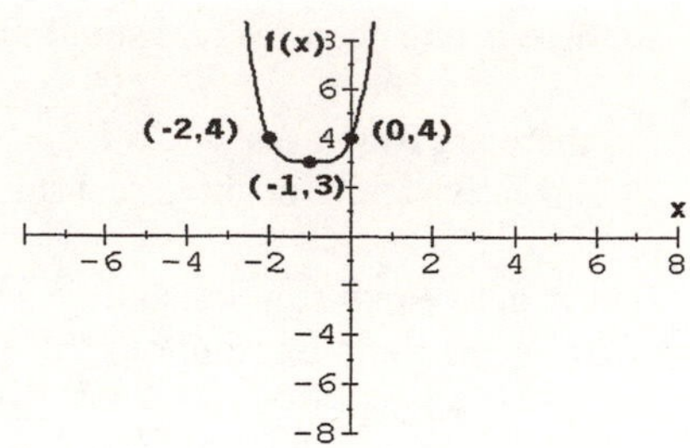

3. $f(x) = (x+1)^3$

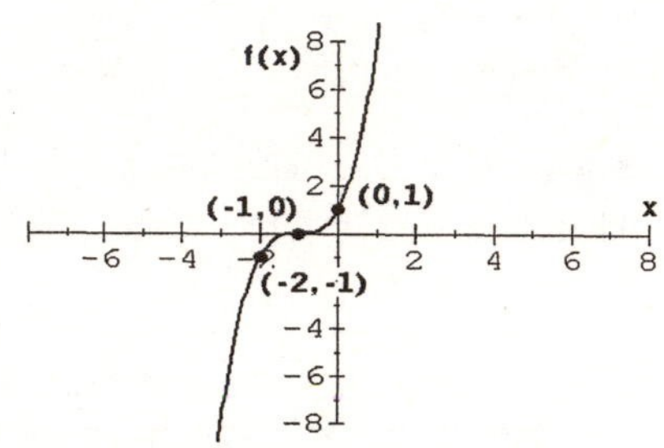

11. $f(x) = (x-2)(x+1)(x+3)$

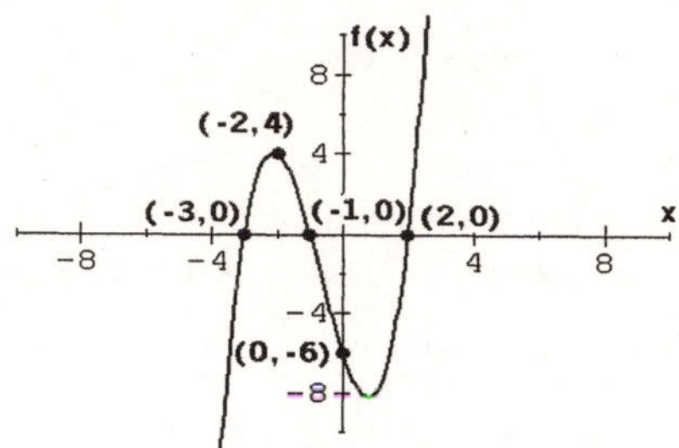

5. $f(x) = (x+3)^4$

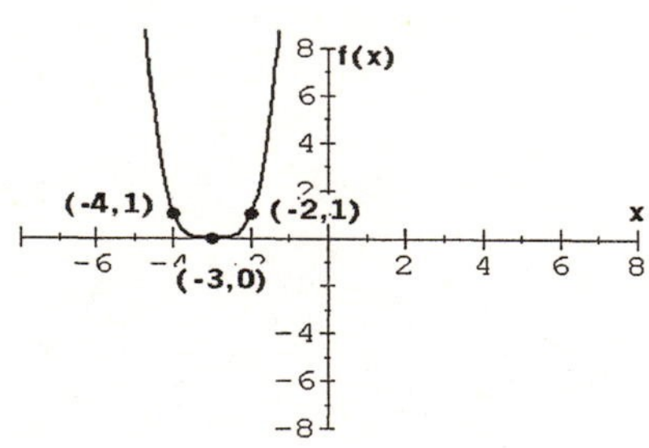

13. $f(x) = x(x+2)(2-x)$

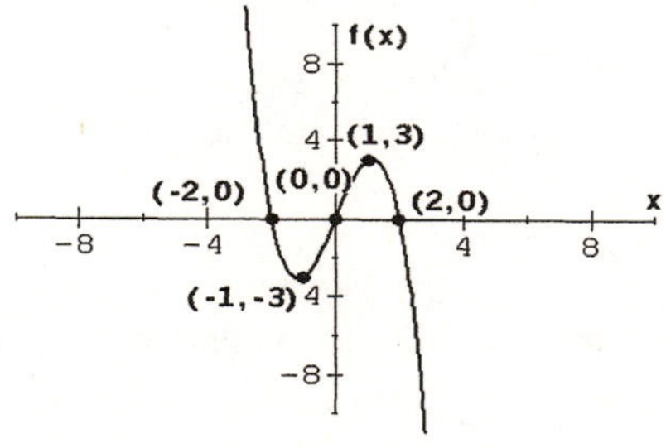

7. $f(x) = -(x-2)^4$

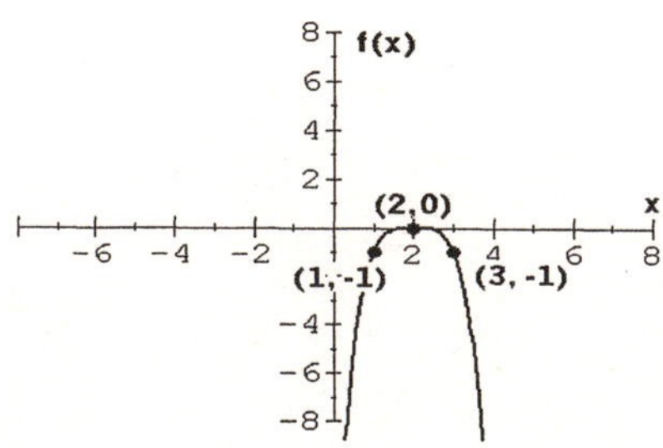

15. $f(x) = -x^2(x-1)(x+1)$

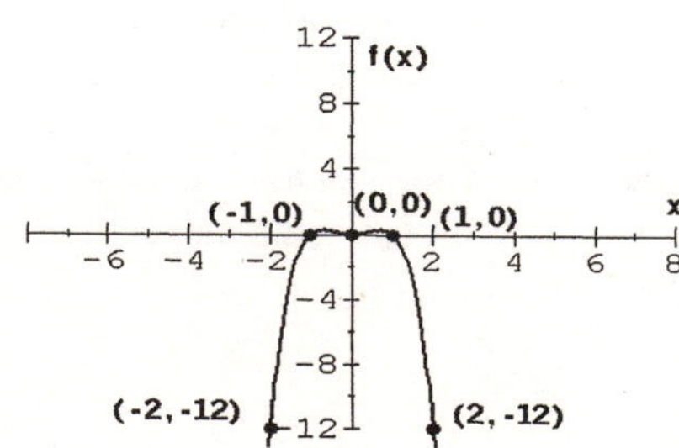

17. $f(x) = (2x - 1)(x - 2)(x - 3)$

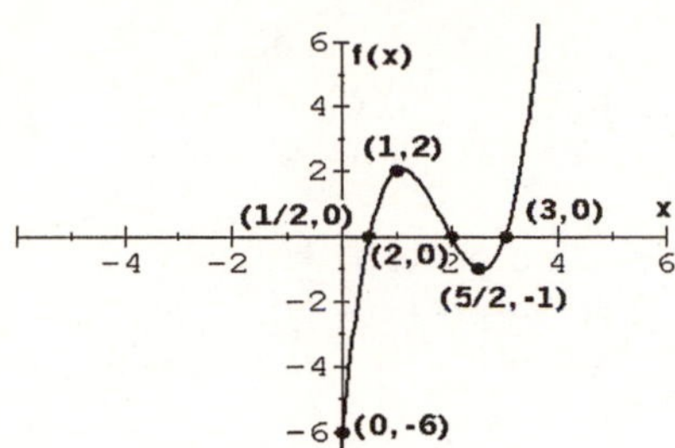

19. $f(x) = (x - 2)(x - 1)(x + 1)(x + 2)$

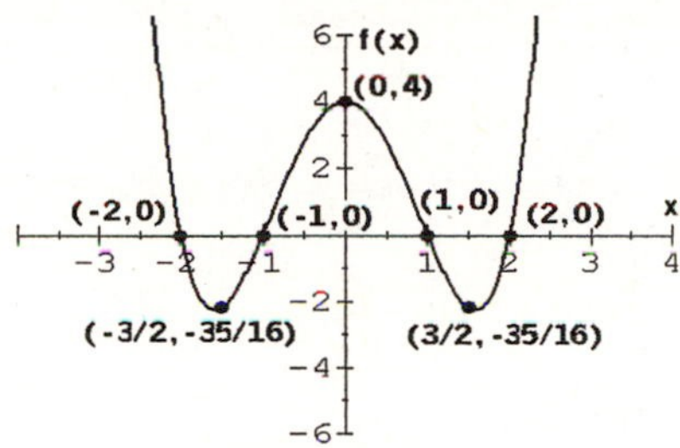

21. $f(x) = x(x - 2)^2(x + 1)$

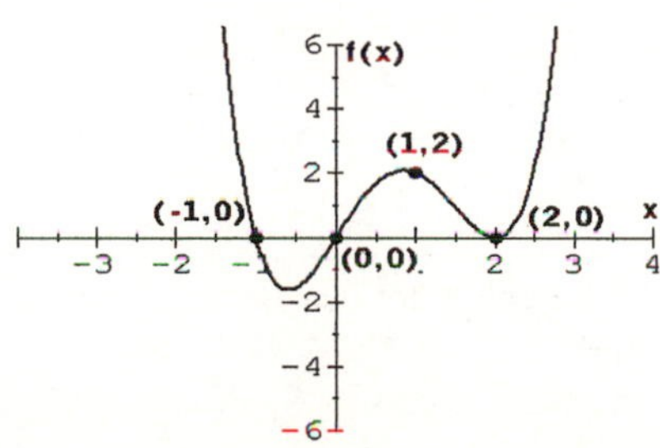

23. $f(x) = -x^3 - x^2 + 6x$
$f(x) = -x(x^2 + x - 6)$
$f(x) = -x(x + 3)(x - 2)$

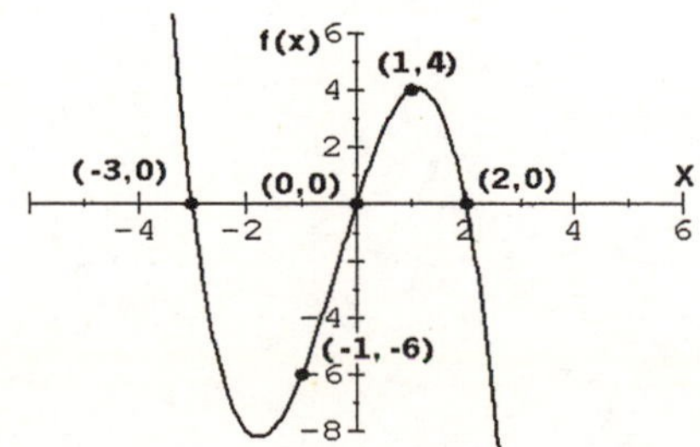

25. $f(x) = x^4 - 5x^3 + 6x^2$
$f(x) = x^2(x^2 - 5x + 6)$
$f(x) = x^2(x - 3)(x - 2)$

27. $f(x) = x^3 + 2x^2 - x - 2$
$f(x) = x^2(x + 2) - 1(x + 2)$
$f(x) = (x + 2)(x^2 - 1)$
$f(x) = (x + 2)(x - 1)(x + 1)$

29. $f(x) = x^3 - 8x^2 + 19x - 12$

1	1	−8	19	−12
		1	−7	12
	1	−7	12	0

$f(x) = (x - 1)(x^2 - 7x + 12)$
$f(x) = (x - 1)(x - 3)(x - 4)$

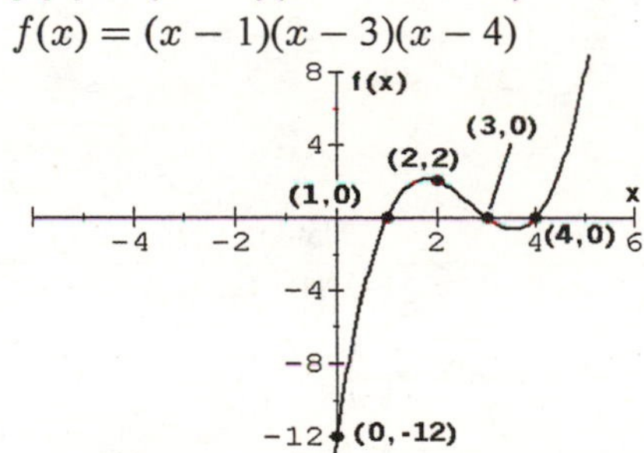

31. $f(x) = 2x^3 - 3x^2 - 3x + 2$

−1	2	−3	−3	2
		−2	5	−2
	2	−5	2	0

$f(x) = (x + 1)(2x^2 - 5x + 2)$
$f(x) = (x + 1)(2x - 1)(x - 2)$

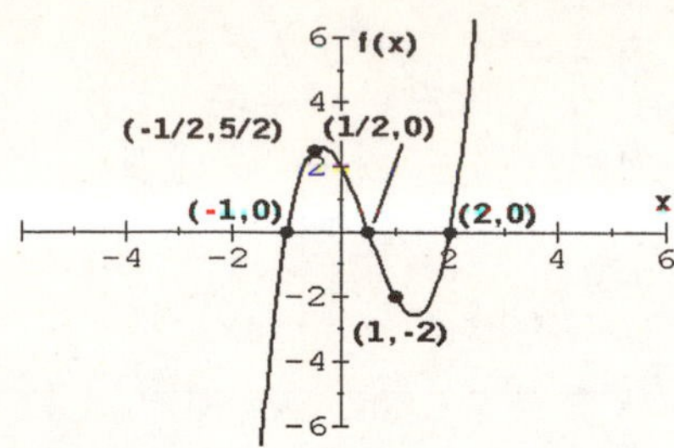

33. $f(x) = x^4 - 5x^2 + 4$
$f(x) = (x^2 - 4)(x^2 - 1)$
$f(x) = (x + 2)(x - 2)(x + 1)(x - 1)$

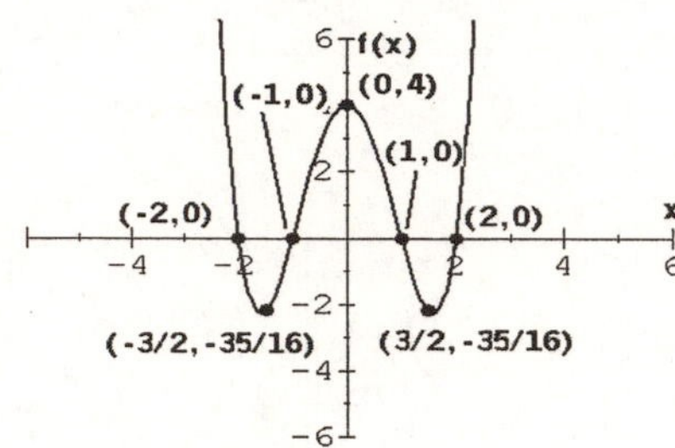

35. $f(x) = (x + 3)(x - 6)(8 - x)$

a. y-intercepts, let $x = 0$
$f(0) = (0 + 3)(0 - 6)(8 - 0)$
$f(0) = (3)(-6)(8) = -144$

b. x-intercepts, let $y = 0$
$0 = (x + 3)(x - 6)(8 - x)$
$x + 3 = 0$ or $x - 6 = 0$ or $8 - x = 0$
$x = -3$ or $x = 6$ or $x = 8$

c.

Interval	Test Value	Sign of $f(x)$	Location of Graph
$x < -3$	$f(-4) = 120$	positive	above x-axis
$-3 < x < 6$	$f(0) = -144$	negative	below x-axis
$6 < x < 8$	$f(7) = 10$	positive	above x-axis
$x > 8$	$f(9) = -36$	negative	below x-axis

$f(x) > 0$ for $\{x | x < -3 \text{ or } 6 < x < 8\}$
$f(x) < 0$ for $\{x | -3 < x < 6 \text{ or } x > 8\}$

37. $f(x) = (x + 3)^4(x - 1)^3$

a. y-intercept, let $x = 0$
$f(0) = (0 + 3)^4(0 - 1)^3$
$f(0) = (81)(-1) = -81$

b. x-intercept, let $y = 0$
$0 = (x + 3)^4(x - 1)^3$
$x + 3 = 0$ or $x - 1 = 0$
$x = -3$ or $x = 1$

c.

Interval	Test Value	Sign of $f(x)$	Location of Graph
$x < -3$	$f(-4) = -125$	negative	below x-axis
$-3 < x < 1$	$f(0) = -81$	negative	below x-axis
$x > 1$	$f(2) = 625$	positive	above x-axis

$f(x) > 0$ for $\{x | x > 1\}$
$f(x) < 0$ for $\{x | x < -3 \text{ or } -3 < x < 1\}$

39. $f(x) = x(x - 6)^2(x + 4)$

a. y-intercept, let $x = 0$
$f(0) = 0(0 - 6)^2(0 + 4)^2$
$f(0) = 0$

b. x-intercept, let $y = 0$
$0 = x(x - 6)^2(x + 4)$
$x = 0$ or $x - 6 = 0$ or $x + 4 = 0$
$x = 0$ or $x = 6$ or $x = -4$

c.

Interval	Test Value	Sign of $f(x)$	Location of Graph
$x < -4$	$f(-5) = 605$	positive	above x-axis
$-4 < x < 0$	$f(-3) = -243$	negative	below x-axis
$0 < x < 6$	$f(1) = 125$	positive	above x-axis
$x > 6$	$f(7) = 77$	positive	above x-axis

$f(x) > 0$ for $\{x | x < -4 \text{ or } 0 < x < 6 \text{ or } x > 6\}$
$f(x) < 0$ for $\{x | -4 < x < 0\}$

41. $f(x) = x^2(2 - x)(x + 3)$

a. y-intercept, let $x = 0$
$f(0) = 0^2(2 - 0)(0 + 3) = 0$

b. x-intercepts, let $y = 0$
$0 = x^2(2 - x)(x + 3)$
$x^2 = 0$ or $2 - x = 0$ or $x + 3 = 0$
$x = 0$ or $x = 2$ or $x = -3$

c.

Interval	Test Value	Sign of $f(x)$	Location of Graph
$x < -3$	$f(-4) = -96$	negative	below x-axis
$-3 < x < 0$	$f(-1) = 6$	positive	above x-axis
$0 < x < 2$	$f(1) = 4$	positive	above x-axis
$x > 2$	$f(3) = -54$	negative	below x-axis

$f(x) > 0$ for $\{x| -3 < x < 0 \text{ or } 0 < x < 2\}$
$f(x) < 0$ for $\{x|x < -3 \text{ or } x > 2\}$

Further Investigations

45a) $x^3 + x - 6 = 0$
There is one variation in sign.
So, there is one positive solution.

Replacing x with $-x$
$(-x)^3 + (-x) - 6$
$-x^3 - x - 6$
There is no variations in sign, so there are no negative solutions.

Therefore, there is 1 positive solution and 2 nonreal complex solutions.

Possible rational solutions are:
$\pm 1, \pm 2, \pm 3$, or ± 6.

	1	0	1	−6
1	1	1	2	−4
2	1	2	5	4

There is an irrational root between 1 and 2.

	1	0	1	−6
1.5	1	1.5	3.25	−1.125
1.6	1	1.6	3.56	−0.304
1.7	1	1.7	3.89	0.613

There is an irrational root between 1.6 and 1.7.

	1	0	1	−6
1.63	1	1.63	3.6569	−0.039
1.64	1	1.64	3.6896	0.051

There is an irrational solution between 1.63 and 1.64. Therefore 1.6 can be used for an approximation to the nearest tenth.

b) $x^3 - 6x - 6 = 0$
There is one variation in sign.
So there is 1 positive solution.

Replacing x with $-x$:
$(-x)^3 - 6(-x) - 6$
$-x^3 + 6x - 6$
There are two variations in sign.
So there are 2 or 0 negative solutions.

Possible rational solutions are:
$\pm 1, \pm 2, \pm 3$ or ± 6

	1	0	−6	−6
−6	1	−6	30	−186
−3	1	−3	3	−15
−2	1	−2	−2	−2
−1	1	−1	−5	−1
1	1	1	−5	−11
2	1	2	−2	−10
3	1	3	3	3

There is an irrational solution between 2 and 3.

	1	0	−6	−6
2.7	1	2.7	1.29	−2.517
2.8	1	2.8	1.84	−0.848
2.9	1	2.9	2.41	0.989

There is an irrational solution between 2.8 and 2.9.

	1	0	− 6	− 6
2.84	1	2.84	2.0656	− 0.134
2.85	1	2.85	2.1225	0.049

There is an irrational solution between 2.84 and 2.85. Therefore, 2.8 can be used for an approximation to the nearest tenth.

c) $x^3 - 27x - 60 = 0$
There is one variation in sign.
So there is 1 positive solution.

Replacing x with $-x$
$(-x)^3 - 27(-x) - 60$
$-x^3 + 27x - 60$
There are two variations in sign.
So there are 2 or 0 negative solutions.
Possible rational solutions are:
$\pm 1, \pm 2, \pm 3, \pm 4, \pm 5, \pm 6, \pm 10,$
$\pm 12, \pm 15, \pm 20, \pm 30, \pm 60$

	1	0	− 27	− 60
− 6	1	− 6	9	− 114
10	1	10	73	670

− 6 is a lower bound.
10 is an upper bound.

	1	0	− 27	− 60
− 5	1	− 5	2	− 50
− 4	1	− 4	− 11	− 16
− 3	1	− 3	− 18	− 6
− 2	1	− 2	− 23	− 14
− 1	1	− 1	− 26	− 34
1	1	1	− 26	− 86
2	1	2	− 23	− 106
3	1	3	− 18	− 114
4	1	4	− 11	− 104
5	1	5	− 2	− 70
6	1	6	9	− 6
10	1	10	73	670

There is an irrational solution between 6 and 10.

	1	0	− 27	− 60
6.0	1	6.0	9	− 6
6.1	1	6.1	10.21	2.281

There is an irrational solution between 6.0 and 6.1.

	1	0	− 27	− 60
6.07	1	6.07	9.84	− 0.241
6.08	1	6.08	9.97	0.596

There is an irrational solution between 6.07 and 6.08.
Therefore, 6.1 can be used for an approximation to the nearest tenth.

d) $x^3 - x^2 - x - 1 = 0$
There is one variation in sign.
So, there is one positive solution.

Replacing x with $-x$:
$(-x)^3 - (-x)^2 - (-x) - 1$
$-x^3 - x^2 + x - 1$
There are two variations in sign,
so there are 2 or 0 negative solutions.

Possible rational solutions are:
± 1

	1	− 1	− 1	− 1
− 1	1	− 2	1	− 2
1	1	0	− 1	− 2
2	1	1	1	1

− 1 is a lower bound.
2 is an upper bound.

Since $f(1) = -2$ and $f(2) = 1$, there is an irrational root between 1 and 2.

	1	− 1	− 1	− 1
1.7	1	0.7	0.19	− 0.68
1.8	1	0.8	0.44	− 0.21
1.9	1	0.9	0.71	0.349

There is an irrational root between 1.8 and 1.9.

	1	− 1	− 1	− 1
1.83	1	0.63	0.52	− 0.05
1.84	1	0.84	0.55	0.003

There is an irrational root
between 1.83 and 1.84.
Therefore 1.8 can be used for
an approximation to the nearest tenth.

e) $x^3 - 2x - 10 = 0$
There is one variation in sign.
So there is 1 positive solution.

Replacing x with $-x$:
$(-x)^3 - 2(-x) - 10$
$-x^3 + 2x - 10$
There are two variations in sign.
So there are 2 or 0 negative solutions.

Possible rational solutions are:
$\pm 1, \pm 2, \pm 5, \pm 10$

	1	0	−2	−10
−5	1	−5	23	−125
−2	1	−2	2	−14
−1	1	−1	−1	−9
1	1	1	−1	−11
2	1	2	2	−6
5	1	5	23	105

There is an irrational solution
between 2 and 5.

	1	0	−2	−10
2	1	2	2	−6
3	1	3	7	11

There is an irrational solution
between 2 and 3.

	1	0	−2	−10
2.4	1	2.4	3.76	−0.976
2.5	1	2.5	4.25	0.625

There is an irrational solution
between 2.4 and 2.5.

	1	0	−2	−10
2.46	1	2.46	4.05	−0.033
2.47	1	2.47	4.10	0.129

There is an irrational solution
between 2.46 and 2.47.
Therefore, 2.5 can be used for
an approximation to the nearest tenth.

f) $x^3 - 5x^2 - 1 = 0$
There is one variation in sign.
So there is 1 positive solution.

Replacing x with $-x$:
$(-x)^3 - 5(-x)^2 - 1$
$-x^3 - 5x^2 - 1$
There are no variations in sign.
So, there are no negative
solutions.

A possible rational solution is
$+1$.

	1	−5	0	−1
1	1	−4	−4	−5
5	1	0	0	−1
6	1	1	6	35

There is an irrational solution
between 5 and 6.

	1	−5	0	−1
5.0	1	0	0	−1
5.1	1	0.1	0.51	1.601

There is an irrational solution
between 5.0 and 5.1.

	1	−5	0	−1
5.03	1	0.03	0.151	−0.241
5.04	1	0.04	0.202	0.016

There is an irrational solution
between 5.03 and 5.04.
Therefore, 5.0 can be used for
an approximation to the nearest tenth.

PROBLEM SET 9.5 Graphing Rational Functions

1. $f(x) = \dfrac{1}{x^2}$

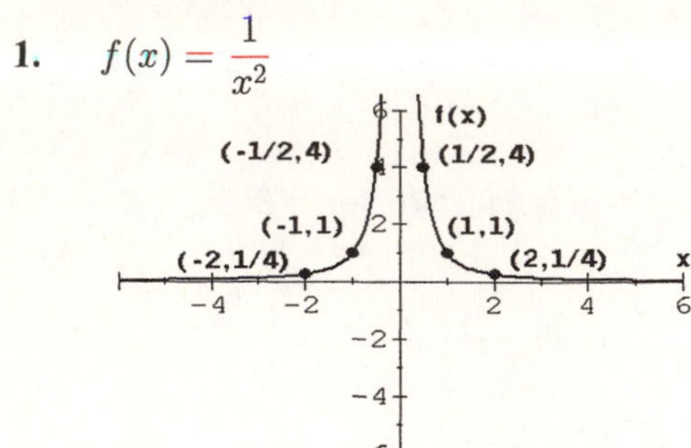

3. $f(x) = \dfrac{-1}{x-3}$

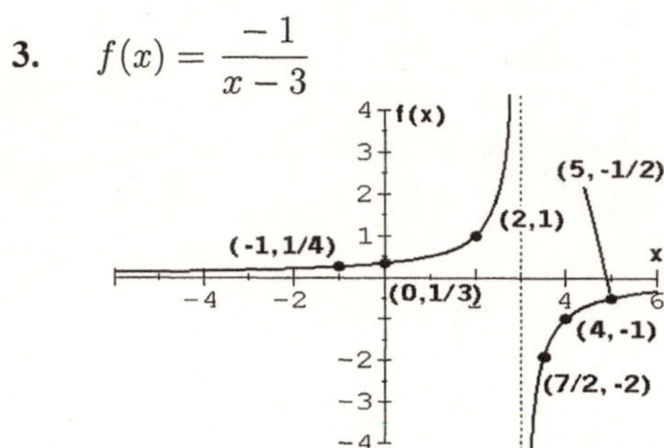

5. $f(x) = \dfrac{-3}{(x+2)^2}$

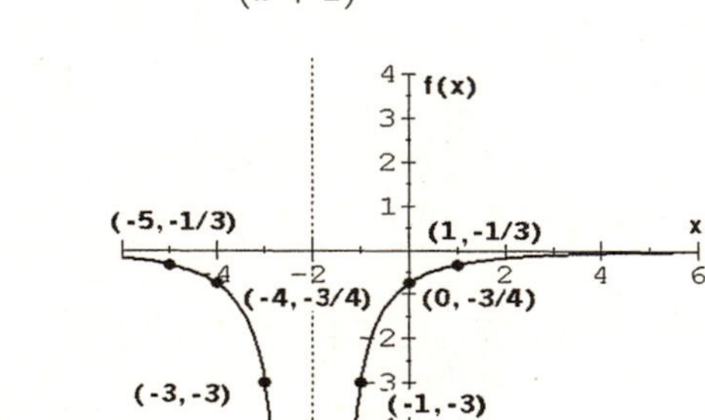

7. $f(x) = \dfrac{2x}{x-1}$

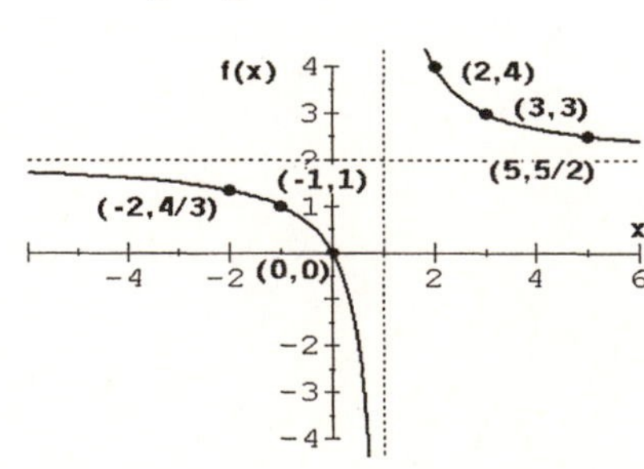

9. $f(x) = \dfrac{-x}{x+1}$

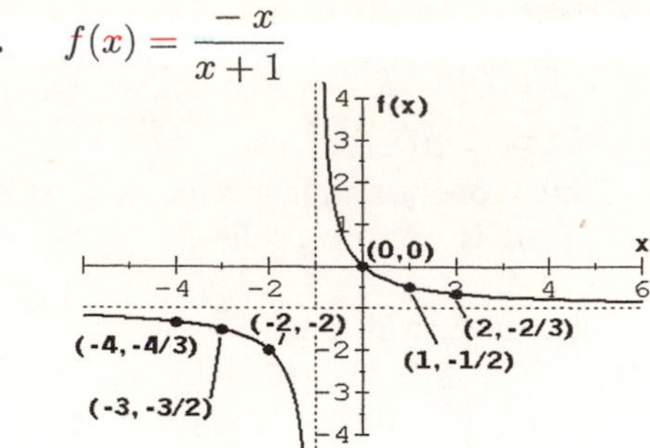

11. $f(x) = \dfrac{-2}{x^2-4}$

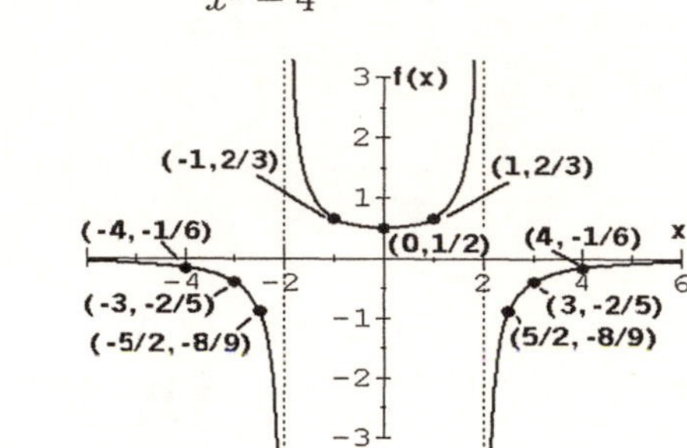

13. $f(x) = \dfrac{3}{(x+2)(x-4)}$

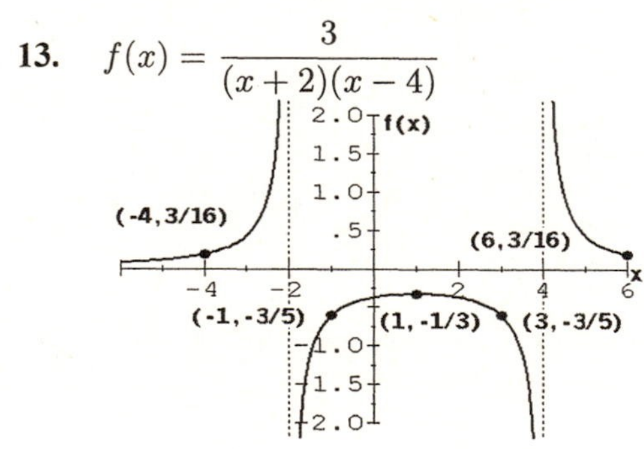

15. $f(x) = \dfrac{-1}{x^2+x-6}$

$$f(x) = \frac{-1}{(x+3)(x-2)}$$

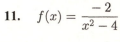

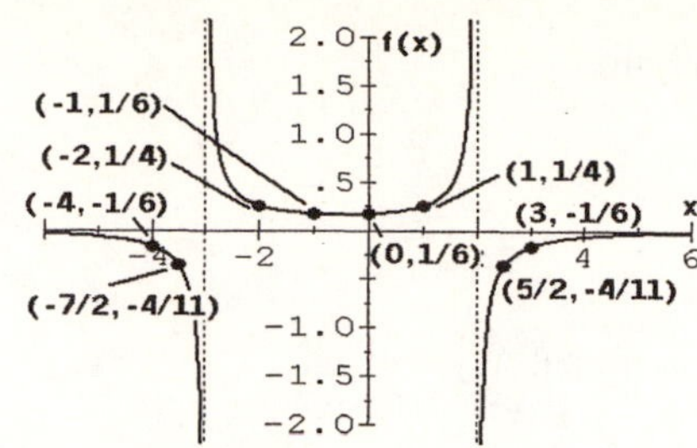

17. $f(x) = \dfrac{2x - 1}{x}$

19. $f(x) = \dfrac{4x^2}{x^2 + 1}$

21. $f(x) = \dfrac{x^2 - 4}{x^2}$

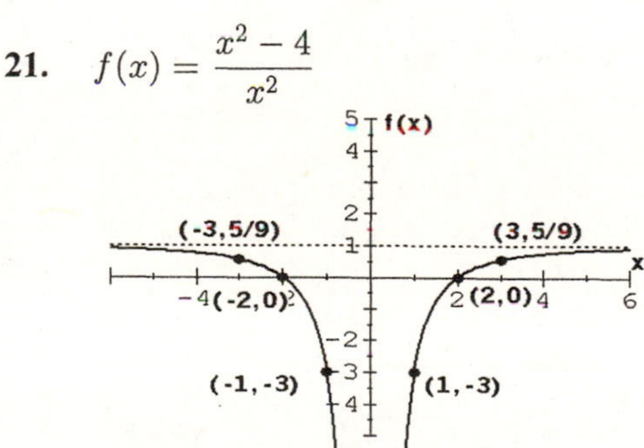

Further Investigations

25a.

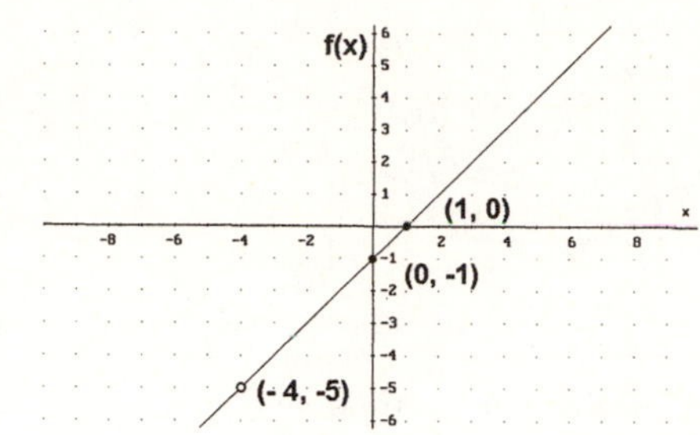

b.

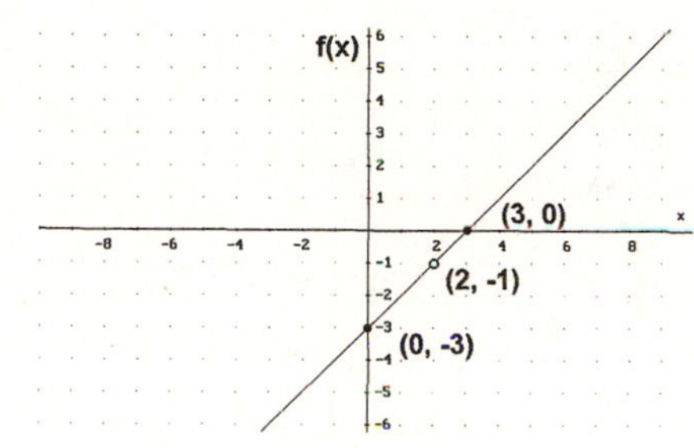

c.

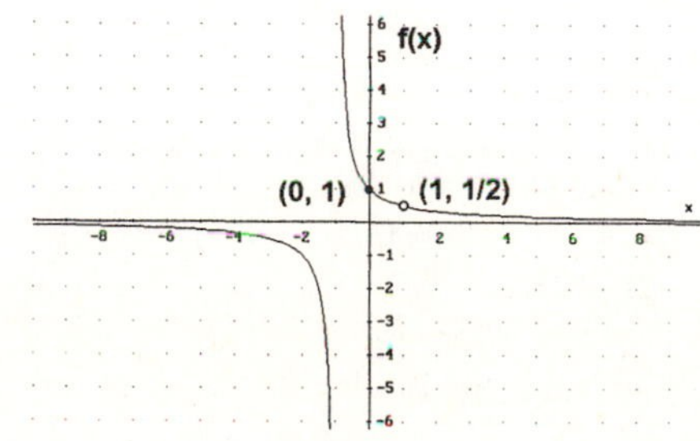

d.

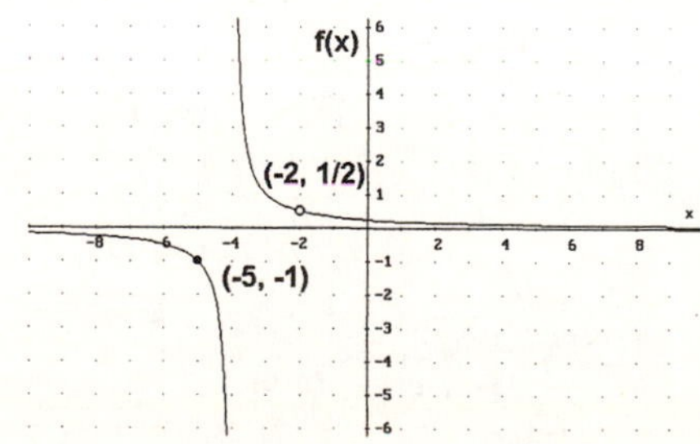

PROBLEM SET 9.6 More on Graphing Rational Functions

1. $f(x) = \dfrac{x^2}{x^2 + x - 2}$

Vertical Asymptotes:
$x^2 + x - 2 = 0$
$(x + 2)(x - 1) = 0$
$x + 2 = 0$ or $x - 1 = 0$
$x = -2$ or $x = 1$

Horizontal Asymptote:
$f(x) = \dfrac{1}{1 + \dfrac{1}{x} - \dfrac{2}{x^2}}$

$f(x) = 1$

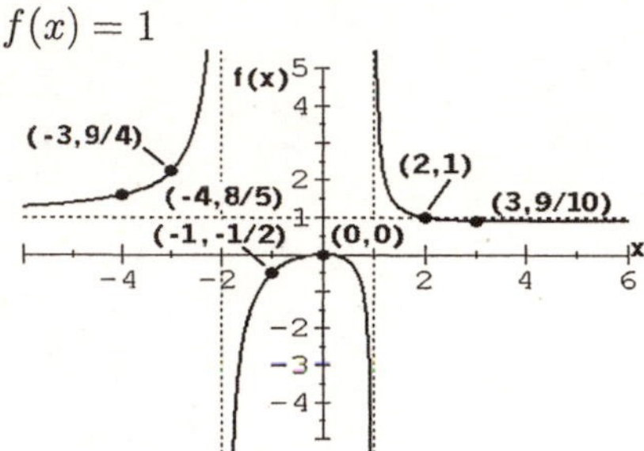

3. $f(x) = \dfrac{2x^2}{x^2 - 2x - 8}$

Vertical Asymptotes:
$x^2 - 2x - 8 = 0$
$(x + 2)(x - 4) = 0$
$x + 2 = 0$ or $x - 4 = 0$
$x = -2$ or $x = 4$

Horizontal Asymptote:

$f(x) = \dfrac{2}{1 - \dfrac{2}{x} - \dfrac{8}{x^2}}$

$f(x) = 2$

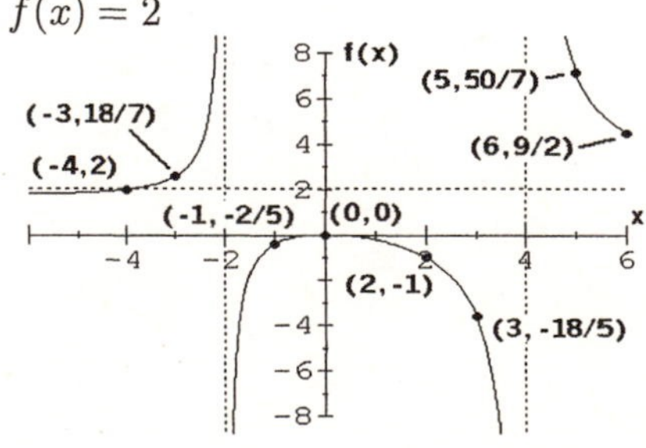

5. $f(x) = \dfrac{-x}{x^2 - 1}$

Vertical Asymptotes:
$x^2 - 1 = 0$
$(x + 1)(x - 1) = 0$
$x + 1 = 0$ or $x - 1 = 0$
$x = -1$ or $x = 1$

Horizontal Asymptote:
$f(x) = \dfrac{\dfrac{-1}{x}}{1 - \dfrac{1}{x^2}}$

$f(x) = 0$

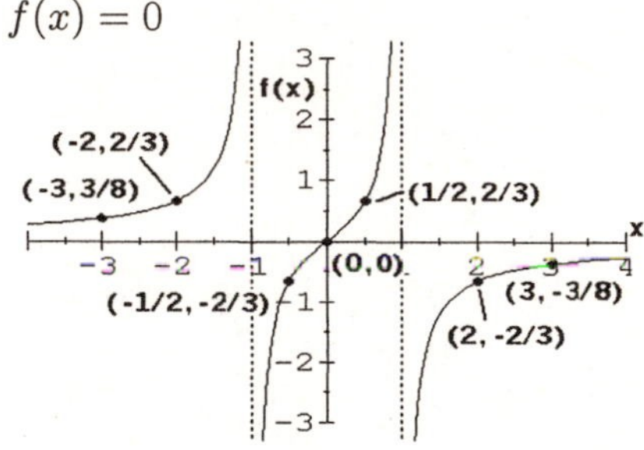

7. $f(x) = \dfrac{x}{x^2 + x - 6}$

Vertical Asymptotes:
$x^2 + x - 6 = 0$
$(x + 3)(x - 2) = 0$
$x + 3 = 0$ or $x - 2 = 0$
$x = -3$ or $x = 2$

Horizontal Asymptote:
$f(x) = \dfrac{\dfrac{1}{x}}{1 + \dfrac{1}{x} - \dfrac{6}{x^2}}$

$f(x) = 0$

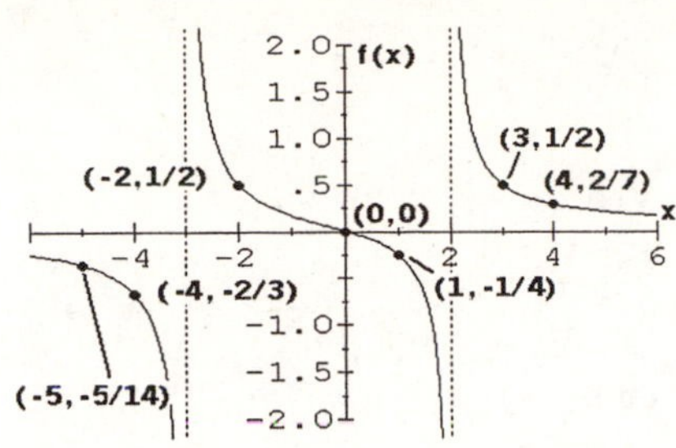

9. $f(x) = \dfrac{x^2}{x^2 - 4x + 3}$

Vertical Asymptotes:
$x^2 - 4x + 3 = 0$
$(x-3)(x-1) = 0$
$x - 3 = 0$ or $x - 1 = 0$
$x = 3$ or $x = 1$

Horizontal Asymptote:
$$f(x) = \frac{1}{1 - \frac{4}{x} + \frac{3}{x^2}}$$
$f(x) = 1$

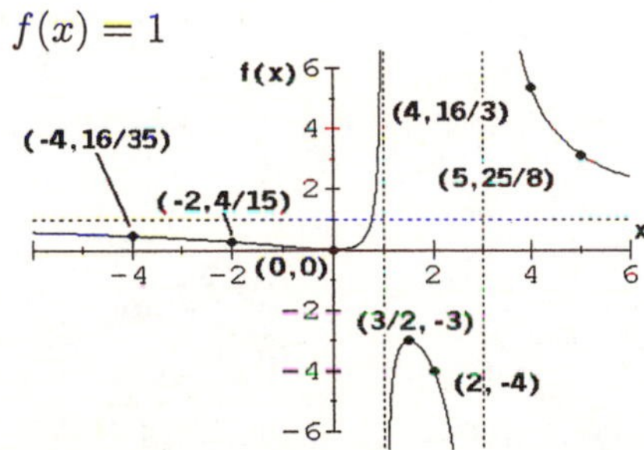

11. $f(x) = \dfrac{x}{x^2 + 2}$

Vertical Asymptotes: none

Horizontal Asymptotes:
$$f(x) = \frac{\frac{1}{x}}{1 + \frac{2}{x^2}}$$
$f(x) = 0$

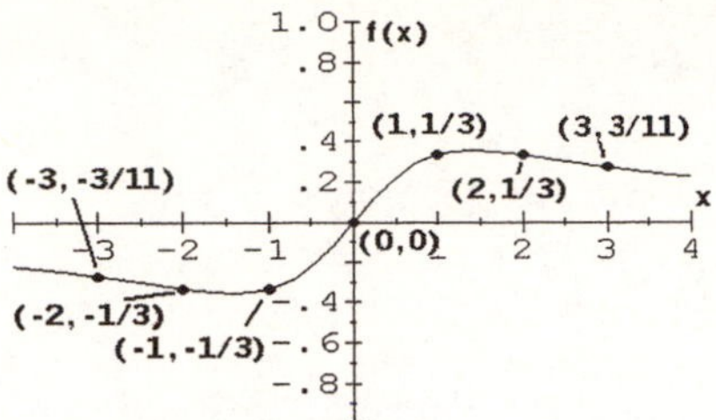

13. $f(x) = \dfrac{-4x}{x^2 + 1}$

Vertical Asymptotes: none
Horizontal Asymptotes:
$$f(x) = \frac{\frac{-4}{x}}{1 + \frac{1}{x^2}}$$
$f(x) = 0$

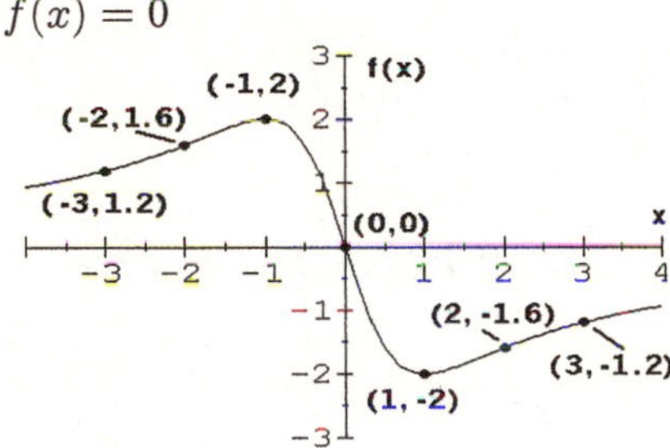

15. $f(x) = \dfrac{x^2 + 2}{x - 1}$

Vertical Asymptotes:
$x - 1 = 0$
$x = 1$

Oblique Asymptote:
$$\frac{x^2 + 2}{x - 1} = x + 1 + \frac{3}{x - 1}$$

Asymptote is $f(x) = x + 1$.

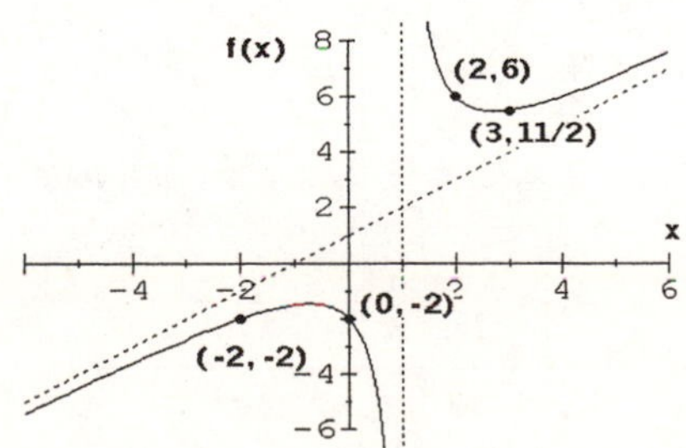

Problem Set 9.6

17. $f(x) = \dfrac{x^2 - x - 6}{x + 1}$

Vertical Asymptotes:
$x + 1 = 0$
$x = -1$

Oblique Asymptote:
$\dfrac{x^2 - x - 6}{x + 1} = x - 2 + \dfrac{-4}{x + 1}$

Asymptote is $f(x) = x - 2$.

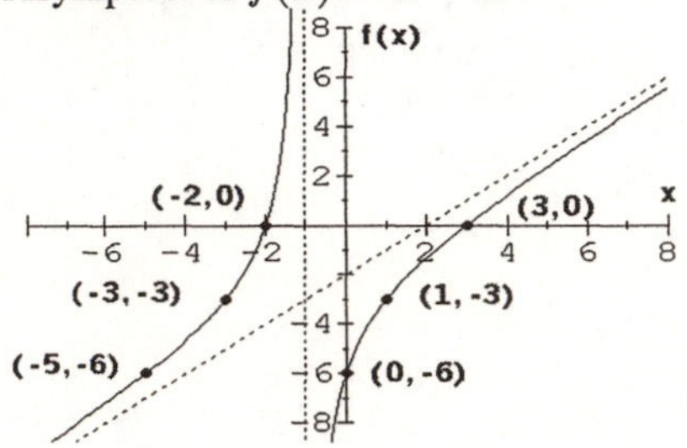

19. $f(x) = \dfrac{x^2 + 1}{1 - x}$

Vertical Asymptotes:
$1 - x = 0$
$x = 1$

Oblique Asymptote:
$\dfrac{x^2 + 1}{1 - x} = -x - 1 + \dfrac{2}{1 - x}$

Asymptote is $f(x) = -x - 1$.

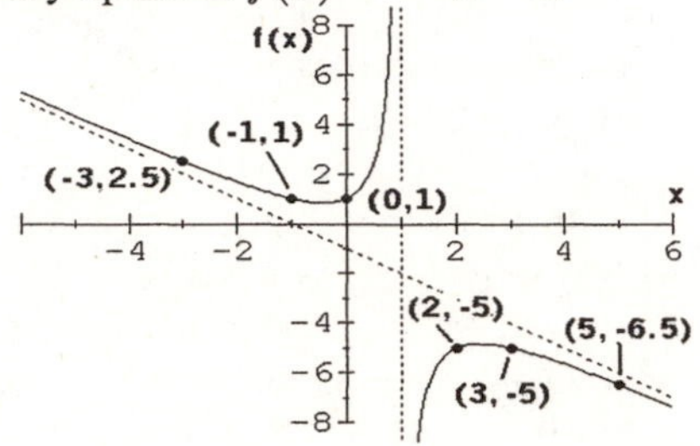

CHAPTER 9 Review Problem Set

1. $(3x^3 - 4x^2 + 6x - 2) \div (x - 1)$

1	3	−4	6	−2
		3	−1	5
	3	−1	5	3

Q: $3x^2 - x + 5$
R: 3

2. $(5x^3 + 7x^2 - 9x + 10) \div (x + 2)$

−2	5	7	−9	10
		−10	6	6
	5	−3	−3	16

Q: $5x^2 - 3x - 3$
R: 16

3. $(-2x^4 + x^3 - 2x^2 - x - 1) \div (x + 4)$

−4	−2	1	−2	−1	−1
		8	−36	152	−604
	−2	9	−38	151	−605

Q: $-2x^3 + 9x^2 - 38x + 151$ R: -605

4. $(-3x^4 - 5x^2 + 9) \div (x - 3)$

3	−3	0	−5	0	9
		−9	−27	−96	−288
	−3	−9	−32	−96	−279

Q: $-3x^3 - 9x^2 - 32x - 96$
R: -279

5. $f(x) = 4x^5 - 3x^3 + x^2 - 1$
$f(1) = 4(1)^5 - 3(1)^3 + (1)^2 - 1$
$f(1) = 4 - 3 + 1 - 1 = 1$

6. $f(x) = 4x^3 - 7x^2 + 6x - 8$

−3	4	−7	6	−8
		−12	57	−189
	4	−19	63	−197

$f(-3) = -197$

7. $f(x) = -x^4 + 9x^2 - x - 2$

-2	-1	0	9	-1	-2
		2	-4	-10	22
	-1	2	5	-11	20

$f(-2) = 20$

8. $f(x) = x^4 - 9x^3 + 9x^2 - 10x + 16$

8	1	-9	9	-10	16
		8	-8	8	-16
	1	-1	1	-2	0

$f(8) = 0$

9.

-2	2	1	-7	-2
		-4	6	2
	2	-3	-1	0

Since $f(-2) = 0$, $(x + 2)$ is a factor of $(2x^3 + x^2 - 7x - 2)$.

10.

3	1	5	-7	-1	3
		3	24	51	150
	1	8	17	50	153

$f(3) = 153$
So $(x - 3)$ is not a factor of $x^4 + 5x^3 - 7x^2 - x + 3$.

11.

4	1	0	0	0	0	-1024
		4	16	64	256	1024
	1	4	16	64	256	0

Since $f(4) = 0$, $(x - 4)$ is a factor of $(x^5 - 1024)$.

12.

-1	1	0	0	0	0	1
		-1	1	-1	1	-1
	1	-1	1	-1	1	0

$f(-1) = 0$
So $(x + 1)$ is a factor of $x^5 + 1$.

13. $x^3 - 3x^2 - 13x + 15 = 0$
c: $\pm 1, \pm 3, \pm 5, \pm 15$
d: ± 1
$\frac{c}{d}$: $\pm 1, \pm 3, \pm 5, \pm 15$

1	1	-3	-13	15
		1	-2	-15
	1	-2	-15	0

$(x - 1)(x^2 - 2x - 15) = 0$
$(x - 1)(x - 5)(x + 3) = 0$
$x - 1 = 0$ or $x - 5 = 0$ or $x + 3 = 0$
$x = 1$ or $x = 5$ or $x = -3$
The solution set is $\{-3, 1, 5\}$.

14. $8x^3 + 26x^2 - 17x - 35 = 0$
c: $\pm 1, \pm 5, \pm 7, \pm 35$
d: $\pm 1, \pm 2, \pm 4, \pm 8$
$\frac{c}{d}$: $\pm 1, \pm 5, \pm 7, \pm 35, \pm \frac{1}{2}, \pm \frac{5}{2},$

$\pm \frac{7}{2}, \pm \frac{35}{2}, \pm \frac{1}{4}, \pm \frac{5}{4}, \pm \frac{7}{4},$

$\pm \frac{35}{4}, \pm \frac{1}{8}, \pm \frac{5}{8}, \pm \frac{7}{8}, \pm \frac{35}{8}$

-1	8	26	-17	-35
		-8	-18	35
	8	18	-35	0

$(x + 1)(8x^2 + 18x - 35) = 0$
$(x + 1)(2x + 7)(4x - 5) = 0$
$x + 1 = 0$ or $2x + 7 = 0$ or $4x - 5 = 0$
$x = -1$ or $2x = -7$ or $4x = 5$

$x = -\frac{7}{2}$ or $x = \frac{5}{4}$

The solution set is $\left\{-\frac{7}{2}, -1, \frac{5}{4}\right\}$.

15. $x^4 - 5x^3 + 34x^2 - 82x + 52 = 0$
c: $\pm 1, \pm 2, \pm 4, \pm 13, \pm 26, \pm 52$
d: ± 1
$\frac{c}{d}$: $\pm 1, \pm 2, \pm 4, \pm 13, \pm 26, \pm 52$

1	1	-5	34	-82	52
		1	-4	30	-52
	1	-4	30	-52	0

$(x - 1)(x^3 - 4x^2 + 30x - 52) = 0$
c: $\pm 1, \pm 2, \pm 4, \pm 13, \pm 26, \pm 52$
d: ± 1

$\frac{c}{d}$: $\pm 1, \pm 2, \pm 4, \pm 13, \pm 26, \pm 52$

$$\begin{array}{r|rrrr} 2 & 1 & -4 & 30 & -52 \\ & & 2 & -4 & 52 \\ \hline & 1 & -2 & 26 & 0 \end{array}$$

$(x-1)(x-2)(x^2-2x+26)=0$
$x-1=0$ or $x-2=0$ or $x^2-2x+26=0$
$x=1$ or $x=2$ or

$$x=\frac{-(-2)\pm\sqrt{(-2)^2-4(1)(26)}}{2(1)}$$
$$x=\frac{2\pm\sqrt{4-104}}{2}$$
$$x=\frac{2\pm\sqrt{-100}}{2}$$
$$x=\frac{2\pm 10i}{2}$$
$$x=1\pm 5i$$

The solution set is $\{1, 2, 1\pm 5i\}$.

16. $x^3-4x^2-10x+4=0$
c: $\pm 1, \pm 2, \pm 4$
d: ± 1
$\frac{c}{d}$: $\pm 1, \pm 2, \pm 4$

$$\begin{array}{r|rrrr} -2 & 1 & -4 & -10 & 4 \\ & & -2 & 12 & -4 \\ \hline & 1 & -6 & 2 & 0 \end{array}$$

$(x+2)(x^2-6x+2)=0$
$x+2=0$ or $x^2-6x+2=0$

$$x=-2 \text{ or } x=\frac{6\pm\sqrt{(-6)^2-4(1)(2)}}{2(1)}$$
$$x=\frac{6\pm\sqrt{28}}{2}$$
$$x=\frac{6\pm 2\sqrt{7}}{2}$$
$$x=3\pm\sqrt{7}$$

The solution set is $\{-2, 3\pm\sqrt{7}\}$.

17. $4x^4-3x^3+2x^2+x+4=0$
There are two variations in sign, so there are two or no positive real solutions.

Replacing x with $-x$:
$4(-x)^4-3(-x)^3+2(-x)^2-x+4$
$4x^4+3x^3+2x^2-x+4$
There are two variations in sign, so there are two or no negative real solutions.

So, there are 4 cases.
I: 2 positive and 2 negative real solutions
or
II: 2 positive real solutions and 2 nonreal complex solutions.
or
III: 2 negative real solutions and 2 nonreal complex solutions
or
IV: 4 nonreal complex solutions.

18. $x^5+3x^3+x+7=0$
There are no variations in sign.
So there are no positive solutions.

Replace x with $-x$:
$(-x)^5+3(-x)^3+(-x)+7$
$-x^5-3x^5-x+7$
There is one variation in sign.
So there is 1 negative solution.
Therefore, there is 1 negative and 4 nonreal complex solutions.

19. $f(x)=-(x-2)^3+3$

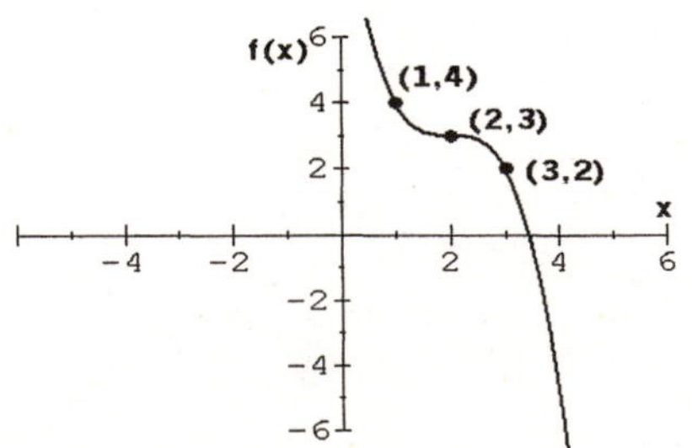

20. $f(x) = (x+3)(x-1)(3-x)$

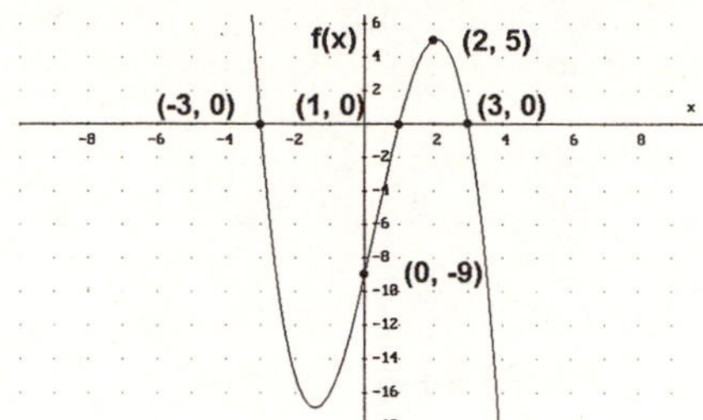

21. $f(x) = x^4 - 4x^2$

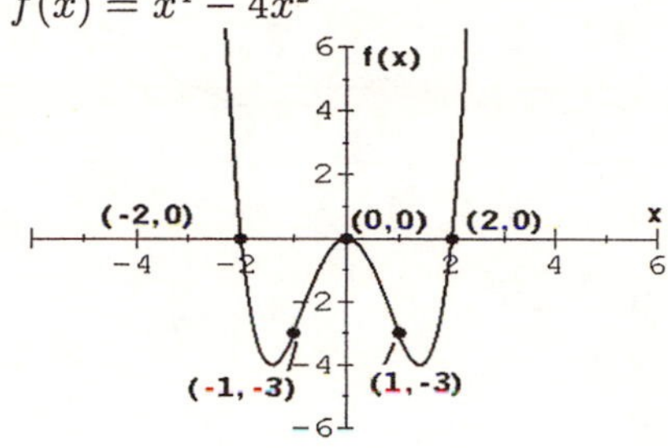

22. $f(x) = x^3 - 4x^2 + x + 6$

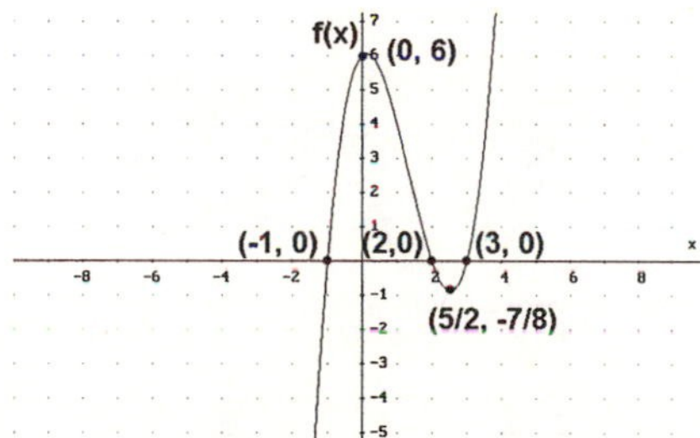

23. $f(x) = \dfrac{2x}{x-3}$

$$f(x) = \frac{\frac{2x}{x}}{\frac{x}{x} - \frac{3}{x}}$$

$$f(x) = \frac{2}{1 - \frac{3}{x}}$$

Horizontal asymptote of $f(x) = \dfrac{2}{1} = 2$.

Vertical asymptote of $x = 3$.

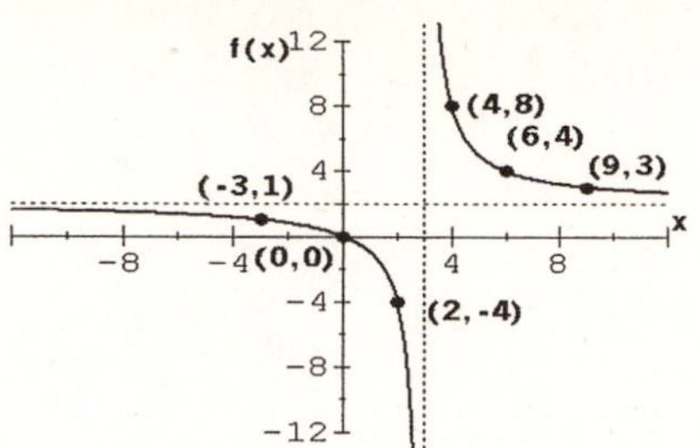

24. $f(x) = \dfrac{-3}{x^2+1}$

$$f(x) = \frac{-\frac{3}{x^2}}{\frac{x^2}{x^2} + \frac{1}{x^2}}$$

$$f(x) = \frac{-\frac{3}{x^2}}{1 + \frac{1}{x^2}}$$

Horizontal asymptote is $f(x) = 0$.
There is no vertical asymptote.

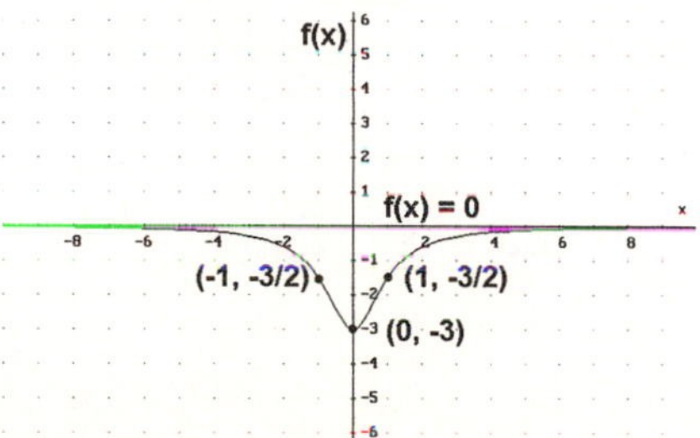

25. $f(x) = \dfrac{-x^2}{x^2 - x - 6}$

$$f(x) = \frac{-x^2}{(x-3)(x+2)}$$

Vertical Asymptotes:
$(x-3)(x+2) = 0$
$x - 3 = 0$ or $x + 2 = 0$
$x = 3$ or $x = -2$

$$f(x) = \frac{-x^2}{x^2 - x - 6} = \frac{-\frac{x^2}{x^2}}{\frac{x^2}{x^2} - \frac{x}{x^2} - \frac{6}{x^2}}$$

$$f(x) = \frac{-1}{1 - \frac{1}{x} - \frac{6}{x^2}}$$

Horizontal asymptote is
$f(x) = -1.$

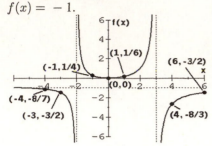

26. $f(x) = \dfrac{x^2+3}{x+1}$
Vertical Asymptotes:
$x + 1 = 0$
$x = -1$

Oblique asymptote :

-1	1	0	3
		-1	1
	1	-1	4

$$f(x) = \frac{x^2+3}{x+1} = x - 1 + \frac{4}{x+1}$$

Oblique asymptote is $f(x) = x - 1$.

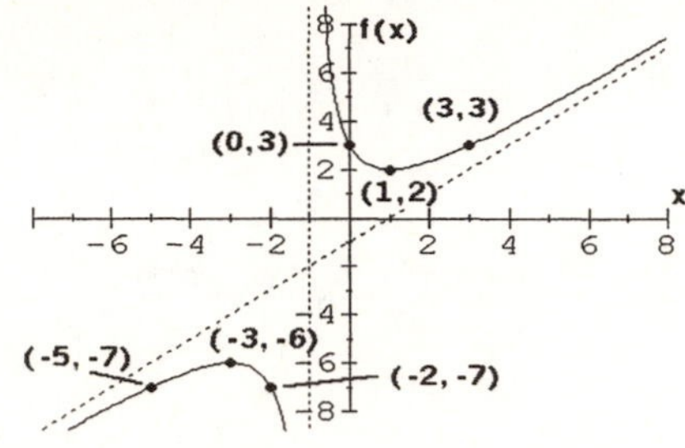

CHAPTER 9 Test

1. $(3x^3 + 5x^2 - 14x - 6) \div (x+3)$

-3	3	5	-14	-6
		-9	12	6
	3	-4	-2	0

Q: $3x^2 - 4x - 2$
R: 0

2. $(4x^4 - 7x^2 - x + 4) \div (x-2)$

2	4	0	-7	-1	4
		8	16	18	34
	4	8	9	17	38

Q: $4x^3 + 8x^2 + 9x + 17$
R: 38

3. $f(x) = x^5 - 8x^4 + 9x^3 - 13x^2 - 9x - 10$

7	1	-8	9	-13	-9	-10
		7	-7	14	7	-14
	1	-1	2	1	-2	-24

$f(7) = -24$

4. $f(x) = 3x^4 + 20x^3 - 6x^2 + 9x + 19$

-7	3	20	-6	9	19
		-21	7	-7	-14
	3	-1	1	2	5

$f(-7) = 5$

5. $f(x) = x^5 - 35x^3 - 32x + 15$

6	1	0	-35	0	-32	15
		6	36	6	36	24
	1	6	1	6	4	39

$f(6) = 39$

6. $f(x) = 3x^3 - 11x^2 - 22x - 20$

5	3	-11	-22	-20
		15	20	-10
	3	4	-2	-30

Since $f(5) = -30$, $(x-5)$ is not a factor of $3x^3 - 11x^2 - 22x - 20$.

7. $f(x) = 5x^3 + 9x^2 - 9x - 17$

-2	5	9	-9	-17
		-10	2	14
	5	-1	-7	-3

Since $f(-2) = -3$, $(x+2)$ is not a factor of $(5x^3 + 9x^2 - 9x - 17)$.

8.

-3	1	0	-16	-17	12
		-3	9	21	-12
	1	-3	-7	4	0

Since $f(-3) = 0$, $(x+3)$ is a factor of $x^4 - 16x^2 - 17x + 12$.

9.

6	1	0	-2	3	-12
		6	36	204	1242
	1	6	34	207	1230

Since $f(6) = 1230$, $(x-6)$ is not a factor of $(x^4 - 2x^2 + 3x - 12)$.

10. $x^3 - 13x + 12 = 0$
c: $\pm 1, \pm 2, \pm 3, \pm 4, \pm 6, \pm 12$
d: ± 1
$\frac{c}{d}$: $\pm 1, \pm 2, \pm 3, \pm 4, \pm 6, \pm 12$

1	1	0	-13	12
		1	1	-12
	1	1	-12	0

$(x-1)(x^2 + x - 12) = 0$
$(x-1)(x+4)(x-3) = 0$
$x - 1 = 0$ or $x + 4 = 0$ or $x - 3 = 0$
$x = 1$ or $x = -4$ or $x = 3$
The solution set is $\{-4, 1, 3\}$.

11. $2x^3 + 5x^2 - 13x - 4 = 0$
c: $\pm 1, \pm 2, \pm 4$
d: $\pm 1, \pm 2$
$\frac{c}{d}$: $\pm 1, \pm 2, \pm 4, \pm \frac{1}{2}$

-4	2	5	-13	-4
		-8	12	4
	2	-3	-1	0

$(x+4)(2x^2 - 3x - 1) = 0$
$x + 4 = 0$ or $2x^2 - 3x - 1 = 0$

$x = -4$ or $x = \frac{-(-3) \pm \sqrt{(-3)^2 - 4(2)(-1)}}{2(2)}$

$x = \frac{3 \pm \sqrt{17}}{4}$

The solution set is $\left\{-4, \frac{3 \pm \sqrt{17}}{4}\right\}$.

12. $x^4 - 4x^3 - 5x^2 + 38x - 30 = 0$
c: $\pm 1, \pm 2, \pm 3, \pm 10, \pm 15, \pm 30$
d: ± 1
$\frac{c}{d}$: $\pm 1, \pm 2, \pm 3, \pm 10, \pm 15, \pm 30$

1	1	-4	-5	38	-30
		1	-3	-8	30
	1	-3	-8	30	0

$(x-1)(x^3 - 3x^2 - 8x + 30) = 0$
c, d, and $\frac{c}{d}$ are the same as above.

-3	1	-3	-8	30
		-3	18	-30
	1	-6	10	0

$(x-1)(x+3)(x^2 - 6x + 10) = 0$
$x - 1 = 0$ or $x + 3 = 0$ or $x^2 - 6x + 10 = 0$
$x = 1$ or $x = -3$ or

$x = \frac{-(-6) \pm \sqrt{(-6)^2 - 4(1)(10)}}{2(1)}$

$x = \frac{6 \pm \sqrt{-4}}{2}$

$x = \frac{6 \pm 2i}{2}$

$x = 3 \pm i$
The solution set is $\{-3, 1, 3 \pm i\}$.

13. $2x^3 + 3x^2 - 17x + 12 = 0$
c: $\pm 1, \pm 2, \pm 3, \pm 4, \pm 6, \pm 12$
d: $\pm 1, \pm 2$
$\frac{c}{d}$: $\pm 1, \pm 2, \pm 3, \pm 4, \pm 6, \pm 12, \pm \frac{1}{2}, \pm \frac{3}{2}$

$$\begin{array}{r|rrrr} 1 & 2 & 3 & -17 & 12 \\ & & 2 & 5 & -12 \\ \hline & 2 & 5 & -12 & 0 \end{array}$$

$(x-1)(2x^2+5x-12)=0$
$(x-1)(2x-3)(x+4)=0$
$x-1=0$ or $2x-3=0$ or $x+4=0$
$x=1$ or $2x=3$ or $x=-4$

$x=1$ or $x=\frac{3}{2}$ or $x=-4$

The solution set is $\left\{-4, 1, \frac{3}{2}\right\}$.

14. $3x^3-7x^2-8x+20=0$
c: $\pm 1, \pm 2, \pm 4, \pm 5, \pm 10, \pm 20$
d: $\pm 1, \pm 3$
$\frac{c}{d}$: $\pm 1, \pm 2, \pm 4, \pm 5, \pm 10, \pm 20, \pm \frac{1}{3},$
$\pm \frac{2}{3}, \pm \frac{4}{3}, \pm \frac{5}{3}, \pm \frac{10}{3}, \pm \frac{20}{3}$

$$\begin{array}{r|rrrr} 2 & 3 & -7 & -8 & 20 \\ & & 6 & -2 & -20 \\ \hline & 3 & -1 & -10 & 0 \end{array}$$

$(x-2)(3x^2-x-10)=0$
$(x-2)(3x+5)(x-2)=0$
$x-2=0$ or $3x+5=0$
$x=2$ or $3x=-5$
$x=-\frac{5}{3}$

The solution set is $\left\{-\frac{5}{3}, 2\right\}$.

15. $5x^4+3x^3-x^2-9=0$
There is one variation in sign,
so there is one positive real solution.

Replacing x with $-x$:
$5(-x)^4+3(-x)^3-(-x)^2-9$
$5x^4-3x^3-x^2-9$
There is one variation in sign, so
there is one negative real solution.

Therefore, there is 1 positive, 1 negative, and 2 nonreal complex solutions.

16. $f(x)=3x^3+19x^2-14x$

x-intercepts, let $y=0$
$0=3x^3+19x^2-14x$
$0=x(3x^2+19x-14)$
$0=x(3x-2)(x+7)$
$x=0$ or $3x-2=0$ or $x+7=0$
or $3x=2$ or $x=-7$
$x=\frac{2}{3}$

The x-intercepts are -7, 0, and $\frac{2}{3}$.

17. $f(x)=\frac{5x}{x+3}$
Vertical asymptote is $x=-3$.

18. $f(x)=\frac{5x^2}{x^2-4}$

$$f(x)=\frac{\frac{5x^2}{x^2}}{\frac{x^2}{x^2}-\frac{4}{x^2}}$$

$$f(x)=\frac{5}{1-\frac{4}{x^2}}$$

$f(x)=5$ is the horizontal asymptote.

19. $f(x)=\frac{x^2}{x^2+2}$
Replace x with $-x$:

$$f(-x)=\frac{(-x)^2}{(-x)^2+2}$$

$$f(-x)=\frac{x^2}{x^2+2}$$

$f(x)=f(-x)$
It is symmetric to the y-axis.

Replace y with $-y$:

$$-y=\frac{x^2}{x^2+2}$$

$$y=\frac{-x^2}{x^2+2}$$

It is not symmetric to the x-axis.

Replace x with $-x$ and y with $-y$:

$$-y = \frac{(-x)^2}{(-x)^2 + 2}$$

$$-y = \frac{x^2}{x^2 + 2}$$

$$y = \frac{-x^2}{x^2 + 2}$$

It is not symmetric with respect to the origin.

20. $f(x) = \dfrac{-3x}{x^2 + 1}$

Replace x with $-x$:

$$f(-x) = \frac{-3(-x)}{(-x)^2 + 1}$$

$$f(-x) = \frac{3x}{x^2 + 1}$$

$f(x) \neq f(-x)$

It is not symmetric to the $y-$ axis.

Replace y with $-y$:

$$-y = \frac{-3x}{x^2 + 1}$$

$$y = \frac{3x}{x^2 + 1}$$

It is not symmetric to $x-$ axis.

Replace x with $-x$ and y with $-y$:

$$-y = \frac{-3(-x)}{(-x)^2 + 1}$$

$$y = \frac{-3x}{x^2 + 1}$$

It is symmetric with respect to the origin.

21. $f(x) = (2 - x)(x - 1)(x + 1)$

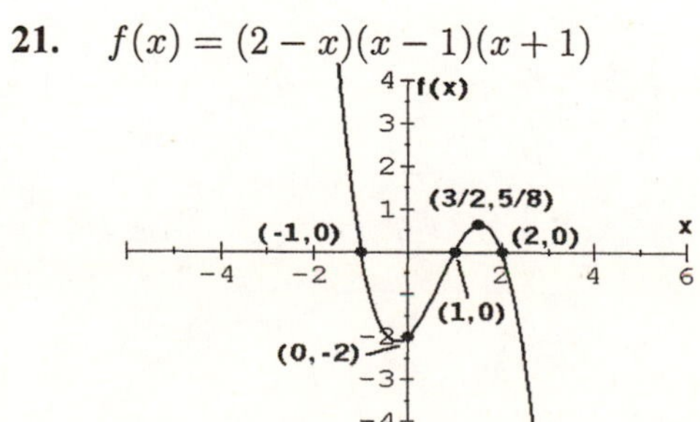

22. $f(x) = -x(x - 3)(x + 2)$

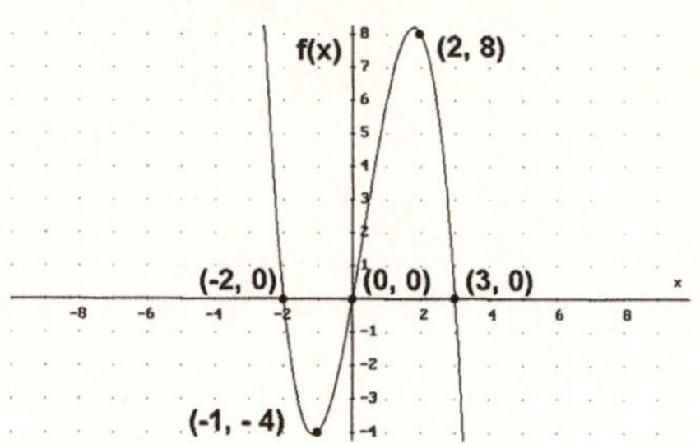

23. $f(x) = \dfrac{-x}{x - 3}$

Vertical asymptote:

$x - 3 = 0$

$x = 3$

Horizontal asymptote:

$$f(x) = \frac{-x}{x - 3}$$

$$f(x) = \frac{-\frac{x}{x}}{\frac{x}{x} - \frac{3}{x}}$$

$$f(x) = \frac{-1}{1 - \frac{3}{x}}$$

Horizontal asymptote is $f(x) = -1$.

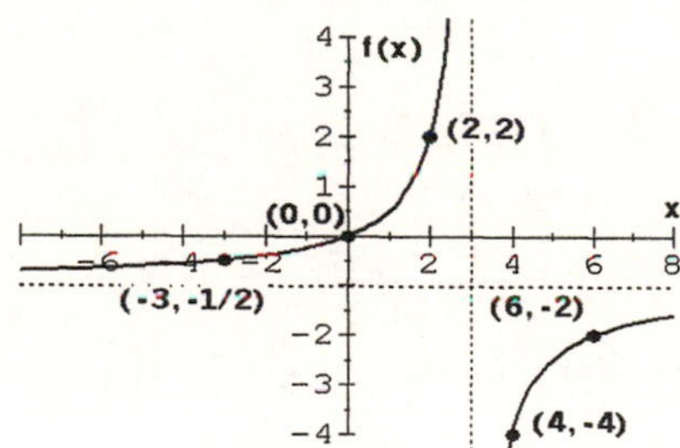

24. $f(x) = \dfrac{-2}{x^2 - 4}$

Vertical asymptotes:

$x^2 - 4 = 0$

$(x - 2)(x + 2) = 0$

$x - 2 = 0$ or $x + 2 = 0$

$x = 2$ or $x = -2$

Horizontal asymptote:

$$f(x)=\frac{-\frac{2}{x^2}}{\frac{x^2}{x^2}-\frac{4}{x^2}}$$

$$f(x)=\frac{-\frac{2}{x^2}}{1-\frac{4}{x^2}}$$

Horizontal asymptote is $f(x)=0$.

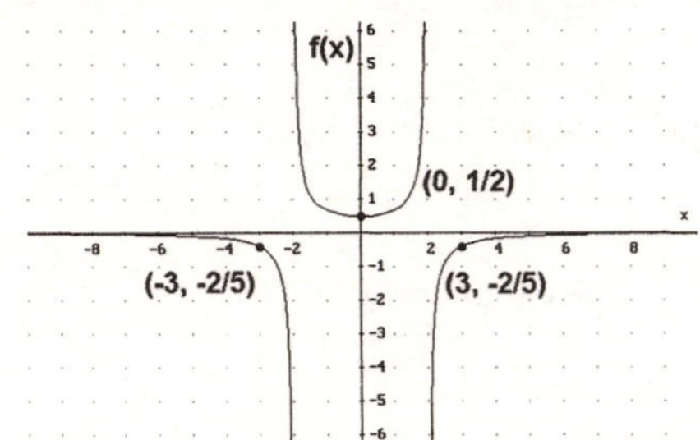

25. $f(x)=\dfrac{4x^2+x+1}{x+1}$

Vertical asymptotes:
$x+1=0$
$x=-1$

Oblique asymptote: $f(x)=\dfrac{4x^2+x+1}{x+1}$

$$\begin{array}{r|rrr} -1 & 4 & 1 & 1 \\ & & -4 & 3 \\ \hline & 4 & -3 & 4 \end{array}$$

$$f(x)=\frac{4x^2+x+1}{x+1}=4x-3+\frac{4}{x+1}$$

Oblique asymptote is $f(x)=4x-3$.

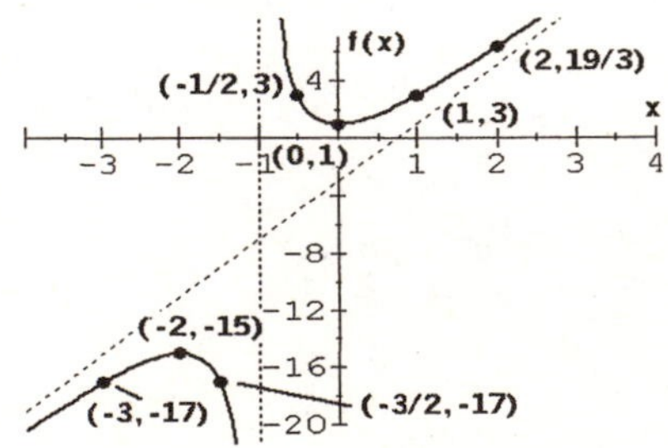

Chapter 10 Exponential and Logarithmic Functions

PROBLEM SET 10.1 Exponents and Exponential Functions

1. $2^x = 64$
$2^x = 2^6$
$x = 6$
The solution set is $\{6\}$.

3. $3^{2x} = 27$
$3^{2x} = 3^3$
$2x = 3$
$x = \frac{3}{2}$
The solution set is $\left\{\frac{3}{2}\right\}$.

5. $\left(\frac{1}{2}\right)^x = \frac{1}{128}$
$\left(\frac{1}{2}\right)^x = \left(\frac{1}{2}\right)^7$
$x = 7$
The solution set is $\{7\}$.

7. $3^{-x} = \frac{1}{243}$
$3^{-x} = \frac{1}{3^5}$
$3^{-x} = 3^{-5}$
$-x = -5$
$x = 5$
The solution set is $\{5\}$.

9. $6^{3x-1} = 36$
$6^{3x-1} = 6^2$
$3x - 1 = 2$
$3x = 3$
$x = 1$
The solution set is $\{1\}$.

11. $\left(\frac{3}{4}\right)^n = \frac{64}{27}$
$\left(\frac{3}{4}\right)^n = \left(\frac{4}{3}\right)^3$
$\left(\frac{3}{4}\right)^n = \left(\frac{3}{4}\right)^{-3}$
$n = -3$
The solution set is $\{-3\}$.

13. $16^x = 64$
$(2^4)^x = 2^6$
$2^{4x} = 2^6$
$4x = 6$
$x = \frac{6}{4} = \frac{3}{2}$
The solution set is $\left\{\frac{3}{2}\right\}$.

15. $27^{4x} = 9^{x+1}$
$(3^3)^{4x} = (3^2)^{x+1}$
$3^{12x} = 3^{2x+2}$
$12x = 2x + 2$
$10x = 2$
$x = \frac{2}{10} = \frac{1}{5}$
The solution set is $\left\{\frac{1}{5}\right\}$.

17. $9^{4x-2} = \frac{1}{81}$
$9^{4x-2} = \frac{1}{9^2}$
$9^{4x-2} = 9^{-2}$
$4x - 2 = -2$
$4x = 0$
$x = 0$
The solution set is $\{0\}$.

19. $10^x = 0.1$
$10^x = 10^{-1}$
$x = -1$
The solution set is $\{-1\}$.

21. $(2^{x+1})(2^x) = 64$

$2^{2x+1} = 2^6$

$2x + 1 = 6$

$2x = 5$

$x = \frac{5}{2}$

The solution set is $\left\{\frac{5}{2}\right\}$.

23. $(27)(3^x) = 9^x$

$(3^3)(3^x) = (3^2)^x$

$3^{x+3} = 3^{2x}$

$x + 3 = 2x$

$3 = x$

The solution set is $\{3\}$.

25. $(4^x)(16^{3x-1}) = 8$

$(2^2)^x(2^4)^{3x-1} = 2^3$

$(2^{2x})(2^{12x-4}) = 2^3$

$2^{14x-4} = 2^3$

$14x - 4 = 3$

$14x = 7$

$x = \frac{7}{14} = \frac{1}{2}$

The solution set is $\left\{\frac{1}{2}\right\}$.

27. $f(x) = 3^x$

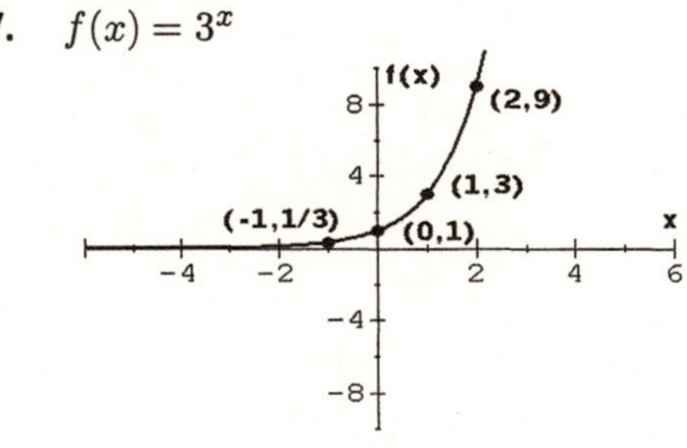

29. $f(x) = \left(\frac{1}{3}\right)^x$

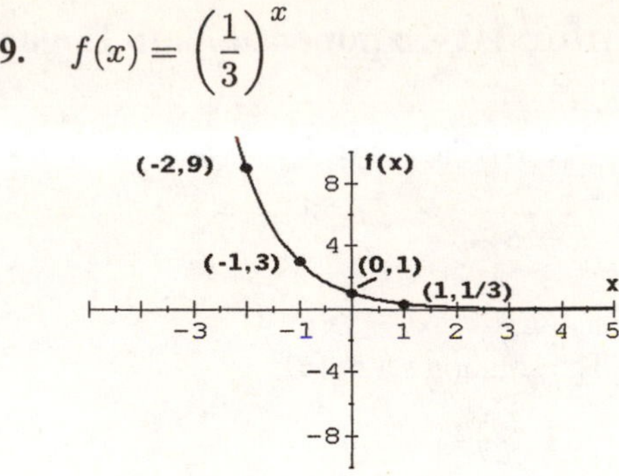

31. $f(x) = \left(\frac{3}{2}\right)^x$

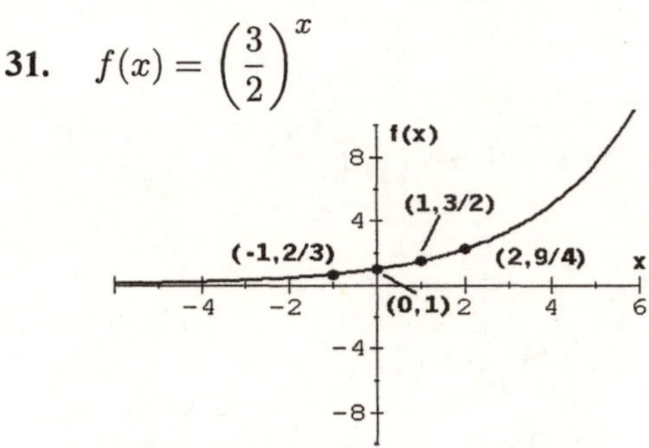

33. $f(x) = 2^x - 3$

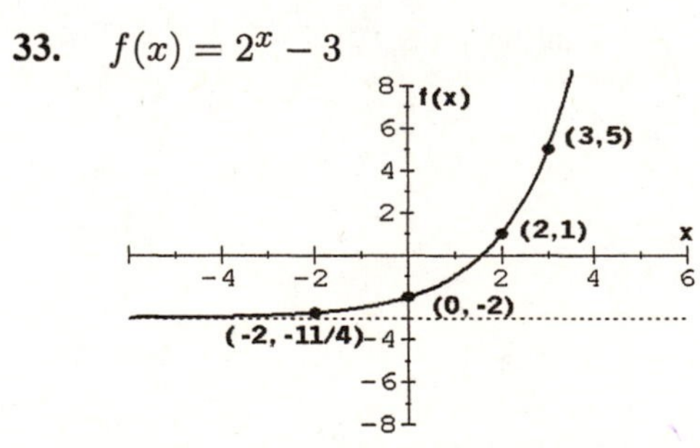

35. $f(x) = 2^{x+2}$

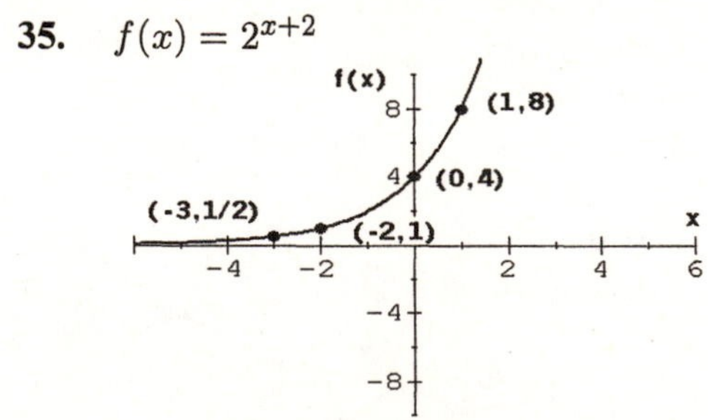

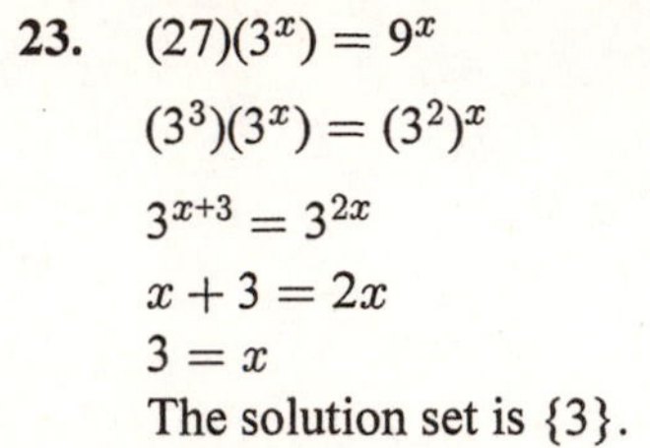

37. $f(x) = -2^x$

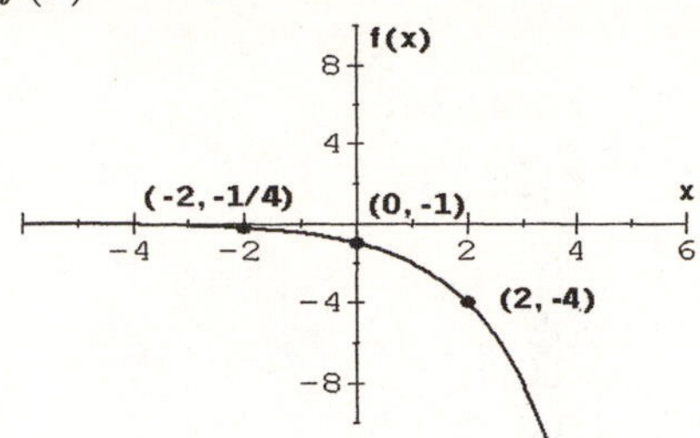

39. $f(x) = 2^{-x-2}$

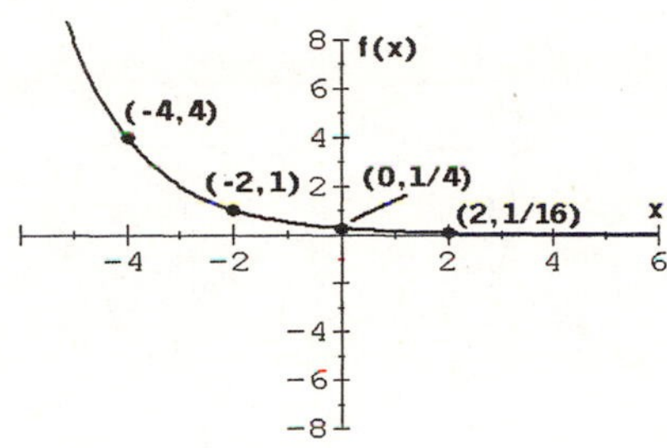

41. $f(x) = 2^{x^2}$

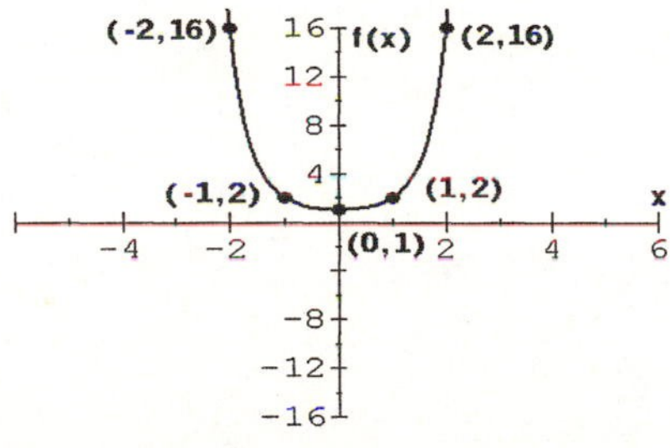

43. $f(x) = 2^{|x|}$

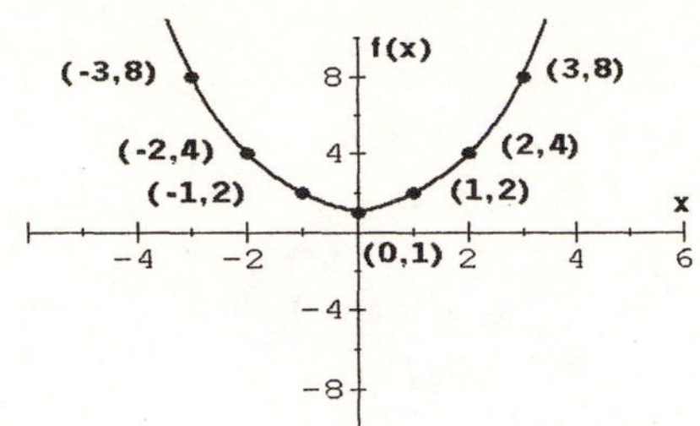

45. $f(x) = 2^x - 2^{-x}$

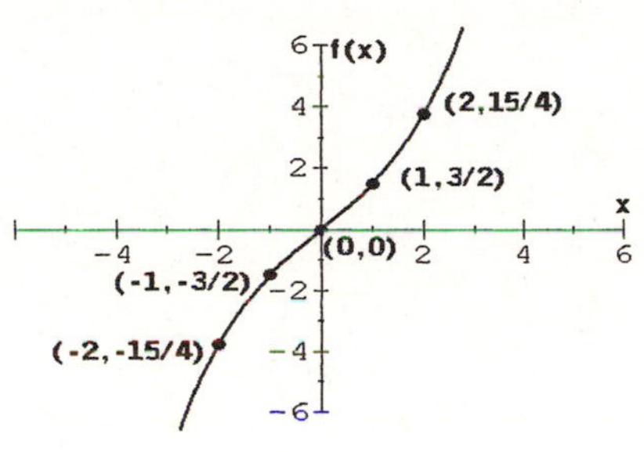

PROBLEM SET 10.2 Applications of Exponential Functions

1. $P = P_0(1.04)^t$

a. $P = 0.77(1.04)^3$
$P = 0.77(1.125)$
$P = \$0.87$

b. $P = 3.43(1.04)^5$
$P = 3.43(1.217)$
$P = \$4.17$

c. $P = 1.99(1.04)^4$
$P = 1.99(1.170)$
$P = \$2.33$

d. $P = 1.05(1.04)^{10}$
$P = 1.05(1.480)$
$P = \$1.55$

e. $P = 18000(1.04)^5$
$P = 18000(1.217)$
$P = \$21{,}900$

f. $P = 120{,}000(1.04)^8$
$P = 120{,}000(1.369)$
$P = \$164{,}228$

g. $P = 500(1.04)^7$
$P = 500(1.316)$
$P = \$658$

3. $A = P\left(1 + \frac{r}{n}\right)^{nt}$
$A = 200\left(1 + \frac{0.06}{1}\right)^{1(6)}$
$A = 200(1.06)^6$
$A = 200(1.419)$
$A = \$283.70$

5. $A = P\left(1 + \frac{r}{n}\right)^{nt}$
$A = 500\left(1 + \frac{0.08}{2}\right)^{2(7)}$
$A = 500(1.04)^{14}$
$A = 500(1.732)$
$A = \$865.84$

7. $A = P\left(1 + \frac{r}{n}\right)^{nt}$
$A = 800\left(1 + \frac{0.09}{4}\right)^{4(9)}$
$A = 800(1.0225)^{36}$
$A = 800(2.228)$
$A = \$1782.25$

9. $A = P\left(1 + \frac{r}{n}\right)^{nt}$
$A = 1500\left(1 + \frac{0.12}{12}\right)^{12(5)}$
$A = 1500(1.01)^{60}$
$A = 1500(1.817)$
$A = \$2725.05$

11. $A = P\left(1 + \frac{r}{n}\right)^{nt}$
$A = 5000\left(1 + \frac{0.085}{1}\right)^{1(15)}$
$A = 5000(1.085)^{15}$
$A = 5000(3.3997)$
$A = \$16,998.71$

13. $A = P\left(1 + \frac{r}{n}\right)^{nt}$
$A = 8000\left(1 + \frac{0.105}{4}\right)^{4(10)}$
$A = 8000(1.02625)^{40}$
$A = 8000(2.819)$
$A = \$22,553.65$

15. $A = Pe^{rt}$
$A = 400e^{0.07(5)}$
$A = 400e^{0.35}$
$A = 400(1.419)$
$A = \$567.63$

17. $A = Pe^{rt}$
$A = 750e^{0.08(8)}$
$A = 750e^{0.64}$
$A = 750(1.896)$
$A = \$1422.36$

19. $A = Pe^{rt}$
$A = 2000e^{0.10(15)}$
$A = 2000e^{1.5}$
$A = 2000(4.482)$
$A = \$8963.38$

21. $A = Pe^{rt}$
$A = 7500e^{0.085(10)}$
$A = 7500e^{0.85}$
$A = 7500(2.3396)$
$A = \$17,547.35$

23. $A = Pe^{rt}$
$A = 15000e^{0.0775(10)}$
$A = 15000e^{0.775}$
$A = 15000(2.171)$
$A = \$32558.88$

25. $A = P\left(1 + \frac{r}{n}\right)^{nt}$
$2700 = 1500\left(1 + \frac{r}{4}\right)^{4(10)}$
$1.8 = \left(1 + \frac{r}{4}\right)^{40}$
$(1.8)^{\frac{1}{40}} = \left[\left(1 + \frac{r}{4}\right)^{40}\right]^{\frac{1}{40}}$
$1.0148 = 1 + \frac{r}{4}$
$0.0148 = \frac{r}{4}$

$0.0592 = r$
$r = 5.9\%$

27. $P(1+r) = Pe^r$
$1+r = e^r$
$1+r = e^{0.0775}$
$1+r = 1.0806$
$r = 0.0806$
$r = 8.06\%$

29. $A = P\left(1+\frac{r}{n}\right)^{nt}$
$A = P\left(1+\frac{0.0825}{4}\right)^4$
$A = P(1.020625)^4$
$A = P(1.08509)$
$A = P\left(1+\frac{0.083}{2}\right)^2$
$A = P(1.0415)^2$
$A = P(1.08472)$
8.25% compounded quarterly will yield more.

31. $Q = Q_0\left(\frac{1}{2}\right)^{\frac{t}{n}}$
$Q = 400\left(\frac{1}{2}\right)^{\frac{87}{29}}$
$Q = 400\left(\frac{1}{2}\right)^3$
$Q = 400\left(\frac{1}{8}\right)$
$Q = 50$
There will be 50 grams after 87 years.

$Q = 400\left(\frac{1}{2}\right)^{\frac{100}{29}}$
$Q = 400\left(\frac{1}{2}\right)^{3.4483}$
$Q = 37$
There will be 37 grams after 100 years.

33. $Q(t) = 1000e^{0.4t}$
$Q(2) = 1000e^{0.4(2)}$
$Q(2) = 1000e^{0.8}$
$Q(2) = 1000(2.2255)$
$Q(2) = 2226$

$Q(3) = 1000e^{0.4(3)}$
$Q(3) = 1000e^{1.2}$
$Q(3) = 1000(3.3201)$
$Q(3) = 3320$

$Q(5) = 1000e^{0.4(5)}$
$Q(5) = 1000e^2$
$Q(5) = 1000(7.3891)$
$Q(5) = 7389$

35. $Q = Q_0e^{0.3t}$
$6640 = Q_0e^{0.3(4)}$
$6640 = Q_0e^{1.2}$
$6640 = Q_0(3.3201)$
$2000 = Q_0$
There were initially 2000 bacteria.

37. $P(a) = 14.7e^{-0.21a}$

a. $P(3.85) = 14.7e^{-0.21(3.85)}$
$P(3.85) = 14.7e^{-0.8085}$
$P(3.85) = 14.7(0.4455)$
$P(3.85) = 6.5$ lbs. per sq. in.

b. $P(1) = 14.7e^{-0.21(1)}$
$P(1) = 14.7e^{-0.21}$
$P(1) = 14.7(0.8106)$
$P(1) = 11.9$ lbs. per sq. in.

c. $1985 \text{ feet} \times \frac{1 \text{ mile}}{5280 \text{ feet}} = 0.376 \text{ miles}$

$P(.376) = 14.7e^{-0.21(.376)}$
$P(.376) = 14.7e^{-0.07896}$
$P(.376) = 14.7(0.9241)$
$P(.376) = 13.6$ lbs. per sq. in.

d. $1090 \text{ feet} \times \frac{1 \text{ mile}}{5280 \text{ feet}} = 0.2064$

$P(0.2064) = 14.7e^{-0.21(0.2064)}$
$P(0.2064) = 14.7e^{-0.0422}$
$P(0.2064) = 14.7(0.9576)$
$P(0.2064) = 14.1$ lbs. per sq. in.

39. $f(x) = e^x + 1$

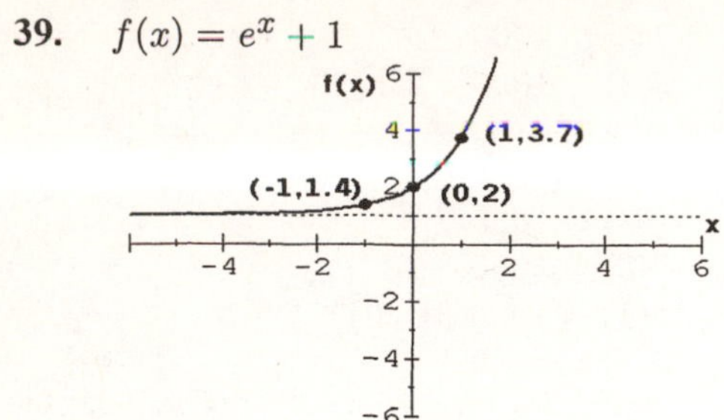

41. $f(x) = 2e^x$

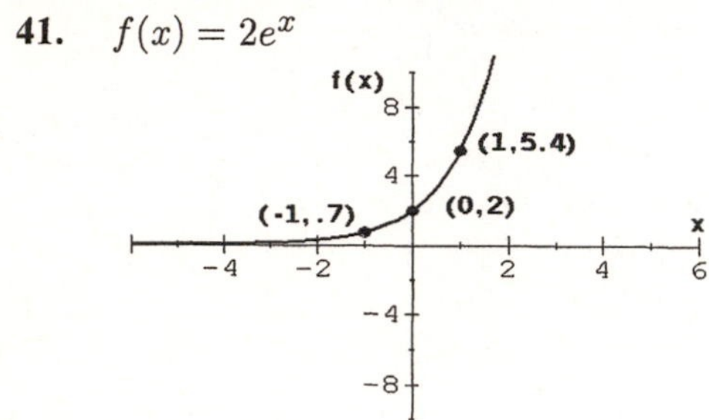

43. $f(x) = e^{2x}$

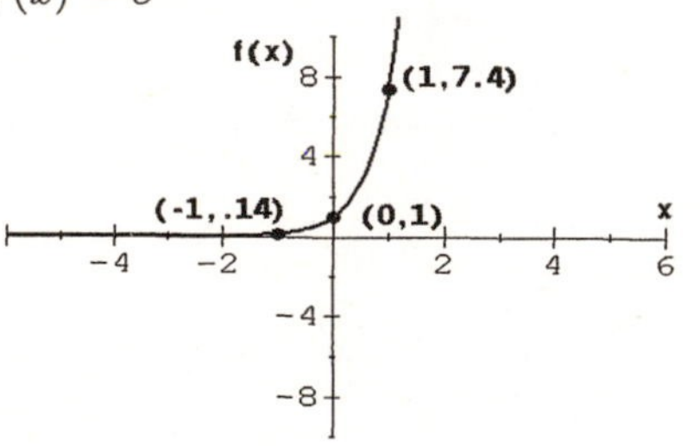

Further Investigations

49. $A = Pe^{rt}$

For 8%

For 5 years

$A = 1000e^{0.08(5)} = 1000e^{0.40}$
$A = 1000(1.492) = 1492$

For 10 years

$A = 1000e^{0.08(10)} = 1000e^{0.8}$
$A = 1000(2.226) = 2226$

For 15 years

$A = 1000e^{0.08(15)} = 1000e^{1.2}$
$A = 1000(3.320) = 3320$

For 20 years

$A = 1000e^{0.08(20)} = 1000e^{1.6}$
$A = 1000(4.953) = 4953$

For 25 years

$A = 1000e^{0.08(25)} = 1000e^{2}$
$A = 1000(7.389) = 7389$

For 10%

For 5 years

$A = 1000e^{1.10(5)} = 1000e^{0.5}$
$A = 1000(1.649) = 1649$

For 10 years

$A = 1000e^{0.10(10)} = 1000e^{1}$
$A = 1000(2.718) = 2718$

For 15 years

$A = 1000e^{0.10(15)} = 1000e^{1.5}$
$A = 1000(4.482) = 4482$

For 20 years

$A = 1000e^{0.10(20)} = 1000e^{2}$
$A = 1000(7.389) = 7389$

For 25 years

$A = 1000e^{0.10(25)} = 1000e^{2.5}$
$A = 1000(12.182) = 12182$

For 12%

For 5 years

$A = 1000e^{0.12(5)} = 1000e^{0.6}$
$A = 1000(1.822) = 1822$

For 10 years

$A = 1000e^{0.12(10)} = 1000e^{1.2}$
$A = 1000(3.320) = 3320$

For 15 years

$A = 1000e^{0.12(15)} = 1000e^{1.8}$
$A = 1000(6.050) = 6050$

For 20 years

$A = 1000e^{0.12(20)} = 1000e^{2.4}$
$A = 1000(11.023) = 11{,}023$

For 25 years

$A = 1000e^{0.12(25)} = 1000e^{3}$
$A = 1000(20.086) = 20{,}086$

For 14%

For 5 years

$A = 1000e^{0.14(5)} = 1000e^{0.7}$
$A = 1000(2.014) = 2014$

For 10 years

$A = 1000e^{0.14(10)} = 1000e^{1.4}$
$A = 1000(4.055) = 4055$

For 15 years

$A = 1000e^{0.14(15)} = 1000^{2.1}$
$A = 1000(8.166) = 8166$

For 20 years

$A = 1000e^{0.14(20)} = 1000e^{2.8}$
$A = 1000(16.445) = 16{,}445$

For 25 years

$A = 1000e^{0.14(25)} = 1000e^{3.5}$

$A = 1000(33.115) = 33,115$

	8%	10%	12%	14%
5 years	\$1492	1649	1822	2014
10 years	2226	2718	3320	4055
15 years	3320	4482	6050	8166
20 years	4953	7389	11023	16445
25 years	7389	12182	20086	33115

51. $A = P\left(1 + \frac{r}{n}\right)^{nt}$

Compounded annually

For 8%

$A = 1000\left(1 + \frac{0.08}{1}\right)^{1(10)} = 1000(1.08)^{10}$

$A = 1000(2.159) = 2159$

For 10%

$A = 1000\left(1 + \frac{0.10}{1}\right)^{1(10)} = 1000(1.1)^{10}$

$A = 1000(2.594) = 2594$

For 12%

$A = 1000\left(1 + \frac{0.12}{1}\right)^{1(10)} = 1000(1.12)^{10}$

$A = 1000(3.106) = 3106$

For 14%

$A = 1000\left(1 + \frac{0.14}{1}\right)^{1(10)} = 1000(1.14)^{10}$

$A = 1000(3.707) = 3707$

Compounded semiannually

For 8%

$A = 1000\left(1 + \frac{0.08}{2}\right)^{2(10)} = 1000(1.04)^{20}$

$A = 1000(2.191) = 2191$

For 10%

$A = 1000\left(1 + \frac{0.10}{2}\right)^{2(10)} = 1000(1.05)^{20}$

$A = 1000(2.653) = 2653$

For 12%

$A = 1000\left(1 + \frac{0.12}{2}\right)^{2(10)} = 1000(1.06)^{20}$

$A = 1000(3.207) = 3207$

For 14%

$A = 1000\left(1 + \frac{0.14}{2}\right)^{2(10)} = 1000(1.07)^{20}$

$A = 1000(3.870) = 3870$

Compounded quarterly

For 8%

$A = 1000\left(1 + \frac{0.08}{4}\right)^{4(10)} = 1000(1.02)^{40}$

$A = 1000(2.208) = 2208$

For 10%

$A = 1000\left(1 + \frac{0.10}{4}\right)^{4(10)} = 1000(1.025)^{40}$

$A = 1000(2.685) = 2685$

For 12%

$A = 1000\left(1 + \frac{0.12}{4}\right)^{4(10)} = 1000(1.03)^{40}$

$A = 1000(3.262) = 3262$

For 14%

$A = 1000\left(1 + \frac{0.14}{4}\right)^{4(10)} = 1000(1.035)^{40}$

$A = 1000(3.959) = 3959$

Compounded monthly

For 8%

$A = 1000\left(1 + \frac{0.08}{12}\right)^{12(10)} = 1000(1.007)^{120}$

$A = 1000(2.220) = 2220$

For 10%

$A = 1000\left(1 + \frac{0.10}{12}\right)^{12(10)} = 1000(1.0083)^{120}$

$A = 1000(2.707) = 2707$

For 12%

$A = 1000\left(1 + \frac{0.12}{12}\right)^{12(10)} = 1000(1.01)^{120}$

$A = 1000(3.300) = 3300$

For 14%

$A = 1000\left(1 + \frac{0.14}{12}\right)^{12(10)} = 1000(1.012)^{120}$

$A = 1000(4.022) = 4022$

Compounded continuously

$A = Pe^{rt}$

For 8%

$A = 1000e^{0.08(10)} = 1000e^{0.8}$

$A = 1000(2.226) = 2226$

Problem Set 10.2

For 10%

$A = 1000e^{0.10(10)} = 1000e^{1}$

$A = 1000(2.718) = 2718$

For 12%

$A = 1000e^{0.12(10)} = 1000e^{1.2}$

$A = 1000(3.320) = 3320$

For 14%

$A = 1000e^{0.14(10)} = 1000e^{1.4}$

$A = 1000(4.055) = 4055$

Compounded	8%	10%	12%	14%
Annually	$2159	2594	3106	3707
Semiannually	2191	2653	3207	3870
Quarterly	2208	2685	3262	3959
Monthly	2220	2707	3300	4022
Countinuosly	2226	2718	3320	4055

53. $f(x) = \dfrac{e^{x} + e^{-x}}{2}$

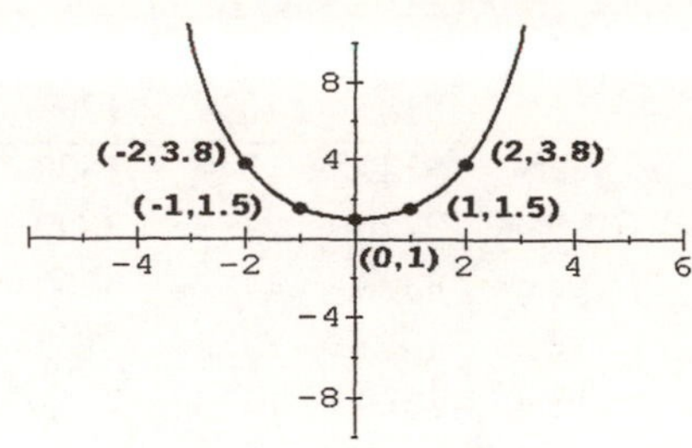

55. $f(x) = \dfrac{e^{x} - e^{-x}}{2}$

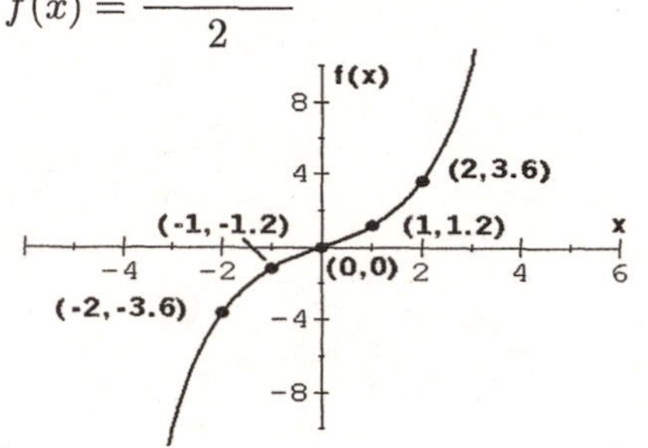

PROBLEM SET 10.3 Inverse Functions

1. Yes, it is one-to-one because no horizontal line intersects the graph in more than one point.

3. No, it is not one-to-one because there is some horizontal line that intersects the graph in more than one point.

5. Yes, it is one-to-one because no horizontal line intersects the graph in more than one point.

7. $f(x) = 5x + 4$
Because the graph is a straight line (slope not equal to 0), it is one-to-one.

9. $f(x) = x^3$
Yes, it is one-to-one because no horizontal line intersects the graph in more than one point.

11. $f(x) = |x| + 1$
No, it is not one-to-one because there is some horizontal line that intersects the graph in more than one point.

13. $f(x) = -x^4$
No, it is not one-to-one because there is some horizontal line that intersects the graph in more than one point.

15. Domain of $f: \{1, 2, 5\}$
Range of $f: \{5, 9, 21\}$
$f^{-1} = \{(5, 1), (9, 2), (21, 5)\}$

Domain of $f^{-1}: \{5, 9, 21\}$
Range of $f^{-1}: \{1, 2, 5\}$

17. Domain of $f: \{0, 2, -1, -2\}$
Range of $f: \{0, 8, -1, -8\}$
$f^{-1} = \{(0, 0), (8, 2), (-1, -1), (-8, -2)\}$

Domain of $f^{-1}: \{0, 8, -1, -8\}$
Range of $f^{-1}: \{0, 2, -1, -2\}$

19. $f(x) = 5x - 9, \qquad g(x) = \dfrac{x+9}{5}$

The set of real numbers is the domain and range of both functions.

$$(f \circ g)(x) = 5\left(\frac{x+9}{5}\right) - 9$$
$$= x + 9 - 9 = x$$

$$(g \circ f)(x) = \frac{(5x-9)+9}{5}$$
$$= \frac{5x}{x} = x$$

Since $(f \circ g)(x) = (g \circ f)(x) = x$ they are inverse functions.

21. $f(x) = -\dfrac{1}{2}x + \dfrac{5}{6}, \; g(x) = -2x + \dfrac{5}{3}$

The set of real numbers is the domain and range of both functions.

$$(f \circ g)(x) = -\frac{1}{2}\left(-2x + \frac{5}{3}\right) + \frac{5}{6}$$
$$= x - \frac{5}{6} + \frac{5}{6} = x$$

$$(g \circ f)(x) = -2\left(-\frac{1}{2}x + \frac{5}{6}\right) + \frac{5}{3}$$
$$= x - \frac{5}{3} + \frac{5}{3} = x$$

Since $(f \circ g)(x) = (g \circ f)(x) = x$ they are inverse functions.

23. $f(x) = \dfrac{1}{x-1}$ for $x > 1$

$g(x) = \dfrac{x+1}{x}$ for $x > 0$

The domain of g equals the range of f, namely, all real numbers greater than 0.

$$(f \circ g)(x) = \frac{1}{\frac{x+1}{x} - 1} = \frac{1}{\frac{x+1-x}{x}}$$
$$= \frac{1}{\frac{1}{x}} = x$$

The domain of f equals the range of g, namely, all real numbers greater than 1.

$$(g \circ f)(x) = \frac{\frac{1}{x-1} + 1}{\frac{1}{x-1}} = \frac{\frac{1+x-1}{x-1}}{\frac{1}{x-1}}$$

$$= \frac{\frac{x}{x-1}}{\frac{1}{x-1}} = \frac{x}{x-1} \cdot \frac{x-1}{1} = x$$

Since $(f \circ g)(x) = (g \circ f)(x) = x$ they are inverse functions.

25. $f(x) = \sqrt{2x-4}$ for $x \geq 2$

$g(x) = \dfrac{x^2+4}{2}$ for $x \geq 0$

The domain of g equals the range of f, namely, all real numbers greater than or equal to 0.

$$(f \circ g)(x) = \sqrt{2\left(\frac{x^2+4}{2}\right) - 4}$$
$$= \sqrt{x^2 + 4 - 4} = \sqrt{x^2} = x$$

The domain of f equals the range of g, namely, all real numbers greater than or equal to 2.

$$(g \circ f)(x) = \frac{\left(\sqrt{2x-4}\right)^2 + 4}{2}$$
$$= \frac{2x-4+4}{2} = \frac{2x}{2} = x$$

Since $(f \circ g)(x) = (g \circ f)(x) = x$ they are inverse functions.

27. $f(x) = 3x, \qquad g(x) = -\dfrac{1}{3}x$

The set of real numbers is the domain and range of both functions.

$$(f \circ g)(x) = 3\left(-\frac{1}{3}x\right) = -x$$

$(g \circ f)(x) = -\frac{1}{3}(3x) = -x$

Since $(f \circ g)(x) = (g \circ f)(x) = -x$
they are not inverse functions.

29. $f(x) = x^3, \quad g(x) = \sqrt[3]{x}$

The set of real numbers is the domain and range of both functions.

$(f \circ g)(x) = \left(\sqrt[3]{x}\right)^3 = x$
$(g \circ f)(x) = \sqrt[3]{x^3} = x$

Since $(f \circ g)(x) = (g \circ f)(x) = x$
they are inverse functions.

31. $f(x) = x, \; g(x) = \frac{1}{x}$

$(f \circ g)(x) = \frac{1}{x}$

They are not inverse functions.

33. $f(x) = x^2 - 3$ for $x \geq 0$,
$g(x) = \sqrt{x+3}$ for $x \geq -3$

The domain of g equals the range of f, namely, all real numbers greater than or equal to -3.

$(f \circ g)(x) = \left(\sqrt{x+3}\right)^2 - 3 = x+3 - 3 = x$

The domain of f equals the range of g, namely,all real numbers greater than or equal to 0.

$(g \circ f)(x) = \sqrt{x^2 + 3 - 3} = \sqrt{x^2} = x$

Since $(f \circ g)(x) = (g \circ f)(x) = x$
they are inverse functions.

35. $f(x) = \sqrt{x+1}, \;\; g(x) = x^2 - 1$ for $x \geq 0$

The domain of g equals the range of f, namely, all real numbers greater than or equal to 0.

$(f \circ g)(x) = \sqrt{x^2 - 1 + 1} = \sqrt{x^2} = x$

The domain of f equals the range of g, namely,all real numbers greater than or equal to -1.

$(g \circ f)(x) = \left(\sqrt{x+1}\right)^2 - 1 = x+1 - 1 = x$

Since $(f \circ g)(x) = (g \circ f)(x) = x$
they are inverse functions.

37. $f(x) = x - 4$
$y = x - 4$

To find the inverse function interchange x and y, and solve for y.

$x = y - 4$
$x + 4 = y$
$f^{-1}(x) = x + 4$

The set of real numbers is the domain and range of both functions.

$(f \circ f^{-1})(x) = (x+4) - 4 = x$
$(f^{-1} \circ f)(x) = (x-4) + 4 = x$

39. $f(x) = -3x - 4$
$y = -3x - 4$

To find the inverse function interchange x and y, and solve for y.

$x = -3y - 4$
$x + 4 = -3y$
$\frac{x+4}{-3} = y$
$f^{-1}(x) = \frac{-x-4}{3}$

The set of real numbers is the domain and range of both functions.

$(f \circ f^{-1})(x) = -3\left(\frac{-x-4}{3}\right) - 4$
$= x + 4 - 4 = x$

$(f^{-1} \circ f)(x) = \frac{-(-3x-4) - 4}{4}$
$= \frac{3x}{3} = x$

41. $f(x) = \frac{3}{4}x - \frac{5}{6}$
$y = \frac{3}{4}x - \frac{5}{6}$

To find the inverse function interchange x and y, and solve for y.

$$x = \frac{3}{4}y - \frac{5}{6}$$
$$12x = 9y - 10$$
$$12x + 10 = 9y$$
$$y = \frac{12x + 10}{9}$$
$$f^{-1}(x) = \frac{12x + 10}{9} = \frac{4}{3}x + \frac{10}{9}$$

The set of real numbers is the domain and range of both functions.

$$(f \circ f^{-1})(x) = \frac{3}{4}\left(\frac{4}{3}x + \frac{10}{9}\right) - \frac{5}{6}$$
$$= x + \frac{5}{6} - \frac{5}{6} = x$$

$$(f^{-1} \circ f)(x) = \frac{4}{3}\left(\frac{3}{4}x - \frac{5}{6}\right) + \frac{10}{9}$$
$$= x - \frac{10}{9} + \frac{10}{9} = x$$

43. $f(x) = -\frac{2}{3}x$

$y = -\frac{2}{3}x$

To find the inverse function interchange x and y, and solve for y.

$$x = -\frac{2}{3}y$$
$$-\frac{3}{2}x = y$$
$$f^{-1}(x) = -\frac{3}{2}x$$

The set of real numbers is the domain and range of both functions.

$$(f \circ f^{-1})(x) = -\frac{2}{3}\left(-\frac{3}{2}x\right) = x$$
$$(f^{-1} \circ f)(x) = -\frac{3}{2}\left(-\frac{2}{3}x\right) = x$$

45. $f(x) = \sqrt{x}$ for $x \geq 0$

$y = \sqrt{x}$

To find the inverse function interchange x and y, and solve for y.

$$x = \sqrt{y}$$
$$x^2 = y$$
$$f^{-1}(x) = x^2 \text{ for } x \geq 0$$

The domain and range for both functions is the same namely, all real numbers greater than or equal to 0.

$$(f \circ f^{-1})(x) = \sqrt{x^2} = x$$
$$(f^{-1} \circ f)(x) = \left(\sqrt{x}\right)^2 = x$$

47. $f(x) = x^2 + 4$ for $x \geq 0$

$y = x^2 + 4$

To find the inverse function interchange x and y, and solve for y.

$$x = y^2 + 4$$
$$x - 4 = y^2$$
$$y = \sqrt{x - 4}$$

(Note the negative root is discarded.)

$f^{-1}(x) = \sqrt{x - 4}$ for $x \geq 4$

The domain of f^{-1} equals the range of f, namely, all real numbers greater than or equal to 4.

$$(f \circ f^{-1})(x) = \left(\sqrt{x - 4}\right)^2 + 4$$
$$= x - 4 + 4 = x$$

The domain of f equals the range of f^{-1}, namely,all real numbers greater than or equal to 0.

$$(f^{-1} \circ f)(x) = \sqrt{x^2 + 4 - 4}$$
$$= \sqrt{x^2} = x$$

49. $f(x) = 1 + \frac{1}{x}$ for $x > 0$

$y = 1 + \frac{1}{x}$ for $x > 0$

To find the inverse function interchange x and y, and solve for y.

$$x = 1 + \frac{1}{y}$$
$$x - 1 = \frac{1}{y}$$

$y(x-1)=1$

$y=\dfrac{1}{x-1}$

$f^{-1}(x)=\dfrac{1}{x-1}$ for $x>1$

The domain of f^{-1} equals the range of f, namely, all real numbers greater than 1.

$(f\circ f^{-1})(x)=1+\dfrac{1}{\dfrac{1}{x-1}}=1+x-1=x$

The domain of f equals the range of f^{-1}, namely,all real numbers greater than 0.

$(f^{-1}\circ f)(x)=\dfrac{1}{1+\dfrac{1}{x}-1}=\dfrac{1}{\dfrac{1}{x}}=x$

51. $f(x)=3x$
$y=3x$

To find the inverse function interchange x and y, and solve for y.

$x=3y$

$\frac{1}{3}x=y$

$f^{-1}(x)=\frac{1}{3}x$

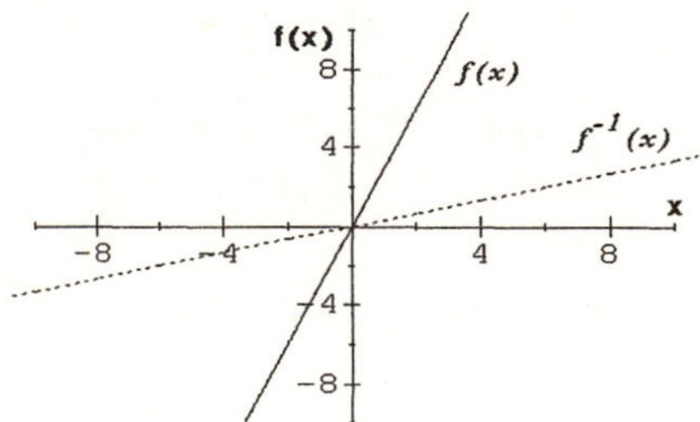

53. $f(x)=2x+1$
$y=2x+1$

To find the inverse function interchange x and y, and solve for y.

$x=2y+1$

$x-1=2y$

$\frac{1}{2}x-\frac{1}{2}=y$

$f^{-1}(x)=\frac{1}{2}x-\frac{1}{2}$

55. $f(x)=\dfrac{2}{x-1}$ for $x>1$

$y=\dfrac{2}{x-1}$

To find the inverse function interchange x and y, and solve for y.

$x=\dfrac{2}{y-1}$

$(y-1)(x)=2$

$y-1=\dfrac{2}{x}$

$y=\dfrac{2}{x}+1$

$f^{-1}(x)=\dfrac{2}{x}+1$ for $x>0$

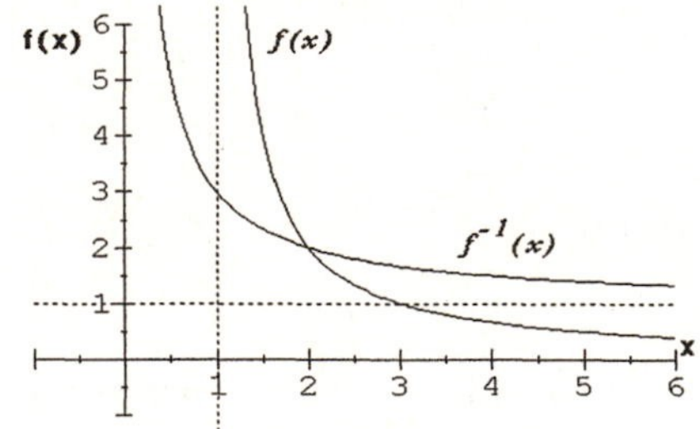

57. $f(x)=x^2-4$ for $x\geq 0$
$y=x^2-4$

To find the inverse function interchange x and y, and solve for y.

$x=y^2-4$
$x+4=y^2$
$y=\sqrt{x+4}$
(Note : discard the negative root.)

$f^{-1}(x) = \sqrt{x+4}$ for $x \geq -4$

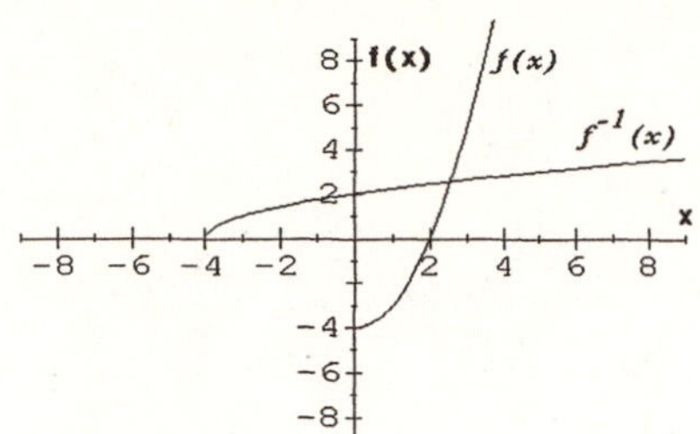

59. $f(x) = x^2 + 1$
Increasing on $[0, \infty)$ and decreasing on $(-\infty, 0]$

61. $f(x) = -3x + 1$
Decreasing on $(-\infty, \infty)$

63. $f(x) = -(x+2)^2 - 1$
Increasing on $(-\infty, -2]$ and decreasing on $[-2, \infty)$

65. $f(x) = -2x^2 - 16x - 35$
Increasing on $(-\infty, -4]$ and decreasing on $[-4, \infty)$

Further Investigations

71a. $f(x) = 3x - 9$

$$f(f^{-1}(x)) = 3\left[f^{-1}(x)\right] - 9 = x$$
$$3\left[f^{-1}(x)\right] = x + 9$$
$$f^{-1}(x) = \frac{x+9}{3}$$

b. $f(x) = -2x + 6$

$$f(f^{-1}(x)) = -2\left[f^{-1}(x)\right] + 6 = x$$
$$-2\left[f^{-1}(x)\right] = x - 6$$
$$f^{-1}(x) = \frac{x-6}{-2} = \frac{6-x}{2}$$

c. $f(x) = -x + 1$

$$f(f^{-1}(x)) = -\left[f^{-1}(x)\right] + 1 = x$$
$$-\left[f^{-1}(x)\right] = x - 1$$
$$f^{-1}(x) = \frac{x-1}{-1} = -x + 1$$
$$f^{-1}(x) = 1 - x$$

d. $f(x) = 2x$

$$f(f^{-1}(x)) = 2\left[f^{-1}(x)\right] = x$$
$$f^{-1}(x) = \frac{x}{2}$$

e. $f(x) = -5x$

$$f(f^{-1}(x)) = -5\left[f^{-1}(x)\right] = x$$
$$f^{-1}(x) = -\frac{x}{5}$$
$$f^{-1}(x) = -\frac{1}{5}x$$

f. $f(x) = x^2 + 6$ for $x \geq 0$

$$f(f^{-1}(x)) = \left[f^{-1}(x)\right]^2 + 6 = x$$
$$\left[f^{-1}(x)\right]^2 = x - 6$$
$$f^{-1}(x) = \sqrt{x-6} \text{ for } x \geq 6$$

PROBLEM SET 10.4 Logarithms

1. $2^7 = 128$
$\log_2 128 = 7$

3. $5^3 = 125$
$\log_5 125 = 3$

5. $10^3 = 1000$
$\log_{10} 1000 = 3$

7. $2^{-2} = \frac{1}{4}$
$\log_2 \frac{1}{4} = -2$

9. $10^{-1} = 0.1$
$\log_{10} 0.1 = -1$

Problem Set 10.4

11. $\log_3 81 = 4$
$3^4 = 81$

13. $\log_4 64 = 3$
$4^3 = 64$

15. $\log_{10} 10000 = 4$
$10^4 = 10000$

17. $\log_2 \left(\frac{1}{16}\right) = -4$
$2^{-4} = \frac{1}{16}$

19. $\log_{10} 0.001 = -3$
$10^{-3} = 0.001$

21. $\log_2 16 = x$
$2^x = 16$
$2^x = 2^4$
$x = 4$

23. $\log_3 81 = x$
$3^x = 81$
$3^x = 3^4$
$x = 4$

25. $\log_6 216 = x$
$6^x = 216$
$6^x = 6^3$
$x = 3$

27. $\log_7 \sqrt{7} = x$
$7^x = \sqrt{7}$
$7^x = 7^{\frac{1}{2}}$
$x = \frac{1}{2}$

29. $\log_{10} 1 = x$
$10^x = 1$
$10^x = 10^0$
$x = 0$

31. $\log_{10} 0.1 = x$
$10^x = 0.1$
$10^x = 10^{-1}$
$x = -1$

33. $10^{\log_{10} 5} = x$
$5 = x$

35. $\log_2 \left(\frac{1}{32}\right) = x$
$2^x = \frac{1}{32}$
$2^x = \frac{1}{2^5}$
$2^x = 2^{-5}$
$x = -5$

37. $\log_5 (\log_2 32) = x$
$\log_5 (\log_2 2^5) = x$
$\log_5 5 = x$
$5^x = 5$
$5^x = 5^1$
$x = 1$

39. $\log_{10} (\log_7 7) = x$
$\log_{10} 1 = x$
$10^x = 1$
$10^x = 10^0$
$x = 0$

41. $\log_7 x = 2$
$7^2 = x$
$49 = x$
The solution set is $\{49\}$.

43. $\log_8 x = \frac{4}{3}$
$8^{\frac{4}{3}} = x$
$(8^{\frac{1}{3}})^4 = x$
$(2)^4 = x$
$16 = x$
The solution set is $\{16\}$.

45. $\log_9 x = \frac{3}{2}$
$9^{\frac{3}{2}} = x$
$(9^{\frac{1}{2}})^3 = x$
$3^3 = x$
$27 = x$
The solution set is $\{27\}$.

47. $\log_4 x = -\frac{3}{2}$

$4^{-\frac{3}{2}} = x$

$\frac{1}{4^{\frac{3}{2}}} = x$

$\frac{1}{(4^{\frac{1}{2}})^3} = x$

$\frac{1}{2^3} = x$

$\frac{1}{8} = x$

The solution set is $\left\{\frac{1}{8}\right\}$.

49. $\log_x 2 = \frac{1}{2}$

$x^{\frac{1}{2}} = 2$

$(x^{\frac{1}{2}})^2 = 2^2$

$x = 4$

The solution set is $\{4\}$.

51. $\log_2 35$

$\log_2 (5 \bullet 7)$

$\log_2 5 + \log_2 7$

$2.3219 + 2.8074$

5.1293

53. $\log_2 125$

$\log_2 5^3$

$3 \log_2 5$

$3(2.3219)$

6.9657

55. $\log_2 \sqrt{7}$

$\log_2 7^{\frac{1}{2}}$

$\frac{1}{2} \log_2 7$

$\frac{1}{2}(2.8074)$

1.4037

57. $\log_2 175$

$\log_2 (5^2 \bullet 7)$

$\log_2 5^2 + \log_2 7$

$2 \log_2 5 + \log_2 7$

$2(2.3219) + 2.8074$

$4.6438 + 2.8074$

7.4512

59. $\log_2 80$

$\log_2 (16 \bullet 5)$

$\log_2 16 + \log_2 5$

$\log_2 2^4 + 2.3219$

$4 + 2.3219$

6.3219

61. $\log_8 \left(\frac{5}{11}\right)$

$\log_8 5 - \log_8 11$

$0.7740 - 1.1531$

-0.3791

63. $\log_8 \sqrt{11}$

$\log_8 11^{\frac{1}{2}}$

$\frac{1}{2} \log_8 11$

$\frac{1}{2}(1.1531)$

0.5766

65. $\log_8 88$

$\log_8 (8 \bullet 11)$

$\log_8 8 + \log_8 11$

$1 + 1.1531$

2.1531

67. $\log_8 \left(\frac{25}{11}\right)$

$\log_8 25 - \log_8 11$

$\log_8 5^2 - 1.1531$

$2 \log_8 5 - 1.1531$

$2(0.7740) - 1.1531$

$1.5480 - 1.1531$

0.3949

69. $\log_b xyz$

$\log_b x + \log_b y + \log_b z$

71. $\log_b \left(\frac{y}{z}\right)$

$\log_b y - \log_b z$

73. $\log_b y^3 z^4$
$\log_b y^3 + \log_b z^4$
$3 \log_b y + 4 \log_b z$

75. $\log_b \left(\dfrac{x^{\frac{1}{2}} y^{\frac{1}{3}}}{z^4}\right)$

$\log_b x^{\frac{1}{2}} + \log_b y^{\frac{1}{3}} - \log_b z^4$
$\dfrac{1}{2} \log_b x + \dfrac{1}{3} \log_b y - 4 \log_b z$

77. $\log_b \sqrt[3]{x^2 z}$
$\log_b x^{\frac{2}{3}} z^{\frac{1}{3}}$
$\log_b x^{\frac{2}{3}} + \log_b z^{\frac{1}{3}}$
$\dfrac{2}{3} \log_b x + \dfrac{1}{3} \log_b z$

79. $\log_b \left(x \sqrt{\dfrac{x}{y}}\right)$

$\log_b \dfrac{x \bullet x^{\frac{1}{2}}}{y^{\frac{1}{2}}}$

$\log_b \dfrac{x^{\frac{3}{2}}}{y^{\frac{1}{2}}}$

$\log_b x^{\frac{3}{2}} - \log_b y^{\frac{1}{2}}$
$\dfrac{3}{2} \log_b x - \dfrac{1}{2} \log_b y$

81. $2 \log_b x - 4 \log_b y$
$\log_b x^2 - \log_b y^4$
$\log_b \left(\dfrac{x^2}{y^4}\right)$

83. $\log_b x - (\log_b y - \log_b z)$
$\log_b x - \log_b y + \log_b z$
$\log_b x + \log_b z - \log_b y$
$\log_b (xz) - \log_b y$
$\log_b \left(\dfrac{xz}{y}\right)$

85. $2 \log_b x + 4 \log_b y - 3 \log_b z$
$\log_b x^2 + \log_b y^4 - \log_b z^3$
$\log_b (x^2 y^4) - \log_b z^3$
$\log_b \left(\dfrac{x^2 y^4}{z^3}\right)$

87. $\dfrac{1}{2} \log_b x - \log_b x + 4 \log_b y$
$\log_b \sqrt{x} - \log_b x + \log_b y^4$
$\log_b \sqrt{x} + \log_b y^4 - \log_b x$
$\log_b (y^4 \sqrt{x}) - \log_b x$
$\log_b \left(\dfrac{y^4 \sqrt{x}}{x}\right)$

89. $\log_3 x + \log_3 4 = 2$
$\log_3 4x = 2$
$3^2 = 4x$
$9 = 4x$
$\dfrac{9}{4} = x$
The solution set is $\left\{\dfrac{9}{4}\right\}$.

91. $\log_{10} x + \log_{10} (x - 21) = 2$
$\log_{10} x(x - 21) = 2$
$10^2 = x(x - 21)$
$100 = x^2 - 21x$
$x^2 - 21x - 100 = 0$
$(x - 25)(x + 4) = 0$
$x - 25 = 0$ or $x + 4 = 0$
$x = 25$ or $x = -4$
Discard the root $x = -4$.
The solution set is $\{25\}$.

93. $\log_2 x + \log_2 (x - 3) = 2$
$\log_2 x(x - 3) = 2$
$2^2 = x(x - 3)$
$4 = x^2 - 3x$
$x^2 - 3x - 4 = 0$
$(x - 4)(x + 1) = 0$
$x - 4 = 0$ or $x + 1 = 0$
$x = 4$ or $x = -1$
Discard the root $x = -1$.
The solution set is $\{4\}$.

95. $\log_3(x + 3) + \log_3(x + 5) = 1$
$\log_3[(x + 3)(x + 5)] = 1$
$3^1 = (x + 3)(x + 5)$
$3 = x^2 + 8x + 15$
$0 = x^2 + 8x + 12$

$0 = (x + 6)(x + 2)$
$x + 6 = 0$ or $x + 2 = 0$
$x = -6$ or $x = -2$
Discard the root $x = -6$.
The solution set is $\{-2\}$.

97. $\log_2 3 + \log_2(x + 4) = 3$
$\log_2 3(x + 4) = 3$
$2^3 = 3(x + 4)$
$8 = 3x + 12$
$-4 = 3x$
$-\frac{4}{3} = x$
The solution set is $\left\{-\frac{4}{3}\right\}$.

99. $\log_{10}(2x - 1) - \log_{10}(x - 2) = 1$
$\log_{10}\left(\frac{2x - 1}{x - 2}\right) = 1$
$10^1 = \frac{2x - 1}{x - 2};\ x \neq 2$
$(x - 2)(10) = (x - 2)\left(\frac{2x - 1}{x - 2}\right)$
$10x - 20 = 2x - 1$
$8x = 19$
$x = \frac{19}{8}$
The solution set is $\left\{\frac{19}{8}\right\}$.

101. $\log_5(3x - 2) = 1 + \log_5(x - 4)$
$\log_5(3x - 2) - \log_5(x - 4) = 1$
$\log_5\left(\frac{3x - 2}{x - 4}\right) = 1$
$5^1 = \frac{3x - 2}{x - 4};\ x \neq 4$
$5(x - 4) = 3x - 2$
$5x - 20 = 3x - 2$
$2x = 18$
$x = 9$
The solution set is $\{9\}$.

103. $\log_2(x - 1) - \log_2(x + 3) = 2$
$\log_2\left(\frac{x - 1}{x + 3}\right) = 2$
$2^2 = \frac{x - 1}{x + 3};\ x \neq -3$
$4 = \frac{x - 1}{x + 3};$
$4(x + 3) = x - 1$
$4x + 12 = x - 1$
$3x = -13$
$x = -\frac{13}{3}$
Discard the root $x = -\frac{13}{3}$.
The solution set is $\emptyset$.

105. $\log_8(x + 7) + \log_8 x = 1$
$\log_8(x + 7)(x) = 1$
$8^1 = (x + 7)(x)$
$8 = x^2 + 7x$
$x^2 + 7x - 8 = 0$
$(x + 8)(x - 1) = 0$
$x + 8 = 0$ or $x - 1 = 0$
$x = -8$ or $x = 1$
Discard the root $x = -8$.
The solution set is $\{1\}$.

107. $\log_b\left(\frac{r}{s}\right) = \log_b r - \log_b s$
Let $m = \log_b r$ and $n = \log_b s$.
$m = \log_b r$ becomes $b^m = r$.
$n = \log_b s$ becomes $b^n = s$.
Then the quotient $\frac{r}{s}$ becomes
$\frac{r}{s} = \frac{b^m}{b^n} = b^{m-n}$.
Changing $\frac{r}{s} = b^{m-n}$ to logarithmic form
$\log_b\left(\frac{r}{s}\right) = m - n$
$\log_b\left(\frac{r}{s}\right) = \log_b r - \log_b s$

PROBLEM SET 10.5 Logarithmic Functions

1. log 7.24
0.8597

3. log 52.23
1.7179

5. log 3214.1
3.5071

7. log 0.729
-0.1373

9. log 0.00034
-3.4685

11. $\log x = 2.6143$
$x = 411.43$

13. $\log x = 4.9547$
$x = 90{,}095$

15. $\log x = 1.9006$
$x = 79.543$

17. $\log x = -1.3148$
$x = 0.048440$

19. $\log x = -2.1928$
$x = 0.0064150$

21. ln 5
1.6094

23. ln 32.6
3.4843

25. ln 430
6.0638

27. ln 0.46
-0.7765

29. ln 0.0314
-3.4609

31. $\ln x = 0.4721$
$x = 1.6034$

33. $\ln x = 1.1425$
$x = 3.1346$

35. $\ln x = 4.6873$
$x = 108.56$

37. $\ln x = -0.7284$
$x = 0.48268$

39. $\ln x = -3.3244$
$x = 0.035994$

41. **(a)** $f(x) = \log x$

x	0.1	0.5	1	2	4	8	10
$\log x$	-1	$-.3$	0	.3	.6	.9	1

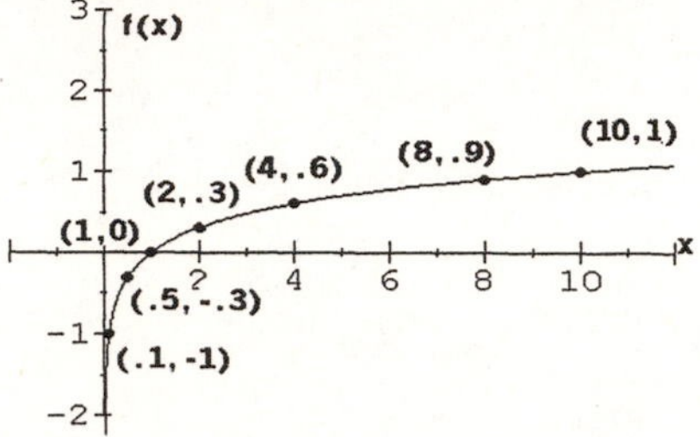

(b) $f(x) = 10^x$

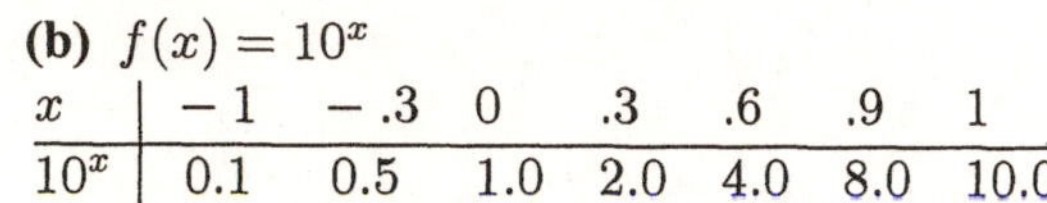

x	-1	$-.3$	0	.3	.6	.9	1
10^x	0.1	0.5	1.0	2.0	4.0	8.0	10.0

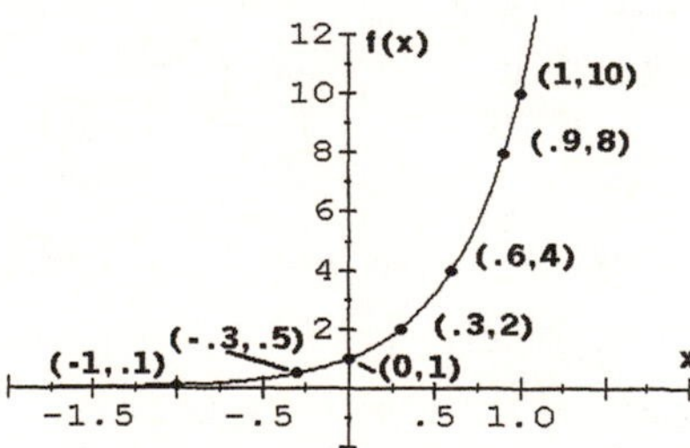

43. $y = \log_{\frac{1}{2}} x$

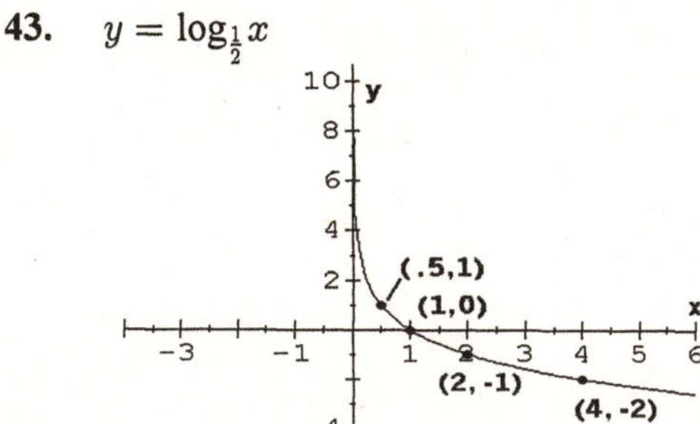

45. $f(x) = \log_3 x \qquad g(x) = 3^x$

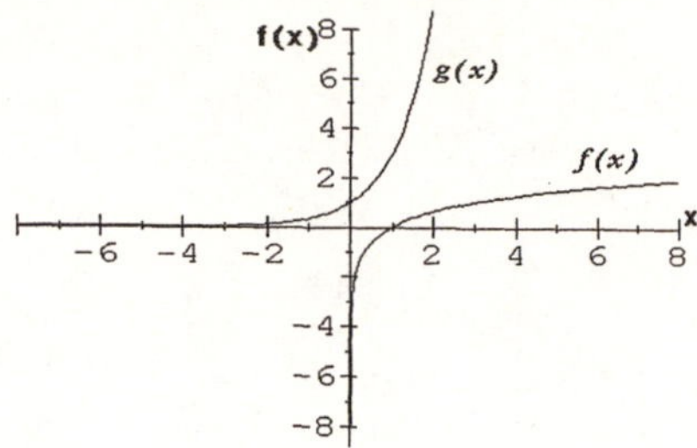

47. $f(x) = 3 + \log_2 x$

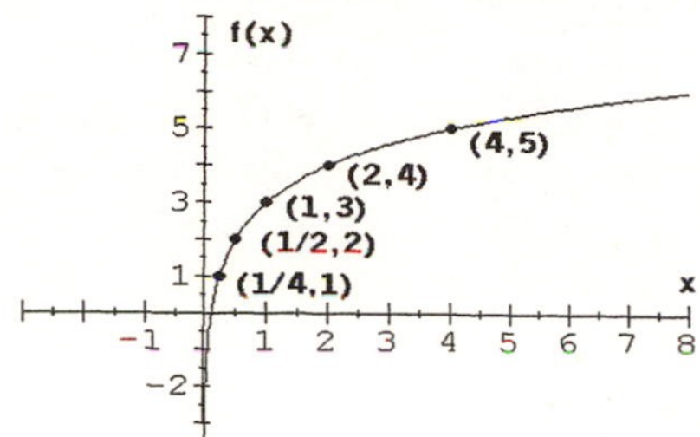

49. $f(x) = \log_2(x + 3)$

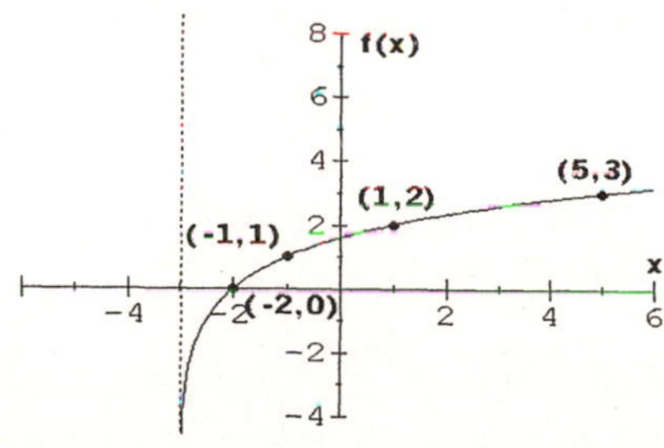

51. $f(x) = \log_2 2x$

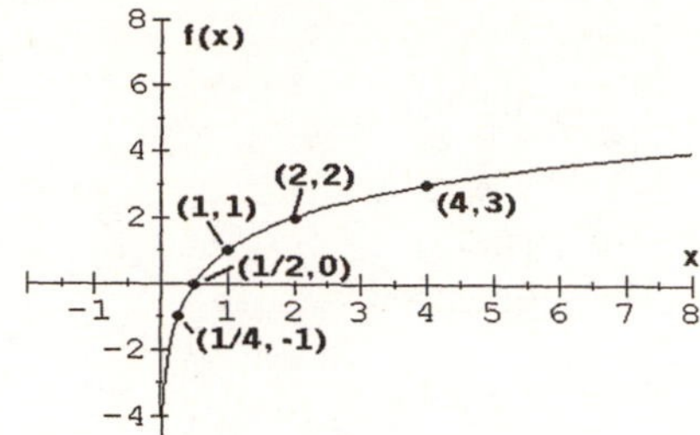

53. $f(x) = 2\log_2 x$

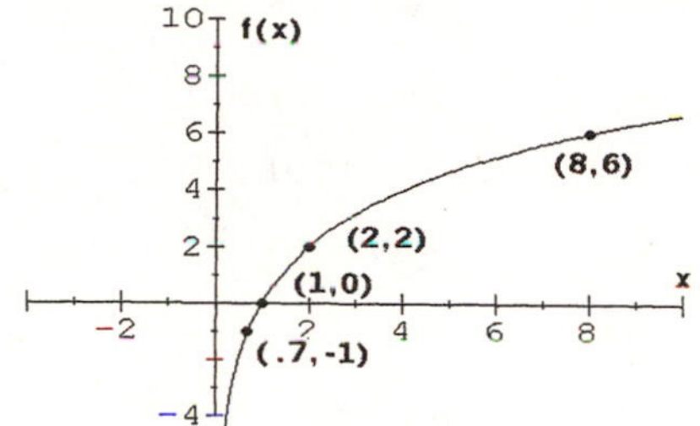

55. $\dfrac{\ln 2}{\ln 7} = \dfrac{0.6931}{1.946} = 0.36$

57. $\dfrac{\ln 5}{2\ln 3} = \dfrac{1.6094}{2(1.0986)} = \dfrac{1.6094}{2.1972} = 0.73$

59. $\dfrac{\ln 2}{0.03} = \dfrac{0.6931}{0.03} = 23.10$

61. $\dfrac{\log 5}{3\log 1.07} = \dfrac{0.6990}{3(0.0294)} = \dfrac{0.6990}{0.0882} = 7.93$

PROBLEM SET 10.6 Exponential Equations, Logarithmic Equations, and Problem Solving

1. $3^x = 13$
$\log 3^x = \log 13$
$x \log 3 = \log 13$
$x(0.4771) = 1.1139$
$x = 2.33$
The solution set is $\{2.33\}$.

3. $4^n = 35$
$\log 4^n = \log 35$
$n \log 4 = \log 35$
$n(0.6021) = 1.5441$
$n = 2.56$
The solution set is $\{2.56\}$.

5. $2^x + 7 = 50$
$2^x = 43$
$\log 2^x = \log 43$
$x \log 2 = \log 43$
$x(0.3010) = 1.6335$
$x = 5.43$
The solution set is $\{5.43\}$.

Problem Set 10.6

7. $3^{x-2} = 11$
$\log 3^{x-2} = \log 11$
$(x-2)(\log 3) = \log 11$
$(x-2)(0.4771) = 1.0414$
$0.4771x - 0.9542 = 1.0414$
$0.4771x = 1.9956$
$x = 4.18$
The solution set is $\{4.18\}$.

9. $5^{3t+1} = 9$
$\log 5^{3t+1} = \log 9$
$(3t+1)(\log 5) = \log 9$
$(3t+1)(.6990) = 0.9542$
$2.097t + .6990 = 0.9542$
$2.097t = 0.2552$
$t = 0.12$
The solution set is $\{0.12\}$.

11. $e^x = 27$
$\ln e^x = \ln 27$
$x(\ln e) = \ln 27$
$x = \ln 27$
$x = 3.30$
The solution set is $\{3.30\}$.

13. $e^{x-2} = 13.1$
$\ln e^{x-2} = \ln 13.1$
$x - 2 = 2.57$
$x = 4.57$
The solution set is $\{4.57\}$.

15. $3e^x - 1 = 17$
$3e^x = 18$
$e^x = 6$
$\ln e^x = \ln 6$
$x = 1.79$
The solution set is $\{1.79\}$.

17. $5^{2x+1} = 7^{x+3}$
$\log 5^{2x+1} = \log 7x^{x+3}$
$(2x+1)(\log 5) = (x+3)(\log 7)$
$(2x+1)(.6990) = (x+3)(0.8451)$
$1.398x + 0.6990 = 0.8451x + 2.5353$
$0.5529x = 1.8363$
$x = 3.32$
The solution set is $\{3.32\}$.

19. $3^{2x+1} = 2^{3x+2}$
$\log 3^{2x+1} = \log 2^{3x+2}$
$(2x+1)(\log 3) = (3x+2)(\log 2)$
$(2x+1)(0.4771) = (3x+2)(0.3010)$
$0.9542x + 0.4771 = 0.9030x + 0.6020$
$0.0512x = 0.1249$
$x = 2.44$
The solution set is $\{2.44\}$.

21. $\log x + \log(x+21) = 2$
$\log x(x+21) = 2$
$10^2 = x(x+21)$
$100 = x^2 + 21x$
$0 = x^2 + 21x - 100$
$0 = (x+25)(x-4)$
$x + 25 = 0$ or $x - 4 = 0$
$x = -25$ or $x = 4$
Discard the root $x = -25$.
The solution set is $\{4\}$.

23. $\log(3x-1) = 1 + \log(5x-2)$
$\log(3x-1) - \log(5x-2) = 1$
$\log \dfrac{3x-1}{5x-2} = 1$
$10^1 = \dfrac{3x-1}{5x-2}$
$10(5x-2) = 3x - 1$
$50x - 20 = 3x - 1$
$47x = 19$
$x = \dfrac{19}{47}$
The solution set is $\left\{\dfrac{19}{47}\right\}$.

25. $\log(x+1) = \log 3 - \log(2x-1)$
$\log(x+1) + \log(2x-1) = \log 3$
$\log(x+1)(2x-1) = \log 3$
$(x+1)(2x-1) = 3$
$2x^2 + x - 1 = 3$
$2x^2 + x - 4 = 0$
$x = \dfrac{-1 \pm \sqrt{1^2 - 4(2)(-4)}}{2(2)}$
$x = \dfrac{-1 \pm \sqrt{33}}{4}$
Discard the root $x = \dfrac{-1 - \sqrt{33}}{4}$.

The solution set is $\left\{\frac{-1+\sqrt{33}}{4}\right\}$.

27. $\log(x+2) - \log(2x+1) = \log x$
$\log(x+2) = \log x + \log(2x+1)$
$\log(x+2) = \log x(2x+1)$
$x+2 = x(2x+1)$
$x+2 = 2x^2 + x$
$2 = 2x^2$
$x^2 = 1$
$x = \pm\sqrt{1} = \pm 1$
Discard the root $x = -1$.
The solution set is $\{1\}$.

29. $\ln(2t+5) = \ln 3 + \ln(t-1)$
$\ln(2t+5) = \ln 3(t-1)$
$2t+5 = 3(t-1)$
$2t+5 = 3t-3$
$8 = t$
The solution set is $\{8\}$.

31. $\log\sqrt{x} = \sqrt{\log x}$

$\log x^{\frac{1}{2}} = (\log x)^{\frac{1}{2}}$

$\frac{1}{2}(\log x) = (\log x)^{\frac{1}{2}}$

$2\left[\frac{1}{2}(\log x)\right] = 2(\log x)^{\frac{1}{2}}$

$\log x = 2(\log x)^{\frac{1}{2}}$

$\log x - 2(\log x)^{\frac{1}{2}} = 0$

Let $u = (\log x)^{\frac{1}{2}}$.
$u^2 - 2u = 0$
$u(u-2) = 0$
$u = 0$ or $u - 2 = 0$
$u = 0$ or $u = 2$
$(\log x)^{\frac{1}{2}} = 0$ or $(\log x)^{\frac{1}{2}} = 2$
$[(\log x)^{\frac{1}{2}}]^2 = 0^2$ or $[(\log x)^{\frac{1}{2}}]^2 = 2^2$
$\log x = 0$ or $\log x = 4$
$10^0 = x$ or $10^4 = x$
$1 = x$ or $10{,}000 = x$
The solution set is $\{1, 10000\}$.

33. $\log_2 40 = \frac{\log 40}{\log 2} = \frac{1.6020}{0.3010} = 5.322$

35. $\log_3 16 = \frac{\log 16}{\log 3} = \frac{1.2041}{0.4771} = 2.524$

37. $\log_4 1.6 = \frac{\log 1.6}{\log 4} = \frac{0.2041}{0.6021} = 0.339$

39. $\log_5 0.26 = \frac{\log 0.26}{\log 5} = \frac{-0.5850}{0.6990} = -0.837$

41. $\log_7 500 = \frac{\log 500}{\log 7} = \frac{2.6990}{0.8451} = 3.194$

43. $A = P\left(1+\frac{r}{n}\right)^{nt}$

$1000 = 750\left(1+\frac{0.12}{4}\right)^{4t}$

$1.3333 = (1.03)^{4t}$
$\log 1.3333 = \log(1.03)^{4t}$
$0.1249 = 4t\log(1.03)$
$0.1249 = 4t(0.0128)$
$0.1249 = 0.0512t$
$2.4 = t$
It will take 2.4 years.

45. $A = Pe^{rt}$
$4000 = 2000e^{0.13t}$
$2 = e^{0.13t}$
$\ln 2 = \ln e^{0.13t}$
$0.6931 = 0.13t$
$5.3 = t$
It will take 5.3 years.

47. $A = Pe^{rt}$
$900 = 500e^{r(10)}$
$1.8 = e^{10r}$
$\ln 1.8 = \ln e^{10r}$
$0.5878 = 10r$
$0.05878 = r$
$5.9\% = r$
The interest rate would need to be 5.9%.

49. $Q = Q_0e^{0.34t}$
$4000 = 400e^{0.34t}$
$10 = e^{0.34t}$
$\ln 10 = \ln e^{0.34t}$
$2.3026 = 0.34t$

$6.8 = t$
It will take 6.8 hours.

51. $P(a) = 14.7e^{-0.21a}$
$11.53 = 14.7e^{-0.21a}$
$0.7844 = e^{-0.21a}$
$\ln 0.7844 = \ln e^{-0.21a}$
$-0.2428 = -0.21a$
$1.156 = a$
$1.156 \text{ miles} \times \dfrac{5280 \text{ feet}}{1 \text{ mile}} = 6104 \text{ feet}$
Cheyenne is approximately 6100 feet above sea level.

53. $Q(t) = Q_0e^{0.4t}$
$2000 = 500e^{0.4t}$
$4 = e^{0.4t}$
$\ln 4 = \ln e^{0.4t}$
$1.3863 = 0.4t$
$3.5 = t$
It will take 3.5 hours.

55. $R = \log \dfrac{I}{I_0}$
$R = \log \dfrac{5{,}000{,}000 I_0}{I_0}$
$R = \log 5{,}000{,}000$
$R = 6.7$

57. For $R = 7.3$
$I = (10^{7.3})I_0$

For $R = 6.4$
$I = (10^{6.4})I_0$
$\dfrac{(10^{7.3})I_0}{(10^{6.4})I_0} = 10^{7.3-6.4} = 10^{0.9}$
$10^{0.9} = 7.9$

It is approximately 8 times more intense.

Further Investigations

63. Let $x = \log_a r$.
$a^x = r$
$\log_b a^x = \log_b r$
$x \log_b a = \log_b r$
$x = \dfrac{\log_b r}{\log_b a}$
a, b, and r are positive numbers with $a \neq 1$ and $b \neq 1$.

65. $\dfrac{5^x - 5^{-x}}{2} = 3$
$5^x - 5^{-x} = 6$
$5^x - \dfrac{1}{5^x} = 6$
$5^x\left(5^x - \dfrac{1}{5^x}\right) = 5^x(6)$
$5^{2x} - 1 = 6(5^x)$
$(5^x)^2 - 6(5^x) - 1 = 0$

Let $u = 5^x$.
$u^2 - 6u - 1 = 0$
$u = \dfrac{-(-6) \pm \sqrt{(-6)^2 - 4(1)(-1)}}{2(1)}$
$u = \dfrac{6 \pm \sqrt{40}}{2}$
$u = 6.162$ or $u = -0.1623$
Substituting $u = 5^x$
$5^x = 6.162$ or $5^x = -0.1623$ no solution
$\log 5^x = \log 6.162$
$x(\log 5) = \log 6.162$
$x(0.6990) = 0.7897$
$x = 1.13$
The solution set is $\{1.13\}$.

67. $y = \dfrac{e^x - e^{-x}}{2}$
$2y = e^x - e^{-x}$
$e^x(2y) = e^x(e^x - e^{-x})$
$2y(e^x) = (e^x)^2 - 1$
$0 = (e^x)^2 - 2y(e^x) - 1 = 0$
Let $u = e^x$.
$u^2 - 2y(u) - 1 = 0$
$u = \dfrac{-(-2y) \pm \sqrt{(-2y)^2 - 4(1)(-1)}}{2(1)}$
$u = \dfrac{2y \pm \sqrt{4y^2 + 4}}{2}$

$$u = \frac{2y \pm \sqrt{4(y^2+1)}}{2}$$
$$u = \frac{2y \pm 2\sqrt{y^2+1}}{2}$$
$$u = y + \sqrt{y^2+1}$$

Substituting $u = e^x$

$e^x = y + \sqrt{y^2+1}$

$\ln e^x = \ln(y + \sqrt{y^2+1})$

$x = \ln(y + \sqrt{y^2+1})$

CHAPTER 10 Review Problem Set

1. $8^{\frac{5}{3}}$

$(8^{\frac{1}{3}})^5 = 2^5 = 32$

2. $-25^{\frac{3}{2}}$

$-(25^{\frac{1}{2}})^3$

$-(5)^3$

-125

3. $(-27)^{\frac{4}{3}}$

$[(-27)^{\frac{1}{3}}]^4 = (-3)^4 = 81$

4. $\log_6 216 = x$

$6^x = 216$

$6^x = 6^3$

$x = 3$

5. $\log_7\left(\frac{1}{49}\right) = x$

$7^x = \frac{1}{49}$

$7^x = \frac{1}{7^2}$

$7^x = 7^{-2}$

$x = -2$

6. $\log_2 \sqrt[3]{2} = x$

$2^x = \sqrt[3]{2}$

$2^x = 2^{\frac{1}{3}}$

$x = \frac{1}{3}$

7. $\log_2\left(\frac{\sqrt[4]{32}}{2}\right) = x$

$2^x = \frac{\sqrt[4]{32}}{2}$

$2^x = \frac{\sqrt[4]{2^5}}{2}$

$2^x = \frac{2^{\frac{5}{4}}}{2^1}$

$2^x = 2^{\frac{5}{4}-1}$

$2^x = 2^{\frac{1}{4}}$

$x = \frac{1}{4}$

8. $\log_{10} 0.00001 = x$

$10^x = 0.00001$

$10^x = 10^{-5}$

$x = -5$

9. $\ln e = x$

$e^x = e$

$x = 1$

10. $7^{\log_7 12} = 12$

11. $\log_{10} 2 + \log_{10} x = 1$

$\log_{10} 2x = 1$

$10^1 = 2x$

$5 = x$

The solution set is $\{5\}$.

12. $\log_3 x = -2$

$3^{-2} = x$

$\frac{1}{9} = x$

The solution set is $\left\{\frac{1}{9}\right\}$.

13. $4^x = 128$

$(2^2)^x = 2^7$

$2^{2x} = 2^7$
$2x = 7$
$x = \frac{7}{2}$
The solution set is $\left\{\frac{7}{2}\right\}$.

14. $3^t = 42$
$\log 3^t = \log 42$
$t(\log 3) = 1.62325$
$t(0.47712) = 1.62325$
$t = 3.40$
The solution set is $\{3.40\}$.

15. $\log_2 x = 3$
$2^3 = x$
$8 = x$
The solution set is $\{8\}$.

16. $\left(\frac{1}{27}\right)^{3x} = 3^{2x-1}$
$(3^{-3})^{3x} = 3^{2x-1}$
$3^{-9x} = 3^{2x-1}$
$-9x = 2x - 1$
$-11x = -1$
$x = \frac{-1}{-11} = \frac{1}{11}$
The solution set is $\left\{\frac{1}{11}\right\}$.

17. $2e^x = 14$
$e^x = 7$
$\ln e^x = \ln 7$
$x = 1.95$
The solution set is $\{1.95\}$.

18. $2^{2x+1} = 3^{x+1}$
$\log 2^{2x+1} = \log 3^{x+1}$
$(2x+1)\log 2 = (x+1)\log 3$
$(2x+1)(0.3010) = (x+1)(0.4771)$
$0.6020x + 0.3010 = 0.4771x + 0.4771$
$0.1249x = 0.1761$
$x = 1.41$
The solution set is $\{1.41\}$.

19. $\ln(x+4) - \ln(x+2) = \ln x$
$\ln(x+4) = \ln x + \ln(x+2)$
$\ln(x+4) = \ln x(x+2)$
$x + 4 = x(x+2)$
$x + 4 = x^2 + 2x$
$0 = x^2 + x - 4$
$x = \frac{-1 \pm \sqrt{1^2 - 4(1)(-4)}}{2(1)}$
$x = \frac{-1 \pm \sqrt{17}}{2}$
$x = \frac{-1 + \sqrt{17}}{2}$ or $x = \frac{-1 - \sqrt{17}}{2}$
$x = 1.56$ or $x = -2.56$
Discard the root $x = -2.56$.
The solution set is $\{1.56\}$.

20. $\log x + \log(x - 15) = 2$
$\log x(x - 15) = 2$
$10^2 = x(x - 15)$
$100 = x^2 - 15x$
$0 = x^2 - 15x - 100$
$0 = (x - 20)(x + 5)$
$x - 20 = 0$ or $x + 5 = 0$
$x = 20$ or $x = -5$
Discard the root $x = -5$.
The solution set is $\{20\}$.

21. $\log(\log x) = 2$
$10^2 = \log x$
$100 = \log x$
$10^{100} = x$
The solution set is $\{10^{100}\}$.

22. $\log(7x - 4) - \log(x - 1) = 1$
$\log\frac{(7x-4)}{(x-1)} = 1$
$10^1 = \frac{7x-4}{x-1}$
$10(x - 1) = 7x - 4$
$10x - 10 = 7x - 4$
$3x = 6$
$x = 2$
The solution set is $\{2\}$.

23. $\ln(2t - 1) = \ln 4 + \ln(t - 3)$
$\ln(2t - 1) = \ln 4(t - 3)$
$2t - 1 = 4(t - 3)$
$2t - 1 = 4t - 12$

$11 = 2t$

$\frac{11}{2} = t$

The solution set is $\left\{\frac{11}{2}\right\}$.

24. $64^{2t+1} = 8^{-t+2}$

$(2^6)^{2t+1} = (2^3)^{-t+2}$

$2^{12t+6} = 2^{-3t+6}$

$12t + 6 = -3t + 6$

$15t = 0$

$t = 0$

The solution set is $\{0\}$.

25. $\log\left(\frac{7}{3}\right)$

$\log 7 - \log 3$

$0.8451 - 0.4771$

0.3680

26. $\log 21$

$\log 7(3)$

$\log 7 + \log 3$

$0.8451 + 0.4771$

1.3222

27. $\log 27$

$\log 3^3$

$3 \log 3$

$3(0.4771)$

1.4313

28. $\log 7^{\frac{2}{3}}$

$\frac{2}{3}\log 7$

$\frac{2}{3}(0.8451)$

0.5634

29a. $\log_b\left(\frac{x}{y^2}\right)$

$\log_b x - \log_b y^2$

$\log_b x - 2\log_b y$

b. $\log_b \sqrt[4]{xy^2}$

$\log_b (x^{\frac{1}{4}}y^{\frac{2}{4}})$

$\log_b x^{\frac{1}{4}} + \log_b y^{\frac{2}{4}}$

$\log_b x^{\frac{1}{4}} + \log_b y^{\frac{1}{2}}$

$\frac{1}{4}\log_b x + \frac{1}{2}\log_b y$

c. $\log_b\left(\frac{\sqrt{x}}{y^3}\right)$

$\log_b \sqrt{x} - \log_b y^3$

$\log_b x^{\frac{1}{2}} - \log_b y^3$

$\frac{1}{2}\log_b x - 3\log_b y$

30a) $3\log_b x + 2\log_b y$

$\log_b x^3 + \log_b y^2$

$\log_b x^3y^2$

b) $\frac{1}{2}\log_b y - 4\log_b x$

$\log_b \sqrt{y} - \log_b x^4$

$\log_b \frac{\sqrt{y}}{x^4}$

c) $\frac{1}{2}(\log_b x + \log_b y) - 2\log_b z$

$\frac{1}{2}(\log_b xy) - \log_b z^2$

$\log_b \sqrt{xy} - \log_b z^2$

$\log_b \frac{\sqrt{xy}}{z^2}$

31. $\log_2 3 = \frac{\log 3}{\log 2} = \frac{0.47712}{0.30103} = 1.585$

32. $\log_3 2 = \frac{\log 2}{\log 3} = \frac{0.30103}{0.477121} = 0.631$

33. $\log_4 191 = \frac{\log 191}{\log 4} = \frac{2.28103}{0.60206} = 3.789$

34. $\log_2 0.23 = \frac{\log 0.23}{\log 2} = -\frac{0.63827}{0.30103} = -2.120$

35. **(a)** $f(x) = \left(\frac{3}{4}\right)^x$

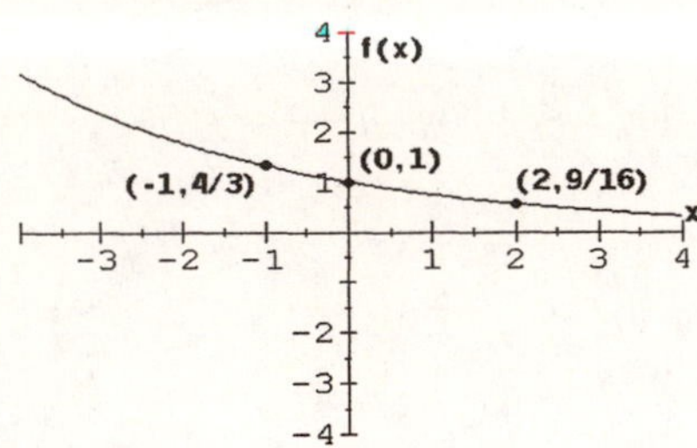

(b) $f(x) = \left(\frac{3}{4}\right)^x + 2$

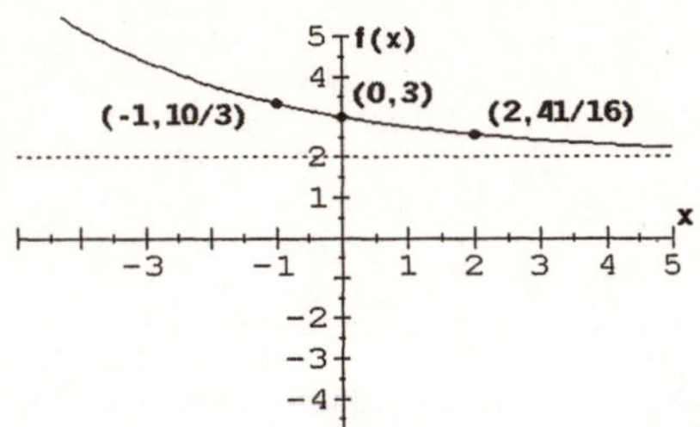

(c) $f(x) = \left(\frac{3}{4}\right)^{-x}$

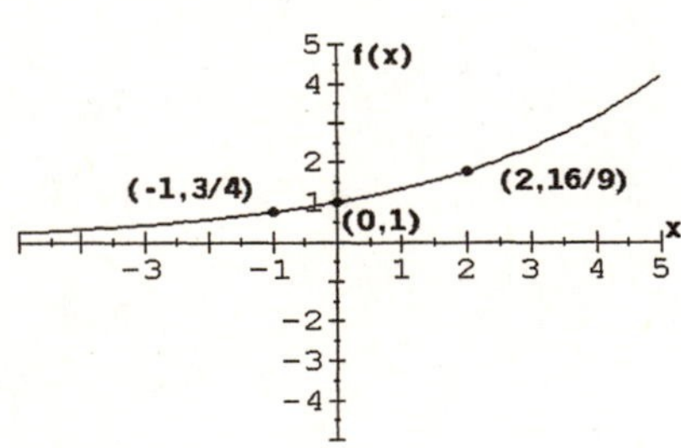

36. **(a)** $f(x) = 2^x$

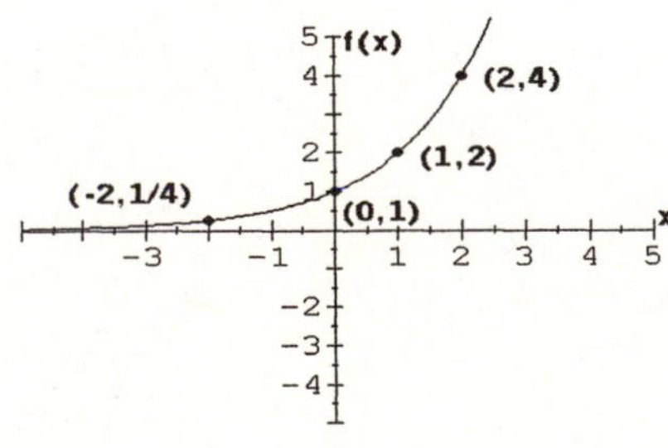

(b) $f(x) = 2^{x+2}$

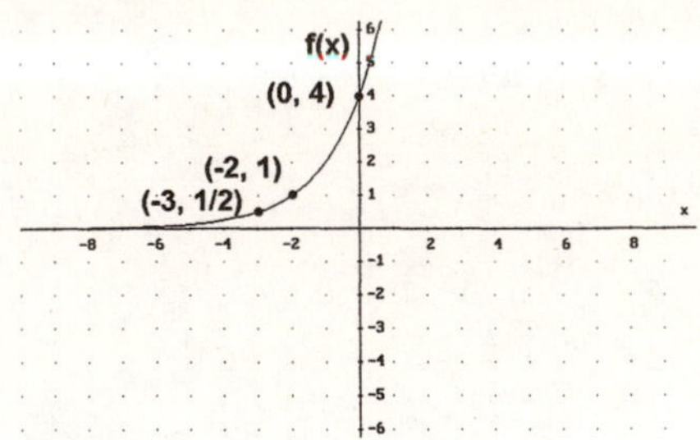

(c) $f(x) = -2^x$

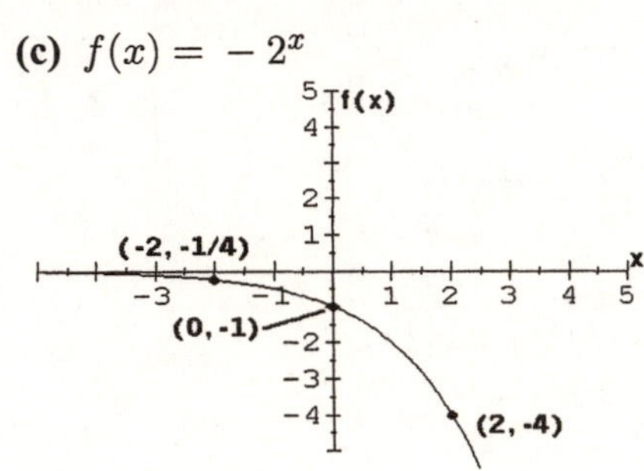

37. **(a)** $f(x) = e^{x-1}$

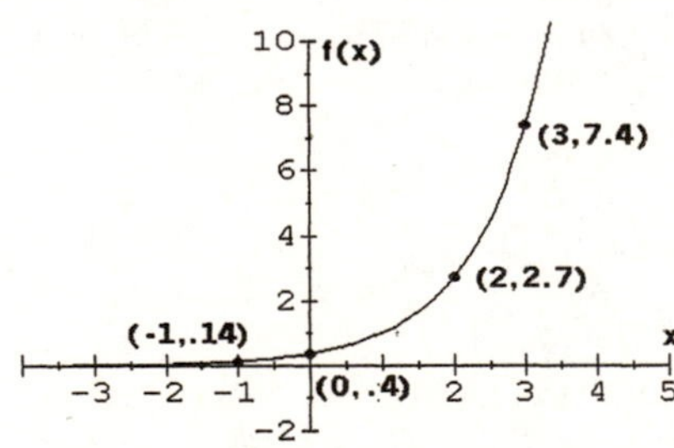

(b) $f(x) = e^x - 1$

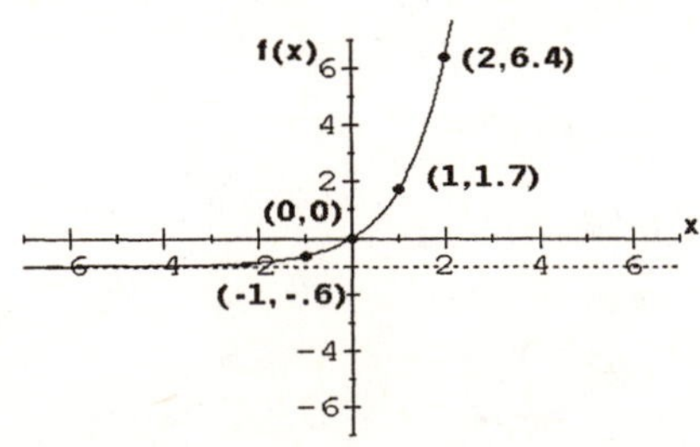

(c) $f(x) = e^{-x+1}$

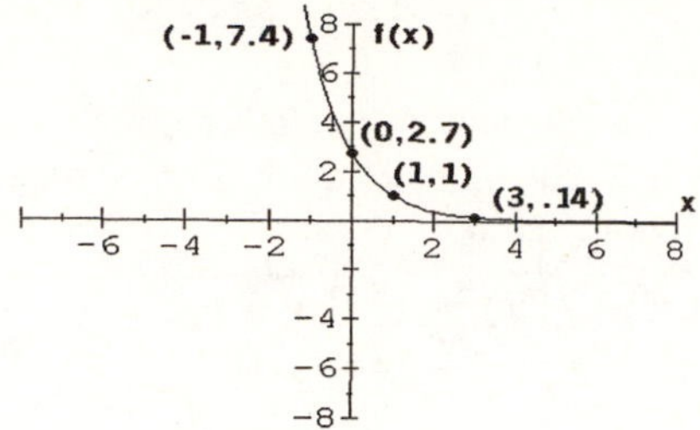

38. **(a)** $f(x) = -1 + \log x$

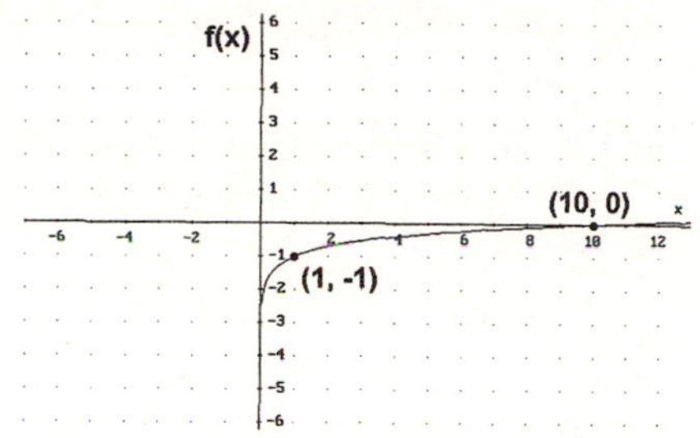

(b) $f(x) = \log(x-1)$

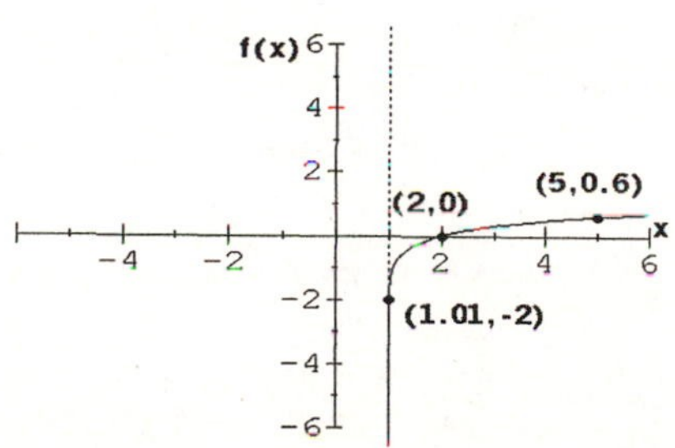

(c) $f(x) = -1 - \log x$

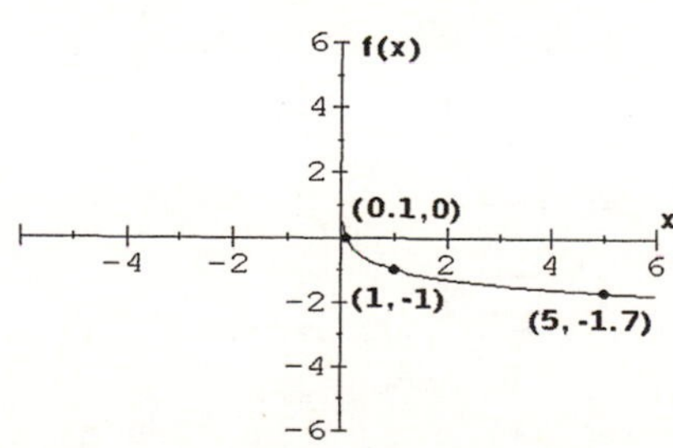

39. $f(x) = 3^x - 3^{-x}$

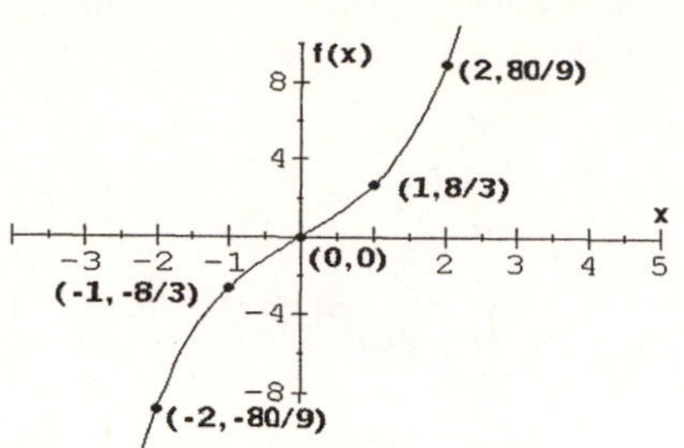

40. $f(x) = e^{-x^2/2}$

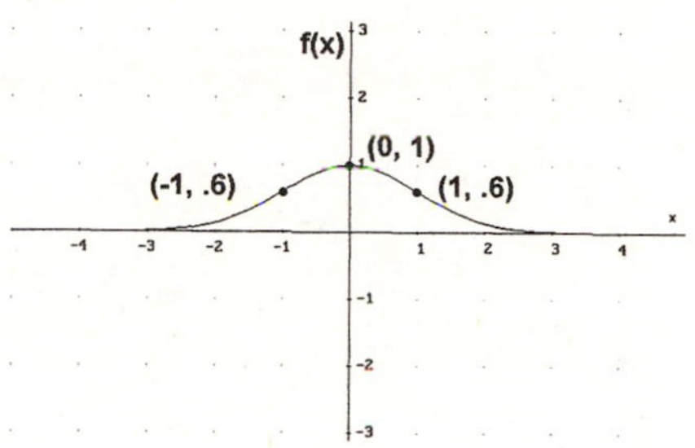

41. $f(x) = \log_2(x-3)$

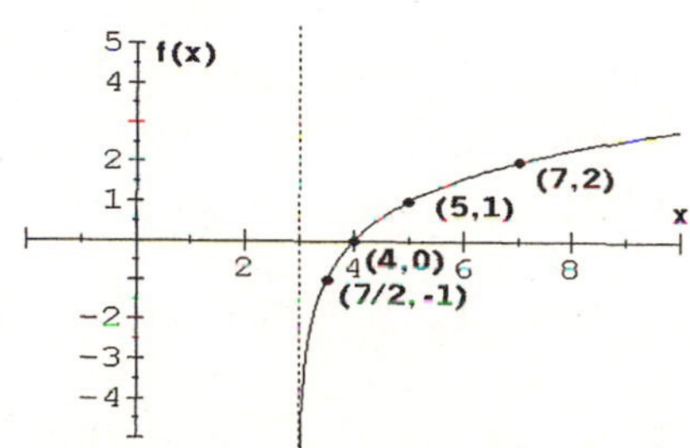

42. $f(x) = 3\log_3 x$

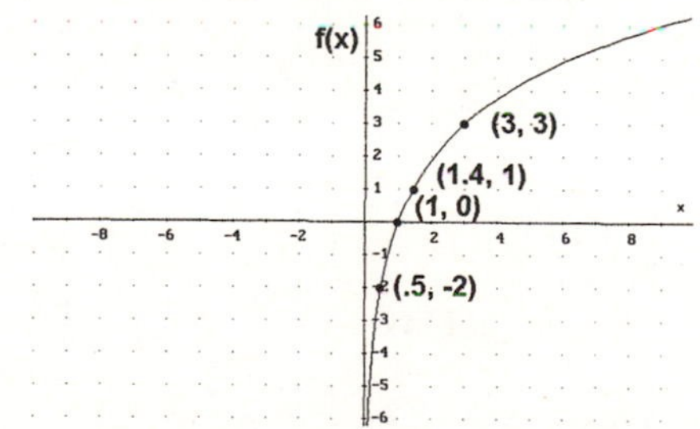

43. $A = P\left(1 + \frac{r}{n}\right)^{nt}$

$$A = 750\left(1 + \frac{0.11}{4}\right)^{4(10)}$$

$A = 750(1 + 0.0275)^{40}$
$A = 750(1.0275)^{40}$
$A = 750(2.959874)$
$A = \$2219.91$

44. $A = P\left(1 + \frac{r}{n}\right)^{nt}$

$A = 1250\left(1 + \frac{0.09}{12}\right)^{12(15)}$

$A = 1250(1.0075)^{180}$
$A = 1250(3.83804)$
$A = \$4797.55$

45. $A = P\left(1 + \frac{r}{n}\right)^{nt}$

$A = 2500\left(1 + \frac{0.095}{2}\right)^{2(20)}$

$A = 2500(1 + 0.0475)^{40}$
$A = 2500(1.0475)^{40}$
$A = 2500(6.399724)$
$A = \$15,999.31$

46. $f(x) = 7x - 1$ and $g(x) = \frac{x+1}{7}$
The set of real numbers is the domain and range of both functions

$(f \circ g)(x) = 7\left(\frac{x+1}{7}\right) - 1$
$= x + 1 - 1 = x$

$(g \circ f)(x) = \frac{7x - 1 + 1}{7} = \frac{7x}{7} = x$
Yes, they are inverse functions.

47. $f(x) = -\frac{2}{3}x$ and $g(x) = \frac{3}{2}x$

$(f \circ g)(x) = -\frac{2}{3}\left(\frac{3}{2}x\right) = -x$

They are not inverses.

48. $f(x) = x^2 - 6$ for $x \geq 0$
$g(x) = \sqrt{x+6}$ for $x \geq -6$

The domain of g equals the range of f, namely, all real numbers greater than or equal to -6.

$(f \circ g)(x) = \left(\sqrt{x+6}\right)^2 - 6$
$= x + 6 - 6 = x$

The domain of f equals the range of g, namely, all real numbers greater than or equal to 0.

$(g \circ f)(x) = \sqrt{x^2 - 6 + 6}$
$= \sqrt{x^2} = x$

Yes, they are inverse functions.

49. $f(x) = 2 - x^2$ for $x \geq 0$
$g(x) = \sqrt{2 - x}$ for $x \leq 2$

The domain of g equals the range of f, namely, all real numbers less than or equal to 2.

$(f \circ g)(x) = 2 - \left(\sqrt{2-x}\right)^2$
$= 2 - (2 - x) = x$

The domain of f equals the range of g, namely, all real numbers greater than or equal to 0.

$(g \circ f)(x) = \sqrt{2 - (2 - x^2)}$
$= \sqrt{2 - 2 + x^2}$
$= \sqrt{x^2} = x$

Yes, the functions are inverses.

50. $f(x) = 4x + 5$
$y = 4x + 5$

Interchange x and y and solve for y.

$x = 4y + 5$
$x - 5 = 4y$

$y = \frac{x-5}{4}$

$f^{-1}(x) = \frac{x-5}{4}$

The domain of f^{-1} equals the range of f, namely, all real numbers.

$(f \circ f^{-1})(x) = 4\left(\frac{x-5}{4}\right) + 5$
$= x - 5 + 5 = x$

The domain of f equals the range of f^{-1}, namely, all real numbers.

$$(f^{-1} \circ f)(x) = \frac{4x+5-5}{4} = \frac{4x}{4} = x$$

51. $f(x) = -3x - 7$

$y = -3x - 7$

To find the inverse interchange x and y, and solve y.

$x = -3y - 7$

$y = \dfrac{x+7}{-3}$

$f^{-1}(x) = \dfrac{-x-7}{3}$

The domain of f^{-1} equals the range of f, namely, all real numbers.

$$(f \circ f^{-1})(x) = -3\left(\frac{-x-7}{3}\right) - 7 = x + 7 - 7 = x$$

The domain of f equals the range of f^{-1}, namely, all real numbers.

$$(f^{-1} \circ f)(x) = \frac{-(-3x-7)-7}{3} = \frac{3x+7-7}{3} = x$$

52. $f(x) = \dfrac{5}{6}x - \dfrac{1}{3}$

$y = \dfrac{5}{6}x - \dfrac{1}{3}$

Interchange x and y and solve for y.

$x = \dfrac{5}{6}y - \dfrac{1}{3}$

$6x = 5y - 2$

$6x + 2 = 5y$

$\dfrac{6x+2}{5} = y$

$f^{-1}(x) = \dfrac{6x+2}{5}$

The domain of f^{-1} equals the range of f, namely, all real numbers.

$$(f \circ f^{-1})(x) = \frac{5}{6}\left(\frac{6x+2}{5}\right) - \frac{1}{3} = \frac{6x+2}{6} - \frac{1}{3} = x + \frac{1}{3} - \frac{1}{3} = x$$

The domain of f equals the range of f^{-1}, namely, all real numbers.

$$(f^{-1} \circ f)(x) = \frac{6\left(\frac{5}{6}x - \frac{1}{3}\right) + 2}{5} = \frac{5x-2+2}{5} = \frac{5x}{5} = x$$

53. $f(x) = -2 - x^2$ for $x \geq 0$

Range: $f(x) \leq -2$

$y = -2 - x^2$

To find the inverse interchange x and y and then solve for y.

$x = -2 - y^2$

$x + 2 = -y^2$

$-x - 2 = y^2$

Since the domain of the inverse function is $x \leq -2$ and the range if $f^{-1}(x) \geq 0$, we want only the positive root of y^2.

$y = \sqrt{-x-2}$

$f^{-1}(x) = \sqrt{-x-2}$

The domain of f^{-1} equals the range of f, namely, all real numbers less than or equal to -2.

$$(f \circ f^{-1})(x) = -2 - (\sqrt{-x-2})^2 = -2 - (-x-2) = -2 + x + 2 = x$$

The domain of f equals the range of f^{-1}, namely, all real numbers greater than or equal to 0.

$$(f^{-1} \circ f)(x) = \sqrt{-(-2-x^2)-2}$$
$$= \sqrt{2+x^2-2}$$
$$= \sqrt{x^2} = x$$

54. $-2x^2 + 16x - 32 = 0$
$-2(x^2 - 8x + 16) = 0$
$-2(x-4)(x-4) = 0$
Critical value $x = 4$

$f(0) = -32$
$f(1) = -18$

Since $0 < 1$ and $f(0) < f(1)$,
$f(x)$ is increasing on $(-\infty, 4]$.

$f(5) = -2$
$f(6) = -8$

Since $5 < 6$ and $f(5) > f(6)$,
$f(x)$ is decreasing on $[4, \infty)$.

55. $f(x) = 2\sqrt{x-3}$
Increasing on $[3, \infty)$.

56. $A = P\left(1 + \frac{r}{n}\right)^{nt}$

$200 = 100\left(1 + \frac{0.14}{1}\right)^{1t}$
$2 = (1.14)^t$
$\log 2 = \log(1.14)^t$
$0.3010 = t \log(1.14)$
$0.3010 = t(0.0569)$
$5.3 = t$
It will take approximately 5.3 years.

57. $A = P\left(1 + \frac{r}{n}\right)^{nt}$

$3500 = 1000\left(1 + \frac{0.105}{4}\right)^{4t}$
$3.5 = (1 + .02625)^{4t}$
$3.5 = (1.02625)^{4t}$
$\log 3.5 = \log(1.02625)^{4t}$
$0.544068 = 4t \log(1.02625)$
$0.544068 = 4t(0.011253)$
$0.544068 = 0.045012t$
$12.1 = t$
It will take 12.1 years.

58. $A = Pe^{rt}$
$1000 = 500e^{r(8)}$
$2 = e^{8r}$
$\ln 2 = \ln e^{8r}$
$0.6931 = 8r$
$0.087 = r$
$8.7\% = r$
The rate will be approximately 8.7%.

59. $P(t) = P_0e^{0.02t}$

$P(10) = 50000e^{0.02(10)}$
$P(10) = 50000e^{0.2}$
$P(10) = 50000(1.2214)$
$P(10) = 61{,}070$

$P(15) = 50000e^{0.02(15)}$
$P(15) = 50000e^{0.3}$
$P(15) = 50000(1.34986)$
$P(15) = 67{,}493$

$P(20) = 50000e^{0.02(20)}$
$P(20) = 50000e^{0.4}$
$P(20) = 50000(1.49182)$
$P(20) = 74{,}591$

60. $Q = Qe^{0.29t}$
$2000 = 500e^{0.29t}$
$4 = e^{0.29t}$
$\ln 4 = \ln e^{0.29t}$
$1.3863 = 0.29t$
$4.8 = t$
It will take approximately 4.8 hours.

61. $Q = Q_0\left(\frac{1}{2}\right)^{\frac{t}{n}}$

$Q = 750\left(\frac{1}{2}\right)^{\frac{100}{40}}$
$Q = 750(0.5)^{2.5}$
$Q = 750(0.1767767)$
$Q = 133$
There will be 133 grams
left after 100 days.

62. $R = \log \dfrac{I}{I_0}$

$R = \log \dfrac{125{,}000{,}000 I_0}{I_0}$

$R = \log 125{,}000{,}000$

$R = 8.1$

The Richter number is 8.1.

CHAPTER 10 Test

1. $\log_3 \sqrt{3} = x$

$3^x = \sqrt{3}$

$3^x = 3^{\frac{1}{2}}$

$x = \dfrac{1}{2}$

2. $\log_2(\log_2 4)$

$\log_2 2 = x$

$2^x = 2^1$

$x = 1$

3. $-2 + \ln e^3$

$-2 + 3$

1

4. $\log_2(0.5) = x$

$2^x = 0.5$

$2^x = \dfrac{1}{2}$

$2^x = 2^{-1}$

$x = -1$

5. $4^x = \dfrac{1}{64}$

$4^x = \dfrac{1}{4^3}$

$4^x = 4^{-3}$

$x = -3$

The solution set is $\{-3\}$.

6. $9^x = \dfrac{1}{27}$

$(3^2)^x = \dfrac{1}{3^3}$

$3^{2x} = 3^{-3}$

$2x = -3$

$x = -\dfrac{3}{2}$

The solution set is $\left\{-\dfrac{3}{2}\right\}$.

7. $2^{3x-1} = 128$

$2^{3x-1} = 2^7$

$3x - 1 = 7$

$3x = 8$

$x = \dfrac{8}{3}$

The solution set is $\left\{\dfrac{8}{3}\right\}$.

8. $\log_9 x = \dfrac{5}{2}$

$9^{\frac{5}{2}} = x$

$(9^{\frac{1}{2}})^5 = x$

$3^5 = x$

$243 = x$

The solution set is $\{243\}$.

9. $\log x + \log(x + 48) = 2$

$\log x(x + 48) = 2$

$10^2 = x(x + 48)$

$100 = x^2 + 48x$

$0 = x^2 + 48x - 100$

$0 = (x + 50)(x - 2)$

$x + 50 = 0$ or $x - 2 = 0$

$x = -50$ or $x = 2$

Discard the root $x = -50$.

The solution set is $\{2\}$.

10. $\ln x = \ln 2 + \ln(3x - 1)$

$\ln x = \ln 2(3x - 1)$

$e^{\ln x} = e^{\ln 2(3x-1)}$

$x = 2(3x - 1)$
$x = 6x - 2$
$-5x = -2$
$x = \frac{-2}{-5} = \frac{2}{5}$
The solution set is $\left\{\frac{2}{5}\right\}$.

11. $\log_3 100$
$\log_3 (5^2 \bullet 4)$
$\log_3 5^2 + \log_3 4$
$2 \log_3 5 + \log_3 4$
$2(1.4650) + 1.2619$
$2.9300 + 1.2619$
4.1919

12. $\log_3 1.25$
$\log_3 \left(\frac{1}{4}\right)(5)$
$\log_3 \left(\frac{1}{4}\right) + \log_3 5$
$\log_3 4^{-1} + 1.4650$
$-1 \log_3 4 + 1.4650$
$-1.2619 + 1.4650$
0.2031

13. $\log_3 \sqrt{5}$
$\log_3 5^{\frac{1}{2}}$
$\frac{1}{2} \log_3 5$
$\frac{1}{2}(1.4650)$
0.7325

14. $f(x) = -3x - 6$
$y = -3x - 6$
Interchange x and y and solve for y.
$x = -3y - 6$
$x + 6 = -3y$
$\frac{x+6}{-3} = y$
$y = \frac{-x-6}{3}$
$f^{-1}(x) = \frac{-x-6}{3}$

15. $e^x = 176$
$\ln e^x = \ln 176$
$x = 5.17$
The solution set is $\{5.17\}$.

16. $2^{x-6} = 314$
$\log 2^{x-2} = \log 314$
$(x-2)\log 2 = \log 314$
$x - 2 = \frac{\log 314}{\log 2}$
$x - 2 = \frac{2.4969}{0.3010}$
$x - 2 = 8.29$
$x = 10.29$
The solution set is $\{10.29\}$.

17. $\log_5 632 = \frac{\log 632}{\log 5} = \frac{2.800717}{0.698970} = 4.0069$

18. $f(x) = \frac{2}{3}x - \frac{3}{5}$
$y = \frac{2}{3}x - \frac{3}{5}$
Interchange x and y and solve for y.
$x = \frac{2}{3}y - \frac{3}{5}$
$x + \frac{3}{5} = \frac{2}{3}y$
$\frac{3}{2}\left(x + \frac{3}{5}\right) = \frac{3}{2}\left(\frac{2}{3}y\right)$
$\frac{3}{2}x + \frac{9}{10} = y$
$f^{-1}(x) = \frac{3}{2}x + \frac{9}{10}$

19. $A = P\left(1 + \frac{r}{n}\right)^{nt}$
$A = 3500\left(1 + \frac{0.075}{4}\right)^{4(8)}$
$A = 3500(1.01875)^{32}$
$A = 3500(1.812024)$
$A = \$6342.08$

20. $A = P\left(1 + \frac{r}{n}\right)^{nt}$

$12500 = 5000\left(1 + \frac{0.07}{1}\right)^{1t}$

$2.5 = (1.07)^t$

$\log 2.5 = \log 1.07^t$

$\log 2.5 = t \log 1.07$

$t = \frac{\log 2.5}{\log 1.07}$

$t = \frac{0.39794}{0.02938}$

$t = 13.5$

It will take 13.5 years.

21. $Q(t) = Q_0 e^{0.23t}$

$2400 = 400e^{0.23t}$

$6 = e^{0.23t}$

$\ln 6 = \ln e^{0.23t}$

$1.79176 = 0.23t$

$7.8 = t$

It will take 7.8 hours.

22. $Q = Q_0\left(\frac{1}{2}\right)^{\frac{r}{n}}$

$Q = 7500\left(\frac{1}{2}\right)^{\frac{32}{50}}$

$Q = 7500\left(\frac{1}{2}\right)^{0.64}$

$Q = 7500(0.64171)$

$Q = 4813$ grams

23. $f(x) = e^x - 2$

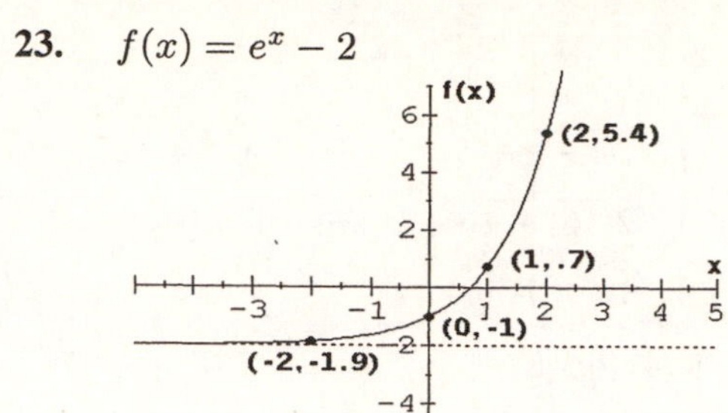

24. $f(x) = -3^{-x}$

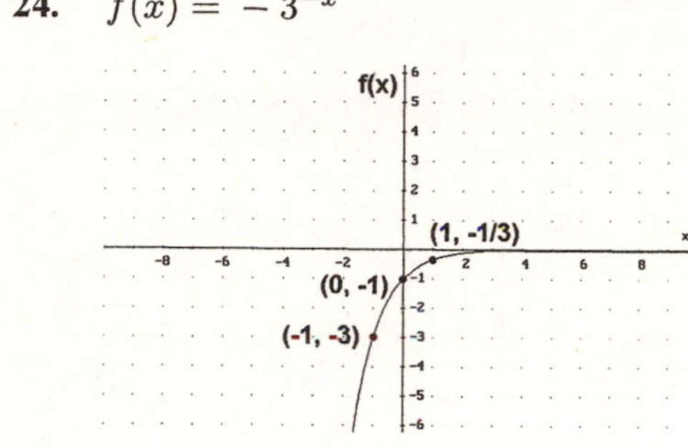

25. $f(x) = \log_2(x - 2)$

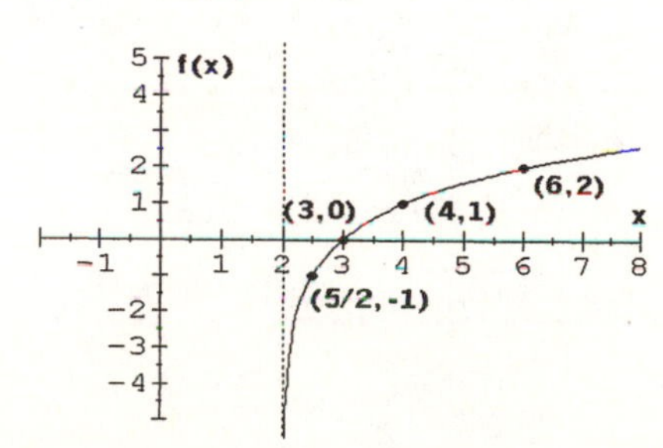

CHAPTER 10 Cumulative Review Chapters 1-10

1. $-5(x-1) - 3(2x+4) + 3(3x-1);$

$-5x + 5 - 6x - 12 + 9x - 3$

$-2x - 10$ for $x = -2$

$-2(-2) - 10$

$4 - 10 = -6$

2. $\frac{14a^3b^2}{7a^2b}$ for $a = -1, b = 4$

$2ab$

$2(-1)(4)$

-8

3. $\frac{2}{n} - \frac{3}{2n} + \frac{5}{3n}$; for $n = 4$

$\frac{2}{4} - \frac{3}{2(4)} + \frac{5}{3(4)}$

$\frac{1}{2} - \frac{3}{8} + \frac{5}{12}$; LCD = 24

$\frac{1}{2} \bullet \frac{12}{12} - \frac{3}{8} \bullet \frac{3}{3} + \frac{5}{12} \bullet \frac{2}{2}$

$\frac{12}{24} - \frac{9}{24} + \frac{10}{24} = \frac{13}{24}$

4. $4\sqrt{2x-y}+5\sqrt{3x+y}$ for $x = 16, y = 16$

$4\sqrt{2(16)-16}+5\sqrt{3(16)+16}$

$4\sqrt{16}+5\sqrt{64}$

$4(4)+5(8)$

$16+40=56$

5. $\frac{3}{x-2} - \frac{5}{x+3}$; for $x = 3$

$\frac{3}{3-2} - \frac{5}{3+3}$

$\frac{3}{1} - \frac{5}{6}$; LCD $= 6$

$\frac{18}{6} - \frac{5}{6}$

$\frac{13}{6}$

6. $(-5\sqrt{6})(3\sqrt{12})$

$-15\sqrt{72}$

$-15\sqrt{36}\sqrt{2}$

$-15(6)\sqrt{2}$

$-90\sqrt{2}$

7. $(2\sqrt{x}-3)(\sqrt{x}+4)$

$2\sqrt{x^2}+8\sqrt{x}-3\sqrt{x}-12$

$2x+5\sqrt{x}-12$

8. $(3\sqrt{2}-\sqrt{6})(\sqrt{2}+4\sqrt{6})$

$3\sqrt{4}+12\sqrt{12}-\sqrt{12}-4\sqrt{36}$

$3(2)+11\sqrt{12}-4(6)$

$6+11(2\sqrt{3})-24$

$-18+22\sqrt{3}$

9. $(2x-1)(x^2+6x-4)$

$2x^3+12x^2-8x-x^2-6x+4$

$2x^3+11x^2-14x+4$

10. $\frac{x^2-x}{x+5} \bullet \frac{x^2+5x+4}{x^4-x^2}$

$\frac{x(x-1)}{x+5} \bullet \frac{(x+4)(x+1)}{x^2(x+1)(x-1)}$

$\frac{x+4}{x(x+5)}$

11. $\frac{16x^2y}{24xy^3} \div \frac{9xy}{8x^2y^2}$

$\frac{16x^2y}{24xy^3} \bullet \frac{8x^2y^2}{9xy}$

$\frac{16(8)x^4y^3}{24(9)x^2y^4}$

$\frac{16x^2}{27y}$

12. $\frac{x+3}{10} + \frac{2x+1}{15} - \frac{x-2}{18}$; LCD $= 90$

$\frac{9}{9} \bullet \frac{(x+3)}{10} + \frac{6}{6} \bullet \frac{(2x+1)}{15} - \frac{5}{5} \bullet \frac{(x-2)}{18}$

$\frac{9(x+3)+6(2x+1)-5(x-2)}{90}$

$\frac{9x+27+12x+6-5x+10}{90}$

$\frac{16x+43}{90}$

13. $\frac{7}{12ab} - \frac{11}{15a^2}$; LCD $= 60a^2b$

$\frac{7}{12ab} \bullet \frac{5a}{5a} - \frac{11}{15a^2} \bullet \frac{4b}{4b}$

$\frac{35a}{60a^2b} - \frac{44b}{60a^2b}$

$\frac{35a-44b}{60a^2b}$

14. $\dfrac{8}{x^2-4x}+\dfrac{2}{x}$

$\dfrac{8}{x(x-4)}+\dfrac{2}{x}$; LCD $= x(x-4)$

$\dfrac{8}{x(x-4)}+\dfrac{2}{x}\bullet\dfrac{(x-4)}{(x-4)}$

$\dfrac{8+2(x-4)}{x(x-4)}$

$\dfrac{8+2x-8}{x(x-4)}$

$\dfrac{2x}{x(x-4)}$

$\dfrac{2}{x-4}$

15.

$$
\begin{array}{r|l}
 & \phantom{8x^3-{}}2x^2-x-4 \\
\hline
4x-1 & 8x^3-6x^2-15x+4 \\
 & 8x^3-2x^2 \\
\hline
 & -4x^2-15x \\
 & -4x^2+x \\
\hline
 & -16x+4 \\
 & -16x+4 \\
\hline
 & 0
\end{array}
$$

$2x^2-x-4$

16. $\dfrac{\dfrac{5}{x^2}-\dfrac{3}{x}}{\dfrac{1}{y}+\dfrac{2}{y^2}}$

$\dfrac{x^2y^2}{x^2y^2}\bullet\dfrac{\left(\dfrac{5}{x^2}-\dfrac{3}{x}\right)}{\left(\dfrac{1}{y}+\dfrac{2}{y^2}\right)}$

$\dfrac{x^2y^2\left(\dfrac{5}{x^2}\right)-x^2y^2\left(\dfrac{3}{x}\right)}{x^2y^2\left(\dfrac{1}{y}\right)+x^2y^2\left(\dfrac{2}{y^2}\right)}$

$\dfrac{5y^2-3xy^2}{x^2y+2x^2}$

17. $\dfrac{\dfrac{2}{x}-3}{\dfrac{3}{y}+4}$

$\dfrac{xy}{xy}\bullet\dfrac{\dfrac{2}{x}-3}{\dfrac{3}{y}+4}$

$\dfrac{xy\left(\dfrac{2}{x}-3\right)}{xy\left(\dfrac{3}{y}+4\right)}$

$\dfrac{xy\left(\dfrac{2}{x}\right)-xy(3)}{xy\left(\dfrac{3}{7}\right)+xy(4)}$

$\dfrac{xy\left(\dfrac{2}{x}\right)-xy(3)}{xy\left(\dfrac{3}{y}\right)+xy(4)}$

$\dfrac{2y-3xy}{3x+4xy}$

18. $\dfrac{2-\dfrac{1}{n-2}}{3+\dfrac{4}{n+3}}$

$\dfrac{(n-2)(n+3)}{(n-2)(n+3)}\bullet\dfrac{\left(2-\dfrac{1}{n-2}\right)}{\left(3+\dfrac{4}{n+3}\right)}$

$\dfrac{2(n-2)(n+3)-(n+3)}{3(n-2)(n+3)+4(n-2)}$

$\dfrac{(n+3)[2(n-2)-1]}{(n-2)[3(n+3)+4]}$

$\dfrac{(n+3)(2n-4-1)}{(n-2)(3n+9+4)}$

$$\frac{(n+3)(2n-5)}{(n-2)(3n+13)}$$

19. $\dfrac{3a}{2-\frac{1}{a}} - 1$

$\dfrac{a}{a} \bullet \dfrac{3a}{\left(2-\frac{1}{a}\right)} - 1$

$\dfrac{3a^2}{2a-1} - 1$

$\dfrac{3a^2}{2a-1} - 1 \bullet \dfrac{(2a-1)}{2a-1}$

$\dfrac{3a^2-1(2a-1)}{2a-1}$

$\dfrac{3a^2-2a+1}{2a-1}$

20. $20x^2+7x-6$
$(5x-2)(4x+3)$

21. $16x^3+54$
$2(8x^3+27)$
$2(2x+3)(4x^2-6x+9)$

22. $4x^4-25x^2+36$
$(4x^2-9)(x^2-4)$
$(2x+3)(2x-3)(x+2)(x-2)$

23. $12x^3-52x^2-40x$
$4x(3x^2-13x-10)$
$4x(3x+2)(x-5)$

24. $xy-6x+3y-18$
$x(y-6)+3(y-6)$
$(y-6)(x+3)$

25. $10+9x-9x^2$
$(5-3x)(2+3x)$

26. $\left(\dfrac{2}{3}\right)^{-4} = \left(\dfrac{3}{2}\right)^{4} = \dfrac{3^4}{2^4} = \dfrac{81}{16}$

27. $\dfrac{3}{\left(\frac{4}{3}\right)^{-1}}$

$3\left(\dfrac{4}{3}\right)^{1} = 4$

28. $\sqrt[3]{-\dfrac{27}{64}} = -\dfrac{3}{4}$

29. $-\sqrt{0.09} = -0.3$

30. $(27)^{-\frac{4}{3}} = \dfrac{1}{27^{\frac{4}{3}}} = \dfrac{1}{(27^{\frac{1}{3}})^4} = \dfrac{1}{3^4} = \dfrac{1}{81}$

31. $4^0+4^{-1}+4^{-2}$

$1+\dfrac{1}{4}+\dfrac{1}{4^2}$

$1+\dfrac{1}{4}+\dfrac{1}{16}$

$\dfrac{16}{16}+\dfrac{4}{16}+\dfrac{1}{16}$

$\dfrac{21}{16}$

32. $\left(\dfrac{3^{-1}}{2^{-3}}\right)^{-2} = \dfrac{3^2}{2^6} = \dfrac{9}{64}$

33. $(2^{-3}-3^{-2})^{-1}$

$\left(\dfrac{1}{2^3}-\dfrac{1}{3^2}\right)^{-1}$

$\left(\dfrac{1}{8}-\dfrac{1}{9}\right)^{-1}$

$\left(\dfrac{9}{72}-\dfrac{8}{72}\right)^{-1}$

$\left(\dfrac{1}{72}\right)^{-1}$

$\left(\dfrac{72}{1}\right)^{1} = 72$

34. $\log_2 64 = x$
$2^x = 64$
$2^x = 2^6$
$x = 6$

35. $\log_3\left(\frac{1}{9}\right) = x$

$3^x = \frac{1}{9}$

$3^x = \frac{1}{3^2}$

$3^x = 3^{-2}$

$x = -2$

36. $(-3x^{-1}y^2)(4x^{-2}y^{-3})$

$-12x^{-3}y^{-1}$

$\frac{-12}{x^3y}$

37. $\frac{48x^{-4}y^2}{6xy}$

$8x^{-4-1}y^{2-1}$

$8x^{-5}y^1$

$\frac{8y}{x^5}$

38. $\left(\frac{27a^{-4}b^{-3}}{-3a^{-1}b^{-4}}\right)^{-1}$

$(-9a^{-3}b^1)^{-1}$

$(-9)^{-1}a^3b^{-1}$

$-\frac{a^3}{9b}$

39. $\sqrt{80}$

$\sqrt{16}\sqrt{5}$

$4\sqrt{5}$

40. $-2\sqrt{54}$

$-2\sqrt{9}\sqrt{6}$

$-2(3)\sqrt{6}$

$-6\sqrt{6}$

41. $\sqrt{\frac{75}{81}} = \frac{\sqrt{75}}{\sqrt{81}} = \frac{\sqrt{25}\sqrt{3}}{9} = \frac{5\sqrt{3}}{9}$

42. $\frac{4\sqrt{6}}{3\sqrt{8}} = \frac{4}{3}\sqrt{\frac{6}{8}} = \frac{4}{3}\sqrt{\frac{3}{4}}$

$\frac{4\sqrt{3}}{3\sqrt{4}} = \frac{4\sqrt{3}}{3(2)} = \frac{2\sqrt{3}}{3}$

43. $\sqrt[3]{56}$

$\sqrt[3]{8}\sqrt[3]{7}$

$2\sqrt[3]{7}$

44. $\frac{\sqrt[3]{3}}{\sqrt[3]{4}} = \frac{\sqrt[3]{3}}{\sqrt[3]{4}} \bullet \frac{\sqrt[3]{2}}{\sqrt[3]{2}} = \frac{\sqrt[3]{6}}{\sqrt[3]{8}} = \frac{\sqrt[3]{6}}{2}$

45. $4\sqrt{52x^3y^2}$

$4\sqrt{4x^2y^2}\sqrt{13x}$

$4(2xy)\sqrt{13x}$

$8xy\sqrt{13x}$

46. $\sqrt{\frac{2x}{3y}} = \frac{\sqrt{2x}}{\sqrt{3y}} = \frac{\sqrt{2x}}{\sqrt{3y}} \bullet \frac{\sqrt{3y}}{\sqrt{3y}} = \frac{\sqrt{6xy}}{3y}$

47. $-3\sqrt{24} + 6\sqrt{54} - \sqrt{6}$

$-3\sqrt{4}\sqrt{6} + 6\sqrt{9}\sqrt{6} - \sqrt{6}$

$-3(2)\sqrt{6} + 6(3)\sqrt{6} - \sqrt{6}$

$-6\sqrt{6} + 18\sqrt{6} - \sqrt{6}$

$11\sqrt{6}$

48. $\frac{\sqrt{8}}{3} - \frac{3\sqrt{18}}{4} - \frac{5\sqrt{50}}{2}$

$\frac{2\sqrt{2}}{3} - \frac{9\sqrt{2}}{4} - \frac{25\sqrt{2}}{2}$; LCD = 12

$\frac{4}{4} \bullet \frac{(2\sqrt{2})}{3} - \frac{3}{3} \bullet \frac{(9\sqrt{2})}{4} - \frac{6}{6} \bullet \frac{(25\sqrt{2})}{2}$

$\frac{8\sqrt{2}}{12} - \frac{27\sqrt{2}}{12} - \frac{150\sqrt{2}}{12}$

$\frac{-169\sqrt{2}}{12} = -\frac{169\sqrt{2}}{12}$

49. $8\sqrt[3]{3} - 6\sqrt[3]{24} - 4\sqrt[3]{81}$

$8\sqrt[3]{3} - 6\sqrt[3]{8}\sqrt[3]{3} - 4\sqrt[3]{27}\sqrt[3]{3}$

$8\sqrt[3]{3} - 6(2)\sqrt[3]{3} - 4(3)\sqrt[3]{3}$

$8\sqrt[3]{3} - 12\sqrt[3]{3} - 12\sqrt[3]{3}$

$-16\sqrt[3]{3}$

50. $\dfrac{\sqrt{3}}{\sqrt{6} - 2\sqrt{2}}$

$\dfrac{\sqrt{3}}{(\sqrt{6} - 2\sqrt{2})} \bullet \dfrac{(\sqrt{6} + 2\sqrt{2})}{(\sqrt{6} + 2\sqrt{2})}$

$\dfrac{\sqrt{18} + 2\sqrt{6}}{\sqrt{36} - 4\sqrt{4}} = \dfrac{3\sqrt{2} + 2\sqrt{6}}{6 - 4(2)}$

$\dfrac{3\sqrt{2} + 2\sqrt{6}}{6 - 8} = \dfrac{3\sqrt{2} + 2\sqrt{6}}{-2}$

$\dfrac{-3\sqrt{2} - 2\sqrt{6}}{2}$

51. $\dfrac{3\sqrt{5} - \sqrt{3}}{2\sqrt{3} + \sqrt{7}}$

$\dfrac{(3\sqrt{5} - \sqrt{3})}{(2\sqrt{3} + \sqrt{7})} \bullet \dfrac{(2\sqrt{3} - \sqrt{7})}{(2\sqrt{3} - \sqrt{7})}$

$\dfrac{6\sqrt{15} - 3\sqrt{35} - 2\sqrt{9} + \sqrt{21}}{4\sqrt{9} - \sqrt{49}}$

$\dfrac{6\sqrt{15} - 3\sqrt{35} - 2(3) + \sqrt{21}}{4(3) - 7}$

$\dfrac{6\sqrt{15} - 3\sqrt{35} - 6 + \sqrt{21}}{5}$

52. $\dfrac{(0.00016)(300)(0.028)}{0.064}$

$\dfrac{(1.6)(10)^{-4}(3)(10)^{2}(2.8)^{-2}}{(6.4)(10)^{-2}}$

$(2.1)(10)^{-4+2-2-(-2)}$

$(2.1)(10)^{-2}$

0.021

53. $\dfrac{0.00072}{0.0000024}$

$\dfrac{(7.2)(10)^{-4}}{(2.4)(10)^{-6}} = 3(10)^{-4-(-6)}$

$3(10)^{2} = 300$

54. $\sqrt{0.00000009}$

$\sqrt{(9)(10)^{-8}} = \sqrt{9}\sqrt{(10)^{-8}}$

$3(10)^{-4} = 0.0003$

55. $(5 - 2i)(4 + 6i)$

$20 + 30i - 8i - 12i^2$

$20 + 22i + 12$

$32 + 22i$

56. $(-3 - i)(5 - 2i)$

$-15 + 6i - 5i + 2i^2$

$-15 + i - 2$

$-17 + i$

57. $\dfrac{5}{4i} = \dfrac{5}{4i} \bullet \dfrac{(-i)}{(-i)} = \dfrac{-5i}{-4i^2} = \dfrac{-5i}{4}$

$0 - \dfrac{5}{4}i$

58. $\dfrac{-1 + 6i}{7 - 2i}$

$\dfrac{(-1 + 6i)}{(7 - 2i)} \bullet \dfrac{(7 + 2i)}{(7 + 2i)}$

$\dfrac{-7 - 2i + 42i + 12i^2}{49 - 4i^2}$

$\dfrac{-7 + 40i - 12}{49 + 4}$

$\dfrac{-19 + 40i}{53} = -\dfrac{19}{53} + \dfrac{40}{53}i$

59. $(2, -3)$ and $(-1, 7)$

$m = \dfrac{7 - (-3)}{-1 - 2} = \dfrac{10}{-3} = -\dfrac{10}{3}$

60. $4x - 7y = 9$
$-7y = -4x + 9$
$\left(-\frac{1}{7}\right)(-7y) = \left(-\frac{1}{7}\right)(-4x + 9)$
$y = \frac{4}{7}x - \frac{9}{7}$
$m = \frac{4}{7}$

61. $(4, 5)$ and $(-2, 1)$
$d = \sqrt{(-2-4)^2 + (1-5)^2}$
$d = \sqrt{(-6)^2 + (-4)^2}$
$d = \sqrt{36 + 16} = \sqrt{52}$
$d = \sqrt{4}\sqrt{13} = 2\sqrt{13}$

62. $(3, -1)$ and $(7, 4)$
$m = \frac{4-(-1)}{7-3} = \frac{5}{4}$
$y - 4 = \frac{5}{4}(x - 7)$
$4(y - 4) = 4\left[\frac{5}{4}(x - 7)\right]$
$4y - 16 = 5(x - 7)$
$4y - 16 = 5x - 35$
$-16 = 5x - 4y - 35$
$19 = 5x - 4y$
$5x - 4y = 19$

63. $3x - 4y = 6; (-3, -2)$
$-4y = -3x + 6$
$\left(-\frac{1}{4}\right)(-4y) = \left(-\frac{1}{4}\right)(-3x + 6)$
$y = \frac{3}{4}x - \frac{3}{2}$

Perpendicular lines have slopes that are negative reciprocals.

$m = -\frac{4}{3}; (-3, -2)$
$y - (-2) = -\frac{4}{3}[x - (-3)]$
$y + 2 = -\frac{4}{3}(x + 3)$
$3(y + 2) = 3\left[-\frac{4}{3}(x + 3)\right]$
$3y + 6 = -4(x + 3)$
$3y + 6 = -4x - 12$
$4x + 3y + 6 = -12$
$4x + 3y = -18$

64. $x^2 + 4x + y^2 - 12y + 31 = 0$
$x^2 + 4x + 4 + y^2 - 12y + 36 = -31 + 4 + 36$
$(x + 2)^2 + (y - 6)^2 = 9$
center $(-2, 6)$, $r = \sqrt{9} = 3$

65. $y = x^2 + 10x + 21$
$y = x^2 + 10x + 25 - 25 + 21$
$y = (x + 5)^2 - 4$
vertex: $(-5, -4)$

66. $x^2 + 4y^2 = 16$
If $y = 0$,
$x^2 + 4(0)^2 = 16$
$x^2 = 16$
$x = \pm\sqrt{16} = \pm 4$
$(4, 0), (-4, 0)$
length of major axis $= 8$

67. 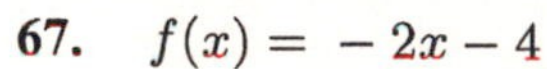
$f(x) = -2x - 4$

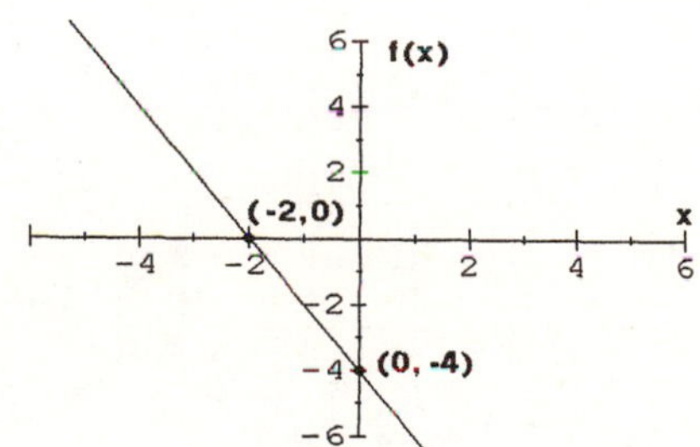

68. $f(x) = -2x^2 - 2$

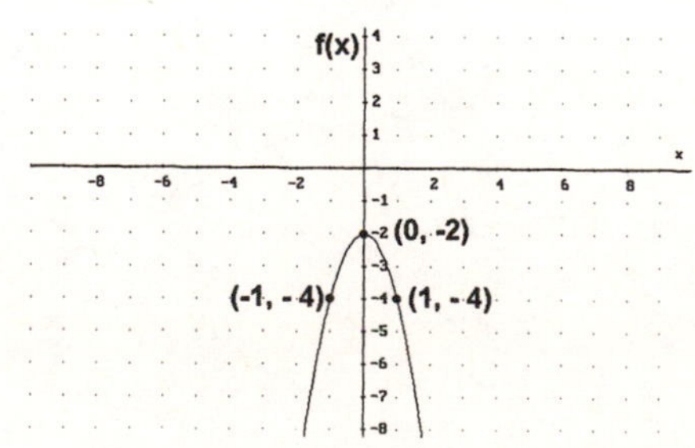

69. $f(x) = x^2 - 2x - 2$

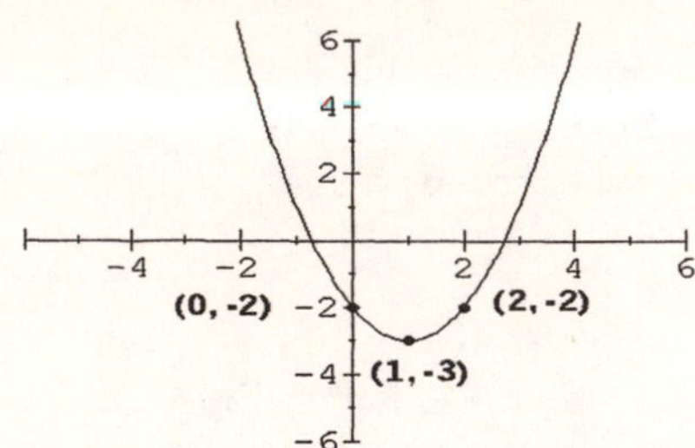

70. $f(x) = \sqrt{x+1} + 2$

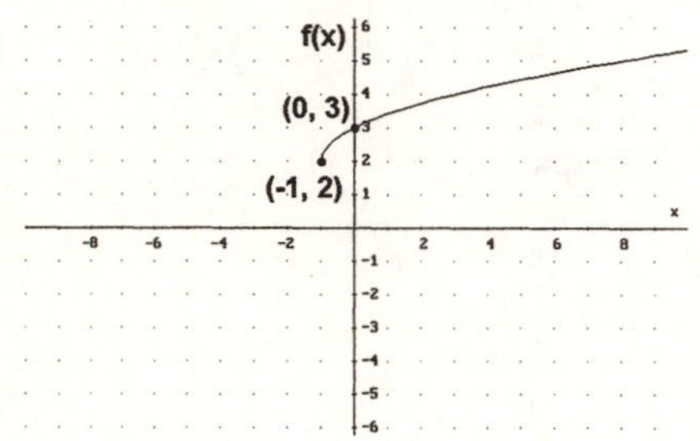

71. $f(x) = 2x^2 + 8x + 9$

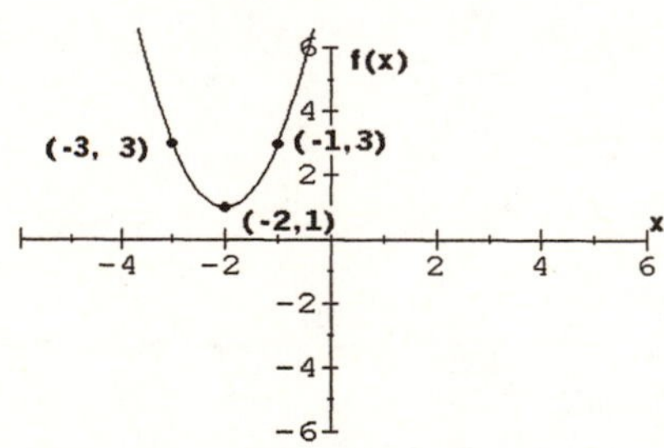

72. $f(x) = -|x-2| + 1$

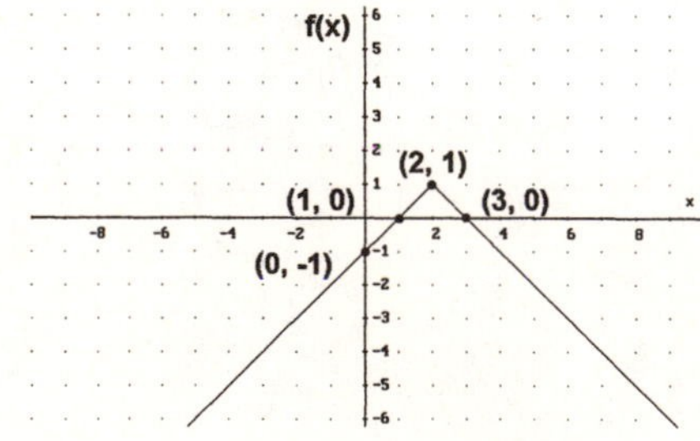

73. $f(x) = 2^x + 2$

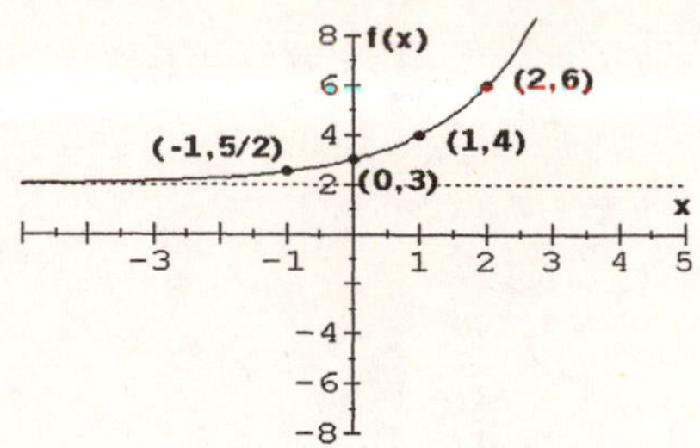

74. $f(x) = \log_2(x-2)$

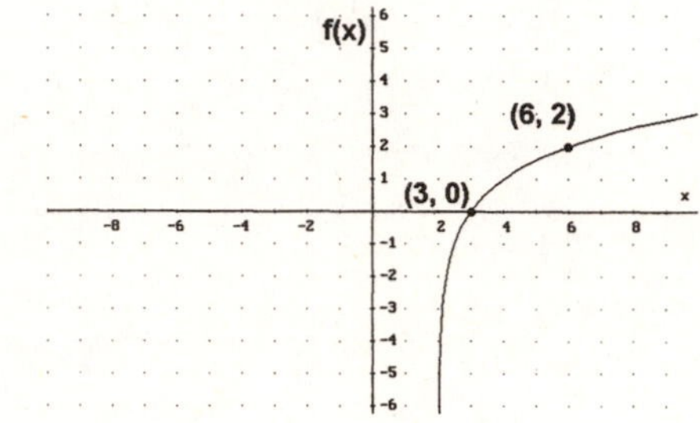

75. $f(x) = -x(x+1)(x-2)$

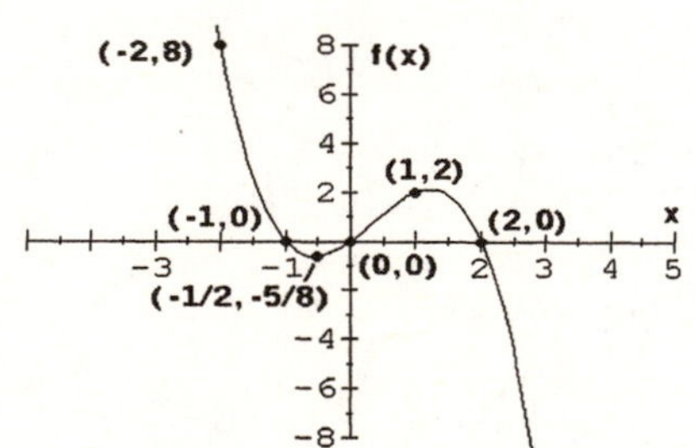

76. $f(x) = \dfrac{-x}{x+2}$

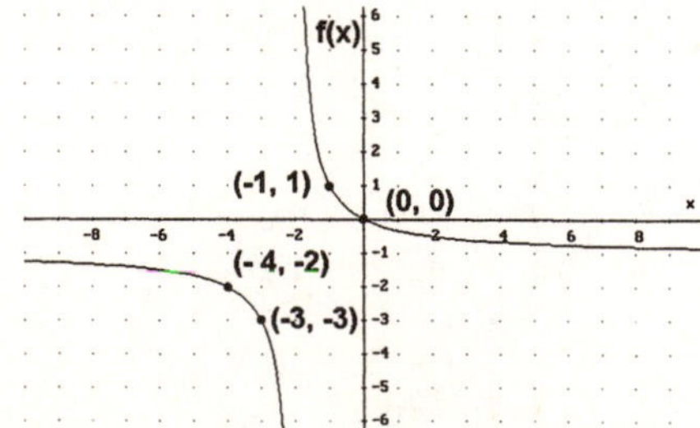

77. $f(x) = x - 3; \quad g(x) = 2x^2 - x - 1$

$$\begin{aligned}(g \circ f)(x) &= g(f(x)) = g(x-3)\\ &= 2(x-3)^2 - (x-3) - 1\\ &= 2(x^2 - 6x + 9) - x + 3 - 1\\ &= 2x^2 - 12x + 18 - x + 2\\ &= 2x^2 - 13x + 20\end{aligned}$$

$$\begin{aligned}(f \circ g)(x) &= f(g(x)) = f(2x^2 - x - 1)\\ &= (2x^2 - x - 1) - 3\\ &= 2x^2 - x - 4\end{aligned}$$

78. $f(x) = 3x - 7$
$y = 3x - 7$

Interchange x and y, then solve for y.

$$\begin{aligned}x &= 3y - 7\\ x + 7 &= 3y\\ \frac{1}{3}(x+7) &= \frac{1}{3}(3y)\\ \frac{x+7}{3} &= y\\ f^{-1}(x) &= \frac{x+7}{3}\end{aligned}$$

79. $f(x) = -\frac{1}{2}x + \frac{2}{3}$
$y = -\frac{1}{2}x + \frac{2}{3}$

Interchange x and y, then solve for y.

$$\begin{aligned}x &= -\frac{1}{2}y + \frac{2}{3}\\ 6(x) &= 6\left(-\frac{1}{2}y + \frac{2}{3}\right)\\ 6x &= -3y + 4\\ 3y + 6x &= 4\\ \frac{1}{3}\left(3y + 6x\right) &= \left(\frac{1}{3}\right)4\\ y + 2x &= \frac{4}{3}\\ y &= -2x + \frac{4}{3}\\ f^{-1}(x) &= -2x + \frac{4}{3}\end{aligned}$$

80.
$$\begin{aligned}y &= kx\\ 2 &= k\left(-\frac{2}{3}\right)\\ 2 &= -\frac{2}{3}k\\ -\frac{3}{2}(2) &= -\frac{3}{2}\left(-\frac{2}{3}k\right)\\ -3 &= k\end{aligned}$$

81.
$$\begin{aligned}y &= \frac{k}{x^2}\\ 4 &= \frac{k}{(3)^2}\\ 4 &= \frac{k}{9}\\ 36 &= k\\ y &= \frac{36}{(6)^2}\\ y &= 1\end{aligned}$$

82.
$$\begin{aligned}V &= \frac{k}{P}\\ 15 &= \frac{k}{20}\\ 300 &= k\\ V &= \frac{300}{P}\\ V &= \frac{300}{25} = 12 \text{ cubic centimeters}\end{aligned}$$

83.
$$\begin{aligned}3(2x-1) - 2(5x+1) &= 4(3x+4)\\ 6x - 3 - 10x - 2 &= 12x + 16\\ -4x - 5 &= 12x + 16\\ -16x &= 21\\ x &= -\frac{21}{16}\end{aligned}$$

The solution set is $\left\{-\frac{21}{16}\right\}$.

84.
$$\begin{aligned}n + \frac{3n-1}{9} - 4 &= \frac{3n+1}{3}\\ 9\left(n + \frac{3n-1}{9} - 4\right) &= 9\left(\frac{3n+1}{3}\right)\\ 9n + 3n - 1 - 36 &= 3(3n+1)\\ 12n - 37 &= 9n + 3\\ 3n &= 40\end{aligned}$$

$n = \frac{40}{3}$

The solution set is $\left\{\frac{40}{3}\right\}$.

85. $0.92 + 0.9(x - 0.3) = 2x - 5.95$
$100[0.92 + 0.9(x - .3)] = 100(2x - 5.95)$
$92 + 90(x - .3) = 200x - 595$
$92 + 90x - 27 = 200x - 595$
$-110x = -660$
$x = 6$
The solution set is $\{6\}$.

86. $|4x - 1| = 11$
$4x - 1 = 11$ or $4x - 1 = -11$
$4x = 12$ or $4x = -10$
$x = 3$ or $x = -\frac{10}{4} = -\frac{5}{2}$
The solution set is $\left\{-\frac{5}{2}, 3\right\}$.

87. $3x^2 = 7x$
$3x^2 - 7x = 0$
$x(3x - 7) = 0$
$x = 0$ or $3x - 7 = 0$
$x = 0$ or $3x = 7$
$x = 0$ or $x = \frac{7}{3}$
The solution set is $\left\{0, \frac{7}{3}\right\}$.

88. $x^3 - 36x = 0$
$x(x^2 - 36) = 0$
$x(x + 6)(x - 6) = 0$
$x = 0$ or $x + 6 = 0$ or $x - 6 = 0$
$x = 0$ or $x = -6$ or $x = 6$
The solution set is $\{-6, 0, 6\}$.

89. $30x^2 + 13x - 10 = 0$
$(6x + 5)(5x - 2) = 0$
$6x + 5 = 0$ or $5x - 2 = 0$
$6x = -5$ or $5x = 2$
$x = -\frac{5}{6}$ or $x = \frac{2}{5}$
The solution set is $\left\{-\frac{5}{6}, \frac{2}{5}\right\}$.

90. $8x^3 + 12x^2 - 36x = 0$
$4x(2x^2 + 3x - 9) = 0$
$4x(2x - 3)(x + 3) = 0$
$4x = 0$ or $2x - 3 = 0$ or $x + 3 = 0$
$x = 0$ or $2x = 3$ or $x = -3$
$x = 0$ or $x = \frac{3}{2}$ or $x = -3$
The solution set is $\left\{-3, 0, \frac{3}{2}\right\}$.

91. $x^4 + 8x^2 - 9 = 0$
$(x^2 + 9)(x^2 - 1) = 0$
$x^2 + 9 = 0$ or $x^2 - 1 = 0$
$x^2 = -9$ or $x^2 = 1$
$x = \pm\sqrt{-9}$ or $x = \pm\sqrt{1}$
$x = \pm 3i$ or $x = \pm 1$
The solution set is $\{\pm 3i, \pm 1\}$.

92. $(n + 4)(n - 6) = 11$
$n^2 - 2n - 24 = 11$
$n^2 - 2n - 35 = 0$
$(n - 7)(n + 2) = 0$
$n - 7 = 0$ or $n + 5 = 0$
$n = 7$ or $n = -5$
The solution set is $\{-5, 7\}$.

93. $2 - \frac{3x}{x - 4} = \frac{14}{x + 7}$; $x \neq 4, x \neq -7$
$(x\text{-}4)(x\text{+}7)\left(2 - \frac{3x}{x\text{-}4}\right) = (x\text{-}4)(x\text{+}7)\left(\frac{14}{x\text{+}7}\right)$
$2(x - 4)(x + 7) - 3x(x + 7) = 14(x - 4)$
$2(x^2 + 3x - 28) - 3x^2 - 21x = 14x - 56$
$2x^2 + 6x - 56 - 3x^2 - 21x = 14x - 56$
$-x^2 - 15x - 56 = 14x - 56$
$0 = x^2 + 29x$
$0 = x(x + 29)$
$x = 0$ or $x + 29 = 0$
$x = 0$ or $x = -29$
The solution set is $\{-29, 0\}$.

94. $\dfrac{2n}{6n^2+7n-3}-\dfrac{n-3}{3n^2+11n-4}=\dfrac{5}{2n^2+11n+12}$

$\dfrac{2n}{(3n-1)(2n+3)}-\dfrac{n-3}{(3n-1)(n+4)}=\dfrac{5}{(2n+3)(n+4)}$

$(3n-1)(2n+3)(n+4)\left[\dfrac{2n}{(3n-1)(2n+3)}-\dfrac{n-3}{(3n-1)(n+4)}\right]=(3n-1)(2n+3)(n+4)\left[\dfrac{5}{(2n+3)(n+4)}\right]$

$2n(n+4)-(2n+3)(n-3)=5(3n-1)$

$2n^2+8n-(2n^2-3n-9)=15n-5$

$2n^2+8n-2n^2+3n+9=15n-5$

$11n+9=15n-5$

$-4n=-14$

$n=\dfrac{-14}{-4}=\dfrac{7}{2}$

The solution set is $\left\{\dfrac{7}{2}\right\}$.

95. $\sqrt{3y}-y=-6$
$\sqrt{3y}=y-6$
$(\sqrt{3y})^2=(y-6)^2$
$3y=y^2-12y+36$
$0=y^2-15y+36$
$0=(y-12)(y-3)$
$y-12=0$ or $y-3=0$
$y=12$ or $y=3$
Checking $y=12$
$\sqrt{3(12)}-12\stackrel{?}{=}-6$
$\sqrt{36}-12\stackrel{?}{=}-6$
$6-12\stackrel{?}{=}-6$
$-6=-6$
Checking $y=3$
$\sqrt{3(3)}-3\stackrel{?}{=}-6$
$\sqrt{9}-3\stackrel{?}{=}-6$
$3-3\stackrel{?}{=}-6$
$0\neq-6$
The solution set is $\{12\}$.

96. $\sqrt{x+19}-\sqrt{x+28}=-1$
$\sqrt{x+19}=\sqrt{x+28}-1$
$(\sqrt{x+19})^2=(\sqrt{x+28}-1)^2$
$x+19=(\sqrt{x+28})^2+2(\sqrt{x+28})(-1)+(-1)^2$
$x+19=x+28-2\sqrt{x+28}+1$
$-10=-2\sqrt{x+28}$
$5=\sqrt{x+28}$
$(5)^2=(\sqrt{x+28})^2$
$25=x+28$
$-3=x$
Check
$\sqrt{-3+19}-\sqrt{-3+28}\stackrel{?}{=}-1$
$\sqrt{16}-\sqrt{25}\stackrel{?}{=}-1$
$4-5\stackrel{?}{=}-1$
$-1=-1$
The solution set is $\{-3\}$.

97. $(3x-1)^2 = 45$

$3x - 1 = \pm\sqrt{45}$

$3x - 1 = \pm 3\sqrt{5}$

$3x = 1 \pm 3\sqrt{5}$

$x = \dfrac{1 \pm 3\sqrt{5}}{3}$

The solution set is $\left\{\dfrac{1 \pm 3\sqrt{5}}{3}\right\}$.

98. $(2x+5)^2 = -32$

$2x + 5 = \pm\sqrt{-32}$

$2x + 5 = \pm 4i\sqrt{2}$

$2x = -5 \pm 4i\sqrt{2}$

$x = \dfrac{-5 \pm 4i\sqrt{2}}{2}$

The solution set is $\left\{\dfrac{-5 \pm 4i\sqrt{2}}{2}\right\}$.

99. $2x^2 - 3x + 4 = 0$

$x = \dfrac{-(-3) \pm \sqrt{(-3)^2 - 4(2)(4)}}{2(2)}$

$x = \dfrac{3 \pm \sqrt{9-32}}{4}$

$x = \dfrac{3 \pm \sqrt{-23}}{4}$

$x = \dfrac{3 \pm i\sqrt{23}}{4}$

The solution set is $\left\{\dfrac{3 \pm i\sqrt{23}}{4}\right\}$.

100. $3n^2 - 6n + 2 = 0$

$n = \dfrac{-(-6) \pm \sqrt{(-6)^2 - 4(3)(2)}}{2(3)}$

$n = \dfrac{6 \pm \sqrt{36-24}}{6}$

$n = \dfrac{6 \pm \sqrt{12}}{6} = \dfrac{6 \pm 2\sqrt{3}}{6}$

$n = \dfrac{2(3 \pm \sqrt{3})}{6} = \dfrac{3 \pm \sqrt{3}}{3}$

The solution set is $\left\{\dfrac{3 \pm \sqrt{3}}{3}\right\}$.

101. $\dfrac{5}{n-3} - \dfrac{3}{n+3} = 1; n \neq 3, n \neq -3$

$(n\text{-}3)(n\text{+}3)\left(\dfrac{5}{n-3} - \dfrac{3}{n+3}\right) = (n\text{-}3)(n\text{+}3)(1)$

$5(n+3) - 3(n-3) = n^2 - 9$

$5n + 15 - 3n + 9 = n^2 - 9$

$2n + 24 = n^2 - 9$

$0 = n^2 - 2n - 33$

$n = \dfrac{-(-2) \pm \sqrt{(-2)^2 - 4(1)(-33)}}{2(1)}$

$n = \dfrac{2 \pm \sqrt{4+132}}{2}$

$n = \dfrac{2 \pm \sqrt{136}}{2}$

$n = \dfrac{2 \pm 2\sqrt{34}}{2}$

$n = \dfrac{2(1 \pm \sqrt{34})}{2}$

$n = 1 \pm \sqrt{34}$

The solution set is $\{1 \pm \sqrt{34}\}$.

102. $12x^4 - 19x^2 + 5 = 0$

$(4x^2 - 5)(3x^2 - 1) = 0$

$4x^2 - 5 = 0$ or $3x^2 - 1 = 0$

$4x^2 = 5$ or $3x^2 = 1$

$x^2 = \dfrac{5}{4}$ or $x^2 = \dfrac{1}{3}$

$x = \pm\sqrt{\dfrac{5}{4}}$ or $x = \pm\sqrt{\dfrac{1}{3}}$

$x = \pm\dfrac{\sqrt{5}}{2}$ or $x = \pm\dfrac{1}{\sqrt{3}} \bullet \dfrac{\sqrt{3}}{\sqrt{3}}$

$x = \pm\dfrac{\sqrt{5}}{2}$ or $x = \pm\dfrac{\sqrt{3}}{3}$

The solution set is $\left\{\pm\dfrac{\sqrt{5}}{2}, \pm\dfrac{\sqrt{3}}{3}\right\}$.

103. $2x^2 + 5x + 5 = 0$

$$x = \frac{-5 \pm \sqrt{(5)^2 - 4(2)(5)}}{2(2)}$$

$$x = \frac{-5 \pm \sqrt{25 - 40}}{4}$$

$$x = \frac{-5 \pm \sqrt{-15}}{4}$$

$$x = \frac{-5 \pm i\sqrt{15}}{4}$$

The solution set is $\left\{\frac{-5 \pm i\sqrt{15}}{4}\right\}$.

104. $x^3 - 4x^2 - 25x + 28 = 0$
possible rational roots:
c: $\pm 1, \pm 2, \pm 4, \pm 7, \pm 14, \pm 28$
d: ± 1
$\frac{c}{d}$: $\pm 1, \pm 2, \pm 4, \pm 7, \pm 14, \pm 28$

1	1	−4	−25	28
		1	−3	−28
	1	−3	−28	0

$x = 1$
$x^2 - 3x - 28 = 0$
$(x - 7)(x + 4) = 0$
$x - 7 = 0$ or $x + 4 = 0$
$x = 7$ or $x = -4$
The solution set is $\{-4, 1, 7\}$.

105. $6x^3 - 19x^2 + 9x + 10 = 0$
Factors of 6: $\pm 1, \pm 2, \pm 3, \pm 6$
Factors of 10: $\pm 1, \pm 2, \pm 5, \pm 10$
Possible rational roots:
$\pm 1, \pm 2, \pm 5, \pm 10 \pm \frac{5}{3}, \pm \frac{10}{3}, \pm \frac{5}{6},$

$\pm \frac{5}{2}, \pm \frac{1}{3}, \pm \frac{2}{3}, \pm \frac{1}{6}, \pm \frac{1}{2}$

2	6	−19	9	10
		12	−14	−10
	6	−7	−5	0

$(x - 2)(6x^2 - 7x - 5) = 0$
$(x - 2)(3x - 5)(2x + 1) = 0$
$x - 2 = 0$ or $3x - 5 = 0$ or $2x + 1 = 0$
$x = 2$ or $3x = 5$ or $2x = -1$
$x = 2$ or $x = \frac{5}{3}$ or $x = -\frac{1}{2}$
The solution set is $\left\{-\frac{1}{2}, \frac{5}{3}, 2\right\}$.

106. $16^x = 64$

$(2^4)^x = 2^6$

$2^{4x} = 2^6$

$4x = 6$

$x = \frac{6}{4} = \frac{3}{2}$

The solution set is $\left\{\frac{3}{2}\right\}$.

107. $\log_3 x = 4$
$3^4 = x$
$81 = x$
The solution set is $\{81\}$.

108. $\log_{10} x + \log_{10} 25 = 2$
$\log_{10} 25x = 2$
$10^2 = 25x$
$100 = 25x$
$4 = x$
The solution set is $\{4\}$.

109. $\ln (3x - 4) - \ln (x + 1) = \ln 2$

$\ln \frac{3x - 4}{x + 1} = \ln 2$

$\frac{3x - 4}{x + 1} = 2;\ x \neq -1$
$3x - 4 = 2(x + 1)$
$3x - 4 = 2x + 2$
$x - 4 = 2$
$x = 6$
The solution set is $\{6\}$.

110. $27^{4x} = 9^{x+1}$

$(3^3)^{4x} = (3^2)^{x+1}$

$3^{12x} = 3^{2x+2}$

$12x = 2x + 2$
$10x = 2$

$x = \frac{2}{10} = \frac{1}{5}$

The solution set is $\left\{\frac{1}{5}\right\}$.

111. $-5(y-1)+3 > 3y-4-4y$

$-5y+5+3 > -y-4$

$-5y+8 > -y-4$

$-4y > -12$

$-\frac{1}{4}(-4y) < -\frac{1}{4}(-12)$

$y < 3$

The solution set is $(-\infty, 3)$.

112. $0.06x + 0.08(250 - x) \geq 19$

$100[0.06x + 0.08(250 - x)] \geq 100(19)$

$6x + 8(250 - x) \geq 1900$

$6x + 2000 - 8x \geq 1900$

$-2x \geq -100$

$-\frac{1}{2}(-2x) \leq -\frac{1}{2}(-100)$

$x \leq 50$

The solution set is $(-\infty, 50]$.

113. $|5x-2| > 13$

$5x-2 < -13$ or $5x-2 > 13$

$5x < -11$ or $5x > 15$

$x < -\frac{11}{5}$ or $x > 3$

The solution set is

$\left(-\infty, -\frac{11}{5}\right) \cup (3, \infty)$.

114. $|6x+2| < 8$

$-8 < 6x+2 < 8$

$-10 < 6x < 6$

$-\frac{10}{6} < x < \frac{6}{6}$

$-\frac{5}{3} < x < 1$

The solution set is $\left(-\frac{5}{3}, 1\right)$.

115. $\frac{x-2}{5} - \frac{3x-1}{4} \leq \frac{3}{10}$

$20\left(\frac{x-2}{5} - \frac{3x-1}{4}\right) \leq 20\left(\frac{3}{10}\right)$

$20\left(\frac{x-2}{5}\right) - 20\left(\frac{3x-1}{4}\right) \leq 6$

$4(x-2) - 5(3x-1) \leq 6$

$4x - 8 - 15x + 5 \leq 6$

$-11x - 3 \leq 6$

$-11x \leq 9$

$x \geq -\frac{9}{11}$

The solution set is $\left[-\frac{9}{11}, \infty\right)$.

116. $(x-2)(x+4) \leq 0$

$(x-2)(x+4) = 0$

$x-2=0$ or $x+4=0$

$x=2$ or $x=-4$

	-4		2
Test Point	-5	0	3
$x-2$:	negative	negative	positive
$x+4$:	negative	positive	positive
product:	positive	negative	positive

The solution set is $[-4, 2]$.

117. $(3x-1)(x-4) > 0$

$(3x-1)(x-4) = 0$

$3x-1=0$ or $x-4=0$

$3x=1$ or $x=4$

$x = \frac{1}{3}$

	$\frac{1}{3}$		4
Test Point	0	1	5
$3x-1$:	negative	positive	positive
$x-4$:	negative	negative	positive
product:	positive	negative	positive

The solution set is

$\left(-\infty, \frac{1}{3}\right) \cup (4, \infty)$.

118. $x(x+5) < 24$
$x^2 + 5x < 24$
$x^2 + 5x - 24 < 0$
$x^2 + 5x - 24 = 0$
$(x+8)(x-3) = 0$
$x + 8 = 0$ or $x - 3 = 0$
$x = -8$ or $x = 3$

	-8		3
Test Point	-9	0	4
$x+8$:	negative	positive	positive
$x-3$:	negative	negative	positive
product:	positive	negative	positive

The solution set is $(-8, 3)$.

119. $\frac{x-3}{x-7} \geq 0$

$x - 3 = 0$ or $x - 7 = 0$
$x = 3$ or $x = 7$

	3		7
Test Point	2	5	8
$x-3$:	negative	positive	positive
$x-7$:	negative	negative	positive
quotient:	positive	negative	negative

The solution set is $(-\infty, 3] \cup (7, \infty)$.

120. $\frac{2x}{x+3} > 4; x \neq -3$

$\frac{2x}{x+3} - 4 > 0$

$\frac{2x}{x+3} - \frac{4(x+3)}{(x+3)} > 0$

$\frac{2x - 4(x+3)}{x+3} > 0$

$\frac{2x - 4x - 12}{x+3} > 0$

$\frac{-2x - 12}{x+3} > 0$

$-2x - 12 = 0$ or $x + 3 = 0$
$-2x = 12$ or $x = -3$
$x = -6$ or $x = -3$

	-6		-3
Test Point	-7	-5	0
$-2x-12$:	positive	negative	negative
$x+3$:	negative	negative	positive
quotient:	negative	positive	negative

The solution set is $(-6, -3)$.

121. Let x = 1st odd integer,
$x + 2$ = 2nd odd integer and
$x + 4$ = 3rd odd integer.

$x + x + 2 + x + 4 = 57$
$3x + 6 = 57$
$3x = 51$
$x = 17$
The integers are 17, 19, and 21.

122. Let x = number of nickels,
y = number of dimes and
z = number of quarters.

$$\begin{pmatrix} x + y + z = 63 \\ y = x + 6 \\ z = 2x + 1 \end{pmatrix}$$

$x + y + z = 63$
$x + x + 6 + 2x + 1 = 63$
$4x + 7 = 63$
$4x = 56$
$x = 14$
$y = x + 6$
$y = 14 + 6$
$y = 20$

$z = 2x + 1$
$z = 2(14) + 1$
$z = 28 + 1$
$z = 29$
There are 14 nickels, 20 dimes, and 29 quarters.

123. Let $x =$ angle and
$180 - x =$ supplementary angle.

$x = \frac{1}{3}(180 - x) + 4$
$x = 60 - \frac{1}{3}x + 4$
$\frac{4}{3}x = 64$
$\frac{3}{4}\left(\frac{4}{3}x\right) = \frac{3}{4}(64)$
$x = 48°$
The angles are 48° and 132°.

124. Let $x =$ selling price.
$300 + (50\%)(x) = x$
$300 + 0.5x = x$
$10(300 + .5x) = 10x$
$3000 + 5x = 10x$
$3000 = 5x$
$600 = x$
The selling price should be \$600.

125. Let $x =$ money invested at 8% and
$x + 300 =$ money invested at 9%.

$0.08x + 0.09(x + 300) = 316$
$100[0.08x + 0.09(x + 300)] = 100(316)$
$8x + 9(x + 300) = 31600$
$8x + 9x + 2700 = 31600$
$17x = 28900$
$x = 1700$
There is \$1700 invested at 8%
and \$2000 invested at 9%.

126.

	Rate	Time	Distance
East	$x + 10$	4.5	$4.5(x + 10)$
West	x	4.5	$4.5x$

$4.5(x + 10) + 4.5x = 639$
$10[4.5(x + 10) + 4.5x] = 10(639)$
$45(x + 10) + 45x = 6390$
$45x + 450 + 45x = 6390$
$90x = 5940$
$x = 66$
The westbound train is traveling
at 66 mph and the eastbound train
is traveling at 76 mph.

127. Let $x =$ amount drained.
$(50\%)(10 - x) + (100\%)(x) = 70\%(10)$
$0.5(10 - x) + 1x = 0.7(10)$
$10[0.5(10 - x) + 1x] = 10[0.7(10)]$
$5(10 - x) + 10x = 7(10)$
$50 - 5x + 10x = 70$
$5x = 20$
$x = 4$
There needs to be 4 quarts drained.

128. Let $x =$ score on the 4th day.
$\frac{70 + 73 + 76 + x}{4} \leq 72$
$\frac{219 + x}{4} \leq 72$
$4\left(\frac{219 + x}{4}\right) \leq 4(72)$
$219 + x \leq 288$
$x \leq 69$
The score must be 69 or less.

129. Let $x =$ number.
$x^3 = 9x$
$x^3 - 9x = 0$
$x(x^2 - 9) = 0$
$x(x - 3)(x + 3) = 0$
$x = 0$ or $x - 3 = 0$ or $x + 3 = 0$
$x = 0$ or $x = 3$ or $x = -3$
The number is -3, 0, or 3.

130. Let $x =$ width of strip.
width $= 8 - 2x$
length $= 14 - 2x$

$(8 - 2x)(14 - 2x) = 72$
$112 - 16x - 28x + 4x^2 = 72$
$4x^2 - 44x + 40 = 0$
$4(x^2 - 11x + 10) = 0$
$4(x - 10)(x - 1) = 0$
$x - 10 = 0$ or $x - 1 = 0$
$x = 10$ or $x = 1$
Discard the root $x = 10$.
The width of the strip is 1 inch.

131. Let x = one part and
$2450 - x$ = other part.

$$\frac{x}{2450 - x} = \frac{3}{4}$$
$$4x = 3(2450 - x)$$
$$4x = 7350 - 3x$$
$$7x = 7350$$
$$x = 1050$$

One part is \$1050 and the other part is \$1400.

132.

	Time in Hours	Rate
Sue	x	$\frac{1}{x}$
Dean	2	$\frac{1}{2}$
Together	$\frac{6}{5}$	$\frac{5}{6}$

$$\frac{1}{x} + \frac{1}{2} = \frac{5}{6}$$
$$6x\left(\frac{1}{x} + \frac{1}{2}\right) = 6x\left(\frac{5}{6}\right)$$
$$6 + 3x = 5x$$
$$6 = 2x$$
$$3 = x$$

It would take Sue 3 hours.

133.

	Price/ share	Number of shares	Cost
Bought	$\frac{300}{x}$	x	300
Sold	$\frac{300}{x-10}$	$x - 10$	300

$$\frac{300}{x} + 5 = \frac{300}{x - 10};\ x \neq 0, x \neq 10$$
$$x(x - 10)\left(\frac{300}{x} + 5\right) = x(x - 10)\left(\frac{300}{x - 10}\right)$$
$$300(x - 10) + 5x(x - 10) = 300x$$
$$300x - 3000 + 5x^2 - 50x = 300x$$
$$5x^2 - 50x - 3000 = 0$$
$$5(x^2 - 10x - 600) = 0$$
$$5(x - 30)(x + 20) = 0$$
$$x - 30 = 0 \quad \text{or} \quad x + 20 = 0$$
$$x = 30 \quad \text{or} \quad x = -20$$

Discard the root $x = -20$.
He originally bought 30 shares at \$10 per share.

134. Let x = units digit
and y = tens digit.

$$\begin{pmatrix} x = 2y + 1 \\ x + y = 10 \end{pmatrix}$$
$$x + y = 10$$
$$2y + 1 + y = 10$$
$$3y + 1 = 10$$
$$3y = 9$$
$$y = 3$$
$$x = 2(3) + 1$$
$$x = 6 + 1$$
$$x = 7$$

The number is 37.

135. Let x = smallest angle,
y = middle angle and
z = largest angle.

$$\begin{pmatrix} x + y = z - 40 \\ x + z = 2y \\ x + y + z = 180 \end{pmatrix}$$

$$\begin{pmatrix} x + y - z = -40 \\ x - 2y + z = 0 \\ x + y + z = 180 \end{pmatrix}$$

$$\begin{bmatrix} 1 & 1 & -1 & -40 \\ 1 & -2 & 1 & 0 \\ 1 & 1 & 1 & 180 \end{bmatrix}$$

-1(row 1) + row 2
-1(row 1) + row 3

$$\begin{bmatrix} 1 & 1 & -1 & -40 \\ 0 & -3 & 2 & 40 \\ 0 & 0 & 2 & 220 \end{bmatrix}$$
$$2z = 220$$
$$z = 110$$
$$-3y + 2z = 40$$
$$-3y + 2(110) = 40$$
$$-3y + 220 = 40$$
$$-3y = -180$$
$$y = 60$$
$$x + y - z = -40$$
$$x + 60 - 110 = -40$$
$$x - 50 = -40$$
$$x = 10$$

The angles are 10°, 60°, and 110°.

Chapter 11 Systems of Equations

PROBLEM SET 11.1 Systems of Two Linear Equations in Two Variables

1. $\begin{pmatrix} x - y = 1 \\ 2x + y = 8 \end{pmatrix}$

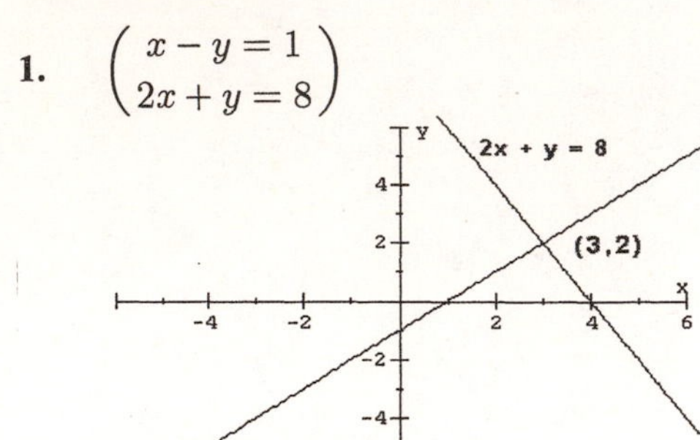

The system is consistent.
Checking $x = 3$, $y = 2$;

$(3) - (2) \stackrel{?}{=} 1$
$1 = 1$

$2(3) + (2) \stackrel{?}{=} 8$
$6 + 2 \stackrel{?}{=} 8$
$8 = 8$

The solution set is $\{(3, 2)\}$.

3. $\begin{pmatrix} 4x + 3y = -5 \\ 2x - 3y = -7 \end{pmatrix}$

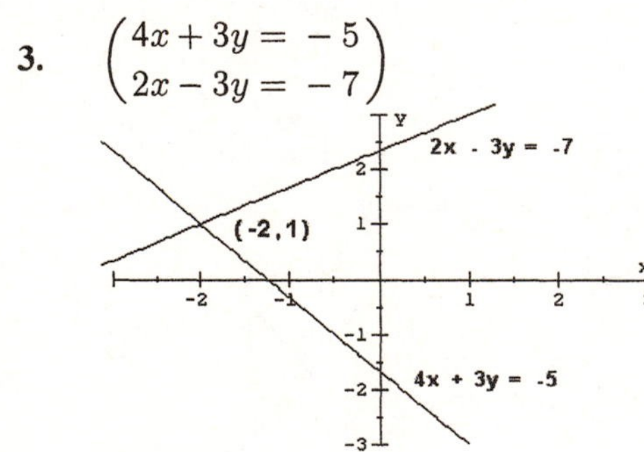

The system is consistent.
Checking $x = -2$, $y = 1$;

$4(-2) + 3(1) \stackrel{?}{=} -5$
$-8 + 3 \stackrel{?}{=} -5$
$-5 = -5$

$2(-2) - 3(1) \stackrel{?}{=} -7$
$-4 - 3 \stackrel{?}{=} -7$
$-7 = -7$

The solution set is $\{(-2, 1)\}$.

5. $\begin{pmatrix} \frac{1}{2}x + \frac{1}{4}y = 9 \\ 4x + 2y = 72 \end{pmatrix}$

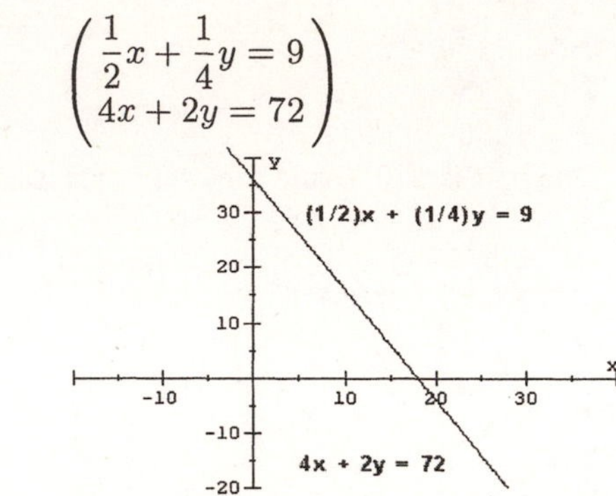

The system is dependent.
Let $x = k$.
$4k + 2y = 72$
$2y = -4k + 72$
$y = -2k + 36$
The solution set is $\{(k, -2k + 36)\}$, where k is any real number.

7. $\begin{pmatrix} \frac{1}{2}x - \frac{1}{3}y = 3 \\ x + 4y = -8 \end{pmatrix}$

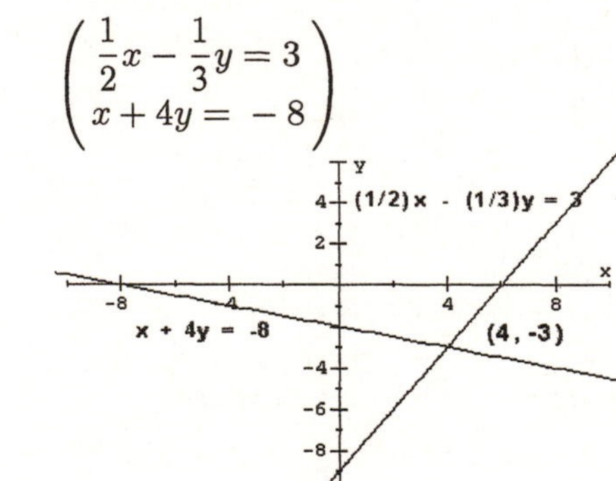

The system is consistent.
Checking $x = 4$, $y = -3$;

$\frac{1}{2}(4) - \frac{1}{3}(-3) \stackrel{?}{=} 3$
$2 + 1 \stackrel{?}{=} 3$
$3 = 3$

$(4) + 4(-3) \stackrel{?}{=} -8$
$4 - 12 \stackrel{?}{=} -8$
$-8 = -8$

The solution set is $\{(4, -3)\}$.

9. $\begin{pmatrix} x - \dfrac{y}{2} = -4 \\ 8x - 4y = -1 \end{pmatrix}$

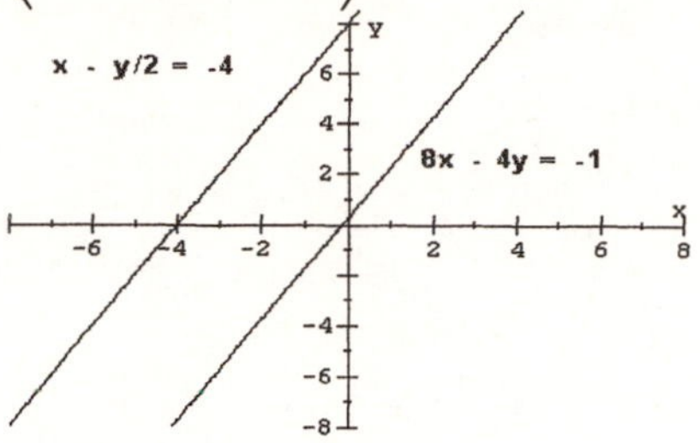

The system is inconsistent.
The solution set is $\emptyset$.

11. $\begin{pmatrix} x + y = 16 \\ y = x + 2 \end{pmatrix}$

Substitute $x + 2$ for y in the 1st equation.

$x + y = 16$
$x + (x + 2) = 16$
$2x + 2 = 16$
$2x = 14$
$x = 7$

$y = x + 2$
$y = 7 + 2$
$y = 9$

The solution set is $\{(7, 9)\}$.

13. $\begin{pmatrix} x = 3y - 25 \\ 4x + 5y = 19 \end{pmatrix}$

Substitute $3y - 25$ for x in the 2nd equation.

$4x + 5y = 19$
$4(3y - 25) + 5y = 19$
$12y - 100 + 5y = 19$
$17y = 119$
$y = 7$

$x = 3y - 25$
$x = 3(7) - 25$
$x = -4$

The solution set is $\{(-4, 7)\}$.

15. $\begin{pmatrix} y = \dfrac{2}{3}x - 1 \\ 5x - 7y = 9 \end{pmatrix}$

Substitute $\dfrac{2}{3}x - 1$ for y in the 2nd equation.

$5x - 7y = 9$
$5x - 7\left(\dfrac{2}{3}x - 1\right) = 9$
$5x - \dfrac{14}{3}x + 7 = 9$
$15x - 14x + 21 = 27$
$x + 21 = 27$
$x = 6$
$y = \dfrac{2}{3}x - 1$
$y = \dfrac{2}{3}(6) - 1$
$y = 3$

The solution set is $\{(6, 3)\}$.

17. $\begin{pmatrix} a = 4b + 13 \\ 3a + 6b = -33 \end{pmatrix}$

Substitute $4b + 13$ for a in the 2nd equation.

$3a + 6b = -33$
$3(4b + 13) + 6b = -33$
$12b + 39 + 6b = -33$
$18b = -72$
$b = -4$

$a = 4b + 13$
$a = 4(-4) + 13$
$a = -3$

Therefore, $a = -3$ and $b = -4$.

19. $\begin{pmatrix} 2x - 3y = 4 \\ y = \dfrac{2}{3}x - \dfrac{4}{3} \end{pmatrix}$

Substitute $\dfrac{2}{3}x - \dfrac{4}{3}$ for y in the 1st equation.

$2x - 3y = 4$
$2x - 3\left(\dfrac{2}{3}x - \dfrac{4}{3}\right) = 4$
$2x - 2x + 4 = 4$
$0 = 0$

The system is dependent. Let $x = k$, then $y = \dfrac{2}{3}k - \dfrac{4}{3}$.

The solution set is $\left\{\left(k, \dfrac{2}{3}k - \dfrac{4}{3}\right)\right\}$ where k is any real number.

21. $\begin{pmatrix} u = t - 2 \\ t + u = 12 \end{pmatrix}$

Substitute $t - 2$ for u in the 2nd equation.

$t + u = 12$

$t + t - 2 = 12$
$2t = 14$
$t = 7$
$u = t - 2$
$u = 7 - 2$
$u = 5$
Therefore, $u = 5$ and $t = 7$.

23. $\begin{pmatrix} 4x + 3y = -7 \\ 3x - 2y = 16 \end{pmatrix}$

$4x + 3y = -7$
$3y = -4x - 7$
$y = -\frac{4}{3}x - \frac{7}{3}$
Substitute into the 2nd equation.
$3x - 2y = 16$
$3x - 2\left(-\frac{4}{3}x - \frac{7}{3}\right) = 16$
$3x + \frac{8}{3}x + \frac{14}{3} = 16$
$9x + 8x + 14 = 48$
$17x = 34$
$x = 2$
$y = -\frac{4}{3}x - \frac{7}{3}$
$y = -\frac{4}{3}(2) - \frac{7}{3}$
$y = -5$
The solution set is $\{(2, -5)\}$.

25. $\begin{pmatrix} 5x - y = 4 \\ y = 5x + 9 \end{pmatrix}$
Substitute $5x + 9$ for y in the 1st equation.
$5x - y = 4$
$5x - (5x + 9) = 4$
$5x - 5x - 9 = 4$
$-9 \neq 4$
The system is inconsistent.
The solution set is $\emptyset$.

27. $\begin{pmatrix} 4x - 5y = 3 \\ 8x + 15y = -24 \end{pmatrix}$
$4x - 5y = 3$
$4x = 5y + 3$
$x = \frac{5}{4}y + \frac{3}{4}$
Substitute into the 2nd equation.
$8x + 15y = -24$
$8\left(\frac{5}{4}y + \frac{3}{4}\right) + 15y = -24$
$10y + 6 + 15y = -24$
$25y = -30$
$y = -\frac{6}{5}$
$x = \frac{5}{4}y + \frac{3}{4}$
$x = \frac{5}{4}\left(-\frac{6}{5}\right) + \frac{3}{4}$
$x = -\frac{3}{4}$
The solution set is $\left\{\left(-\frac{3}{4}, -\frac{6}{5}\right)\right\}$.

29. $\begin{pmatrix} 3x + 2y = 1 \\ 5x - 2y = 23 \end{pmatrix}$

$$\begin{array}{l} 3x + 2y = 1 \\ \underline{5x - 2y = 23} \\ 8x \qquad = 24 \end{array}$$
$x = 3$
$3x + 2y = 1$
$3(3) + 2y = 1$
$9 + 2y = 1$
$2y = -8$
$y = -4$
The solution set is $\{(3, -4)\}$.

31. $\begin{pmatrix} x - 3y = -22 \\ 2x + 7y = 60 \end{pmatrix}$

Multiply equation 1 by -2 and then add the result to equation 2.
$$\begin{array}{r} -2x + 6y = 44 \\ \underline{2x + 7y = 60} \\ 13y = 104 \end{array}$$
$y = 8$
$x - 3y = -22$
$x - 3(8) = -22$
$x - 24 = -22$

$x = 2$
The solution set is $\{(2, 8)\}$.

33. $\begin{pmatrix} 4x - 5y = 21 \\ 3x + 7y = -38 \end{pmatrix}$
Multiply equation 1 by -3,
multiply equation 2 by 4,
and then add the resulting equations.

$$\begin{aligned} -12x + 15y &= -63 \\ 12x + 28y &= -152 \\ \hline 43y &= -215 \\ y &= -5 \end{aligned}$$

$4x - 5y = 21$
$4x - 5(-5) = 21$
$4x = -4$
$x = -1$
The solution set is $\{(-1, -5)\}$.

35. $\begin{pmatrix} 5x - 2y = 19 \\ 5x - 2y = 7 \end{pmatrix}$
Multiply equation 1 by -1 and then add the result to equation 2.

$$\begin{aligned} -5x + 2y &= -19 \\ 5x - 2y &= 7 \\ \hline 0 &\neq -12 \end{aligned}$$

The system is inconsistent.
The solution set is $\emptyset$.

37. $\begin{pmatrix} 5a + 6b = 8 \\ 2a - 15b = 9 \end{pmatrix}$
Multiply equation 1 by 5,
multiply equation 2 by 2,
and then add the resulting equations.

$$\begin{aligned} 25a + 30b &= 40 \\ 4a - 30b &= 18 \\ \hline 29a \quad &= 58 \\ a &= \frac{58}{29} = 2 \end{aligned}$$

$5a + 6b = 8$
$5(2) + 6b = 8$
$10 + 6b = 8$
$6b = -2$
$b = -\frac{2}{6} = -\frac{1}{3}$
Therefore, $a = 2$ and $b = -\frac{1}{3}$.

39. $\begin{pmatrix} \frac{2}{3}s + \frac{1}{4}t = -1 \\ \frac{1}{2}s - \frac{1}{3}t = -7 \end{pmatrix}$

Multiply equation 1 by 12 and multiply equation 2 by 6.
$\begin{pmatrix} 8s + 3t = -12 \\ 3s - 2t = -42 \end{pmatrix}$

Multiply equation 1 by 2,
multiply equation 2 by 3,
and then add the resulting equations.

$$\begin{aligned} 16s + 6t &= -24 \\ 9s - 6t &= -126 \\ \hline 25s \quad &= -150 \\ s &= -6 \end{aligned}$$

$3s - 2t = -42$
$3(-6) - 2t = -42$
$-18 - 2t = -42$
$-2t = -24$
$t = 12$
Therefore, $s = -6$ and $t = 12$.

41. $\begin{pmatrix} \frac{x}{2} - \frac{2y}{5} = -\frac{23}{60} \\ \frac{2x}{3} + \frac{y}{4} = -\frac{1}{4} \end{pmatrix}$

Multiply equation 1 by 60 and multiply equation 2 by 12.
$\begin{pmatrix} 30x - 24y = -23 \\ 8x + 3y = -3 \end{pmatrix}$

Multiply equation 2 by 8 and then add the result to equation 1.

$$\begin{aligned} 30x - 24y &= -23 \\ 64x + 24y &= -24 \\ \hline 94x \quad &= -47 \\ x &= -\frac{47}{94} = -\frac{1}{2} \end{aligned}$$

$8x + 3y = -3$
$8\left(-\frac{1}{2}\right) + 3y = -3$
$-4 + 3y = -3$
$3y = 1$

$y = \frac{1}{3}$

The solution set is $\left\{\left(-\frac{1}{2}, \frac{1}{3}\right)\right\}$.

43. $\begin{pmatrix} \frac{2x}{3} + \frac{1y}{2} = \frac{1}{6} \\ 4x + 6y = -1 \end{pmatrix}$

Multiply equation 1 by 6,
multiply equation 2 by -1,
and then add the resulting equations.

$$\begin{array}{r} 4x + 3y = 1 \\ -4x - 6y = 1 \\ \hline -3y = 2 \\ y = -\frac{2}{3} \end{array}$$

$4x + 6y = -1$

$4x + 6\left(-\frac{2}{3}\right) = -1$

$4x - 4 = -1$

$4x = 3$

$x = \frac{3}{4}$

The solution set is $\left\{\left(\frac{3}{4}, -\frac{2}{3}\right)\right\}$.

45. $\begin{pmatrix} 5x - y = -22 \\ 2x + 3y = -2 \end{pmatrix}$

Multiply equation 1 by 3 and then add the result to equation 2.

$$\begin{array}{r} 15x - 3y = -66 \\ 2x + 3y = -2 \\ \hline 17x \quad\quad = -68 \\ x = -4 \end{array}$$

$5x - y = -22$

$5(-4) - y = -22$

$-20 - y = -22$

$-y = -2$

$y = 2$

The solution set is $\{(-4, 2)\}$.

47. $\begin{pmatrix} x = 3y - 10 \\ x = -2y + 15 \end{pmatrix}$

Substitute $3y - 10$ for x in the 2nd equation.

$x = -2y + 15$

$3y - 10 = -2y + 15$

$5y = 25$

$y = 5$

$x = -2y + 15$

$x = -2(5) + 15$

$x = -10 + 15$

$x = 5$

The solution set is $\{(5, 5)\}$.

49. $\begin{pmatrix} 3x - 5y = 9 \\ 6x - 10y = -1 \end{pmatrix}$

Multiply equation 1 by -2 and then add the result to equation 2.

$$\begin{array}{r} -6x + 10y = -18 \\ 6x - 10y = -1 \\ \hline 0 \neq -19 \end{array}$$

The system is inconsistent.
The solution set is $\emptyset$.

51. $\begin{pmatrix} \frac{1}{2}x - \frac{2}{3}y = 22 \\ \frac{1}{2}x + \frac{1}{4}y = 0 \end{pmatrix}$

Multiply equation 1 by 6 and multiply equation 2 by 4.

$\begin{pmatrix} 3x - 4y = 132 \\ 2x + y = 0 \end{pmatrix}$

Multiply equation 2 by 4 and then add the result to equation 1.

$$\begin{array}{r} 3x - 4y = 132 \\ 8x + 4y = 0 \\ \hline 11x \quad\quad = 132 \\ x = 12 \end{array}$$

$2x + y = 0$

$2(12) + y = 0$

$y = -24$

The solution set is $\{(12, -24)\}$.

53. $\begin{pmatrix} t = 2u + 2 \\ 9u - 9t = -45 \end{pmatrix}$

Substitue $2u + 2$ for t in the 2nd equation.

$9u - 9t = -45$

$9u - 9(2u + 2) = -45$

$9u - 18u - 18 = -45$

$-9u = -27$

$u = 3$

$t = 2u + 2$
$t = 2(3) + 2$
$t = 6 + 2$
$t = 8$
Therefore, $t = 8$ and $u = 3$.

55. $\begin{pmatrix} x + y = 1000 \\ 0.12x + 0.14y = 136 \end{pmatrix}$

Multiply equation 2 by 100.
$\begin{pmatrix} x + y = 1000 \\ 12x + 14y = 13600 \end{pmatrix}$

Multiply equation 1 by -12 and then add the result to equation 2.

$$\begin{array}{r} -12x - 12y = -12000 \\ 12x + 14y = 13600 \\ \hline 2y = 1600 \\ y = 800 \end{array}$$

$x + y = 1000$
$x + 800 = 1000$
$x = 200$
The solution set is $\{(200, 800)\}$.

57. $\begin{pmatrix} y = 2x \\ 0.09x + 0.12y = 132 \end{pmatrix}$
Substitute $2x$ for y in the 2nd equation.
$0.09x + 0.12y = 132$
$0.09x + 0.12(2x) = 132$
$0.09x + 0.24x = 132$
$0.33x = 132$
$x = 400$

$y = 2x$
$y = 2(400)$
$y = 800$
The solution set is $\{(400, 800)\}$.

59. $\begin{pmatrix} x + y = 10.5 \\ 0.5x + 0.8y = 7.35 \end{pmatrix}$

Multiply equation 1 by 10 and multiply equation 2 by 100.
$\begin{pmatrix} 10x + 10y = 105 \\ 50x + 80y = 735 \end{pmatrix}$

Multiply equation 1 by -5 and then add the result to equation 2.

$$\begin{array}{r} -50x - 50y = -525 \\ 50x + 80y = 735 \\ \hline 30y = 210 \\ y = 7 \end{array}$$

$10x + 10y = 105$
$10x + 10(7) = 105$
$10x + 70 = 105$
$10x = 35$
$x = 3.5$
The solution set is $\{(3.5, 7)\}$.

61. Let $x =$ one number
and $y =$ other number.

$\begin{pmatrix} x + y = 53 \\ x - y = 19 \end{pmatrix}$

$$\begin{array}{r} x + y = 53 \\ x - y = 19 \\ \hline 2x \quad = 72 \\ x = 36 \end{array}$$

$x + y = 53$
$36 + y = 53$
$y = 17$
The numbers are 17 and 36.

63. Let x represent the smaller angle and y represent the larger angle.

$\begin{pmatrix} y = 15 + 4x \\ x + y = 90 \end{pmatrix}$
Substitute $15 + 4x$ for y in the 2nd equation.
$x + y = 90$
$x + 15 + 4x = 90$
$5x = 75$
$x = 15$
$y = 15 + 4x$
$y = 15 + 4(15)$
$y = 15 + 60$
$y = 75$
The angles are 15° and 75°.

65. Let $t =$ tens digit
and $u =$ units digit.

$\begin{pmatrix} t = 3u + 1 \\ u + t = 9 \end{pmatrix}$

Substitute $3u + 1$ for t in the 2nd equation.
$u + t = 9$
$u + 3u + 1 = 9$
$4u + 1 = 9$
$4u = 8$
$u = 2$

$t = 3u + 1$
$t = 3(2) + 1$
$t = 6 + 1$
$t = 7$
The number is 72.

67. Let u represent the units digit and t represent the tens digit.

value of original number $= 10t + u$
value of number with digits reversed $= 10u + t$

Sum of the digits is 7.
$t + u = 7$

$$\begin{pmatrix} t + u = 7 \\ 10u + t = 9 + 10t + u \end{pmatrix}$$

Simplify the 2nd equation.
$$\begin{pmatrix} t + u = 7 \\ -9t + 9u = 9 \end{pmatrix}$$

Divide the 2nd equation by 9 and then add the result to equation 1.
$$\begin{array}{r} t + u = 7 \\ -t + u = 1 \\ \hline 2u = 8 \\ u = 4 \end{array}$$
$t + u = 7$
$t + 4 = 7$
$t = 3$
The original number is 34.

69. Let $x =$ number of double rooms and $y =$ number of single rooms.

$$\begin{pmatrix} x + y = 23 \\ 32x + 26y = 688 \end{pmatrix}$$

Multiply equation 1 by -26 and then add the result to equation 2.

$$\begin{array}{rcr} -26x - 26y & = & -598 \\ 32x + 26y & = & 688 \\ \hline 6x & = & 90 \\ & x = 15 & \end{array}$$
$x + y = 23$
$15 + y = 23$
$y = 8$
There were 8 single rooms and 15 double rooms.

71. Let s represent the number of student tickets and n the number of nonstudent tickets.

$$\begin{pmatrix} 3s + 5n = 10,000 \\ s + n = 3000 \end{pmatrix}$$

Multiply equation 2 by -3 and then add the result to equation 1.
$$\begin{array}{rcr} 3s + 5n & = & 10000 \\ -3s - 3n & = & -9000 \\ \hline 2n & = & 1000 \\ & n = 500 & \end{array}$$
$s + n = 3000$
$s + 500 = 3000$
$s = 2500$
There were 2500 student tickets and 500 nonstudent tickets.

73. Let $x =$ money invested at 11% and $y =$ money invested at 9%.

$$\begin{pmatrix} x = 3y \\ 0.11x + 0.09y = 210 \end{pmatrix}$$
Substitute $3y$ for x in the 2nd equation.

$0.11(3y) + 0.09y = 210$
$0.33y + 0.09y = 210$
$0.42y = 210$
$y = 500$
$x = 3y$
$x = 3(500)$
$x = 1500$
She invested \$1,500 at 11% and \$500 invested at 9%.

75. Let x represent the rate of the kayak in still water and y the rate of the current.

	Rate	Time	Distance
upstream	$x-y$	4	$4(x-y)$
downstream	$2x+y$	1	$1(x+y)$

$$\begin{pmatrix} 4(x-y)=20 \\ 1(2x+y)=19 \end{pmatrix}$$

Divide the 1st equation by 4 and solve equation 2 for y. Substitute into equation 1.

$x-y=5$
$y=-2x+19$
$x-(-2x+19)=5$
$x+2x-19=5$
$3x=24$
$x=8$

$8-y=5$
$-y=-3$
$y=3$

The rate of the current is 3 mph.

77. Let $x =$ price of tennis ball and $y =$ price of golf ball.

$$\begin{pmatrix} 4x+3y=10.25 \\ 2x+5y=11.25 \end{pmatrix}$$

Multiply the 2nd equation by -2 and then add the result to equation 1.

$$\begin{array}{rl} 4x+3y= & 10.25 \\ -4x-10y= & -22.50 \\ \hline -7y= & -12.25 \\ y= & 1.75 \end{array}$$

$4x+3y=10.25$
$4x+3(1.75)=10.25$
$4x=5$
$x=1.25$

One tennis ball costs \$1.25 and one golf ball costs \$1.75.

79. Let f represent the number of five-dollar bills and t the number of ten-dollar bills.

$$\begin{pmatrix} f=12+t \\ 5f+10t=330 \end{pmatrix}$$

Substitute $12+t$ for f in the 2nd equation.

$5f+10t=330$
$5(12+t)+10t=330$
$60+5t+10t=330$
$60+15t=330$
$15t=270$
$t=18$
$f=12+t$
$f=12+18$
$f=30$

There were 18 ten-dollar bills and 30 five-dollar bills.

Further Investigations

85. $$\begin{pmatrix} \frac{1}{x}+\frac{2}{y}=\frac{7}{12} \\ \frac{3}{x}-\frac{2}{y}=\frac{5}{12} \end{pmatrix}$$

$$\begin{array}{l} \frac{1}{x}+\frac{2}{y}=\frac{7}{12} \\ \frac{3}{x}-\frac{2}{y}=\frac{5}{12} \\ \hline \frac{4}{x}\qquad=\frac{12}{12} \end{array}$$

$\frac{4}{x}=1$
$4=x$

$\frac{1}{x}+\frac{2}{y}=\frac{7}{12}$
$\frac{1}{4}+\frac{2}{y}=\frac{7}{12}$
$\frac{2}{y}=\frac{4}{12}$
$4y=24$
$y=6$

The solution set is $\{(4, 6)\}$.

87. $$\begin{pmatrix} \frac{3}{x}-\frac{2}{y}=\frac{13}{6} \\ \frac{2}{x}+\frac{3}{y}=0 \end{pmatrix}$$

Multiply equation 1 by 3,
multiply equation 2 by 2,
and then add the resulting equations.

$$\frac{9}{x} - \frac{6}{y} = \frac{13}{2}$$
$$\frac{4}{x} + \frac{6}{y} = 0$$
$$\frac{13}{x} \quad = \frac{13}{2}$$
$$x = 2$$
$$\frac{2}{x} + \frac{3}{y} = 0$$
$$\frac{2}{2} + \frac{3}{y} = 0$$
$$\frac{3}{y} = -1$$
$$-y = 3$$
$$y = -3$$

The solution set is $\{(2, -3)\}$.

89. $\begin{pmatrix} \frac{5}{x} - \frac{2}{y} = 23 \\ \frac{4}{x} + \frac{3}{y} = \frac{23}{2} \end{pmatrix}$

Multiply equation 1 by 3,
multiply equation 2 by 2,
and then add the resulting equations.

$$\frac{15}{x} - \frac{6}{y} = 69$$
$$\frac{8}{x} + \frac{6}{y} = 23$$
$$\frac{23}{x} \quad = 92$$
$$23 = 92x$$
$$\frac{23}{92} = x$$
$$x = \frac{1}{4}$$
$$\frac{5}{x} - \frac{2}{y} = 23$$
$$\frac{5}{\frac{1}{4}} - \frac{2}{y} = 23$$
$$20 - \frac{2}{y} = 23$$
$$-\frac{2}{y} = 3$$
$$-2 = 3y$$
$$-\frac{2}{3} = y$$

The solution set is $\left\{\left(\frac{1}{4}, -\frac{2}{3}\right)\right\}$.

91. Answers vary.

PROBLEM SET 11.2 Systems of Three Linear Equations in Three Variables

1. $\begin{pmatrix} 2x - 3y + 4z = 10 \\ 5y - 2z = -16 \\ 3z = 9 \end{pmatrix}$

$$3z = 9$$
$$z = 3$$

Substitute 3 for z in the 2nd equation.

$$5y - 2z = -16$$
$$y - 2(3) = -16$$
$$5y - 6 = -16$$
$$5y = -10$$
$$y = -2$$

Substitute -2 for y and 3 for z in the 1st equation.

$$2x - 3y + 4z = 10$$
$$2x - 3(-2) + 4(3) = 10$$
$$2x + 6 + 12 = 10$$
$$2x = -8$$
$$x = -4$$

The solution set is $\{(-4, -2, 3)\}$.

3. $\begin{pmatrix} x + 2y - 3z = 2 \\ 3y - z = 13 \\ 3y + 5z = 25 \end{pmatrix}$

Multiply the 2nd equation by -1 and then add the result to equation 3.

$$-3y + z = -13$$
$$3y + 5z = 25$$
$$6z = 12$$
$$z = 2$$

Substitute 2 for z in the 2nd equation.
$3y - z = 13$
$3y - 2 = 13$
$3y = 15$
$y = 5$
Substitute -2 for x and
5 for y in the 1st equation.
$x + 2y - 3z = 2$
$x + 2(5) - 3(2) = 2$
$x + 10 - 6 = 2$
$x + 4 = 2$
$x = -2$
The solution set is $\{(-2, 5, 2)\}$.

5. $\begin{pmatrix} 3x + 2y - 2z = 14 \\ x - 6y = 16 \\ 2x + 5z = -2 \end{pmatrix}$

Multiply the 2nd equation by -2 and then add the result to equation 3.

$$\begin{array}{r} -2x + 12z = -32 \\ 2x + \ \ 5z = -\ \ 2 \\ \hline 17z = -34 \\ z = -2 \end{array}$$

Substitute -2 for z in the 2nd equation.
$x - 6z = 16$
$x - 6(-2) = 16$
$x + 12 = 16$
$x = 4$
Substitute 4 for x and
-2 for z in the 1st equation.
$3x + 2y - 2z = 14$
$3(4) + 2y - 2(-2) = 14$
$12 + 2y + 4 = 14$
$2y + 16 = 14$
$2y = -2$
$y = -1$
The solution set is $\{(4, -1, -2)\}$.

7. $\begin{pmatrix} x - 2y + 3z = 7 \\ 2x + y + 5z = 17 \\ 3x - 4y - 2z = 1 \end{pmatrix}$

Multiply equation 1 by -2 and add to equation 2 to replace equation 2.
Multiply equation 1 by -3 and add to equation 3 to replace equation 3.

$\begin{pmatrix} x - 2y + 3z = 7 \\ 5y - z = 3 \\ 2y - 11z = -20 \end{pmatrix}$

Multiply equation 2 by -11 and add to equation 3 to replace equation 3.

$\begin{pmatrix} x - 2y + 3z = 7 \\ 5y - z = 3 \\ -53y = -53 \end{pmatrix}$

$-53y = -53$
$y = 1$
Substitute 1 for y in the 2nd equation.
$5y - z = 3$
$5(1) - z = 3$
$5 - z = 3$
$-z = -2$
$z = 2$
Substitute 1 for y and
2 for z in the 1st equation.
$x - 2y + 3z = 7$
$x - 2(1) + 3(2) = 7$
$x - 2 + 6 = 7$
$x + 4 = 7$
$x = 3$
The solution set is $\{(3, 1, 2)\}$.

9. $\begin{pmatrix} 2x - y + z = 0 \\ 3x - 2y + 4z = 11 \\ 5x + y - 6z = -32 \end{pmatrix}$

Add the 1st and 3rd equations.

$$\begin{array}{l} 2x - y + \ \ z = 0 \\ 5x + y - 6z = -32 \\ \hline 7x \qquad - 5z = -32 \end{array}$$

Multiply the 3rd equation by 2 and then add the result to equation 2.

$$\begin{array}{r} 3x - 2y + \ \ 4z = \quad 11 \\ 10x + 2y - 12z = -64 \\ \hline 13x \qquad - 8z = -53 \end{array}$$

$\begin{pmatrix} 7x - 5z = -32 \\ 13x - 8z = -53 \end{pmatrix}$

Multiply equation 1 by -8,
multiply equation 2 by 5,
and then add the resulting equations.

$$\begin{array}{rl} -56x + 40z = & 256 \\ 65x - 40z = & -265 \\ \hline 9x \quad = & -9 \end{array}$$

$x = -1$

Substitute -1 for x in the 1st equation.

$7x - 5z = -32$

$7(-1) - 5z = -32$

$-7 - 5z = -32$

$-5z = -25$

$z = 5$

Substitute -1 for x and 5 for z in the 1st equation.

$2x - y + z = 0$

$2(-1) - y + 5 = 0$

$-2 - y + 5 = 0$

$-y + 3 = 0$

$-y = -3$

$y = 3$

The solution set is $\{(-1, 3, 5)\}$.

11. $\begin{pmatrix} 3x + 2y - z = -11 \\ 2x - 3y + 4z = 11 \\ 5x + y - 2z = -17 \end{pmatrix}$

Multiply equation 1 by 4 and add to equation 2 to replace equation 2. Multiply equation 1 by -2 and add to equation 3 to replace equation 3.

$\begin{pmatrix} 3x + 2y - z = 11 \\ 14x + 5y = -33 \\ -x - 3y = 5 \end{pmatrix}$

Multiply equation 3 by 14 and add to equation 2 to replace equation 2.

$\begin{pmatrix} 3x + 2y - z = -11 \\ -37y = 37 \\ -x - 3y = 5 \end{pmatrix}$

$-37y = 37$

$y = -1$

Substitute -1 for y in the 3rd equation.

$-x - 3y = 5$

$-x - 3(-1) = 5$

$-x + 3 = 5$

$-x = 2$

$x = -2$

Substitute -2 for x and -1 for y in the 1st equation.

$3x + 2y - z = -11$

$3(-2) + 2(-1) - z = -11$

$-6 - 2 - z = -11$

$-8 - z = -11$

$-z = -3$

$z = 3$

The solution set is $\{(-2, -1, 3)\}$.

13. $\begin{pmatrix} 2x + 3y - 4z = -10 \\ 4x - 5y + 3z = 2 \\ 2y + z = 8 \end{pmatrix}$

Multiply the 1st equation by -2 and then add the result to equation 2.

$$\begin{array}{rl} -4x - 6y + 8z = & 20 \\ 4x - 5y + 3z = & 2 \\ \hline -11y + 11z = & 22 \end{array}$$

$-y + z = 2$

$\begin{pmatrix} -y + z = 2 \\ 2y + z = 8 \end{pmatrix}$

Multiply the 1st equation by -1 and then add the result to equation 2.

$$\begin{array}{rl} y - z = & -2 \\ 2y + z = & 8 \\ \hline 3y \quad = & 6 \end{array}$$

$y = 2$

Substitute 2 for y in the 1st equation.

$-y + z = 2$

$-2 + z = 2$

$z = 4$

Substitute 2 for y and 4 for z in the 1st equation.

$2x + 3y - 4z = -10$

$2x + 3(2) - 4(4) = -10$

$2x + 6 - 16 = -10$

$2x - 10 = -10$

$2x = 0$

$x = 0$

The solution set is $\{(0, 2, 4)\}$.

15. $\begin{pmatrix} 3x + 2y - 2z = 14 \\ 2x - 5y + 3z = 7 \\ 4x - 3y + 7z = 5 \end{pmatrix}$

Multiply equation 1 by 2,
multiply equation 2 by -3,
and then add the resulting equations.

$$\begin{array}{r} 6x + 4y - 4z = 28 \\ -6x + 15y - 9z = -21 \\ \hline 19y - 13z = 7 \end{array}$$

Multiply the 2nd equation by -2 and then add the result to equation 3.

$$\begin{array}{r} -4x + 10y - 6z = -14 \\ 4x - 3y + 7z = 5 \\ \hline 7y + z = -9 \end{array}$$

$$\begin{pmatrix} 19y - 13z = 7 \\ 7y + z = -9 \end{pmatrix}$$

Multiply the 2nd equation by 13 and then add the result to equation 1.

$$\begin{array}{r} 91y + 13z = -117 \\ 19y - 13z = 7 \\ \hline 110y = -110 \end{array}$$

$y = -1$

Substitute -1 for y in the 2nd equation.

$7y + z = -9$
$7(-1) + z = -9$
$-7 + z = -9$
$z = -2$

Substitute -1 for y and -2 for z in the 1st equation.

$3x + 2y - 2z = 14$
$3x + 2(-1) - 2(-2) = 14$
$3x - 2 + 4 = 14$
$3x + 2 = 14$
$3x = 12$
$x = 4$

The solution set is $\{(4, -1, -2)\}$.

17. $\begin{pmatrix} 2x - 3y + 4z = -12 \\ 4x + 2y - 3z = -13 \\ 6x - 5y + 7z = -31 \end{pmatrix}$

Multiply the 1st equation by -2 and then add the result to equation 2.

$$\begin{array}{r} -4x + 6y - 8z = 24 \\ 4x + 2y - 3z = -13 \\ \hline 8y - 11z = 11 \end{array}$$

Multiply the 1st equation by -3 and then add the result to equation 3.

$$\begin{array}{r} -6x + 9y - 12z = 36 \\ 6x - 5y + 7z = -31 \\ \hline 4y - 5z = 5 \end{array}$$

$$\begin{pmatrix} 8y - 11z = 11 \\ 4y - 5z = 5 \end{pmatrix}$$

Multiply the 2nd equation by -2 and then add the result to equation 1.

$$\begin{array}{r} 8y - 11z = 11 \\ -8y + 10z = -10 \\ \hline -z = 1 \end{array}$$

$z = -1$

Substitute -1 for z in the 2nd equation.

$4y - 5z = 5$
$4y - 5(-1) = 5$
$4y + 5 = 5$
$4y = 0$
$y = 0$

Substitute 0 for y and -1 for z in the 1st equation.

$2x - 3y + 4z = -12$
$2x - 3(0) + 4(-1) = -12$
$2x - 0 - 4 = -12$
$2x - 4 = -12$
$2x = -8$
$x = -4$

The solution set is $\{(-4, 0, -1)\}$.

19. $\begin{pmatrix} 5x - 3y - 6z = 22 \\ x - y + z = -3 \\ -3x + 7y - 5z = 23 \end{pmatrix}$

Multiply equation 2 by -5 and add to equation 1 to replace equation 1.
Multiply equation 2 by 3 and add to equation 3 to replace equation 3.

$$\begin{pmatrix} 2y - 11z = 37 \\ x - y + z = -3 \\ 4y - 2z = 14 \end{pmatrix}$$

Multiply equation 1 by -2 and add to equation 3 to replace equation 3.

$$\begin{pmatrix} 2y - 11z = 37 \\ x - y + z = -3 \\ 20z = -60 \end{pmatrix}$$

$20z = -60$
$z = -3$
Substitute -3 for z in the 1st equation.
$2y - 11z = 37$
$2y - 11(-3) = 37$
$2y + 33 = 37$
$2y = 4$
$y = 2$
Substitute 2 for y and
-3 for z in the 2nd equation.
$x - y + z = -3$
$x - 2 + (-3) = -3$
$x - 5 = -3$
$x = 2$
The solution set is $\{(2, 2, -3)\}$.

21. Let x represent the number of pounds of almonds, y the number of pounds of pecans, and z the number of pounds of peanuts.

$$\begin{pmatrix} x + y + z = 20 \\ 3.50x + 4y + 2z = (2.70)(20) \\ z = 3y \end{pmatrix}$$

Substitute $3y$ for z in the
1st and 2nd equations.
$x + y + (3y) = 20$
$x + 4y = 20$

$3.50x + 4y + 2(3y) = 54$
$3.50x + 4y + 6y = 54$
$3.50x + 10y = 54$

$-5(x + 4y) = -5(20)$
$2(3.50x + 10y) = 2(54)$

$$\begin{array}{rcr} -5x - 20y &=& -100 \\ 7x + 20y &=& 108 \\ \hline 2x \qquad &=& 8 \end{array}$$

$x = 4$
$x + 4y = 20$
$4 + 4y = 20$
$4y = 16$
$y = 4$

$x + y + z = 20$
$4 + 4 + z = 20$
$8 + z = 20$
$z = 12$

The mix should have 4 pounds of almonds, 4 pounds of pecans and 12 pounds of peanuts.

23. Let x = number of nickels,
y = number of dimes and
z = number of quarters.

$$\begin{pmatrix} x + y + z = 42 \\ x + y = z - 2 \\ 0.05x + 0.10y + 0.25z = 7.15 \end{pmatrix}$$

$$\begin{pmatrix} x + y + z = 42 \\ x + y - z = -2 \\ 0.05x + 0.10y + 0.25z = 7.15 \end{pmatrix}$$

Multiply equation 1 by -1 and add to equation 2 to replace equation 2.

$$\begin{pmatrix} x + y + z = 42 \\ -2z = -44 \\ 0.05x + 0.10y + 0.25z = 7.15 \end{pmatrix}$$

$-2z = -44$
$z = 22$
Substitute 22 for z in the
1st and 2nd equations.

$$\begin{pmatrix} x + y + 22 = 42 \\ 0.05x + 0.10y + 0.25(22) = 7.15 \end{pmatrix}$$

$$\begin{pmatrix} x + y = 20 \\ 0.05x + 0.10y = 1.65 \end{pmatrix}$$

Multiply equation 1 by -0.05 and add to equation 2 to replace equation 2.

$$\begin{pmatrix} x + y = 20 \\ 0.05y = 0.65 \end{pmatrix}$$

$0.05y = 0.65$
$y = 13$
Substitute 13 for y in the 1st equation.
$x + y = 20$
$x + 13 = 20$
$x = 7$
There are 7 nickels, 13 dimes and 22 quarters.

25. Let x represent the smallest angle, y the middle angle, and z the largest angle.

$$\begin{pmatrix} x+y+z=180 \\ z=2x \\ x+z=2y \end{pmatrix}$$

Substitute $2x$ for z in the 1st and 3rd equations.

$$\begin{pmatrix} x+y+2x=180 \\ x+2x=2y \end{pmatrix}$$

$$\begin{pmatrix} 3x+y=180 \\ 3x-2y=0 \end{pmatrix}$$

Multiply the 2nd equation by -1 and then add the result to equation 1.

$$\begin{array}{r} 3x+\ \ y=180 \\ -3x+2y=\ \ \ 0 \\ \hline 3y=180 \\ y=60 \end{array}$$

Substitute 60 for y in the 1st equation.

$3x+y=180$
$3x+60=180$
$3x=120$
$x=40$

Substitute 40 for x in the 2nd equation.

$z=2x$
$z=2(40)$
$z=80$

The angles measure 40°, 60°, and 80°.

27. Let $x=$ money invested at 12%, $y=$ money invested at 13% and $z=$ money invested at 14%.

$$\begin{pmatrix} x+y+z=3000 \\ 0.12x+0.13y+0.14z=400 \\ x+y=z \end{pmatrix}$$

Substitute z for $x+y$ in the 1st equation.

$x+y+z=3000$
$z+z=3000$
$2z=3000$
$z=1500$

Substitute 1500 for z in the 1st and 2nd equations.

$$\begin{pmatrix} x+y+1500=3000 \\ 0.12x+0.13y+0.14(1500)=400 \end{pmatrix}$$

$$\begin{pmatrix} x+y=1500 \\ 0.12x+0.13y=190 \end{pmatrix}$$

Multiply equation 1 by -0.12 and add to equation 2 to replace equation 2.

$$\begin{pmatrix} x+y=1500 \\ 0.01y=10 \end{pmatrix}$$

$0.01y=10$
$y=1000$

Substitute 1000 for y in the 1st equation.

$x+y=1500$
$x+1000=1500$
$x=500$

There is \$500 is invested at 12%, \$1,000 is invested at 13% and \$1,500 is invested at 14%.

29. Let x represent the number of Type A birdhouses, y the number of Type B, and z the number of Type C.

$$\begin{pmatrix} 0.1x+0.2y+0.1z=35 \\ 0.4x+0.4y+0.3z=95 \\ 0.2x+0.1y+0.3z=62.5 \end{pmatrix}$$

Multiply each equation by 10.

$$\begin{pmatrix} x+2y+z=350 \\ 4x+4y+3z=950 \\ 2x+y+3z=625 \end{pmatrix}$$

Multiply the 1st equation by -2 and then add the result to equation 3.

$$\begin{array}{r} -2x-4y-2z=\ -700 \\ 2x+\ \ y+3z=\ \ \ \ 625 \\ \hline -3y+\ \ z=\ \ -75 \end{array}$$

Multiply the 1st equation by -4 and then add the result to equation 2.

$$\begin{array}{r} -4z-8y-4z=-1400 \\ 4x+4y+3z=\ \ \ \ 950 \\ \hline -4y-\ \ z=\ -450 \end{array}$$

$$\begin{array}{r} -3y+z=\ \ -75 \\ -4y-z=-450 \\ \hline -7y\ \ \ \ \ \ =-525 \\ y=75 \end{array}$$

$-3y+z=-75$
$-3(75)+z=-75$
$-225+z=-75$
$z=150$
$x+2y+z=350$
$x+2(75)+150=350$
$x+150+150=350$
$x+300=350$
$x=50$
Each week the company makes 50 Type A birdhouses, 75 Type B birdhouses and 150 Type C birdhouses.

PROBLEM SET 11.3 Matrix Approach to Solving Linear Systems

1. Yes

3. Yes

5. No

7. No

9. Yes

11. $\left[\begin{array}{cc|c} 1 & -3 & 14 \\ 3 & 2 & -13 \end{array}\right]$

-3(row 1) + (row 2) replace (row 2)
$\left[\begin{array}{cc|c} 1 & -3 & 14 \\ 0 & 11 & -55 \end{array}\right]$

$\frac{1}{11}$(row 2)
$\left[\begin{array}{cc|c} 1 & -3 & 14 \\ 0 & 1 & -5 \end{array}\right]$

3(row 2) + (row 1) replace (row 1)
$\left[\begin{array}{cc|c} 1 & 0 & -1 \\ 0 & 1 & -5 \end{array}\right]$

The solution set is $\{(-1,\ -5)\}$.

13. $\left[\begin{array}{cc|c} 3 & -4 & 33 \\ 1 & 7 & -39 \end{array}\right]$

Exchange row 1 and row 2.
$\left[\begin{array}{cc|c} 1 & 7 & -39 \\ 3 & -4 & 33 \end{array}\right]$

-3(row 1) + (row 2) replace (row 2)
$\left[\begin{array}{cc|c} 1 & 7 & -39 \\ 0 & -25 & 150 \end{array}\right]$

$-\frac{1}{25}$(row 2) replace (row 2)
$\left[\begin{array}{cc|c} 1 & 7 & -39 \\ 0 & 1 & -6 \end{array}\right]$

-7(row 2) + (row 1) replace (row 1)
$\left[\begin{array}{cc|c} 1 & 0 & 3 \\ 0 & 1 & -6 \end{array}\right]$

The solution set is $\{(3,\ -6)\}$.

15. $\left[\begin{array}{cc|c} 1 & -6 & -2 \\ 2 & -12 & 5 \end{array}\right]$

-2(row 1) + (row 2) replace (row 2)
$\left[\begin{array}{cc|c} 1 & -6 & -2 \\ 0 & 0 & 9 \end{array}\right]$

$0 \neq 9$
The system is inconsistent.
The solution set is $\emptyset$.

17. $\left[\begin{array}{cc|c} 3 & -5 & 39 \\ 2 & 7 & -67 \end{array}\right]$

(row 1) − (row 2) replace (row 1)
$\left[\begin{array}{cc|c} 1 & -12 & 106 \\ 2 & 7 & -67 \end{array}\right]$

-2(row 1) + (row 2) replace (row 2)
$\left[\begin{array}{cc|c} 1 & -12 & 106 \\ 0 & 31 & -279 \end{array}\right]$

$\frac{1}{31}$(row 2) replace (row 2)
$\left[\begin{array}{cc|c} 1 & -12 & 106 \\ 0 & 1 & -9 \end{array}\right]$

12(row 2) + (row 1) replace (row 1)

$$\left[\begin{array}{cc|c} 1 & 0 & -2 \\ 0 & 1 & -9 \end{array}\right]$$

The solution set is $\{(-2, -9)\}$.

19. $$\left[\begin{array}{ccc|c} 1 & -2 & -3 & -6 \\ 3 & -5 & -1 & 4 \\ 2 & 1 & 2 & 2 \end{array}\right]$$

-3(row 1) + (row 2) replace (row 2)
-2(row 1) + (row 3) replace (row 3)

$$\left[\begin{array}{ccc|c} 1 & -2 & -3 & -6 \\ 0 & 1 & 8 & 22 \\ 0 & 5 & 8 & 14 \end{array}\right]$$

2(row 2) + (row 1) replace (row 1)
-5(row 2) + (row 3) replace (row 3)

$$\left[\begin{array}{ccc|c} 1 & 0 & 13 & 38 \\ 0 & 1 & 8 & 22 \\ 0 & 0 & -32 & -96 \end{array}\right]$$

$-\frac{1}{32}$(row 3)

$$\left[\begin{array}{ccc|c} 1 & 0 & 13 & 38 \\ 0 & 1 & 8 & 22 \\ 0 & 0 & 1 & 3 \end{array}\right]$$

-8(row 3) + (row 2) replace (row 2)
-13(row 3) + (row 1) replace (row 1)

$$\left[\begin{array}{ccc|c} 1 & 0 & 0 & -1 \\ 0 & 1 & 0 & -2 \\ 0 & 0 & 1 & 3 \end{array}\right]$$

The solution set is $\{(-1, -2, 3)\}$.

21. $$\left[\begin{array}{ccc|c} -2 & -5 & 3 & 11 \\ 1 & 3 & -3 & -12 \\ 3 & -2 & 5 & 31 \end{array}\right]$$

Exchange rows 1 and 2.

$$\left[\begin{array}{ccc|c} 1 & 3 & -3 & -12 \\ -2 & -5 & 3 & 11 \\ 3 & -2 & 5 & 31 \end{array}\right]$$

2(row 1) + (row 2) replace (row 2)
-3(row 1) + (row 3) replace (row 3)

$$\left[\begin{array}{ccc|c} 1 & 3 & -3 & -12 \\ 0 & 1 & -3 & -13 \\ 0 & -11 & 14 & 67 \end{array}\right]$$

-3(row 2) + (row 1) replace (row 1)
11(row 2) + (row 3) replace (row 3)

$$\left[\begin{array}{ccc|c} 1 & 0 & 6 & 27 \\ 0 & 1 & -3 & -13 \\ 0 & 0 & -19 & -76 \end{array}\right]$$

$-\frac{1}{19}$(row 3)

$$\left[\begin{array}{ccc|c} 1 & 0 & 6 & 27 \\ 0 & 1 & -3 & -13 \\ 0 & 0 & 1 & 4 \end{array}\right]$$

-6(row 3) + (row 1) replace (row 1)
3(row 3) + (row 2) replace (row 2)

$$\left[\begin{array}{ccc|c} 1 & 0 & 0 & 3 \\ 0 & 1 & 0 & -1 \\ 0 & 0 & 1 & 4 \end{array}\right]$$

The solution set is $\{(3, -1, 4)\}$.

23. $$\left[\begin{array}{ccc|c} 1 & -3 & -1 & 2 \\ 3 & 1 & -4 & -18 \\ -2 & 5 & 3 & 2 \end{array}\right]$$

-3(row 1) + (row 2) replace (row 2)
2(row 1) + (row 3) replace (row 3)

$$\left[\begin{array}{ccc|c} 1 & -3 & -1 & 2 \\ 0 & 10 & -1 & -24 \\ 0 & -1 & 1 & 6 \end{array}\right]$$

$\frac{1}{10}$(row 2)

$$\left[\begin{array}{ccc|c} 1 & -3 & -1 & 2 \\ 0 & 1 & -\frac{1}{10} & -\frac{12}{5} \\ 0 & -1 & 1 & 6 \end{array}\right]$$

3(row 2) + (row 1) replace (row 1)
(row 2) + (row 3) replace (row 3)

$$\left[\begin{array}{ccc|c} 1 & 0 & -\frac{13}{10} & -\frac{26}{5} \\ 0 & 1 & -\frac{1}{10} & -\frac{12}{5} \\ 0 & 0 & \frac{9}{10} & \frac{18}{5} \end{array}\right]$$

$\frac{10}{9}$(row 3)

$$\left[\begin{array}{ccc|c} 1 & 0 & -\frac{13}{10} & -\frac{26}{5} \\ 0 & 1 & -\frac{1}{10} & -\frac{12}{5} \\ 0 & 0 & 1 & 4 \end{array}\right]$$

$\frac{1}{10}$(row 3) + (row 2) replace (row 2)

$\frac{13}{10}$(row 3) + (row 1) replace (row 1)

$$\left[\begin{array}{ccc|c} 1 & 0 & 0 & 0 \\ 0 & 1 & 0 & -2 \\ 0 & 0 & 1 & 4 \end{array}\right]$$

The solution set is $\{(0,\ -2,\ 4)\}$.

25. $$\left[\begin{array}{ccc|c} 1 & -1 & 2 & 1 \\ -3 & 4 & -1 & 4 \\ -1 & 2 & 3 & 6 \end{array}\right]$$

3(row 1) + (row 2) replace (row 2)
(row 1) + (row 3) replace (row 3)

$$\left[\begin{array}{ccc|c} 1 & -1 & 2 & 1 \\ 0 & 1 & 5 & 7 \\ 0 & 1 & 5 & 7 \end{array}\right]$$

(row 2) − (row 3) replace (row 3)

$$\left[\begin{array}{ccc|c} 1 & -1 & 2 & 1 \\ 0 & 1 & 5 & 7 \\ 0 & 0 & 0 & 0 \end{array}\right]$$

The row of zeros indicates a dependent system.

$y + 5z = 7$
$y = 7 - 5z$

$x - y + 2z = 1$
$x - (7 - 5z) + 2z = 1$
$x - 7 + 7z = 1$
$x = 8 - 7z$

The solution set is $\{(8 - 7k,\ 7 - 5k,\ k)\}$, where k is any real number.

27. $$\left[\begin{array}{ccc|c} -2 & 1 & 5 & -5 \\ 3 & 8 & -1 & -34 \\ 1 & 2 & 1 & -12 \end{array}\right]$$

Interchange row 1 and row 3.

$$\left[\begin{array}{ccc|c} 1 & 2 & 1 & -12 \\ 3 & 8 & -1 & -34 \\ -2 & 1 & 5 & -5 \end{array}\right]$$

− 3(row 1) + (row 2) replace (row 2)
2(row 1) + (row 3) replace (row 3)

$$\left[\begin{array}{ccc|c} 1 & 2 & 1 & -12 \\ 0 & 2 & -4 & 2 \\ 0 & 5 & 7 & -29 \end{array}\right]$$

$\frac{1}{2}$(row 2)

$$\left[\begin{array}{ccc|c} 1 & 2 & 1 & -12 \\ 0 & 1 & -2 & 1 \\ 0 & 5 & 7 & -29 \end{array}\right]$$

− 2(row 2) + (row 1) replace (row 1)
− 5(row 2) + (row 3) replace (row 3)

$$\left[\begin{array}{ccc|c} 1 & 0 & 5 & -14 \\ 0 & 1 & -2 & 1 \\ 0 & 0 & 17 & -34 \end{array}\right]$$

$\frac{1}{17}$(row 3)

$$\left[\begin{array}{ccc|c} 1 & 0 & 5 & -14 \\ 0 & 1 & -2 & 1 \\ 0 & 0 & 1 & -2 \end{array}\right]$$

2(row 3) + (row 2) replace (row 2)
− 5(row 3) + (row 1) replace (row 1)

$$\left[\begin{array}{ccc|c} 1 & 0 & 0 & -4 \\ 0 & 1 & 0 & -3 \\ 0 & 0 & 1 & -2 \end{array}\right]$$

The solution set is $\{(-4,\ -3,\ -2)\}$.

29. $\left[\begin{array}{ccc|c} 2 & 3 & -1 & 7 \\ 3 & 4 & 5 & -2 \\ 5 & 1 & 3 & 13 \end{array}\right]$

-1(row 1) + (row 2) replace (row 1)

$\left[\begin{array}{ccc|c} 1 & 1 & 6 & -9 \\ 3 & 4 & 5 & -2 \\ 5 & 1 & 3 & 13 \end{array}\right]$

-3(row 1) + (row 2) replace (row 2)
-5(row 1) + (row 3) replace (row 3)

$\left[\begin{array}{ccc|c} 1 & 1 & 6 & -9 \\ 0 & 1 & -13 & 25 \\ 0 & -4 & -27 & 58 \end{array}\right]$

(row 1) − (row 2) replace (row 1)
4(row 2) + (row 3) replace (row 3)

$\left[\begin{array}{ccc|c} 1 & 0 & 19 & -34 \\ 0 & 1 & -13 & 25 \\ 0 & 0 & -79 & 158 \end{array}\right]$

$-\frac{1}{79}$(row 3)

$\left[\begin{array}{ccc|c} 1 & 0 & 19 & -34 \\ 0 & 1 & -13 & 25 \\ 0 & 0 & 1 & -2 \end{array}\right]$

-19(row 3) + (row 1) replace (row 1)
13(row 3) + (row 2) replace (row 2)

$\left[\begin{array}{ccc|c} 1 & 0 & 0 & 4 \\ 0 & 1 & 0 & -1 \\ 0 & 0 & 1 & -2 \end{array}\right]$

The solution set is $\{(4, -1, -2)\}$.

31. $\left[\begin{array}{cccc|c} 1 & -3 & -2 & 1 & -3 \\ -2 & 7 & 1 & -2 & -1 \\ 3 & -7 & -3 & 3 & -5 \\ 5 & 1 & 4 & -2 & 18 \end{array}\right]$

2(row 1) + (row 2) replace (row 2)
-3(row 1) + (row 3) replace (row 3)
-5(row 1) + (row 4) replace (row 4)

$\left[\begin{array}{cccc|c} 1 & -3 & -2 & 1 & -3 \\ 0 & 1 & -3 & 0 & -7 \\ 0 & 2 & 3 & 0 & 4 \\ 0 & 16 & 14 & -7 & 33 \end{array}\right]$

3(row 2) + (row 1) replace (row 1)
-2(row 2) + (row 3) replace (row 3)
-16(row 2) + (row 4) replace (row 4)

$\left[\begin{array}{cccc|c} 1 & 0 & -11 & 1 & -24 \\ 0 & 1 & -3 & 0 & -7 \\ 0 & 0 & 9 & 0 & 18 \\ 0 & 0 & 62 & -7 & 145 \end{array}\right]$

$\frac{1}{9}$(row 3)

$\left[\begin{array}{cccc|c} 1 & 0 & -11 & 1 & -24 \\ 0 & 1 & -3 & 0 & -7 \\ 0 & 0 & 1 & 0 & 2 \\ 0 & 0 & 62 & -7 & 145 \end{array}\right]$

11(row 3) + (row 1) replace (row 1)
3(row 3) + (row 2) replace (row 2)
-62(row 3) + (row 4) replace (row 4)

$\left[\begin{array}{cccc|c} 1 & 0 & 0 & 1 & -2 \\ 0 & 1 & 0 & 0 & -1 \\ 0 & 0 & 1 & 0 & 2 \\ 0 & 0 & 0 & -7 & 21 \end{array}\right]$

$-\frac{1}{7}$(row 4)

$\left[\begin{array}{cccc|c} 1 & 0 & 0 & 1 & -2 \\ 0 & 1 & 0 & 0 & -1 \\ 0 & 0 & 1 & 0 & 2 \\ 0 & 0 & 0 & 1 & -3 \end{array}\right]$

$-$(row 4) + (row 1) replace (row 1)

$\left[\begin{array}{cccc|c} 1 & 0 & 0 & 0 & 1 \\ 0 & 1 & 0 & 0 & -1 \\ 0 & 0 & 1 & 0 & 2 \\ 0 & 0 & 0 & 1 & -3 \end{array}\right]$

The solution set is $\{(1, -1, 2, -3)\}$.

33. $\left[\begin{array}{cccc|c} 1 & 3 & -1 & 2 & -2 \\ 2 & 7 & 2 & -1 & 19 \\ -3 & -8 & 3 & 1 & -7 \\ 4 & 11 & -2 & -3 & 19 \end{array}\right]$

-2(row 1) + (row 2) replace (row 2)
3(row 1) + (row 3) replace (row 3)
-4(row 1) + (row 4) replace (row 4)

$$\left[\begin{array}{cccc|c} 1 & 3 & -1 & 2 & -2 \\ 0 & 1 & 4 & -5 & 23 \\ 0 & 1 & 0 & 7 & -13 \\ 0 & -1 & 2 & -11 & 27 \end{array}\right]$$

-3(row 2) + (row 1) replace (row 1)
(row 2) − (row 3) replace (row 3)
(row 3) + (row 4) replace (row 4)

$$\left[\begin{array}{cccc|c} 1 & 0 & -13 & 17 & -71 \\ 0 & 1 & 4 & -5 & 23 \\ 0 & 0 & 4 & -12 & 36 \\ 0 & 0 & 2 & -4 & 14 \end{array}\right]$$

$\frac{1}{4}$(row 3)
-2(row 4) + (row 3) replace (row 4)

$$\left[\begin{array}{cccc|c} 1 & 0 & -13 & 17 & -71 \\ 0 & 1 & 4 & -5 & 23 \\ 0 & 0 & 1 & -3 & 9 \\ 0 & 0 & 0 & -4 & 8 \end{array}\right]$$

$-4x_4 = 8$
$x_4 = -2$

$x_3 - 3x_4 = 9$
$x_3 - 3(-2) = 9$
$x_3 + 6 = 9$
$x_3 = 3$

$x_2 + 4x_3 - 5x_4 = 23$
$x_2 + 4(3) - 5(-2) = 23$
$x_2 + 12 + 10 = 23$
$x_2 + 22 = 23$
$x_2 = 1$

$x_1 - 13x_3 + 17x_4 = -71$
$x_1 - 13(3) + 17(-2) = -71$
$x_1 - 39 - 34 = -71$
$x_1 - 73 = -71$
$x_1 = 2$
The solution set is $\{(2, 1, 3, -2)\}$.

35. $$\left[\begin{array}{cccc|c} 1 & 0 & 0 & 0 & -2 \\ 0 & 1 & 0 & 0 & 4 \\ 0 & 0 & 1 & 0 & -3 \\ 0 & 0 & 0 & 1 & 0 \end{array}\right]$$

The solution set is $\{(-2, 4, -3, 0)\}$.

37. $$\left[\begin{array}{cccc|c} 1 & 0 & 0 & 0 & -8 \\ 0 & 1 & 0 & 0 & 5 \\ 0 & 0 & 1 & 0 & -2 \\ 0 & 0 & 0 & 0 & 1 \end{array}\right]$$

The last row represents
$0x_1 + 0x_2 + 0x_3 + 0x_4 = 1$
$0 = 1$
This is false, so the system is inconsistent and the solution is $\emptyset$.

39. $$\left[\begin{array}{cccc|c} 1 & 0 & 0 & 3 & 5 \\ 0 & 1 & 0 & 0 & -1 \\ 0 & 0 & 1 & 4 & 2 \\ 0 & 0 & 0 & 0 & 0 \end{array}\right]$$

The bottom row at all zeros represents an identity. From the other row we get
$x_3 + 4x_4 = 2$
$x_2 = -1$
$x_1 + 3x_4 = 5$

Let $x_4 = k$.
Then $x_3 + 4k = 2$,
$x_3 = -4k + 2$ and
$x_1 + 3k = 5$
$x_1 = -3k + 5$
The solution set is
$\{(-3k + 5, -1, -4k + 2, k)\}$
where k is any real number.

41. $$\left[\begin{array}{cccc|c} 1 & 3 & 0 & 0 & 9 \\ 0 & 0 & 1 & 0 & 2 \\ 0 & 0 & 0 & 1 & -3 \\ 0 & 0 & 0 & 0 & 0 \end{array}\right]$$

$x_4 = -3$
$x_3 = 2$
$x_1 + 3x_2 = 9$
$x_1 = 9 - 3x_2$

Let $k = x_2$.
The solution set is
$\{(9 - 3k, k, 2, -3)\}$
where k is any real number.

45. $\left[\begin{array}{ccc|c} 1 & -2 & 3 & 4 \\ 3 & -5 & -1 & 7 \end{array}\right]$

-3(row 1) + (row 2) replace (row 2)

$\left[\begin{array}{ccc|c} 1 & -2 & 3 & 4 \\ 0 & 1 & -10 & -5 \end{array}\right]$

2(row 2) + (row 1) replace (row 1)

$\left[\begin{array}{ccc|c} 1 & 0 & -17 & -6 \\ 0 & 1 & -10 & -5 \end{array}\right]$

$y - 10z = -5$
$y = 10z - 5$
$x - 17z = -6$
$x = 17z - 6$
Let $z = k$.
The solution set is $\{(17k - 6, 10k - 5, k)\}$, where k is a real number.

47. $\left[\begin{array}{ccc|c} 2 & -4 & 3 & 8 \\ 3 & 5 & -1 & 7 \end{array}\right]$

$-\frac{3}{2}$(row 1) + (row 2) replace (row 2)

$\left[\begin{array}{ccc|c} 2 & -4 & 3 & 8 \\ 0 & 11 & -\frac{11}{2} & -5 \end{array}\right]$

$\frac{4}{11}$(row 2) + (row 1) replace (row 1)

$\left[\begin{array}{ccc|c} 2 & 0 & 1 & \frac{68}{11} \\ 0 & 11 & -\frac{11}{2} & -5 \end{array}\right]$

$11y - \frac{11}{2}z = -5$
$11y = \frac{11}{2}z - 5$
$y = \frac{1}{2}z - \frac{5}{11}$
$2x + z = \frac{68}{11}$
$2x = -z + \frac{68}{11}$
$x = -\frac{1}{2}z + \frac{34}{11}$
Let $z = k$.
The solution set is $\left\{\left(-\frac{1}{2}k + \frac{34}{11}, \frac{1}{2}k - \frac{5}{11}, k\right)\right\}$, where k is any real number.

49. $\left[\begin{array}{ccc|c} 1 & -2 & 4 & 9 \\ 2 & -4 & 8 & 3 \end{array}\right]$

-2(row 1) + (row 2) replace (row 2)

$\left[\begin{array}{ccc|c} 1 & -26 & 4 & 9 \\ 0 & 0 & 0 & -15 \end{array}\right]$

$0x + 0y + 0z = -15$
$0 \neq -15$
Therefore the solution set is $\emptyset$.

PROBLEM SET 11.4 Determinants

1. $\begin{vmatrix} 4 & 3 \\ 2 & 7 \end{vmatrix} = (4)(7) - (2)(3) = 22$

3. $\begin{vmatrix} -3 & 2 \\ 7 & 5 \end{vmatrix} = -15 - 14 = -29$

5. $\begin{vmatrix} 2 & -3 \\ 8 & -2 \end{vmatrix} = (2)(-2) - (8)(-3) = 20$

7. $\begin{vmatrix} -2 & -3 \\ -1 & -4 \end{vmatrix} = 8 - 3 = 5$

9. $\begin{vmatrix} \frac{1}{2} & \frac{1}{3} \\ -3 & -6 \end{vmatrix}$

$= \left(\frac{1}{2}\right)(-6) - (-3)\left(\frac{1}{3}\right) = -2$

Problem Set 11.4

11. $\begin{vmatrix} \frac{1}{2} & \frac{2}{3} \\ \frac{3}{4} & -\frac{1}{3} \end{vmatrix} = -\frac{1}{6} - \frac{1}{2} = -\frac{2}{3}$

13. $\begin{vmatrix} 1 & 2 & -1 \\ 3 & 1 & 2 \\ 2 & 4 & 3 \end{vmatrix}$

-3(row 1) + (row 2) replace (row 2)
-2(row 1) + (row 3) replace (row 3)

$\begin{vmatrix} 1 & 2 & -1 \\ 0 & -5 & 5 \\ 0 & 0 & 5 \end{vmatrix}$

Expand about column 1.

$D = 1(-1)^{1+1}\begin{vmatrix} -5 & 5 \\ 0 & 5 \end{vmatrix}$
$D = 1(-25 - 0) = -25$

15. $\begin{vmatrix} 1 & -4 & 1 \\ 2 & 5 & -1 \\ 3 & 3 & 4 \end{vmatrix}$

-2(row 1) + (row 2) replace (row 2)
-3(row 1) + (row 3) replace (row 3)

$\begin{vmatrix} 1 & -4 & 1 \\ 0 & 13 & -3 \\ 0 & 15 & 1 \end{vmatrix}$

Expand about column 1.

$D = 1(-1)^{1+1}\begin{vmatrix} 13 & -3 \\ 15 & 1 \end{vmatrix}$
$D = 13 - (-45) = 58$

17. $\begin{vmatrix} 6 & 12 & 3 \\ -1 & 5 & 1 \\ -3 & 6 & 2 \end{vmatrix}$

6(row 2) + (row 1) replace (row 1)
-3(row 2) + (row 3) replace (row 3)

$\begin{vmatrix} 0 & 42 & 9 \\ -1 & 5 & 1 \\ 0 & -9 & -1 \end{vmatrix}$

Expand about column 1.

$D = (-1)(-1)^{2+1}\begin{vmatrix} 42 & 9 \\ -9 & -1 \end{vmatrix}$
$D = 1(-42 + 81) = 39$

19. $\begin{vmatrix} 2 & -1 & 3 \\ 0 & 3 & 1 \\ 1 & -2 & -1 \end{vmatrix}$

-2(row 3) + (row 1) replace (row 1)

$\begin{vmatrix} 0 & 3 & 5 \\ 0 & 3 & 1 \\ 1 & -2 & -1 \end{vmatrix}$

Expand about column 1.

$D = 1(-1)^{3+1}\begin{vmatrix} 3 & 5 \\ 3 & 1 \end{vmatrix}$
$D = 3 - 15 = -12$

21. $\begin{vmatrix} -3 & -2 & 1 \\ 5 & 0 & 6 \\ 2 & 1 & -4 \end{vmatrix}$

2(row 3) + (row 1) replace (row 1)

$\begin{vmatrix} 1 & 0 & -7 \\ 5 & 0 & 6 \\ 2 & 1 & -4 \end{vmatrix}$

Expand about column 2.

$D = 1(-1)^{3+2}\begin{vmatrix} 1 & -7 \\ 5 & 6 \end{vmatrix}$
$D = (-1)(6 + 35) = -41$

23. $\begin{vmatrix} 3 & -4 & -2 \\ 5 & -2 & 1 \\ 1 & 0 & 0 \end{vmatrix}$

Expand about row 3.

$D = 1(-1)^{3+1}\begin{vmatrix} -4 & -2 \\ -2 & 1 \end{vmatrix}$
$D = -4 - 4 = -8$

25. $\begin{vmatrix} 24 & -1 & 4 \\ 40 & 2 & 0 \\ -16 & 6 & 0 \end{vmatrix}$

Expand about column 3.

$D = 4(-1)^{1+3}\begin{vmatrix} 40 & 2 \\ -16 & 6 \end{vmatrix}$

$D = 4(240 + 32) = 1088$

27. $\begin{vmatrix} 2 & 3 & -4 \\ 4 & 6 & -1 \\ -6 & 1 & -2 \end{vmatrix}$

-2(row 1) + (row 2) replace (row 2)

$\begin{vmatrix} 2 & 3 & -4 \\ 0 & 0 & 7 \\ -6 & 1 & -2 \end{vmatrix}$

Expand about row 2.

$D = 7(-1)^{2+3}\begin{vmatrix} 2 & 3 \\ -6 & 1 \end{vmatrix}$

$D = -7[2 - (-18)]$

$D = -7(20) = -140$

29. $\begin{vmatrix} 1 & -2 & 3 & 2 \\ 2 & -1 & 0 & 4 \\ -3 & 4 & 0 & -2 \\ -1 & 1 & 1 & 5 \end{vmatrix}$

-3(row 4) + (row 1) replace (row 1)

$\begin{vmatrix} 4 & -5 & 0 & -13 \\ 2 & -1 & 0 & 4 \\ -3 & 4 & 0 & -2 \\ -1 & 1 & 1 & 5 \end{vmatrix}$

Expand about column 3.

$D = 1(-1)^{4+3}\begin{vmatrix} 4 & -5 & -13 \\ 2 & -1 & 4 \\ -3 & 4 & -2 \end{vmatrix}$

-5(row 2) + (row 1) replace (row 1)
4(row 2) + (row 3) replace (row 3)

$D = -1\begin{vmatrix} -6 & 0 & -33 \\ 2 & -1 & 4 \\ 5 & 0 & 14 \end{vmatrix}$

Expand about column 2.

$D = -1\left[(-1)(-1)^{2+2}\begin{vmatrix} -6 & -33 \\ 5 & 14 \end{vmatrix}\right]$

$D = (-1)(-1)(-84 + 165) = 81$

31. $\begin{vmatrix} 3 & -1 & 2 & 3 \\ 1 & 0 & 2 & 1 \\ 2 & 3 & 0 & 1 \\ 5 & 2 & 4 & -5 \end{vmatrix}$

3(row 1) + (row 3) replace (row 3)
2(row 1) + (row 4) replace (row 4)

$\begin{vmatrix} 3 & -1 & 2 & 3 \\ 1 & 0 & 2 & 1 \\ 11 & 0 & 6 & 10 \\ 11 & 0 & 8 & 1 \end{vmatrix}$

Expand about column 2.

$D = -1(-1)^{1+2}\begin{vmatrix} 1 & 2 & 1 \\ 11 & 6 & 10 \\ 11 & 8 & 1 \end{vmatrix}$

$D = (1)\begin{vmatrix} 1 & 2 & 1 \\ 11 & 6 & 10 \\ 11 & 8 & 1 \end{vmatrix}$

-11(row 1) + (row 2) replace (row 2)
-11(row 1) + (row 3) replace (row 3)

$D = \begin{vmatrix} 1 & 2 & 1 \\ 0 & -16 & -1 \\ 0 & -14 & -10 \end{vmatrix}$

Expand about column 1.

$D = 1(-1)^{1+1}\begin{vmatrix} -16 & -1 \\ -14 & -10 \end{vmatrix}$

$D = \begin{vmatrix} -16 & -1 \\ -14 & -10 \end{vmatrix} = 160 - 14 = 146$

33. Property 11.3

35. Property 11.2

37. Property 11.4

39. Property 11.3

41. Property 11.5

Further Investigations

47. and 49. Each of these properties can be verified 3 × 3 matrices by evaluating the determinants of the general 3 × 3 matrices.

PROBLEM SET 11.5 Cramer's Rule

1. $D = \begin{vmatrix} 2 & -1 \\ 3 & 2 \end{vmatrix} = 4 - (-3) = 7$

$D_x = \begin{vmatrix} -2 & -1 \\ 11 & 2 \end{vmatrix} = -4 - (-11) = 7$

$D_y = \begin{vmatrix} 2 & -2 \\ 3 & 11 \end{vmatrix} = 22 - (-6) = 28$

$x = \dfrac{D_x}{D} = \dfrac{7}{7} = 1$

$y = \dfrac{D_y}{D} = \dfrac{28}{7} = 4$

The solution set is $\{(1, 4)\}$.

3. $D = \begin{vmatrix} 5 & 2 \\ 3 & -4 \end{vmatrix} = -20 - 6 = -26$

$D_x = \begin{vmatrix} 5 & 2 \\ 29 & -4 \end{vmatrix} = -20 - 58 = -78$

$D_y = \begin{vmatrix} 5 & 5 \\ 3 & 29 \end{vmatrix} = 145 - 15 = 130$

$x = \dfrac{D_x}{D} = \dfrac{-78}{-26} = 3$

$y = \dfrac{D_y}{D} = \dfrac{130}{-26} = -5$

The solution set is $\{(3, -5)\}$.

5. $D = \begin{vmatrix} 5 & -4 \\ -1 & 2 \end{vmatrix} = 10 - 4 = 6$

$D_x = \begin{vmatrix} 14 & -4 \\ -4 & 2 \end{vmatrix} = 28 - 16 = 12$

$D_y = \begin{vmatrix} 5 & 14 \\ -1 & -4 \end{vmatrix} = -20 + 14 = -6$

$x = \dfrac{D_x}{D} = \dfrac{12}{6} = 2$

$y = \dfrac{D_y}{D} = \dfrac{-6}{6} = -1$

The solution set is $\{(2, -1)\}$.

7. $D = \begin{vmatrix} -2 & 1 \\ 6 & -3 \end{vmatrix} = 6 - 6 = 0$

Because the equations are not multiples, the system is inconsistent. The solution set is $\emptyset$.

9. $D = \begin{vmatrix} -4 & 3 \\ 4 & -6 \end{vmatrix} = 24 - 12 = 12$

$D_x = \begin{vmatrix} 3 & 3 \\ -5 & -6 \end{vmatrix} = -18 + 15 = -3$

$D_y = \begin{vmatrix} -4 & 3 \\ 4 & -5 \end{vmatrix} = 20 - 12 = 8$

$x = \dfrac{D_x}{D} = \dfrac{-3}{12} = -\dfrac{1}{4}$

$y = \dfrac{D_y}{D} = \dfrac{8}{12} = \dfrac{2}{3}$

The solution set is $\left\{\left(-\frac{1}{4}, \frac{2}{3}\right)\right\}$.

11. $D=\begin{vmatrix} 9 & -1 \\ 8 & 1 \end{vmatrix}=9-(-8)=17$

$D_x=\begin{vmatrix} -2 & -1 \\ 4 & 1 \end{vmatrix}=-2-(-4)=2$

$D_y=\begin{vmatrix} 9 & -2 \\ 8 & 4 \end{vmatrix}=36-(-16)=52$

$x=\frac{D_x}{D}=\frac{2}{17}$

$y=\frac{D_y}{D}=\frac{52}{17}$

The solution set is $\left\{\left(\frac{2}{17}, \frac{52}{17}\right)\right\}$.

13. $D=\begin{vmatrix} -\frac{2}{3} & \frac{1}{2} \\ \frac{1}{3} & -\frac{3}{2} \end{vmatrix}=1-\frac{1}{6}=\frac{5}{6}$

$D_x=\begin{vmatrix} -7 & \frac{1}{2} \\ 6 & -\frac{3}{2} \end{vmatrix}=\frac{21}{2}-3=\frac{15}{2}$

$D_y=\begin{vmatrix} -\frac{2}{3} & -7 \\ \frac{1}{3} & 6 \end{vmatrix}=-4+\frac{7}{3}=-\frac{5}{3}$

$x=\frac{D_x}{D}=\frac{\frac{15}{2}}{\frac{5}{6}}=9$

$y=\frac{D_y}{D}=\frac{-\frac{5}{3}}{\frac{5}{6}}=-2$

The solution set is $\{(9, -2)\}$.

15. $\begin{pmatrix} 2x+7y=-1 \\ x=2 \end{pmatrix}$

$D=\begin{vmatrix} 2 & 7 \\ 1 & 0 \end{vmatrix}=0-7==-7$

$D_x=\begin{vmatrix} -1 & 7 \\ 2 & 0 \end{vmatrix}=0-14=-14$

$D_y=\begin{vmatrix} 2 & -1 \\ 1 & 2 \end{vmatrix}=4-(-1)=5$

$x=\frac{D_x}{D}=\frac{-14}{-7}=2$

$y=\frac{D_y}{D}=\frac{5}{-7}$

The solution set is $\left\{\left(2, -\frac{5}{7}\right)\right\}$.

17. $D=\begin{vmatrix} 1 & -1 & 2 \\ 2 & 3 & -4 \\ -1 & 2 & -1 \end{vmatrix}$

Expand about row 1.

$D=1(-1)^{1+1}\begin{vmatrix} 3 & -4 \\ 2 & -1 \end{vmatrix}$

$-1(-1)^{1+2}\begin{vmatrix} 2 & -4 \\ -1 & -1 \end{vmatrix}$

$+2(-1)^{1+3}\begin{vmatrix} 2 & 3 \\ -1 & 2 \end{vmatrix}$

$D=(-3+5)+(-2-4)+2(4+3)$
$D=5-6+14=13$

$D_x=\begin{vmatrix} -8 & -1 & 2 \\ 18 & 3 & -4 \\ 7 & 2 & -1 \end{vmatrix}$

Expand about row 1.

$D_x=-8(-1)^{1+1}\begin{vmatrix} 3 & -4 \\ 2 & -1 \end{vmatrix}$

$-1(-1)^{1+2}\begin{vmatrix} 18 & -4 \\ 7 & -1 \end{vmatrix}$

$+2(-1)^{1+3}\begin{vmatrix} 18 & 3 \\ 7 & 2 \end{vmatrix}$

$D_x=-8(-3+8)+(-18+28)+2(36-21)$
$D_x=-8(5)+10+2(15)=0$

$$D_y = \begin{vmatrix} 1 & -8 & 2 \\ 2 & 18 & -4 \\ -1 & 7 & -1 \end{vmatrix}$$

Expand about row 1.

$$D_y = 1(-1)^{1+1}\begin{vmatrix} 18 & -4 \\ 7 & -1 \end{vmatrix}$$

$$-8(-1)^{1+2}\begin{vmatrix} 2 & -4 \\ -1 & -1 \end{vmatrix}$$

$$+2(-1)^{1+3}\begin{vmatrix} 2 & 18 \\ -1 & 7 \end{vmatrix}$$

$D_y = (-18+28)+8(-2-4)-2(14+18)$
$D_y = 10+8(-6)+2(32) = 26$

$$D_z = \begin{vmatrix} 1 & -1 & -8 \\ 2 & 3 & 18 \\ -1 & 2 & 7 \end{vmatrix}$$

Expand abour row 1.

$$D_z = 1(-1)^{1+1}\begin{vmatrix} 3 & 18 \\ 2 & 7 \end{vmatrix}$$

$$-1(-1)^{1+2}\begin{vmatrix} 2 & 18 \\ -1 & 7 \end{vmatrix}$$

$$-8(-1)^{1+3}\begin{vmatrix} 2 & 3 \\ -1 & 2 \end{vmatrix}$$

$D_z = (21-36)+1(14+18)-8(4+3)$
$D_z = -15+32-8(7) = -39$

$$x = \frac{D_x}{D} = \frac{0}{13} = 0$$

$$y = \frac{D_y}{D} = \frac{26}{13} = 2$$

$$z = \frac{D_z}{D} = \frac{-39}{13} = -3$$

The solution set is $\{(0, 2, -3)\}$.

19. $\begin{pmatrix} 2x-3y+z = -7 \\ -3x+y-z = -7 \\ x-2y-5z = -45 \end{pmatrix}$

$$D = \begin{vmatrix} 2 & -3 & 1 \\ -3 & 1 & -1 \\ 1 & -2 & -5 \end{vmatrix}$$

(row 1) + (row 2) replace (row 2)
5(row 1) + (row 3) replace (row 3)

$$D = \begin{vmatrix} 2 & -3 & 1 \\ -1 & -2 & 0 \\ 11 & -17 & 0 \end{vmatrix}$$

Expand about column 3.

$$D = 1(-1)^{1+3}\begin{vmatrix} -1 & -2 \\ 11 & -17 \end{vmatrix}$$

$D = 17-(-22) = 39$

$$D_x = \begin{vmatrix} -7 & -3 & 1 \\ -7 & 1 & -1 \\ -45 & -2 & -5 \end{vmatrix}$$

(row 1) + (row 2) replace (row 2)
5(row 1) + (row 3) replace (row 3)

$$D_x = \begin{vmatrix} -7 & -3 & 1 \\ -14 & -2 & 0 \\ -80 & -17 & 0 \end{vmatrix}$$

Expand about column 3.

$$D_x = 1(-1)^{1+3}\begin{vmatrix} -14 & -2 \\ -80 & -17 \end{vmatrix}$$

$D_x = -14(-17)-(-80)(-2) = 78$

$$D_y = \begin{vmatrix} 2 & -7 & 1 \\ -3 & -7 & -1 \\ 1 & -45 & -5 \end{vmatrix}$$

(row 1) + (row 2) replace (row 2)
5(row 1) + (row 3) replace (row 3)

$$D_y = \begin{vmatrix} 2 & -7 & 1 \\ -1 & -14 & 0 \\ 11 & -80 & 0 \end{vmatrix}$$

Expand about column 3.

$$D_y = 1(-1)^{1+3}\begin{vmatrix} -1 & -14 \\ 11 & -80 \end{vmatrix}$$

$D_y = -1(-80)-(11)(-14) = 234$

$$D_z = \begin{vmatrix} 2 & -3 & -7 \\ -3 & 1 & -7 \\ 1 & -2 & -45 \end{vmatrix}$$

3(row 2) + (row 1) replace (row 1)
2(row 2) + (row 3) replace (row 3)

$$D_z = \begin{vmatrix} -7 & 0 & -28 \\ -3 & 1 & -7 \\ -5 & 0 & -59 \end{vmatrix}$$

Expand about column 2.

$$D_z = 1(-1)^{2+2}\begin{vmatrix} -7 & -28 \\ -5 & -59 \end{vmatrix}$$

$$D_z = -7(-59) - (-5)(-28) = 273$$

$$x = \frac{D_x}{D} = \frac{78}{39} = 2$$

$$y = \frac{D_y}{D} = \frac{234}{39} = 6$$

$$z = \frac{D_z}{D} = \frac{273}{39} = 7$$

The solution set is $\{(2, 6, 7)\}$.

21. $D = \begin{vmatrix} 4 & 5 & -2 \\ 7 & -1 & 2 \\ 3 & 1 & 4 \end{vmatrix}$

5(row 2) + (row 1) replace (row 1)
(row 3) + (row 2) replace (row 2)

$$D = \begin{vmatrix} 39 & 0 & 8 \\ 10 & 0 & 6 \\ 3 & 1 & 4 \end{vmatrix}$$

Expand about column 2.

$$D = 1(-1)^{3+2}\begin{vmatrix} 39 & 8 \\ 10 & 6 \end{vmatrix}$$

$$D = -1(234 - 80) = -154$$

$$D_x = \begin{vmatrix} -14 & 5 & -2 \\ 42 & -1 & 2 \\ 28 & 1 & 4 \end{vmatrix}$$

5(row 2) + (row 1) replace (row 1)
(row 3) + (row 2) replace (row 2)

$$D_x = \begin{vmatrix} 39 & 0 & 8 \\ 10 & 0 & 6 \\ 3 & 1 & 4 \end{vmatrix}$$

Expand about column 2.

$$D_x = 1(-1)^{3+2}\begin{vmatrix} 196 & 8 \\ 70 & 6 \end{vmatrix}$$

$$D_x = -1(1176 - 560) = -616$$

$$D_y = \begin{vmatrix} 4 & -14 & -2 \\ 7 & 42 & 2 \\ 3 & 28 & 4 \end{vmatrix}$$

2(row 1) + (row 3) replace (row 3)
(row 1) + (row 2) replace (row 2)

$$D_y = \begin{vmatrix} 4 & -14 & -2 \\ 11 & 28 & 0 \\ 11 & 0 & 0 \end{vmatrix}$$

Expand about column 3.

$$D_y = -2(-1)^{1+3}\begin{vmatrix} 11 & 28 \\ 11 & 0 \end{vmatrix}$$

$$D_y = -2(0 - 308) = 616$$

$$D_z = \begin{vmatrix} 4 & 5 & -14 \\ 7 & -1 & 42 \\ 3 & 1 & 28 \end{vmatrix}$$

5(row 2) + (row 1) replace (row 1)
(row 3) + (row 2) replace (row 2)

$$D_z = \begin{vmatrix} 39 & 0 & 8 \\ 10 & 0 & 6 \\ 3 & 1 & 4 \end{vmatrix}$$

Expand about column 2.

$$D_z = 1(-1)^{3+2}\begin{vmatrix} 39 & 196 \\ 10 & 70 \end{vmatrix}$$

$$D_z = -1(2730 - 1960) = -770$$

$$x = \frac{D_x}{D} = \frac{-616}{-154} = 4$$

$$y = \frac{D_y}{D} = \frac{616}{-154} = -4$$

$$z = \frac{D_z}{D} = \frac{-770}{-154} = 5$$

The solution set is $\{(4, -4, 5)\}$.

23. $\begin{pmatrix} 2x - y + 3z = -17 \\ 3y + z = 5 \\ x - 2y - z = -3 \end{pmatrix}$

$$D = \begin{vmatrix} 2 & -1 & 3 \\ 0 & 3 & 1 \\ 1 & -2 & -1 \end{vmatrix}$$

-2(row 3) + (row 1) replace (row 1)

$$D = \begin{vmatrix} 0 & 3 & 5 \\ 0 & 3 & 1 \\ 1 & -2 & -1 \end{vmatrix}$$

Expand about column 1.

$$D = 1(-1)^{3+1}\begin{vmatrix} 3 & 5 \\ 3 & 1 \end{vmatrix}$$

$$D = 3 - 15 = -12$$

$$D_x = \begin{vmatrix} -17 & -1 & 3 \\ 5 & 3 & 1 \\ -3 & -2 & -1 \end{vmatrix}$$

(row 3) + (row 2) replace (row 2)
3(row 3) + (row 1) replace (row 1)

$$D_x = \begin{vmatrix} -26 & -7 & 0 \\ 2 & 1 & 0 \\ -3 & -2 & -1 \end{vmatrix}$$

Expand about column 3.

$$D_x = -1(-1)^{3+3}\begin{vmatrix} -26 & -7 \\ 2 & 1 \end{vmatrix}$$

$$D_x = -[-26 - (-14)] = 12$$

$$D_y = \begin{vmatrix} 2 & -17 & 3 \\ 0 & 5 & 1 \\ 1 & -3 & -1 \end{vmatrix}$$

-2(row 3) + (row 1) replace (row 1)

$$D_y = \begin{vmatrix} 0 & -11 & 5 \\ 0 & 5 & 1 \\ 1 & -3 & -1 \end{vmatrix}$$

Expand about column 1.

$$D_y = (1)(-1)^{3+1}\begin{vmatrix} -11 & 5 \\ 5 & 1 \end{vmatrix}$$

$$D_y = -11 - 25 = -36$$

$$D_z = \begin{vmatrix} 2 & -1 & -17 \\ 0 & 3 & 5 \\ 1 & -2 & -3 \end{vmatrix}$$

-2(row 3) + (row 1) replace (row 1)

$$D_z = \begin{vmatrix} 0 & 3 & -11 \\ 0 & 3 & 5 \\ 1 & -2 & -3 \end{vmatrix}$$

Expand about column 1.

$$D_z = 1(-1)^{3+1}\begin{vmatrix} 3 & -11 \\ 3 & 5 \end{vmatrix}$$

$$D_z = 15 - (-33) = 48$$

$$x = \frac{D_x}{D} = \frac{12}{-12} = -1$$

$$y = \frac{D_y}{D} = \frac{-36}{-12} = 3$$

$$z = \frac{D_z}{D} = \frac{48}{-12} = -4$$

The solution set is $\{(-1, 3, -4)\}$.

25. $D = \begin{vmatrix} 1 & 3 & -4 \\ 2 & -1 & 1 \\ 4 & 5 & -7 \end{vmatrix}$

5(row 2) + (row 3) replace (row 3)
3(row 2) + (row 1) replace (row 1)

$$D = \begin{vmatrix} 7 & 0 & -1 \\ 2 & -1 & 1 \\ 14 & 0 & -2 \end{vmatrix}$$

Expand about column 2.

$$D = -1(-1)^{2+2}\begin{vmatrix} 7 & -1 \\ 14 & -2 \end{vmatrix}$$

$$D = -[-14 - (-14)] = 0$$

$$D_x = \begin{vmatrix} -1 & 3 & -4 \\ 2 & -1 & 1 \\ 0 & 5 & -7 \end{vmatrix}$$

5(row 2) + (row 3) replace (row 3)
3(row 2) + (row 1) replace (row 1)

$$D_x = \begin{vmatrix} 5 & 0 & -1 \\ 2 & -1 & 1 \\ 10 & 0 & -2 \end{vmatrix}$$

Expand about column 2.

$$D_x = -1(-1)^{2+2}\begin{vmatrix} 5 & -1 \\ 10 & -2 \end{vmatrix}$$

$$D_x = -[-10-(-10)] = 0$$

$$D_y = \begin{vmatrix} 1 & -1 & -4 \\ 2 & 2 & 1 \\ 4 & 0 & -7 \end{vmatrix}$$

2(row 1) + (row 2) replace (row 2)

$$D_y = \begin{vmatrix} 1 & -1 & -4 \\ 4 & 0 & -7 \\ 4 & 0 & -7 \end{vmatrix}$$

Expand about column 2.

$$D_y = -1(-1)^{1+2}\begin{vmatrix} 4 & -7 \\ 4 & -7 \end{vmatrix}$$

$$D_y = -28-(-28) = 0$$

$$D_z = \begin{vmatrix} 1 & 3 & -1 \\ 2 & -1 & 2 \\ 4 & 5 & 0 \end{vmatrix}$$

2(row 1) + (row 2) replace (row 2)

$$D_z = \begin{vmatrix} 1 & 3 & -1 \\ 4 & 5 & 0 \\ 4 & 5 & 0 \end{vmatrix}$$

Expand about column 3.

$$D_z = -1(-1)^{1+3}\begin{vmatrix} 4 & 5 \\ 4 & 5 \end{vmatrix}$$

$$D_z = -(20-20) = 0$$

Equations are dependent.
The solution set has infinitely many solutions.

27. $$D = \begin{vmatrix} 3 & -2 & -3 \\ 1 & 2 & 3 \\ -1 & 4 & -6 \end{vmatrix}$$

(row 2) + (row 1) replace (row 1)

$$D = \begin{vmatrix} 4 & 0 & 0 \\ 1 & 2 & 3 \\ -1 & 4 & -6 \end{vmatrix}$$

Expand about row 1.

$$D = 4(-1)^{1+1}\begin{vmatrix} 2 & 3 \\ 4 & -6 \end{vmatrix}$$

$$D = 4[-12-12] = -96$$

$$D_x = \begin{vmatrix} -5 & -2 & -3 \\ -3 & 2 & 3 \\ 8 & 4 & -6 \end{vmatrix}$$

(row 2) + (row 1) replace (row 1)

$$D_x = \begin{vmatrix} -8 & 0 & 0 \\ -3 & 2 & 3 \\ 8 & 4 & -6 \end{vmatrix}$$

Expand about row 1.

$$D_x = -8(-1)^{1+1}\begin{vmatrix} 2 & 3 \\ 4 & -6 \end{vmatrix}$$

$$D_x = -8(-12-12) = 192$$

$$D_y = \begin{vmatrix} 3 & -5 & -3 \\ 1 & -3 & 3 \\ -1 & 8 & -6 \end{vmatrix}$$

3(row 3) + (row 1) replace (row 1)
(row 3) + (row 2) replace (row 2)

$$D_y = \begin{vmatrix} 0 & 19 & -21 \\ 0 & 5 & -3 \\ -1 & 8 & -6 \end{vmatrix}$$

Expand about column 1.

$$D_y = -1(-1)^{3+1}\begin{vmatrix} 19 & -21 \\ 5 & -3 \end{vmatrix}$$

$$D_y = -[-57-(-105)] = -48$$

$$D_z = \begin{vmatrix} 3 & -2 & -5 \\ 1 & 2 & -3 \\ -1 & 4 & 8 \end{vmatrix}$$

3(row 3) + (row 1) replace (row 1)
(row 3) + (row 2) replace (row 2)

$$D_z = \begin{vmatrix} 0 & 10 & 19 \\ 0 & 6 & 5 \\ -1 & 4 & 8 \end{vmatrix}$$

Expand about column 1.

$$D_z = -1(-1)^{3+1}\begin{vmatrix} 10 & 19 \\ 6 & 5 \end{vmatrix}$$
$$D_z = -[50-114] = 64$$

$$x = \frac{D_x}{D} = \frac{192}{-96} = -2$$

$$y = \frac{D_y}{D} = \frac{-48}{-96} = \frac{1}{2}$$

$$z = \frac{D_z}{D} = \frac{64}{-96} = -\frac{2}{3}$$

The solution set is $\left\{\left(-2, \frac{1}{2}, -\frac{2}{3}\right)\right\}$.

29. $D = \begin{vmatrix} 1 & -2 & 3 \\ -2 & 4 & -3 \\ 5 & -6 & 6 \end{vmatrix}$

(row 1) + (row 2) replace (row 2)
− 2(row 1) + (row 3) replace (row 3)

$$D = \begin{vmatrix} 1 & -2 & 3 \\ -1 & 2 & 0 \\ 3 & -2 & 0 \end{vmatrix}$$

Expand about column 3.

$$D = 3(-1)^{1+3}\begin{vmatrix} -1 & 2 \\ 3 & -2 \end{vmatrix}$$
$$D = 3(2-6) = -12$$

$$D_x = \begin{vmatrix} 1 & -2 & 3 \\ -3 & 4 & -3 \\ 10 & -6 & 6 \end{vmatrix}$$

(row 1) + (row 2) replace (row 2)
− 2(row 1) + (row 3) replace (row 3)

$$D_x = \begin{vmatrix} 1 & -2 & 3 \\ -2 & 2 & 0 \\ 8 & -2 & 0 \end{vmatrix}$$

Expand about column 3.

$$D_x = 3(-1)^{1+3}\begin{vmatrix} -2 & 2 \\ 8 & -2 \end{vmatrix}$$
$$D_x = 3(4-16) = 3(-12) = -36$$

$$D_y = \begin{vmatrix} 1 & 1 & 3 \\ -2 & -3 & -3 \\ 5 & 10 & 6 \end{vmatrix}$$

(row 1) + (row 2) replace (row 2)
− 2(row 1) + (row 3) replace (row 3)

$$D_y = \begin{vmatrix} 1 & 1 & 3 \\ -2 & -3 & -3 \\ 5 & 10 & 6 \end{vmatrix}$$

Expand about column 3.

$$D_y = 3(-1)^{1+3}\begin{vmatrix} -1 & -2 \\ 3 & 8 \end{vmatrix}$$
$$D_y = 3[-8-(-6)] = 3(-2) = -6$$

$$D_z = \begin{vmatrix} 1 & -2 & 1 \\ -2 & 4 & -3 \\ 5 & -6 & 10 \end{vmatrix}$$

2(row 1) + (row 2) replace (row 2)

$$D_z = \begin{vmatrix} 1 & -2 & 1 \\ 0 & 0 & -1 \\ 5 & -6 & 10 \end{vmatrix}$$

Expand about row 2.

$$D_z = -1(-1)^{2+3}\begin{vmatrix} 1 & -2 \\ 5 & -6 \end{vmatrix}$$
$$D_z = -6-(-10) = 4$$

$$x = \frac{D_x}{D} = \frac{-36}{-12} = 3$$

$$y = \frac{D_y}{D} = \frac{-6}{-12} = \frac{1}{2}$$

$$z = \frac{D_z}{D} = \frac{4}{-12} = -\frac{1}{3}$$

The solution set is $\left\{\left(3, \frac{1}{2}, -\frac{1}{3}\right)\right\}$.

31. $D = \begin{vmatrix} -1 & -1 & 3 \\ -2 & 1 & 7 \\ 3 & 4 & -5 \end{vmatrix}$

− 2(row 1) + (row 2) replace (row 2)
3(row 1) + (row 3) replace (row 3)

$$D = \begin{vmatrix} -1 & -1 & 3 \\ 0 & 3 & 1 \\ 0 & 1 & 4 \end{vmatrix}$$

Expand about column 1.

$$D = -1(-1)^{1+1}\begin{vmatrix} 3 & 1 \\ 1 & 4 \end{vmatrix}$$
$$D = -(12 - 1) = -11$$

$$D_x = \begin{vmatrix} -2 & -1 & 3 \\ 14 & 1 & 7 \\ 12 & 4 & -5 \end{vmatrix}$$

-2(column2)+(column 1) replace (column1)
3(column 2) + (column 3) replace (column 3)

$$D_x = \begin{vmatrix} 0 & -1 & 0 \\ 12 & 1 & 10 \\ 4 & 4 & 7 \end{vmatrix}$$

Expand about row 1.

$$D_x = -1(-1)^{1+2}\begin{vmatrix} 12 & 10 \\ 4 & 7 \end{vmatrix}$$
$$D_x = 84 - 40 = 44$$

$$D_y = \begin{vmatrix} -1 & -2 & 3 \\ -2 & 14 & 7 \\ 3 & 12 & -5 \end{vmatrix}$$

-2(row 1) + (row 2) replace (row 2)
3(row 1) + (row 3) replace (row 3)

$$D_y = \begin{vmatrix} -1 & -2 & 3 \\ 0 & 18 & 1 \\ 0 & 6 & 4 \end{vmatrix}$$

Expand about column 1.

$$D_y = -1(-1)^{1+1}\begin{vmatrix} 18 & 1 \\ 6 & 4 \end{vmatrix}$$
$$D_y = -(72 - 6) = -66$$

$$D_z = \begin{vmatrix} -1 & -1 & -2 \\ -2 & 1 & 14 \\ 3 & 4 & 12 \end{vmatrix}$$

(row 1) + (row 2) replace (row 2)
4(row 1) + (row 3) replace (row 3)

$$D_z = \begin{vmatrix} -1 & -1 & -2 \\ -3 & 0 & 12 \\ -1 & 0 & 4 \end{vmatrix}$$

Expand about column 2.

$$D_z = -1(-1)^{1+2}\begin{vmatrix} -3 & 12 \\ -1 & 4 \end{vmatrix}$$
$$D_z = -12 - (-12) = 0$$

$$x = \frac{D_x}{D} = \frac{44}{-11} = -4$$

$$y = \frac{D_y}{D} = \frac{-66}{-11} = 6$$

$$z = \frac{D_z}{D} = \frac{0}{-11} = 0$$

The solution set is $\{(-4, 6, 0)\}$.

Further Investigations

35. (a) If the constant terms are all zero, then $D_x = 0$, $D_y = 0$, and $D_z = 0$.

If $D \neq 0$ then

$$x = \frac{D_x}{D} = \frac{0}{D} = 0$$

$$y = \frac{D_y}{D} = \frac{0}{D} = 0$$

$$z = \frac{D_z}{D} = \frac{0}{D} = 0$$

Hence the solution is $(0, 0, 0)$.

(b) If $D = 0$ and the constants are all zero, then at least one equation is a multiple or linear combination of the other equations. Therefore the system would be dependent.

37. $$D = \begin{vmatrix} 2 & -1 & 1 \\ 3 & 2 & 5 \\ 4 & -7 & 1 \end{vmatrix}$$

2(row 1) + (row 2) replace (row 2)
-7(row 1) + (row 3) replace (row 3)

$$D = \begin{vmatrix} 2 & -1 & 1 \\ 7 & 0 & 7 \\ -10 & 0 & -6 \end{vmatrix}$$

Expand about column 1.

$$D = -1(-1)^{1+2}\begin{vmatrix} 7 & 7 \\ -10 & -6 \end{vmatrix}$$
$$D = -42 - (-70) = 28$$

Since $D \neq 0$, the solution set is $\{(0, 0, 0)\}$.

39. $D=\begin{vmatrix} 2 & -1 & 2 \\ 1 & 2 & 1 \\ 1 & -3 & 1 \end{vmatrix}$

$-$ (column 1)+(column 3) replace (column 3)

$D=\begin{vmatrix} 2 & -1 & 0 \\ 1 & 2 & 0 \\ 1 & -3 & 0 \end{vmatrix}=0$

Since $D=0$, the system has infinitely many solutions.

PROBLEM SET 11.6 Partial Fractions

OPTIONAL -- See Page 521

CHAPTER 11 Review Problem Set

1. $\begin{pmatrix} 3x-y=16 \\ 5x+7y=-34 \end{pmatrix}$

$3x-y=16$
$-y=-3x+16$
$y=3x-16$
Substitute $3x-16$ for y in equation 2.
$5x-7y=-34$
$5x+7(3x-16)=-34$
$5x+21x-112=-34$
$26x=78$
$x=3$
$y=3x-16$
$y=3(3)-16$
$y=9-16$
$y=-7$
The solution set is $\{(3,\ -7)\}$.

2. $\begin{pmatrix} 6x+5y=-21 \\ x-4y=11 \end{pmatrix}$

$x-4y=11$
$x=4y+11$
Substitute $4y+11$ for x in equation 1.
$6x+5y=-21$
$6(4y+11)+5y=-21$
$24y+66+5y=-21$
$29y=-87$
$y=-3$
$x=4y+11$
$x=4(-3)+11$
$x=-12+11$
$x=-1$
The sollution set is $\{(-1,\ -3)\}$.

3. $\begin{pmatrix} 2x-3y=12 \\ 3x+5y=-20 \end{pmatrix}$

$2x-3y=12$
$2x=3y+12$
$x=\frac{3}{2}y+6$
Substitute $\frac{3}{2}y+6$ for x in equation 2.
$3\left(\frac{3}{2}y+6\right)+5y=-20$
$\frac{9}{2}y+18+5y=-20$
$\frac{19}{2}y=-38$
$y=\frac{2}{19}(-38)=-4$
$x=\frac{3}{2}y+6$
$x=\frac{3}{2}(-4)+6$
$x=-6+6$
$x=0$
The solution set is $\{(0,\ -4)\}$.

4. $\begin{pmatrix} 5x+8y=1 \\ 4x+7y=-2 \end{pmatrix}$

$5x=-8y+1$
$x=\frac{-8y+1}{5}$

Substitute $\frac{-8y+1}{5}$ for x in equation 2.

$$4\left(\frac{-8y+1}{5}\right)+7y=-2$$
$$\frac{-32y+4}{5}+7y=-2$$
$$-32y+4+35y=-10$$
$$3y=-14$$
$$y=-\frac{14}{3}$$
$$x=\frac{-8y+1}{5}$$
$$x=\frac{1}{5}\left[-8\left(-\frac{14}{3}\right)+1\right]$$
$$x=\frac{1}{5}\left(\frac{112}{3}+\frac{3}{3}\right)$$
$$x=\frac{1}{5}\left(\frac{115}{3}\right)$$
$$x=\frac{23}{3}$$

The solution set is $\left\{\left(\frac{23}{3},\ -\frac{14}{3}\right)\right\}$.

5. $\begin{pmatrix} 4x-3y=34 \\ 3x+2y=0 \end{pmatrix}$

Multiply equation 1 by 2,
multiply equation 2 by 3,
and then add the resulting equations.

$$\begin{array}{l} 8x-6y=68 \\ \underline{9x+6y=\ \ 0} \\ 17x\qquad=68 \\ \qquad x=4 \end{array}$$

$$4x-3y=34$$
$$4(4)-3y=34$$
$$16-3y=34$$
$$-3y=18$$
$$y=-6$$

The solution set is $\{(4,\ -6)\}$.

6. $\begin{pmatrix} \frac{1}{2}x-\frac{2}{3}y=1 \\ \frac{3}{4}x+\frac{1}{6}y=-1 \end{pmatrix}$

Multiply equation 1 by 6.

Multiply equation 2 by 12.

$\begin{pmatrix} 3x-4y=6 \\ 9x+2y=-12 \end{pmatrix}$

Multiply equation 2 by 2 and then add the result to equation 1.

$$\begin{array}{l} 3x-4y=6 \\ \underline{18x+4y=-24} \\ 21x\qquad=-18 \\ \qquad x=-\frac{18}{21}=-\frac{6}{7} \end{array}$$

$$3x-4y=6$$
$$3\left(-\frac{6}{7}\right)-4y=6$$
$$-\frac{18}{7}-4y=6$$
$$-4y=\frac{42}{7}+\frac{18}{7}$$
$$-4y=\frac{60}{7}$$
$$y=-\frac{1}{4}\left(\frac{60}{7}\right)$$
$$y=-\frac{15}{7}$$

The solution set is $\left\{\left(-\frac{6}{7},\ -\frac{15}{7}\right)\right\}$.

7. $\begin{pmatrix} 2x-y+3z=-19 \\ 3x+2y-4z=21 \\ 5x-4y-z=-8 \end{pmatrix}$

Multiply equation 1 by 2 and add to equation 2 to replace equation 2. Multiply equation 1 by -4 and add to equation 3 to replace equation 3.

$\begin{pmatrix} 2x-y+3z=-19 \\ 7x+2z=-17 \\ -3x-13z=68 \end{pmatrix}$

Multiply equation 2 by 3 and multiply equation 3 by 7, then add the resulting equations to replace equation 3.

$\begin{pmatrix} 2x-y+3z=-19 \\ 7x+2z=-17 \\ -85z=425 \end{pmatrix}$

$-85z=425$
$z=-5$
$7x+2z=-17$
$7x+2(-5)=-17$
$7x-10=-17$
$7x=-7$
$x=-1$
$2x-y+3z=-19$
$2(-1)-y+3(-5)=-19$
$-2-y-15=-19$
$-y=-2$
$y=2$
The solution set is $\{(-1,\ 2,\ -5)\}$.

8. $\begin{pmatrix} 3x+2y-4z=4 \\ 5x+3y-z=2 \\ 4x-2y+3z=11 \end{pmatrix}$

Multiply equation 2 by -4 and add to equation 1 to replace equation 1. Multiply equation 2 by 3 and add to equation 3 to replace equation 3.

$\begin{pmatrix} -17x-10y=-4 \\ 5x+3y-z=2 \\ 19x+7y=17 \end{pmatrix}$

Multiply equation 1 by 19 and multiply equation 3 by 17, then add the resulting equations to replace equation 3.

$\begin{pmatrix} -17x-10y=-4 \\ 5x+3y-z=2 \\ -71y=213 \end{pmatrix}$

$-71y=213$
$y=-3$
$-17x-10y=-4$
$-17x-10(-3)=-4$
$-17x+30=-4$
$-17x=-34$
$x=2$
$5x+3y-z=2$
$5(2)+3(-3)-z=2$
$10-9-z=2$
$1-z=2$
$-z=1$
$z=-1$
The solution set is $\{(2,\ -3,\ -1)\}$.

9. $\begin{pmatrix} x-3y=17 \\ -3x+2y=-23 \end{pmatrix}$

$\left[\begin{array}{cc|c} 1 & -3 & 17 \\ -3 & 2 & -23 \end{array}\right]$

3(row 1) + (row 2) replace (row 2)

$\left[\begin{array}{cc|c} 1 & -3 & 17 \\ 0 & -7 & 28 \end{array}\right]$

$-\frac{1}{7}$(row 2)

$\left[\begin{array}{cc|c} 1 & -3 & 17 \\ 0 & 1 & -4 \end{array}\right]$

3(row 2) + (row 1) replace (row 1)

$\left[\begin{array}{cc|c} 1 & 0 & 5 \\ 0 & 1 & -4 \end{array}\right]$

The solution set is $\{(5,\ -4)\}$.

10. $\begin{pmatrix} 2x+3y=25 \\ 3x-5y=-29 \end{pmatrix}$

$\left[\begin{array}{cc|c} 2 & 3 & 25 \\ 3 & -5 & -29 \end{array}\right]$

-3(row 1) + 2(row 2) replace (row 2)

$\left[\begin{array}{cc|c} 2 & 3 & 25 \\ 0 & -19 & -133 \end{array}\right]$

$-19y=-133$
$y=7$
$2x+3y=25$
$2x+3(7)=25$
$2x+21=25$
$2x=4$
$x=2$
The solution set is $\{(2,\ 7)\}$.

11. $\begin{pmatrix} x-2y+z=-7 \\ 2x-3y+4z=-14 \\ -3x+y-2z=10 \end{pmatrix}$

$\left[\begin{array}{ccc|c} 1 & -2 & 1 & -7 \\ 2 & -3 & 4 & -14 \\ -3 & 1 & -2 & 10 \end{array}\right]$

-2(row 1) + (row 2) replace (row 2)
3(row 1) + (row 3) replace (row 3)

$$\begin{bmatrix} 1 & -2 & 1 & | & -7 \\ 0 & 1 & 2 & | & 0 \\ 0 & -5 & 1 & | & -11 \end{bmatrix}$$

2(row 2) + (row 1) replace (row 1)
5(row 2) + (row 3) replace (row 3)

$$\begin{bmatrix} 1 & 0 & 5 & | & -7 \\ 0 & 1 & 2 & | & 0 \\ 0 & 0 & 11 & | & -11 \end{bmatrix}$$

$\frac{1}{11}$(row 3)

$$\begin{bmatrix} 1 & 0 & 5 & | & -7 \\ 0 & 1 & 2 & | & 0 \\ 0 & 0 & 1 & | & -1 \end{bmatrix}$$

-2(row 3) + (row 2) replace (row 2)
-5(row 3) + (row 1) replace (row 1)

$$\begin{bmatrix} 1 & 0 & 0 & | & -2 \\ 0 & 1 & 0 & | & 2 \\ 0 & 0 & 1 & | & -1 \end{bmatrix}$$

The solution set is $\{(-2, 2, -1)\}$.

12. $\begin{pmatrix} -2x - 7y + z = 9 \\ x + 3y - 4z = -11 \\ 4x + 5y - 3z = -11 \end{pmatrix}$

$$\begin{bmatrix} -2 & -7 & 1 & | & 9 \\ 1 & 3 & -4 & | & -11 \\ 4 & 5 & -3 & | & -11 \end{bmatrix}$$

2(row 2) + (row 1) replace (row 1)
-4(row 2) + (row 3) replace (row 3)

$$\begin{bmatrix} 0 & -1 & -7 & | & -13 \\ 1 & 3 & -4 & | & -11 \\ 0 & -7 & 13 & | & 33 \end{bmatrix}$$

-7(row 1) + (row 3) replace (row 3)

$$\begin{bmatrix} 0 & -1 & -7 & | & -13 \\ 1 & 3 & -4 & | & -11 \\ 0 & 0 & 62 & | & 124 \end{bmatrix}$$

$62z = 124$
$z = 2$
$-y - 7z = -13$
$-y - 7(2) = -13$
$-y - 14 = -13$
$-y = 1$
$y = -1$
$x + 3y - 4z = -11$
$x + 3(-1) - 4(2) = -11$
$x - 3 - 8 = -11$
$x - 11 = -11$
$x = 0$
The solution set is $\{(0, -1, 2)\}$.

13. $\begin{pmatrix} 5x + 3y = -18 \\ 4x - 9y = -3 \end{pmatrix}$

$D = \begin{vmatrix} 5 & 3 \\ 4 & -9 \end{vmatrix} = -45 - 12 = -57$

$D_x = \begin{vmatrix} -18 & 3 \\ -3 & -9 \end{vmatrix} = 162 - (-9) = 171$

$D_y = \begin{vmatrix} 5 & -18 \\ 4 & -3 \end{vmatrix} = -15 - (-72) = 57$

$x = \frac{D_x}{D} = \frac{171}{-57} = -3$

$y = \frac{D_y}{D} = \frac{57}{-57} = -1$

The solution set is $\{(-3, -1)\}$.

14. $\begin{pmatrix} 0.2x + 0.3y = 2.6 \\ 0.5x - 0.1y = 1.4 \end{pmatrix}$

$D = \begin{vmatrix} 0.2 & 0.3 \\ 0.5 & -0.1 \end{vmatrix}$

$D = 0.2(-0.1) - (0.5)(0.3) = -0.17$

$D_x = \begin{vmatrix} 2.6 & 0.3 \\ 1.4 & -0.1 \end{vmatrix}$

$D_x = 2.6(-0.1) - 1.4(0.3) = -0.68$

$D_y = \begin{vmatrix} 0.2 & 2.6 \\ 0.5 & 1.4 \end{vmatrix}$

$D_y = 0.2(1.4) - 0.5(2.6) = -1.02$

$x = \frac{D_x}{D} = \frac{-0.68}{-0.17} = 4$

$$y = \frac{D_y}{D} = \frac{-1.02}{-0.17} = 6$$

The solution set is $\{(4, 6)\}$.

15. $\begin{pmatrix} 2x - 3y - 3z = 25 \\ 3x + y + 2z = -5 \\ 5x - 2y - 4z = 32 \end{pmatrix}$

$$D = \begin{vmatrix} 2 & -3 & -3 \\ 3 & 1 & 2 \\ 5 & -2 & -4 \end{vmatrix}$$

2(row 2) + (row 3) replace (row 3)

$$D = \begin{vmatrix} 2 & -3 & -3 \\ 3 & 1 & 2 \\ 11 & 0 & 0 \end{vmatrix}$$

Expand about row 3.

$$D = 11(-1)^{3+1}\begin{vmatrix} -3 & -3 \\ 1 & 2 \end{vmatrix}$$

$$D = 11[-6 - (-3)] = -33$$

$$D_x = \begin{vmatrix} 25 & -3 & -3 \\ -5 & 1 & 2 \\ 32 & -2 & -4 \end{vmatrix}$$

2(row 2) + (row 3) replace (row 3)

$$D_x = \begin{vmatrix} 25 & -3 & -3 \\ -5 & 1 & 2 \\ 22 & 0 & 0 \end{vmatrix}$$

Expand about row 3.

$$D_x = 22(-1)^{3+1}\begin{vmatrix} -3 & -3 \\ 1 & 2 \end{vmatrix}$$

$$D_x = 22[-6 - (-3)] = -66$$

$$D_y = \begin{vmatrix} 2 & 25 & -3 \\ 3 & -5 & 2 \\ 5 & 3 & -4 \end{vmatrix}$$

2(row 2) + (row 3) replace (row 3)

$$D_y = \begin{vmatrix} 2 & 25 & -3 \\ 3 & -5 & 2 \\ 11 & 22 & 0 \end{vmatrix}$$

Expand about column 3.

$$Dy = -3(-1)^{1+3}\begin{vmatrix} 3 & -5 \\ 11 & 22 \end{vmatrix}$$

$$+ 2(-1)^{2+3}\begin{vmatrix} 2 & 25 \\ 11 & 22 \end{vmatrix}$$

$$D_y = -3[66 - (-55)] - 2[44 - 275]$$

$$D_y = -3(121) - 2(-231) = 99$$

$$D_z = \begin{vmatrix} 2 & -3 & 25 \\ 3 & 1 & -5 \\ 5 & -2 & 32 \end{vmatrix}$$

3(row 2) + (row 1) replace (row 1)
2(row 2) + (row 3) replace (row 3)

$$D_z = \begin{vmatrix} 11 & 0 & 10 \\ 3 & 1 & -5 \\ 11 & 0 & 22 \end{vmatrix}$$

Expand about column 2.

$$D_z = 1(-1)^{2+2}\begin{vmatrix} 11 & 10 \\ 11 & 22 \end{vmatrix}$$

$$D_z = 11(22) - 11(10) = 132$$

$$x = \frac{D_x}{D} = \frac{-66}{-33} = 2$$

$$y = \frac{D_y}{D} = \frac{99}{-33} = -3$$

$$z = \frac{D_z}{D} = \frac{132}{-33} = -4$$

The solution set is $\{(2, -3, -4)\}$.

16. $\begin{pmatrix} 3x - y + z = -10 \\ 6x - 2y + 5z = -35 \\ 7x + 3y - 4z = 19 \end{pmatrix}$

$$D = \begin{vmatrix} 3 & -1 & 1 \\ 6 & -2 & 5 \\ 7 & 3 & -4 \end{vmatrix}$$

-2(row 1) + (row 2) replace (row 2)

$$D = \begin{vmatrix} 3 & -1 & 1 \\ 0 & 0 & 3 \\ 7 & 3 & -4 \end{vmatrix}$$

Expand about row 2.

$$D = 3(-1)^{2+3}\begin{vmatrix} 3 & -1 \\ 7 & 3 \end{vmatrix}$$

$$D = -3[9-(-7)] = -48$$

$$D_x = \begin{vmatrix} -10 & -1 & 1 \\ -35 & -2 & 5 \\ 19 & 3 & -4 \end{vmatrix}$$

-2(row 1) + (row 2) replace (row 2)
3(row 1) + (row 3) replace (row 3)

$$D_x = \begin{vmatrix} -10 & -1 & 1 \\ -15 & 0 & 3 \\ -11 & 0 & -1 \end{vmatrix}$$

Expand about column 2.

$$D_x = -1(-1)^{1+2}\begin{vmatrix} -15 & 3 \\ -11 & -1 \end{vmatrix}$$

$$D_x = 15-(-33) = 48$$

$$D_y = \begin{vmatrix} 3 & -10 & 1 \\ 6 & -35 & 5 \\ 7 & 19 & -4 \end{vmatrix}$$

-5(row 1) + (row 2) replace (row 2)
4(row 1) + (row 3) replace (row 3)

$$D_y = \begin{vmatrix} 3 & -10 & 1 \\ -9 & 15 & 0 \\ 19 & -21 & 0 \end{vmatrix}$$

Expand about column 3.

$$D_y = 1(-1)^{1+3}\begin{vmatrix} -9 & 15 \\ 19 & -21 \end{vmatrix}$$

$$D_y = [-9(-21)-(19)15] = -96$$

$$D_z = \begin{vmatrix} 3 & -1 & -10 \\ 6 & -2 & -35 \\ 7 & 3 & 19 \end{vmatrix}$$

-2(row 1) + (row 2) replace (row 2)

$$D_z = \begin{vmatrix} 3 & -1 & -10 \\ 0 & 0 & -15 \\ 7 & 3 & 19 \end{vmatrix}$$

Expand about row 2.

$$D_z = -15(-1)^{2+3}\begin{vmatrix} 3 & -1 \\ 7 & 3 \end{vmatrix}$$

$$D_z = 15[9-(-7)] = 240$$

$$x = \frac{D_x}{D} = \frac{48}{-48} = -1$$

$$y = \frac{D_y}{D} = \frac{-96}{-48} = 2$$

$$z = \frac{D_z}{D} = \frac{240}{-48} = -5$$

The solution set is $\{(-1, 2, -5)\}$.

17. $\begin{pmatrix} 4x+7y=-15 \\ 3x-2y=25 \end{pmatrix}$

Multiply equation 1 by 2,
multiply equation 2 by 7,
and then add the resulting equations.

$$\begin{aligned} 8x+14y &= -30 \\ 21x-14y &= 175 \\ \hline 29x \quad &= 145 \\ x &= 5 \end{aligned}$$

$4x+7y=-15$
$4(5)+7y=-15$
$20+7y=-15$
$7y=-35$
$y=-5$
The solution set is $\{(5, -5)\}$.

18. $\begin{pmatrix} \frac{3}{4}x-\frac{1}{2}y=-15 \\ \frac{2}{3}x+\frac{1}{4}y=-5 \end{pmatrix}$

$\begin{pmatrix} 3x-2y=-60 \\ 8x+3y=-60 \end{pmatrix}$

Multiply equation 1 by 4 and multiply equation 2 by 12, then add the resulting equations to replace equation 2.

$\begin{pmatrix} 3x-2y=-60 \\ 25x=-300 \end{pmatrix}$

$25x=-300$
$x=-12$
$3x-2y=-60$

$3(-12) - 2y = -60$
$-36 - 2y = -60$
$-2y = -24$
$y = 12$
The solution set is $\{(-12, 12)\}$.

19. $\begin{pmatrix} x + 4y = 3 \\ 3x - 2y = 1 \end{pmatrix}$

$x = -4y + 3$

Substitute $-4y + 3$ for x in equation 2.
$3x - 2y = 1$
$3(-4y + 3) - 2y = 1$
$-12y + 9 - 2y = 1$
$-14y = -8$

$y = \frac{-8}{-14} = \frac{4}{7}$

$x = -4x + 3$

$x = -4\left(\frac{4}{7}\right) + 3$

$x = \frac{5}{7}$

The solution set is $\left\{\left(\frac{5}{7}, \frac{4}{7}\right)\right\}$.

20. $\begin{pmatrix} 7x - 3y = -49 \\ y = \frac{3}{5}x - 1 \end{pmatrix}$

Substitute $\frac{3}{5}x - 1$ for y in equation 1.

$7x - 3y = -47$

$7x - 3\left(\frac{3}{5}x - 1\right) = -49$

$7x - \frac{9}{5}x + 3 = -49$

$\frac{26}{5}x = -52$

$x = -52\left(\frac{5}{26}\right)$

$x = -10$

$y = \frac{3}{5}x - 1$

$y = \frac{3}{5}(-10) - 1$

$y = -6 - 1$

$y = -7$

The solution set is $\{(-10, -7)\}$.

21. $\begin{pmatrix} x - y - z = 4 \\ -3x + 2y + 5z = -21 \\ 5x - 3y - 7z = 30 \end{pmatrix}$

Multiply equation 1 by 3 and add to equation 2 to replace equation 2.
Multiply equation 1 by -5 and add to equation 3 to replace equation 3.

$\begin{pmatrix} x - y - z = 4 \\ -y + 2z = -9 \\ 2y - 2z = 10 \end{pmatrix}$

Add equation 2 to equation 3 to replace equation 3.

$\begin{pmatrix} x - y - z = 4 \\ -y + 2z = -9 \\ y = 1 \end{pmatrix}$

$y = 1$
$-y + 2z = -9$
$-1 + 2z = -9$
$2z = -8$
$z = -4$
$x - y - z = 4$
$x - 1 - (-4) = 4$
$x + 3 = 4$
$x = 1$
The solution set is $\{(1, 1, -4)\}$.

22. $\begin{pmatrix} 2x - y + z = -7 \\ -5x + 2y - 3z = 17 \\ 3x + y + 7z = -5 \end{pmatrix}$

$$\left[\begin{array}{ccc|c} 2 & -1 & 1 & -7 \\ -5 & 2 & -3 & 17 \\ 3 & 1 & 7 & -5 \end{array}\right]$$

2(row 1) + (row 2) replace (row 2)
(row 1) + (row 3) replace (row 3)

$$\left[\begin{array}{ccc|c} 2 & -1 & 1 & -7 \\ -1 & 0 & -1 & 3 \\ 5 & 0 & 8 & -12 \end{array}\right]$$

5(row 2) + (row 3) replace (row 3)

$$\left[\begin{array}{ccc|c} 2 & -1 & 1 & -7 \\ -1 & 0 & -1 & 3 \\ 0 & 0 & 3 & 3 \end{array}\right]$$

$3z = 3$
$z = 1$
$-x - z = 3$
$-x = 4$
$x = -4$
$2x - y + z = -7$
$2(-4) - y + 1 = -7$
$-8 - y + 1 = -7$
$-y - 7 = -7$
$-y = 0$
$y = 0$

The solution set is $\{(-4, 0, 1)\}$.

23. $\begin{pmatrix} 3x - 2y - 5z = 2 \\ -4x + 3y + 11z = 3 \\ 2x - y + z = -1 \end{pmatrix}$

Multiply equation 3 by -2 and add to equation 1 to replace equation 1. Multiply equation 3 by 3 and add to equation 2 to replace equation 2.

$$\begin{pmatrix} -x - 7z = 4 \\ 2x + 14z = 0 \\ 2x - y + z = -1 \end{pmatrix}$$

Multiply equation 1 by 2 and add to equation 2 to replace equation 2.

$$\begin{pmatrix} -x - 7z = 4 \\ 0 = 8 \\ 2x - y + z = -1 \end{pmatrix}$$

Since $0 \neq 8$, the system is inconsistent. The solution set is $\emptyset$.

24. $\begin{pmatrix} 7x - y + z = -4 \\ -2x + 9y - 3z = -50 \\ x - 5y + 4z = 42 \end{pmatrix}$

$$\left[\begin{array}{ccc|c} 7 & -1 & 1 & -4 \\ -2 & 9 & -3 & -50 \\ 1 & -5 & 4 & 42 \end{array}\right]$$

3(row 1) + (row 2) replace (row 2)
-4(row 1) + (row 3) replace (row 3)

$$\left[\begin{array}{ccc|c} 7 & -1 & 1 & -4 \\ 19 & 6 & 0 & -62 \\ -27 & -1 & 0 & 58 \end{array}\right]$$

6(row 3) + (row 2) replace (row 2)

$$\left[\begin{array}{ccc|c} 7 & -1 & 1 & -4 \\ -143 & 0 & 0 & 286 \\ -27 & -1 & 0 & 58 \end{array}\right]$$

$-143x = 286$
$x = -2$

$-27x - y = 58$
$-27(-2) - y = 58$
$54 - y = 58$
$-y = 4$
$y = -4$

$7x - y + z = -4$
$7(-2) - (-4) + z = -4$
$-14 + 4 + z = -4$
$-10 + z = -4$
$z = 6$

The solution set is $\{(-2, -4, 6)\}$.

25. $D = \begin{vmatrix} -2 & 6 \\ 3 & 8 \end{vmatrix} = -16 - 18 = -34$

26. $D = \begin{vmatrix} 5 & -4 \\ 7 & -3 \end{vmatrix} = 5(-3) - 7(-4) = 13$

27. $D = \begin{vmatrix} 2 & 3 & -1 \\ 3 & 4 & -5 \\ 6 & 4 & 2 \end{vmatrix}$

-5(row 1) + (row 2) replace (row 2)
2(row 1) + (row 3) replace (row 3)

$$D = \begin{vmatrix} 2 & 3 & -1 \\ -7 & -11 & 0 \\ 10 & 10 & 0 \end{vmatrix}$$

Expand about column 3.

$$D = -1(-1)^{1+3}\begin{vmatrix} -7 & -11 \\ 10 & 10 \end{vmatrix}$$

$$D = -[-70 - (-110)] = -40$$

28. $D = \begin{vmatrix} 3 & -2 & 4 \\ 1 & 0 & 6 \\ 3 & -3 & 5 \end{vmatrix}$

Expand about row 2.

$$D = 1(-1)^{2+1}\begin{vmatrix} -2 & 4 \\ -3 & 5 \end{vmatrix}$$

$$+6(-1)^{2+3}\begin{vmatrix} 3 & -2 \\ 3 & -3 \end{vmatrix}$$

$D = -[-10-(-12)] - 6[-9-(-6)]$
$D = -2 - 6(-3) = 16$

29. $D = \begin{vmatrix} 5 & 4 & 3 \\ 2 & -7 & 0 \\ 3 & -2 & 0 \end{vmatrix}$

Expand about column 3.

$D = 3(-1)^{1+3}\begin{vmatrix} 2 & -7 \\ 3 & -2 \end{vmatrix}$

$D = 3[-4-(-21)] = 3(17) = 51$

30. $\begin{vmatrix} 5 & -4 & 2 & 1 \\ 3 & 7 & 6 & -2 \\ 2 & 1 & -5 & 0 \\ 3 & -2 & 4 & 0 \end{vmatrix}$

2(row 1) + (row 2) replace (row 2)

$\begin{vmatrix} 5 & -4 & 2 & 1 \\ 13 & -1 & 10 & 0 \\ 2 & 1 & -5 & 0 \\ 3 & -2 & 4 & 0 \end{vmatrix}$

(row 2) + (row 3) replace (row 3)
$-$ 2(row 2) + (row 4) replace (row 4)

$\begin{vmatrix} 5 & -4 & 2 & 1 \\ 13 & -1 & 10 & 0 \\ 15 & 0 & 5 & 0 \\ -23 & 0 & -16 & 0 \end{vmatrix}$

Expand about column 4.

$$D = 1(-1)^{1+4}\begin{vmatrix} 13 & -1 & 10 \\ 15 & 0 & 5 \\ -23 & 0 & -16 \end{vmatrix}$$

$$D = -\begin{vmatrix} 13 & -1 & 10 \\ 15 & 0 & 5 \\ -23 & 0 & -16 \end{vmatrix}$$

Expand about column 2.

$D = -(-1)(-1)^{1+2}\begin{vmatrix} 15 & 5 \\ -23 & -16 \end{vmatrix}$

$D = -[-240-(-115)] = 125$

31. Let u = unit digit
and t = tens digit.

$$\begin{pmatrix} u+t=9 \\ 10u+t=u+10t-45 \end{pmatrix}$$

$$\begin{pmatrix} u+t=9 \\ 9u-9t=-45 \end{pmatrix}$$

$u+t=9$
$u=9-t$
$9u-9t=-45$
$9(9-t)-9t=-45$
$81-9t-9t=-45$
$81-18t=-45$
$-18t=-126$
$t=7$
$u=9-t$
$u=9-7$
$u=2$
The original number is 72.

32. Let x = money invested at 12%
and y = money invested at 10%.

$$\begin{pmatrix} x+y=2500 \\ 0.12x=0.10y+102 \end{pmatrix}$$

$$\begin{pmatrix} x+y=2500 \\ 0.12x-0.10y=102 \end{pmatrix}$$

Multiply equation 1 by 0.10 and add to equation 2 to replace equation 2.

$$\begin{pmatrix} x+y=2500 \\ 0.22x=352 \end{pmatrix}$$

$0.22x=352$
$x=1600$

$1600 + y = 2500$
$y = 900$
She invested \$1600 at 12%
and \$900 at 10%.

33. Let $n =$ number of nickels,
$d =$ number of dimes and
$q =$ number of quarters.

$$\begin{pmatrix} 0.05n + 0.10d + 0.25q = 17.70 \\ d = 2n - 8 \\ q = n + d + 2 \end{pmatrix}$$

$$\begin{pmatrix} 5n + 10d + 25q = 1770 \\ -2n + d = -8 \\ -n - d + q = 2 \end{pmatrix}$$

Multiply equation 3 by -25 and add to equation 1 to replace equation 1.

$$\begin{pmatrix} 30n + 35d = 1720 \\ -2n + d = -8 \\ -n - d + q = 2 \end{pmatrix}$$

Multiply equation 2 by 15 and add to equation 1 to replace equation 1.

$$\begin{pmatrix} 50d = 1600 \\ -2n + d = -8 \\ -n - d + q = 2 \end{pmatrix}$$

$50d = 1600$
$d = 32$
$-2n + d = -8$
$-2n + 32 = -8$
$-2n = -40$
$n = 20$

$-n - d + q = 2$
$-20 - 32 + q = 2$
$-52 + q = 2$
$q = 54$

There are 20 nickels, 32 dimes,
and 54 quarters.

34. Let a, b, and c represent the measures of the angeles respectively.

$$\begin{pmatrix} c = 4a + 10 \\ a + c = 3b \\ a + b + c = 180 \end{pmatrix}$$

$$\begin{pmatrix} -4a + c = 10 \\ a - 3b + c = 0 \\ a + b + c = 180 \end{pmatrix}$$

Multiply equation 3 by 3 and add to equation 2 to replace equation 2.

$$\begin{pmatrix} -4a + c = 10 \\ 4a + 4c = 540 \\ a + b + c = 180 \end{pmatrix}$$

Add equation 1 to equation 2 to replace equation 1.

$$\begin{pmatrix} 5c = 550 \\ 4a + 4c = 540 \\ a + b + c = 180 \end{pmatrix}$$

$5c = 550$
$c = 110$
$4a + 4c = 540$
$4a + 4(110) = 540$
$4a + 440 = 540$
$4a = 100$
$a = 25$
$a + b + c = 180$
$25 + b + 110 = 180$
$b + 135 = 180$
$b = 45$
The measures of the angles are
25°, 45°, and 110°.

CHAPTER 11 Test

1. III

2. I

3. III

4. II

5. $D = \begin{vmatrix} -2 & 4 \\ -5 & 6 \end{vmatrix} = -12 - (-20) = 8$

6. $D = \begin{vmatrix} \frac{1}{2} & \frac{1}{3} \\ \frac{3}{4} & -\frac{2}{3} \end{vmatrix}$

$D = \frac{1}{2}\left(-\frac{2}{3}\right) - \left(\frac{3}{4}\right)\left(\frac{1}{3}\right)$

$D = -\frac{1}{3} - \frac{1}{4} = -\frac{7}{12}$

7. $D = \begin{vmatrix} -1 & 2 & 1 \\ 3 & 1 & -2 \\ 2 & -1 & 1 \end{vmatrix}$

3(row 1) + (row 2) replace (row 2)
2(row 1) + (row 3) replace (row 3)

$D = \begin{vmatrix} -1 & 2 & 1 \\ 0 & 7 & 1 \\ 0 & 3 & 3 \end{vmatrix}$

Expand about column 1.

$D = -1(-1)^{1+1}\begin{vmatrix} 7 & 1 \\ 3 & 3 \end{vmatrix}$

$D = -(21 - 3) = -18$

8. $D = \begin{vmatrix} 2 & 4 & -5 \\ -4 & 3 & 0 \\ -2 & 6 & 1 \end{vmatrix}$

Expand about column 3.

$D = -5(-1)^{1+3}\begin{vmatrix} -4 & 3 \\ -2 & 6 \end{vmatrix}$

$+1(-1)^{3+3}\begin{vmatrix} 2 & 4 \\ -4 & 3 \end{vmatrix}$

$D = -5[-24 - (-6)] + [6 - (-16)]$
$D = -5(-18) + 22 = 112$

9. The system is dependent, because the equations are multiple. Hence, there are an infinite number of solutions.

10. $\begin{pmatrix} 3x - 2y = -14 \\ 7x + 2y = -6 \end{pmatrix}$

Add equation 1 to equation 2 to replace equation 2.

$\begin{pmatrix} 3x - 2y = -14 \\ 10x = -20 \end{pmatrix}$

$10x = -20$
$x = -2$
$3x - 2y = -14$
$3(-2) - 2y = -14$
$-6 - 2y = -14$
$-2y = -8$
$y = 4$

The solution set is $\{(-2, 4)\}$.

11. $\begin{pmatrix} 4x - 5y = 17 \\ y = -3x + 8 \end{pmatrix}$

Substitute $-3x + 8$ for y in equation 1.

$4x - 5y = 17$
$4x - 5(-3x + 8) = 17$
$4x + 15x - 40 = 17$
$19x = 57$
$x = 3$
$y = -3x + 8$
$y = -3(3) + 8$
$y = -9 + 8$
$y = -1$

The solution set is $\{(3, -1)\}$.

12. $\begin{pmatrix} \frac{3}{4}x - \frac{1}{2}y = -21 \\ \frac{2}{3}x + \frac{1}{6}y = -4 \end{pmatrix}$

$\begin{pmatrix} 3x - 2y = -84 \\ 4x + y = -24 \end{pmatrix}$

Multiply equation 2 by 2 and add to equation 1 to replace equation 1.

$\begin{pmatrix} 11x = -132 \\ 4x + y = -24 \end{pmatrix}$

$11x = -132$
$x = -12$

13. $\begin{pmatrix} 4x - y = 7 \\ 3x + 2y = 2 \end{pmatrix}$

$$D = \begin{vmatrix} 4 & -1 \\ 3 & 2 \end{vmatrix} = 8 - (-3) = 11$$

$$D_y = \begin{vmatrix} 4 & 7 \\ 3 & 2 \end{vmatrix} = 8 - 21 = -13$$

$$y = \frac{D_y}{D} = -\frac{13}{11}$$

14. $\left[\begin{array}{ccc|c} 1 & 1 & -4 & 3 \\ 0 & 1 & 4 & 5 \\ 0 & 0 & 3 & 6 \end{array}\right]$

$\frac{1}{3}$(row 3)

$$\left[\begin{array}{ccc|c} 1 & 1 & -4 & 3 \\ 0 & 1 & 4 & 5 \\ 0 & 0 & 1 & 2 \end{array}\right]$$

-4(row 3) + (row 2) replace (row 2)
4(row 3) + (row 1) replace (row 1)

$$\left[\begin{array}{ccc|c} 1 & 1 & 0 & 11 \\ 0 & 1 & 0 & -3 \\ 0 & 0 & 1 & 2 \end{array}\right]$$

$-$(row 2) + (row 1) replace (row 1)

$$\left[\begin{array}{ccc|c} 1 & 0 & 0 & 14 \\ 0 & 1 & 0 & -3 \\ 0 & 0 & 1 & 2 \end{array}\right]$$

$x = 14$

15. $\left[\begin{array}{ccc|c} 1 & 2 & -3 & 4 \\ 0 & 1 & 2 & 5 \\ 0 & 0 & 2 & -8 \end{array}\right]$

$2z = -8$
$z = -4$
$y + 2z = 5$
$y + 2(-4) = 5$
$y - 8 = 5$
$y = 13$

16. $\begin{pmatrix} x + 3y - z = 5 \\ 2x - y - z = 7 \\ 5x + 8y - 4z = 22 \end{pmatrix}$

Multiply equation 2 by -4 and add to equation 3 to replace equation 3. Multiply equation 2 by -1 and add to equation 1 to replace equation 1.

$$\begin{pmatrix} -x + 4y = -2 \\ 2x - y - z = 7 \\ -3x + 12y = -6 \end{pmatrix}$$

Multiply equation 1 by -3 and add to equation 3 to replace equation 3.

$$\begin{pmatrix} -x + 4y = -2 \\ 2x - y - z = 7 \\ 0 = 0 \end{pmatrix}$$

Since $0 = 0$ is an identity, the system is dependent. Hence, there are an infinite number of solutions.

17. $\begin{pmatrix} 3x - y - 2z = 1 \\ 4x + 2y + z = 5 \\ 6x - 2y - 4z = 9 \end{pmatrix}$

Multiply equation 1 by -2 and add to equation 3 to replace equation 3.

$$\begin{pmatrix} 3x - y - 2z = 1 \\ 4x + 2y + z = 5 \\ 0 = 7 \end{pmatrix}$$

Since $0 \neq 7$, the system is inconsistent. There is no solution.

18. $\begin{pmatrix} 5x - 3y - 2z = -1 \\ 4y + 7z = 3 \\ 4z = -12 \end{pmatrix}$

$4z = -12$
$z = -3$
$4y + 7z = 3$
$4y + 7(-3) = 3$
$4y - 21 = 3$
$4y = 24$
$y = 6$
$5x - 3y - 2z = -1$
$5x - 3(6) - 2(-3) = -1$
$5x - 18 + 6 = -1$
$5x - 12 = -1$
$5x = 11$
$x = \frac{11}{5}$

The solution set is $\left\{\left(\frac{11}{5}, 6, -3\right)\right\}$.

19. $\begin{pmatrix} x-2y+z=0 \\ y-3z=-1 \\ 2y+5z=-2 \end{pmatrix}$

Multiply equation 2 by -2 and add to equation 3 to replace equation 3.

$\begin{pmatrix} x-2y+z=0 \\ y-3z=-1 \\ 11z=0 \end{pmatrix}$

$11z=0$
$z=0$
$y-3z=-1$
$y-3(0)=-1$
$y=-1$
$x-2y+z=0$
$x-2(-1)+0=0$
$x+2=0$
$x=-2$
The solution set is $\{(-2,\ -1,\ 0)\}$.

20. $\begin{pmatrix} x-4y+z=12 \\ -2x+3y-z=-11 \\ 5x-3y+2z=17 \end{pmatrix}$

Add equation 1 to equation 2 to replace equation 1.
Multiply equation 2 by 2 and add to equation 3 to replace equation 3.

$\begin{pmatrix} -x-y=1 \\ -2x+3y-z=-11 \\ x+3y=-5 \end{pmatrix}$

Multiply equation 1 by 3 and add to equation 3 to replace equation 3.

$\begin{pmatrix} -x-y=1 \\ -2x+3y-z=-11 \\ -2x=-2 \end{pmatrix}$

$-2x=-2$
$x=1$

21. $\begin{pmatrix} x-3y+z=-13 \\ 3x+5y-z=17 \\ 5x-2y+2z=-13 \end{pmatrix}$

Add equation 1 to equation 2 to replace equation 2.

Multiply equation 1 by -2 and add to equation 3 to replace equation 3.

$\begin{pmatrix} x-3y+z=-13 \\ 4x+2y=4 \\ 3x+4y=13 \end{pmatrix}$

Multiply equation 2 by -3 and multiply equation 3 by 4, then add the resulting equations to replace equation 3.

$\begin{pmatrix} x-3y+z=-13 \\ 4x+2y=4 \\ 10y=40 \end{pmatrix}$

$10y=40$
$y=4$

22. Let x = amount of 30% solution and y = amount of 70% solution.

$\begin{pmatrix} x+y=8 \\ 0.30x+0.70y=0.40(8) \end{pmatrix}$

$x=8-y$

Substitute $8-y$ for x in equation 2.
$0.30(8-y)+0.70y=3.2$
$2.40-0.30y+0.70y=3.2$
$0.40y=0.8$
$y=2$
There should be 2 liters of 30% solution.

23. Let n = number of nickels, d = number of dimes and q = number of quarters.

$\begin{pmatrix} 0.05n+0.10d+0.25q=7.25 \\ n+d+q=43 \\ q=3n+1 \end{pmatrix}$

$\begin{pmatrix} 5n+10d+25q=725 \\ n+d+q=43 \\ -3n+q=1 \end{pmatrix}$

Multiply equation 2 by -10 and add to equation 1 to replace equation 1.

$\begin{pmatrix} -5n+15q=295 \\ n+d+q=43 \\ -3n+q=1 \end{pmatrix}$

Multiply equation 1 by $-\frac{3}{5}$ and add to equation 3 to replace equation 3.

$$\begin{pmatrix} -5n+15q=295 \\ n+d+q=43 \\ -8q=-176 \end{pmatrix}$$

$-8q=-176$
$q=22$
There are 22 quarters.

24. Let $x=$ number of batches of cream puffs, $y=$ number of batches of eclairs and $z=$ number of batches of danish rolls.

$$\begin{pmatrix} 0.2x+0.5y+0.4z=7.0 \\ 0.3x+0.1y+0.2z=3.9 \\ 0.1x+0.5y+0.3z=5.5 \end{pmatrix}$$

Multiply each equation by 10.

$$\begin{pmatrix} 2x+5y+4z=70 \\ 3x+y+2z=39 \\ x+5y+3z=55 \end{pmatrix}$$

Multiply equation 2 by -5 and add to equation 3 to replace equation 3.

$$\begin{pmatrix} 2x+5y+4z=70 \\ 3x+y+2z=39 \\ -14x-7z=-140 \end{pmatrix}$$

Multiply equation 3 by $-\frac{1}{7}$.

$$\begin{pmatrix} 2x+5y+4z=70 \\ 3x+y+2z=39 \\ 2x+z=20 \end{pmatrix}$$

Multiply equation 2 by -5 and add to equation 1 to replace equation 1.

$$\begin{pmatrix} -13x-6z=-125 \\ 3x+y+2z=39 \\ 2x+z=20 \end{pmatrix}$$

Multiply equation 3 by 6 and add to equation 1 to replace equation 1.

$$\begin{pmatrix} -x=-5 \\ 3x+y+2z=39 \\ 2x+z=20 \end{pmatrix}$$

$-x=-5$
$x=5$
$2x+z=20$
$2(5)+z=20$
$10+z=20$
$z=10$
$3x+y+2z=39$
$3(5)+y+2(10)=39$
$15+y+20=39$
$y=4$
There are 5 batches of cream puffs, 4 batches of eclairs and 10 batches of danish rolls.

25. Let $x=$ measure of the largest angle, $y=$ measure of the other angle and $z=$ measure of thesmallest angle.

$$\begin{pmatrix} x=20+y+z \\ x-z=65 \\ x+y+z=180 \end{pmatrix}$$

$$\begin{pmatrix} x-y-z=20 \\ x \quad -z=65 \\ x+y+z=180 \end{pmatrix}$$

Add equation 1 to equation 3 to replace equation 3.

$$\begin{pmatrix} x-y-z=20 \\ x \quad -z=65 \\ 2x=200 \end{pmatrix}$$

$2x=200$
$x=100$
$x-z=65$
$100-z=65$
$-z=-35$
$z=35$
$x-y-z=20$
$100-y-35=20$
$65-y=20$
$-y=-45$
$y=45$
The measures of the angles of the triangle are $35°$, $45°$ and $100°$.

Chapter 12 Algebra of Matrices

PROBLEM SET 12.1 Algebra of 2x2 Matrices

1. $A+B=\begin{bmatrix}1 & -2\\ 3 & 4\end{bmatrix}+\begin{bmatrix}2 & -3\\ 5 & -1\end{bmatrix}$

$A+B=\begin{bmatrix}3 & -5\\ 8 & 3\end{bmatrix}$

3. $3C=\begin{bmatrix}0 & 18\\ -12 & 6\end{bmatrix}$ $D=\begin{bmatrix}-2 & 3\\ 5 & -4\end{bmatrix}$

$3C+D=\begin{bmatrix}0-2 & 18+3\\ -12+5 & 6-4\end{bmatrix}$

$3C+D=\begin{bmatrix}-2 & 21\\ -7 & 2\end{bmatrix}$

5. $4A-3B=4\begin{bmatrix}1 & -2\\ 3 & 4\end{bmatrix}-3\begin{bmatrix}2 & -3\\ 5 & -1\end{bmatrix}$

$4A-3B=\begin{bmatrix}4 & -8\\ 12 & 16\end{bmatrix}-\begin{bmatrix}6 & -9\\ 15 & -3\end{bmatrix}$

$4A-3B=\begin{bmatrix}-2 & 1\\ -3 & 19\end{bmatrix}$

7. $A-B=\begin{bmatrix}1-2 & -2-(-3)\\ 3-5 & 4-(-1)\end{bmatrix}$

$A-B=\begin{bmatrix}-1 & 1\\ -2 & 5\end{bmatrix}$

$(A-B)-C=\begin{bmatrix}-1-0 & 1-6\\ -2-(-4) & 5-2\end{bmatrix}$

$(A-B)-C=\begin{bmatrix}-1 & -5\\ 2 & 3\end{bmatrix}$

9. $2D-4E=2\begin{bmatrix}-2 & 3\\ 5 & -4\end{bmatrix}-4\begin{bmatrix}2 & 5\\ 7 & 3\end{bmatrix}$

$2D-4E=\begin{bmatrix}-4 & 6\\ 10 & -8\end{bmatrix}-\begin{bmatrix}8 & 20\\ 28 & 12\end{bmatrix}$

$2D-4E=\begin{bmatrix}-12 & -14\\ -18 & -20\end{bmatrix}$

11. $D+E=\begin{bmatrix}-2+2 & 3+5\\ 5+7 & -4+3\end{bmatrix}$

$D+E=\begin{bmatrix}0 & 8\\ 12 & -1\end{bmatrix}$

$B-(D+E)=\begin{bmatrix}2-0 & -3-8\\ 5-12 & -1-(-1)\end{bmatrix}$

$B-(D+E)=\begin{bmatrix}2 & -11\\ -7 & 0\end{bmatrix}$

13. $AB=\begin{bmatrix}1 & -1\\ 2 & -2\end{bmatrix}\begin{bmatrix}3 & -4\\ -1 & 2\end{bmatrix}$

$AB=\begin{bmatrix}1(3)-1(-1) & 1(-4)-1(2)\\ 2(3)-2(-1) & 2(-4)-2(2)\end{bmatrix}$

$AB=\begin{bmatrix}4 & -6\\ 8 & -12\end{bmatrix}$

$BA=\begin{bmatrix}3 & -4\\ -1 & 2\end{bmatrix}\begin{bmatrix}1 & -1\\ 2 & -2\end{bmatrix}$

$BA=\begin{bmatrix}3(1)-4(2) & 3(-1)-4(-2)\\ -1(1)+2(2) & -1(-1)+2(-2)\end{bmatrix}$

$BA=\begin{bmatrix}-5 & 5\\ 3 & -3\end{bmatrix}$

15. $AB=\begin{bmatrix}1 & -3\\ -4 & 6\end{bmatrix}\begin{bmatrix}7 & -3\\ 4 & 5\end{bmatrix}$

$AB=\begin{bmatrix}1(7)-3(4) & 1(-3)-3(5)\\ -4(7)+6(4) & -4(-3)+6(5)\end{bmatrix}$

$AB=\begin{bmatrix}-5 & -18\\ -4 & 42\end{bmatrix}$

$$BA=\begin{bmatrix}7 & -3\\4 & 5\end{bmatrix}\begin{bmatrix}1 & -3\\-4 & 6\end{bmatrix}$$

$$BA=\begin{bmatrix}7(1)-3(-4) & 7(-3)-3(6)\\4(1)+5(-4) & 4(-3)+5(6)\end{bmatrix}$$

$$BA=\begin{bmatrix}19 & -39\\-16 & 18\end{bmatrix}$$

17. $$AB=\begin{bmatrix}2 & -4\\1 & -2\end{bmatrix}\begin{bmatrix}1 & -2\\-3 & 6\end{bmatrix}$$

$$AB=\begin{bmatrix}2(1)-4(-3) & 2(-2)-4(6)\\1(1)-2(-3) & 1(-2)-2(6)\end{bmatrix}$$

$$AB=\begin{bmatrix}14 & -28\\7 & -14\end{bmatrix}$$

$$BA=\begin{bmatrix}1 & -2\\-3 & 6\end{bmatrix}\begin{bmatrix}2 & -4\\1 & -2\end{bmatrix}$$

$$BA=\begin{bmatrix}1(2)-2(1) & 1(-4)-2(-2)\\-3(2)+6(1) & -3(-4)+6(-2)\end{bmatrix}$$

$$BA=\begin{bmatrix}0 & 0\\0 & 0\end{bmatrix}$$

19. $$AB=\begin{bmatrix}-3 & -2\\-4 & -1\end{bmatrix}\begin{bmatrix}2 & -1\\4 & 5\end{bmatrix}$$

$$AB=\begin{bmatrix}-3(2)-2(4) & -3(-1)-2(5)\\-4(2)-1(4) & -4(-1)-1(5)\end{bmatrix}$$

$$AB=\begin{bmatrix}-14 & -7\\-12 & -1\end{bmatrix}$$

$$BA=\begin{bmatrix}2 & -1\\4 & 5\end{bmatrix}\begin{bmatrix}-3 & -2\\-4 & -1\end{bmatrix}$$

$$BA=\begin{bmatrix}2(-3)\text{-}1(-4) & 2(-2)\text{-}1(-1)\\4(-3)\text{+}5(-4) & 4(-2)\text{+}5(-1)\end{bmatrix}$$

$$BA=\begin{bmatrix}-2 & -3\\-32 & -13\end{bmatrix}$$

21. $$AB=\begin{bmatrix}2 & -1\\-5 & 3\end{bmatrix}\begin{bmatrix}3 & 1\\5 & 2\end{bmatrix}$$

$$AB=\begin{bmatrix}2(3)-1(5) & 2(1)-1(2)\\-5(3)+3(5) & -5(1)+3(2)\end{bmatrix}$$

$$AB=\begin{bmatrix}1 & 0\\0 & 1\end{bmatrix}$$

$$BA=\begin{bmatrix}3 & 1\\5 & 2\end{bmatrix}\begin{bmatrix}2 & -1\\-5 & 3\end{bmatrix}$$

$$BA=\begin{bmatrix}3(2)+1(-5) & 3(-1)+1(3)\\5(2)+2(-5) & 5(-1)+2(3)\end{bmatrix}$$

$$BA=\begin{bmatrix}1 & 0\\0 & 1\end{bmatrix}$$

23. $$AB=\begin{bmatrix}\frac{1}{2} & -\frac{1}{3}\\\frac{1}{3} & \frac{1}{4}\end{bmatrix}\begin{bmatrix}4 & -6\\6 & -4\end{bmatrix}$$

$$AB=\begin{bmatrix}\frac{1}{2}(4)-\frac{1}{3}(6) & \frac{1}{2}(-6)-\frac{1}{3}(-4)\\\frac{1}{3}(4)+\frac{1}{4}(6) & \frac{1}{3}(-6)+\frac{1}{4}(-4)\end{bmatrix}$$

$$AB=\begin{bmatrix}0 & -\frac{5}{3}\\\frac{17}{6} & -3\end{bmatrix}$$

$$BA=\begin{bmatrix}4 & -6\\6 & -4\end{bmatrix}\begin{bmatrix}\frac{1}{2} & -\frac{1}{3}\\\frac{1}{3} & \frac{1}{4}\end{bmatrix}$$

$$BA=\begin{bmatrix}4\left(\frac{1}{2}\right)-6\left(\frac{1}{3}\right) & 4\left(-\frac{1}{3}\right)-6\left(\frac{1}{4}\right)\\6\left(\frac{1}{2}\right)-4\left(\frac{1}{3}\right) & 6\left(-\frac{1}{3}\right)-4\left(\frac{1}{4}\right)\end{bmatrix}$$

$$BA=\begin{bmatrix}0 & -\frac{17}{6}\\\frac{5}{3} & -3\end{bmatrix}$$

Problem Set 12.1

25. $AB = \begin{bmatrix} 5 & 6 \\ 2 & 3 \end{bmatrix} \begin{bmatrix} 1 & -2 \\ -\frac{2}{3} & \frac{5}{3} \end{bmatrix}$

$$AB = \begin{bmatrix} 5(1)+6\left(-\frac{2}{3}\right) & 5(-2)+6\left(\frac{5}{3}\right) \\ 2(1)+3\left(-\frac{2}{3}\right) & 2(-2)+3\left(\frac{5}{3}\right) \end{bmatrix}$$

$$AB = \begin{bmatrix} 1 & 0 \\ 0 & 1 \end{bmatrix}$$

$$BA = = \begin{bmatrix} 1 & -2 \\ -\frac{2}{3} & \frac{5}{3} \end{bmatrix} \begin{bmatrix} 5 & 6 \\ 2 & 3 \end{bmatrix}$$

$$BA = \begin{bmatrix} 1(5)-2(2) & 1(6)-2(3) \\ -\frac{2}{3}(5)+\frac{5}{3}(2) & -\frac{2}{3}(6)+\frac{5}{3}(3) \end{bmatrix}$$

$$BA = \begin{bmatrix} 1 & 0 \\ 0 & 1 \end{bmatrix}$$

27. $AB = \begin{bmatrix} -2 & 3 \\ 5 & 4 \end{bmatrix} \begin{bmatrix} 0 & 1 \\ 1 & 0 \end{bmatrix}$

$$AB = \begin{bmatrix} -2(0)+3(1) & -2(1)+3(0) \\ 5(0)+4(1) & 5(1)+4(0) \end{bmatrix}$$

$$AB = \begin{bmatrix} 3 & -2 \\ 4 & 5 \end{bmatrix}$$

$$BA = \begin{bmatrix} 0 & 1 \\ 1 & 0 \end{bmatrix} \begin{bmatrix} -2 & 3 \\ 5 & 4 \end{bmatrix}$$

$$BA = \begin{bmatrix} 0(-2)+1(5) & 0(3)+1(4) \\ 1(-2)+0(5) & 1(3)+0(4) \end{bmatrix}$$

$$BA = \begin{bmatrix} 5 & 4 \\ -2 & 3 \end{bmatrix}$$

29. $AD = \begin{bmatrix} -2 & 3 \\ 5 & 4 \end{bmatrix} \begin{bmatrix} 1 & 1 \\ 1 & 1 \end{bmatrix}$

$$AD = \begin{bmatrix} -2(1)+3(1) & -2(1)+3(1) \\ 5(1)+4(1) & 5(1)+4(1) \end{bmatrix}$$

$$AD = \begin{bmatrix} 1 & 1 \\ 9 & 9 \end{bmatrix}$$

$$DA = \begin{bmatrix} 1 & 1 \\ 1 & 1 \end{bmatrix} \begin{bmatrix} -2 & 3 \\ 5 & 4 \end{bmatrix}$$

$$DA = \begin{bmatrix} 1(-2)+1(5) & 1(3)+1(4) \\ 1(-2)+1(5) & 1(3)+1(4) \end{bmatrix}$$

$$DA = \begin{bmatrix} 3 & 7 \\ 3 & 7 \end{bmatrix}$$

31. $AB = \begin{bmatrix} 2 & 4 \\ 5 & -3 \end{bmatrix} \begin{bmatrix} -2 & 3 \\ -1 & 2 \end{bmatrix}$

$$AB = \begin{bmatrix} 2(-2)+4(-1) & 2(3)+4(2) \\ 5(-2)-3(-1) & 5(3)-3(2) \end{bmatrix}$$

$$AB = \begin{bmatrix} -8 & 14 \\ -7 & 9 \end{bmatrix}$$

$$(AB)C = \begin{bmatrix} -8 & 14 \\ -7 & 9 \end{bmatrix} \begin{bmatrix} 2 & 1 \\ 3 & 7 \end{bmatrix}$$

$$(AB)C = \begin{bmatrix} -8(2)+14(3) & -8(1)+14(7) \\ -7(2)+9(3) & -7(1)+9(7) \end{bmatrix}$$

$$(AB)C = \begin{bmatrix} 26 & 90 \\ 13 & 56 \end{bmatrix}$$

$$BC = \begin{bmatrix} -2 & 3 \\ -1 & 2 \end{bmatrix} \begin{bmatrix} 2 & 1 \\ 3 & 7 \end{bmatrix}$$

$$BC = \begin{bmatrix} -2(2)+3(3) & -2(1)+3(7) \\ -1(2)+2(3) & -1(1)+2(7) \end{bmatrix}$$

$$BC = \begin{bmatrix} 5 & 19 \\ 4 & 13 \end{bmatrix}$$

$$A(BC) = \begin{bmatrix} 2 & 4 \\ 5 & -3 \end{bmatrix} \begin{bmatrix} 5 & 19 \\ 4 & 13 \end{bmatrix}$$

$$A(BC) = \begin{bmatrix} 2(5)+4(4) & 2(19)+4(13) \\ 5(5)-3(4) & 5(19)-3(13) \end{bmatrix}$$

$$A(BC) = \begin{bmatrix} 26 & 90 \\ 13 & 56 \end{bmatrix}$$

33. $(A+B)C=$

$$\left(\begin{bmatrix}2 & 4\\5 & -3\end{bmatrix}+\begin{bmatrix}-2 & 3\\-1 & 2\end{bmatrix}\right)\begin{bmatrix}2 & 1\\3 & 7\end{bmatrix}$$

$(A+B)C=$

$$\begin{bmatrix}0 & 7\\4 & -1\end{bmatrix}\begin{bmatrix}2 & 1\\3 & 7\end{bmatrix}=\begin{bmatrix}21 & 49\\5 & -3\end{bmatrix}$$

$AC+BC=$

$$\begin{bmatrix}2 & 4\\5 & -3\end{bmatrix}\begin{bmatrix}2 & 1\\3 & 7\end{bmatrix}+\begin{bmatrix}-2 & 3\\-1 & 2\end{bmatrix}\begin{bmatrix}2 & 1\\3 & 7\end{bmatrix}$$

$$AC+BC=\begin{bmatrix}16 & 30\\1 & -16\end{bmatrix}+\begin{bmatrix}5 & 19\\4 & 13\end{bmatrix}$$

$$AC+BC=\begin{bmatrix}21 & 49\\5 & -3\end{bmatrix}$$

$(A+B)C=AC+BC$

35. $$A+B=\begin{bmatrix}a_{11} & a_{12}\\a_{21} & a_{22}\end{bmatrix}+\begin{bmatrix}b_{11} & b_{12}\\b_{21} & b_{22}\end{bmatrix}$$

$$A+B=\begin{bmatrix}a_{11}+b_{11} & a_{12}+b_{12}\\a_{21}+b_{21} & a_{22}+b_{22}\end{bmatrix}$$

$$B+A=\begin{bmatrix}b_{11} & b_{12}\\b_{21} & b_{22}\end{bmatrix}\begin{bmatrix}a_{11} & a_{12}\\a_{21} & a_{22}\end{bmatrix}$$

$$B+A=\begin{bmatrix}b_{11}+a_{11} & b_{12}+a_{12}\\b_{21}+a_{21} & b_{22}+a_{22}\end{bmatrix}$$

37. $$A+(-A)=\begin{bmatrix}a_{11} & a_{12}\\a_{21} & a_{22}\end{bmatrix}+\begin{bmatrix}-a_{11} & -a_{12}\\-a_{21} & -a_{22}\end{bmatrix}$$

$$A+(-A)=\begin{bmatrix}0 & 0\\0 & 0\end{bmatrix}$$

39. $$(k+l)A=\begin{bmatrix}ka_{11}+la_{11} & ka_{12}+la_{12}\\ka_{21}+la_{21} & ka_{22}+la_{22}\end{bmatrix}$$

$$kA+lA=\begin{bmatrix}ka_{11} & ka_{12}\\ka_{21} & ka_{22}\end{bmatrix}+\begin{bmatrix}la_{11} & la_{12}\\la_{21} & la_{22}\end{bmatrix}$$

$$kA+lA=\begin{bmatrix}ka_{11}+la_{11} & ka_{12}+la_{12}\\ka_{21}+la_{21} & ka_{22}+la_{22}\end{bmatrix}$$

41. The element of the first row and the first column of $(AB)C$ is
$a_{11}b_{11}c_{11}+a_{12}b_{21}c_{11}+a_{11}b_{12}c_{21}+a_{12}b_{22}c_{21}$

Rearranging terms
$a_{11}b_{11}c_{11}+a_{11}b_{12}c_{21}+a_{12}b_{21}c_{11}+a_{12}b_{22}c_{21}$

Now factoring
$a_{11}(b_{11}c_{11}+b_{12}c_{21})+a_{12}(b_{21}c_{11}+b_{22}c_{21})$

The same term would be obtained as the element in the first row and first column of the product $A(BC)$.

$$\begin{bmatrix}a_{11} & a_{12}\\a_{21} & a_{22}\end{bmatrix}\begin{bmatrix}b_{11}c_{11}+b_{12}c_{21} & b_{11}c_{12}+b_{12}c_{22}\\b_{21}c_{11}+b_{22}c_{21} & b_{21}c_{12}+b_{21}c_{22}\end{bmatrix}$$

The same would be true for the other elements in the product.

43. We will verify this for the element in the first row and first column. The same will follow for the other elements.

$$(A+B)C=\begin{bmatrix}a_{11}+b_{11} & a_{12}+b_{12}\\a_{21}+b_{21} & a_{22}+a_{22}\end{bmatrix}\begin{bmatrix}c_{11} & c_{12}\\c_{21} & c_{22}\end{bmatrix}$$

$$(A+B)C=\begin{bmatrix}a_{11}c_{11}+b_{11}c_{11}+a_{12}c_{21}+b_{12}c_{21} & \cdots\\ \cdots & \cdots\end{bmatrix}$$

$$AC = \begin{bmatrix} a_{11}c_{11} + a_{12}c_{21} & \cdots \\ \cdots & \cdots \end{bmatrix}$$

$$BC = \begin{bmatrix} b_{11}c_{11} + b_{12}c_{21} & \cdots \\ \cdots & \cdots \end{bmatrix}$$

$$AC + BC = \begin{bmatrix} a_{11}c_{11} + a_{12}c_{21} + b_{11}c_{11} + b_{12}c_{21} & \cdots \\ \cdots & \cdots \end{bmatrix}$$

Further Investigations

49. $A^2 = AA = \begin{bmatrix} 1(1) - 1(2) & 1(-1) - 1(3) \\ 2(1) + 3(2) & 2(-1) + 3(3) \end{bmatrix} = \begin{bmatrix} -1 & -4 \\ 8 & 7 \end{bmatrix}$

$A^3 = A^2A = \begin{bmatrix} -1(1) - 4(2) & -1(-1) - 4(3) \\ 8(1) + 7(2) & 8(-1) + 7(3) \end{bmatrix} = \begin{bmatrix} -9 & -11 \\ 22 & 13 \end{bmatrix}$

PROBLEM SET 12.2 Multiplicative Inverses

1. $|A| = (5)(3) - 2(7) = 1$

$A^{-1} = \frac{1}{1}\begin{bmatrix} 3 & -7 \\ -2 & 5 \end{bmatrix}$

$A^{-1} = \begin{bmatrix} 3 & -7 \\ -2 & 5 \end{bmatrix}$

3. $|A| = 15 - 16 = -1$

$A^{-1} = \frac{1}{-1}\begin{bmatrix} 5 & -8 \\ -2 & 3 \end{bmatrix}$

$A^{-1} = \begin{bmatrix} -5 & 8 \\ 2 & -3 \end{bmatrix}$

5. $|A| = (-1)(4) - (3)(2) = -10$

$A^{-1} = \frac{1}{-10}\begin{bmatrix} 4 & -2 \\ -3 & -1 \end{bmatrix}$

$A^{-1} = \begin{bmatrix} -\frac{2}{5} & \frac{1}{5} \\ \frac{3}{10} & \frac{1}{10} \end{bmatrix}$

7. $|A| = -12 - (-12) = 0$
Inverse does not exist.

9. $|A| = (-3)(5) - (-4)(2) = -7$

$A^{-1} = \frac{1}{-7}\begin{bmatrix} 5 & -2 \\ 4 & -3 \end{bmatrix}$

$A^{-1} = \begin{bmatrix} -\frac{5}{7} & \frac{2}{7} \\ -\frac{4}{7} & \frac{3}{7} \end{bmatrix}$

11. $|A| = 0 - 5 = -5$

$A^{-1} = \frac{1}{-5}\begin{bmatrix} 3 & -1 \\ -5 & 0 \end{bmatrix}$

$A^{-1} = \begin{bmatrix} -\frac{3}{5} & \frac{1}{5} \\ 1 & 0 \end{bmatrix}$

13. $|A| = (-2)(-4) - (-1)(-3) = 5$

$A^{-1} = \frac{1}{5}\begin{bmatrix} -4 & 3 \\ 1 & -2 \end{bmatrix}$

$A^{-1} = \begin{bmatrix} -\frac{4}{5} & \frac{3}{5} \\ \frac{1}{5} & -\frac{2}{5} \end{bmatrix}$

15. $|A| = -12 - (-15) = 3$

$$A^{-1} = \frac{1}{3}\begin{bmatrix} 6 & -5 \\ 3 & -2 \end{bmatrix}$$

$$A^{-1} = \begin{bmatrix} 2 & -\frac{5}{3} \\ 1 & -\frac{2}{3} \end{bmatrix}$$

17. $|A| = (1)(-1) - (1)(1) = -2$

$$A^{-1} = \frac{1}{-2}\begin{bmatrix} -1 & -1 \\ -1 & 1 \end{bmatrix}$$

$$A^{-1} = \begin{bmatrix} \frac{1}{2} & \frac{1}{2} \\ \frac{1}{2} & -\frac{1}{2} \end{bmatrix}$$

19. $AB = \begin{bmatrix} 4 & 3 \\ 2 & 5 \end{bmatrix}\begin{bmatrix} 3 \\ 6 \end{bmatrix}$

$$AB = \begin{bmatrix} 4(3) + 3(6) \\ 2(3) + 5(6) \end{bmatrix} = \begin{bmatrix} 30 \\ 36 \end{bmatrix}$$

21. $AB = \begin{bmatrix} -3 & -4 \\ 2 & 1 \end{bmatrix}\begin{bmatrix} 4 \\ -3 \end{bmatrix}$

$$AB = \begin{bmatrix} -3(4) - 4(-3) \\ 2(4) + 1(-3) \end{bmatrix} = \begin{bmatrix} 0 \\ 5 \end{bmatrix}$$

23. $AB = \begin{bmatrix} -4 & 2 \\ 7 & -5 \end{bmatrix}\begin{bmatrix} -1 \\ -4 \end{bmatrix}$

$$AB = \begin{bmatrix} -4(-1) + 2(-4) \\ 7(-1) - 5(-4) \end{bmatrix} = \begin{bmatrix} -4 \\ 13 \end{bmatrix}$$

25. $AB = \begin{bmatrix} -2 & -3 \\ -5 & -6 \end{bmatrix}\begin{bmatrix} 5 \\ -2 \end{bmatrix}$

$$AB = \begin{bmatrix} -2(5) - 3(-2) \\ -5(5) - 6(-2) \end{bmatrix} = \begin{bmatrix} -4 \\ -13 \end{bmatrix}$$

27. $A = \begin{bmatrix} 2 & 3 \\ 1 & 2 \end{bmatrix}$

$|A| = 4 - 3 = 1$

$$A^{-1} = \frac{1}{1}\begin{bmatrix} 2 & -3 \\ -1 & 2 \end{bmatrix}$$

$$A^{-1} = \begin{bmatrix} 2 & -3 \\ -1 & 2 \end{bmatrix}$$

$$A^{-1}B = \begin{bmatrix} 2 & -3 \\ -1 & 2 \end{bmatrix}\begin{bmatrix} 13 \\ 8 \end{bmatrix}$$

$$A^{-1}B = \begin{bmatrix} 2(13) - 3(8) \\ -1(13) + 2(8) \end{bmatrix} = \begin{bmatrix} 2 \\ 3 \end{bmatrix}$$

The solution set is $\{(2, 3)\}$.

29. $|A| = 4(2) - (-3)(-3) = -1$

$$A^{-1} = \frac{1}{-1}\begin{bmatrix} 2 & 3 \\ 3 & 4 \end{bmatrix}$$

$$A^{-1} = \begin{bmatrix} -2 & -3 \\ -3 & -4 \end{bmatrix}$$

$$A^{-1}B = \begin{bmatrix} -2 & -3 \\ -3 & -4 \end{bmatrix}\begin{bmatrix} -23 \\ 16 \end{bmatrix}$$

$$A^{-1}B = \begin{bmatrix} -2(-23) - 3(16) \\ -3(-23) - 4(16) \end{bmatrix} = \begin{bmatrix} -2 \\ 5 \end{bmatrix}$$

The solution set is $\{(-2, 5)\}$.

31. $A = \begin{bmatrix} 1 & -7 \\ 6 & 5 \end{bmatrix}$

$|A| = 5 - (-42) = 47$

$$A^{-1} = \frac{1}{47}\begin{bmatrix} 5 & 7 \\ -6 & 1 \end{bmatrix}$$

$$A^{-1}B = \frac{1}{47}\begin{bmatrix} 5 & 7 \\ -6 & 1 \end{bmatrix}\begin{bmatrix} 7 \\ -5 \end{bmatrix}$$

$$A^{-1}B = \frac{1}{47}\begin{bmatrix} 5(7) + 7(-5) \\ -6(7) + 1(-5) \end{bmatrix}$$

$$A^{-1}B = \frac{1}{47}\begin{bmatrix} 0 \\ -47 \end{bmatrix} = \begin{bmatrix} 0 \\ -1 \end{bmatrix}$$

The solution set is $\{(0, -1)\}$.

33. $A = \begin{bmatrix} 3 & -5 \\ 4 & -3 \end{bmatrix} \quad B = \begin{bmatrix} 2 \\ -1 \end{bmatrix}$

$|A| = -9 - (-20) = 11$

$$A^{-1} = \frac{1}{11}\begin{bmatrix} -3 & 5 \\ -4 & 3 \end{bmatrix}$$

$$A^{-1}B = \frac{1}{11}\begin{bmatrix} -3 & 5 \\ -4 & 3 \end{bmatrix}\begin{bmatrix} 2 \\ -1 \end{bmatrix}$$

$$A^{-1}B = \frac{1}{11}\begin{bmatrix} -3(2) + 5(-1) \\ -4(2) + 3(-1) \end{bmatrix}$$

$$A^{-1}B = \frac{1}{11}\begin{bmatrix} -11 \\ -11 \end{bmatrix} = \begin{bmatrix} -1 \\ -1 \end{bmatrix}$$

The solution set is $\{(-1, -1)\}$.

35. $A = \begin{bmatrix} 3 & 1 \\ 9 & -5 \end{bmatrix} \quad B = \begin{bmatrix} 19 \\ 1 \end{bmatrix}$

$|A| = -15 - 9 = -24$

$$A^{-1} = \frac{1}{-24}\begin{bmatrix} -5 & -1 \\ -9 & 3 \end{bmatrix}$$

$$A^{-1}B = -\frac{1}{24}\begin{bmatrix} -5 & -1 \\ -9 & 3 \end{bmatrix}\begin{bmatrix} 19 \\ 1 \end{bmatrix}$$

$$A^{-1}B = -\frac{1}{24}\begin{bmatrix} -5(19) - 1(1) \\ -9(19) + 3(1) \end{bmatrix}$$

$$A^{-1}B = -\frac{1}{24}\begin{bmatrix} -96 \\ -168 \end{bmatrix} = \begin{bmatrix} 4 \\ 7 \end{bmatrix}$$

The solution set is $\{(4, 7)\}$.

37. $|A| = 3(-18) - (30)(2) = -114$

$$A^{-1} = \frac{1}{-114}\begin{bmatrix} -18 & -2 \\ -30 & 3 \end{bmatrix}$$

$$A^{-1} = \begin{bmatrix} \frac{3}{19} & \frac{1}{57} \\ \frac{5}{19} & -\frac{1}{38} \end{bmatrix}$$

$$A^{-1}B = \begin{bmatrix} \frac{3}{19} & \frac{1}{57} \\ \frac{5}{19} & -\frac{1}{38} \end{bmatrix}\begin{bmatrix} 0 \\ -19 \end{bmatrix}$$

$$A^{-1}B = \begin{bmatrix} -\frac{1}{3} \\ \frac{1}{2} \end{bmatrix}$$

The solution set is $\left\{\left(-\frac{1}{3}, \frac{1}{2}\right)\right\}$.

39. $A = \begin{bmatrix} \frac{1}{3} & \frac{3}{4} \\ \frac{2}{3} & \frac{1}{5} \end{bmatrix} \quad B = \begin{bmatrix} 12 \\ -2 \end{bmatrix}$

$$|A| = \frac{1}{15} - \frac{1}{2} = -\frac{13}{30}$$

$$A^{-1} = -\frac{1}{\frac{13}{30}}\begin{bmatrix} \frac{1}{5} & -\frac{3}{4} \\ -\frac{2}{3} & \frac{1}{3} \end{bmatrix}$$

$$A^{-1}B = -\frac{30}{13}\begin{bmatrix} \frac{1}{5} & -\frac{3}{4} \\ -\frac{2}{3} & \frac{1}{3} \end{bmatrix}\begin{bmatrix} 12 \\ -2 \end{bmatrix}$$

$$A^{-1}B = -\frac{30}{13}\begin{bmatrix} \frac{1}{5}(12) - \frac{3}{4}(-2) \\ -\frac{2}{3}(12) + \frac{1}{3}(-2) \end{bmatrix}$$

$$A^{-1}B = -\frac{30}{13}\begin{bmatrix} \frac{39}{10} \\ \frac{-26}{3} \end{bmatrix} = \begin{bmatrix} -9 \\ 20 \end{bmatrix}$$

The solution set is $\{(-9, 20)\}$.

PROBLEM SET 12.3 $m \times n$ Matrices

1. $A+B=\begin{bmatrix} 2 & -1 & 4 \\ -2 & 0 & 5 \end{bmatrix}+\begin{bmatrix} -1 & 4 & -7 \\ 5 & -6 & 2 \end{bmatrix}=\begin{bmatrix} 1 & 3 & -3 \\ 3 & -6 & 7 \end{bmatrix}$

$A-B=\begin{bmatrix} 2 & -1 & 4 \\ -2 & 0 & 5 \end{bmatrix}-\begin{bmatrix} -1 & 4 & -7 \\ 5 & -6 & 2 \end{bmatrix}=\begin{bmatrix} 3 & -5 & 11 \\ -7 & 6 & 3 \end{bmatrix}$

$2A+3B=2\begin{bmatrix} 2 & -1 & 4 \\ -2 & 0 & 5 \end{bmatrix}+3\begin{bmatrix} -1 & 4 & -7 \\ 5 & -6 & 2 \end{bmatrix}$

$2A+3B=\begin{bmatrix} 4 & -2 & 8 \\ -4 & 0 & 10 \end{bmatrix}+\begin{bmatrix} -3 & 12 & -21 \\ 15 & -18 & 6 \end{bmatrix}=\begin{bmatrix} 1 & 10 & -13 \\ 11 & -18 & 16 \end{bmatrix}$

$4A-2B=4\begin{bmatrix} 2 & -1 & 4 \\ -2 & 0 & 5 \end{bmatrix}-2\begin{bmatrix} -1 & 4 & -7 \\ 5 & -6 & 2 \end{bmatrix}$

$4A-2B=\begin{bmatrix} 8 & -4 & 16 \\ -8 & 0 & 20 \end{bmatrix}-\begin{bmatrix} -2 & 8 & -14 \\ 10 & -12 & 4 \end{bmatrix}=\begin{bmatrix} 10 & -12 & 30 \\ -18 & 12 & 16 \end{bmatrix}$

3. $A+B=[2-3 \quad -1-6 \quad 4+9 \quad 12-5]=[-1 \quad -7 \quad 13 \quad 7]$

$A-B=[2-(-3) \quad -1-(-6) \quad 4-9 \quad 12-(-5)]=[5 \quad 5 \quad -5 \quad 17]$

$2A=[4 \quad -2 \quad 8 \quad 24]$

$3B=[-9 \quad -18 \quad 27 \quad -15]$

$2A+3B=[4-9 \quad -2-18 \quad 8+27 \quad 24-15]=[-5 \quad -20 \quad 35 \quad 9]$

$4A=[8 \quad -4 \quad 16 \quad 48]$

$-2B=[6 \quad 12 \quad -18 \quad 10]$

$4A-2B=[8+6 \quad -4+12 \quad 16-18 \quad 48+10]=[14 \quad 8 \quad -2 \quad 58]$

5. $A+B=\begin{bmatrix} 3 & -2 & 1 \\ -1 & 4 & -7 \\ 0 & 5 & 9 \end{bmatrix}+\begin{bmatrix} 5 & -1 & -3 \\ 10 & -2 & 4 \\ 7 & 0 & 12 \end{bmatrix}=\begin{bmatrix} 8 & -3 & -2 \\ 9 & 2 & -3 \\ 7 & 5 & 21 \end{bmatrix}$

$A-B=\begin{bmatrix} 3 & -2 & 1 \\ -1 & 4 & -7 \\ 0 & 5 & 9 \end{bmatrix}-\begin{bmatrix} 5 & -1 & -3 \\ 10 & -2 & 4 \\ 7 & 0 & 12 \end{bmatrix}=\begin{bmatrix} -2 & -1 & 4 \\ -11 & 6 & -11 \\ -7 & 5 & -3 \end{bmatrix}$

$2A+3B=2\begin{bmatrix} 3 & -2 & 1 \\ -1 & 4 & -7 \\ 0 & 5 & 9 \end{bmatrix}+3\begin{bmatrix} 5 & -1 & -3 \\ 10 & -2 & 4 \\ 7 & 0 & 12 \end{bmatrix}$

Problem Set 12.3

$$2A+3B=\begin{bmatrix}6 & -4 & 2\\ -2 & 8 & -14\\ 0 & 10 & 18\end{bmatrix}+\begin{bmatrix}15 & -3 & -9\\ 30 & -6 & 12\\ 21 & 0 & 36\end{bmatrix}=\begin{bmatrix}21 & -7 & -7\\ 28 & 2 & -2\\ 21 & 10 & 54\end{bmatrix}$$

$$4A-2B=4\begin{bmatrix}3 & -2 & 1\\ -1 & 4 & -7\\ 0 & 5 & 9\end{bmatrix}-2\begin{bmatrix}5 & -1 & -3\\ 10 & -2 & 4\\ 7 & 0 & 12\end{bmatrix}$$

$$4A-2B=\begin{bmatrix}12 & -8 & 4\\ -4 & 16 & -28\\ 0 & 20 & 36\end{bmatrix}-\begin{bmatrix}10 & -2 & -6\\ 20 & -4 & 8\\ 14 & 0 & 24\end{bmatrix}=\begin{bmatrix}2 & -6 & 10\\ -24 & 20 & -36\\ -14 & 20 & 12\end{bmatrix}$$

7. $$A+B=\begin{bmatrix}-1+1 & 0+2\\ 2-3 & 3+7\\ -5+6 & -4-5\\ -7+9 & 11-2\end{bmatrix}=\begin{bmatrix}0 & 2\\ -1 & 10\\ 1 & -9\\ 2 & 9\end{bmatrix}$$

$$A-B=\begin{bmatrix}-1-1 & 0-2\\ 2-(-3) & 3-7\\ -5-6 & -4-(-5)\\ -7-9 & 11-(-2)\end{bmatrix}=\begin{bmatrix}-2 & -2\\ 5 & -4\\ -11 & 1\\ -16 & 13\end{bmatrix}$$

$$2A=\begin{bmatrix}-2 & 0\\ 4 & 6\\ -10 & -8\\ -14 & 22\end{bmatrix}\qquad 3B=\begin{bmatrix}3 & 6\\ -9 & 21\\ 18 & -15\\ 27 & -6\end{bmatrix}$$

$$2A+3B=\begin{bmatrix}-2+3 & 0+6\\ 4-9 & 6+21\\ -10+18 & -8-15\\ -14+27 & 22-6\end{bmatrix}=\begin{bmatrix}1 & 6\\ -5 & 27\\ 8 & -23\\ 13 & 16\end{bmatrix}$$

$$4A=\begin{bmatrix}-4 & 0\\ 8 & 12\\ -20 & -16\\ -28 & 44\end{bmatrix}\qquad -2B=\begin{bmatrix}-2 & -4\\ 6 & -14\\ -12 & 10\\ -18 & 4\end{bmatrix}$$

$$4A-2B=\begin{bmatrix}-4-2 & 0-4\\ 8+6 & 12-14\\ -20-12 & -16+10\\ -28-18 & 44+4\end{bmatrix}=\begin{bmatrix}-6 & -4\\ 14 & -2\\ -32 & -6\\ -46 & 48\end{bmatrix}$$

9. $$AB=\begin{bmatrix}2 & -1\\ 0 & -4\\ -5 & 3\end{bmatrix}\begin{bmatrix}5 & -2 & 6\\ -1 & 4 & -2\end{bmatrix}$$

$$AB = \begin{bmatrix} 2(5)+(-1)(-1) & 2(-2)+(-1)(4) & 2(6)+(-1)(-2) \\ 0(5)+(-4)(-1) & 0(-2)+(-4)(4) & 0(6)+(-4)(-2) \\ -5(5)+3(-1) & -5(-2)+3(4) & -5(6)+3(-2) \end{bmatrix}$$

$$AB = \begin{bmatrix} 11 & -8 & 14 \\ 4 & -16 & 8 \\ -28 & 22 & -36 \end{bmatrix}$$

$$BA = \begin{bmatrix} 5 & -2 & 6 \\ -1 & 4 & -2 \end{bmatrix} \begin{bmatrix} 2 & -1 \\ 0 & -4 \\ -5 & 3 \end{bmatrix}$$

$$BA = \begin{bmatrix} 5(2)+(-2)(0)+6(-5) & 5(-1)+(-2)(-4)+6(3) \\ -1(2)+4(0)+(-2)(-5) & -1(-1)+4(-4)+(-2)(3) \end{bmatrix} = \begin{bmatrix} -20 & 21 \\ 8 & -21 \end{bmatrix}$$

11. $$AB = \begin{bmatrix} 2 & -1 & -3 \\ 0 & -4 & 7 \end{bmatrix} \begin{bmatrix} 2 & 1 & -1 & 4 \\ 0 & -2 & 3 & 5 \\ -6 & 4 & -2 & 0 \end{bmatrix}$$

$$AB = \begin{bmatrix} 2(2)\text{ -}1(0)\text{ -}3(-6) & 2(1)\text{ -}1(-2)\text{ -}3(4) & 2(-1)\text{ -}1(3)\text{ -}3(-2) & 2(4)\text{ - }(5)\text{-}3(0) \\ 0(2)\text{ -}4(0)+7(-6) & 0(1)\text{ -}4(-2)+7(4) & 0(-1)\text{ -}4(3)+7(-2) & 0(4)\text{ - }4(5)+7(0) \end{bmatrix}$$

$$AB = \begin{bmatrix} 22 & -8 & 1 & 3 \\ -42 & 36 & -26 & -20 \end{bmatrix}$$

BA does not exist.

13. $$AB = \begin{bmatrix} 1 & -1 & 2 \\ 0 & 1 & -2 \\ 3 & 1 & 4 \end{bmatrix} \begin{bmatrix} 2 & 3 & -1 \\ 4 & 0 & 2 \\ -5 & 1 & -1 \end{bmatrix}$$

$$AB = \begin{bmatrix} 1(2)+(-1)(4)+2(-5) & 1(3)+(-1)(0)+2(1) & 1(-1)+(-1)(2)+2(-1) \\ 0(2)+1(4)+(-2)(-5) & 0(3)+1(0)+(-2)(1) & 0(-1)+1(2)+(-2)(-1) \\ 3(2)+1(4)+4(-5) & 3(3)+1(0)+4(1) & 3(-1)+1(2)+4(-1) \end{bmatrix}$$

$$AB = \begin{bmatrix} -12 & 5 & -5 \\ 14 & -2 & 4 \\ -10 & 13 & -5 \end{bmatrix}$$

$$BA = \begin{bmatrix} 2 & 3 & -1 \\ 4 & 0 & 2 \\ -5 & 1 & -1 \end{bmatrix} \begin{bmatrix} 1 & -1 & 2 \\ 0 & 1 & -2 \\ 3 & 1 & 4 \end{bmatrix}$$

$$BA = \begin{bmatrix} 2(1)+3(0)+(-1)(3) & 2(-1)+3(1)+(-1)(1) & 2(2)+3(-2)+(-1)(4) \\ 4(1)+0(0)+2(3) & 4(-1)+0(1)+(2)(1) & 4(2)+0(-2)+2(4) \\ (-5)(1)+1(0)+(-1)(3) & -5(-1)+1(1)+(-1)(1) & (-5)(2)+1(-2)+(-1)(4) \end{bmatrix}$$

$$BA = \begin{bmatrix} -1 & 0 & -6 \\ 10 & -2 & 16 \\ -8 & 5 & -16 \end{bmatrix}$$

15. $$AB = [2 \quad -1 \quad 3 \quad 4]\begin{bmatrix} -1 \\ -3 \\ 2 \\ -4 \end{bmatrix} = [2(-1) - 1(-3) + 3(2) + 4(-4)] = [-9]$$

$$BA = \begin{bmatrix} -1 \\ -3 \\ 2 \\ -4 \end{bmatrix}[2 \quad -1 \quad 3 \quad 4]$$

$$BA = \begin{bmatrix} -1(2) & -1(-1) & -1(3) & -1(4) \\ -3(2) & -3(-1) & -3(3) & -3(4) \\ 2(2) & 2(-1) & 2(3) & 2(4) \\ -4(2) & -4(-1) & -4(3) & -4(4) \end{bmatrix} = \begin{bmatrix} -2 & 1 & -3 & -4 \\ -6 & 3 & -9 & -12 \\ 4 & -2 & 6 & 8 \\ -8 & 4 & -12 & -16 \end{bmatrix}$$

17. AB does not exist since the number of columns of A, 1, does not equal the number of rows of B, 3.

$$BA = \begin{bmatrix} 3 & -2 \\ 1 & 0 \\ -1 & 4 \end{bmatrix}\begin{bmatrix} 2 \\ -7 \end{bmatrix} = \begin{bmatrix} 3(2) + (-2)(-7) \\ 1(2) + 0(-7) \\ -1(2) + 4(-7) \end{bmatrix} = \begin{bmatrix} 20 \\ 2 \\ -30 \end{bmatrix}$$

19. $$AB = \begin{bmatrix} 3 \\ -4 \\ 2 \end{bmatrix}[3 \quad -4] = \begin{bmatrix} 3(3) & 3(-4) \\ -4(3) & -4(-4) \\ 2(3) & 2(-4) \end{bmatrix} = \begin{bmatrix} 9 & -12 \\ -12 & 16 \\ 6 & -8 \end{bmatrix}$$

BA does not exist.

21. $$\left[\begin{array}{cc|cc} 1 & 3 & 1 & 0 \\ 4 & 2 & 0 & 1 \end{array}\right]$$

-4(row 1) + (row 2) replace (row 2)

$$\left[\begin{array}{cc|cc} 1 & 3 & 1 & 0 \\ 0 & -10 & -4 & 1 \end{array}\right]$$

$-\frac{1}{10}$(row 2) replace (row 2)

$$\left[\begin{array}{cc|cc} 1 & 3 & 1 & 0 \\ 0 & 1 & \frac{2}{5} & -\frac{1}{10} \end{array}\right]$$

21. Continued

-3(row 2) + (row 1) replace (row 1)

$$\left[\begin{array}{cc|cc} 1 & 0 & -\frac{1}{5} & \frac{3}{10} \\ 0 & 1 & \frac{2}{5} & -\frac{1}{10} \end{array}\right]$$

Multiplicative inverse: $\begin{bmatrix} -\frac{1}{5} & \frac{3}{10} \\ \frac{2}{5} & -\frac{1}{10} \end{bmatrix}$

23. $$\left[\begin{array}{cc|cc} 2 & 1 & 1 & 0 \\ 7 & 4 & 0 & 1 \end{array}\right]$$

$-\frac{7}{2}$(row 1) + (row 2) replace (row 2)

$$\left[\begin{array}{cc|cc} 2 & 1 & 1 & 0 \\ 0 & \frac{1}{2} & -\frac{7}{2} & 1 \end{array}\right]$$

$\frac{1}{2}$(row 1) replace (row 1)

$$\left[\begin{array}{cc|cc} 1 & \frac{1}{2} & \frac{1}{2} & 0 \\ 0 & \frac{1}{2} & -\frac{7}{2} & 1 \end{array}\right]$$

$-$ (row 2) + (row 1) replace (row 1)

$$\left[\begin{array}{cc|cc} 1 & 0 & 4 & -1 \\ 0 & \frac{1}{2} & -\frac{7}{2} & 1 \end{array}\right]$$

2(row 2) replace (row 2)

$$\left[\begin{array}{cc|cc} 1 & 0 & 4 & -1 \\ 0 & 1 & -7 & 2 \end{array}\right]$$

$$A^{-1} = \begin{bmatrix} 4 & -1 \\ -7 & 2 \end{bmatrix}$$

25. $$\left[\begin{array}{cc|cc} -2 & 1 & 1 & 0 \\ 3 & -4 & 0 & 1 \end{array}\right]$$

(row 1) + (row 2) replace (row 1)

$$\left[\begin{array}{cc|cc} 1 & -3 & 1 & 1 \\ 3 & -4 & 0 & 1 \end{array}\right]$$

$-$ 3(row 1) + (row 2) replace (row 2)

$$\left[\begin{array}{cc|cc} 1 & -3 & 1 & 1 \\ 0 & 5 & -3 & -2 \end{array}\right]$$

$\frac{1}{5}$(row 2) replace (row 2)

$$\left[\begin{array}{cc|cc} 1 & -3 & 1 & 1 \\ 0 & 1 & -\frac{3}{5} & -\frac{2}{5} \end{array}\right]$$

3(row 2) + (row 1) replace (row 1)

$$\left[\begin{array}{cc|cc} 1 & 0 & -\frac{4}{5} & -\frac{1}{5} \\ 0 & 1 & -\frac{3}{5} & -\frac{2}{5} \end{array}\right]$$

Multiplicative inverse: $\begin{bmatrix} -\frac{4}{5} & -\frac{1}{5} \\ -\frac{3}{5} & -\frac{2}{5} \end{bmatrix}$

27. $$\left[\begin{array}{ccc|ccc} 1 & 2 & 3 & 1 & 0 & 0 \\ 1 & 3 & 4 & 0 & 1 & 0 \\ 1 & 4 & 3 & 0 & 0 & 1 \end{array}\right]$$

$-$ (row 1) + (row 2) replace (row 2)
$-$ (row 1) + (row 3) replace (row 3)

$$\left[\begin{array}{ccc|ccc} 1 & 2 & 3 & 1 & 0 & 0 \\ 0 & 1 & 1 & -1 & 1 & 0 \\ 0 & 2 & 0 & -1 & 0 & 1 \end{array}\right]$$

$-$ 2(row 2) + (row 1) replace (row 1)
$-$ 2(row 2) + (row 3) replace (row 3)

$$\left[\begin{array}{ccc|ccc} 1 & 0 & 1 & 3 & -2 & 0 \\ 0 & 1 & 1 & -1 & 1 & 0 \\ 0 & 0 & -2 & 1 & -2 & 1 \end{array}\right]$$

$-\frac{1}{2}$(row 3) replace (row 3)

$$\left[\begin{array}{ccc|ccc} 1 & 0 & 1 & 3 & -2 & 0 \\ 0 & 1 & 1 & -1 & 1 & 0 \\ 0 & 0 & 1 & -\frac{1}{2} & 1 & -\frac{1}{2} \end{array}\right]$$

$-$ (row 3) + (row 2) replace (row 2)
$-$ (row 3) + (row 1) replace (row 1)

$$\left[\begin{array}{ccc|ccc} 1 & 0 & 0 & \frac{7}{2} & -3 & \frac{1}{2} \\ 0 & 1 & 0 & -\frac{1}{2} & 0 & \frac{1}{2} \\ 0 & 0 & 1 & -\frac{1}{2} & 1 & -\frac{1}{2} \end{array}\right]$$

$$A^{-1} = \begin{bmatrix} \frac{7}{2} & -3 & \frac{1}{2} \\ -\frac{1}{2} & 0 & \frac{1}{2} \\ -\frac{1}{2} & 1 & -\frac{1}{2} \end{bmatrix}$$

29. $$\left[\begin{array}{ccc|ccc} 1 & -2 & 1 & 1 & 0 & 0 \\ -2 & 5 & 3 & 0 & 1 & 0 \\ 3 & -5 & 7 & 0 & 0 & 1 \end{array}\right]$$

2(row 1) + (row 2) replace (row 2)
− 3(row 1) + (row 3) replace (row 3)

$$\left[\begin{array}{ccc|ccc} 1 & -2 & 1 & 1 & 0 & 0 \\ 0 & 1 & 5 & 2 & 1 & 0 \\ 0 & 1 & 4 & -3 & 0 & 1 \end{array}\right]$$

2(row 2) + (row 1) replace (row 1)
(row 2) − (row 3) replace (row 3)

$$\left[\begin{array}{ccc|ccc} 1 & 0 & 11 & 5 & 2 & 0 \\ 0 & 1 & 5 & 2 & 1 & 0 \\ 0 & 0 & 1 & 5 & 1 & -1 \end{array}\right]$$

− 11(row 3) + (row 1) replace (row 1)
− 5(row 3) + (row 2) replace (row 2)

$$\left[\begin{array}{ccc|ccc} 1 & 0 & 0 & -50 & -9 & 11 \\ 0 & 1 & 0 & -23 & -4 & 5 \\ 0 & 0 & 1 & 5 & 1 & -1 \end{array}\right]$$

Multiplicative inverse: $\begin{bmatrix} -50 & -9 & 11 \\ -23 & -4 & 5 \\ 5 & 1 & -1 \end{bmatrix}$

31. $$\left[\begin{array}{ccc|ccc} 2 & 3 & -4 & 1 & 0 & 0 \\ 3 & -1 & -2 & 0 & 1 & 0 \\ 1 & -4 & 2 & 0 & 0 & 1 \end{array}\right]$$

$\frac{1}{2}$(row 1) replace (row 1)

$$\left[\begin{array}{ccc|ccc} 1 & \frac{3}{2} & -2 & \frac{1}{2} & 0 & 0 \\ 3 & -1 & -2 & 0 & 1 & 0 \\ 1 & -4 & 2 & 0 & 0 & 1 \end{array}\right]$$

− 3(row 1) + (row 2) replace (row 2)
− (row 1) + (row 3) replace (row 3)

$$\left[\begin{array}{ccc|ccc} 1 & \frac{3}{2} & -2 & \frac{1}{2} & 0 & 0 \\ 0 & -\frac{11}{2} & 4 & -\frac{3}{2} & 1 & 0 \\ 0 & -\frac{11}{2} & 4 & -\frac{1}{2} & 0 & 1 \end{array}\right]$$

− (row 2) + (row 3) replace (row 3)

$$\left[\begin{array}{ccc|ccc} 1 & \frac{3}{2} & -2 & \frac{1}{2} & 0 & 0 \\ 0 & -\frac{11}{2} & 4 & -\frac{3}{2} & 1 & 0 \\ 0 & 0 & 0 & 1 & -1 & 1 \end{array}\right]$$

Because row 3 has 0, 0, 0, the inverse does not exist.

33. $$\left[\begin{array}{ccc|ccc} 1 & 2 & 3 & 1 & 0 & 0 \\ -3 & -4 & 3 & 0 & 1 & 0 \\ 2 & 4 & -1 & 0 & 0 & 1 \end{array}\right]$$

3(row 1) + (row 2) replace (row 2)
− 2(row 1) + (row 3) replace (row 3)

$$\left[\begin{array}{ccc|ccc} 1 & 2 & 3 & 1 & 0 & 0 \\ 0 & 2 & 12 & 3 & 1 & 0 \\ 0 & 0 & -7 & -2 & 0 & 1 \end{array}\right]$$

$\frac{1}{2}$(row 2) replace (row 2)
$-\frac{1}{7}$(row 3) replace (row 3)

$$\left[\begin{array}{ccc|ccc} 1 & 2 & 3 & 1 & 0 & 0 \\ 0 & 1 & 6 & \frac{3}{2} & \frac{1}{2} & 0 \\ 0 & 0 & 1 & \frac{2}{7} & 0 & -\frac{1}{7} \end{array}\right]$$

− 2(row 2) + (row 1) replace (row 1)
− 6(row 3) + (row 2) replace (row 2)

$$\left[\begin{array}{ccc|ccc} 1 & 0 & -9 & -2 & -1 & 0 \\ 0 & 1 & 0 & -\frac{3}{14} & \frac{1}{2} & \frac{6}{7} \\ 0 & 0 & 1 & \frac{2}{7} & 0 & -\frac{1}{7} \end{array}\right]$$

9(row 3) + (row 1) replace (row 1)

$$\left[\begin{array}{ccc|ccc} 1 & 0 & 0 & \frac{4}{7} & -1 & -\frac{9}{7} \\ 0 & 1 & 0 & -\frac{3}{14} & \frac{1}{2} & \frac{6}{7} \\ 0 & 0 & 1 & \frac{2}{7} & 0 & -\frac{1}{7} \end{array}\right]$$

Multiplicative inverse: $\begin{bmatrix} \frac{4}{7} & -1 & -\frac{9}{7} \\ -\frac{3}{14} & \frac{1}{2} & \frac{6}{7} \\ \frac{2}{7} & 0 & -\frac{1}{7} \end{bmatrix}$

35. $$\left[\begin{array}{ccc|ccc} 2 & 0 & 0 & 1 & 0 & 0 \\ 0 & 4 & 0 & 0 & 1 & 0 \\ 0 & 0 & 10 & 0 & 0 & 1 \end{array}\right]$$

$\frac{1}{2}$(row 1) replace (row 1)

$\frac{1}{4}$(row 2) replace (row 2)

$\frac{1}{10}$(row 3) replace (row 3)

$$\left[\begin{array}{ccc|ccc} 1 & 0 & 0 & \frac{1}{2} & 0 & 0 \\ 0 & 1 & 0 & 0 & \frac{1}{4} & 0 \\ 0 & 0 & 1 & 0 & 0 & \frac{1}{10} \end{array}\right]$$

$$A^{-1} = \begin{bmatrix} \frac{1}{2} & 0 & 0 \\ 0 & \frac{1}{4} & 0 \\ 0 & 0 & \frac{1}{10} \end{bmatrix}$$

37. $A = \begin{bmatrix} 2 & 1 \\ 7 & 4 \end{bmatrix} \qquad B = \begin{bmatrix} -4 \\ -13 \end{bmatrix}$

$A^{-1} = \begin{bmatrix} 4 & -1 \\ -7 & 2 \end{bmatrix}$ (from problem #23)

$$A^{-1}B = \begin{bmatrix} 4 & -1 \\ -7 & 2 \end{bmatrix}\begin{bmatrix} -4 \\ -13 \end{bmatrix} = \begin{bmatrix} -3 \\ 2 \end{bmatrix}$$

The solution set is $\{(-3, 2)\}$.

39. $A = \begin{bmatrix} -2 & 1 \\ 3 & -4 \end{bmatrix} \qquad B = \begin{bmatrix} 1 \\ -14 \end{bmatrix}$

$A^{-1} = \begin{bmatrix} -\frac{4}{5} & -\frac{1}{5} \\ -\frac{3}{5} & -\frac{2}{5} \end{bmatrix}$ (from problem #25)

$$A^{-1}B = \begin{bmatrix} -\frac{4}{5} & -\frac{1}{5} \\ -\frac{3}{5} & -\frac{2}{5} \end{bmatrix}\begin{bmatrix} 1 \\ -14 \end{bmatrix}$$

$$A^{-1}B = \begin{bmatrix} -\frac{4}{5}(1) - \frac{1}{5}(-14) \\ -\frac{3}{5}(1) - \frac{2}{5}(-14) \end{bmatrix}$$

$$A^{-1}B = \begin{bmatrix} 2 \\ 5 \end{bmatrix}$$

The solution set is $\{(2, 5)\}$.

41. $A = \begin{bmatrix} 1 & 2 & 3 \\ 1 & 3 & 4 \\ 1 & 4 & 3 \end{bmatrix} \qquad B = \begin{bmatrix} -2 \\ -3 \\ -6 \end{bmatrix}$

$A^{-1} = \begin{bmatrix} \frac{7}{2} & -3 & \frac{1}{2} \\ -\frac{1}{2} & 0 & \frac{1}{2} \\ -\frac{1}{2} & 1 & -\frac{1}{2} \end{bmatrix}$ (from problem #27)

$$A^{-1}B = \begin{bmatrix} \frac{7}{2} & -3 & \frac{1}{2} \\ -\frac{1}{2} & 0 & \frac{1}{2} \\ -\frac{1}{2} & 1 & -\frac{1}{2} \end{bmatrix}\begin{bmatrix} -2 \\ -3 \\ -6 \end{bmatrix}$$

$$A^{-1}B = \begin{bmatrix} -1 \\ -2 \\ 1 \end{bmatrix}$$

The solution set is $\{(-1, -2, 1)\}$.

43. A^{-1} is found in problem #29.

$$A^{-1}B = \begin{bmatrix} -50 & -9 & 11 \\ -23 & -4 & 5 \\ 5 & 1 & -1 \end{bmatrix}\begin{bmatrix} -3 \\ 34 \\ 14 \end{bmatrix}$$

$$A^{-1}B = \begin{bmatrix} -50(-3) - 9(34) + 11(14) \\ -23(-3) - 4(34) + 5(14) \\ 5(-3) + 1(34) - 1(14) \end{bmatrix}$$

$$A^{-1}B = \begin{bmatrix} -2 \\ 3 \\ 5 \end{bmatrix}$$

The solution set is $\{(-2, 3, 5)\}$.

45. A^{-1} is found in problem #33.

$$A^{-1}B = \begin{bmatrix} \frac{4}{7} & -1 & -\frac{9}{7} \\ -\frac{3}{14} & \frac{1}{2} & \frac{6}{7} \\ \frac{2}{7} & 0 & -\frac{1}{7} \end{bmatrix} \begin{bmatrix} 2 \\ 0 \\ 4 \end{bmatrix}$$

$$A^{-1}B = \begin{bmatrix} -4 \\ 3 \\ 0 \end{bmatrix}$$

The solution set is $\{(-4, 3, 0)\}$.

47a. $$A^{-1}B = \begin{bmatrix} -\frac{5}{24} & \frac{1}{6} & -\frac{7}{24} \\ \frac{7}{24} & \frac{1}{6} & \frac{5}{24} \\ \frac{11}{24} & -\frac{1}{6} & \frac{1}{24} \end{bmatrix} \begin{bmatrix} 7 \\ 1 \\ -1 \end{bmatrix}$$

$$A^{-1}B = \begin{bmatrix} -\frac{5}{24}(7) + \frac{1}{6}(1) - \frac{7}{24}(-1) \\ \frac{7}{24}(7) + \frac{1}{6}(1) + \frac{5}{24}(-1) \\ \frac{11}{24}(7) - \frac{1}{6}(1) + \frac{1}{24}(-1) \end{bmatrix}$$

$$A^{-1}B = \begin{bmatrix} -1 \\ 2 \\ 3 \end{bmatrix}$$

$A^{-1}B = \{(-1, 2, 3)\}$
The solution set is $\{(-1, 2, 3)\}$.

b. $$A^{-1}B = \begin{bmatrix} -\frac{5}{24} & \frac{1}{6} & -\frac{7}{24} \\ \frac{7}{24} & \frac{1}{6} & \frac{5}{24} \\ \frac{11}{24} & -\frac{1}{6} & \frac{1}{24} \end{bmatrix} \begin{bmatrix} -7 \\ 5 \\ 1 \end{bmatrix}$$

$$A^{-1}B = \begin{bmatrix} -\frac{5}{24}(-7) + \frac{1}{6}(5) - \frac{7}{24}(1) \\ \frac{7}{24}(-7) + \frac{1}{6}(5) + \frac{5}{24}(1) \\ \frac{11}{24}(-7) - \frac{1}{6}(5) + \frac{1}{24}(1) \end{bmatrix}$$

$$A^{-1}B = \begin{bmatrix} 2 \\ -1 \\ -4 \end{bmatrix}$$

$A^{-1}B = \{(2, -1, -4)\}$
The solution set is $\{(2, -1, -4)\}$.

c. $$A^{-1}B = \begin{bmatrix} -\frac{5}{24} & \frac{1}{6} & -\frac{7}{24} \\ \frac{7}{24} & \frac{1}{6} & \frac{5}{24} \\ \frac{11}{24} & -\frac{1}{6} & \frac{1}{24} \end{bmatrix} \begin{bmatrix} -9 \\ -8 \\ 19 \end{bmatrix}$$

$$A^{-1}B = \begin{bmatrix} -\frac{5}{24}(-9)+\frac{1}{6}(-8) - \frac{7}{24}(19) \\ \frac{7}{24}(-9)+\frac{1}{6}(-8)+\frac{5}{24}(19) \\ \frac{11}{24}(-9) - \frac{1}{6}(-8)+\frac{1}{24}(19) \end{bmatrix}$$

$$A^{-1}B = \begin{bmatrix} -5 \\ 0 \\ -2 \end{bmatrix}$$

$A^{-1}B = \{(-5, 0, -2)\}$
The solution set is $\{(-5, 0, -2)\}$.

d. $$A^{-1}B = \begin{bmatrix} -\frac{5}{24} & \frac{1}{6} & -\frac{7}{24} \\ \frac{7}{24} & \frac{1}{6} & \frac{5}{24} \\ \frac{11}{24} & -\frac{1}{6} & \frac{1}{24} \end{bmatrix} \begin{bmatrix} -1 \\ -13 \\ -17 \end{bmatrix}$$

$$A^{-1}B = \begin{bmatrix} -\frac{5}{24}(-1) + \frac{1}{6}(-13) - \frac{7}{24}(-17) \\ \frac{7}{24}(-1) + \frac{1}{6}(-13) + \frac{5}{24}(-17) \\ \frac{11}{24}(-1) - \frac{1}{6}(-13) + \frac{1}{24}(-17) \end{bmatrix}$$

$$A^{-1}B = \begin{bmatrix} 3 \\ -6 \\ 1 \end{bmatrix}$$

$A^{-1}B = \{(3, -6, 1)\}$
The solution set is $\{(3, -6, 1)\}$.

e. $A^{-1}B = \begin{bmatrix} -\frac{5}{24} & \frac{1}{6} & -\frac{7}{24} \\ \frac{7}{24} & \frac{1}{6} & \frac{5}{24} \\ \frac{11}{24} & -\frac{1}{6} & \frac{1}{24} \end{bmatrix} \begin{bmatrix} -2 \\ 0 \\ -2 \end{bmatrix}$

$A^{-1}B = \begin{bmatrix} -\frac{5}{24}(-2)+\frac{1}{6}(0)-\frac{7}{24}(-2) \\ \frac{7}{24}(-2)+\frac{1}{6}(0)+\frac{5}{24}(-2) \\ \frac{11}{24}(-2)-\frac{1}{6}(0)+\frac{1}{24}(-2) \end{bmatrix}$

$A^{-1}B = \begin{bmatrix} 1 \\ -1 \\ -1 \end{bmatrix}$

$A^{-1}B = \{(1, -1, -1)\}$

The solution set is $\{(1, -1, -1)\}$.

Further Investigations

51a. $\begin{bmatrix} 2 & -3 \\ -1 & 2 \end{bmatrix} \begin{bmatrix} 68 & 77 & 78 & 23 & 29 & 85 & 41 \\ 40 & 51 & 49 & 15 & 19 & 52 & 27 \end{bmatrix}$

$\begin{bmatrix} 16 & 1 & 9 & 1 & 1 & 14 & 1 \\ 12 & 25 & 20 & 7 & 9 & 19 & 13 \end{bmatrix}$

16 12 1 25 9 20 1 7 1 9 14 19 1 13
P L A Y I T A G A I N S A M

b. $\begin{bmatrix} 2 & -3 \\ -1 & 2 \end{bmatrix} \begin{bmatrix} 62 & 78 & 64 & 19 & 93 & 93 & 88 \\ 40 & 47 & 36 & 11 & 57 & 56 & 57 \end{bmatrix} = \begin{bmatrix} 4 & 15 & 20 & 5 & 15 & 18 & 5 \\ 18 & 16 & 8 & 3 & 21 & 19 & 26 \end{bmatrix}$

4 18 15 16 20 8 5 3 15 21 18 19 5 26
D R O P T H E C O U R S E Z

c. $\begin{bmatrix} 2 & -3 \\ -1 & 2 \end{bmatrix} \begin{bmatrix} 64 & 58 & 63 & 21 & 75 & 63 & 38 & 118 \\ 36 & 37 & 36 & 13 & 47 & 36 & 23 & 72 \end{bmatrix} = \begin{bmatrix} 20 & 5 & 18 & 3 & 9 & 18 & 7 & 20 \\ 8 & 16 & 9 & 5 & 19 & 9 & 8 & 26 \end{bmatrix}$

20 8 5 16 18 9 3 5 9 19 18 9 7 8 20 26
T H E P R I C E I S R I G H T Z

d. $\begin{bmatrix} 2 & -3 \\ -1 & 2 \end{bmatrix} \begin{bmatrix} 61 & 115 & 93 & 36 & 78 & 68 & 77 & 60 & 47 & 84 & 21 \\ 38 & 69 & 57 & 20 & 49 & 40 & 51 & 37 & 26 & 51 & 11 \end{bmatrix} =$

$\begin{bmatrix} 8 & 23 & 15 & 12 & 9 & 16 & 1 & 9 & 16 & 15 & 9 \\ 15 & 23 & 21 & 4 & 20 & 12 & 25 & 14 & 5 & 18 & 1 \end{bmatrix}$

8 15 23 23 15 21 12 4 9 20 16 12 1 25 9 14 16 5 15 18 9 1
H O W W O U L D I T P L A Y I N P E O R I A

PROBLEM SET 12.4 Systems of Linear Inequalities: Linear Programming

1. $\begin{pmatrix} x+y>3 \\ x-y>1 \end{pmatrix}$

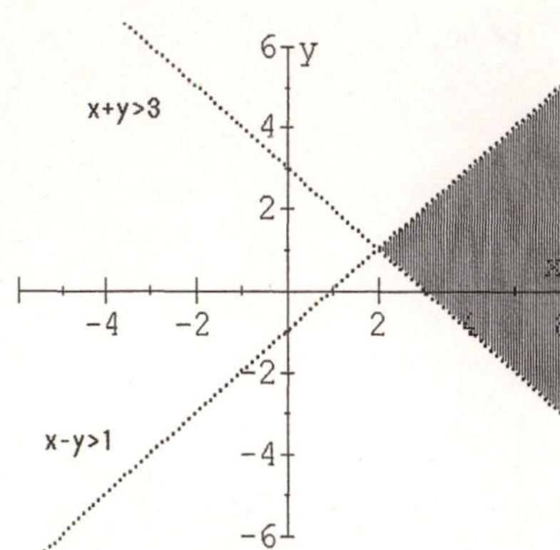

3. $\begin{pmatrix} x-2y\le 4 \\ x+2y>4 \end{pmatrix}$

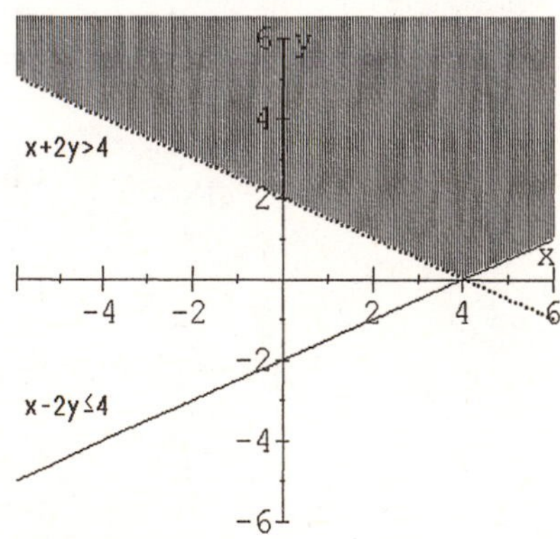

5. $\begin{pmatrix} 2x+3y\le 6 \\ 3x-2y\le 6 \end{pmatrix}$

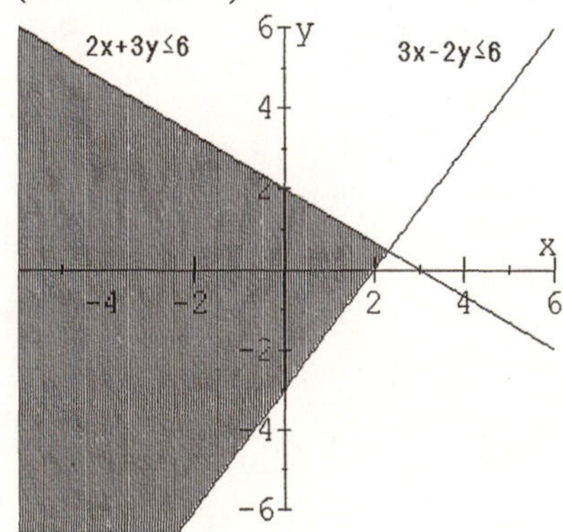

7. $\begin{pmatrix} 2x-y\ge 4 \\ x+3y<3 \end{pmatrix}$

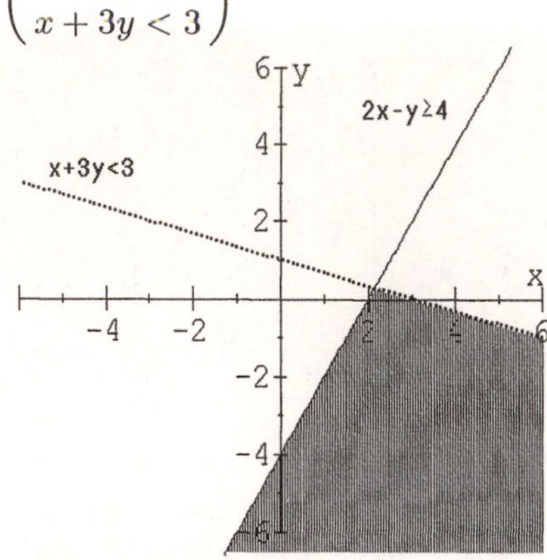

9. $\begin{pmatrix} x+2y> -2 \\ x-y< -3 \end{pmatrix}$

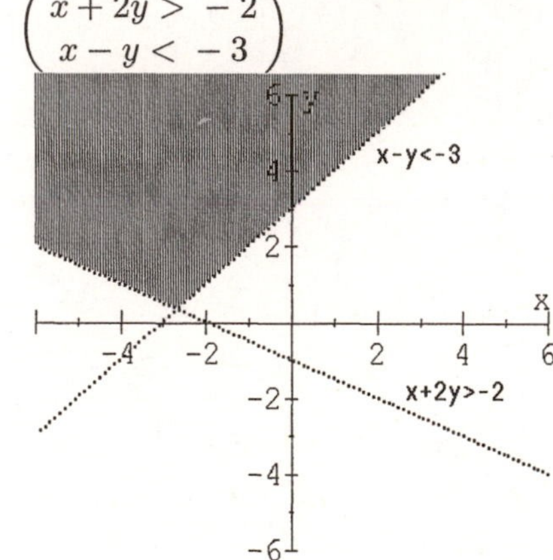

11. $\begin{pmatrix} y>x-4 \\ y<x \end{pmatrix}$

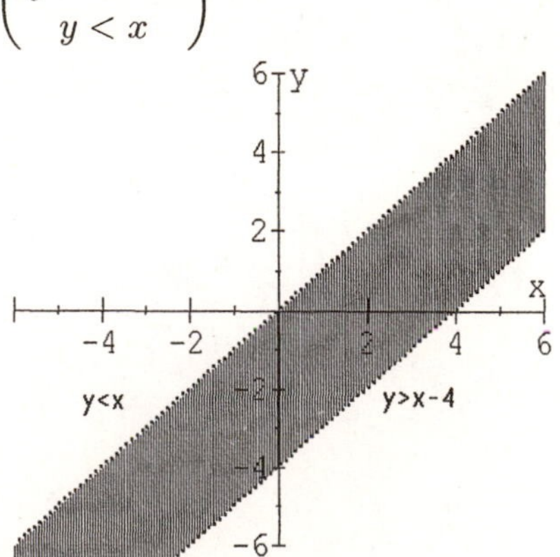

13. $\begin{pmatrix} x - y > 2 \\ x - y > -1 \end{pmatrix}$

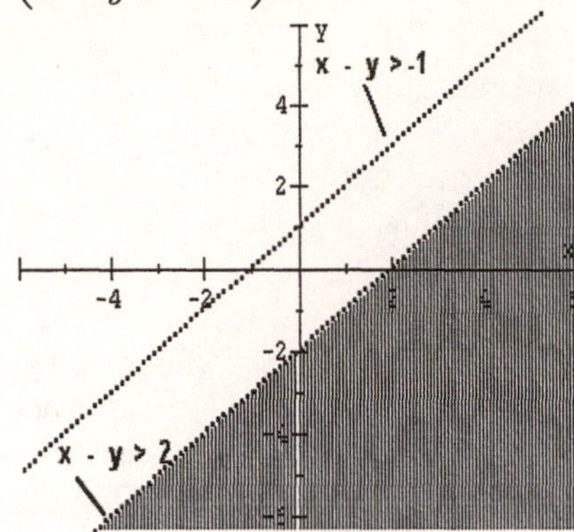

15. $\begin{pmatrix} y \geq x \\ x > -1 \end{pmatrix}$

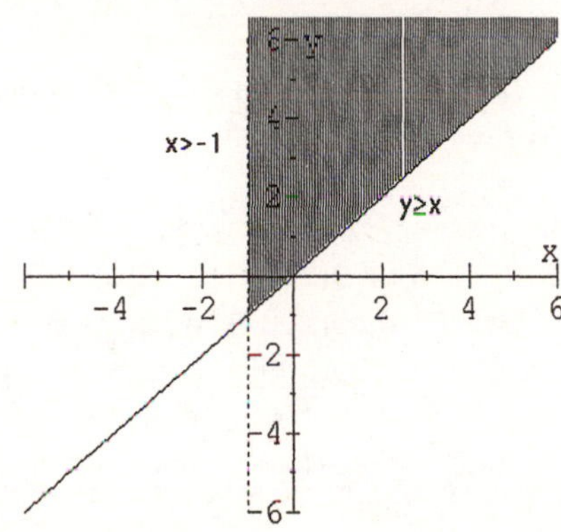

17. $\begin{pmatrix} y < x \\ y > x + 3 \end{pmatrix}$

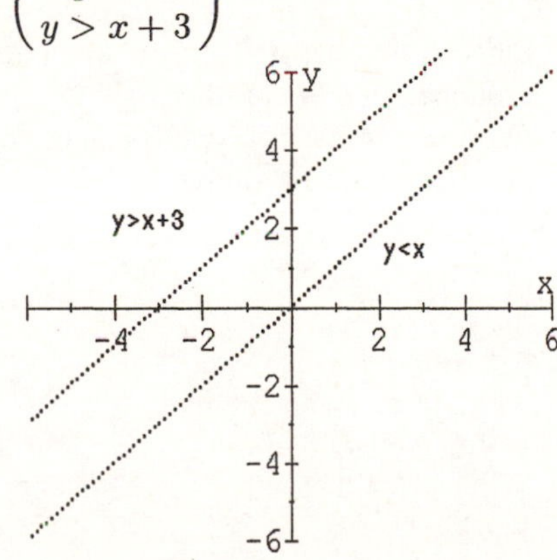

19. $\begin{pmatrix} y > -2 \\ x > 1 \end{pmatrix}$

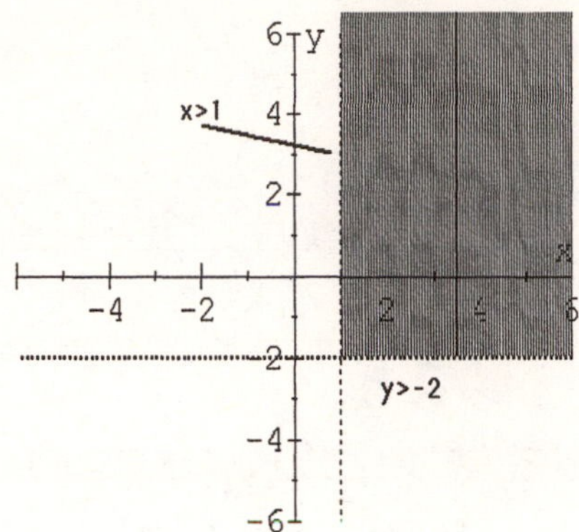

21. $\begin{pmatrix} x \geq 0 \\ y \geq 0 \\ x + y \leq 4 \\ 2x + y \leq 6 \end{pmatrix}$

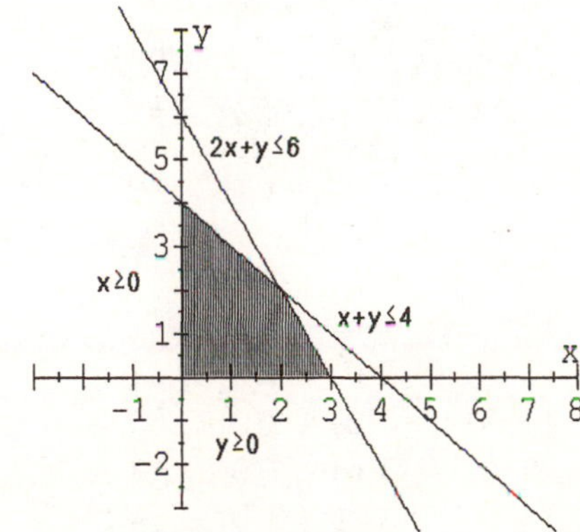

23. $\begin{pmatrix} x \geq 0 \\ y \geq 0 \\ 2x + y \leq 4 \\ 2x - 3y \leq 6 \end{pmatrix}$

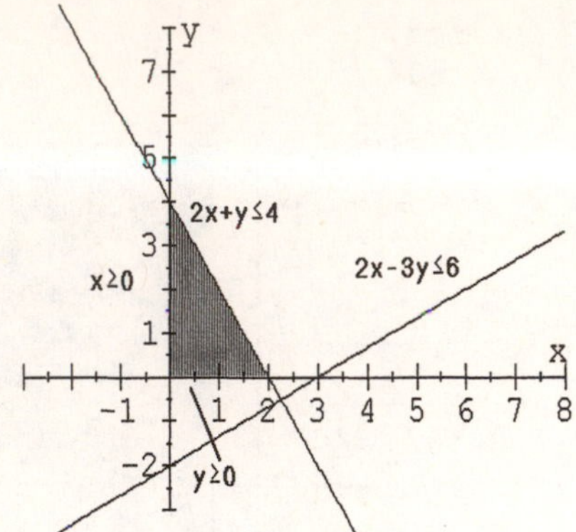

25.

vertices	$f(x, y) = 3x + 5y$
(1, 1)	$3(1) + 5(1) = 8$
(2, 4)	$3(2) + 5(4) = 26$
(4, 8)	$3(4) + 5(8) = 52$
(5, 2)	$3(5) + 5(2) = 25$

The minimum is 8 and the maximum is 52.

27.

vertices	$f(x, y) = x + 4y$
(0, 0)	$(0) + 4(0) = 0$
(0, 7)	$(0) + 4(7) = 28$
(5, 4)	$(5) + 4(4) = 21$
(6, 2)	$(6) + 4(2) = 14$

The minimum is 0 and the maximum is 28.

29.

vertices	$f(x, y) = 3x + 7y$
(0, 3)	$3(0) + 7(3) = 21$
(0, 9)	$3(0) + 7(9) = 63$
(6, 0)	$3(6) + 7(0) = 18$
(4, 0)	$3(4) + 7(0) = 25$

The maximum is 63.

31.

vertices	$f(x, y) = 40x + 55y$
(0, 0)	$40(0) + 55(0) = 0$
(0, 6)	$40(0) + 55(6) = 330$
(3, 4)	$40(3) + 55(4) = 340$
(5, 0)	$40(5) + 55(0) = 200$

The maximum at (3, 4) is $f(3, 4) = 340$.

33.

vertices	$f(x, y) = 0.2x + 0.5y$
(0, 12)	$0.2(0) + 0.5(12) = 6$
(5, 2)	$0.2(5) + 0.5(2) = 2$
(10, 0)	$0.2(10) + 0.5(0) = 2$

The minimum is 2.

35.

vertices	$f(x, y) = 9x + 2y$
(0, 0)	$9(0) + 2(0) = 0$
(0, 4)	$9(0) + 2(4) = 8$
(5, 8)	$9(5) + 2(8) = 61$
(10, 4)	$9(10) + 2(4) = 98$
(10, 0)	$9(10) + 2(0) = 90$

The maximum at (10, 4) is $f(10, 4) = 98$.

37. Let x represent the amount invested in the conservative stock and y the amount invested in the speculative stock.

$x + y \leq 10,000$
$x \geq 2000$
$y \leq 6000$
$y \leq x$
$x \geq 0$
$y \geq 0$

$$\begin{matrix}\text{Total}\\ \text{return}\end{matrix} = \begin{matrix}\text{conservative}\\ \text{return}\end{matrix} + \begin{matrix}\text{speculative}\\ \text{return}\end{matrix}$$

vertices	$f(x, y) = 0.09x + 0.12y$
(2000, 0)	0.09(2000)+0.12(0) = 180
(2000, 2000)	0.09(2000)+0.12(2000) = 420
(5000, 5000)	0.09(5000)+0.12(5000) = 1050
(10000, 0)	0.09(10000)+0.12(0) = 900

The maximum is 1050.
Thus, \$5000 should be invested at 9% and \$5000 at 12%.

39. Let x = number of Type A calculators produced in one month.
Let y = number of Type B calculators produced in one month.

$f(x, y) = 12x + 10y - (9x + 8y)$
$f(x, y) = 3x + 2y$
$x + y \leq 500$
$200 \leq x \leq 300$
$100 \leq y \leq 250$

vertices	$f(x, y) = 3x + 2y$
(200, 100)	$3(200) + 2(100) = 800$
(200, 250)	$3(200) + 2(250) = 1100$
(250, 250)	$3(250) + 2(250) = 1250$
(300, 200)	$3(300) + 2(200) = 1300$
(300, 100)	$3(300) + 2(100) = 1100$

There are 300 Type A calculators and 200 Type B calculators that should be produced in one month.

41. Let x be the number of units of product A and y the number of units of product B.

$x + y \leq 40$ Machine I is available for no more than 40 hours.

$2x + y \leq 40$ Machine II is available for no more than 40 hours

$x + 3y \leq 60$ Machine III is available for no more than 60 hours

$x \geq 0$
$y \geq 0$

Total profit = profit from Product A + profit from Product B

vertices	$f(x, y) = 2.75x + 3.50y$
(0, 0)	$2.75(0) + 3.50(0) = 0$
(0, 20)	$2.75(0) + 3.50(20) = 70$
(12, 16)	$2.75(12) + 3.50(16) = 89$
(20, 0)	$2.75(20) + 3.50(0) = 55$

The maximum is 89.
12 units of Product A and 16 units of Product B should be produced.

CHAPTER 12 Review Problem Set

1. $A + B = \begin{bmatrix} 2+5 & -4-1 \\ -3+0 & 8+2 \end{bmatrix} = \begin{bmatrix} 7 & -5 \\ -3 & 10 \end{bmatrix}$

2. $B - A = \begin{bmatrix} 5-2 & -1-(-4) \\ 0-(-3) & 2-8 \end{bmatrix} = \begin{bmatrix} 3 & 3 \\ 3 & -6 \end{bmatrix}$

3. $C - F = \begin{bmatrix} 3-1 & -1-(-2) \\ -2-4 & 4-(-4) \\ 5-7 & -6-(-8) \end{bmatrix} = \begin{bmatrix} 2 & 1 \\ -6 & 8 \\ -2 & 2 \end{bmatrix}$

4. $2A = \begin{bmatrix} 4 & -8 \\ -6 & 16 \end{bmatrix} \qquad 3B = \begin{bmatrix} 15 & -3 \\ 0 & 6 \end{bmatrix}$

$$2A + 3B = \begin{bmatrix} 4+15 & -8-3 \\ -6+0 & 16+6 \end{bmatrix} = \begin{bmatrix} 19 & -11 \\ -6 & 22 \end{bmatrix}$$

5. $3C = \begin{bmatrix} 9 & -3 \\ -6 & 12 \\ 15 & -18 \end{bmatrix} \qquad -2F = \begin{bmatrix} -2 & 4 \\ -8 & 8 \\ -14 & 16 \end{bmatrix}$

$$3C - 2F = \begin{bmatrix} 9-2 & -3+4 \\ -6-8 & 12+8 \\ 15-14 & -18+16 \end{bmatrix} = \begin{bmatrix} 7 & 1 \\ -14 & 20 \\ 1 & -2 \end{bmatrix}$$

6. $$CD = \begin{bmatrix} 3(-2)-1(5) & 3(-1)-1(0) & 3(4)-1(-3) \\ -2(-2)+4(5) & -2(-1)+4(0) & -2(4)+4(-3) \\ 5(-2)-6(5) & 5(-1)-6(0) & 5(4)-6(-3) \end{bmatrix} = \begin{bmatrix} -11 & -3 & 15 \\ 24 & 2 & -20 \\ -40 & -5 & 38 \end{bmatrix}$$

7. $$DC = \begin{bmatrix} -2 & -1 & 4 \\ 5 & 0 & -3 \end{bmatrix}\begin{bmatrix} 3 & -1 \\ -2 & 4 \\ 5 & -6 \end{bmatrix}$$

$$DC = \begin{bmatrix} -2(3)-1(-2)+4(5) & -2(-1)-1(4)+4(-6) \\ 5(3)+0(-2)-3(5) & 5(-1)+0(4)-3(-6) \end{bmatrix} = \begin{bmatrix} 16 & -26 \\ 0 & 13 \end{bmatrix}$$

8. $$DC = \begin{bmatrix} -2(3)-1(-2)+4(5) & -2(-1)-1(4)+4(-6) \\ 5(3)-3(5) & 5(-1)-3(-6) \end{bmatrix} = \begin{bmatrix} 16 & -26 \\ 0 & 13 \end{bmatrix}$$

$$AB = \begin{bmatrix} 2(5) & 2(-1)-4(2) \\ -3(5) & -3(-1)+8(2) \end{bmatrix} = \begin{bmatrix} 10 & -10 \\ -15 & 19 \end{bmatrix}$$

$$DC + AB = \begin{bmatrix} 16+10 & -26-10 \\ 0-15 & 13+19 \end{bmatrix} = \begin{bmatrix} 26 & -36 \\ -15 & 32 \end{bmatrix}$$

9. $$DE = \begin{bmatrix} -2 & -1 & 4 \\ 5 & 0 & -3 \end{bmatrix}\begin{bmatrix} 1 \\ -3 \\ -7 \end{bmatrix} = \begin{bmatrix} -2(1)-1(-3)+4(-7) \\ 5(1)+0(-3)-3(-7) \end{bmatrix} = \begin{bmatrix} -27 \\ 26 \end{bmatrix}$$

10. EF does not exist.

11. $$AB = \begin{bmatrix} 2 & -4 \\ -3 & 8 \end{bmatrix}\begin{bmatrix} 5 & -1 \\ 0 & 2 \end{bmatrix} = \begin{bmatrix} 2(5)-4(0) & 2(-1)-4(2) \\ -3(5)+8(0) & -3(-1)+8(2) \end{bmatrix} = \begin{bmatrix} 10 & -10 \\ -15 & 19 \end{bmatrix}$$

$$BA = \begin{bmatrix} 5 & -1 \\ 0 & 2 \end{bmatrix}\begin{bmatrix} 2 & -4 \\ -3 & 8 \end{bmatrix} = \begin{bmatrix} 5(2)-1(-3) & 5(-4)-1(8) \\ 0(2)+2(-3) & 0(-4)+2(8) \end{bmatrix} = \begin{bmatrix} 13 & -28 \\ -6 & 16 \end{bmatrix}$$

Therefore AB $\neq$ BA.

12. $$C + F = \begin{bmatrix} 4 & -3 \\ 2 & 0 \\ 12 & -14 \end{bmatrix} \qquad D(C+F) = \begin{bmatrix} 38 & -50 \\ -16 & 27 \end{bmatrix}$$

$$DC = \begin{bmatrix} 16 & -26 \\ 0 & 13 \end{bmatrix} \qquad DF = \begin{bmatrix} 22 & -24 \\ -16 & 14 \end{bmatrix} \qquad DC + DF = \begin{bmatrix} 38 & -50 \\ -16 & 27 \end{bmatrix}$$

Therefore, D(C + F) = DC + DF

13. $$C + F = \begin{bmatrix} 3+1 & -1-2 \\ -2+4 & 4-4 \\ 5+7 & -6-8 \end{bmatrix} = \begin{bmatrix} 4 & -3 \\ 2 & 0 \\ 12 & -14 \end{bmatrix}$$

$$(C+F)D = \begin{bmatrix} 4 & -3 \\ 2 & 0 \\ 12 & -14 \end{bmatrix}\begin{bmatrix} -2 & -1 & 4 \\ 5 & 0 & -3 \end{bmatrix}$$

$$(C+F)D = \begin{bmatrix} 4(-2)-3(5) & 4(-1)-3(0) & 4(4)-3(-3) \\ 2(-2)+0(5) & 2(-1)+0(0) & 2(4)+0(-3) \\ 12(-2)-14(5) & 12(-1)-14(0) & 12(4)-14(-3) \end{bmatrix} = \begin{bmatrix} -23 & -4 & 25 \\ -4 & -2 & 8 \\ -94 & -12 & 90 \end{bmatrix}$$

$$CD = \begin{bmatrix} 3 & -1 \\ -2 & 4 \\ 5 & -6 \end{bmatrix}\begin{bmatrix} -2 & -1 & 4 \\ 5 & 0 & -3 \end{bmatrix} = \begin{bmatrix} -11 & -3 & 15 \\ 24 & 2 & -20 \\ -40 & -5 & 38 \end{bmatrix} \text{(From \#6 in this section)}$$

$$FD = \begin{bmatrix} 1 & -2 \\ 4 & -4 \\ 7 & -8 \end{bmatrix}\begin{bmatrix} -2 & -1 & 4 \\ 5 & 0 & -3 \end{bmatrix}$$

$$FD = \begin{bmatrix} 1(-2)-2(5) & 1(-1)-2(0) & 1(4)-2(-3) \\ 4(-2)-4(5) & 4(-1)-4(0) & 4(4)-4(-3) \\ 7(-2)-8(5) & 7(-1)-8(0) & 7(4)-8(-3) \end{bmatrix} = \begin{bmatrix} -12 & -1 & 10 \\ -28 & -4 & 28 \\ -54 & -7 & 52 \end{bmatrix}$$

$$CD+FD = \begin{bmatrix} -11-12 & -3-1 & 15+10 \\ 24-28 & 2-4 & -20+28 \\ -40-54 & -5-7 & 38+52 \end{bmatrix} = \begin{bmatrix} -23 & -4 & 25 \\ -4 & -2 & 8 \\ -94 & -12 & 90 \end{bmatrix}$$

Therefore, (C + F)D = CD + FD

14. $D = \begin{vmatrix} 9 & 5 \\ 7 & 4 \end{vmatrix} = 36 - 35 = 1$

$$A^{-1} = \frac{1}{1}\begin{bmatrix} 4 & -5 \\ -7 & 9 \end{bmatrix} = \begin{bmatrix} 4 & -5 \\ -7 & 9 \end{bmatrix}$$

15. $|A| = 27 - 28 = -1$

$$A^{-1} = \frac{1}{-1}\begin{bmatrix} 3 & -4 \\ -7 & 9 \end{bmatrix} = \begin{bmatrix} -3 & 4 \\ 7 & -9 \end{bmatrix}$$

16. $D = \begin{vmatrix} -2 & 1 \\ 2 & 3 \end{vmatrix} = -6 - 2 = -8$

$$A^{-1} = \frac{1}{-8}\begin{bmatrix} 3 & -1 \\ -2 & -2 \end{bmatrix} = \begin{bmatrix} -\frac{3}{8} & \frac{1}{8} \\ \frac{1}{4} & \frac{1}{4} \end{bmatrix}$$

17. $|A| = -12 - (-12) = 0$
Inverse does not exist.

18. $D = \begin{vmatrix} -1 & -3 \\ -4 & -5 \end{vmatrix} = 5 - 12 = -7$

$$A^{-1} = \frac{1}{-7}\begin{bmatrix} -5 & 3 \\ 4 & -1 \end{bmatrix} = \begin{bmatrix} \frac{5}{7} & -\frac{3}{7} \\ -\frac{4}{7} & \frac{1}{7} \end{bmatrix}$$

19. $|A| = 0 - (-21) = 21$

$$A^{-1} = \frac{1}{21}\begin{bmatrix} 6 & 3 \\ -7 & 0 \end{bmatrix} = \begin{bmatrix} \frac{2}{7} & \frac{1}{7} \\ -\frac{1}{3} & 0 \end{bmatrix}$$

20. $\left[\begin{array}{ccc|ccc} 1 & -2 & 1 & 1 & 0 & 0 \\ 2 & -5 & 2 & 0 & 1 & 0 \\ -3 & 7 & 5 & 0 & 0 & 1 \end{array}\right]$

$-2(\text{row } 1) + (\text{row } 2)$ replace (row 2)
$3(\text{row } 1) + (\text{row } 3)$ replace (row 3)

$$\left[\begin{array}{ccc|ccc} 1 & -2 & 1 & 1 & 0 & 0 \\ 0 & -1 & 0 & -2 & 1 & 0 \\ 0 & 1 & 8 & 3 & 0 & 1 \end{array}\right]$$

$-(\text{row } 2)$ replace (row 2)

$$\left[\begin{array}{ccc|ccc} 1 & -2 & 1 & 1 & 0 & 0 \\ 0 & 1 & 0 & 2 & -1 & 0 \\ 0 & 1 & 8 & 3 & 0 & 1 \end{array}\right]$$

$2(\text{row } 2) + (\text{row } 1)$ replace (row 1)
$-(\text{row } 2) + (\text{row } 3)$ replace (row 3)

$$\left[\begin{array}{ccc|ccc} 1 & 0 & 1 & 5 & -2 & 0 \\ 0 & 1 & 0 & 2 & -1 & 0 \\ 0 & 0 & 8 & 1 & 1 & 1 \end{array}\right]$$

$\frac{1}{8}(\text{row } 3)$ replace (row 3)

$$\left[\begin{array}{ccc|ccc} 1 & 0 & 1 & 5 & -2 & 0 \\ 0 & 1 & 0 & 2 & -1 & 0 \\ 0 & 0 & 1 & \frac{1}{8} & \frac{1}{8} & \frac{1}{8} \end{array}\right]$$

$-(\text{row } 3) + (\text{row } 1)$ replace (row 1)

$$\left[\begin{array}{ccc|ccc} 1 & 0 & 0 & \frac{39}{8} & -\frac{17}{8} & -\frac{1}{8} \\ 0 & 1 & 0 & 2 & -1 & 0 \\ 0 & 0 & 1 & \frac{1}{8} & \frac{1}{8} & \frac{1}{8} \end{array}\right]$$

$$A^{-1} = \begin{bmatrix} \frac{39}{8} & -\frac{17}{8} & -\frac{1}{8} \\ 2 & -1 & 0 \\ \frac{1}{8} & \frac{1}{8} & \frac{1}{8} \end{bmatrix}$$

21. $$\left[\begin{array}{ccc|ccc} 1 & 3 & -2 & 1 & 0 & 0 \\ 4 & 13 & -7 & 0 & 1 & 0 \\ 5 & 16 & -8 & 0 & 0 & 1 \end{array}\right]$$

$-4(\text{row } 1) + (\text{row } 2)$ replace (row 2)
$-5(\text{row } 1) + (\text{row } 3)$ replace (row 3)

$$\left[\begin{array}{ccc|ccc} 1 & 3 & -2 & 1 & 0 & 0 \\ 0 & 1 & 1 & -4 & 1 & 0 \\ 0 & 1 & 2 & -5 & 0 & 1 \end{array}\right]$$

$-3(\text{row } 2) + (\text{row } 1)$ replace (row 1)
$-(\text{row } 2) + (\text{row } 3)$ replace (row 3)

$$\left[\begin{array}{ccc|ccc} 1 & 0 & -5 & 13 & -3 & 0 \\ 0 & 1 & 1 & -4 & 1 & 0 \\ 0 & 0 & 1 & -1 & -1 & 1 \end{array}\right]$$

$-(\text{row } 3) + (\text{row } 2)$ replace (row 2)
$5(\text{row } 3) + (\text{row } 1)$ replace (row 1)

$$\left[\begin{array}{ccc|ccc} 1 & 0 & 0 & 8 & -8 & 5 \\ 0 & 1 & 0 & -3 & 2 & -1 \\ 0 & 0 & 1 & -1 & -1 & 1 \end{array}\right]$$

$$A^{-1} = \begin{bmatrix} 8 & -8 & 5 \\ -3 & 2 & -1 \\ -1 & -1 & 1 \end{bmatrix}$$

22. $$\left[\begin{array}{ccc|ccc} -2 & 4 & 7 & 1 & 0 & 0 \\ 1 & -3 & 5 & 0 & 1 & 0 \\ 1 & -5 & 22 & 0 & 0 & 1 \end{array}\right]$$

$-\frac{1}{2}(\text{row } 1)$ replace (row 1)

$$\left[\begin{array}{ccc|ccc} 1 & -2 & -\frac{7}{2} & -\frac{1}{2} & 0 & 0 \\ 1 & -3 & 5 & 0 & 1 & 0 \\ 1 & -5 & 22 & 0 & 0 & 1 \end{array}\right]$$

$-(\text{row } 1) + (\text{row } 2)$ replace (row 2)
$-(\text{row } 1) + (\text{row } 3)$ replace (row 3)

$$\left[\begin{array}{ccc|ccc} 1 & -2 & -\frac{7}{2} & -\frac{1}{2} & 0 & 0 \\ 0 & -1 & \frac{17}{2} & \frac{1}{2} & 1 & 0 \\ 0 & -3 & \frac{51}{2} & \frac{1}{2} & 0 & 1 \end{array}\right]$$

$-(\text{row } 2)$ replace (row 2)

$$\left[\begin{array}{ccc|ccc} 1 & -2 & -\frac{7}{2} & \frac{1}{2} & 0 & 0 \\ 0 & 1 & -\frac{17}{2} & -\frac{1}{2} & -1 & 0 \\ 0 & -3 & \frac{51}{2} & \frac{1}{2} & 0 & 1 \end{array}\right]$$

$2(\text{row } 2) + (\text{row } 1)$ replace (row 1)
$3(\text{row } 2) + (\text{row } 3)$ replace (row 3)

$$\left[\begin{array}{ccc|ccc} 1 & 0 & -\frac{41}{2} & -\frac{1}{2} & -2 & 0 \\ 0 & 1 & -\frac{17}{2} & \frac{1}{2} & -1 & 0 \\ 0 & 0 & 0 & -1 & -3 & 1 \end{array}\right]$$

Inverse does not exist.

23. $\left[\begin{array}{ccc|ccc} -1 & 2 & 3 & 1 & 0 & 0 \\ 2 & -5 & -7 & 0 & 1 & 0 \\ -3 & 5 & 11 & 0 & 0 & 1 \end{array}\right]$

$-$(row 1) replace (row 1)
2(row 1) + (row 2) replace (row 2)
$-$ 3(row 1) + (row 3) replace (row 3)

$$\left[\begin{array}{ccc|ccc} 1 & -2 & -3 & -1 & 0 & 0 \\ 0 & -1 & -1 & 2 & 1 & 0 \\ 0 & -1 & 2 & -3 & 0 & 1 \end{array}\right]$$

$-$(row 2) replace (row 2)
$-$ (row 2) + (row 3) replace (row 3)
$-$ 2(row 2) + (row 1) replace (row 1)

$$\left[\begin{array}{ccc|ccc} 1 & 0 & -1 & -5 & -2 & 0 \\ 0 & 1 & 1 & -2 & -1 & 0 \\ 0 & 0 & 3 & -5 & -1 & 1 \end{array}\right]$$

$\frac{1}{3}$(row 3) replace (row 3)

$$\left[\begin{array}{ccc|ccc} 1 & 0 & -1 & -5 & -2 & 0 \\ 0 & 1 & 1 & -2 & -1 & 0 \\ 0 & 0 & 1 & -\frac{5}{3} & -\frac{1}{3} & \frac{1}{3} \end{array}\right]$$

$-$ (row 3) + (row 2) replace (row 2)
(row 3) + (row 1) replace (row 1)

$$\left[\begin{array}{ccc|ccc} 1 & 0 & 0 & -\frac{20}{3} & -\frac{7}{3} & \frac{1}{3} \\ 0 & 1 & 0 & -\frac{1}{3} & -\frac{2}{3} & -\frac{1}{3} \\ 0 & 0 & 1 & -\frac{5}{3} & -\frac{1}{3} & \frac{1}{3} \end{array}\right]$$

$$A^{-1} = \begin{bmatrix} -\frac{20}{3} & -\frac{7}{3} & \frac{1}{3} \\ -\frac{1}{3} & -\frac{2}{3} & -\frac{1}{3} \\ -\frac{5}{3} & -\frac{1}{3} & \frac{1}{3} \end{bmatrix}$$

24. A^{-1} found in problem #14.

$$A^{-1}B = \begin{bmatrix} 4 & -5 \\ -7 & 9 \end{bmatrix}\begin{bmatrix} 12 \\ 10 \end{bmatrix}$$

$$A^{-1}B = \begin{bmatrix} 4(12) - 5(10) \\ -7(12) + 9(10) \end{bmatrix} = \begin{bmatrix} -2 \\ 6 \end{bmatrix}$$

The solution set is $\{(-2, 6)\}$.

25. A^{-1} found in problem #16.

$$A^{-1}B = \begin{bmatrix} -\frac{3}{8} & \frac{1}{8} \\ \frac{1}{4} & \frac{1}{4} \end{bmatrix}\begin{bmatrix} -9 \\ 5 \end{bmatrix}$$

$$A^{-1}B = \begin{bmatrix} -\frac{3}{8}(-9) + \frac{1}{8}(5) \\ \frac{1}{4}(-9) + \frac{1}{4}(5) \end{bmatrix}$$

$$A^{-1}B = \begin{bmatrix} 4 \\ -1 \end{bmatrix}$$

The solution set is $\{(4, -1)\}$.

26. A^{-1} found in problem #20.

$$A^{-1}B = \begin{bmatrix} \frac{39}{8} & -\frac{17}{8} & -\frac{1}{8} \\ 2 & -1 & 0 \\ \frac{1}{8} & \frac{1}{8} & \frac{1}{8} \end{bmatrix}\begin{bmatrix} 7 \\ 17 \\ -32 \end{bmatrix}$$

$$A^{-1}B = \begin{bmatrix} \frac{39}{8}(7) - \frac{17}{8}(17) - \frac{1}{8}(-32) \\ 2(7) - 1(17) + 0 \\ \frac{1}{8}(7) + \frac{1}{8}(17) + \frac{1}{8}(-32) \end{bmatrix}$$

$$A^{-1}B = \begin{bmatrix} 2 \\ -3 \\ -1 \end{bmatrix}$$

The solution set is $\{(2, -3, -1)\}$.

27. A^{-1} found in problem #21.

$$A^{-1}B = \begin{bmatrix} 8 & -8 & 5 \\ -3 & 2 & -1 \\ -1 & -1 & 1 \end{bmatrix}\begin{bmatrix} -7 \\ -21 \\ -23 \end{bmatrix}$$

$$A^{-1}B = \begin{bmatrix} 8(-7)-8(-21)+5(-23) \\ -3(-7)+2(-21)-1(-23) \\ -1(-7)-1(-21)+1(-23) \end{bmatrix}$$

$$A^{-1}B = \begin{bmatrix} -3 \\ 2 \\ 5 \end{bmatrix}$$

The solution set is $\{(-3,\ 2,\ -5)\}$.

28. A^{-1} found in problem #23.

$$A^{-1}B = \begin{bmatrix} -\frac{20}{3} & -\frac{7}{3} & \frac{1}{3} \\ -\frac{1}{3} & -\frac{2}{3} & -\frac{1}{3} \\ -\frac{5}{3} & -\frac{1}{3} & \frac{1}{3} \end{bmatrix}\begin{bmatrix} 22 \\ -51 \\ 71 \end{bmatrix}$$

$$A^{-1}B = \begin{bmatrix} -\frac{20}{3}(22)-\frac{7}{3}(-51)+\frac{1}{3}(71) \\ -\frac{1}{3}(22)-\frac{2}{3}(-51)-\frac{1}{3}(71) \\ -\frac{5}{3}(22)-\frac{1}{3}(-51)+\frac{1}{3}(71) \end{bmatrix}$$

$$A^{-1}B = \begin{bmatrix} -4 \\ 3 \\ 4 \end{bmatrix}$$

The solution set is $\{(-4,\ 3,\ 4)\}$.

29. $\begin{pmatrix} 3x-4y \ge 0 \\ 2x+3y \le 0 \end{pmatrix}$

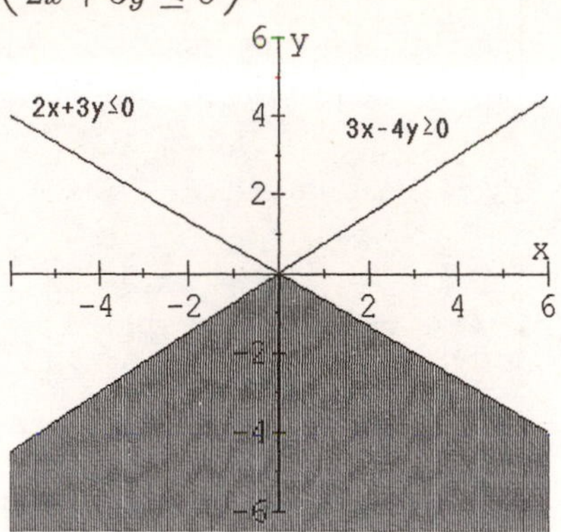

30. $\begin{pmatrix} 3x-2y < 6 \\ 2x-3y < 6 \end{pmatrix}$

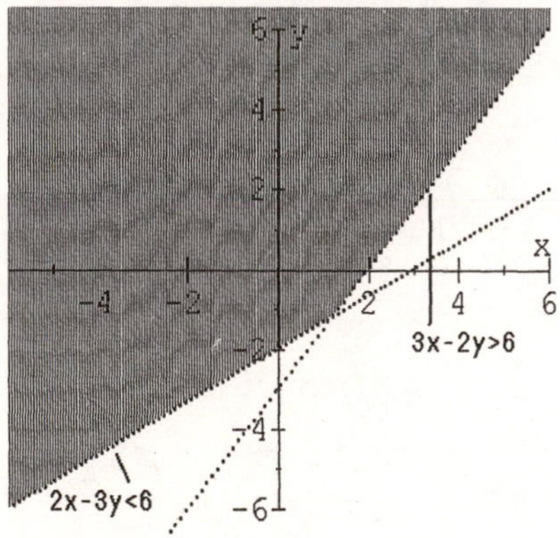

31. $\begin{pmatrix} x-4y < 4 \\ 2x+y \ge 2 \end{pmatrix}$

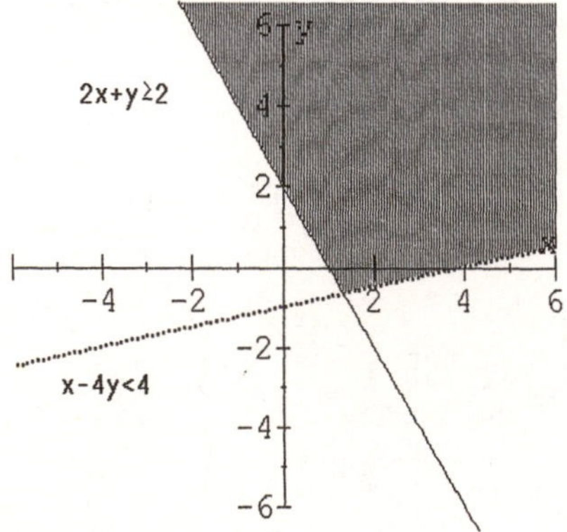

32. $\begin{pmatrix} x \geq 0 \\ y \geq 0 \\ x + 2y \leq 4 \\ 2x - y \leq 4 \end{pmatrix}$

33.

vertices	$f(x, y) = 8x + 5y$
(0, 0)	$8(0) + 5(0) = 0$
(1, 4)	$8(1) + 5(4) = 28$
(4, 1)	$8(4) + 5(1) = 37$
(4, 0)	$8(4) + 5(0) = 32$

The maximum at (4, 1) is $f(4, 1) = 37$.

34.

vertices	$f(x, y) = 2x + 7y$
(0, 8)	56
(2, 7)	53
(6, 3)	33
(8, 0)	16
(0, 0)	0

The maximum is 56.

35.

vertices	$f(x, y) = 7x + 5y$
(0, 0)	$7(0) + 5(0) = 0$
(0, 8)	$7(0) + 5(8) = 40$
(2, 7)	$7(2) + 5(7) = 49$
(6, 3)	$7(6) + 5(3) = 57$
(8, 0)	$7(8) + 5(0) = 56$

The maximum at (6, 3) is $f(6, 3) = 57$.

36.

vertices	$f(x, y) = 150x + 200y$
(0, 8)	1600
(2, 7)	1700
(6, 3)	1500
(8, 0)	1200
(0, 0)	0

The maximum is 1700.

37. Let x = number of one-gallon freezers produced.
Let y = number of two-gallon freezers produced.

$f(x, y) = 4.50x + 5.25y$
$x \geq 75$
$y \geq 100$
$x + y \leq 250$

vertices	$f(x, y) = 4.50x + 5.25y$
(75, 175)	$4.50(75) + 5.25(175) = 1256.25$
(75, 100)	$4.50(75) + 5.25(100) = 862.50$
(150, 100)	$4.50(150) + 5.25(100) = 1200.00$

Maximum when 75 one-gallon freezers and 175 two-gallon freezers are produced.

CHAPTER 12 Test

1. $AB = \begin{bmatrix} -1 & 3 \\ 4 & -2 \end{bmatrix}\begin{bmatrix} 3 & -2 \\ 4 & -1 \end{bmatrix} = \begin{bmatrix} -1(3) + 3(4) & -1(-2) + 3(-1) \\ 4(3) - 2(4) & 4(-2) - 2(-1) \end{bmatrix} = \begin{bmatrix} 9 & -1 \\ 4 & -6 \end{bmatrix}$

2. $BA = \begin{bmatrix} 3(-1) - 2(4) & 3(3) - 2(-2) \\ 4(-1) - 1(4) & 4(3) - 1(-2) \end{bmatrix} = \begin{bmatrix} -11 & 13 \\ -8 & 14 \end{bmatrix}$

3. $DE = \begin{bmatrix} 2 & -1 \\ 3 & -2 \\ 6 & 5 \end{bmatrix}\begin{bmatrix} 2 & -1 & 4 \\ 5 & 1 & -3 \end{bmatrix}$

$$DE = \begin{bmatrix} 2(2)-1(5) & 2(-1)-1(1) & 2(4)-1(-3) \\ 3(2)-2(5) & 3(-1)-2(1) & 3(4)-2(-3) \\ 6(2)+5(5) & 6(-1)+5(1) & 6(4)+5(-3) \end{bmatrix} = \begin{bmatrix} -1 & -3 & 11 \\ -4 & -5 & 18 \\ 37 & -1 & 9 \end{bmatrix}$$

4. BC does not exist.

5. $$EC = \begin{bmatrix} 2 & -1 & 4 \\ 5 & 1 & -3 \end{bmatrix}\begin{bmatrix} -3 \\ 5 \\ -6 \end{bmatrix} = \begin{bmatrix} 2(-3)-1(5)+4(-6) \\ 5(-3)+1(5)-3(-6) \end{bmatrix} = \begin{bmatrix} -35 \\ 8 \end{bmatrix}$$

6. $$2A = \begin{bmatrix} -2 & 6 \\ 8 & -4 \end{bmatrix} \qquad 2A - B = \begin{bmatrix} -2-3 & 6-(-2) \\ 8-4 & -4-(-1) \end{bmatrix} = \begin{bmatrix} -5 & 8 \\ 4 & -3 \end{bmatrix}$$

7. $$3D = \begin{bmatrix} 6 & -3 \\ 9 & -6 \\ 18 & 15 \end{bmatrix} 2F = \begin{bmatrix} -2 & 12 \\ 4 & -10 \\ 6 & 8 \end{bmatrix}$$

$$3D + 2F = \begin{bmatrix} 6-2 & -3+12 \\ 9+4 & -6-10 \\ 18+6 & 15+8 \end{bmatrix} = \begin{bmatrix} 4 & 9 \\ 13 & -16 \\ 24 & 23 \end{bmatrix}$$

8. $$-3A = \begin{bmatrix} 3 & -9 \\ -12 & 6 \end{bmatrix} \qquad -2B = \begin{bmatrix} -6 & 4 \\ -8 & 2 \end{bmatrix}$$

$$-3A - 2B = \begin{bmatrix} 3-6 & -9+4 \\ -12-8 & 6+2 \end{bmatrix} = \begin{bmatrix} -3 & -5 \\ -20 & 8 \end{bmatrix}$$

9. $$EF = \begin{bmatrix} 2 & -1 & 4 \\ 5 & 1 & -3 \end{bmatrix}\begin{bmatrix} -1 & 6 \\ 2 & -5 \\ 3 & 4 \end{bmatrix}$$

$$EF = \begin{bmatrix} 2(-1)-1(2)+4(3) & 2(6)-1(-5)+4(4) \\ 5(-1)+1(2)-3(3) & 5(6)+1(-5)-3(4) \end{bmatrix} = \begin{bmatrix} 8 & 33 \\ -12 & 13 \end{bmatrix}$$

10. $$AB = \begin{bmatrix} 9 & -1 \\ 4 & -6 \end{bmatrix}$$

$$EF = \begin{bmatrix} 8 & 33 \\ -12 & 13 \end{bmatrix}$$

$$AB - EF = \begin{bmatrix} 9-8 & -1-33 \\ 4-(-12) & -6-13 \end{bmatrix}$$

$$AB - EF = \begin{bmatrix} 1 & -34 \\ 16 & -19 \end{bmatrix}$$

11. $|A| = -9-(-10) = 1$

$$A^{-1} = \frac{1}{1}\begin{bmatrix} -3 & 2 \\ -5 & 3 \end{bmatrix} = \begin{bmatrix} -3 & 2 \\ -5 & 3 \end{bmatrix}$$

12. $$D = \begin{vmatrix} -2 & 5 \\ 3 & -7 \end{vmatrix} = 14 - 15 = -1$$

$$A^{-1} = \frac{1}{-1}\begin{bmatrix} -7 & -5 \\ -3 & -2 \end{bmatrix} = \begin{bmatrix} 7 & 5 \\ 3 & 2 \end{bmatrix}$$

13. $|A| = 8 - 6 = 2$

$$A^{-1} = \frac{1}{2}\begin{bmatrix} 8 & 3 \\ 2 & 1 \end{bmatrix} = \begin{bmatrix} 4 & \frac{3}{2} \\ 1 & \frac{1}{2} \end{bmatrix}$$

14. $D = \begin{vmatrix} 3 & 5 \\ 1 & 4 \end{vmatrix} = 12 - 5 = 7$

$$A^{-1} = \frac{1}{7}\begin{bmatrix} 4 & -5 \\ -1 & 3 \end{bmatrix} = \begin{bmatrix} \frac{4}{7} & -\frac{5}{7} \\ -\frac{1}{7} & \frac{3}{7} \end{bmatrix}$$

15. $$\left[\begin{array}{ccc|ccc} -2 & 2 & 3 & 1 & 0 & 0 \\ 1 & -1 & 0 & 0 & 1 & 0 \\ 0 & 1 & 4 & 0 & 0 & 1 \end{array}\right]$$

$-\frac{1}{2}$(row 1) replace (row 1)

$$\left[\begin{array}{ccc|ccc} 1 & -1 & -\frac{3}{2} & -\frac{1}{2} & 0 & 0 \\ 1 & -1 & 0 & 0 & 1 & 0 \\ 0 & 1 & 4 & 0 & 0 & 1 \end{array}\right]$$

$-$ (row 1) + (row 2) replace (row 2)

$$\left[\begin{array}{ccc|ccc} 1 & -1 & -\frac{3}{2} & -\frac{1}{2} & 0 & 0 \\ 0 & 0 & \frac{3}{2} & \frac{1}{2} & 1 & 0 \\ 0 & 1 & 4 & 0 & 0 & 1 \end{array}\right]$$

Interchange row 2 and row 3.

$$\left[\begin{array}{ccc|ccc} 1 & -1 & -\frac{3}{2} & -\frac{1}{2} & 0 & 0 \\ 0 & 1 & 4 & 0 & 0 & 1 \\ 0 & 0 & \frac{3}{2} & \frac{1}{2} & 1 & 0 \end{array}\right]$$

(row 2) + (row 1) replace (row 1)

$$\left[\begin{array}{ccc|ccc} 1 & 0 & \frac{5}{2} & -\frac{1}{2} & 0 & 1 \\ 0 & 1 & 4 & 0 & 0 & 1 \\ 0 & 0 & \frac{3}{2} & \frac{1}{2} & 1 & 0 \end{array}\right]$$

$\frac{2}{3}$(row 3) replace (row 3)

$$\left[\begin{array}{ccc|ccc} 1 & 0 & \frac{5}{2} & -\frac{1}{2} & 0 & 1 \\ 0 & 1 & 4 & 0 & 0 & 1 \\ 0 & 0 & 1 & \frac{1}{3} & \frac{2}{3} & 0 \end{array}\right]$$

$-$ 4(row 3) + (row 2) replace (row 2)
$-\frac{5}{2}$(row 3) + (row 1) replace (row 1)

$$\left[\begin{array}{ccc|ccc} 1 & 0 & 0 & -\frac{4}{3} & -\frac{5}{3} & 1 \\ 0 & 1 & 0 & -\frac{4}{3} & -\frac{8}{3} & 1 \\ 0 & 0 & 1 & \frac{1}{3} & \frac{2}{3} & 0 \end{array}\right]$$

$$A^{-1} = \begin{bmatrix} -\frac{4}{3} & -\frac{5}{3} & 1 \\ -\frac{4}{3} & -\frac{8}{3} & 1 \\ \frac{1}{3} & \frac{2}{3} & 0 \end{bmatrix}$$

16. $$\left[\begin{array}{ccc|ccc} 1 & -2 & 4 & 1 & 0 & 0 \\ 0 & 1 & 3 & 0 & 1 & 0 \\ 0 & 0 & 1 & 0 & 0 & 1 \end{array}\right]$$

2(row 2) + (row 1) replace (row 1)

$$\left[\begin{array}{ccc|ccc} 1 & 0 & 10 & 1 & 2 & 0 \\ 0 & 1 & 3 & 0 & 1 & 0 \\ 0 & 0 & 1 & 0 & 0 & 1 \end{array}\right]$$

$-$ 3(row 3) + (row 2) replace (row 2)
$-$ 10(row 3) + (row 1) replace (row 1)

$$\left[\begin{array}{ccc|ccc} 1 & 0 & 0 & 1 & 2 & -10 \\ 0 & 1 & 0 & 0 & 1 & -3 \\ 0 & 0 & 1 & 0 & 0 & 1 \end{array}\right]$$

$$A^{-1} = \begin{bmatrix} 1 & 2 & -10 \\ 0 & 1 & -3 \\ 0 & 0 & 1 \end{bmatrix}$$

17. $A = \begin{bmatrix} 3 & -2 \\ 5 & -3 \end{bmatrix}$ $\qquad B = \begin{bmatrix} 48 \\ 76 \end{bmatrix}$

$|A| = -9 - (-10) = 1$

$$A^{-1} = \frac{1}{1}\begin{bmatrix} -3 & 2 \\ -5 & 3 \end{bmatrix} = \begin{bmatrix} -3 & 2 \\ -5 & 3 \end{bmatrix}$$

$$A^{-1}B = \begin{bmatrix} -3 & 2 \\ -5 & 3 \end{bmatrix}\begin{bmatrix} 48 \\ 76 \end{bmatrix}$$

$$A^{-1}B = \begin{bmatrix} -3(48) + 2(76) \\ -5(48) + 3(76) \end{bmatrix} = \begin{bmatrix} 8 \\ -12 \end{bmatrix}$$

The solution set is $\{(8, -12)\}$.

18. $$A^{-1}B = \begin{bmatrix} 4 & \frac{3}{2} \\ 1 & \frac{1}{2} \end{bmatrix}\begin{bmatrix} 36 \\ -100 \end{bmatrix}$$

$$A^{-1}B = \begin{bmatrix} 4(36) + \frac{3}{2}(-100) \\ 1(36) + \frac{1}{2}(-100) \end{bmatrix} = \begin{bmatrix} -6 \\ -14 \end{bmatrix}$$

The solution set is $\{(-6, -14)\}$.

19. $A = \begin{bmatrix} 3 & 5 \\ 1 & 4 \end{bmatrix}$ $\quad B = \begin{bmatrix} 92 \\ 61 \end{bmatrix}$

$$|A| = 12 - 5 = 7$$

$$A^{-1} = \frac{1}{7}\begin{bmatrix} 4 & -5 \\ -1 & 3 \end{bmatrix}$$

$$A^{-1}B = \frac{1}{7}\begin{bmatrix} 4 & -5 \\ -1 & 3 \end{bmatrix}\begin{bmatrix} 92 \\ 61 \end{bmatrix}$$

$$A^{-1}B = \frac{1}{7}\begin{bmatrix} 4(92) - 5(61) \\ -1(92) + 3(61) \end{bmatrix}$$

$$A^{-1}B = \frac{1}{7}\begin{bmatrix} 63 \\ 91 \end{bmatrix} = \begin{bmatrix} 9 \\ 13 \end{bmatrix}$$

The solution set is $\{(9, 13)\}$.

20. $$A^{-1}B = \begin{bmatrix} -\frac{10}{9} & \frac{7}{9} & -\frac{5}{9} \\ \frac{4}{9} & -\frac{1}{9} & \frac{2}{9} \\ -\frac{13}{9} & \frac{10}{9} & -\frac{11}{9} \end{bmatrix}\begin{bmatrix} 1 \\ 3 \\ -2 \end{bmatrix}$$

$$A^{-1}B = \begin{bmatrix} -\frac{10}{9}(1) + \frac{7}{9}(3) - \frac{5}{9}(-2) \\ \frac{4}{9}(1) - \frac{1}{9}(3) + \frac{2}{9}(-2) \\ -\frac{13}{9}(1) + \frac{10}{9}(3) - \frac{11}{9}(-2) \end{bmatrix}$$

$$A^{-1}B = \begin{bmatrix} \frac{7}{3} \\ -\frac{1}{3} \\ \frac{13}{3} \end{bmatrix}$$

The solution set is $\left\{\left(\frac{7}{3}, -\frac{1}{3}, \frac{13}{3}\right)\right\}$.

21. $$A^{-1}B = \begin{bmatrix} -\frac{5}{24} & \frac{1}{6} & -\frac{7}{24} \\ \frac{7}{24} & \frac{1}{6} & \frac{5}{24} \\ \frac{11}{24} & -\frac{1}{6} & \frac{1}{24} \end{bmatrix}\begin{bmatrix} 3 \\ 3 \\ 3 \end{bmatrix}$$

$$A^{-1}B = \begin{bmatrix} -\frac{5}{24}(3) + \frac{1}{6}(3) - \frac{7}{24}(3) \\ \frac{7}{24}(3) + \frac{1}{6}(3) + \frac{5}{24}(3) \\ \frac{11}{24}(3) - \frac{1}{6}(3) + \frac{1}{24}(3) \end{bmatrix}$$

$$A^{-1}B = \begin{bmatrix} -1 \\ 2 \\ 1 \end{bmatrix}$$

The solution set is $\{(-1, 2, 1)\}$.

22. $\begin{pmatrix} 2x - y > 4 \\ x + 3y < 3 \end{pmatrix}$

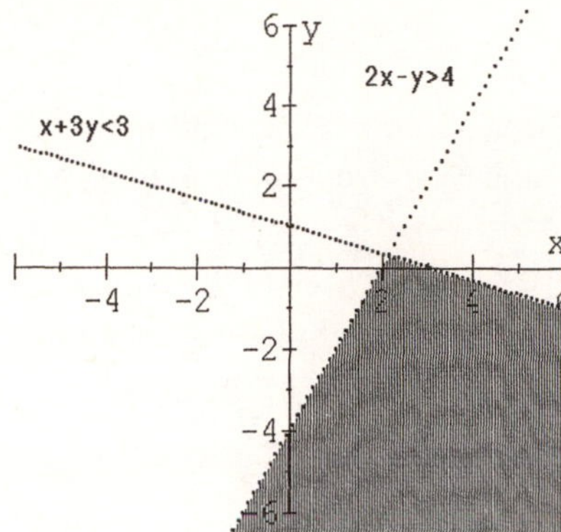

23. $\begin{pmatrix} 2x - 3y \leq 6 \\ x + 4y > 4 \end{pmatrix}$

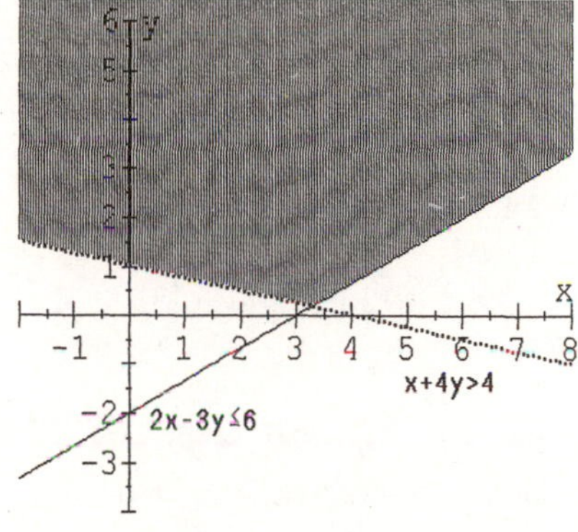

24. $\begin{pmatrix} y \leq 2x - 2 \\ y \geq x + 1 \end{pmatrix}$

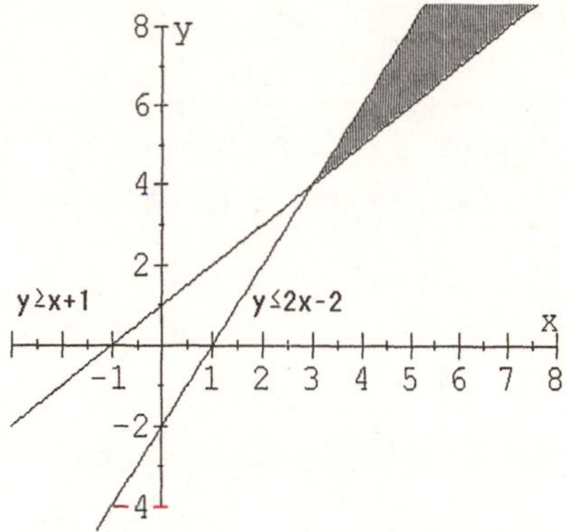

25.

vertices	$f(x, y) = 500x + 350y$
(0, 0)	$500(0) + 350(0) = 0$
(0, 8)	$500(0) + 350(8) = 2800$
(2, 7)	$500(2) + 350(7) = 3450$
(6, 3)	$500(6) + 350(3) = 4050$
(8, 0)	$500(8) + 350(0) = 4000$

The maximum at (6, 3) is $f(6, 3) = 4050$.

Chapter 13 Conic Sections

PROBLEM SET 13.1 Circles

1. Center at (2, 3) and $r = 5$
$(x-2)^2 + (y-3)^2 = 5^2$
$x^2 - 4x + 4 + y^2 - 6y + 9 = 25$
$x^2 + y^2 - 4x - 6y - 12 = 0$

3. Center at $(-1, -5)$ and $r = 3$
$(x+1)^2 + (y+5)^2 = 3^2$
$x^2 + 2x + 1 + y^2 + 10y + 25 = 9$
$x^2 + y^2 + 2x + 10y + 17 = 0$

5. Center at (3, 0) and $r = 3$
$(x-3)^2 + (y-0)^2 = 3^2$
$x^2 - 6x + 9 + y^2 = 9$
$x^2 + y^2 - 6x = 0$

7. Center at the origin and $r = 7$
$(x-0)^2 + (y-0)^2 = 7^2$
$x^2 + y^2 - 49 = 0$

9. Center could be at $(-3, 4)$ or $(-3, -4)$ and $r = 4$. If the center is at $(-3, 4)$, then the equation is as follows:
$(x+3)^2 + (y-4)^2 = 4^2$
$x^2 + 6x + 9 + y^2 - 8y + 16 = 16$
$x^2 + y^2 + 6x - 8y + 9 = 0$

If the center is at $(-3, -4)$, then the equation is as follows
$(x+3)^2 + (y+4)^2 = 4^2$
$x^2 + 6x + 9 + y^2 + 8y + 16 = 16$
$x^2 + y^2 + 6x + 8y + 9 = 0$

11. The center would be at $(-6, -6)$ and $r = 6$.
$(x+6)^2 + (y+6)^2 = 6^2$
$x^2 + 12x + 36 + y^2 + 12y + 36 = 36$
$x^2 + y^2 + 12x + 12y + 36 = 0$

13. The circle passes through (2, 0) and (6, 0). The perpendicualr bisector of the chord from (2, 0) to (6, 0) passes through $(4, 0)$ and is perpendicular to the x-axis, so its equation is $x = 4$.

Because the circle is tangent to the y-axis, the radius goes from $(4, y)$ to $(0, y)$ and therefore $r = 4$.

The distance from $(2, 0)$ to the center $(4, y)$ is given by the following equation:
$4 = \sqrt{(4-2)^2 + (y-0)^2}$
$4 = \sqrt{2^2 + y^2}$
$16 = 4 + y^2$
$12 = y^2$
$y = \pm 2\sqrt{3}$

<u>Case 1</u>: Center at $(4, 2\sqrt{3})$, $r = 4$
$(x-4)^2 + (y - 2\sqrt{3})^2 = 4^2$
$x^2 - 8x + 16 + y^2 - 4\sqrt{3}y + 12 = 16$
$x^2 + y^2 - 8x - 4\sqrt{3}y + 12 = 0$

<u>Case 2</u>: Center at $(4, -2\sqrt{3})$, $r = 4$
$(x-4)^2 + (y + 2\sqrt{3})^2 = 4^2$
$x^2 - 8x + 16 + y^2 + 4\sqrt{3}y + 12 = 16$
$x^2 + y^2 - 8x + 4\sqrt{3}y + 12 = 0$

15. $(x-5)^2 + (y-7)^2 = 25$
Center: $(5, 7)$, $r = \sqrt{25} = 5$

17. $(x+1)^2 + (y+8)^2 = 12$
Center: $(-1, -8)$, $r = \sqrt{12} = 2\sqrt{3}$

19. $3(x-10)^2 + 3(y+5)^2 = 9$
$\frac{1}{3}\left[3(x-10)^2 + 3(y+5)^2\right] = \frac{1}{3}(9)$
$(x-10)^2 + (y+5)^2 = 3$
Center: $(10, -5)$, $r = \sqrt{3}$

21. $x^2 + y^2 - 6x - 10y + 30 = 0$
$x^2 - 6x + y^2 - 10y = -30$
$x^2 - 6x + 9 + y^2 - 10y + 25 = -30+9+25$
$(x-3)^2 + (y-5)^2 = 4$
Center at (3, 5) and $r = 2$.

23. $x^2 + y^2 + 10x + 14y + 73 = 0$
$x^2 + 10x + y^2 + 14y = -73$
$x^2 + 10x + 25 + y^2 + 14y + 49 = -73 + 25 + 49$
$(x + 5)^2 + (y + 7)^2 = 1$
Center at $(-5, -7)$ and $r = 1$.

25. $x^2 + y^2 - 10x = 0$
$x^2 - 10x + y^2 = 0$
$x^2 - 10x + 25 + y^2 = 25$
$(x - 5)^2 + (y + 0)^2 = 25$
Center at $(5, 0)$ and $r = 5$.

27. $x^2 + y^2 - 5y - 1 = 0$
$x^2 + y^2 - 5y = 1$
$x^2 + y^2 - 5y + \frac{25}{4} = 1 + \frac{25}{4}$
$(x - 0)^2 + \left(y - \frac{5}{2}\right)^2 = \frac{29}{4}$
Center at $\left(0, \frac{5}{2}\right)$
and $r = \sqrt{\frac{29}{4}} = \frac{\sqrt{29}}{\sqrt{4}} = \frac{\sqrt{29}}{2}$

29. $x^2 + y^2 = 8$
Center at $(0, 0)$ and $r = \sqrt{8} = 2\sqrt{2}$.

31. $4x^2 + 4y^2 - 4x - 8y - 11 = 0$
Divide both sides of the equation by 4.
$x^2 + y^2 - x - 2y - \frac{11}{4} = 0$
$x^2 - x + y^2 - 2y = \frac{11}{4}$
$x^2 - x + \frac{1}{4} + y^2 - 2y + 1 = \frac{11}{4} + \frac{1}{4} + 1$
$\left(x - \frac{1}{2}\right)^2 + (y - 1)^2 = 4$
Center at $\left(\frac{1}{2}, 1\right)$ and $r = 2$.

33. $x^2 + y^2 - 2x + 3y - 12 = 0$
$x^2 - 2x + y^2 + 3y = 12$
$x^2 - 2x + 1 + y^2 + 3y + \frac{9}{4} = 12 + 1 + \frac{9}{4}$
$(x - 1)^2 + \left(y + \frac{3}{2}\right)^2 = \frac{61}{4}$
Center at $\left(1, -\frac{3}{2}\right)$

The slope from the center to $(4, 1)$ is:
$$m = \frac{1 - \left(-\frac{3}{2}\right)}{4 - 1} = \frac{\frac{5}{2}}{3} = \frac{5}{6}$$
Slope of the tangent line is $-\frac{6}{5}$.
$y - 1 = -\frac{6}{5}\left(x - 4\right)$
$5(y - 1) = -6(x - 4)$
$5y - 5 = -6x + 24$
$6x + 5y = 29$

35. Center at $(-3, -4)$ and passes through $(0, 0)$.

$(x - (-3))^2 + (y - (-4))^2 = r^2$
$(x + 3)^2 + (y + 4)^2 = r^2$
When $x = 0$ and $y = 0$,
$(0 + 3)^2 + (0 + 4)^2 = r^2$
$9 + 16 = r^2$
$25 = r^2$
$r = 5$
$(x + 3)^2 + (y + 4)^2 = 25$
$x^2 + 6x + 9 + y^2 + 8y + 16 = 25$
$x^2 + y^2 + 6x + 8y = 0$

37. Centers on the line $2x + 3y = 10$ and tangent to both axis. When tangent to both axis, one case would be the coordinates of the center would be equal and in the other case the coordinates of the center would be opposite each other.
Case 1: Let the coordinates of the center be (a, a). Since the center lies on the line $2x + 3y = 10$, the point (a, a) must satisfy the equation.
$2a + 3a = 10$
$5a = 10$
$a = 2$
So center is at $(2, 2)$ and hence $r = 2$.
$(x - 2)^2 + (y - 2)^2 = 2^2$
$x^2 - 4x + 4 + y^2 - 4y + 4 = 4$
$x^2 + y^2 - 4x - 4y + 4 = 0$
Case 2: Let the coordinates of the center be $(a, -a)$.
$2a + 3(-a) = 10$
$-a = 10$
$a = -10$

The center is at $(-10, 10)$ and hence $r = 10$.
$(x+10)^2 + (y-10)^2 = 10^2$
$x^2 + 20x + 100 + y^2 - 20y + 100 = 100$
$x^2 + y^2 + 20x - 20y + 100 = 0$

39. $x^2 + y^2 + 8x + 4y - 30 = 0$
$x^2 + 8x + y^2 + 4y = 30$
$x^2 + 8x + 16 + y^2 + 4y + 4 = 30 + 16 + 4$
$(x+4)^2 + (y+2)^2 = 50$
Center at $(-4, -2)$
Slope from the center to the midpoint of the chord is

$$m = \frac{-2-4}{-4-(-1)} = \frac{-6}{-3} = 2.$$

So slope of the chord is $-\frac{1}{2}$.

$y - 4 = -\frac{1}{2}[x - (-1)]$
$2y - 8 = -(x+1)$
$2y - 8 = -x - 1$
$x + 2y = 7$

41. $(1, 2)$, $(-3, -8)$, and $(-9, 6)$
When $x = 1$ and $y = 2$,
$1^2 + 2^2 + D + 2E + F = 0$
$D + 2E + F = -5.$

When $x = -3$ and $y = -8$,
$(-3)^2 + (-8)^2 - 3D - 8E + F = 0$
$-3D - 8E + F = -73.$

When $x = -9$ and $y = 6$,
$(-9)^2 + 6^2 - 9D + 6E + F = 0$
$-9D + 6E + F = -117.$

$D + 2E + F = -5$
$-3D - 8E + F = -73$
$-9D + 6E + F = -117$

3(equation 1) + (equation 2) replaces equation 2
9(equation 1) + (equation 3) replaces equation 3

$D + 2E + F = -5$
$-2E + 4F = -88$
$24E + 10F = -162$
12(equation 2) + (equation 3) replaces equation 3

$D + 2E + F = -5$
$-2E + 4F = -88$
$58F = -1218$

$58F = -1218$
$F = -21$

$-2E + 4F = -88$
$-2E + 4(-21) = -88$
$-2E = -4$
$E = 2$

$D + 2E + F = -5$
$D + 2(2) - 21 = -5$
$D = 12$

$x^2 + y^2 + 12x + 2y - 21 = 0$

Further Investigations

47.

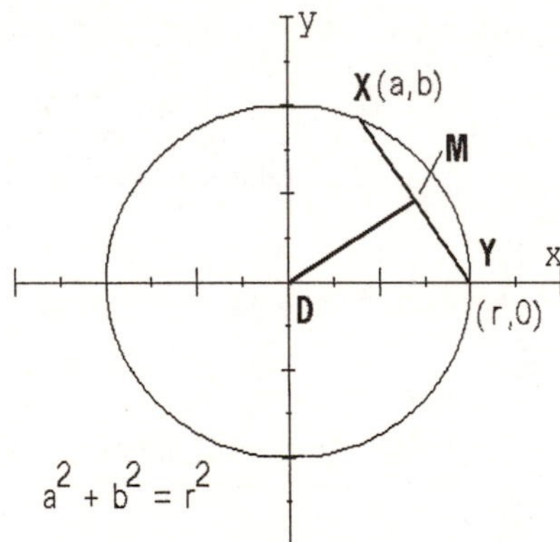

Show segment DM is perpendicular to chord XY where M is the midpoint of chord XY and D is the center of the circle.
To show $\overline{DM}$ is perpendicular to $\overline{XY}$ show that the product of their slopes is -1.

$$\text{Midpoint of } \overline{XY} = M\left(\frac{a+r}{2}, \frac{b+0}{2}\right) = M\left(\frac{a+r}{2}, \frac{b}{2}\right)$$

$D = (0, 0)$

$$\text{Slope of } \overline{DM} = \frac{\frac{b}{2} - 0}{\frac{a+r}{2} - 0} = \frac{\frac{b}{2}}{\frac{a+r}{2}} = \frac{b}{a+r}$$

$$\text{Slope of } \overline{XY} = \frac{b-0}{a-r} = \frac{b}{a-r}$$

Product of slopes:

$$\left(\frac{b}{a+r}\right)\left(\frac{b}{a-r}\right)=\frac{b^2}{a^2-r^2}$$

Remember: $a^2+b^2=r^2$
Substituting a^2+b^2 for r^2:

$$\left(\frac{b}{a+r}\right)\left(\frac{b}{a-r}\right)=\frac{b^2}{a^2-(a^2+b^2)}$$

$$=\frac{b^2}{a^2-a^2-b^2}$$

$$=\frac{b^2}{-b^2}=-1$$

Product of the slopes is -1 which implies that the line segments are perpendicular. So segment DM is perpendicular to chord XY.

PROBLEM SET 13.2 Parabolas

1. $y^2=8x$
$4p=8$
$p=2$
vertex: $(0, 0)$
focus: $(2, 0)$
directrix: $x=-2$

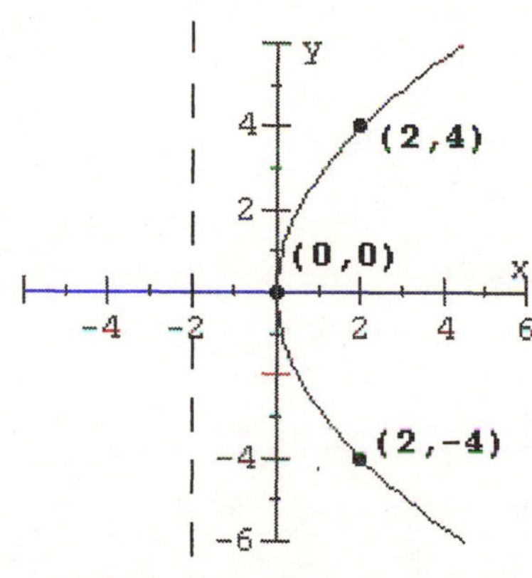

3. $x^2=-12y$

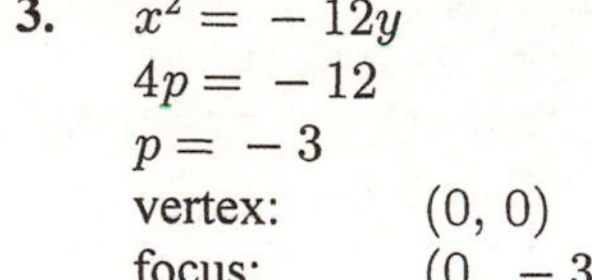

$4p=-12$
$p=-3$
vertex: $(0, 0)$
focus: $(0, -3)$
directrix: $y=3$

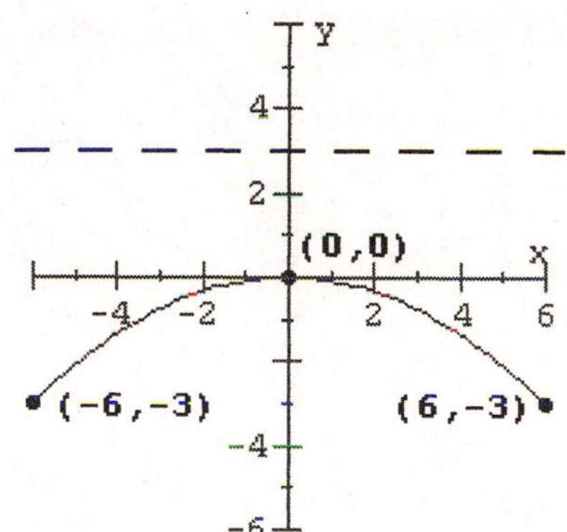

5. $y^2=-2x$

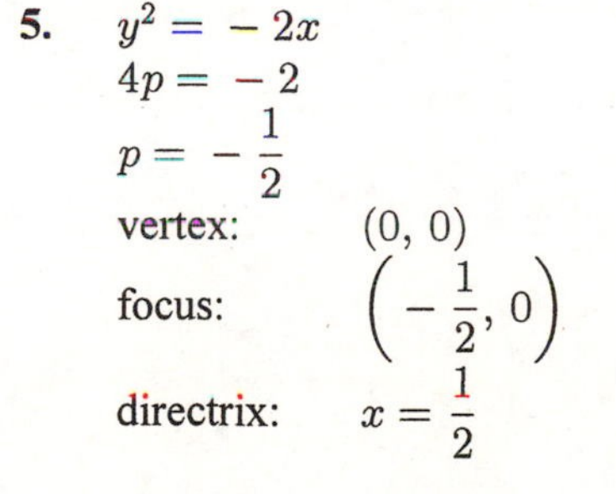

$4p=-2$
$p=-\frac{1}{2}$
vertex: $(0, 0)$
focus: $\left(-\frac{1}{2}, 0\right)$
directrix: $x=\frac{1}{2}$

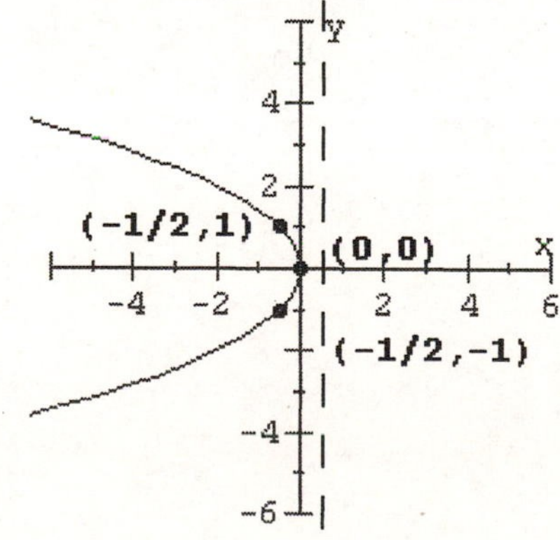

Problem Set 13.2

7. $x^2 = 6y$

$4p = 6$

$p = \frac{6}{4} = \frac{3}{2}$

vertex: $(0, 0)$

focus: $\left(0, \frac{3}{2}\right)$

directrix: $y = -\frac{3}{2}$

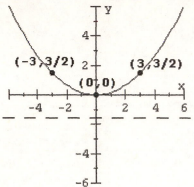

9. $x^2 = 12(y + 1)$

$4p = 12$

$p = 3$

vertex: $(0, -1)$

focus: $(0, 2)$

directrix: $y = -4$

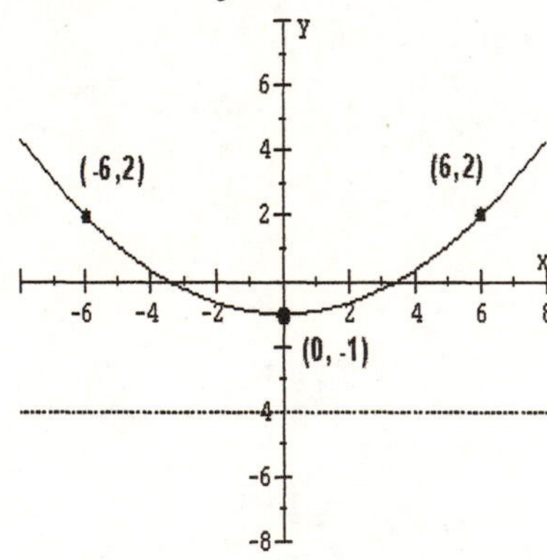

11. $y^2 = -8(x - 3)$

$4p = -8$

$p = -2$

vertex: $(3, 0)$

focus: $(1, 0)$

directrix: $x = 5$

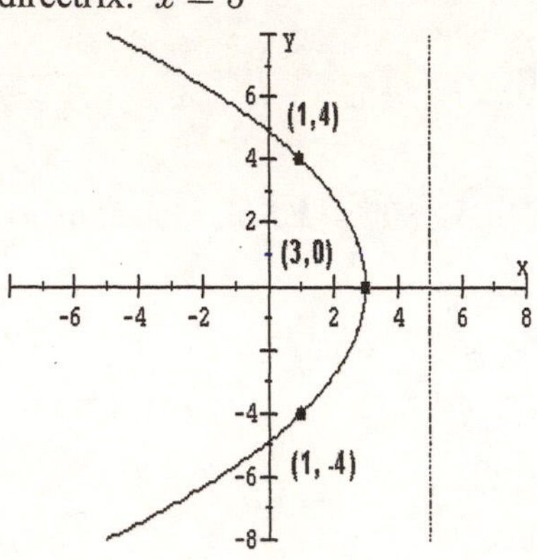

13. $x^2 - 4y + 8 = 0$

$x^2 = 4y - 8$

$x^2 = 4(y - 2)$

$4p = 4$

$p = 1$

vertex: $(0, 2)$

focus: $(0, 3)$

directrix: $y = 1$

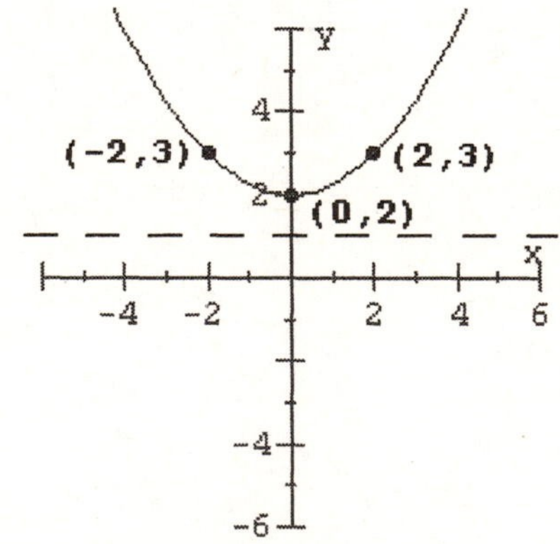

15. $x^2 + 8y + 16 = 0$
$x^2 = -8y - 16$
$x^2 = -8(y + 2)$
$4p = -8$
$p = -2$
vertex: $(0, -2)$
focus: $(0, -4)$
directrix: $y = -2 - (-2) = 0$

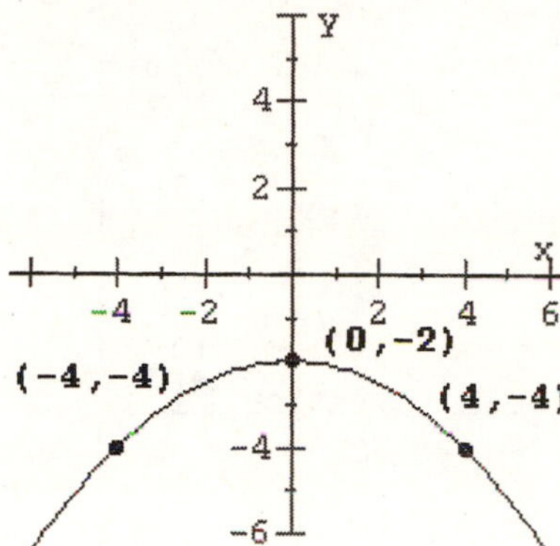

17. $y^2 - 12x + 24 = 0$
$y^2 = 12x - 24$
$y^2 = 12(x - 2)$
$4p = 12$
$p = 3$
vertex: $(2, 0)$
focus: $(5, 0)$
directrix: $x = -1$

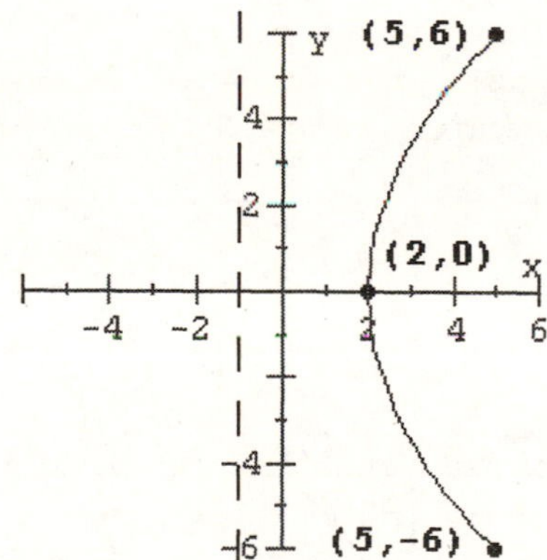

19. $(x - 2)^2 = -4(y + 2)$
$4p = -4$
$p = -1$
vertex: $(2, -2)$
focus: $(2, -3)$
directrix: $y = -1$

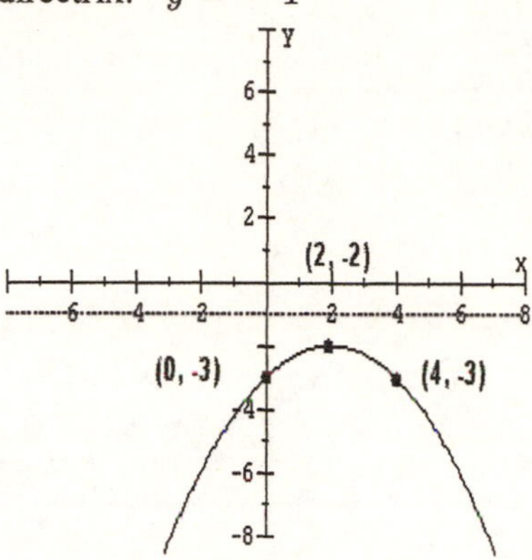

21. $(y + 4)^2 = -8(x + 2)$
$4p = -8$
$p = -2$
vertex: $(-2, -4)$
focus: $(-4, -4)$
directrix: $x = 0$

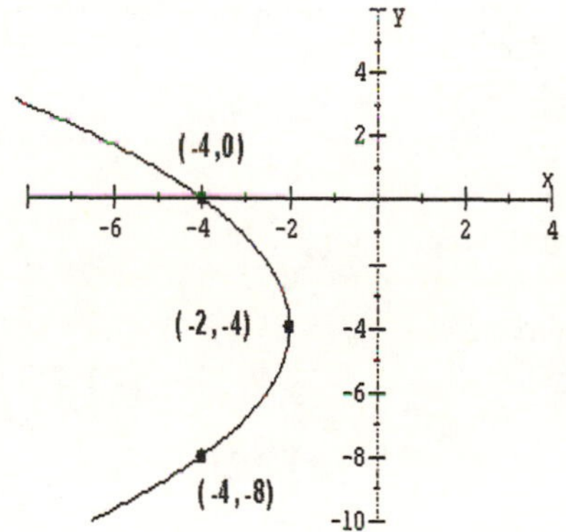

Problem Set 13.2

23. $x^2 - 2x - 4y + 9 = 0$
$x^2 - 2x = 4y - 9$
$x^2 - 2x + 1 = 4y - 9 + 1$
$(x-1)^2 = 4y - 8$
$(x-1)^2 = 4(y-2)$
$4p = 4$
$p = 1$
vertex: $(1, 2)$
focus: $(1, 3)$
directrix: $y = 2 - 1 = 1$

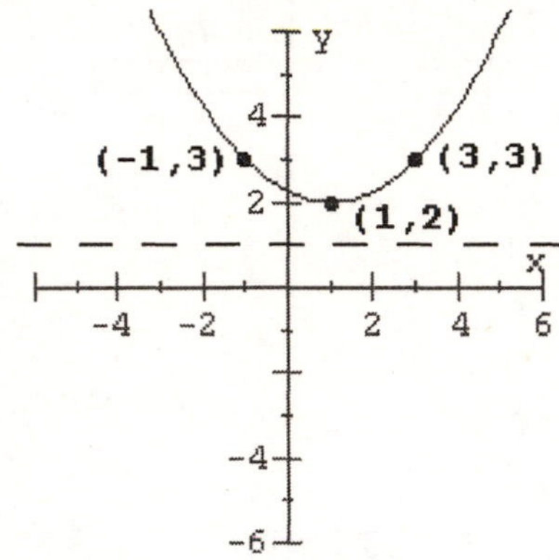

25. $x^2 + 6x + 8y + 1 = 0$
$z^2 + 6x = -8y - 1$
$x^2 + 6x + 9 = -8y - 1 + 9$
$(x+3)^2 = -8y + 8$
$(x+3)^2 = -8(y-1)$
$4p = -8$
$p = -2$
vertex: $(-3, 1)$
focus: $(-3, -1)$
directrix: $y = 3$

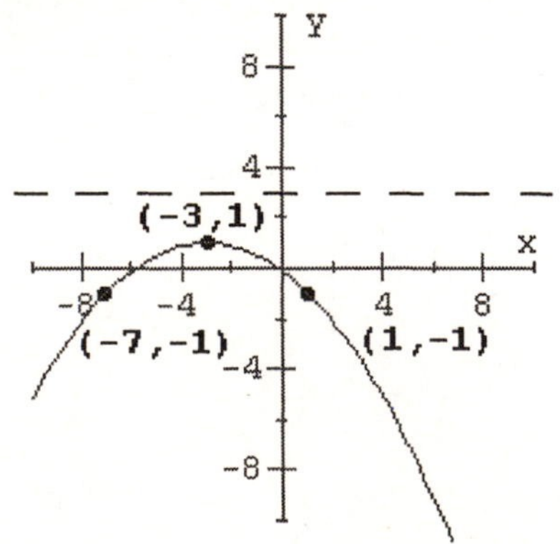

27. $y^2 - 2y + 12x - 35 = 0$
$y^2 - 2y = -12x + 35$
$y^2 - 2y + 1 = -12x + 35 + 1$
$(y-1)^2 = -12x + 36$
$(y-1)^2 = -12(x-3)$
$4p = -12$
$p = -3$
vertex: $(3, 1)$
focus: $(0, 1)$
directrix: $x = 3 - (-3) = 6$

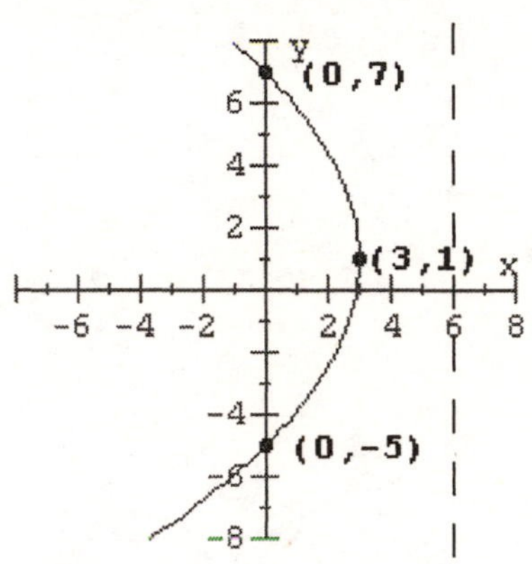

29. $y^2 + 6y - 4x + 1 = 0$
$y^2 + 6y = 4x - 1$
$y^2 + 6y + 9 = 4x - 1 + 9$
$(y+3)^2 = 4x + 8$
$(y+3)^2 = 4(x+2)$
$4p = 4$
$p = 1$
vertex: $(-2, -3)$
focus: $(-1, -3)$
directrix: $x = -3$

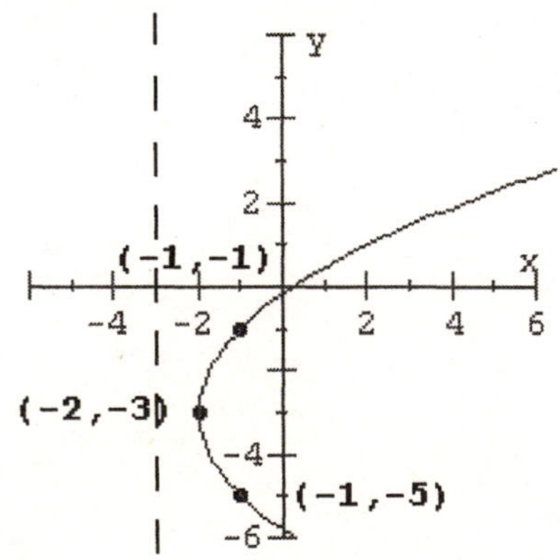

31. Focus $(0, 3)$, directrix $y = -3$

$h = 0$

$$\begin{aligned} k + p &= 3 \\ k - p &= -3 \\ \hline 2k &= 0 \\ k &= 0 \end{aligned}$$

$k + p = 3$
$0 + p = 3$
$p = 3$

$(x - 0)^2 = 4(3)(y - 0)$
$x^2 = 12y$

33. Focus $(-1, 0)$, directrix $x = 1$
The parabola is of the general form $y^2 = 4px$ and $p = -1$
$y^2 = 4(-1)x$
$y^2 = -4x$

35. Focus $(0, 1)$, directrix $y = 7$

$h = 0$

$$\begin{aligned} k + p &= 1 \\ k - p &= 7 \\ \hline 2k &= 8 \\ k &= 4 \end{aligned}$$

$k + p = 1$
$4 + p = 1$
$p = -3$

$(x - 0)^2 = 4(-3)(y - 4)$
$x^2 = -12(y - 4)$
$x^2 = -12y + 48$
$x^2 + 12y - 48 = 0$

37. Focus $(3, 4)$, directrix $y = -2$
The vertex is midway between the focus and the directrix at $(3, 1)$.
$p = 3$
$(x - h)^2 = 4p(y - k)$
$(x - 3)^2 = 4(3)(y - 1)$
$x^2 - 6x + 9 = 12y - 12$
$x^2 - 6x - 12y + 21 = 0$

39. Focus $(-4, 5)$, directrix $x = 0$

$$\begin{aligned} h + p &= -4 \\ h - p &= 0 \\ \hline 2h &= -4 \\ h &= -2 \end{aligned}$$

$k = 5$

$h + p = -4$
$-2 + p = -4$
$p = -2$

$(y - k)^2 = 4p(x - h)$
$(y - 5)^2 = 4(-2)(x + 2)$
$y^2 - 10y + 25 = -8(x + 2)$
$y^2 - 10y + 25 = -8x - 16$
$y^2 - 10y + 8x + 41 = 0$

41. Vertex $(0, 0)$, symmetric with respect to the x-axis and contains the point $(-3, 5)$.
$y^2 = 4px$
Using point $(-3, 5)$ for x and y
$5^2 = 4p(-3)$
$25 = -12p$

$$\frac{-25}{12} = p$$

$$y^2 = 4\left(\frac{-25}{12}\right)x$$

$$y^2 = \frac{-25}{3}x$$

43. Vertex $(0, 0)$, Focus $\left(\frac{5}{2}, 0\right)$

$$p = \frac{5}{2}$$

$$y^2 = 4\left(\frac{5}{2}\right)x$$

$y^2 = 10x$

45. Vertex $(7, 3)$, Focus $(7, 5)$
symmetry with respect to the line $x = 7$.
p equals the distance between the focus and the directrix $p = 2$.
$(x - h)^2 = 4p(y - k)$
$(x - 7)^2 = 4(2)(y - 3)$
$x^2 - 14x + 49 = 8y - 24$
$x^2 - 14x - 8y + 73 = 0$

47. Vertex $(8, -3)$, Focus $(11, -3)$
symmetry with respect to the line $y = -3$.

$h = 8$ $\qquad k = -3$
$h + p = 11$
$8 + p = 11$
$p = 3$

$(y+3)^2 = 4(3)(x-8)$
$y^2+6y+9 = 12(x-8)$
$y^2+6y+9 = 12x-96$
$y^2+6y-12x+105 = 0$

49. Vertex $(-9, 1)$, symmetry with respect to the line $x = -9$, contains the point $(-8, 0)$.

$(x-h)^2 = 4p(y-k)$
$[x-(-9)]^2 = 4p(y-1)$
$(x+9)^2 = 4p(y-1)$

Use the point $(-8, 0)$ for x and y.
$(-8+9)^2 = 4p(0-1)$
$1^2 = 4p(-1)$
$1 = -4p$
$-\frac{1}{4} = p$
$(x+9)^2 = 4\left(-\frac{1}{4}\right)(y-1)$
$x^2+18x+81 = -y+1$
$x^2+18x+y+80 = 0$

51. Vertex $(0, 10)$
$(x-0)^2 = 4p(y-10)$
$x^2 = 4p(y-10)$
Substitute the points
$(150, 40)$ to find p
$150^2 = 4p(40-10)$
$22500 = 120p$
$187.5 = p$
$(x-0)^2 = 4(187.5)(y-10)$
$x^2 = 750(y-10)$

53. Vertex $(0, 100)$
$(x-h)^2 = 4p(y-k)$
$(x-0)^2 = 4p(y-100)$
$x^2 = 4p(y-100)$
Using the point $(10, 0)$ for x and y
$10^2 = 4p(0-100)$
$100 = -400p$
$-\frac{1}{4} = p$
$(x-0)^2 = 4\left(-\frac{1}{4}\right)(y-100)$
$x^2 = -y+100$
Find x when $y = 50$
$x^2 = -50+100$
$x^2 = 50$
$x = \sqrt{50} = 5\sqrt{2}$

The width of the arch is
$2\left(5\sqrt{2}\right) = 10\sqrt{2}$ feet.

55. Set up the arch on a coordinate system with the vertex on the y-axis. Then the points $(-100, 0)$, $(100, 0)$, $(-60, 40)$ and $(60, 40)$ are on the parabola.
$x^2 = 4p(y-k)$

Substituting $(100, 0)$ and $(60, 40)$
$100^2 = 4p(0-k)$
$10000 = -4pk$

$60^2 = 4p(40-k)$
$3600 = 160p-4pk$

Substitute $10000 = -4pk$
$3600 = 160p+10000$
$-6400 = 160p$
$-40 = p$

$10000 = -4(-40)k$
$10000 = 160k$
$62.5 = k$
The arch must be 62.5 feet high in the center.

PROBLEM SET 13.3 Ellipses

1. $\frac{x^2}{4}+\frac{y^2}{1} = 1$
$a^2 = 4$ $\qquad$ $b^2 = 1$
$c^2 = a^2-b^2$
$c^2 = 4-1$
$c^2 = 3$, thus $c = \sqrt{3}$
vertices: $(\pm 2, 0)$
endpoints of minor axis: $(0, \pm 1)$
foci: $(\pm\sqrt{3}, 0)$

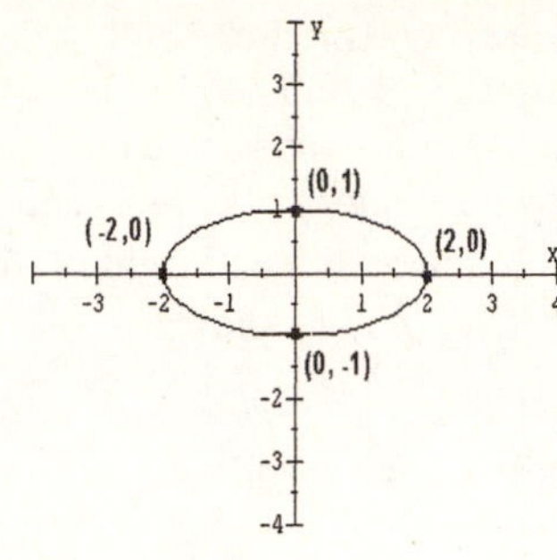

3. $\dfrac{x^2}{4} + \dfrac{y^2}{9} = 1$

$a^2 = 4 \qquad\qquad b^2 = 9$
$c^2 = b^2 - a^2 = 9 - 4 = 5$
$c^2 = 5,\ \text{thus } c = \sqrt{5}$

vertices: $(0,\ \pm 3)$
endpoints of minor axis: $(\pm 2,\ 0)$
foci: $(0,\ \pm\sqrt{5})$

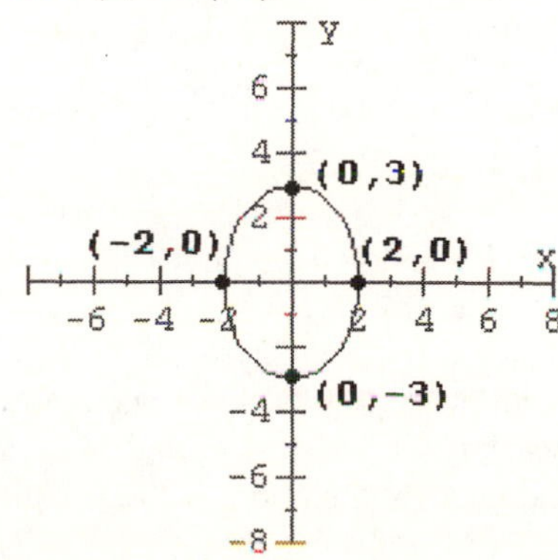

5. $9x^2 + 3y^2 = 27$

$\dfrac{x^2}{3} + \dfrac{y^2}{9} = 1$

$a^2 = 3 \qquad\qquad b^2 = 9$
$c^2 = b^2 - a^2 = 9 - 3 = 6$
$c^2 = 6,\ \text{thus } c = \sqrt{6}$
vertices: $(0,\ \pm 3)$
endpoints of minor axis: $(\pm\sqrt{3},\ 0)$
foci: $(0,\ \pm\sqrt{6})$

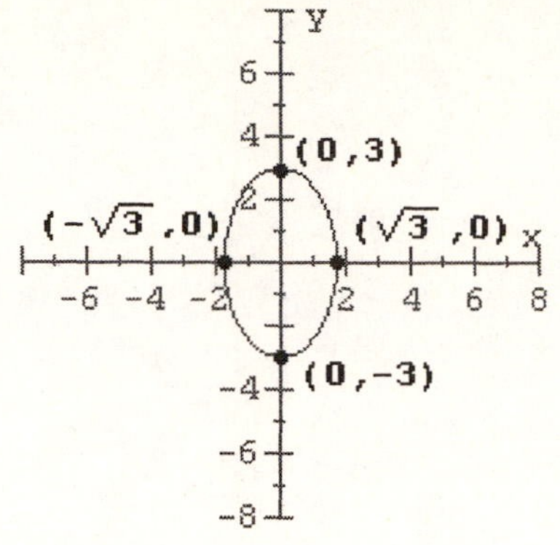

7. $2x^2 + 5y^2 = 50$

$\dfrac{x^2}{25} + \dfrac{y^2}{10} = 1$

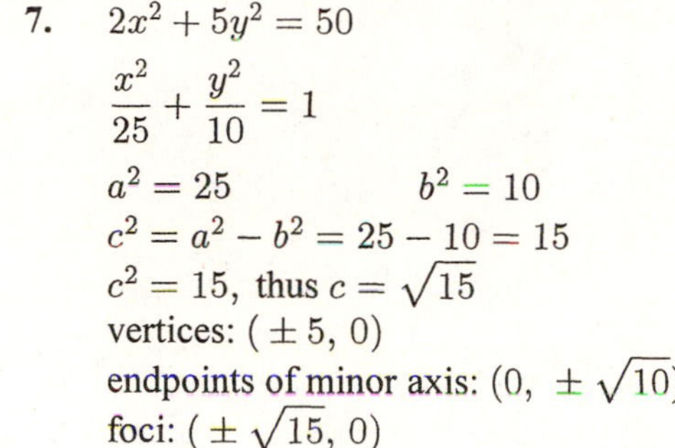

$a^2 = 25 \qquad\qquad b^2 = 10$
$c^2 = a^2 - b^2 = 25 - 10 = 15$
$c^2 = 15,\ \text{thus } c = \sqrt{15}$
vertices: $(\pm 5,\ 0)$
endpoints of minor axis: $(0,\ \pm\sqrt{10})$
foci: $(\pm\sqrt{15},\ 0)$

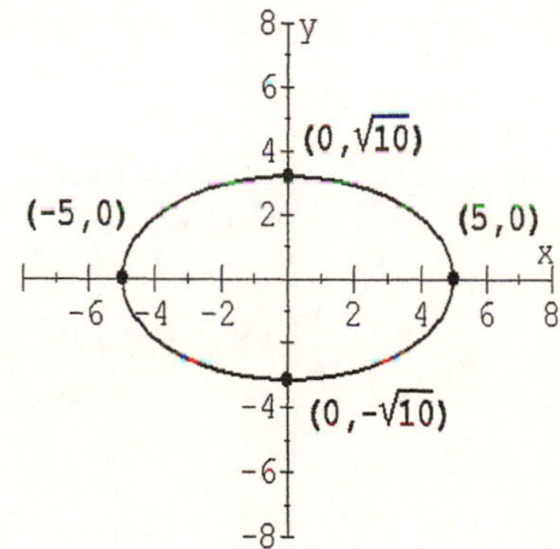

9. $12x^2 + y^2 = 36$

$\dfrac{x^2}{3} + \dfrac{y^2}{36} = 1$

$a^2 = 3 \qquad\qquad b^2 = 36$
$c^2 = b^2 - a^2 = 36 - 3 = 33$
$c^2 = 33,\ \text{thus } c = \sqrt{33}$
vertices: $(0,\ \pm 6)$
endpoints of minor axis: $(\pm\sqrt{3},\ 0)$
foci: $(0,\ \pm\sqrt{33})$

Problem Set 13.3

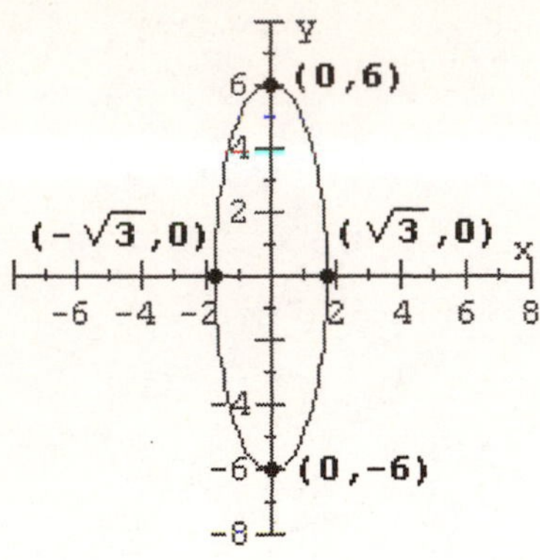

11. $7x^2 + 11y^2 = 77$

$\dfrac{x^2}{11} + \dfrac{y^2}{7} = 1$

$a^2 = 11 \qquad b^2 = 7$

$c^2 = a^2 - b^2 = 11 - 7 = 4$

$c^2 = 4,$ thus $c = 2$

vertices: $(\pm\sqrt{11},\ 0)$

endpoints of minor axis: $(0,\ \pm\sqrt{7})$

foci: $(\pm 2,\ 0)$

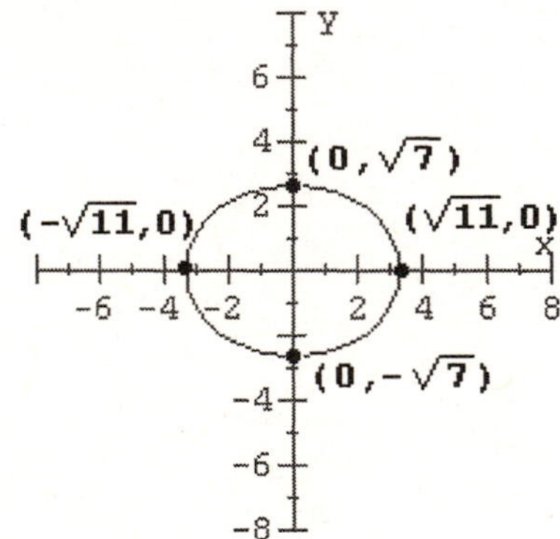

13. $\dfrac{(x-2)^2}{9} + \dfrac{(y-1)^2}{4} = 1$

$a^2 = 9; \qquad b^2 = 4$

$a = 3; \qquad b = 2$

$h = 2, \qquad k = 1$

$c^2 = a^2 - b^2 = 9 - 4 = 5$

thus $c = \sqrt{5}$

Center at $(2,\ 1)$

Vertices are 3 units left and right of the center at $(-1, 1)$ and $(5, 1)$.
Endpoints of minor axis are 2 units above and below the center at $(2,\ -1)$ and $(2, 3)$.
Foci are $\sqrt{5}$ units left and right of the center at $(2+\sqrt{5},\ 1)$ and $(2-\sqrt{5}, 1)$.

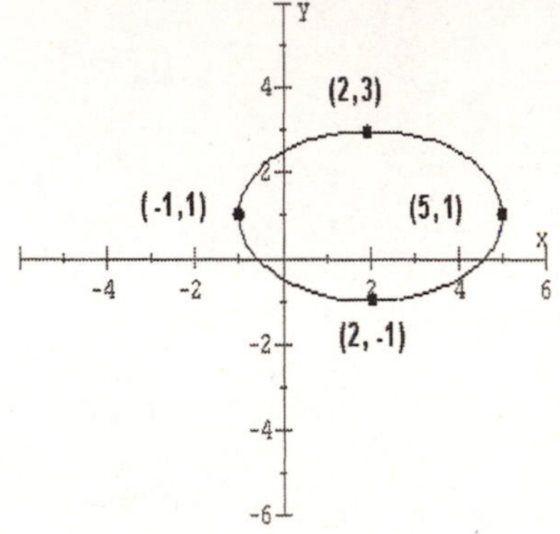

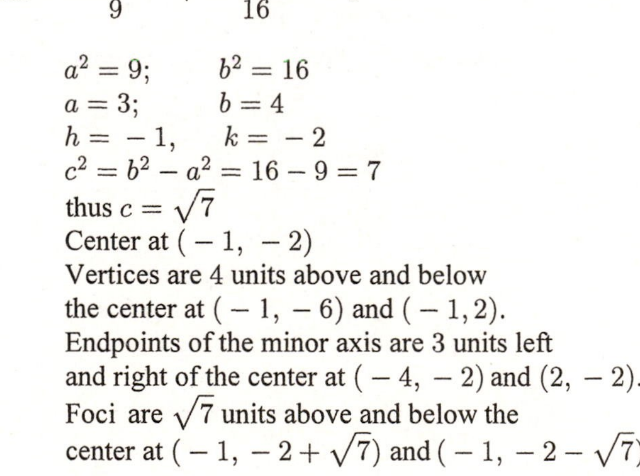

15. $\dfrac{(x+1)^2}{9} + \dfrac{(y+2)^2}{16} = 1$

$a^2 = 9; \qquad b^2 = 16$

$a = 3; \qquad b = 4$

$h = -1, \qquad k = -2$

$c^2 = b^2 - a^2 = 16 - 9 = 7$

thus $c = \sqrt{7}$

Center at $(-1,\ -2)$

Vertices are 4 units above and below the center at $(-1,\ -6)$ and $(-1, 2)$.
Endpoints of the minor axis are 3 units left and right of the center at $(-4,\ -2)$ and $(2,\ -2)$.
Foci are $\sqrt{7}$ units above and below the center at $(-1,\ -2+\sqrt{7})$ and $(-1,\ -2-\sqrt{7})$.

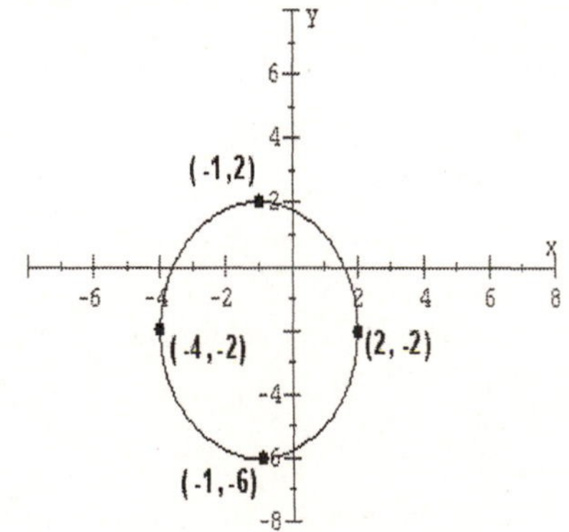

17. $4x^2 - 8x + 9y^2 - 36y + 4 = 0$
$4(x^2 - 2x) + 9(y^2 - 4y) = -4$
$4(x^2 - 2x+1)+9(y^2 - 4y+4) = -4+4+36$
$4(x-1)^2 + 9(y-2)^2 = 36$

$$\frac{(x-1)^2}{9} + \frac{(y-2)^2}{4} = 1$$

$h = 1,\ k = 2,\ a = 3,\ b = 2$
$c^2 = a^2 - b^2 = 9 - 4 = 5$
$c^2 = 5$, thus $c = \sqrt{5}$
Center is at $(1, 2)$.
Vertices are 3 units left and right of the center at $(-2, 2)$ and $(4, 2)$.
Endpoints of minor axis are 2 units above and below the center at $(1, 4)$ and $(1, 0)$.
Foci are $\sqrt{5}$ units left and right of the center at $(1 - \sqrt{5}, 2)$ and $(1 + \sqrt{5}, 2)$.

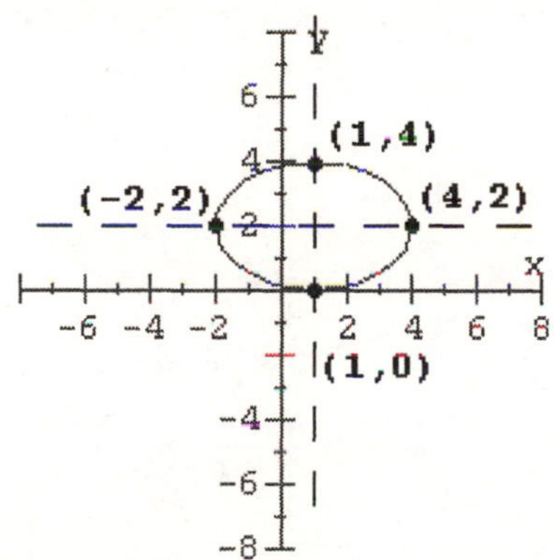

19. $4x^2 + 16x + y^2 + 2y + 1 = 0$
$4(x^2 + 4x) + (y+1)^2 = 0$
$4(x^2 + 4x + 4) + (y+1)^2 = 0 + 16$
$4(x+2)^2 + (y+1)^2 = 16$

$$\frac{(x+2)^2}{4} + \frac{(y+1)^2}{16} = 1$$

$h = -2,\ k = -1,\ a = 2,\ b = 4$
$c^2 = b^2 - a^2$
$c^2 = 16 - 4$
$c^2 = 12$, thus $c = \sqrt{12} = 2\sqrt{3}$
Center is at $(-2, -1)$.
Vertices are 4 units above and below the center at $(-2, 3)$ and $(-2, -5)$.
Endpoints of minor axis are 2 units left and right of the center at $(-4, -1)$ and $(0, -1)$.

Foci are $2\sqrt{3}$ units above and below the center at $(-2, -1 - 2\sqrt{3})$ and $(-2, -1 + 2\sqrt{3})$.

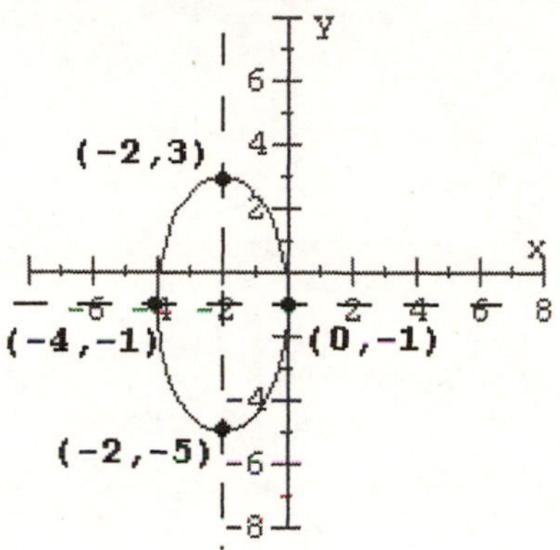

21. $x^2 - 6x + 4y^2 + 5 = 0$
$x^2 - 6x + 4y^2 = -5$
$(x^2 - 6x + 9) + 4y^2 = -5 + 9$
$(x-3)^2 + 4y^2 = 4$

$$\frac{(x-3)^2}{4} + \frac{y^2}{1} = 1$$

$h = 3,\ k = 0,\ a = 2,\ b = 1$
$c^2 = a^2 - b^2$
$c^2 = 4 - 1$
$c^2 = 3$, thus $c = \sqrt{3}$
Center is at $(3, 0)$.
Vertices are 2 units left and right of the center at $(1, 0)$ and $(5, 0)$.
Endpoints of minor axis are 1 unit above and below the center at $(3, -1)$ and $(3, 1)$.
Foci are $\sqrt{3}$ units left and right of the center at $(3 - \sqrt{3}, 0)$ and $(3 + \sqrt{3}, 0)$.

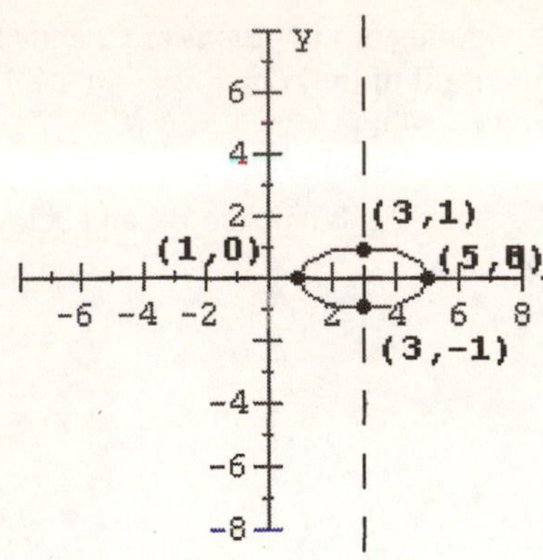

23. $9x^2 - 72x + 2y^2 + 4y + 128 = 0$
$9(x^2 - 8x) + 2(y^2 + 2y) = -128$
$9(x^2 - 8x+16)+2(y^2+2y+1) = -128+144+2$
$9(x - 4)^2 + 2(y + 1)^2 = 18$
$\frac{(x-4)^2}{2} + \frac{(y+1)^2}{9} = 1$
$h = 4,\ k = -1,\ a = \sqrt{2},\ b = 3$
$c^2 = b^2 - a^2$
$c^2 = 9 - 2$
$c^2 = 7$, thus $c = \sqrt{7}$
Center is at $(4, -1)$.
Vertices are 3 units above and below the center at $(4, 2)$ and $(4, -4)$.
Endpoints of minor axis are $\sqrt{2}$ units left and right of the center at $(4 - \sqrt{2}, -1)$ and $(4 + \sqrt{2}, -1)$.
Foci are $\sqrt{7}$ units above and below the center at $(4, -1 - \sqrt{7})$ and $(4, -1 + \sqrt{7})$.

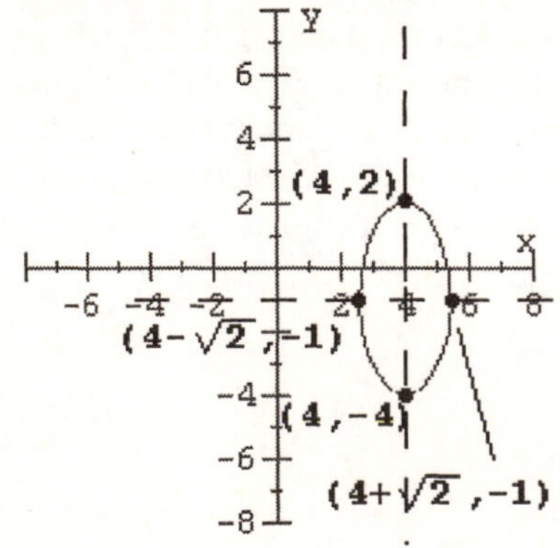

25. $2x^2 + 12x + 11y^2 - 88y + 172 = 0$
$2(x^2 + 6x) + 11(y^2 - 8y) = -172$
$2(x^2+6x+9)+11(y^2-8y+16) = -172+18+176$
$2(x + 3)^2 + 11(y - 4)^2 = 22$
$\frac{(x+3)^2}{11} + \frac{(y-4)^2}{2} = 1$
$h = -3,\ k = 4,\ a = \sqrt{11},\ b = \sqrt{2}$
$c^2 = a^2 - b^2$
$c^2 = 11 - 2$
$c^2 = 9$, thus $c = 3$

Center at $(-3, 4)$
Vertices are $\sqrt{11}$ units left and right of the center at $(-3 - \sqrt{11}, 4)$ and $(-3 + \sqrt{11}, 4)$.
Endpoints of minor axis are $\sqrt{2}$ units above and below the center at $(-3, 4 - \sqrt{2})$ and $(-3, 4 + \sqrt{2})$.
Foci are 3 units left and right of the center at $(-6, 4)$ and $(0, 4)$.

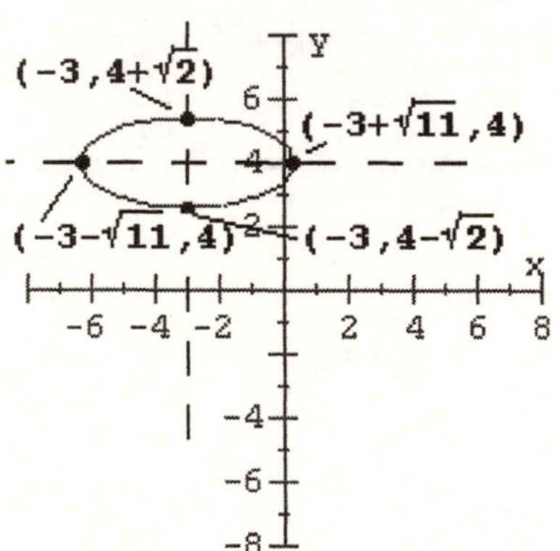

27. Vertices $(\pm 5, 0)$, foci $(\pm 3, 0)$
$a^2 = 25$, thus $a = 5$
$c^2 = 9$, thus $c = 3$
$c^2 = a^2 - b^2$
$9 = 25 - b^2$
$b^2 = 16$
$\frac{x^2}{25} + \frac{y^2}{16} = 1$

$16x^2 + 25y^2 = 400$

29. Vertices $(0, \pm 6)$, foci $(0, \pm 5)$
Note: the vertices are on the y-axis.
$a = 6 \qquad c = 5$
$a^2 = 36 \qquad c^2 = 25$
$c^2 = a^2 - b^2$
$25 = 36 - b^2$
$-11 = -b^2$
$b^2 = 11$
$$\frac{x^2}{b^2} + \frac{y^2}{a^2} = 1$$
$$\frac{x^2}{11} + \frac{y^2}{36} = 1$$
$36x^2 + 11y^2 = 396$

31. Vertices $(\pm 3, 0)$
Length of minor axis is 2.
$a^2 = 9 \qquad b^2 = \frac{1}{2}(2) = 1$
$$\frac{x^2}{9} + \frac{y^2}{1} = 1$$
$x^2 + 9y^2 = 9$

33. Foci $(0, \pm 2)$ Length of minor axis $= 3$.
Note: foci on y-axis
$c = 2 \qquad 2a = 3$
$c^2 = 4a = \frac{3}{2}; \qquad a^2 = \frac{9}{4}$
$c^2 = b^2 - a^2$
$$4 = b^2 - \frac{9}{4}$$
$$\frac{25}{4} = b^2$$
$$\frac{x^2}{a^2} + \frac{y^2}{b^2} = 1$$
$$\frac{x^2}{\frac{9}{4}} + \frac{y^2}{\frac{25}{4}} = 1$$
$$\frac{4x^2}{9} + \frac{4y^2}{25} = 1$$
$100x^2 + 36y^2 = 225$

35. Vertices $(0, \pm 5)$
Contains the point $(3, 2)$
$b^2 = 25$
$$\frac{x^2}{a^2} + \frac{y^2}{25} = 1$$
Substitute $(3, 2)$.
$$\frac{3^2}{a^2} + \frac{2^2}{25} = 1$$
$$\frac{9}{a^2} + \frac{4}{25} = 1$$
$$\frac{9}{a^2} = \frac{21}{25}$$
$21a^2 = 25(9)$
$21a^2 = 225$
$$a^2 = \frac{225}{21} = \frac{75}{7}$$
$$\frac{x^2}{\frac{75}{7}} + \frac{y^2}{25} = 1$$
$$\frac{7x^2}{75} + \frac{y^2}{25} = 1$$
$7x^2 + 3y^2 = 75$

37. Vertices $(5, 1)$ and $(-3, 1)$
Foci $(3, 1)$ and $(-1, 1)$
Center is midway between the vertices at $(1, 1)$.
a is the distance between a vertex and the center, so $a = 4$.
c is the distance between a focus and the center, so $c = 2$

$a = 4 \qquad c = 2$
$a^2 = 16 \qquad c^2 = 4$
$c^2 = a^2 - b^2$
$4 = 16 - b^2$
$-12 = -b^2$
$b^2 = 12$

$$\frac{(x-h)^2}{a^2}+\frac{(y-k)^2}{b^2}=1$$

$$\frac{(x-1)^2}{16}+\frac{(y-1)^2}{12}=1$$

$3(x-1)^2+4(y-1)^2=48$
$3(x^2-2x+1)+4(y^2-2y+1)=48$
$3x^2-6x+3+4y^2-8y+4=48$
$3x^2-6x+4y^2-8y-41=0$

39. Center $(0,\ 1)$ and focus at $(-4,\ 1)$, length of minor axis is 6.

$2b=6 \qquad c=4$
$b=3 \qquad c^2=16$
$b^2=9$

$c^2=a^2-b^2$
$16=a^2-9$
$25=a^2$

$$\frac{(x-0)^2}{25}+\frac{(y-1)^2}{9}=1$$

$9x^2+25(y-1)^2=225$
$9x^2+25(y^2-2y+1)=225$
$9x^2+25y^2-50y+25=225$
$9x^2+25y^2-50y-200=0$

41. This is an ellipse with foci $(2,\ 0)$ and $(-2,\ 0)$ on the x-axis. Furthermore, $c=2$. The center is midway between the foci, at $(0,\ 0)$. The sum of the distance is $2a$ so,
$2a=8$
$a=4$
$c^2=a^2-b^2$
$2^2=4^2-b^2$
$4=16-b^2$
$-12=-b^2$
$b^2=12$

$$\frac{x^2}{a^2}+\frac{y^2}{b^2}=1$$

$$\frac{x^2}{16}+\frac{y^2}{12}=1$$

$3x^2+4y^2=48$

43. Orientate the ellipse on a coordinate system with the center at $(0,\ 0)$.
The vertices would be $(-15,\ 0)$ and $(15,\ 0)$.
Endpoints of the minor axis would be $(0,\ -10)$ and $(0, 10)$.

$a^2=225 \qquad b^2=100$

$$\frac{x^2}{225}+\frac{y^2}{100}=1$$

Find y, when $x=10$.

$$\frac{10^2}{225}+\frac{y^2}{100}=1$$

$$\frac{100}{225}+\frac{y^2}{100}=1$$

$$\frac{4}{9}+\frac{y^2}{100}=1$$

$$\frac{y^2}{100}=\frac{5}{9}$$

$9y^2=500$

$$y^2=\frac{500}{9}$$

$$y=\pm\sqrt{\frac{500}{9}}=\pm\frac{10\sqrt{5}}{3}$$

The height would be $\frac{10\sqrt{5}}{3}$ feet.

PROBLEM SET 13.4 Hyperbolas

1. $\dfrac{x^2}{9} - \dfrac{y^2}{4} = 1$

$a^2 = 9$, thus $a = 3$
$b^2 = 4$, thus $b = 2$
$c^2 = a^2 + b^2 = 4 + 9 = 13$
$c^2 = 13$, thus $c = \sqrt{13}$
vertices: $(\pm 3, 0)$
foci: $(\pm \sqrt{13}, 0)$
asymptotes: $y = \pm \dfrac{b}{a}x$; $\quad y = \pm \dfrac{2}{3}x$

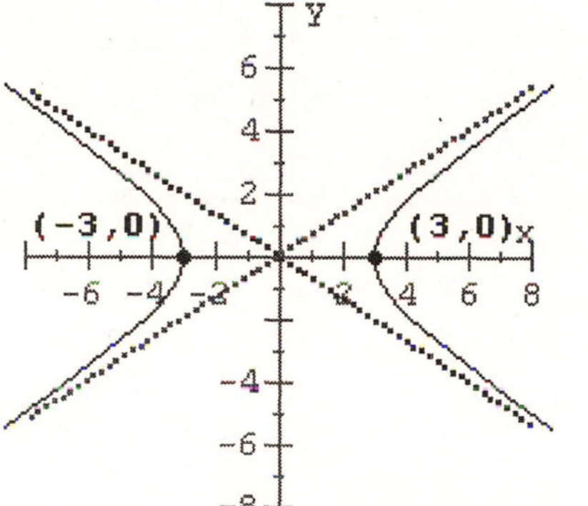

3. $\dfrac{y^2}{4} - \dfrac{x^2}{9} = 1$

$a^2 = 9$ so $a = 3$
$b^2 = 4$ so $b = 2$
$c^2 = a^2 + b^2 = 9 + 4 = 13$
$c^2 = 13$, thus $c = \sqrt{13}$
vertices: $(0, \pm 2)$
foci: $(0, \pm \sqrt{13})$
asymptotes: $y = \pm \dfrac{b}{a}x$, $\quad y = \pm \dfrac{2}{3}x$

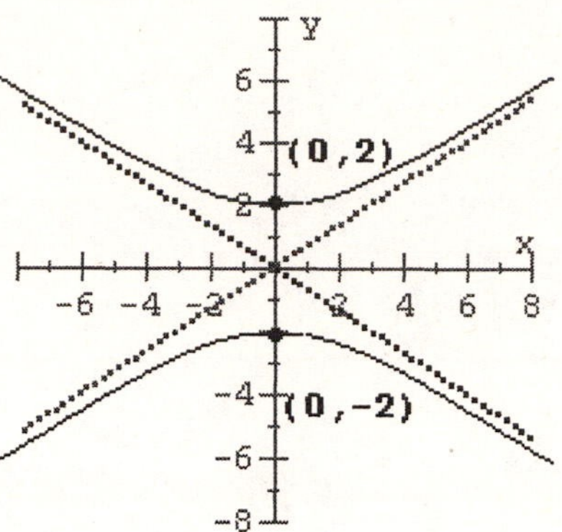

5. $9y^2 - 16x^2 = 144$

$\dfrac{y^2}{16} - \dfrac{x^2}{9} = 1$
$b^2 = 16$, thus $b = 4$
$a^2 = 9$, thus $a = 3$
$c^2 = a^2 + b^2 = 9 + 16 = 25$
$c^2 = 25$, thus $c = 5$

vertices: $(0, \pm 4)$
foci: $(0, \pm 5)$
asymptotes: $y = \pm \dfrac{b}{a}x$, $\quad y = \pm \dfrac{4}{3}x$

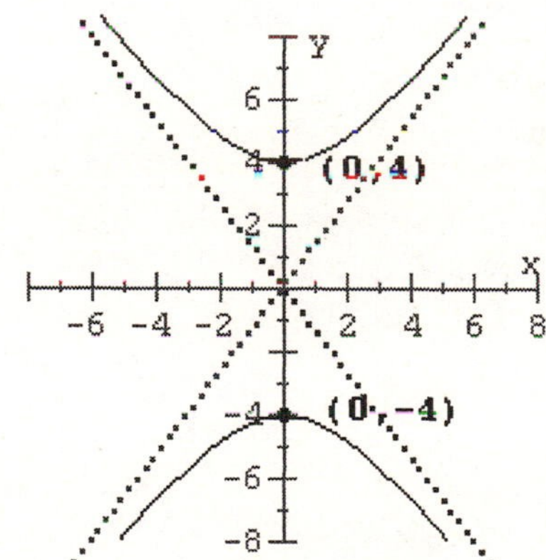

Problem Set 13.4

7. $x^2 - y^2 = 9$

$\frac{x^2}{9} - \frac{y^2}{9} = 1$

$a^2 = 9 \quad a = 3$
$b^2 = 9 \quad b = 3$
$c^2 = a^2 + b^2 = 9 + 9 = 18$
$c^2 = 18$, thus $c = \sqrt{18} = 3\sqrt{2}$
vertices: $(\pm 3, 0)$
foci: $(\pm 3\sqrt{2}, 0)$
asymptotes: $y = \pm \frac{3}{3}x = \pm x$

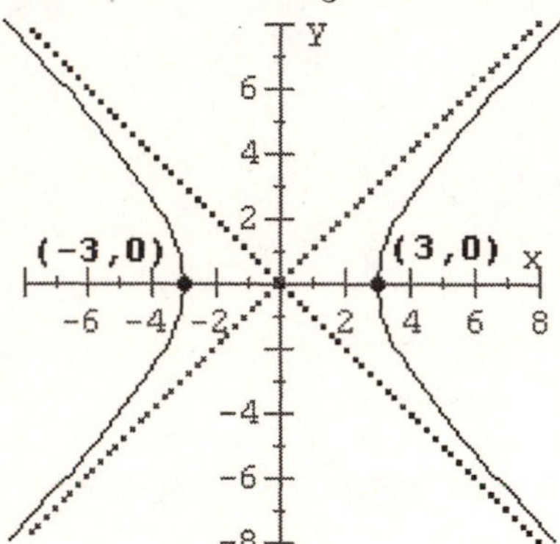

9. $5y^2 - x^2 = 25$

$\frac{y^2}{5} - \frac{x^2}{25} = 1$

$a^2 = 25 \quad a = 5$
$b^2 = 5 \quad b = \sqrt{5}$
$c^2 = a^2 + b^2 = 25 + 5 = 30$
$c^2 = 30$, thus $c = \sqrt{30}$
vertices: $(0, \pm\sqrt{5})$
foci: $(0, \pm\sqrt{30})$
asymptotes: $y = \pm\frac{b}{a}x$, $y = \pm\frac{\sqrt{5}}{5}x$

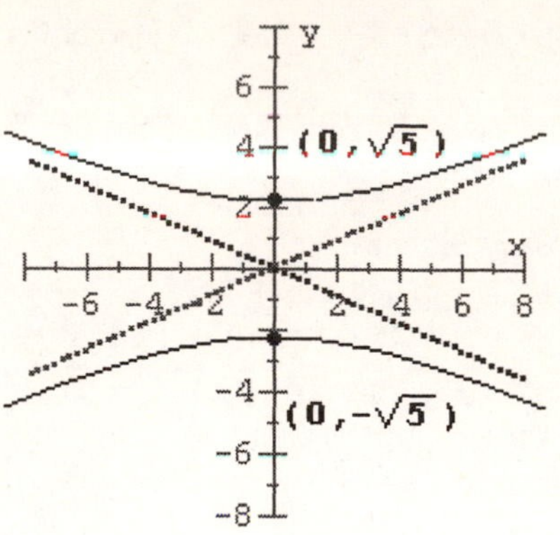

11. $y^2 - 9x^2 = -9$

$\frac{-y^2}{9} + \frac{x^2}{1} = 1$

$\frac{x^2}{1} - \frac{y^2}{9} = 1$

$a^2 = 1 \quad a = 1$
$b^2 = 9 \quad b = 3$
$c^2 = a^2 + b^2 = 1 + 9 = 10$
$c^2 = 10$, thus $c = \sqrt{10}$

vertices: $(\pm 1, 0)$
foci: $(\pm\sqrt{10}, 0)$
asymptotes: $y = \pm\frac{3}{1}x = \pm 3x$

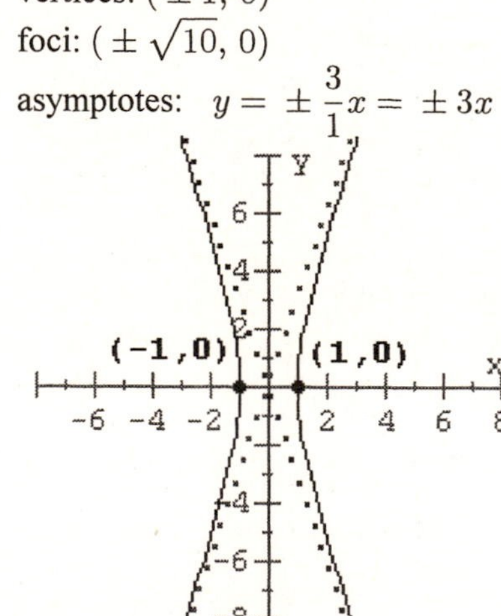

13. $4x^2 - 24x - 9y^2 - 18y - 9 = 0$
$4(x^2 - 6x) - 9(y^2 + 2y) = 9$
$4(x^2 - 6x+9) - 9(y^2 + 2y+1) = 9 + 36 - 9$
$4(x-3)^2 - 9(y+1)^2 = 36$
$$\frac{(x-3)^2}{9} - \frac{(y+1)^2}{4} = 1$$
$a^2 = 9$, thus $a = 3$
$b^2 = 4$, thus $b = 2$
$c^2 = a^2 + b^2 = 4 + 9 = 13$
$c^2 = 13$, thus $c = \sqrt{13}$

Center at $(3, -1)$
Vertices are 3 units left and right of the center at $(0, -1)$ and $(6, -1)$.
Foci are $\sqrt{13}$ units left and right of the center at $(3 - \sqrt{13}, -1)$ and $(3+\sqrt{13}, -1)$.

Asymptotes: Using $a = 3$ and $b = 2$, the slopes are $\pm \frac{b}{a} = \pm \frac{2}{3}$.
The lines going through the center $(3, -1)$ with slopes $\pm \frac{2}{3}$ are:
$[y - (-1)] = \pm \frac{2}{3}(x-3)$

$y + 1 = \pm \frac{2}{3}(x-3)$

$y + 1 = \frac{2}{3}(x-3)$, $y + 1 = -\frac{2}{3}(x-3)$
$3y + 3 = 2x - 6$, $3y + 3 = -2x + 6$
$2x - 3y = 9$, $2x + 3y = 3$

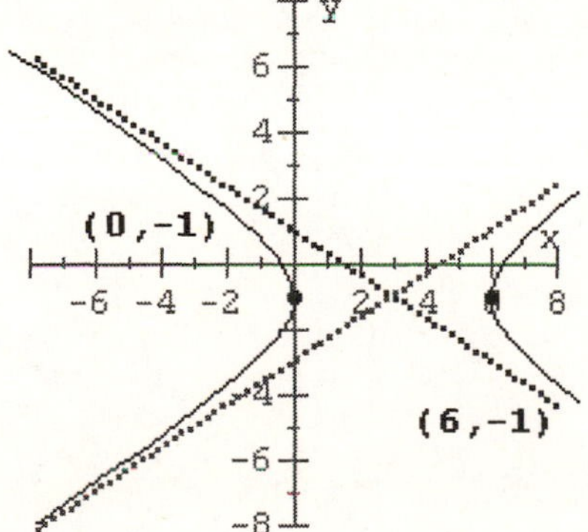

15. $y^2 - 4y - 4x^2 - 24x - 36 = 0$
$y^2 - 4y - 4(x^2 + 6x) = 36$
$y^2 - 4y+4 - 4(x^2+6x+9) = 36+4 - 36$
$(y-2)^2 - 4(x+3)^2 = 4$
$$\frac{(y-2)^2}{4} - \frac{(x+3)^2}{1} = 1$$

Center is at $(-3, 2)$.

$a^2 = 1 \qquad a = 1$
$b^2 = 4 \qquad b = 2$
$c^2 = a^2 + b^2 = 4 + 1 = 5$
$c^2 = 5$, thus $c = \sqrt{5}$

Vertices are 2 units above and below the center at $(-3, 0)$ and $(-3, 4)$.
Foci are $\sqrt{5}$ units above and below the center at $(-3, 2 - \sqrt{5})$ and $(-3, 2 + \sqrt{5})$.

Asymptotes: Using $a = 1$ and $b = 2$, the slopes are $\pm \frac{b}{a} = \pm \frac{2}{1} = \pm 2$.

$y - 2 = \pm 2(x+3)$
$y - 2 = 2(x+3)$, $y - 2 = -2(x+3)$
$y - 2 = 2x + 6$, $y - 2 = -2x - 6$
$2x - y = -8$, $2x + y = -4$

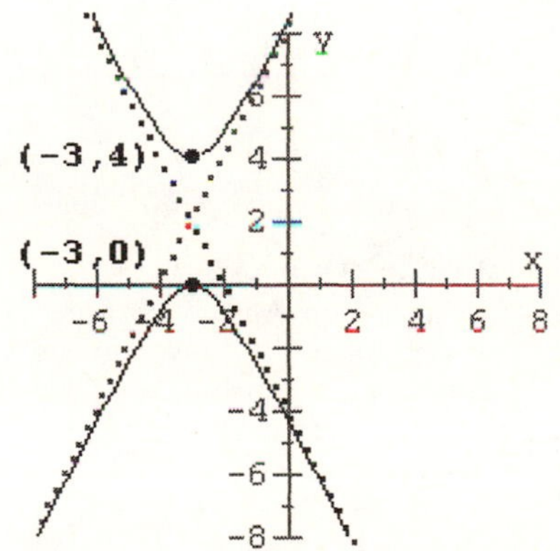

17. $2x^2 - 8x - y^2 + 4 = 0$
$2(x^2 - 4x) - y^2 = -4$
$2(x^2 - 4x + 4) - y^2 = -4 + 8$
$2(x-2)^2 - y^2 = 4$

$$\frac{(x-2)^2}{2} - \frac{y^2}{4} = 1$$

$a^2 = 2$, thus $a = \sqrt{2}$
$b^2 = 4$, thus $b = 2$
$c^2 = a^2 + b^2$
$c^2 = 2 + 4$
$c^2 = 6$, thus $c = \sqrt{6}$

Center at $(2, 0)$
Vertices are $\sqrt{2}$ units left and right of the Center at $(2 - \sqrt{2}, 0)$ and $(2 + \sqrt{2}, 0)$.
Foci are $\sqrt{6}$ units left and right of the center at $(2 - \sqrt{6}, 0)$ and $(2 + \sqrt{6}, 0)$.

Asymptotes: Using $a = \sqrt{2}$ and $b = 2$, the slopes are $\pm\frac{b}{a} = \pm\frac{2}{\sqrt{2}} = \pm\sqrt{2}$.
$y - 0 = \pm\sqrt{2}(x-2)$
$y - 0 = \sqrt{2}(x-2)$, $\quad y - 0 = -\sqrt{2}(x-2)$
$y = \sqrt{2}x - 2\sqrt{2}$, $\quad y = -\sqrt{2}x + 2\sqrt{2}$

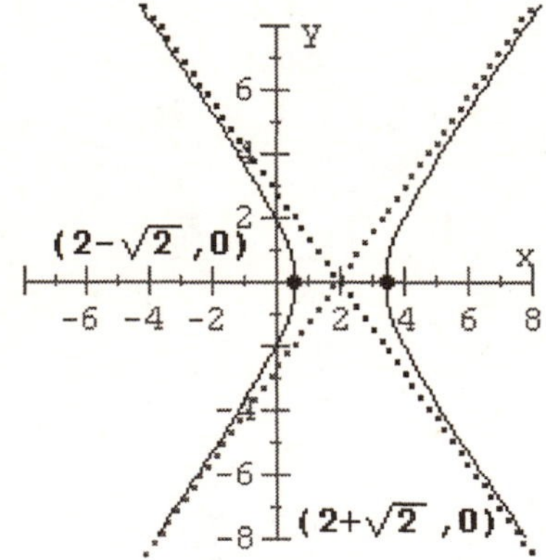

19. $y^2 + 10y - 9x^2 + 16 = 0$
$y^2 + 10y + 25 - 9x^2 = -16 + 25$
$(y+5)^2 - 9x^2 = 9$

$$\frac{(y+5)^2}{9} - \frac{x^2}{1} = 1$$

$a^2 = 1 \quad a = 1$
$b^2 = 9 \quad b = 3$
$c^2 = a^2 + b^2$
$c^2 = 9 + 1$
$c^2 = 10$, thus $c = \sqrt{10}$

Center at $(0, -5)$
Vertices are 3 units above and below the center at $(0, -8)$ and $(0, -2)$.
Foci are $\sqrt{10}$ units above and below the center at $(0, -5 - \sqrt{10})$ and $(0, -5 + \sqrt{10})$.

Asymptotes: Using $a = 1$ and $b = 3$, the slopes are $\pm\frac{b}{a} = \pm\frac{3}{1} = \pm 3$

$y - (-5) = \pm 3(x - 0)$
$y + 5 = \pm 3x$
$y + 5 = -3x$, $\quad y + 5 = 3x$
$3x + y = -5$, $\quad 3x - y = 5$

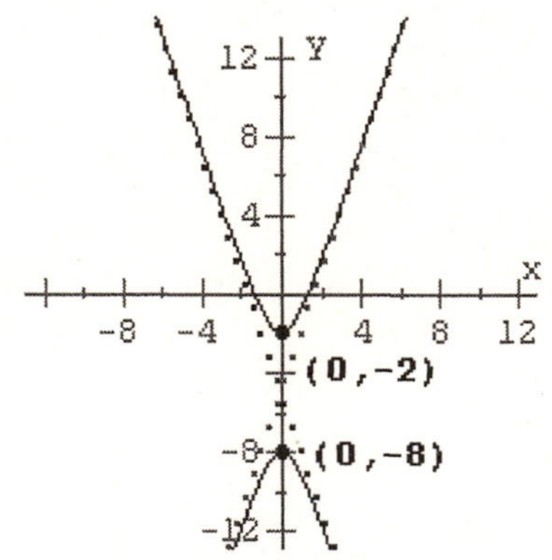

21. $x^2 + 4x - y^2 - 4y - 1 = 0$
$(x^2 + 4x) - (y^2 + 4y) = 1$
$(x^2 + 4x + 4) - (y^2 + 4y + 4) = 1 + 4 - 4$
$(x + 2)^2 - (y + 2)^2 = 1$

$a^2 = 1 \qquad a = 1$
$b^2 = 1 \qquad b = 1$
$c^2 = a^2 + b^2 = 1 + 1 = 2$
$c^2 = 2$, thus $c = \sqrt{2}$

Center at $(-2, -2)$
Vertices are 1 unit left and right of the center at $(-3, -2)$ and $(-1, -2)$.

Foci are $\sqrt{2}$ units left and right of the center at $(-2 - \sqrt{2}, -2)$ and $(-2 + \sqrt{2}, -2)$.

Asymptotes: Using $a = 1$ and $b = 1$, the slopes are $\pm\frac{b}{a} = \pm 1$.
$y - (-2) = \pm 1[x - (-2)]$
$y + 2 = \pm 1(x + 2)$

$y + 2 = -x - 2, \quad y + 2 = x + 2$

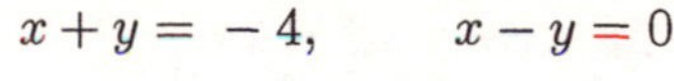
$x + y = -4, \qquad x - y = 0$

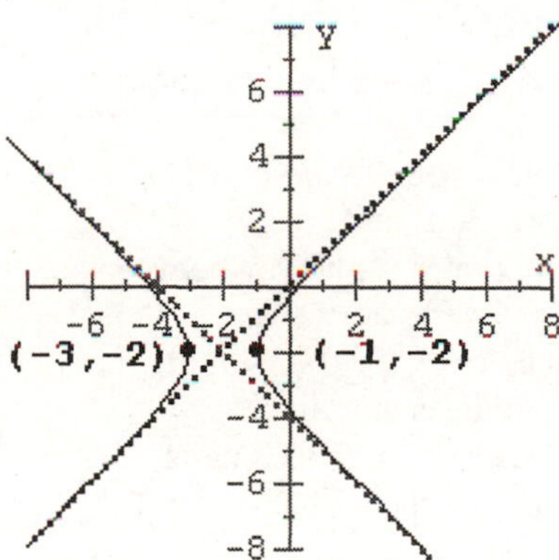

23. Vertices $(\pm 2, 0)$, foci $(\pm 3, 0)$
$a = 2 \qquad c = 3$
$c^2 = a^2 + b^2$
$3^2 = 2^2 + b^2$
$9 = 4 + b^2$
$5 = b^2$, thus $b = \sqrt{5}$
$$\frac{x^2}{4} - \frac{y^2}{5} = 1$$
$5x^2 - 4y^2 = 20$

25. Vertices $(0, \pm 3)$, foci $(0, \pm 5)$
$b = 3 \qquad c = 5$
$c^2 = a^2 + b^2$
$5^2 = a^2 + 3^2$
$25 = a^2 + 9$
$16 = a^2$, thus $a = 4$
$$\frac{y^2}{b^2} - \frac{x^2}{a^2} = 1$$
$$\frac{y^2}{9} - \frac{x^2}{16} = 1$$
$16y^2 - 9x^2 = 144$

27. Vertices $(\pm 1, 0)$, contains the point $(2, 3)$
$a = 1$
$$\frac{x^2}{1} - \frac{y^2}{b^2} = 1$$
Substitute 2 for x and 3 for y.
$$\frac{2^2}{1} - \frac{3^2}{b^2} = 1$$
$$4 - \frac{9}{b^2} = 1$$
$$-\frac{9}{b^2} = -3$$
$-9 = -3b^2$
$3 = b^2$
$$\frac{x^2}{1} - \frac{y^2}{3} = 1$$
$3x^2 - y^2 = 3$

29. Vertices $(0, \pm\sqrt{3})$, length of the conjugate axis is 4.
$b = \sqrt{3}$
$2a = 4$
$a = 2$

$$\frac{y^2}{b^2} - \frac{x^2}{a^2} = 1$$

$$\frac{y^2}{(\sqrt{3})^2} - \frac{x^2}{2^2} = 1$$

$$\frac{y^2}{3} - \frac{x^2}{4} = 1$$

$4y^2 - 3x^2 = 12$

31. Foci $(\pm\sqrt{23}, 0)$
Length of the transverse axis is 8.
$c = \sqrt{23}$ $\quad 2a = 8$
$a = 4$
$c^2 = a^2 + b^2$
$(\sqrt{23})^2 = 4^2 + b^2$
$23 = 16 + b^2$
$7 = b^2$, thus $b = \sqrt{7}$

$$\frac{x^2}{16} - \frac{y^2}{7} = 1$$

$7x^2 - 16y^2 = 112$

33. Vertices $(6, -3)$ and $(2, -3)$
Foci $(7, -3)$ and $(1, -3)$
The center is midway between the vertices, hence is at $(4, -3)$.
The vertex $(6, -3)$ is 2 units from the center, hence $a = 2$.
The focus $(7, -3)$ is 3 units from the center, hence $c = 3$.

$c^2 = a^2 + b^2$
$3^2 = 2^2 + b^2$
$9 = 4 + b^2$
$5 = b^2$, thus $b = \sqrt{5}$

$$\frac{(x-h)^2}{a^2} - \frac{(y-k)^2}{b^2} = 1$$

$$\frac{(x-4)^2}{2^2} - \frac{[y-(-3)]^2}{(\sqrt{5})^2} = 1$$

$$\frac{(x-4)^2}{4} - \frac{(y+3)^2}{5} = 1$$

$5(x-4)^2 - 4(y+3)^2 = 20$
$5(x^2 - 8x + 16) - 4(y^2 + 6y + 9) = 20$
$5x^2 - 40x + 80 - 4y^2 - 24y - 36 = 20$
$5x^2 - 40x - 4y^2 - 24y + 24 = 0$

35. Vertices $(-3, 7)$ and $(-3, 3)$
Foci $(-3, 9)$ and $(-3, 1)$
The center is midway between the vertices, so the center is at $(-3, 5)$ and hence $b = 2$ and $c = 4$.

$c^2 = a^2 + b^2$
$4^2 = a^2 + 2^2$
$16 = a^2 + 4$
$12 = a^2$, thus $a = \sqrt{12} = 2\sqrt{3}$

$$\frac{(y-k)^2}{b^2} - \frac{(x-h)^2}{a^2} = 1$$

$$\frac{(y-5)^2}{4} - \frac{(x+3)^2}{12} = 1$$

$3(y-5)^2 - (x+3)^2 = 12$
$3(y^2 - 10y + 25) - (x^2 + 6x + 9) = 12$
$3y^2 - 30y + 75 - x^2 - 6x - 9 = 12$
$3y^2 - 30y - x^2 - 6x + 54 = 0$

37. Vertices $(0,0)$ and $(4,0)$
Foci $(5,0)$ and $(-1,0)$
The center is midway between the vertices at $(2, 0)$.
The vertex $(0, 0)$ is 2 units from the center, hence $a = 2$.
The focus $(5, 0)$ is 3 units from the center, hence $c = 3$.

$c^2 = a^2 + b^2$
$3^2 = 2^2 + b^2$
$9 = 4 + b^2$
$5 = b^2$, thus $b = \sqrt{5}$

$$\frac{(x-h)^2}{a^2} - \frac{(y-k)^2}{b^2} = 1$$

$$\frac{(x-2)^2}{4} - \frac{(y-0)^2}{5} = 1$$

$5(x-2)^2 - 4y^2 = 20$
$5(x^2 - 4x + 4) - 4y^2 = 20$
$5x^2 - 20x + 20 - 4y^2 - 20 = 0$
$5x^2 - 20x - 4y^2 = 0$

39. Circle
41. Straight Line
43. Ellipse
45. Hyperbola
47. Parabola

PROBLEM SET 13.5 Systems Involving Nonlinear Equations

1. $\begin{pmatrix} x^2 + y^2 = 5 \\ x + 2y = 5 \end{pmatrix}$
$x = -2y + 5$
$(-2y+5)^2 + y^2 = 5$
$4y^2 - 20y + 25 + y^2 = 5$
$5y^2 - 20y + 20 = 0$
$5(y^2 - 4y + 4) = 0$
$y^2 - 4y + 4 = 0$
$(y-2)^2 = 0$
$y = 2$
When $y = 2$,
$x = -2y + 5$
$x = -2(2) + 5 = 1$
The solution set is $\{(1, 2)\}$.

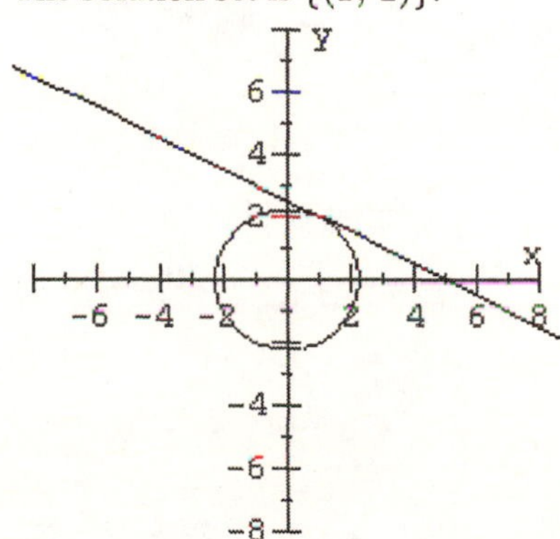

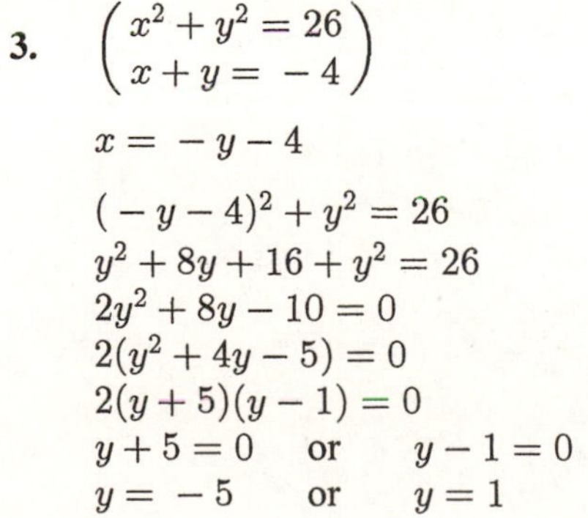

3. $\begin{pmatrix} x^2 + y^2 = 26 \\ x + y = -4 \end{pmatrix}$

$x = -y - 4$

$(-y-4)^2 + y^2 = 26$
$y^2 + 8y + 16 + y^2 = 26$
$2y^2 + 8y - 10 = 0$
$2(y^2 + 4y - 5) = 0$
$2(y+5)(y-1) = 0$
$y + 5 = 0$ or $y - 1 = 0$
$y = -5$ or $y = 1$

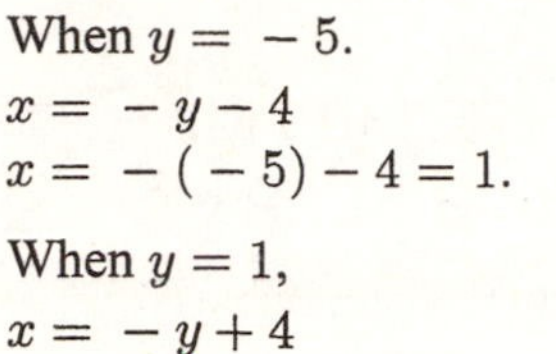

When $y = -5$.
$x = -y - 4$
$x = -(-5) - 4 = 1.$

When $y = 1$,
$x = -y + 4$
$x = -1 - 4 = -5.$

The solution set is $\{(1, -5), (-5, 1)\}$.

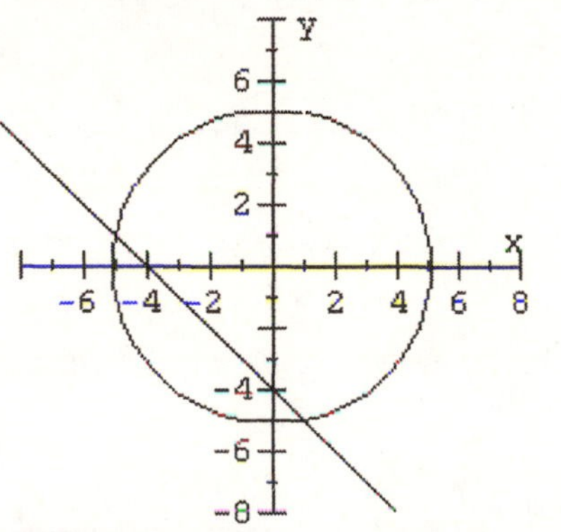

5. $\begin{pmatrix} x^2 + y^2 = 2 \\ x - y = 4 \end{pmatrix}$

$x = y + 4$

$(y+4)^2 + y^2 = 2$
$y^2 + 8y + 16 + y^2 = 2$
$2y^2 + 8y + 14 = 0$
$2(y^2 + 4y + 7) = 0$
$y^2 + 4y + 7 = 0$

$$y = \frac{-4 \pm \sqrt{4^2 - 4(1)(7)}}{2(1)}$$

$$y = \frac{-4 \pm \sqrt{-12}}{2}$$

$$y = \frac{-4 \pm 2i\sqrt{3}}{2}$$

$y = -2 \pm i\sqrt{3}$

When $y = -2 - i\sqrt{3}$,
$x = y + 4$
$x = -2 - i\sqrt{3} + 4$
$x = 2 - i\sqrt{3}$.
When $y = -2 + i\sqrt{3}$,
$x = y + 4$
$x = -2 + i\sqrt{3} + 4$
$x = 2 + i\sqrt{3}$.
The solution set is
$\left\{\left(2 + i\sqrt{3},\ -2 + i\sqrt{3}\right),\right.$
$\left.\left(2 - i\sqrt{3},\ -2 - i\sqrt{3}\right)\right\}$.
There are no real solutions.

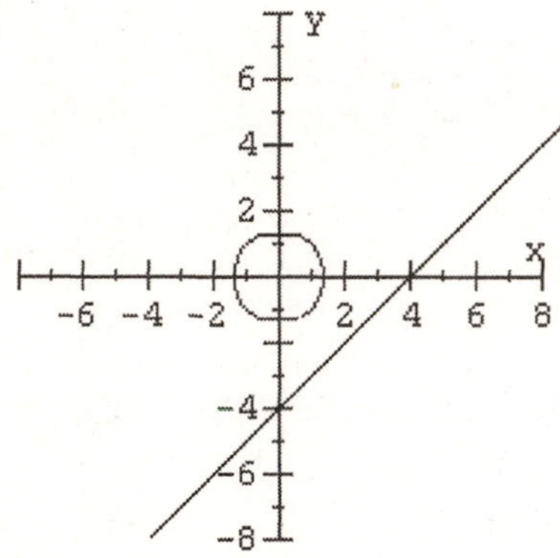

7. $\begin{pmatrix} y = x^2 + 6x + 7 \\ 2x + y = -5 \end{pmatrix}$

$2x + x^2 + 6x + 7 = -5$
$x^2 + 8x + 12 = 0$
$(x + 6)(x + 2) = 0$
$x + 6 = 0$ or $x + 2 = 0$
$x = -6$ or $x = -2$

When $x = -6$,
$y = (-6)^2 + 6(-6) + 7$
$y = 7$.
When $x = -2$,
$y = (-2)^2 + 6(-2) + 7$
$y = -1$.
The solution set is $\{(-6, 7), (-2, -1)\}$.

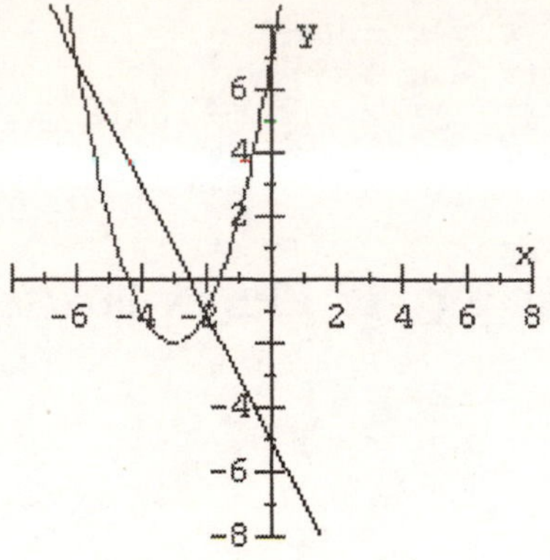

9. $\begin{pmatrix} 2x + y = -2 \\ y = x^2 + 4x + 7 \end{pmatrix}$

$2x + x^2 + 4x + 7 = -2$
$x^2 + 6x + 9 = 0$
$(x + 3)^2 = 0$
$x + 3 = 0$
$x = -3$
When $x = -3$,
$2x + y = -2$
$2(-3) + y = -2$
$-6 + y = -2$
$y = 4$.
The solution set is $\{(-3, 4)\}$.

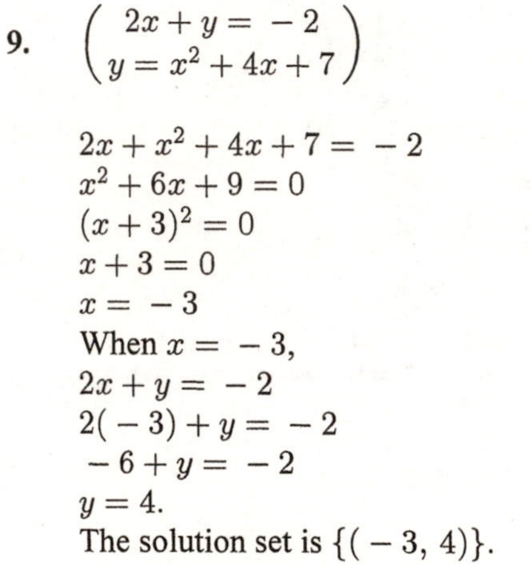

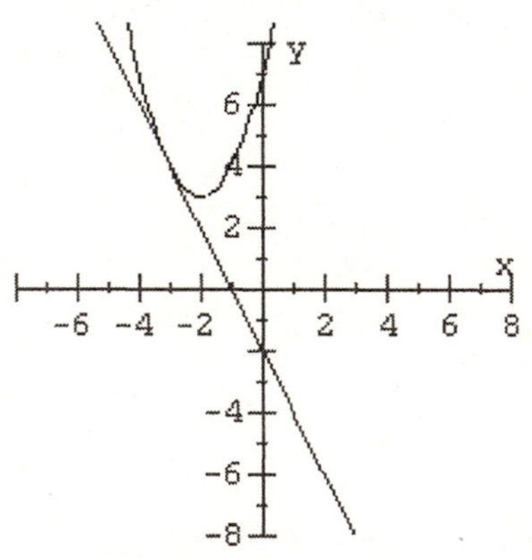

11. $\begin{pmatrix} y = x^2 - 3 \\ x + y = -4 \end{pmatrix}$

$x + x^2 - 3 = -4$
$x^2 + x + 1 = 0$

$$x = \frac{-1 \pm \sqrt{1^2 - 4(1)(1)}}{2(1)}$$
$$x = \frac{-1 \pm \sqrt{-3}}{2}$$
$$x = \frac{-1 \pm i\sqrt{3}}{2}$$

When $x = \dfrac{-1 - i\sqrt{3}}{2}$,

$x + y = -4$

$\dfrac{-1 - i\sqrt{3}}{2} + y = -4$

$y = -4 - \left(\dfrac{-1 - i\sqrt{3}}{2}\right)$

$y = \dfrac{-8 + 1 + i\sqrt{3}}{2} = \dfrac{-7 + i\sqrt{3}}{2}.$

When $x = \dfrac{-1 + i\sqrt{3}}{2}$,

$x + y = -4$

$\dfrac{-1 + i\sqrt{3}}{2} + y = -4$

$y = -4 - \left(\dfrac{-1 + i\sqrt{3}}{2}\right)$

$y = \dfrac{-8 + 1 - i\sqrt{3}}{2} = \dfrac{-7 - i\sqrt{3}}{2}.$

The solution set is

$$\left\{\left(\frac{-1 + i\sqrt{3}}{2}, \frac{-7 - i\sqrt{3}}{2}\right), \left(\frac{-1 - i\sqrt{3}}{2}, \frac{-7 + i\sqrt{3}}{2}\right)\right\}.$$

There are no real solutions.

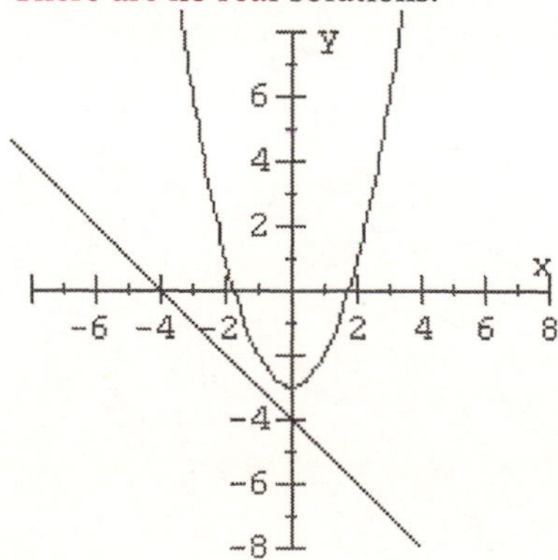

13. $\begin{pmatrix} x^2 + 2y^2 = 9 \\ x - 4y = -9 \end{pmatrix}$

$x = 4y - 9$

$(4y - 9)^2 + 2y^2 = 9$
$16y^2 - 72y + 81 + 2y^2 = 9$
$18y^2 - 72y + 72 = 0$
$18(y^2 - 4y + 4) = 0$
$y^2 - 4y + 4 = 0$
$(y - 2)^2 = 0$
$y - 2 = 0$
$y = 2$
$x = 4y - 9$
$x = 4(2) - 9 = -1$

The solution set is $\{(-1, 2)\}$.

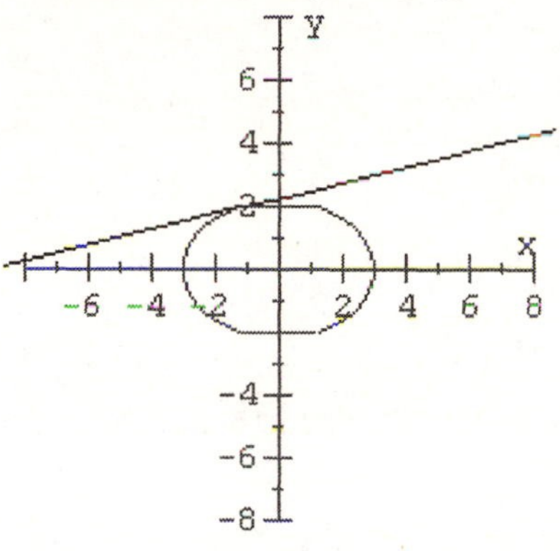

15. $\begin{pmatrix} x + y = -3 \\ x^2 + 2y^2 - 12y - 18 = 0 \end{pmatrix}$

$x = -y - 3$

$(-y - 3)^2 + 2y^2 - 12y - 18 = 0$
$y^2 + 6y + 9 + 2y^2 - 12y - 18 = 0$
$3y^2 - 6y - 9 = 0$
$3(y^2 - 2y - 3) = 0$
$3(y - 3)(y + 1) = 0$
$y - 3 = 0$ or $y + 1 = 0$
$y = 3$ or $y = -1$

When $y = 3$,
$x = -3 - 3$
$x = -6$

When $y = -1$,
$x = -(-1) - 3$
$x = -2$

The solution set is $\{(-6, 3), (-2, -1)\}$.

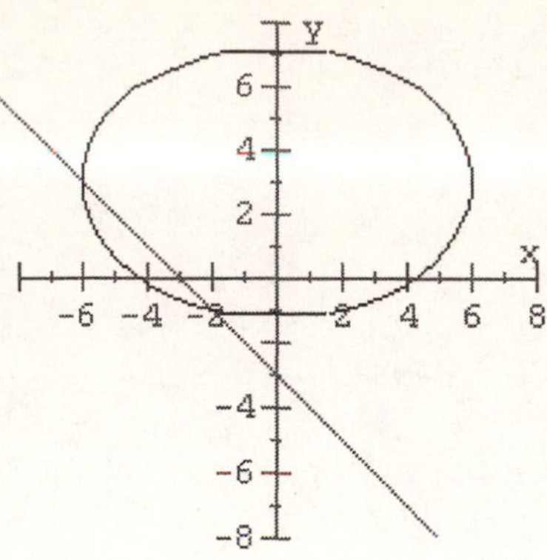

17. $\begin{pmatrix} x - y = 2 \\ x^2 - y^2 = 16 \end{pmatrix}$

$x = y + 2$

$(y+2)^2 - y^2 = 16$

$y^2 + 4y + 4 - y^2 = 16$

$4y + 4 = 16$

$4y = 12$

$y = 3$

$x = 3 + 2$

$x = 5$

The solution set is$\{(5,\ 3)\}$.

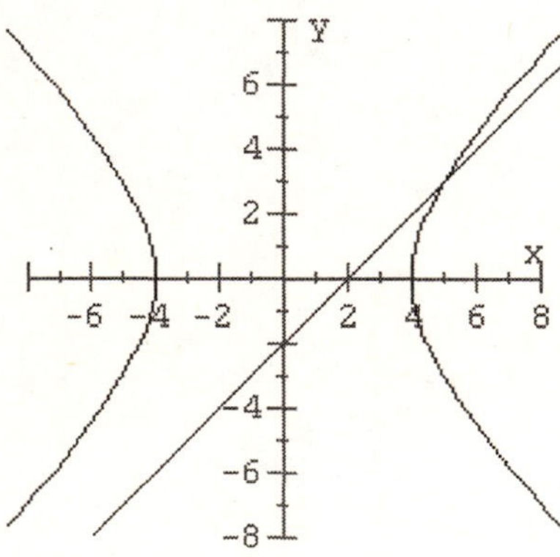

19. $\begin{pmatrix} y = -x^2 + 3 \\ y = x^2 + 1 \end{pmatrix}$

$x^2 + 1 = -x^2 + 3$

$2x^2 = 2$

$x^2 = 1$

$x = \pm\sqrt{1} = \pm 1$

When $x = -1$,

$y = (-1)^2 + 1$

$y = 1 + 1 = 2$

When $x = 1$,

$y = 1^2 + 1 = 2$

The solution set is $\{(1,2),\ (-1,2)\}$.

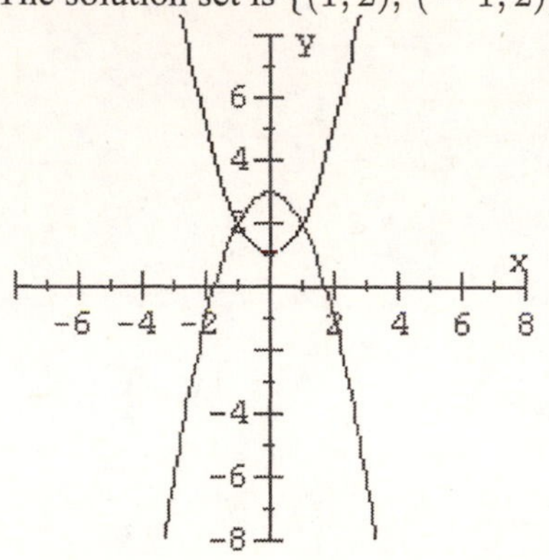

21. $\begin{pmatrix} y = x^2 + 2x - 1 \\ y = x^2 + 4x + 5 \end{pmatrix}$

$x^2 + 2x - 1 = x^2 + 4x + 5$

$2x - 1 = 4x + 5$

$-6 = 2x$

$-3 = x$

When $x = -3$,

$y = x^2 + 2x - 1$

$y = (-3)^2 + 2(-3) - 1$

$y = 2.$

The solution set is $\{(-3,2)\}$.

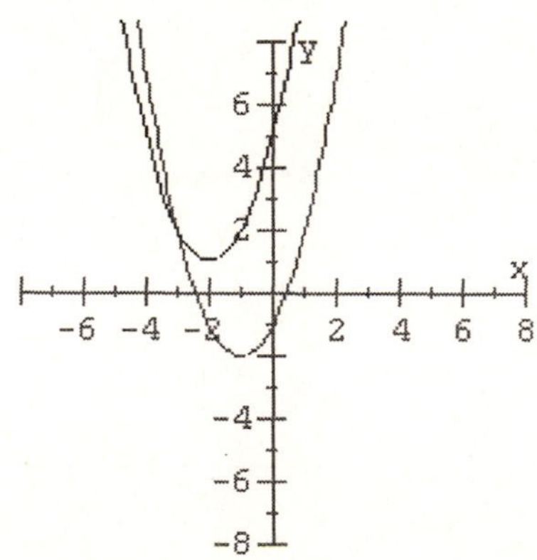

23. $\begin{pmatrix} x^2 - y^2 = 4 \\ x^2 + y^2 = 4 \end{pmatrix}$

Add the equations.

$2x^2 = 8$

$x^2 = 4$
$x = \pm 2$
When $x = -2$,
$x^2 + y^2 = 4$
$(-2)^2 + y^2 = 4$
$4 + y^2 = 4$
$y^2 = 0$
$y = 0.$
When $x = 2$,
$x^2 + y^2 = 4$
$2^2 + y^2 = 4$
$4 + y^2 = 4$
$y^2 = 0$
$y = 0.$
The solution set is $\{(2, 0), (-2, 0)\}$.

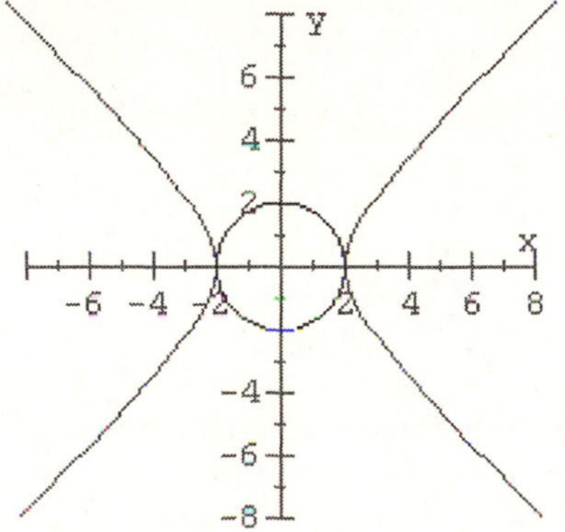

25. $\begin{pmatrix} 8y^2 - 9x^2 = 6 \\ 8x^2 - 3y^2 = 7 \end{pmatrix}$

$3(8y^2 - 9x^2 = 6)$
$8(-3y^2 + 8x^2 = 7)$

Add the equations.

$$\begin{array}{r} 24y^2 - 27x^2 = 18 \\ -24y^2 + 64x^2 = 56 \\ \hline 37x^2 = 74 \\ x^2 = 2 \\ x = \pm\sqrt{2} \end{array}$$

When $x = -\sqrt{2}$,
$8x^2 - 3y^2 = 7$
$8(-\sqrt{2})^2 - 3y^2 = 7$
$8(2) - 3y^2 = 7$
$16 - 3y^2 = 7$
$-3y^2 = -9$
$y^2 = 3$
$y = \pm\sqrt{3}$
$(-\sqrt{2}, \sqrt{3})$ and $(-\sqrt{2}, -\sqrt{3})$

When $x = \sqrt{2}$,
$8x^2 - 3y^2 = 7$
$8(\sqrt{2})^2 - 3y^2 = 7$
$8(2) - 3y^2 = 7$
$16 - 3y^2 = 7$
$-3y^2 = -9$
$y^2 = 3$
$y = \pm\sqrt{3}$
$(\sqrt{2}, \sqrt{3})$ and $(\sqrt{2}, -\sqrt{3})$

The solution set is
$\left\{\left(\sqrt{2}, \sqrt{3}\right), \left(\sqrt{2}, -\sqrt{3}\right), \left(-\sqrt{2}, \sqrt{3}\right), \left(-\sqrt{2}, -\sqrt{3}\right)\right\}$.

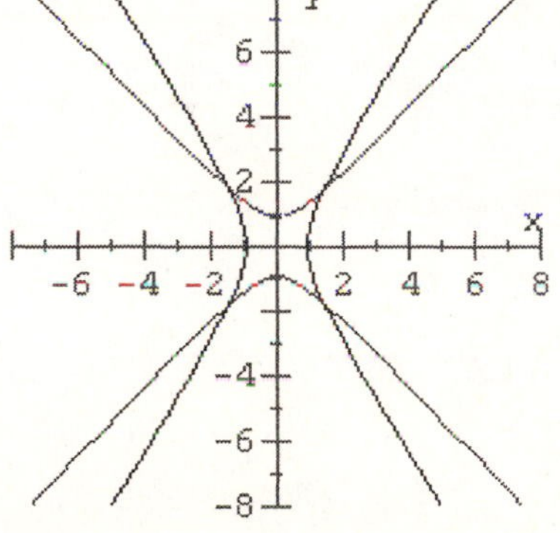

27. $\begin{pmatrix} 2x^2 - 3y^2 = -1 \\ 2x^2 + 3y^2 = 5 \end{pmatrix}$

Add the equations.
$4x^2 = 4$
$x^2 = 1$
$x = \pm 1$

When $x = 1$,
$2x^2 + 3y^2 = 5$
$2(1)^2 + 3y^2 = 5$
$2 + 3y^2 = 5$
$3y^2 = 3$
$y^2 = 1$
$y = \pm 1.$

When $x = -1$,
$2(-1)^2 + 3y^2 = 5$
$2 + 3y^2 = 5$
$3y^3 = 3$
$y^2 = 1$
$y = \pm 1.$
The solution set is
$\{(1,1), (1,-1), (-1,1), (-1,-1)\}$.

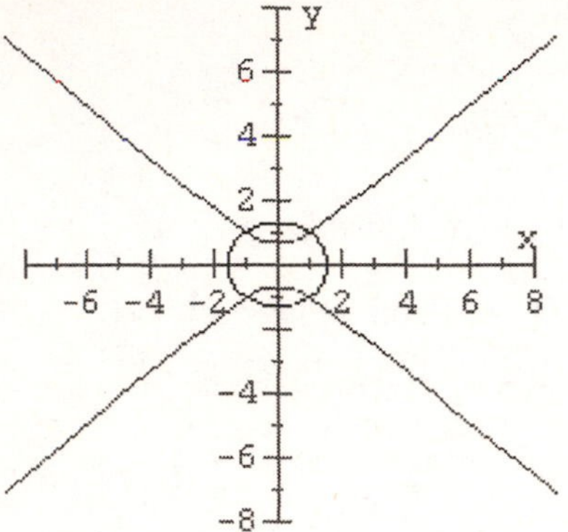

$x = \frac{3}{y}$
$x = \frac{3}{2}$.
The solution set is $\left\{\left(2, \frac{3}{2}\right), \left(\frac{3}{2}, 2\right)\right\}$.

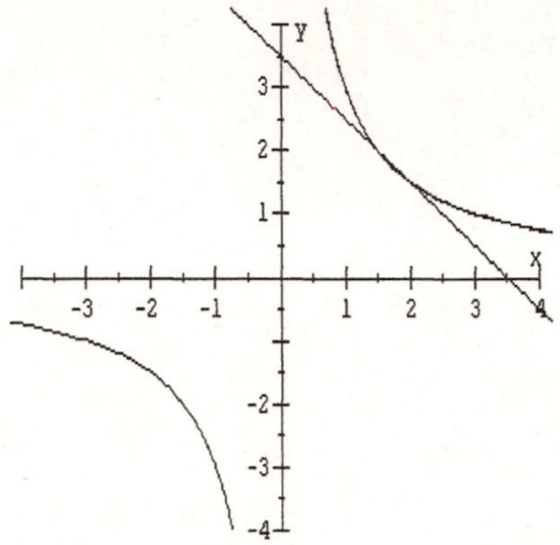

29. $\begin{pmatrix} xy = 3 \\ 2x + 2y = 7 \end{pmatrix}$

$x = \frac{3}{y}$

$2\left(\frac{3}{y}\right) + 2y = 7$

$\frac{6}{y} + 2y = 7$

$6 + 2y^2 = 7y$
$2y^2 - 7y + 6 = 0$
$(2y - 3)(y - 2) = 0$
$2y - 3 = 0$ or $y - 2 = 0$
$2y = 3$ or $y = 2$
$y = \frac{3}{2}$

When $y = \frac{3}{2}$,
$x = \frac{3}{y}$
$x = \frac{3}{\frac{3}{2}} = 2.$
When $y = 2$,

31. $\begin{pmatrix} y = \log_3(x-6) - 3 \\ y = -\log_3 x \end{pmatrix}$

$\log_3(x-6) - 3 = -\log_3 x$
$\log_3(x-6) + \log_3 x = 3$
$\log_3(x-6)x = 3$
$(x-6)x = 3^3$
$x^2 - 6x = 27$
$x^2 - 6x - 27 = 0$
$(x-9)(x+3) = 0$
$x - 9 = 0$ or $x + 3 = 0$
$x = 9$ or $x = -3$
Discard the root $x = -3$.

When $x = 9$,
$y = -\log_3 x$
$y = -\log_3 9$
$y = -2.$
The solution set is $\{(9, -2)\}$.

33. $\begin{pmatrix} y = e^x - 1 \\ y = 2e^{-x} \end{pmatrix}$

$e^x - 1 = 2e^{-x}$
$e^x - 1 = \frac{2}{e^x}$
$e^{2x} - e^x = 2$

$e^{2x} - e^x - 2 = 0$
$(e^x - 2)(e^x + 1) = 0$
$e^x - 2 = 0 \quad \text{or} \quad e^x + 1 = 0$
$e^x = 2 \quad \text{or} \quad e^x = -1$
$e^x = -1$ has no solution
$e^x = 2$
$\ln e^x = \ln 2$
$x(\ln e) = \ln 2$
$x = \ln 2$

When $x = \ln 2$,
$y = e^x - 1$
$y = e^{\ln 2} - 1$
$y = 2 - 1$
$y = 1.$
The solution set is $\{(\ln 2, 1)\}$.

35. $\begin{pmatrix} y = x^3 \\ y = x^3 + 2x^2 + 5x - 3 \end{pmatrix}$

$x^3 = x^3 + 2x^2 + 5x - 3$
$0 = 2x^2 + 5x - 3$
$0 = (2x - 1)(x + 3)$
$2x - 1 = 0 \quad \text{or} \quad x + 3 = 0$
$2x = 1 \quad \text{or} \quad x = -3$
$x = \frac{1}{2}$

When $x = \frac{1}{2}$,
$y = x^3$
$y = \left(\frac{1}{2}\right)^3 = \frac{1}{8}.$

When $x = -3$,
$y = x^3$
$y = (-3)^3 = -27.$

The solution set is
$\left\{\left(\frac{1}{2}, \frac{1}{8}\right), (-3, -27)\right\}.$

CHAPTER 13 Review Problem Set

1. $x^2 + 2y^2 = 32$, Ellipse

$\frac{x^2}{32} + \frac{y^2}{16} = 1$

$a^2 = 32$, thus $a = 4\sqrt{2}$
$b^2 = 16$, thus $b = 4$
$c^2 = a^2 - b^2$
$c^2 = 32 - 16$
$c^2 = 16$, thus $c = 4$
Vertices: $(\pm 4\sqrt{2}, 0)$
Endpoints of minor axis: $(0, \pm 4)$
Foci: $(\pm 4, 0)$

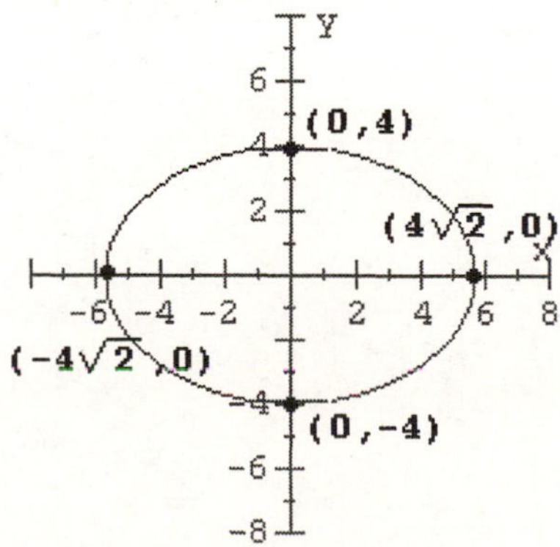

2. $y^2 = -12x$, Parabola
$4p = -12$
$p = -3$
Vertex: $(0, 0)$
Focus: $(-3, 0)$
Directrix: $x = 3$

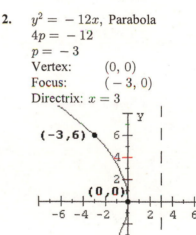

3. $3y^2 - x^2 = 9$, Hyperbola

$\frac{y^2}{3} - \frac{x^2}{9} = 1$

$a^2 = 9$, thus $a = 3$
$b^2 = 3$, thus $b = \sqrt{3}$
$c^2 = a^2 + b^2 = 9 + 3 = 12$
$c^2 = 12$, thus $c = \sqrt{12} = 2\sqrt{3}$
Vertices: $(0, \pm\sqrt{3})$
Endpoints of conjugate axis: $(\pm 3, 0)$
Foci: $(0, \pm 2\sqrt{3})$
Asymptotes: $y = \pm\frac{b}{a}x$, $y = \pm\frac{\sqrt{3}}{3}x$

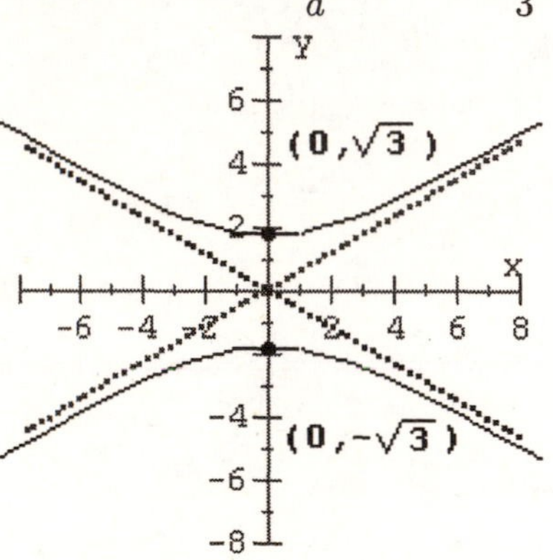

4. $2x^2 - 3y^2 = 18$, Hyperbola

$\frac{x^2}{9} - \frac{y^2}{6} = 1$

$a^2 = 9$, thus $a = 3$
$b^2 = 6$, thus $b = \sqrt{6}$
$c^2 = a^2 + b^2 = 9 + 6 = 15$
$c^2 = 15$, thus $c = \sqrt{15}$

Vertices: $(\pm 3, 0)$
Endpoints of conjugate axis: $(0, \pm\sqrt{6})$
Foci: $(\pm\sqrt{15}, 0)$
Asymptotes: $y = \pm\frac{b}{a}x$, $y = \pm\frac{\sqrt{6}}{3}x$

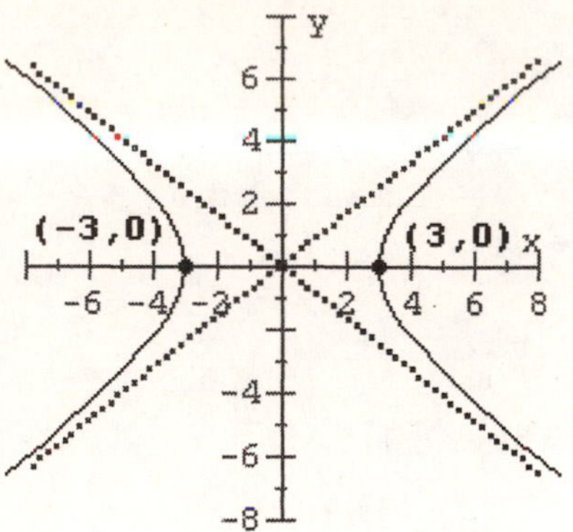

5. $5x^2 + 2y^2 = 20$, Ellipse

$\frac{x^2}{4} + \frac{y^2}{10} = 1$

$a^2 = 4$, thus $a = 2$
$b^2 = 10$, thus $b = \sqrt{10}$
$c^2 = b^2 - a^2 = 10 - 4 = 6$
$c^2 = 6$, thus $c = \sqrt{6}$
Vertices: $(0, \pm\sqrt{10})$
Endpoints of minor axis: $(\pm 2, 0)$
Foci: $(0, \pm\sqrt{6})$

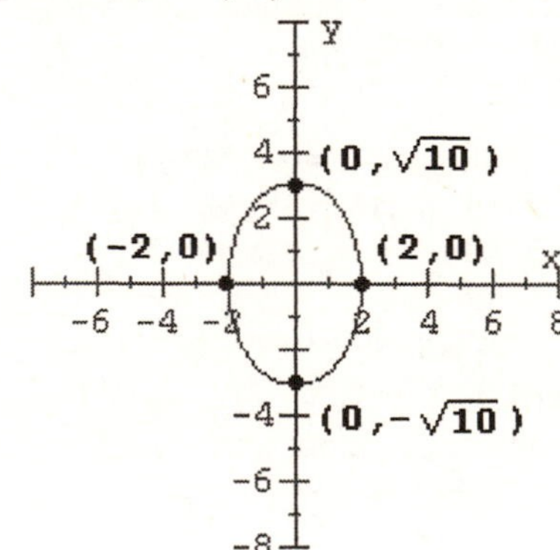

6. $x^2 = 2y$, Parabola
$4p = 2$
$p = \frac{1}{2}$
Vertex: (0, 0)
Focus: $\left(0, \frac{1}{2}\right)$
Directrix: $y = -\frac{1}{2}$

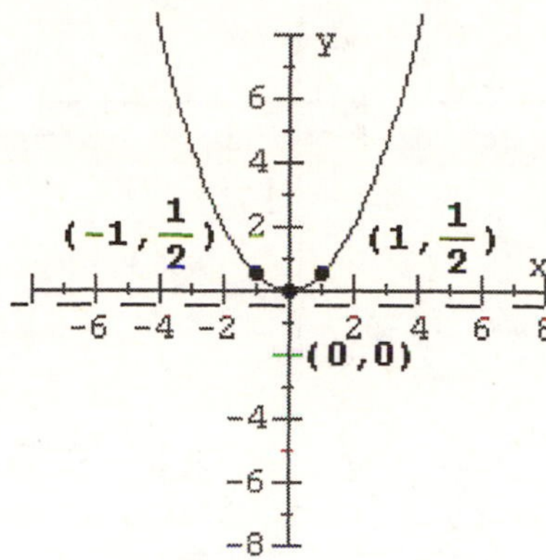

7. $x^2 + y^2 = 10$, Circle
Center: (0, 0)
$r = \sqrt{10}$

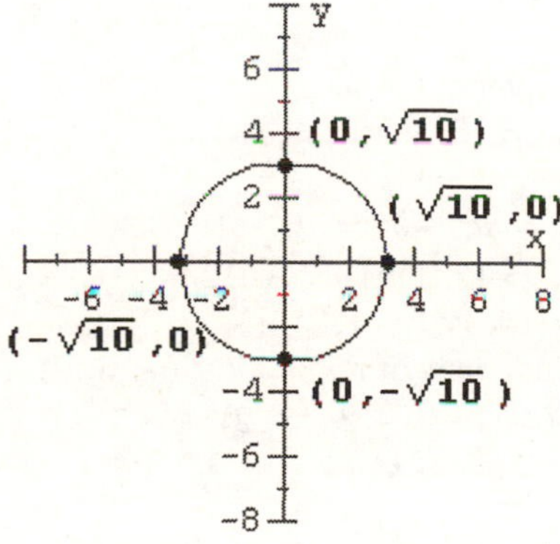

8. $x^2 - 8x - 2y^2 + 4y + 10 = 0$
$x^2 - 8x - 2(y^2 - 2y) = -10$
$x^2 - 8x+16 - 2(y^2 - 2y+1) = -10+16 - 2$
$(x-4)^2 - 2(y-1)^2 = 4$
$\frac{(x-4)^2}{4} - \frac{(y-1)^2}{2} = 1$, Hyberbola

$a^2 = 4$, thus $a = 2$
$b^2 = 2$, thus $b = \sqrt{2}$
$c^2 = a^2 + b^2 = 4 + 2 = 6$
$c^2 = 6$, thus $c = \sqrt{6}$
Center: $(-4, 1)$
Vertices: (6, 1), (2, 1)
Foci: $(4 + \sqrt{6}, 1)$, $(4 - \sqrt{6}, 1)$
Endpoints of conjugate axis
are $(4, 1+\sqrt{2})$, $(4, 1 - \sqrt{2})$.
Asymptotes: $y - k = \pm \frac{b}{a}(x - h)$
$y - 1 = \pm \frac{\sqrt{2}}{2}(x - 4)$
$y - 1 = \frac{\sqrt{2}}{2}(x - 4)$
$2y - 2 = \sqrt{2}x - 4\sqrt{2}$
$\sqrt{2}x - 2y = -2 + 4\sqrt{2}$
AND
$y - 1 = -\frac{\sqrt{2}}{2}(x - 4)$
$2y - 2 = -\sqrt{2}x + 4\sqrt{2}$
$\sqrt{2}x + 2y = 2 + 4\sqrt{2}$

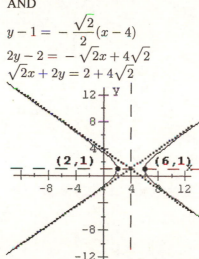

9. $9x^2 - 54x + 2y^2 + 8y + 71 = 0$
$9(x^2 - 6x) + 2(y^2 + 4y) = -71$
$9(x^2 - 6x+9)+2(y^2+4y+4) = -71+81+8$
$9(x-3)^2 + 2(y+2)^2 = 18$
$\frac{(x-3)^2}{2} + \frac{(y+2)^2}{9} = 1$, Ellipse
$a^2 = 2$, thus $a = \sqrt{2}$
$b^2 = 9$, thus $b = 3$
$c^2 = b^2 - a^2$
$c^2 = 9 - 2$
$c^2 = 7$, thus $c = \sqrt{7}$
Center is at $(3, -2)$.

Vertices are 3 units above and below the center at $(3,\ -5)$ and $(3,\ 1)$.
Endpoints of minor axis are $\sqrt{2}$ units left and right of the center at $(3-\sqrt{2},\ -2)$ and $(3+\sqrt{2},\ -2)$.
Foci are $\sqrt{7}$ units above and below the center at $(3,\ -5-\sqrt{7})$ and $(3,\ -5+\sqrt{7})$.

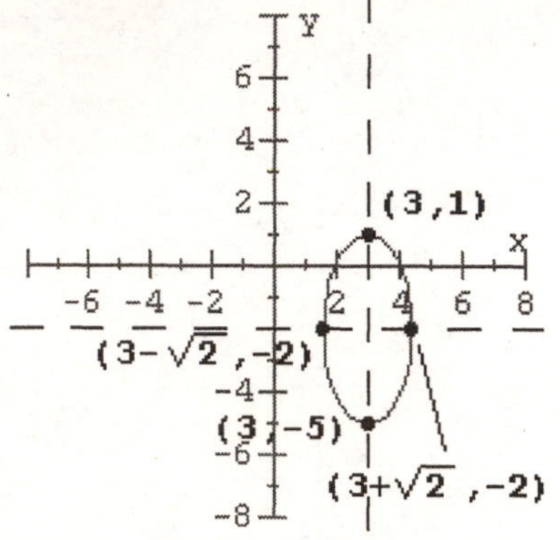

10. $y^2-2y+4x+9=0$
$y^2-2y+1=\ -4x-8$
$(y-1)^2=\ -4(x+2)$, Parabola
$4p=\ -4$
$p=\ -1$
Vertex: $(-2,\ 1)$
Focus: $(-3,\ 1)$
Directrix: $x=\ -1$

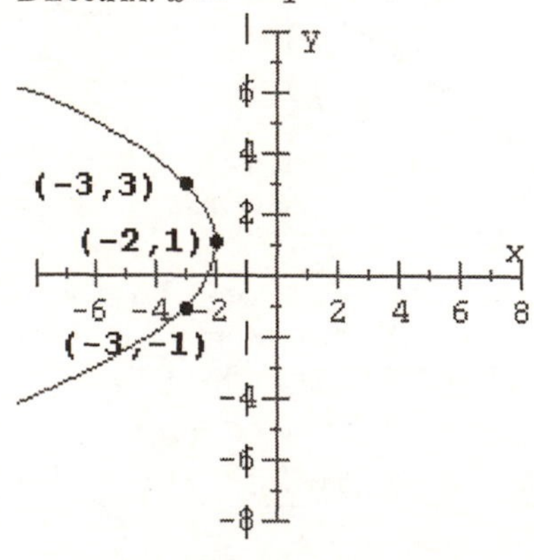

11. $x^2+2x+8y+25=0$
$x^2+2x+1=\ -8y-25+1$
$(x+1)^2=\ -8y-24$
$(x+1)^2=\ -8(y+3)$, Parabola
$4p=\ -8$
$p=\ -2$

Vertex $(-1,\ -3)$
Focus is 2 units below the vertex: $(-1,\ -5)$.
Directrix passes through a point 2 units above the vertex so the equation of the directrix is $y=\ -1$.

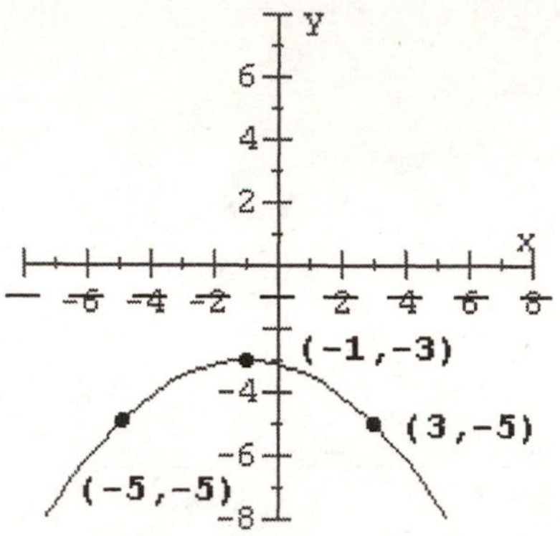

12. $x^2+10x+4y^2-16y+25=0$
$x^2+10x+4(y^2-4y)=\ -25$
$x^2+10x+25+4(y^2-4y+4)=\ -25+25+16$
$(x+5)^2+4(y-2)^2=16$

$$\frac{(x+5)^2}{16}+\frac{(y-2)^2}{4}=1,\text{ Ellipse}$$

Center at $(-5,\ 2)$
$a^2=16$, thus $a=4$
$b^2=4$, thus $b=2$
$c^2=a^2-b^2=16-4=12$
$c^2=12$, thus $c=\sqrt{12}=2\sqrt{3}$
Vertices: $(-1,2)$, $(-9,\ 2)$
Endpoints of minor axis: $(-5,\ 0)$, $(-5,\ 4)$
Foci: $(-5+2\sqrt{3},\ 2)$, $(-5-2\sqrt{3},\ 2)$

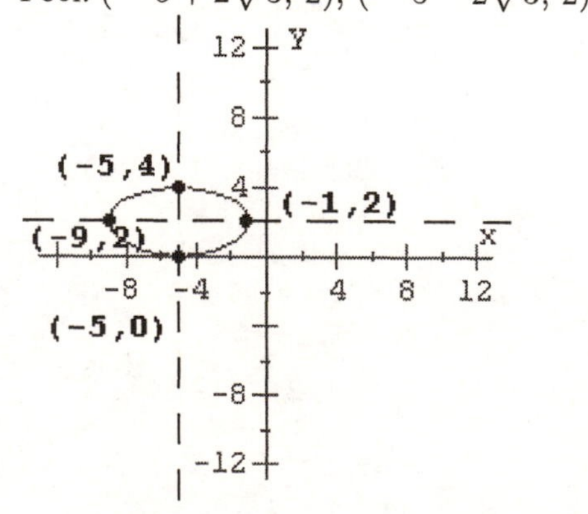

13. $3y^2 + 12y - 2x^2 - 8x - 8 = 0$
$3(y^2 + 4y) - 2(x^2 + 4x) = 8$
$3(y^2{+}4y{+}4) - 2(x^2{+}4x{+}4) = 8{+}12 - 8$
$3(y+2)^2 - 2(x+2)^2 = 12$

$$\frac{(y+2)^2}{4} - \frac{(x+2)^2}{6} = 1, \text{ Hyperbola}$$

$a^2 = 6$, thus $a = \sqrt{6}$
$b^2 = 4$, thus $b = 2$
$c^2 = a^2 + b^2 = 6 + 4 = 10$
$c^2 = 10$, thus $c = \sqrt{10}$
Center: $(-2, -2)$
Vertices are 2 units above and below the center. $(-2, -4)$, $(-2, 0)$
Endpoints of conjugate axis are $\sqrt{6}$ units left and right of the center.
$(-2 - \sqrt{6}, -2)$, $(-2 + \sqrt{6}, -2)$
Foci are $\sqrt{10}$ units above and below the center. $(-2, -2 - \sqrt{10})$, $(-2, -2 + \sqrt{10})$
Asymptotes: $y - k = \pm\frac{b}{a}(x - h)$
$y + 2 = \pm\frac{2}{\sqrt{6}}(x + 2)$
$\sqrt{6}y + 2\sqrt{6} = \pm 2(x + 2)$
$\sqrt{6}y + 2\sqrt{6} = -2(x + 2)$
$\sqrt{6}y + 2\sqrt{6} = -2x - 4$
$2x + \sqrt{6}y = -4 - 2\sqrt{6}$
AND
$\sqrt{6}y + 2\sqrt{6} = 2(x + 2)$
$\sqrt{6}y + 2\sqrt{6} = 2x + 4$
$2x - \sqrt{6}y = -4 + 2\sqrt{6}$

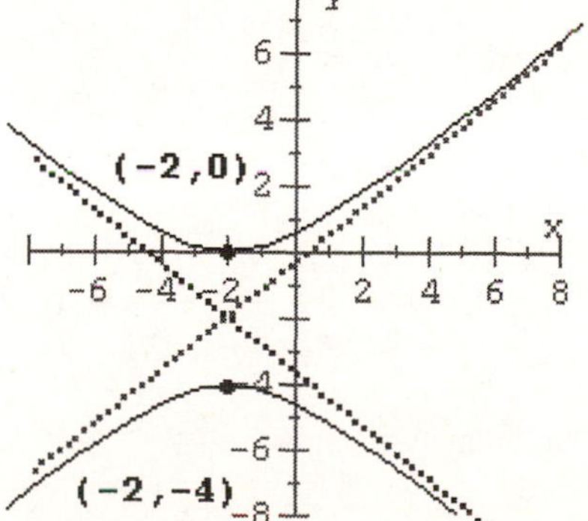

14. $x^2 - 6x + y^2 + 4y - 3 = 0$
$x^2 - 6x + 9 + y^2 + 4y + 4 = 3 + 9 + 4$
$(x-3)^2 + (y+2)^2 = 16$, Circle
Center $(3, -2)$, $r = 4$

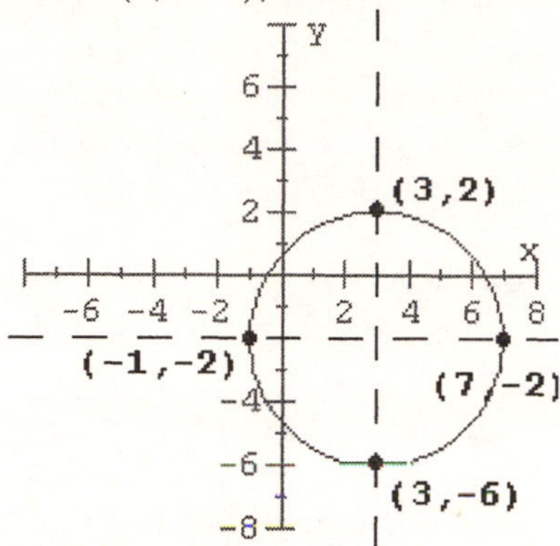

15. Circle: Center $(-8, 3)$, $r = \sqrt{5}$
$(x+8)^2 + (y-3)^2 = (\sqrt{5})^2$
$x^2 + 16x + 64 + y^2 - 6y + 9 = 5$
$x^2 + y^2 + 16x - 6y + 68 = 0$

16. Parabola: Vertex $(0,0)$, Focus $(-5,0)$
directrix $x = 5$
$p = -5$
$y^2 = 4(-5)x$
$y^2 = -20x$

17. Ellipse: Vertices $(0, \pm 4)$,
foci $(0, \pm\sqrt{15})$
$b = 4 \qquad c = \sqrt{15}$
$c^2 = b^2 - a^2$
$(\sqrt{15})^2 = 4^2 - a^2$
$15 = 16 - a^2$
$-1 = -a^2$
$a^2 = 1$, thus $a = 1$

$$\frac{x^2}{1} + \frac{y^2}{16} = 1$$

$16x^2 + y^2 = 16$

18. Hyperbola: Vertices: $(\pm\sqrt{2},\ 0)$
Length of conjugate axis $= 10$
Center $(0,\ 0)$

$2b = 10$
$b = 5$

$$\frac{x^2}{2} - \frac{y^2}{25} = 1$$

$25x^2 - 2y^2 = 50$

19. Circle : Center $(5,\ -12)$
Passes through the origin.
$(x-5)^2 + (y+12)^2 = r^2$
Substitute $(0,\ 0)$ for x and y.
$(0-5)^2 + (0+12)^2 = r^2$
$25 + 144 = r^2$
$169 = r^2$
$r = 13$
$(x-5)^2 + (y+12)^2 = 13^2$
$x^2 - 10x + 25 + y^2 + 24y + 144 = 169$
$x^2 + y^2 - 10x + 24y = 0$

20. Ellipse: Vertices: $(\pm 2,\ 0)$
Contains the point $(1,\ -2)$
Center at $(0, 0)$

$$\frac{x^2}{4} + \frac{y^2}{b^2} = 1$$

Contains $(1,\ -2)$

$$\frac{1^2}{4} + \frac{(-2)^2}{b^2} = 1$$

$$\frac{1}{4} + \frac{4}{b^2} = 1$$

$$\frac{4}{b^2} = \frac{3}{4}$$

$3b^2 = 16$

$$b^2 = \frac{16}{3}$$

$$\frac{x^2}{4} + \frac{3y^2}{16} = 1$$

$4x^2 + 3y^2 = 16$

21. Parabola: Vertex $(0,\ 0)$, contains $(2,\ 6)$
Symmetric with respect to the y-axis
$x^2 = 4py$
Substitute $(2,\ 6)$ for x and y.
$2^2 = 4p(6)$
$4 = 24p$

$$\frac{1}{6} = p$$

$$x^2 = 4\left(\frac{1}{6}\right)y$$

$$x^2 = \frac{2}{3}y$$

22. Hyperbola: Vertices: $(0,\ \pm 1)$,
Foci: $(0,\ \pm\sqrt{10})$
Center at $(0, 0)$
$b = 1$
$c = \sqrt{10}$
$c^2 = a^2 + b^2$
$10 = a^2 + 1$
$9 = a^2$, hence $a = 3$

$$\frac{y^2}{1} - \frac{x^2}{9} = 1$$

$9y^2 - x^2 = 9$

23. Ellipse: Vertices $(6,\ 1)$ and $(6,\ 7)$
Length of minor axis 2 units.
Center is midway between
vertices at $(6,\ 4)$, hence $b = 3$.
Length of minor axis is the same as $2a$.
$2a = 2$
$a = 1$

$$\frac{(x-6)^2}{1} + \frac{(y-4)^2}{9} = 1$$

$9(x-6)^2 + (y-4)^2 = 9$
$9(x^2 - 12x + 36) + y^2 - 8y + 16 = 9$
$9x^2 - 108x + 324 + y^2 - 8y + 16 = 9$
$9x^2 + y^2 - 108x - 8y + 331 = 0$

24. Parabola: Vertex: $(4,\ -2)$,
Focus: $(6,\ -2)$
$h = 4 \qquad k = -2$

$h + p = 6$
$4 + p = 6$
$p = 2$

$(y+2)^2 = 4(2)(x-4)$
$y^2 + 4y + 4 = 8x - 32$
$y^2 + 4y - 8x + 36 = 0$

25. Hyperbola: Vertices $(-5, -3)$, $(-5, -5)$
Foci $(-5, -2)$, $(-5, -6)$
Center is midway between the vertices at $(-5, -4)$, hence $b = 1$ and $c = 2$.
$c^2 = a^2 + b^2$
$2^2 = a^2 + 1^2$
$4 = a^2 + 1$
$3 = a^2$

$$\frac{(y+4)^2}{1} - \frac{(x+5)^2}{3} = 1$$

$3(y+4)^2 - (x+5)^2 = 3$
$3(y^2 + 8y + 16) - (x^2 + 10x + 25) = 3$
$3y^2 + 24y + 48 - x^2 - 10x - 25 = 3$
$3y^2 + 24y - x^2 - 10x + 20 = 0$

26. Parabola: Vertex $(-6, -3)$
Symmetric with respect to the line $x = -6$.
Contains $(-5, -2)$
$h = -6 \qquad k = -3$
$(x+6)^2 = 4p(y+3)$
Substitute $x = -5$ and $y = -2$.
$(-5+6)^2 = 4p(-2+3)$
$1 = 4p$
$\frac{1}{4} = p$
$(x+6)^2 = 4\left(\frac{1}{4}\right)(y+3)$
$x^2 + 12x + 36 = y + 3$
$x^2 + 12x - y + 33 = 0$

27. Ellipse: Endpoints of minor axis are $(-5, 2)$ and $(-5, -2)$.
Length of major axis is 10 units.
The center is midway between the endpoints of the minor axis at $(-5, 0)$.
Hence, $b = 2$. Length of major axis is the same as $2a$.
$2a = 10$
$a = 5$

$$\frac{(x+5)^2}{5^2} + \frac{(y-0)^2}{2^2} = 1$$

$$\frac{(x+5)^2}{25} + \frac{y^2}{4} = 1$$

$4(x+5)^2 + 25y^2 = 100$
$4(x^2 + 10x + 25) + 25y^2 = 100$
$4x^2 + 40x + 100 + 25y^2 = 100$
$4x^2 + 40x + 25y^2 = 0$

28. Hyperbola: Vertices $(2, 0)$ and $(6, 0)$
Length of conjugate axis $= 8$
$h - a = 2 \qquad k = 0$
$h + a = 6$
$2h = 8$
$h = 4$
$a = 2$
$2b = 8$
$b = 4$

$$\frac{(x-4)^2}{4} - \frac{(y-0)^2}{16} = 1$$

$4(x-4)^2 - y^2 = 16$
$4(x^2 - 8x + 16) - y^2 = 16$
$4x^2 - 32x + 64 - y^2 = 16$
$4x^2 - 32x - y^2 + 48 = 0$

29. $\begin{pmatrix} x^2 + y^2 = 17 \\ x - 4y = -17 \end{pmatrix}$

$x = 4y - 17$
$x^2 + y^2 = 17$
$(4y - 17)^2 + y^2 = 17$
$16y^2 - 136y + 289 + y^2 = 17$
$17y^2 - 136y + 272 = 0$
$17(y^2 - 8y + 16) = 0$
$17(y-4)^2 = 0$
$y - 4 = 0$
$y = 4$
When $y = 4$,
$x = 4(4) - 17$
$x = 16 - 17 = -1$.
The solution set is $\{(-1, 4)\}$.

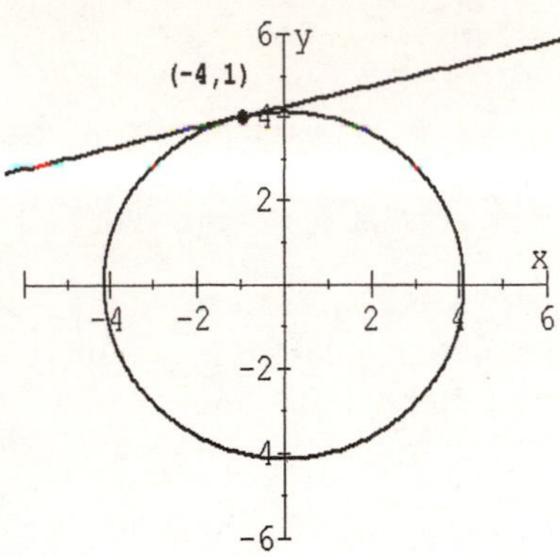

30. $\begin{pmatrix} x^2 - y^2 = 8 \\ 3x - y = 8 \end{pmatrix}$

$y = 3x - 8$
$x^2 - (3x - 8)^2 = 8$
$x^2 - (9x^2 - 48x + 64) = 8$
$x^2 - 9x^2 + 48x - 64 = 8$
$-8x^2 + 48x - 72 = 0$
$-8(x^2 - 6x + 9) = 0$
$-8(x - 3)^2 = 0$
$x - 3 = 0$
$x = 3$
When $x = 3$,
$y = 3(3) - 8$
$y = 9 - 8 = 1.$
The solution set is $\{(3,\ 1)\}$.

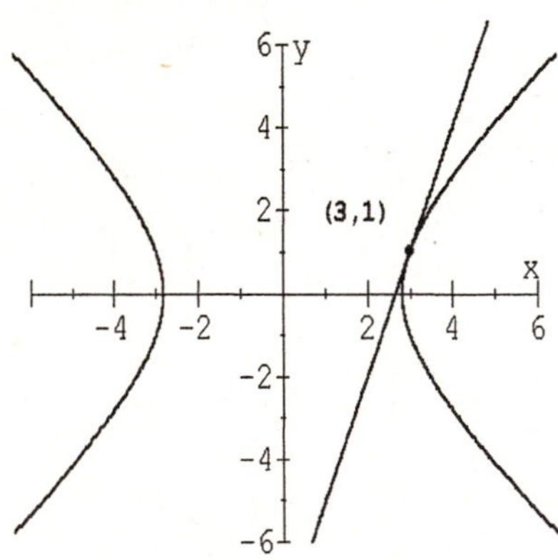

31. $\begin{pmatrix} x - y = 1 \\ y = x^2 + 4x + 1 \end{pmatrix}$

$x - (x^2 + 4x + 1) = 1$
$x - x^2 - 4x - 1 = 1$
$-x^2 - 3x - 2 = 0$
$x^2 + 3x + 2 = 0$
$(x + 1)(x + 2) = 0$
$x + 1 = 0$ or $x + 2 = 0$
$x = -1$ $\quad x = -2$

When $x = -1$,
$y = x^2 + 4x + 1$
$y = (-1)^2 + 4(-1) + 1$
$y = 1 - 4 + 1$
$y = -2$
When $x = -2$,
$y = x^2 + 4x + 1$
$y = (-2)^2 + 4(-2) + 1$
$y = 4 - 8 + 1$
$y = -3$

The solution set is
$\{(-1,\ -2),\ (-2,\ -3)\}$.

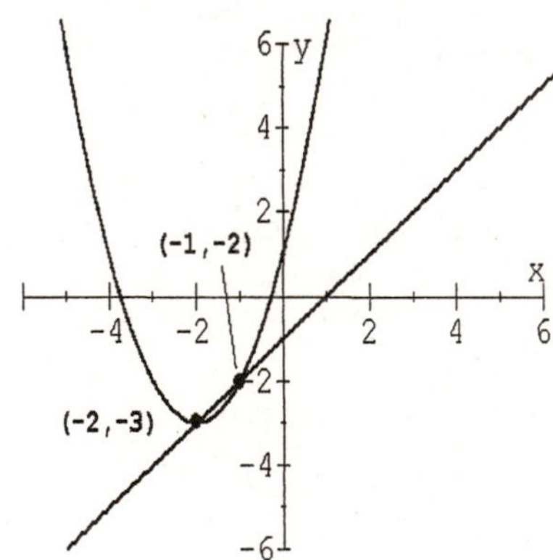

32. $\begin{pmatrix} 4x^2 - y^2 = 16 \\ 9x^2 + 9y^2 = 16 \end{pmatrix}$

9(equation 1) + (equation 2) replace (equation 2)

$\begin{pmatrix} 4x^2 - y^2 = 16 \\ 45x^2 = 160 \end{pmatrix}$

$45x^2 = 160$

$x^2 = \frac{160}{45} = \frac{32}{9}$

$x = \pm\frac{4\sqrt{2}}{3}$

When $x = \frac{4\sqrt{2}}{3}$, $x^2 = \frac{32}{9}$

$4x^2 - y^2 = 16$

$y^2 = 4x^2 - 16$

$y^2 = 4\left(\frac{32}{9}\right) - 16$

$y^2 = -\frac{16}{9}$

$y = \pm\frac{4i}{3}$

When $x = -\frac{4\sqrt{2}}{3}$, $x^2 = \frac{32}{9}$

Similarly $y = \pm\frac{4i}{3}$

The solution set is

$\left\{\left(\frac{4\sqrt{2}}{3}, \frac{4i}{3}\right), \left(\frac{4\sqrt{2}}{3}, -\frac{4i}{3}\right)\right.$

$\left.\left(-\frac{4\sqrt{2}}{3}, \frac{4i}{3}\right), \left(-\frac{4\sqrt{2}}{3}, -\frac{4i}{3}\right)\right\}.$

There are no real solutions.

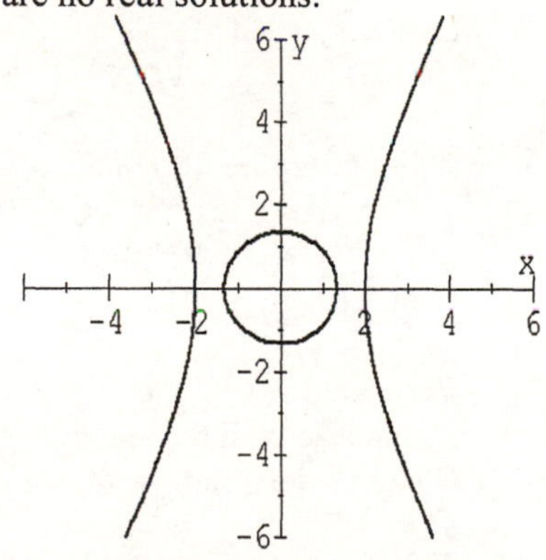

33. $\begin{pmatrix} x^2 + 2y^2 = 8 \\ 2x^2 + 3y^2 = 12 \end{pmatrix}$

$x^2 + 2y^2 = 8$
$x^2 = -2y^2 + 8$

$2x^2 + 3y^2 = 12$
$2(-2y^2 + 8) + 3y^2 = 12$
$-4y^2 + 16 + 3y^2 = 12$
$-y^2 + 16 = 12$
$-y^2 = -4$
$y^2 = 4$
$y = \pm 2$

When $y = -2$,
$x^2 = -2y^2 + 8$
$x^2 = -2(-2)^2 + 8$
$x^2 = -2(4) + 8$
$x^2 = -8 + 8$
$x^2 = 0$
$x = 0.$

When $y = 2$,
$x^2 = -2y^2 + 8$
$x^2 = -2(2)^2 + 8$
$x^2 = -2(4) + 8$
$x^2 = -8 + 8$
$x^2 = 0$
$x = 0.$
The solution set is $\{(0, -2), (0, 2)\}$.

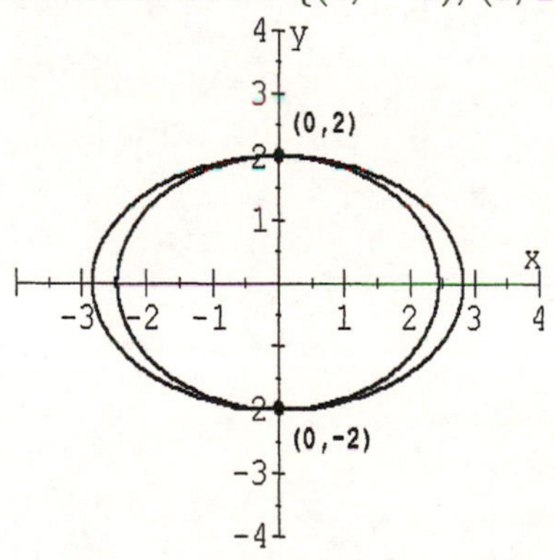

34. $\begin{pmatrix} y^2 - x^2 = 1 \\ 4x^2 + y^2 = 4 \end{pmatrix}$

$y^2 = x^2 + 1$
$4x^2 + x^2 + 1 = 4$
$5x^2 = 3$
$x^2 = \frac{3}{5}$
$x = \pm\sqrt{\frac{3}{5}} = \pm\frac{\sqrt{15}}{5}$

When $x = \frac{\sqrt{15}}{5}$, then $x^2 = \frac{3}{5}$

$y^2 = \frac{3}{5} + 1$
$y^2 = \frac{8}{5}$
$y = \pm\sqrt{\frac{8}{5}} = \pm\frac{2\sqrt{10}}{5}$

When $x = -\frac{\sqrt{15}}{5}$, then $x^2 = \frac{3}{5}$

Same results as above: $y = \pm\frac{2\sqrt{10}}{5}$

The solution set is

$\left\{\left(\frac{\sqrt{15}}{5}, \frac{2\sqrt{10}}{5}\right), \left(\frac{\sqrt{15}}{5}, -\frac{2\sqrt{10}}{5}\right), \left(-\frac{\sqrt{15}}{5}, \frac{2\sqrt{10}}{5}\right), \left(-\frac{\sqrt{15}}{5}, -\frac{2\sqrt{10}}{5}\right)\right\}.$

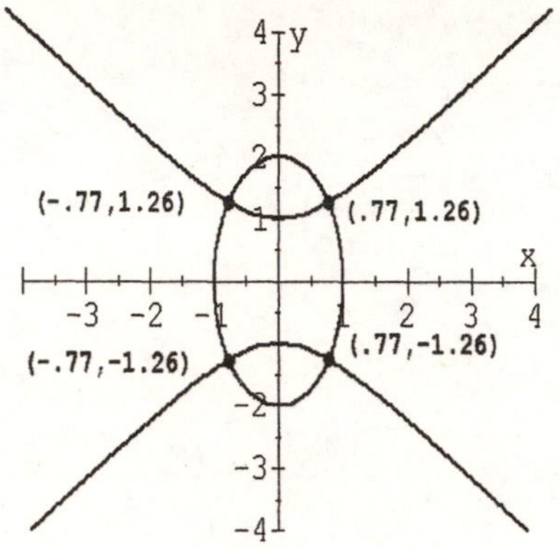

CHAPTER 13 Test

1. $x^2 = -20y$
$4p = -20$
$p = -5$
Focus: $(0, -5)$

2. $y^2 - 4y - 8x - 20 = 0$
$y^2 - 4y + 4 = 8x + 20 + 4$
$(y-2)^2 = 8x + 24$
$(y-2)^2 = 8(x+3)$
Vertex: $(-3, 2)$

3. $2y^2 = 24x$
$y^2 = 12x$
$4p = 12$
$p = 3$
Directrix: $x = -3$

4. $y^2 = 24x$
$4p = 24$
$p = 6$
Focus $(6, 0)$

5. $x^2 + 4x - 12y - 8 = 0$
$x^2 + 4x = = 12y + 8$
$x^2 + 4x + 4 = 12y + 8 + 4$
$(x+2)^2 = 12y + 12$
$(x+2)^2 = 12(y+1)$
Vertex: $(-2, -1)$

6. $x^2 + 6x + y^2 + 18y + 87 = 0$
$x^2+6x+9+y^2+18y+81 = -87 + 9 + 81$
$(x+3)^2 + (y+9)^2 = 3$
Center $(-3, -9)$

7. Vertex at $(0, 0)$
Symmetric with respect to the x-axis.
Contains the point $(-2, 4)$.
$y^2 = 4px$
Substitute $(-2, 4)$ for x and y.
$4^2 = 4p(-2)$
$16 = -8p$
$-2 = p$
$y^2 = 4(-2)x$

$y^2 = -8x$
$y^2 + 8x = 0$

8. Vertex (3, 4), focus (3, 1)
$h = 3 \qquad k = 4$
$k + p = 1$
$4 + p = 1$
$p = -3$
$(x-3)^2 = 4(-3)(y-4)$
$x^2 - 6x + 9 = -12y + 48$
$x^2 - 6x + 12y - 39 = 0$

9. Center $(-1, 6)$, $r = 5$
$(x+1)^2 + (y-6)^2 = 5^2$
$x^2 + 2x + 1 + y^2 - 12y + 36 = 25$
$x^2 + y^2 + 2x - 12y + 12 = 0$

10. $x^2 - 4x + 9y^2 - 18y + 4 = 0$
$x^2 - 4x{+}4{+}9(y^2 - 2y{+}1) = -4 + 4 + 9$
$(x-2)^2 + 9(y-1)^2 = 9$

$$\frac{(x-2)^2}{9} + \frac{(y-1)^2}{1} = 1$$

$a^2 = 9$, thus $a = 3$
Length of major axis $= |2a| = 6$ units

11. $9x^2 + 90x + 4y^2 - 8y + 193 = 0$
$9(x^2 + 10x) + 4(y^2 - 2y) = -193$
$9(x^2{+}10x{+}25){+}4(y^2{-}2y{+}1) = -193{+}225{+}4$
$9(x+5)^2 + 4(y-1)^2 = 36$

$$\frac{(x+5)^2}{4} + \frac{(y-1)^2}{9} = 1$$

Center at $(-5, 1)$
$a^2 = 4$, thus $a = 2$
Endpoints of the minor axis are 2 units left and right of the center, $(-7, 1)$ and $(-3, 1)$.

12. $x^2 + 4y^2 = 16$

$$\frac{x^2}{16} + \frac{y^2}{4} = 1$$

$a^2 = 16$ so $a = 4$
$b^2 = 4$ so $b = 2$
$c^2 = a^2 - b^2 = 16 - 4 = 12$
$c = 2\sqrt{3}$
Foci $(\pm 2\sqrt{3}, 0)$

13. $3x^2 + 30x + y^2 - 16y + 79 = 0$
$3(x^2 + 10x) + y^2 - 16y = -79$
$3(x^2{+}10x{+}25){+}y^2 - 16y{+}64 = -79{+}75{+}64$
$3(x+5)^2 + (y-8)^2 = 60$
Center is at $(-5, 8)$.

14. Vertices: $(0, \pm 10)$
Foci: $(0, \pm 8)$
$b = 10 \qquad c = 8$
$a^2 = b^2 - c^2$
$a^2 = 100 - 64 = 36$

$$\frac{x^2}{36} + \frac{y^2}{100} = 1$$

$25x^2 + 9y^2 = 900$

15. Endpoints of major axis at $(2, -2)$ and $(10, -2)$.
Endpoints of minor axis at $(6, 0)$ and $(6, -4)$

$$\begin{aligned} h - a &= 2 \\ h + a &= 10 \\ \hline 2h &= 12 \\ h &= 6 \end{aligned} \qquad \begin{aligned} k + b &= 0 \\ k - b &= -4 \\ \hline 2k &= -4 \\ k &= -2 \end{aligned}$$

$$\begin{aligned} h + a &= 10 \\ 6 + a &= 10 \\ a &= 4 \end{aligned} \qquad \begin{aligned} k + b &= 0 \\ -2 + b &= 0 \\ b &= 2 \end{aligned}$$

$$\frac{(x-6)^2}{4^2} + \frac{(y+2)^2}{2^2} = 1$$

$$\frac{(x-6)^2}{16} + \frac{(y+2)^2}{4} = 1$$

$(x-6)^2 + 4(y+2)^2 = 16$
$x^2 - 12x + 36 + 4(y^2 + 4y + 4) = 16$
$x^2 - 12x + 36 + 4y^2 + 16y + 16 = 16$
$x^2 - 12x + 4y^2 + 16y + 36 = 0$

16. $4y^2 - 9x^2 = 32$

$$\frac{x^2}{8} - \frac{9x^2}{32} = 1$$

$b = \sqrt{8} = 2\sqrt{2}$

$$a = \sqrt{\frac{32}{9}} = \frac{4\sqrt{2}}{3}$$

$$y = \pm\frac{b}{a}x = \pm\frac{2\sqrt{2}}{\frac{4\sqrt{2}}{3}}x$$

$$y = \pm\frac{3}{2}x$$

17. $y^2 - 6y - 3x^2 - 6x - 3 = 0$
$y^2 - 6y - 3(x^2 + 2x) = 3$
$y^2 - 6y + 9 - 3(x^2 + 2x + 1) = 3 + 9 - 3$
$(y-3)^2 - 3(x+1)^2 = 9$

$$\frac{(y-3)^2}{9} - \frac{(x+1)^2}{3} = 1$$

Center is at $(-1, 3)$.
Vertices are 3 units above and below the center, $(-1, 0)$ and $(-1, -6)$.

18. $5x^2 - 4y^2 = 20$

$$\frac{x^2}{4} - \frac{y^2}{5} = 1$$

$a^2 = 4$
$b^2 = 5$
$c^2 = a^2 + b^2$
$c^2 = 4 + 5 = 9$
$c = 3$
Foci: $(\pm 3, 0)$

19. Vertices: $(\pm 6, 0)$
Foci at $(\pm 4\sqrt{3}, 0)$
$a = 6 \qquad c = 4\sqrt{3}$
$c^2 = a^2 + b^2$
$(4\sqrt{3})^2 = 6^2 + b^2$
$48 = 36 + b^2$
$12 = b^2$

$$\frac{x^2}{36} - \frac{y^2}{12} = 1$$

$x^2 - 3y^2 = 36$

20. Vertices $(0, 4)$ and $(-2, 4)$
Foci $(2, 4)$ and $(-4, 4)$

$h + a = 0 \qquad k = 4$
$h - a = -2$
$2h = -2$
$h = -1$

$-1 + a = 0$
$a = 1$

$h + c = 2$
$-1 + c = 2$
$c = 3$

$c^2 = a^2 + b^2$
$3^2 = 1^2 + b^2$
$9 = 1 + b^2$
$8 = b^2$

$$\frac{(x+1)^2}{1} - \frac{(y-4)^2}{8} = 1$$

$8(x+1)^2 - (y-4)^2 = 8$
$8(x^2 + 2x + 1) - (y^2 - 8y + 16) = 8$
$8x^2 + 16x + 8 - y^2 + 8y - 16 - 8 = 0$
$8x^2 + 16x - y^2 + 8y - 16 = 0$

21. $\begin{pmatrix} x^2 + y^2 = 16 \\ x - 4y = 8 \end{pmatrix}$

By inspection of a graph of the system, the parabola intersects the circle in two points, so there are two real solutions.

22. $\begin{pmatrix} x^2 + 4y^2 = 25 \\ xy = 6 \end{pmatrix}$

$$x = \frac{6}{y}$$

$$\left(\frac{6}{y}\right)^2 + 4y^2 = 25$$

$$\frac{36}{y^2} + 4y^2 = 25$$

$36 + 4y^4 = 25y^2$
$4y^4 - 25y^2 + 36 = 0$
$(4y^2 - 9)(y^2 - 4) = 0$
$4y^2 - 9 = 0$ or $y^2 - 4 = 0$

$y^2 = \frac{9}{4}$ or $y^2 = 4$

$y = \pm\frac{3}{2}$ or $y = \pm 2$

When $y = \frac{3}{2}$, $x = \frac{6}{\frac{3}{2}} = 4$.

When $y = -\frac{3}{2}$, $x = \frac{6}{-\frac{3}{2}} = -4$.

When $y = 2$, $x = \frac{6}{2} = 3$.

When $y = -2$, $x = \frac{6}{-2} = -3$.

The solution set is
$\left\{\left(4, \frac{3}{2}\right), \left(-4, -\frac{3}{2}\right), (3, 2), (-3, -2)\right\}$.

23. $y^2 + 4y + 8x - 4 = 0$
$y^2 + 4y + 4 = -8x + 4 + 4$
$(y + 2)^2 = -8x + 8$
$(y + 2)^2 = -8(x - 1)$
$4p = -8$ so $p = -2$
Vertex: $(1, -2)$
directrix: $x = h - p$
$x = 1 - (-2) = 3$

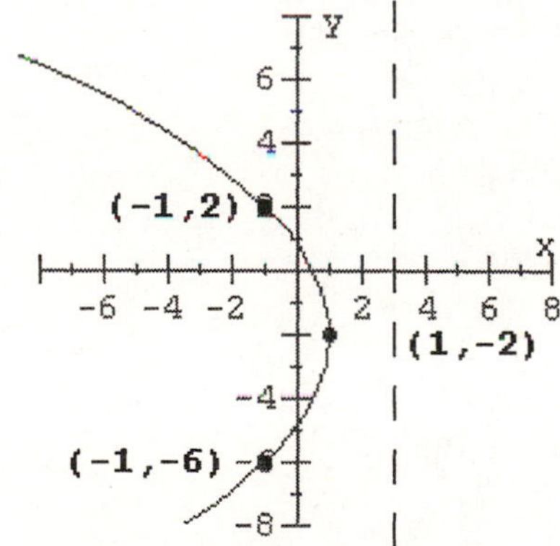

24. $9x^2 - 36x + 4y^2 + 16y + 16 = 0$
$9(x^2 - 4x) + 4(y^2 + 4y) = -16$
$9(x^2\text{-}4x\text{+}4)\text{+}4(y^2\text{+}4y\text{+}4) = -16 + 36 + 16$
$9(x - 2)^2 + 4(y + 2)^2 = 36$

$\frac{(x-2)^2}{4} + \frac{(y+2)^2}{9} = 1$

Center : $(2, -2)$

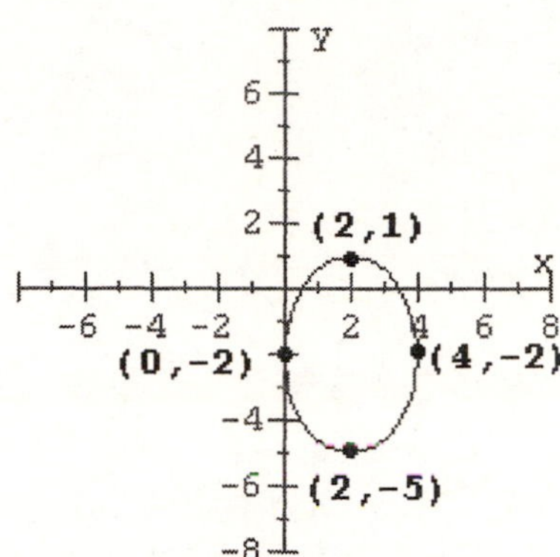

25. $x^2 + 6x - 3y^2 = 0$
$x^2 + 6x + 9 - 3y^2 = 0 + 9$
$(x + 3)^2 - 3y^2 = 9$
$\frac{(x+3)^2}{9} - \frac{(y-0)^2}{3} = 1$

Center: $(-3, 0)$
$a^2 = 9$ so $a = 3$
$b^2 = 3$ so $b = \sqrt{3}$
Vertices: $(-6, 0)$ and $(0, 0)$

Asymptotes: $y = \pm\frac{b}{a}x$
$y = \pm\frac{\sqrt{3}}{3}x$

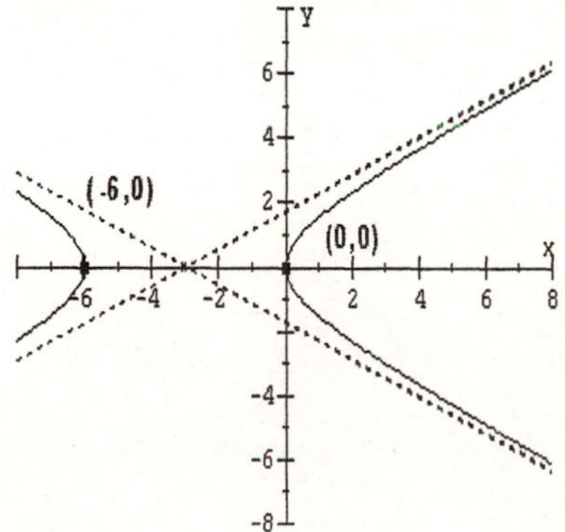

Chapter 14 Sequences and Mathematical Induction

PROBLEM SET 14.1 Arithmetic Sequences

1. $a_n = 3n - 7$
$a_1 = 3(1) - 7 = -4$
$a_2 = 3(2) - 7 = -1$
$a_3 = 3(3) - 7 = 2$
$a_4 = 3(4) - 7 = 5$
$a_5 = 3(5) - 7 = 8$

3. $a_n = -2n + 4$
$a_1 = -2(1) + 4 = 2$
$a_2 = -2(2) + 4 = 0$
$a_3 = -2(3) + 4 = -2$
$a_4 = -2(4) + 4 = -4$
$a_5 = -2(5) + 4 = -6$

5. $a_n = 3n^2 - 1$
$a_1 = 3(1)^2 - 1 = 2$
$a_2 = 3(2)^2 - 1 = 11$
$a_3 = 3(3)^2 - 1 = 26$
$a_4 = 3(4)^2 - 1 = 47$
$a_5 = 3(5)^2 - 1 = 74$

7. $a_n = n(n-1)$
$a_1 = 1(1-1) = 0$
$a_2 = 2(2-1) = 2$
$a_3 = 3(3-1) = 6$
$a_4 = 4(4-1) = 12$
$a_5 = 5(5-1) = 20$

9. $a_n = 2^{n+1}$
$a_1 = 2^{1+1} = 4$
$a_2 = 2^{2+1} = 8$
$a_3 = 2^{3+1} = 16$
$a_4 = 2^{4+1} = 32$
$a_5 = 2^{5+1} = 64$

11. $a_n = -5n - 4$
$a_{15} = -5(15) - 4 = -79$
$a_{30} = -5(30) - 4 = -154$

13. $a_n = (-1)^{n+1}$
$a_{25} = (-1)^{25+1} = 1$
$a_{50} = (-1)^{50+1} = -1$

15. $d = 13 - 11 = 2$
$a_n = a_1 + (n-1)d$
$a_n = 11 + (n-1)(2)$
$a_n = 11 + 2n - 2$
$a_n = 2n + 9$

17. $d = -1 - 2 = -3$
$a_n = a_1 + (n-1)d$
$a_n = 2 + (n-1)(-3)$
$a_n = 2 - 3n + 3$
$a_n = -3n + 5$

19. $d = 2 - \frac{3}{2} = \frac{1}{2}$
$a_n = a_1 + (n-1)d$
$a_n = \frac{3}{2} + (n-1)\left(\frac{1}{2}\right)$
$a_n = \frac{3}{2} + \frac{1}{2}n - \frac{1}{2}$
$a_n = \frac{1}{2}n + 1 = \frac{n+2}{2}$

21. $d = 6 - 2 = 4$
$a_n = a_1 + (n-1)d$
$a_n = 2 + (n-1)4$
$a_n = 2 + 4n - 4$
$a_n = 4n - 2$

23. $d = -6 - (-3) = -3$
$a_n = a_1 + (n-1)d$
$a_n = -3 + (n-1)(-3)$
$a_n = -3 - 3n + 3$
$a_n = -3n$

25. $d = 8 - 3 = 5$
$a_n = a_1 + (n-1)d$
$a_{15} = 3 + (15-1)(5)$
$a_{15} = 73$

27. $d = 26 - 15 = 11$
$a_n = a_1 + (n-1)d$
$a_{30} = 15 + (30-1)(11) = 334$

29. $d = \frac{5}{3} - 1 = \frac{2}{3}$
$a_n = a_1(n-1)d$
$a_{52} = 1 + (52-1)\left(\frac{2}{3}\right)$
$a_{52} = 35$

31. $a_n = a_1 + (n-1)d$

$a_6 = 12$
$12 = a_1 + (6-1)d$
$12 = a_1 + 5d$

$a_{10} = 16$
$16 = a_1 + (10-1)d$
$16 = a_1 + 9d$

$$\begin{array}{rrr} -12 = & -a_1 & -5d \\ 16 = & a_1 & +9d \\ \hline 4 = & & 4d \end{array}$$

$1 = d$
$12 = a_1 + 5d$
$12 = a_1 + 5(1)$
$7 = a_1$

33. $a_n = a_1 + (n-1)d$

$a_3 = 20$
$20 = a_1 + (3-1)d$
$20 = a_1 + 2d$

$a_7 = 32$
$32 = a_1 + (7-1)d$
$32 = a_1 + 6d$

$$\begin{array}{rrr} -20 = & -a_1 & -2d \\ 32 = & a_1 & +6d \\ \hline 12 = & & 4d \end{array}$$

$3 = d$
$20 = a_1 + 2d$
$20 = a_1 + 2(3)$
$14 = a_1$
$a_{25} = 14 + (25-1)(3)$
$a_{25} = 86$

35. $d = 7 - 5 = 2$
$a_n = a_1 + (n-1)d$
$a_{50} = 5 + (50-1)(2)$
$a_{50} = 103$

$S_n = \frac{n(a_1 + a_n)}{2}$

$S_{50} = \frac{50(5 + 103)}{2} = 2700$

37. $d = 6 - 2 = 4$
$a_n = a_1 + (n-1)d$
$a_{40} = 2 + (40-1)(4)$
$a_{40} = 158$

$S_n = \frac{n(a_1 + a_n)}{2}$

$S_{40} = \frac{40(2 + 158)}{2} = 3200$

39. $d = 2 - 5 = -3$
$a_n = a_1 + (n-1)d$
$a_{75} = 5 + (75-1)(-3)$
$a_{75} = -217$

$S_n = \frac{n(a_1 + a_n)}{2}$

$S_{75} = \frac{75(5 - 217)}{2} = -7950$

41. $d = 1 - \frac{1}{2} = \frac{1}{2}$
$a_n = a_1 + (n-1)d$
$a_{50} = \frac{1}{2} + (50-1)\left(\frac{1}{2}\right)$
$a_{50} = 25$

$S_n = \frac{n(a_1 + a_n)}{2}$

$S_{50} = \frac{50\left(\frac{1}{2} + 25\right)}{2} = 637.5$

43. $d = 5 - 1 = 4$
$a_n = a_1 + (n-1)d$
$197 = 1 + (n-1)(4)$
$197 = 1 + 4n - 4$
$200 = 4n$
$50 = n$

$S_n = \frac{n(a_1 + a_n)}{2}$

$S_{50} = \frac{50(1 + 197)}{2} = 4950$

45. $d = 8 - 2 = 6$
$a_n = a_1 + (n-1)d$
$146 = 2 + (n-1)6$
$146 = 2 + 6n - 6$
$150 = 6n$
$25 = n$
$S_n = \dfrac{n(a_1 - a_n)}{2}$
$S_{25} = \dfrac{25(2 + 146)}{2} = 1850$

47. $d = -10 - (-7) = -3$
$a_n = a_1 + (n-1)d$
$-109 = -7 + (n-1)(-3)$
$-109 = -7 - 3n + 3$
$-105 = -3n$
$35 = n$
$S_n = \dfrac{n(a_1 + a_n)}{2}$
$S_{35} = \dfrac{35[-7 + (-109)]}{2} = -2030$

49. $d = -3 - (-5) = 2$
$a_n = a_1 + (n-1)d$
$119 = -5 + (n-1)2$
$119 = -5 + 2n - 2$
$126 = 2n$
$63 = n$
$S_n = \dfrac{n(a_1 + a_n)}{2}$
$S_{63} = \dfrac{63(-5 + 119)}{2} = 3591$

51. $a_1 = 1 \qquad d = 2 \qquad n = 200$
$a_{200} = 1 + (200 - 1)(2)$
$a_{200} = 399$
$S_n = \dfrac{n(a_1 + a_n)}{2}$
$S_{200} = \dfrac{200(1 + 399)}{2} = 40,000$

53. $a_1 = 18 \qquad d = 2$
$a_n = a_1 + (n-1)d$
$482 = 18 + (n-1)2$
$482 = 18 + 2n - 2$
$466 = 2n$
$233 = n$
$S_n = \dfrac{n(a_1 + a_n)}{2}$
$S_{233} = \dfrac{233(18 + 482)}{2} = 58,250$

55. $a_n = 5n - 4$
$a_1 = 5(1) - 4 = 1$
$a_{30} = 5(30) - 4 = 146$
$S_n = \dfrac{n(a_1 + a_n)}{2}$
$S_{30} = \dfrac{30(1 + 146)}{2} = 2205$

57. $a_n = -4n - 1$
$a_1 = -4(1) - 1 = -5$
$a_{25} = -4(25) - 1 = -101$
$S_n = \dfrac{n(a_1 + a_n)}{2}$
$S_{25} = \dfrac{25[-5 + (-101)]}{2} = -1325$

59. $\displaystyle\sum_{i=1}^{45}(5i + 2)$
$a_1 = 5(1) + 2 = 7$
$a_{45} = 5(45) + 2 = 227$
$S_n = \dfrac{n(a_1 + a_n)}{2}$
$S_{45} = \dfrac{45(7 + 227)}{2} = 5265$

61. $\displaystyle\sum_{i=1}^{30}(-2i + 4)$
$a_1 = -2(1) + 4 = 2$
$a_{30} = -2(30) + 4 = -56$
$S_n = \dfrac{n(a_1 + a_n)}{2}$
$S_{30} = \dfrac{30[2 + (-56)]}{2} = -810$

63. $\displaystyle\sum_{i=4}^{32}(3i - 10)$
$a_4 = 3(4) - 10 = 2$
$a_{32} = 3(32) - 10 = 86$
$S_n = \dfrac{n(a_1 + a_n)}{2}$

$$S_{29} = \frac{29(2+86)}{2} = 1276$$

65. $\sum_{i=10}^{20} 4i$

$a_{10} = 4(10) = 40$
$a_{20} = 4(20) = 80$

$$S_n = \frac{n(a_1 + a_n)}{2}$$

$$S_{11} = \frac{11(40+80)}{2} = 660$$

67. $\sum_{i=1}^{5} i^2 = 1^2 + 2^2 + 3^2 + 4^2 + 5^2$

$= 1 + 4 + 9 + 16 + 25 = 55$

69. $\sum_{i=3}^{8} (2i^2 + i) = [2(3)^2 + 3] + [2(4)^2 + 4]$

$+ [2(5)^2 + 5] + [2(6)^2 + 6]$

$+ [2(7)^2 + 7] + [2(8)^2 + 8]$

$= 21+36+55+78+105+136$

$= 431$

Further Investigations

75. $a_n = \begin{cases} 2n+1 & \text{for } n \text{ odd} \\ 2n-1 & \text{for } n \text{ even} \end{cases}$

$a_1 = 2(1) + 1 = 3$
$a_2 = 2(2) - 1 = 3$
$a_3 = 2(3) + 1 = 7$
$a_4 = 2(4) - 1 = 7$
$a_5 = 2(5) + 1 = 11$
$a_6 = 2(6) - 1 = 11$

77. $a_n = \begin{cases} 3n+1 & \text{for } n \leq 3 \\ 4n-3 & \text{for } n > 3 \end{cases}$

$a_1 = 3(1) + 1 = 4$
$a_2 = 3(2) + 1 = 7$
$a_3 = 3(3) + 1 = 10$
$a_4 = 4(4) - 3 = 13$
$a_5 = 4(5) - 3 = 17$
$a_6 = 4(6) - 3 = 21$

79. $\begin{cases} a_1 = 4 \\ a_n = 3a_{n-1} & \text{for } n \geq 2 \end{cases}$

$a_1 = 4$
$a_2 = 3(4) = 12$
$a_3 = 3(12) = 36$
$a_4 = 3(36) = 108$
$a_5 = 3(108) = 324$
$a_6 = 3(324) = 972$

81. $\begin{cases} a_1 = 1 \\ a_2 = 1 \\ a_n = a_{n-2} + a_{n-1} & \text{for } n \geq 3 \end{cases}$

$a_1 = 1$
$a_2 = 1$
$a_3 = 1 + 1 = 2$
$a_4 = 1 + 2 = 3$
$a_5 = 2 + 3 = 5$
$a_6 = 3 + 5 = 8$

83. $\begin{cases} a_1 = 3 \\ a_2 = 1 \\ a_n = (a_{n-1} - a_{n-2})^2 & \text{for } n \geq 3 \end{cases}$

$a_1 = 3$
$a_2 = 1$
$a_3 = (1-3)^2 = 4$
$a_4 = (4-1)^2 = 9$
$a_5 = (9-4)^2 = 25$
$a_6 = (25-9)^2 = 256$

PROBLEM SET 14.2 Geometric Sequences

1. $r = \frac{6}{3} = 2$
$a_n = a_1 r^{n-1}$
$a_n = 3(2)^{n-1}$

3. $r = \frac{9}{3} = 3$
$a_n = a_1 r^{n-1}$
$a_n = 3(3)^{n-1}$
$a_n = 3^n$

5. $r = \frac{\frac{1}{8}}{\frac{1}{4}} = \frac{1}{2}$
$a_n = a_1 r^{n-1}$
$a_n = \frac{1}{4}\left(\frac{1}{2}\right)^{n-1}$
$a_n = \left(\frac{1}{2}\right)^2\left(\frac{1}{2}\right)^{n-1}$
$a_n = \left(\frac{1}{2}\right)^{2+n-1}$
$a_n = \left(\frac{1}{2}\right)^{n+1}$

7. $r = \frac{16}{4} = 4$
$a_n = a_1 r^{n-1}$
$a_n = 4(4)^{n-1}$
$a_n = 4^n$

9. $r = \frac{0.3}{1} = 0.3$
$a_n = a_1 r^{n-1}$
$a_n = 1(0.3)^{n-1}$
$a_n = (0.3)^{n-1}$

11. $r = \frac{-2}{1} = -2$
$a_n = a_1 r^{n-1}$
$a_n = 1(-2)^{n-1}$
$a_n = (-2)^{n-1}$

13. $r = \frac{1}{\frac{1}{2}} = 2$
$a_n = a_1 r^{n-1}$
$a_n = \frac{1}{2}(2)^{n-1}$
$a_8 = \frac{1}{2}(2)^{8-1} = 2^6 = 64$

15. $r = \frac{243}{729} = \frac{1}{3}$
$a_n = a_1 r^{n-1}$
$a_9 = 729\left(\frac{1}{3}\right)^{9-1}$
$a_9 = \frac{1}{9}$

17. $r = \frac{-2}{1} = -2$
$a_n = a_1 r^{n-1}$
$a_n = 1(-2)^{n-1}$
$a_{10} = (-2)^{10-1}$
$a_{10} = (-2)^9$
$a_{10} = -512$

19. $r = \frac{\frac{1}{6}}{\frac{1}{2}} = \frac{2}{6} = \frac{1}{3}$
$a_n = a_1 r^{n-1}$
$a_8 = \frac{1}{2}\left(\frac{1}{3}\right)^{8-1}$
$a_8 = \frac{1}{4374}$

21. $a_5 = \frac{32}{3}$ $\quad r = 2$
$a_n = a_1 r^{n-1}$
$\frac{32}{3} = a_1(2)^{5-1}$
$\frac{32}{3} = 16a_1$
$\frac{2}{3} = a_1$

23. $a_3 = 12$ $\quad a_6 = 96$

$12 = a_1 r^2$ $\quad 96 = a_1 r^5$

$$a_1 = \frac{12}{r^2}$$

$$96 = \frac{12}{r^2}(r^5)$$

$$96 = 12r^3$$
$$8 = r^3$$
$$2 = r$$

25. $r = \frac{2}{1} = 2$

$$S_n = \frac{a_1 r^n - a_1}{r - 1}$$

$$S_{10} = \frac{1(2)^{10} - 1}{2 - 1}$$

$$S_{10} = \frac{2^{10} - 1}{1} = 1023$$

27. $r = \frac{6}{2} = 3$

$$S_n = \frac{a_1 r^n - a_1}{r - 1}$$

$$S_9 = \frac{2(3^9) - 2}{3 - 1}$$

$$S_9 = \frac{39364}{2} = 19,682$$

29. $r = \frac{12}{8} = \frac{3}{2}$

$$S_n = \frac{a_1 r^n - a_1}{r - 1}$$

$$S_8 = \frac{8\left(\frac{3}{2}\right)^8 - 8}{\frac{3}{2} - 1}$$

$$S_8 = \frac{8\left(\frac{6561}{256}\right) - 8}{\frac{1}{2}} = \frac{6305}{16}$$

$$S_8 = 394\frac{1}{16}$$

31. $r = \frac{8}{-4} = -2$

$$S_n = \frac{a_1 r^n - a_1}{r - 1}$$

$$S_{10} = \frac{-4(-2)^{10} - (-4)}{-2 - 1}$$

$$S_{10} = \frac{-4092}{-3} = 1364$$

33. $r = \frac{27}{9} = 3$

$$S = 9 + 27 + 81 + \ldots + 729$$
$$3S = 27 + 81 + \ldots + 729 + 2187$$

Subtract the 1st equation from the 2nd equation.

$$2S = -9 + 2187$$
$$2S = 2178$$
$$S = 1089$$

35. $r = \frac{2}{4} = \frac{1}{2}$

$$\frac{1}{512} = 4\left(\frac{1}{2}\right)^{n-1}$$

$$\frac{1}{2048} = \left(\frac{1}{2}\right)^{n-1}$$

$$\left(\frac{1}{2}\right)^{11} = \left(\frac{1}{2}\right)^{n-1}$$

$$11 = n - 1$$
$$12 = n$$

$$S_{12} = \frac{4\left(\frac{1}{2}\right)^{12} - 4}{\frac{1}{2} - 1}$$

$$S_{12} = \frac{4\left[\left(\frac{1}{2}\right)^{12} - 1\right]}{-\frac{1}{2}}$$

$$S_{12} = -2(4)\left(\frac{1}{4096} - 1\right)$$

$$S_{12} = -8\left(-\frac{4095}{4096}\right) = \frac{32760}{4096}$$

$$S_{12} = 7\frac{511}{512}$$

37. $r = \frac{3}{-1} = -3$

$S = (-1) + 3 + (-9) + ... + (-729)$
$-3S = 3 + (-9) + ... + (-729) + 2187$

Subtract the 1st equation from the 2nd equation.

$-4S = 2187 - (-1)$
$-4S = 2188$
$S = -547$

39. $\sum_{i=1}^{9} 2^{i-3}$

$a_1 = 2^{1-3} = 2^{-2} = \frac{1}{4}$

$$S_9 = \frac{\frac{1}{4}(2)^9 - \frac{1}{4}}{2-1}$$

$$S_9 = \frac{128 - \frac{1}{4}}{1} = 127\frac{3}{4}$$

41. $\sum_{i=2}^{5} (-3)^{i+1}$

$= (-3)^{2+1} + (-3)^{3+1} + (-3)^{4+1} + (-3)^{5+1}$
$= -27 + 81 + (-243) + 729$
$= 540$

43. $\sum_{i=1}^{6} 3\left(\frac{1}{2}\right)^i =$

$$3\left(\frac{1}{2}\right)^1 + 3\left(\frac{1}{2}\right)^2 + 3\left(\frac{1}{2}\right)^3 +$$

$$3\left(\frac{1}{2}\right)^4 + 3\left(\frac{1}{2}\right)^5 + 3\left(\frac{1}{2}\right)^6$$

$$= \frac{3}{2} + \frac{3}{4} + \frac{3}{8} + \frac{3}{16} + \frac{3}{32} + \frac{3}{64}$$

$$= \frac{96}{64} + \frac{48}{64} + \frac{24}{64} + \frac{12}{64} + \frac{6}{64} + \frac{3}{64}$$

$$= \frac{189}{64} = 2\frac{61}{64}$$

45. $r = \frac{1}{2}$

$$S_\infty = \frac{a_1}{1-r}$$

$$S_\infty = \frac{2}{1-\frac{1}{2}} = \frac{2}{\frac{1}{2}} = 4$$

47. $r = \frac{\frac{2}{3}}{1} = \frac{2}{3}$

$$S_\infty = \frac{a_1}{1-r}$$

$$S_\infty = \frac{1}{1-\frac{2}{3}} = \frac{1}{\frac{1}{3}} = 3$$

49. $r = \frac{8}{4} = 2$
Because $|r| > 1$, this infinite sequence has no sum.

51. $r = -\frac{3}{9} = -\frac{1}{3}$

$$S_\infty = \frac{a_1}{1-r}$$

$$S_\infty = \frac{9}{1-\left(-\frac{1}{3}\right)} = \frac{9}{\frac{4}{3}}$$

$$S_\infty = 9\left(\frac{3}{4}\right) = \frac{27}{4}$$

53. $r = \frac{\frac{3}{8}}{\frac{1}{2}} = \frac{3}{4}$

$$S_\infty = \frac{a_1}{1-r}$$

$$S_\infty = \frac{\frac{1}{2}}{1-\frac{3}{4}} = \frac{\frac{1}{2}}{\frac{1}{4}} = 2$$

55. $r = -\frac{4}{8} = -\frac{1}{2}$

$$S_\infty = \frac{a_1}{1-r}$$

$$S_\infty = \frac{8}{1-\left(-\frac{1}{2}\right)} = \frac{8}{\frac{3}{2}}$$

$$S_\infty = 8\left(\frac{2}{3}\right) = \frac{16}{3}$$

57. $0.\overline{3} = 0.3 + 0.03 + 0.003 + ...$

$$r = \frac{0.03}{0.3} = 0.1$$

$$S_\infty = \frac{a_1}{1-r}$$

$$S_\infty = \frac{0.3}{1-0.1} = \frac{0.3}{0.9} = \frac{1}{3}$$

59. $0.\overline{26} = 0.26 + 0.0026 + 0.000026 + ...$

$$r = \frac{0.0026}{0.26} = 0.01$$

$$S_\infty = \frac{a_1}{1-r}$$

$$S_\infty = \frac{0.26}{1-0.01} = \frac{0.26}{0.99} = \frac{26}{99}$$

61. $0.\overline{123} = 0.123 + 0.000123 + 0.000000123 + ...$

$$r = \frac{0.000123}{0.123} = 0.001$$

$$S_\infty = \frac{a_1}{1-r}$$

$$S_\infty = \frac{0.123}{1-0.001} = \frac{0.123}{0.999} = \frac{123}{999} = \frac{41}{333}$$

63. $0.2\overline{6} = 0.2 + [0.06 + 0.006 + ...]$
Consider $0.06 + 0.006 + ...$

$$r = \frac{0.006}{0.06} = 0.1$$

$$S_\infty = \frac{a_1}{1-r}$$

$$S_\infty = \frac{0.06}{1-0.01} = \frac{0.06}{0.9} = \frac{6}{90} = \frac{1}{15}$$

$$0.2\overline{6} = \frac{2}{10} + \frac{1}{15} = \frac{6}{30} + \frac{2}{30} = \frac{8}{30} = \frac{4}{15}$$

65. $0.2\overline{14} = 0.2 + [0.014 + 0.00014 + 0.0000014 + ...]$
Consider $0.014 + 0.00014 + 0.0000014 + ...$

$$r = \frac{0.00014}{0.014} = 0.01$$

$$S_\infty = \frac{a_1}{1-r}$$

$$S_\infty = \frac{0.014}{1-0.01} = \frac{0.014}{0.99} = \frac{14}{990} = \frac{7}{495}$$

Then, $0.2\overline{14} = 0.2 + \frac{7}{495}$

$$= \frac{1}{5} + \frac{7}{495}$$

$$= \frac{99}{495} + \frac{7}{495} = \frac{106}{495}$$

67. $2.\overline{3} = 2 + [0.3 + 0.03 + 0.003 + ...]$
Consider $0.3 + 0.03 + 0.003 + ...$

$$r = \frac{0.03}{0.3} = 0.1$$

$$S_\infty = \frac{a_1}{1-r}$$

$$S_\infty = \frac{0.3}{1-0.1} = \frac{0.3}{0.9} = \frac{3}{9} = \frac{1}{3}$$

$$2.\overline{3} = 2 + \frac{1}{3} = \frac{7}{3}$$

PROBLEM SET 14.3 Another Look at Problem Solving

1. This is an arithmetic sequence with $a_1 = 9500$ and $d = 700$. For 1980 let $n = 1$, then for 2001 $n = 22$.
$a_n = a_1 + (n-1)d$
$a_{22} = 9500 + (22-1)(700) = 24,200$
His salary for 2001 was \$24, 200.

3. $a_1 = 9600 \quad d = 150 \quad n = 14$
$a_n = a_1 + (n-1)d$
$a_n = 9600 + (14-1)(150)$
$a_n = 11,550$
The enrollment in 2003 was $11,500$.

5. Each year's enrollment is $100\% + 10\% = 110\%$ of the previous year's enrollment. This is a geometric sequence with $a_1 = 5000$ and $r = 110\% = 1.1$. For 1999 let $n = 1$, then for 2003, $n = 5$.
$a_n = a_1 r^{n-1}$
$a_5 = 5000(1.1)^{5-1} = 5000(1.1)^4 = 7320.5$
The predicted enrollment for 2003 is 7320.

7. $a_1 = 16,000 \qquad r = \frac{1}{2} \qquad n = 8$
$a_n = a_1 r^{n-1}$
$a_8 = 16000\left(\frac{1}{2}\right)^{8-1}$
$a_8 = 125$
After seven days 125 liters remain.

9. If $\frac{1}{3}$ of the water is removed, then $\frac{2}{3}$ of the water remains. This is a geometric sequence witn $a_1 = \frac{2}{3}(5832) = 3888$ and $r = \frac{2}{3}$.
$a_n = a_1 r^{n-1}$
$a_6 = 3888\left(\frac{2}{3}\right)^{6-1}$
$a_6 = 3888\left(\frac{2}{3}\right)^{5}$
$a_6 = 3888\left(\frac{32}{243}\right)$
$a_6 = 512$
At the end of 6 days, 512 gallons remain.

11. $a_1 = 1 \qquad d = 1 \qquad n = 30$
$a_n = a_1 + (n-1)d$
$a_{31} = 1 + (30-1)(1) = 30$
$S_n = \frac{n(a_1 + a_n)}{2}$
$S_{31} = \frac{30(1+30)}{2} = 465$
465 quarters $= 465(.25) = 116.25$
After thirty days she saves \$116.25.

13. The geometric sequence 1, 2, 4, 8, ... represents the savings, in cents, on a day-by-day basis.
$a_n = a_1 r^{n-1}$
$a_{15} = 1(2)^{15-1} = 2^{14}$
$a_{15} = 16384$ cents $= \$163.84$
$S_n = \frac{a_1 r^n - a_1}{r-1}$
$S_{15} = \frac{1(2)^{15} - 1}{2-1} = \frac{2^{15} - 1}{1}$
$S_{15} = 32767 = 327.67$
Savings for fifteen days is \$327.67.

15. Interest $= 1500 \times 0.12 \times$ (1+2+3+...+10)
$I = 1500 \times 0.12 \times \left[\frac{10(1+10)}{2}\right]$
$I = 1500 \times 0.12 \times 55$
$I = 9,900$
Principal $= 1500 \times 10 = 15,000$
Amount $=$ Interest $+$ Principal
$= 9,900 + 15,000 = 24,900$
The value of all investments is \$24, 900.

17. 16, 48, 80, 112, ...
$d = 48 - 16 = 32$
$a_n = a_1 + (n-1)d$
$a_{11} = 16 + (11-1)(32) = 336$
$S_n = \frac{n(a_1 + a_n)}{2}$

$$S_{11} = \frac{11(16+336)}{2} = 1936$$

It will fall 1936 feet.

19. $a_1 = 60 \qquad r = \frac{1}{2} \qquad n = 7$

After 24 hours there would be 6 time periods and including the initial term $a_1 = 60$, there will be 7 terms.

$$a_n = a_1 r^{n-1}$$

$$a_7 = 60\left(\frac{1}{2}\right)^{7-1} = \frac{60}{64} = \frac{15}{16}$$

After 24 hours, $\frac{15}{16}$ grams remain.

21. For the falling distance, the geometric sequence is 1458, 486, 162, ... with $r = \frac{1}{3}$.

$$S_n = \frac{a_1 r^n - a_1}{r-1}$$

$$S_6 = \frac{1458\left(\frac{1}{3}\right)^6 - 1458}{\frac{1}{3} - 1}$$

$$= \frac{2 - 1458}{-\frac{2}{3}} = 2184$$

For the distance of the rebound, the sequence is 486, 162, ...

$$S_5 = \frac{486\left(\frac{1}{3}\right)^5 - 486}{\frac{1}{3} - 1} = \frac{2 - 486}{-\frac{2}{3}} = 726$$

Total distance $= 2184 + 726 = 2910$ feet.

23. $a_1 = 25 \qquad d = -1 \qquad a_n = 1$

$$a_n = a_1 + (n-1)d$$
$$1 = 25 + (n-1)(-1)$$
$$1 = 25 - n + 1$$
$$-25 = -n$$
$$25 = n$$
$$S_n = \frac{n(a_1 + a_n)}{2}$$
$$S_{25} = \frac{25(25+1)}{2} = 325$$

There are 325 logs in the pile.

25. If $\frac{1}{3}$ of the air is pumped out on each stroke, then $\frac{2}{3}$ of the air remains.

Thus, the geometric sequence $\frac{2}{3}, \frac{4}{9}, \frac{8}{27}, \ldots$ with $r = \frac{2}{3}$ represents the air remaining after each stroke.

$$a_n = a_1 r^{n-1}$$
$$a_7 = \frac{2}{3}\left(\frac{2}{3}\right)^{7-1}$$
$$a_7 = \frac{2}{3}\left(\frac{2}{3}\right)^6$$
$$a_7 = \left(\frac{2}{3}\right)^7 = \frac{128}{2187}$$
$$a_7 = 0.059 = 5.9\%$$

After seven strokes, 5.9% of air remains.

27. $a_1 = 20 \qquad r = \frac{1}{2} \qquad n = 9$

$$a_n = a_1 r^{n-1}$$
$$a_9 = 20\left(\frac{1}{2}\right)^{9-1} = \frac{5}{64}$$

$\frac{5}{64}$ of a gallon of water remains.

PROBLEM SET 14.4 Mathematical Induction

1. Part 1: If $n = 1$, then $a_1 = 1$.

$$S_1 = \frac{1(1+1)}{2} = 1$$

True for $n = 1$.

Part 2: We need to prove that if

$$S_k = \frac{k(k+1)}{2}, \text{ then}$$

$$S_{k+1} = \frac{(k+1)(k+2)}{2}.$$

Proof: $S_{k+1} = S_k + a_{k+1}$

$$= \frac{k(k+1)}{2} + k + 1$$

$$= \frac{k(k+1) + 2(k+1)}{2} = \frac{(k+1)(k+2)}{2}$$

3. Part 1: If $n = 1$, then $a_1 = 3(1) - 1 = 2$

$$S_1 = \frac{1[3(1)+1]}{2} = \frac{1(4)}{2} = 2$$

True for $n = 1$.

Part 2: We need to prove that if

$$S_k = \frac{k(3k+1)}{2}, \text{ then}$$

$$S_{k+1} = \frac{(k+1)[3(k+1)+1]}{2}$$

$$S_{k+1} = \frac{(k+1)(3k+4)}{2}$$

Proof: $S_{k+1} = S_k + a_{k+1}$

$$= \frac{k(3k+1)}{2} + 3(k+1) - 1$$

$$= \frac{k(3k+1)}{2} + 3k + 2$$

$$= \frac{k(3k+1)}{2} + \frac{6k+4}{2}$$

$$= \frac{3k^2 + k + 6k + 4}{2}$$

$$= \frac{3k^2 + 7k + 4}{2} = \frac{(k+1)(3k+4)}{2}$$

5. Part 1: If $n = 1$, then $a_1 = 2^1 = 2$.

$S_1 = 2(2^1 - 1) = 2$

True for $n = 1$.

Part 2: We need to prove that if

$S_k = 2(2^k - 1)$, then

$S_{k+1} = 2(2^{k+1} - 1)$.

Proof: $S_{k+1} = S_k + a_{k+1}$

$= 2(2^k - 1) + 2^{k+1}$

$= 2^{k+1} - 2 + 2^{k+1}$

$= 2 \cdot 2^{k+1} - 2$

$= 2(2^{k+1} - 1)$

7. Part 1: If $n = 1$, then

$a_1 = 1^2 = 1$

$$S_1 = \frac{1(1+1)[2(1)+1]}{6}$$

$$S_1 = \frac{1(2)(3)}{6} = 1$$

True for $n = 1$.

Part 2: We need to prove that if

$$S_k = \frac{k(k+1)(2k+1)}{6}, \text{ then}$$

$$S_{k+1} = \frac{(k+1)(k+2)(2k+3)}{6}$$

Proof: $S_{k+1} = S_k + a_{k+1}$

$$= \frac{k(k+1)(2k+1)}{6} + (k+1)^2$$

$$= \frac{k(k+1)(2k+1)}{6} + \frac{6(k+1)^2}{6}$$

$$= \frac{k(k+1)(2k+1) + 6(k+1)^2}{6}$$

$$= \frac{(k+1)[k(2k+1) + 6(k+1)]}{6}$$

$$= \frac{(k+1)(2k^2 + k + 6k + 6)}{6}$$

$$= \frac{(k+1)(2k^2 + 7k + 6)}{6}$$

$$= \frac{(k+1)(k+2)(2k+3)}{6}$$

9. Part 1: If $n = 1$, then $a_1 = \frac{1}{1(1+1)} = \frac{1}{2}$.

$S_1 = \frac{1}{1+1} = \frac{1}{2}$

True for $n = 1$.

Part 2: We need to prove that if

$S_k = \frac{k}{k+1}$, then

$S_{k+1} = \frac{k+1}{k+2}$.

Proof: $S_{k+1} = S_k + a_{k+1}$

$= \frac{k}{k+1} + \frac{1}{(k+1)(k+2)}$

$= \frac{k(k+2)+1}{(k+1)(k+2)}$

$= \frac{k^2+2k+1}{(k+1)(k+2)}$

$= \frac{(k+1)^2}{(k+1)(k+2)}$

$= \frac{k+1}{k+2}$

11. Part1: If $n = 1$, then

$3^1 \geq 2(1) + 1$

$3 \geq 3$

True

Part 2: We need to prove that if

$3^k \geq 2k + 1$ then

$3^{k+1} \geq 2(k+1) + 1$

$3^{k+1} \geq 2k + 3$

Proof: $3^k \geq 2k + 1$

Multiply both sides of the equation by 3.

$3(3^k) \geq 3(2k+1)$

$3^{k+1} \geq 6k + 3$

Since $k \geq 0$, $6k \geq 2k$

Therefore $6k + 3 \geq 2k + 3$

By transitive property

If $3^{k+1} \geq 6k+3$ and $6k+3 \geq 2k+3$, then $3^{k+1} \geq 2k + 3$

13. Part 1: If $n = 1$, then $n^2 \geq n$

$1^2 \geq 1$

$1 \geq 1$

True

Part 2: We need to prove that if

$k^2 \geq k$, then

$(k+1)^2 \geq k + 1$

Proof: $k^2 \geq k$

Add $2k + 1$ to both sides.

$k^2 + 2k + 1 \geq k + 2k + 1$

$(k+1)^2 \geq 3k + 1$

Furthermore,

$k \geq 0$

$2k \geq 0$

Add $k + 1$ to both sides.

$2k + k + 1 \geq 0 + k + 1$

$3k + 1 \geq k + 1$

Therefore:

$(k+1)^2 \geq 3k + 1 \geq k + 1$

$(k+1)^2 \geq k + 1$

15. Part 1: If $n = 1$, then $4^n - 1$

Becomes $4^1 - 3 = 3$

Which is divisible by 3.

Part 2: We need to prove that if

$4^k - 1$ is divisible by 3, then

$4^{k+1} - 1$ is divisible by 3.

Proof: If $4^k - 1$ is divisible by 3, then for some integer x we have

$4^k - 1 = 3x$

Hence:

$4^k - 1 = 3x$

$4^k = 3x + 1$

Multiply both sides by 4.

$4(4^k) = 4(3x+1)$

$4^{k+1} = 12x + 4$

Subtract 1 from both sides.

$4^{k+1} - 1 = 12x + 4 - 1$

$4^{k+1} - 1 = 12x + 3$

$4^{k+1} - 1 = 3(4x+1)$

Therefore $4^{k+1} - 1$ is divisible by 3.

Problem Set 14.4

17. Part 1: If $n = 1$, then $6^n - 1$ becomes $6^1 - 1 = 5$ which is divisible by 5.

Part 2: We need to prove that if $6^k - 1$ is divisible by 5, then $6^{k+1} - 1$ is divisible by 5.

Proof: If $6^k - 1$ is divisible by 5, then for some integer x we have
$6^k - 1 = 5x$
Therefore:
$6^k - 1 = 5x$
$6^k = 1 + 5x$
Multiply both sides by 6.
$6(6^k) = 6(1 + 5x)$
$6^{k+1} = 6 + 30x$
$6^{k+1} = 1 + 5 + 30x$
$6^{k+1} - 1 = 5(1 + 6x)$

Hence $6^{k+1} - 1$ is divisible by 5.

19. Part 1: If $n = 1$, $n^2 + n = (1)^2 + 1 = 2$ and is divisible by 2.

Part 2: We need to prove that if $k^2 + k$ is divisible by 2, then
$(k+1)^2 + (k+1) =$
$k^2 + 2k + 1 + k + 1 =$
$k^2 + 3k + 2$ is divisible by 2.

Proof: If $k^2 + k$ is divisible by 2, then for some integer x we have
$k^2 + k = 2x$
Since $2k + 2$ is divisible by 2
Then the sum
$k^2 + k + 2k + 2 = 2y$
for some integer y
$k^2 + 3k + 2 = 2y$

Therefore:
$k^2 + 3k + 2$ is divisible by 2.

25. Answers vary.

CHAPTER 14 Review Problem Set

1. $d = 9 - 3 = 6$
$a_n = a_1 + (n-1)d$
$a_n = 3 + (n-1)(6)$
$a_n = 3 + 6n - 6$
$a_n = 6n - 3$

2. $r = \dfrac{1}{\frac{1}{3}} = 3$
$a_n = a_1 r^{n-1}$
$a_n = \frac{1}{3}(3)^{n-1}$
$a_n = (3^{-1})(3^{n-1})$
$a_n = 3^{n-2}$

3. $r = \dfrac{20}{10} = 2$
$a_n = a_1 r^{n-1}$
$a_n = 10(2)^{n-1}$
$a_n = (5 \cdot 2)(2)^{n-1}$
$a_n = 5(2)^n$

4. $d = 2 - 5 = -3$
$a_n = a_1 + (n-1)d$
$a_n = 5 + (n-1)(-3)$
$a_n = 5 - 3n + 3$
$a_n = -3n + 8$

5. $d = -3 - (-5) = 2$
$a_n = a_1 + (n-1)d$
$a_n = -5 + (n-1)(2)$
$a_n = -5 + 2n - 2$
$a_n = 2n - 7$

6. $r = \dfrac{3}{9} = \dfrac{1}{3}$
$a_n = a_1 r^{n-1}$
$a_n = 9\left(\dfrac{1}{3}\right)^{n-1}$
$a_n = 3^2(3^{-1})^{n-1}$
$a_n = 3^2(3)^{-n+1}$
$a_n = 3^{3-n}$

7. $r = \dfrac{2}{-1} = -2$
$a_n = a_1 r^{n-1}$
$a_n = -1(-2)^{n-1}$

8. $d = 15 - 12 = 3$
$a_n = a_1 + (n-1)d$
$a_n = 12 + (n-1)(3)$
$a_n = 12 + 3n - 3$
$a_n = 3n + 9$

9. $d = 1 - \dfrac{2}{3} = \dfrac{1}{3}$
$a_n = a_1 + (n-1)d$
$a_n = \dfrac{2}{3} + (n-1)\left(\dfrac{1}{3}\right)$
$a_n = \dfrac{2}{3} + \dfrac{1}{3}n - \dfrac{1}{3}$
$a_n = \dfrac{1}{3}n + \dfrac{1}{3} = \dfrac{n+1}{3}$

10. $r = \dfrac{4}{1} = 4$
$a_n = a_1 r^{n-1}$
$a_n = 1(4)^{n-1}$
$a_n = 4^{n-1}$

11. $d = 5 - 1 = 4$
$a_n = a_1 + (n-1)d$
$a_{19} = 1 + (19-1)(4)$
$a_{19} = 73$

12. $d = 2 - (-2) = 4$
$a_n = a_1 + (n-1)d$
$a_{28} = -2 + (28-1)(4)$
$a_{28} = 106$

13. $r = \dfrac{4}{8} = \dfrac{1}{2}$
$a_n = a_1 r^{n-1}$
$a_9 = 8\left(\dfrac{1}{2}\right)^{9-1}$
$a_9 = \dfrac{8}{256} = \dfrac{1}{32}$

14. $r = \dfrac{\frac{81}{16}}{\frac{243}{32}} = \left(\dfrac{81}{16}\right)\left(\dfrac{32}{243}\right) = \dfrac{2}{3}$
$a_n = a_1 r^{n-1}$
$a_8 = \dfrac{243}{32}\left(\dfrac{2}{3}\right)^{8-1}$
$a_8 = \left(\dfrac{3^5}{2^5}\right)\left(\dfrac{2^7}{3^7}\right) = \dfrac{2^2}{3^2} = \dfrac{4}{9}$

15. $d = 4 - 7 = -3$
$a_n = a_1 + (n-1)d$
$a_{34} = 7 + (34-1)(-3)$
$a_{34} = -92$

16. $r = \dfrac{16}{-32} = -\dfrac{1}{2}$
$a_n = a_1 r^{n-1}$
$a_{10} = -32\left(-\dfrac{1}{2}\right)^{10-1}$
$a_{10} = -32\left(-\dfrac{1}{512}\right) = \dfrac{1}{16}$

17. $a_n = a_1 + (n-1)d$

$a_5 = -19$ $\quad$ $a_8 = -34$
$-19 = a_1 + 4d$ $\quad$ $-34 = a_1 + 7d$

$$\begin{aligned} 19 &= -a_1 - 4d \\ -34 &= a_1 + 7d \\ \hline -15 &= 3d \\ -5 &= d \end{aligned}$$

18. $a_8 = 37$ $\quad$ $a_{13} = 57$
$37 = a_1 + 7d$ $\quad$ $57 = a_1 + 12d$

$$\begin{aligned} -37 &= -a_1 - 7d \\ 57 &= a_1 + 12d \\ \hline 20 &= 5d \\ 4 &= d \end{aligned}$$

$37 = a_1 + 7(4)$
$37 = a_1 + 28$
$9 = a_1$

$a_{20} = 9 + (20-1)(4)$
$a_{20} = 85$

19. $a_3 = 5$ $\quad a_6 = 135$

$a_n = a_1 r^{n-1}$

$5 = a_1 r^2$ $\quad 135 = a_1 r^5$

$a_1 = \frac{5}{r^2}$

$135 = \left(\frac{5}{r^2}\right) r^5$

$135 = 5r^3$

$27 = r^3$

$3 = r$

$a_1 = \frac{5}{(3)^2} = \frac{5}{9}$

20. $a_2 = \frac{1}{2}$ $\quad a_6 = 8$

$\frac{1}{2} = a_1 r$ $\quad 8 = a_1 r^5$

$a_1 = \frac{1}{2r}$

$8 = \left(\frac{1}{2r}\right) r^5$

$16 = r^4$

$r = \pm 2$

$r = 2$ or $r = -2$

21. $r = \frac{27}{81} = \frac{1}{3}$ $\quad a_1 = 81$

$S_n = \frac{a_1 r^n - a_1}{r - 1}$

$S_9 = \frac{81\left(\frac{1}{3}\right)^9 - 81}{\frac{1}{3} - 1}$

$S_9 = \frac{\frac{1}{243} - 81}{-\frac{2}{3}}$

$S_9 = -\frac{3}{2}\left(\frac{1}{243} - 81\right)$

$S_9 = -\frac{3}{2}\left(\frac{-19682}{243}\right) = 121\frac{40}{81}$

22. $d = 0 - (-3) = 3$

$a_{70} = -3 + (70 - 1)(3)$

$a_{70} = 204$

$S_n = \frac{n(a_1 + a_n)}{2}$

$S_{70} = \frac{70(-3 + 204)}{2} = 7035$

23. $d = 1 - 5 = -4$

$a_{75} = 5 + (75 - 1)(-4) = -291$

$S_n = \frac{n(a_1 + a_n)}{2}$

$S_n = \frac{75(5 - 291)}{2} = -10,725$

24. $a_1 = 2^{5-1} = 2^4 = 16$

$a_2 = 2^{5-2} = 2^3 = 8$

$r = \frac{8}{16} = \frac{1}{2}$

$S_n = \frac{a_1 r^n - a_1}{r - 1}$

$S_{10} = \frac{16\left(\frac{1}{2}\right)^{10} - 16}{\frac{1}{2} - 1} = 31\frac{31}{32}$

25. $a_1 = 7(1) + 1 = 8$

$a_{95} = 7(95) + 1 = 666$

$S_{95} = \frac{95(8 + 666)}{2} = 32,015$

26. $d = 7 - 5 = 2$

$137 = 5 + (n - 1)(2)$

$137 = 5 + 2n - 2$

$134 = 2n$

$67 = n$

$S_n = \frac{n(a_1 + a_n)}{2}$

$S_{67} = \frac{67(5 + 137)}{2} = 4757$

27. $r = \frac{16}{64} = \frac{1}{4}$

$a_n = a_1 r^{n-1}$

$$\frac{1}{64} = 64\left(\frac{1}{4}\right)^{n-1}$$
$$\frac{1}{4^3} = 4^3\left(\frac{1}{4}\right)^{n-1}$$
$$\frac{1}{4^6} = \left(\frac{1}{4}\right)^{n-1}$$
$$\left(\frac{1}{4}\right)^6 = \left(\frac{1}{4}\right)^{n-1}$$
$$6 = n - 1$$
$$7 = n$$
$$S_n = \frac{a_1 r^n - a_1}{r - 1}$$

$$S_7 = \frac{64\left(\frac{1}{4}\right)^7 - 64}{\frac{1}{4} - 1}$$

$$S_7 = \frac{\frac{1}{256} - 64}{-\frac{3}{4}}$$

$$S_7 = -\frac{4}{3}\left(\frac{1}{256} - 64\right) = 85\frac{21}{64}$$

28. $d = 2$
$$384 = 8 + (n-1)(2)$$
$$384 = 8 + 2n - 2$$
$$378 = 2n$$
$$189 = n$$
$$S_{189} = \frac{189(8 + 384)}{2} = 37,044$$

29. $a_1 = 27$ $\quad\quad d = 3$
$$a_n = a_1 + (n-1)d$$
$$276 = 27 + (n-1)(3)$$
$$276 = 27 + 3n - 3$$
$$252 = 3n$$
$$84 = n$$
$$S_n = \frac{n(a_1 + a_n)}{2}$$
$$S_{84} = \frac{84(27 + 276)}{2} = 12,726$$

30. $\sum_{i=1}^{45}(-2i + 5)$
$$a_1 = -2(1) + 5 = 3$$
$$a_{45} = -2(45) + 5 = -85$$
$$S_{45} = \frac{45(3 - 85)}{2} = -1845$$

31. $\sum_{i=1}^{5} i^3 = 1^3 + 2^3 + 3^3 + 4^3 + 5^3$
$$= 1 + 8 + 27 + 64 + 125 = 225$$

32. $\sum_{i=1}^{8} 2^{8-i} = 2^7 + 2^6 + 2^5 + 2^4 + 2^3 + 2^2 + 2^1 + 2^0$
$$= 128 + 64 + 32 + 16 + 8 + 4 + 2 + 1 = 255$$

33. $\sum_{i=4}^{75}(3i - 4)$
$$a_4 = 3(4) - 4 = 8$$
$$a_{75} = 3(75) - 4 = 221$$
$$S_n = \frac{n(a_1 + a_n)}{2}$$
$$S_{72} = \frac{72(8 + 221)}{2} = 8244$$

34. $r = \frac{16}{64} = \frac{1}{4}$
$$S_\infty = \frac{a_1}{1 - r}$$
$$S_\infty = \frac{64}{1 - \frac{1}{4}} = 85\frac{1}{3}$$

35. $0.\overline{36} = 0.36 + 0.0036 + \ldots$
$$r = \frac{0.0036}{0.36} = 0.01$$
$$S_\infty = \frac{a_1}{1 - r}$$
$$S_\infty = \frac{0.36}{1 - 0.01} = \frac{0.36}{0.99} = \frac{4}{11}$$

36. $0.4\overline{5} = 0.4 + (0.05 + 0.005 + \ldots)$
$$r = \frac{0.005}{0.05} = 0.1$$

$$S_\infty = \frac{a_1}{1-r}$$

$$S_\infty = \frac{0.05}{1-0.1} = \frac{0.05}{0.90} = \frac{1}{18}$$

$$0.4\bar{5} = \frac{4}{10} + \frac{1}{18} = \frac{41}{90}$$

37. $a_1 = 3750 \quad d = -250 \quad n = 13$
$a_{13} = 3750 + (13-1)(-250)$
$a_{13} = 750$

The savings account will contain \$750.

38. $a_1 = 0.10 \quad d = 0.10 \quad n = 30$

$a_{30} = 0.10 + (30-1)(0.10) = 3.00$

$$S_{30} = \frac{30(0.10+3.00)}{2} = 46.50$$

In April she will save \$46.50.

39. $r = 2 \quad a_1 = 0.10 \quad n = 15$

$$S_n = \frac{a_1 r^n - a_1}{r-1}$$

$$S_{15} = \frac{0.10(2)^{15} - 0.10}{2-1} = \$3,276.70$$

40. $a_1 = 61,440 \quad r = \frac{3}{4}$ remaining

$$a_7 = 61,440\left(\frac{3}{4}\right)^{7-1}$$

$a_7 = 10,935$

After six days 10,935 gallons remain.

41. Part 1: If $n = 1$, then
$5^1 > 5(1) - 1$
$5 > 4$
True for $n = 1$.

Part 2: We need to prove that if
$5^k > 5k - 1$, then
$5^{k+1} > 5(k+1) - 1$
$5^{k+1} > 5k + 4$

Proof: Assume $5^k > 5k - 1$,
Multiply each side of the inequality by 5.
$5(5^k) > 5(5k-1)$
$5^{k+1} > 25k - 5$
$5^{k+1} > 5k + 20k - 5$
Since $k \geq 1$, $20k > 9$
By transitive property
$5^{k+1} > 5k + 20k - 5$
and $5k + 20k - 5 > 5k + 9 - 5$
Therefore: $5^{k+1} > 5k + 4$

42. Part 1: If $n = 1$, then
$1^3 - 1 + 3 = 3$
which is divisible by 3
True for $n = 1$.

Part 2: We need to prove that if
$k^3 - k + 3$ is divisible by 3, then
$(k+1)^3 - (k+1) + 3 =$
$k^3 + 3k^2 + 3k + 1 - k - 1 + 3 =$
$k^3 + 3k^2 + 2k + 3$ is divisible by 3.

Proof: If $k^3 - k + 3$ is divisible by 3, then for some integer x, we have
$k^3 - k + 3 = 3x$
Since $3k^2 + 3k$ is divisible by 3, the sum
$k^3 - k + 3 + 3k^2 + 3k = 3y$
for some integer y.
Therefore: $k^3 + 3k^2 + 2k + 3$ is divisible by 3.

43. Part 1: If $n = 1$, then

$$a_1 = \frac{1}{1(1+1)(1+2)} = \frac{1}{6}$$

$$S_1 = \frac{1(1+3)}{4(1+1)(1+2)} = \frac{4}{24} = \frac{1}{6}$$

True for $n = 1$.

Part 2: We need to prove that if

$$S_k = \frac{k(k+3)}{4(k+1)(k+2)}, \text{ then}$$

$$S_{k+1} = \frac{(k+1)(k+1+3)}{4(k+1+1)(k+1+2)}$$

$$S_{k+1} = \frac{(k+1)(k+4)}{4(k+2)(k+3)}$$

Proof: $S_{k+1} = S_k + a_{k+1}$

$$= \frac{k(k+3)}{4(k+1)(k+2)} + \frac{1}{(k+1)(k+1+1)(k+1+2)}$$

$$= \frac{k(k+3)}{4(k+1)(k+2)} + \frac{1}{(k+1)(k+2)(k+3)}$$

$$= \frac{k(k+3)(k+3)}{4(k+1)(k+2)(k+3)} + \frac{4}{4(k+1)(k+2)(k+3)}$$

$$= \frac{k(k^2+6k+9)+4}{4(k+1)(k+2)(k+3)}$$

$$= \frac{k^3+6k^2+9k+4}{4(k+1)(k+2)(k+3)}$$

$$= \frac{(k+1)(k+4)(k+1)}{4(k+1)(k+2)(k+3)}$$

$$= \frac{(k+1)(k+4)}{4(k+2)(k+3)}$$

CHAPTER 14 Test

1. $a_{15} = -(15)^2 - 1 = -226$

2. $a_5 = 3(2)^{5-1} = 48$

3. $d = -8 - (-3) = -5$
$a_n = a_1 + (n-1)d$
$a_n = -3 + (n-1)(-5)$
$a_n = -5n + 2$

4. $r = \dfrac{\frac{5}{2}}{5} = \dfrac{1}{2}$
$a_n = a_1 r^{n-1}$
$a_n = 5\left(\dfrac{1}{2}\right)^{n-1}$
$a_n = 5(2^{-1})^{n-1}$
$a_n = 5(2)^{1-n}$

5. $d = 16 - 10 = 6$
$a_n = a_1 + (n-1)d$
$a_n = 10 + (n-1)(6)$
$a_n = 6n + 4$

6. $r = \dfrac{12}{8} = \dfrac{3}{2}$
$a_7 = 8\left(\dfrac{3}{2}\right)^{7-1} = \dfrac{729}{8} = 91\dfrac{1}{8}$

7. $d = 4 - 1 = 3$
$a_n = a_1 + (n-1)d$
$a_{75} = 1 + (75-1)(3)$
$a_{75} = 223$

8. $a_1 = 7 \qquad d = 4$
$243 = 7 + (n-1)(4)$
$243 = 7 + 4n - 4$
$240 = 4n$
$60 = n$

9. $d = 4 - 1 = 3$
$a_{40} = 1 + (40-1)(3) = 118$
$S_{40} = \dfrac{40(1+118)}{2} = 2380$

10. $r = 2 \qquad a_1 = 3$
$S_8 = \dfrac{3(2)^8 - 3}{2-1} = 765$

11. $a_1 = 7(1) - 2 = 5$
$a_{45} = 7(45) - 2 = 313$
$S_{45} = \dfrac{45(5+313)}{2} = 7155$

12. $a_1 = 3(2)^1 = 6 \qquad r = 2$
$S_{10} = \dfrac{6(2)^{10} - 6}{2-1} = 6138$

13. $a_1 = 2 \qquad d = 2 \qquad n = 150$
$a_{150} = 2 + (150-1)(2) = 300$
$S_{150} = \dfrac{150(2+300)}{2} = 22,650$

14. $a_1 = 11 \qquad d = 2$
$193 = 11 + (n-1)(2)$
$193 = 11 + 2n - 2$
$184 = 2n$
$92 = n$
$$S_{92} = \frac{92(11+193)}{2} = 9384$$

15. $\sum_{i=1}^{50}(3i+5)$
$a_1 = 3(1) + 5 = 8$
$a_{50} = 3(50) + 5 = 155$
$$S_{50} = \frac{50(8+155)}{2} = 4075$$

16. $\sum_{i=1}^{10}(-2)^{i-1}$
$a_1 = (-2)^{1-1} = 1 \qquad r = -2$
$$S_{10} = \frac{1(-2)^{10} - 1}{-2-1} = -341$$

17. $r = \frac{\frac{3}{2}}{3} = \frac{1}{2}$
$$S_\infty = \frac{3}{1-\frac{1}{2}} = \frac{3}{\frac{1}{2}} = 6$$

18. $a_1 = 2\left(\frac{1}{3}\right)^{1+1} = \frac{2}{9} \qquad r = \frac{1}{3}$
$$a_2 = 2\left(\frac{1}{3}\right)^{2+1} = \frac{2}{27}$$
$$r = \frac{\frac{2}{27}}{\frac{2}{9}} = \frac{1}{3}$$
$$S_\infty = \frac{\frac{2}{9}}{1-\frac{1}{3}} = \frac{\frac{2}{9}}{\frac{2}{3}} = \frac{1}{3}$$

19. $0.\overline{18} = 0.18 + 0.0018 + ...$
$$r = \frac{0.0018}{0.18} = 0.01$$
$$S_\infty = \frac{0.18}{1-0.01} = \frac{0.18}{0.99} = \frac{2}{11}$$

20. $0.2\overline{6} = 0.2 + [0.06 + 0.006 + ...]$
$$r = \frac{0.006}{0.06} = 0.1$$
$$S_\infty = \frac{0.06}{1-0.1} = \frac{0.06}{0.90} = \frac{6}{90} = \frac{1}{15}$$
$$0.2\overline{6} = \frac{2}{10} + \frac{1}{15} = \frac{4}{15}$$

21. $a_1 = 49,152 \qquad r = \frac{1}{4}$
$$a_8 = 49,152\left(\frac{1}{4}\right)^{8-1} = 3$$
After seven days, 3 liters remain.

22. $a_1 = 0.10 \quad r = 2 \qquad n = 14$
$$S_{15} = \frac{0.10(2)^{14} - 0.10}{2-1} = 1,638.30$$
The amount saved is $\$1,638.30$.

23. $I = 350 \times 0.12 \times (1 + 2 + ... + 10)$
$I = 350 \times 0.12 \times \frac{10(1+10)}{2}$
$I = 350 \times 0.12 \times 55 = 2,310$
$P = 350 \times 10 = 3,500$
$A = P + I$
$A = 3500 + 2310$
$A = 5810$
The amount accumulated after ten years is $\$5,810$.

24. Part 1:If $n = 1$, then
$a_1 = 3(1) - 2 = 1$
$$S_1 = \frac{1[3(1) - 1]}{2} = \frac{2}{2} = 1$$
True for $n = 1$

Part 2: We need to prove that if

$$S_k = \frac{k(3k-1)}{2}, \text{ then}$$

$$S_{k+1} = \frac{(k+1)[3(k+1)-1]}{2}$$

$$S_{k+1} = \frac{(k+1)(3k+2)}{2}$$

Proof: $S_{k+1} = S_k + a_{k+1}$

$$= \frac{k(3k-1)}{2} + 3(k+1) - 2$$

$$= \frac{k(3k-1)}{2} + 3k + 1$$

$$= \frac{3k^2 - k}{2} + \frac{6k+2}{2}$$

$$= \frac{3k^2 - k + 6k + 2}{2}$$

$$= \frac{3k^2 + 5k + 2}{2}$$

$$= \frac{(k+1)(3k+2)}{2}$$

25. Part 1: If $n = 1$, then
$9^1 - 1 = 8$ and is divisible by 8.

Part 2: We need to prove that if
$9^k - 1$ is divisible by 8, then
$9^{k+1} - 1$ is divisible by 8.

Proof: If $9^k - 1$ is divisible by 8, then
for some integer x we have
$9^k - 1 = 8x$
$9^k = 1 + 8x$
Multiply both sides by 9.
$9(9^k) = 9(1 + 8x)$
$9^{k+1} = 9 + 72x$
$9^{k+1} = 1 + 8 + 72x$
$9^{k+1} - 1 = 8 + 72x$
$9^{k+1} - 1 = 8(1 + 9x)$

Hence $9^{k+1} - 1$ is divisible by 8.

Chapter 15 Counting Techniques, Probability, and the Binomial Theorem

PROBLEM SET 15.1 Fundamental Principle of Counting

1. $2 \times 10 = 20$

3. $4 \times 3 \times 2 \times 1 = 24$

5. $7 \times 6 \times 4 = 168$

7. $6 \times 2 \times 4 = 48$

9. $2 \times 3 \times 6 = 36$

11. $20 \times 19 \times 18 = 6840$

13. $6 \times 5 \times 4 \times 3 \times 2 \times 1 = 720$

15. $6 \times 5 \times 4 \times 3 \times 2 \times 1 = 720$

17. $\underline{3} \times \underline{3} \times \underline{2} \times \underline{1} \times \underline{2} = 36$

19. $\underline{4} \times \underline{3} \times \underline{1} \times \underline{2} \times \underline{1} = 24$

21. There are three choices of mailboxes for the first letter, three choices of mailboxes for the second letter, etc.
$3 \times 3 \times 3 \times 3 \times 3 = 3^5 = 243$

23. Impossible

25. $6 \times 6 \times 6 = 216$

27. Count the ways of getting 5 or less.

1st die	2nd die
1	1
1	2
1	3
1	4
2	1
2	2
2	3
3	1
3	2
4	1

There are ten ways to get a sum of 5 or less.
There are $6 \times 6 = 36$ possible outcomes.
Therefore there are $36 - 10 = 26$ outcomes of greater than 5.

29. For 3 digit numbers
$2 \times 3 \times 2 = 12$

For 4 digit number
$4 \times 3 \times 2 \times 1 = 24$

$12 + 24 = 36$ numbers

31. $\underline{4} \times \underline{3} \times \underline{3} \times \underline{2} \times \underline{2} \times \underline{1} \times \underline{1} = 144$

33. There are two choices for each of the ten questions.
$2 \times 2 \times 2 \times 2 \times 2 \times 2 \times 2 \times 2 \times 2 \times 2$
$= 2^{10} = 1024$

35. If the first digit is 4
$\underline{1} \times \underline{3} \times \underline{2} \times \underline{1} \times \underline{3} = 18$

If the first digit is 5
$\underline{1} \times \underline{3} \times \underline{2} \times \underline{1} \times \underline{2} = 12$

$18 + 12 = 30$ numbers

37a. $26 \times 26 \times 9 \times 10 \times 10 \times 10 = 6{,}084{,}000$
b. $26 \times 25 \times 9 \times 10 \times 10 \times 10 = 5{,}850{,}000$
c. $26 \times 26 \times 9 \times 9 \times 8 \times 7 = 3{,}066{,}336$
d. $26 \times 25 \times 9 \times 9 \times 8 \times 7 = 2{,}948{,}400$

PROBLEM SET 15.2 Permutations and Combinations

1. $P(5, 3) = 5 \cdot 4 \cdot 3 = 60$

3. $P(6, 4) = 6 \cdot 5 \cdot 4 \cdot 3 = 360$

5. $C(7, 2) = \frac{P(7, 2)}{2!} = \frac{7 \cdot 6}{2 \cdot 1} = 21$

7. $C(10, 5) = \frac{P(10, 5)}{5!} = \frac{10 \cdot 9 \cdot 8 \cdot 7 \cdot 6}{5 \cdot 4 \cdot 3 \cdot 2 \cdot 1} = 252$

9. $C(15, 2) = \frac{P(15, 2)}{2!} = \frac{15 \cdot 14}{2 \cdot 1} = 105$

11. $C(5, 5) = \frac{P(5, 5)}{5!} = \frac{5 \cdot 4 \cdot 3 \cdot 2 \cdot 1}{5 \cdot 4 \cdot 3 \cdot 2 \cdot 1} = 1$

13. $P(4, 4) = 4 \cdot 3 \cdot 2 \cdot 1 = 24$

15. $C(9, 3) = \frac{P(9, 3)}{3!} = \frac{9 \cdot 8 \cdot 7}{3 \cdot 2 \cdot 1} = 84$

17. **a.** $P(8, 3) = 8 \cdot 7 \cdot 6 = 336$
b. $8 \cdot 8 \cdot 8 = 512$

19. $P(4, 4) \times P(5, 5)$
$4 \cdot 3 \cdot 2 \cdot 1 \times 5 \cdot 4 \cdot 3 \cdot 2 \cdot 1 = 2880$

21. $C(7, 4) \times C(8, 4)$

$\frac{P(7, 4)}{4!} \times \frac{P(8, 4)}{4!}$

$\frac{7 \cdot 6 \cdot 5 \cdot 4}{4 \cdot 3 \cdot 2 \cdot 1} \times \frac{8 \cdot 7 \cdot 6 \cdot 5}{4 \cdot 3 \cdot 2 \cdot 1} = 2450$

23. $C(5, 3) = \frac{P(5, 3)}{3!} = \frac{5 \cdot 4 \cdot 3}{3 \cdot 2 \cdot 1} = 10$

25. $C(5, 2) = \frac{P(5, 2)}{2!} = \frac{5 \cdot 4}{2 \cdot 1} = 10$

27. $\frac{7!}{4!3!} = \frac{7 \cdot 6 \cdot 5 \cdot 4 \cdot 3 \cdot 2 \cdot 1}{4 \cdot 3 \cdot 2 \cdot 1 \cdot 3 \cdot 2 \cdot 1} = 35$

29. $\frac{9!}{3!4!2!} = \frac{9 \cdot 8 \cdot 7 \cdot 6 \cdot 5 \cdot 4 \cdot 3 \cdot 2 \cdot 1}{3 \cdot 2 \cdot 1 \cdot 4 \cdot 3 \cdot 2 \cdot 1 \cdot 2 \cdot 1} = 1260$

31. $\frac{7!}{2!1!1!1!1!1!} = \frac{7 \cdot 6 \cdot 5 \cdot 4 \cdot 3 \cdot 2 \cdot 1}{2 \cdot 1} = 2520$

33. $\frac{6!}{4!2!} = \frac{6 \cdot 5 \cdot 4 \cdot 3 \cdot 2 \cdot 1}{4 \cdot 3 \cdot 2 \cdot 1 \cdot 2 \cdot 1} = 15$

35. $C(10, 5) = \frac{P(10, 5)}{5!}$

$= \frac{10 \cdot 9 \cdot 8 \cdot 7 \cdot 6}{5 \cdot 4 \cdot 3 \cdot 2 \cdot 1} = 252$

The order of the two teams is not important so the number of teams is $\frac{252}{2} = 126$.

37. One defective bulb

$C(4, 1) \times C(9, 2)$

$\frac{P(4, 1)}{1!} \times \frac{P(9, 2)}{2!}$

$4 \times \frac{9 \cdot 8}{2 \cdot 1}$

$4 \times 36 = 144$

At least one defective bulb
First find the number of all possible samples and subtract the number of samples that have no defective bulbs.

$C(13, 3) = \frac{P(13, 3)}{3!}$

$= \frac{13 \cdot 12 \cdot 11}{3 \cdot 2 \cdot 1} = 286$

$C(9, 3) = \frac{P(9, 3)}{3!}$

$= \frac{9 \cdot 8 \cdot 7}{3 \cdot 2 \cdot 1} = 84$

$286 - 84 = 202$

39. $C(6, 4) \times C(2, 2)$

$$\frac{P(6,4)}{4!} \times \frac{P(2,2)}{2!}$$

$$\frac{6 \cdot 5 \cdot 4 \cdot 3}{4 \cdot 3 \cdot 2 \cdot 1} \times \frac{2 \cdot 1}{2 \cdot 1} = 15$$

$C(6, 3) \times C(3, 3)$

$$\frac{P(6,3)}{3!} \times \frac{P(3,3)}{3!}$$

$$\frac{6 \cdot 5 \cdot 4}{3 \cdot 2 \cdot 1} \times \frac{3 \cdot 2 \cdot 1}{3 \cdot 2 \cdot 1} = 20$$

41. Subsets containing A and not B

$$C(5,3) = \frac{P(5,3)}{3!} = \frac{5 \cdot 4 \cdot 3}{3 \cdot 2 \cdot 1} = 10$$

Subsets containing B and not A

$$C(5,3) = \frac{P(5,3)}{3!} = \frac{5 \cdot 4 \cdot 3}{3 \cdot 2 \cdot 1} = 10$$

$10 + 10 = 20$

43. 5 points

$$C(5,2) = \frac{P(5,2)}{2!} = \frac{5 \cdot 4}{2 \cdot 1} = 10$$

6 points

$$C(6,2) = \frac{P(6,2)}{2!} = \frac{6 \cdot 5}{2 \cdot 1} = 15$$

7 points

$$C(7,2) = \frac{P(7,2)}{2!} = \frac{7 \cdot 6}{2 \cdot 1} = 21$$

n points

$$C(n,2) = \frac{P(n,2)}{2!} = \frac{n(n-1)}{2}$$

Further Investigations

47. $$\frac{P(6,6)}{6} = \frac{6 \cdot 5 \cdot 4 \cdot 3 \cdot 2 \cdot 1}{6} = 120$$

49. $$C(n,r) = \frac{P(n,r)}{r!} = \frac{\frac{n!}{(n-r)!}}{r!} =$$

$$\frac{1}{r!} \cdot \frac{n!}{(n-r)!} = \frac{n!}{r!(n-r)!}$$

PROBLEM SET 15.3 Probability

1. $P(E) = \frac{n(E)}{n(S)} = \frac{2}{4} = \frac{1}{2}$

3. $P(E) = \frac{n(E)}{n(S)} = \frac{3}{4}$

5. $P(E) = \frac{n(E)}{n(S)} = \frac{1}{8}$

7. $P(E) = \frac{n(E)}{n(S)} = \frac{7}{8}$

9. $P(E) = \frac{n(E)}{n(S)} = \frac{1}{16}$

11. $P(E) = \frac{n(E)}{n(S)} = \frac{6}{16} = \frac{3}{8}$

13. $P(E) = \frac{n(E)}{n(S)} = \frac{2}{6} = \frac{1}{3}$

15. $P(E) = \frac{n(E)}{n(S)} = \frac{3}{6} = \frac{1}{2}$

17. $P(E) = \frac{n(E)}{n(S)} = \frac{5}{36}$

19. $P(E) = \frac{n(E)}{n(S)} = \frac{6}{36} = \frac{1}{6}$

21. $P(E) = \frac{n(E)}{n(S)} = \frac{11}{36}$

23. $P(E) = \frac{n(E)}{n(S)} = \frac{13}{52} = \frac{1}{4}$

25. $P(E) = \frac{n(E)}{n(S)} = \frac{26}{52} = \frac{1}{2}$

27. $P(E) = \frac{n(E)}{n(S)} = \frac{1}{25}$

29. $P(E) = \frac{n(E)}{n(S)} = \frac{9}{25}$

31. $n(S) = C(5, 2) = 10$

$n(E) = C(4, 1) = 4$

$P(E) = \frac{n(E)}{n(S)} = \frac{4}{10} = \frac{2}{5}$

33. $n(S) = C(5, 2) = 10$

Only one committee with Bill and Carl.

$n(E) = 9$

$P(E) = \frac{n(E)}{n(S)} = \frac{9}{10}$

35. $n(S) = C(8, 5) = 56$

$n(E) = C(6, 3) = 20$

$P(E) = \frac{n(E)}{n(S)} = \frac{20}{56} = \frac{5}{14}$

37. $n(S) = C(8, 5) = 56$

Number of committees
with either Chad or Dominique.

$C(7, 4) + C(7, 4) - C(6, 3)$
$35 + 35 - 20 = 50$

There are 50 committees with Chad
and Dominique but now we need to find
the number of committees with Chad or
Dominique but not both Chad and Dominique.

$n(E) = 50 - C(6, 3) = 50 - 20 = 30$

$P(E) = \frac{n(E)}{n(S)} = \frac{30}{56} = \frac{15}{28}$

39. $n(S) = C(10, 3) = 120$

$n(E) = C(8, 3) = 56$

$P(E) = \frac{n(E)}{n(S)} = \frac{56}{120} = \frac{7}{15}$

41. $n(S) = C(10, 3) = 120$

$n(E) = C(2, 2) \cdot C(8, 1) = 1 \cdot 8 = 8$

$P(E) = \frac{n(E)}{n(S)} = \frac{8}{120} = \frac{1}{15}$

43. $n(S) = P(3, 3) = 6$

$n(E) = 4$

$P(E) = \frac{n(E)}{n(S)} = \frac{4}{6} = \frac{2}{3}$

45. $n(S) = C(5, 4) = 5$

$n(E) = C(4, 4) = 1$

$P(E) = \frac{n(E)}{n(S)} = \frac{1}{5}$

47. $n(S) = P(9, 9) = 362{,}880$

There are two ways, either
math books first or history
books first.

$n(E) = 2 \times P(4, 4) \times P(5, 5)$

$n(E) = 2 \times 24 \times 120 = 5760$

$P(E) = \frac{n(E)}{n(S)} = \frac{5760}{362{,}880} = \frac{1}{63}$

49. $n(S) = P(4, 4) = 24$

$n(E) = 2 \times P(3, 3)$

$n(E) = 2 \times 6 = 12$

$P(E) = \frac{n(E)}{n(S)} = \frac{12}{24} = \frac{1}{2}$

51. $n(S) = C(11, 4) = 330$

$n(E) = C(6, 2) \times C(5, 2)$

$n(E) = 15 \times 10 = 150$

$P(E) = \frac{n(E)}{n(S)} = \frac{150}{330} = \frac{5}{11}$

53. n(S) = C(10, 5) = 252

There are two teams that Ahmed, Bob, and Carl could be on together.

n(E) = 2 × C(7, 2)

n(E) = 2 × 21 = 42

$$P(E) = \frac{n(E)}{n(S)} = \frac{42}{252} = \frac{1}{6}$$

55. $n(S) = 2^9 = 512$

n(E) = C(9, 3) = 84

$$P(E) = \frac{n(E)}{n(S)} = \frac{84}{512} = \frac{21}{128}$$

57. $n(S) = 2^5 = 32$

Getting 0 heads
$n(E_0) = C(5, 0) = 1$

Getting 1 head
$n(E_1) = C(5, 1) = 5$

Getting 2 heads
$n(E_2) = C(5, 2) = 10$

Getting 3 heads
$n(E_3) = C(5, 3) = 10$

For not getting more than 3 heads
n(E) = 1 + 5 + 10 + 10 = 26

$$P(E) = \frac{n(E)}{n(S)} = \frac{26}{32} = \frac{13}{16}$$

59. $$n(S) = \frac{7!}{2!3!1!1!} = 420$$

$$n(E) = \frac{5!}{3!1!1!} = 20$$

$$P(E) = \frac{20}{420} = \frac{1}{21}$$

Further Investigations

63. For each suit there are ten cards that could begin the straight flush and there are 4 suits.

n(E) = 4 × 10 = 40

65. n(E) = 13 × C(4, 3) × 12 × C(4, 2)
= 3744

67. There are ten cards for each of the four suits that could begin the straight flush. So there are 40 cards that could begin the straight flush.

For the second through the fifth cards, the card can be any suit, so there are 4 possibilities for each. This includes the 40 sequences that are straight flushes (of the same suit), so 40 needs to be subtracted.

n(E) = 40 × 4 × 4 × 4 × 4 − 40
= 10,200

$$P(E) = \frac{n(E)}{n(S)} = \frac{10{,}200}{2{,}598{,}960} = \frac{5}{1274}$$

69. $$n(E) = \frac{13 \times C(4, 2) \times 12 \times C(4, 2)}{2!} \times 44$$
= 123,552

71. P(no pairs)
In poker this means no pairs and also no straights and no flushes.

From the previous problem
n(one pair) = 1,098,240
n(two pair) = 123,552
n(3 of a kind) = 54,912
n(full house) = 3,744
n(4 of a kind) = 624
n(straight flushes) = 40
n(flushes) = 5108
n(straight) = 10,200

total 1,296,420

n(poker hands) = 2,598,960
n(no pairs) = 2,598,960 − 1,296,420
= 1,302,540

PROBLEM SET 15.4 Some Properties of Probability: Expected Values

1. E = a sum of 6
$n(E) = 5$
$n(S) = 36$
$P(E) = \frac{5}{36}$

3. E = a sum less than 8
$n(E) = 21$
$n(S) = 36$
$P(E) = \frac{21}{36} = \frac{7}{12}$

5. $n(S) = 6^3 = 216$
$n(E) = 1$
$P(E) = \frac{n(E)}{n(S)} = \frac{1}{216}$

7. $n(S) = 6^3 = 216$
$n(E') = 4$
$n(E) = 216 - 4 = 212$
$P(E) = \frac{n(E)}{n(S)} = \frac{212}{216} = \frac{53}{54}$

9. $n(S) = 2^4 = 16$
$n(E) = 1$
$P(E) = \frac{n(E)}{n(S)} = \frac{1}{16}$

11. E = at least one tail
E′ = no tails or all heads
$n(E') = 1$
$n(S) = 2^4 = 16$
$P(E') = \frac{1}{16}$
$P(E) = 1 - P(E') = 1 - \frac{1}{16} = \frac{15}{16}$

13. E = five tails
$n(E) = 1$
$n(S) = 2^5 = 32$
$P(E) = \frac{1}{32}$

15. E = at least one tail
E′ = no tails - all heads
$n(E') = 1$
$n(S) = 2^5 = 32$
$P(E') = \frac{1}{32}$
$P(E) = 1 - P(E') = 1 - \frac{1}{32} = \frac{31}{32}$

17. E = not getting a double
E′ = getting a double
$n(E') = 6$
$n(S) = 36$
$P(E') = \frac{6}{36} = \frac{1}{6}$
$P(E) = 1 - P(E') = 1 - \frac{1}{6} = \frac{5}{6}$

19. E = not an ace
E′ = an ace
$n(E') = 4$
$n(S) = 52$
$P(E') = \frac{4}{52} = \frac{1}{13}$
$P(E) = 1 - P(E') = 1 - \frac{1}{13} = \frac{12}{13}$

21. E = at least one vowel
E′ = no vowels - all consonants
$n(E') = {}_6C_2 = 15$
$n(S) = {}_9C_2 = 36$
$P(E') = \frac{15}{36} = \frac{5}{12}$
$P(E) = 1 - P(E') = 1 - \frac{5}{12} = \frac{7}{12}$

23. E = at least one man
E′ = no men - all women
$n(E') = {}_7C_3 = 35$
$n(S) = {}_{12}C_3 = 220$
$P(E') = \frac{35}{220} = \frac{7}{44}$
$P(E) = 1 - P(E') = 1 - \frac{7}{44} = \frac{37}{44}$

25. $P(2 \cup \text{odd}) = P(2) + P(\text{odd}) - P(2 \cap \text{odd})$
$= \frac{1}{6} + \frac{3}{6} - \frac{0}{6} = \frac{4}{6} = \frac{2}{3}$

Problem Set 15.4

27. P(odd ∪ multiple of 3) =
P(odd) + P(multiple of 3)
− P(odd ∩ multiple of 3)

$$= \frac{3}{6} + \frac{2}{6} - \frac{1}{6} = \frac{4}{6} = \frac{2}{3}$$

29. P(double ∪ sum of 6) =
P(double) + P(sum of 6)
− P(double ∩ sum of 6)

$$= \frac{6}{36} + \frac{5}{36} - \frac{1}{36} = \frac{10}{36} = \frac{5}{18}$$

31. P(double ∪ sum of 7) =
P(double) + P(sum of 7)
− P(double and sum of 7)

$$= \frac{6}{36} + \frac{6}{36} - \frac{0}{2} = \frac{12}{36} = \frac{1}{3}$$

33. P(at least 2 heads ∪ exactly one tail) =
P(at least 2 heads) + P(exactly one tail)
− P(at least 2 heads ∩ exactly one tail)

$$= \frac{4}{8} + \frac{3}{8} - \frac{3}{8} = \frac{4}{8} = \frac{1}{2}$$

35. P(head ∪ 2)=P(head) + P(2) − P(head ∩ 2)

$$= \frac{6}{12} + \frac{2}{12} - \frac{1}{12} = \frac{7}{12}$$

37a. P(A ∪ R) = P(A) + P(R) − P(A ∩ R)

$$= \frac{60}{1000} + \frac{395}{1000} - \frac{45}{1000}$$

$$= \frac{410}{1000} = \frac{41}{100} = 0.410$$

b. P(A′ ∪ R) = P(A′) + P(R) − P(A′ ∩ R)

$$= \frac{940}{1000} + \frac{395}{1000} - \frac{350}{1000}$$

$$= \frac{985}{1000} = \frac{197}{200} = 0.985$$

c. P(A′ ∪ R′) = P(A′) + P(R′) − P(A′ ∩ R′)

$$= \frac{940}{1000} + \frac{605}{1000} - \frac{590}{1000}$$

$$= \frac{955}{1000} = \frac{191}{200} = 0.955$$

39. P(A ∪ B) = P(A) + P(B) − P(A ∩ B)

$$= \frac{300}{1000} + \frac{400}{1000} - \frac{175}{1000}$$

$$= \frac{525}{1000} = \frac{21}{40} = 0.525$$

41. $E(x) = n \cdot p = 360\left(\frac{1}{6}\right) = 60$

43. $E(x) = n \cdot p = 720\left(\frac{6}{36}\right) = 120$

45. $E(x) = n \cdot p = 144\left(\frac{1}{16}\right) = 9$

47. $E(x) = n \cdot p = 448\left(\frac{1}{8}\right) = 56$

49. Let x = money gained.

x	5	2	1	− 1
$P(x)$	$\frac{2}{36}$	$\frac{4}{36}$	$\frac{6}{36}$	$\frac{24}{36}$

$$E(x) = 5\left(\frac{2}{36}\right) + 2\left(\frac{4}{36}\right) + 1\left(\frac{6}{36}\right) - 1\left(\frac{24}{36}\right)$$

$$= \frac{5}{18} + \frac{2}{9} + \frac{1}{6} - \frac{2}{3} = 0$$

Yes, it is a fair game.

51. Let x = money gained.

x	5	2	− 2
$P(x)$	$\frac{1}{4}$	$\frac{1}{2}$	$\frac{1}{4}$

$$E(x) = 5\left(\frac{1}{4}\right) + 2\left(\frac{1}{2}\right) + 2\left(\frac{1}{4}\right)$$

$$= \frac{5}{4} + \frac{4}{4} - \frac{2}{4} = \frac{7}{4}$$

Yes, it advantageous to him.

53. $E(x) = 0.7(20{,}000) + (0.3)(-10000)$
$= \$11{,}000$

55. $E(x) = 2425(0.02) + (-75)(0.98)$
$= -\$25$

Further Investigations

59. 1 to 7

61. 11 to 5

63. $n(S) = 36$
$n(E) = 4$
Odds in favor are 4 to 32
or 1 to 8.

65. $n(S) = 52$
$n(E) = 26$
Odds against are 26 to 26
or 1 to 1.

67. $n(S) = 7$
$n(E) = 4$
Odds in favor are 4 to 3.

69. $P(E) = \frac{40}{100} = \frac{2}{5}$
$n(S) = 5$
$n(E) = 2$
Odds against are 3 to 2.

71. $n(\text{against}) = 5$
$n(\text{favor}) = 2$
$P(E) = \frac{2}{7}$

73. $n(\text{favor}) = 7$
$n(\text{against}) = 5$
$P(E) = \frac{7}{12}$

PROBLEM SET 15.5 Conditional Probability: Dependent and Independent Events

1. $P(5|\text{odd number}) = \frac{P(5 \cap \text{odd number})}{P(\text{odd number})}$

$= \frac{\frac{1}{6}}{\frac{1}{2}} = \frac{2}{6} = \frac{1}{3}$

3. $P(6|\text{dice are different}) = \frac{P(6 \cap \text{different})}{P(\text{even})}$

$\frac{\frac{4}{36}}{\frac{30}{36}} = \frac{4}{36}\left(\frac{36}{30}\right) = \frac{4}{30} = \frac{2}{15}$

5. $P(J|\text{Face card}) = \frac{P(J \text{ and Face card})}{P(\text{Face card})}$

$\frac{\frac{4}{52}}{\frac{12}{52}} = \left(\frac{4}{52}\right)\left(\frac{52}{12}\right) = \frac{1}{3}$

7. Independent events
$P(5|H) = P(5) = \frac{1}{6}$

9. $P(M|H) = \frac{P(M \cap H)}{P(H)}$

$= \frac{0.2}{0.3} = \frac{2}{3}$

$P(H|M) = \frac{P(M \cap H)}{P(M)}$

$= \frac{0.2}{0.7} = \frac{2}{7}$

11. $P(D|F) = \frac{P(D \cap F)}{P(F)}$

$= \frac{\frac{30}{100}}{\frac{45}{100}} = \frac{30}{45} = \frac{2}{3}$

$P(M|D) = \frac{P(M \cap D)}{P(D)}$

$= \frac{\frac{20}{100}}{\frac{50}{100}} = \frac{20}{50} = \frac{2}{5}$

Problem Set 15.5

13. $P(A|M) = \dfrac{P(A \cap M)}{P(M)}$

$= \dfrac{\frac{5}{100}}{\frac{75}{100}} = \dfrac{5}{75} = \dfrac{1}{15}$

$P(F|A) = \dfrac{P(F \cap A)}{P(A)}$

$= \dfrac{\frac{2}{100}}{\frac{7}{100}} = \dfrac{2}{7}$

15. Dependent, because
$P(F|E) \neq P(F)$

17. Independent

19. The event could happens
in 4 ways.
HHHT, HHTH, HTHH, THHH

$P(E) = 4\left(\frac{1}{2}\right)^4 = 4\left(\frac{1}{16}\right) = \frac{1}{4}$

21. $P(D \cap D \cap D) = P(D) \cdot P(D) \cdot P(D)$

$= \frac{1}{6} \cdot \frac{1}{6} \cdot \frac{1}{6} = \frac{1}{216}$

23. $P(4 \cap 4) = P(4) \cdot P(4|4)$

$= \frac{4}{52} \cdot \frac{3}{51} = \frac{1}{221}$

25. $P(S \cap D) = P(S) \cdot P(D|S)$

$= \frac{13}{52} \cdot \frac{13}{51} = \frac{13}{204}$

Since this can occur
in the other order

P(one spade and one diamond)

$= 2\left(\frac{13}{204}\right) = \frac{13}{102}$

27. $P(S \cap S) = \frac{1}{4} \cdot \frac{1}{4} = \frac{1}{16}$

29. $P(\text{A and K}) = \frac{1}{52} \cdot \frac{1}{52} = \frac{1}{2704}$

Since this can occur
in the other order

$P(E) = 2\left(\frac{1}{2704}\right) = \frac{1}{1352}$

31. If the first card drawn
is a king then $P = \frac{1}{49}$

If the first card is not
a king then $P = \frac{48}{49}\left(\frac{1}{48}\right) = \frac{1}{49}$

$P(E) = \frac{1}{49} + \frac{1}{49} = \frac{2}{49}$

33. $P(R \cap R) = \frac{5}{9} \cdot \frac{5}{9} = \frac{25}{81}$

35. $P(R \cap W) = \frac{5}{9} \cdot \frac{4}{9} = \frac{20}{81}$

37. $P(W \cap W) = \frac{5}{13} \cdot \frac{5}{13} = \frac{25}{169}$

39. $P(R \cap B) = \frac{4}{13} \cdot \frac{4}{13} = \frac{16}{169}$

$P(B \cap R) = \frac{4}{13} \cdot \frac{4}{13} = \frac{16}{169}$

$P(E) = \frac{16}{169} + \frac{16}{169} = \frac{32}{169}$

41. $P(R \cap W) = \frac{1}{3}\left(\frac{2}{2}\right) = \frac{1}{3}$

$P(W \cap R) = \frac{2}{3}\left(\frac{1}{2}\right) = \frac{1}{3}$

$P(E) = \frac{1}{3} + \frac{1}{3} = \frac{2}{3}$

43. $P(W \cap W) = \frac{2}{3}\left(\frac{1}{2}\right) = \frac{1}{3}$

45. $P(R \cap R) = \frac{5}{17}\left(\frac{4}{16}\right) = \frac{5}{68}$

47. $P(R \cap W) = \frac{5}{17}\left(\frac{12}{16}\right) = \frac{15}{68}$

$P(W \cap R) = \frac{12}{17}\left(\frac{5}{16}\right) = \frac{15}{68}$

$P(E) = \frac{15}{68} + \frac{15}{68} = \frac{30}{68} = \frac{15}{34}$

49. $P(W \cap W) = \frac{3}{9}\left(\frac{2}{8}\right) = \frac{1}{12}$

51. $P(B \cap B) = \frac{4}{9}\left(\frac{3}{8}\right) = \frac{1}{6}$

53. $P(B \cap B \cap B) = \frac{1}{9}\left(\frac{1}{9}\right)\left(\frac{1}{9}\right) = \frac{1}{729}$

55. This can happen in 3 orders
WRR, RWR, RRW

$P(E) = 3\left(\frac{5}{9}\right)\left(\frac{3}{9}\right)\left(\frac{3}{9}\right) = \frac{5}{27}$

57. $P(W \cap W \cap W) = \frac{4}{7}\left(\frac{3}{6}\right)\left(\frac{2}{5}\right) = \frac{4}{35}$

59. This can happen in 6 orders.

$P(E) = 6\left(\frac{4 \cdot 1 \cdot 2}{7 \cdot 6 \cdot 5}\right) = \frac{8}{35}$

61. $P(W \cap W) = \frac{4}{7}\left(\frac{1}{3}\right) = \frac{4}{21}$

$P(R \cap R) = \frac{3}{7}\left(\frac{2}{3}\right) = \frac{2}{7}$

$P(R \cap W)$ or $P(W \cap R) =$

$\frac{3}{7}\left(\frac{1}{3}\right) + \frac{4}{7}\left(\frac{2}{3}\right) = \frac{11}{21}$

PROBLEM SET 15.6 Binomial Theorem

1. $(x+y)^8 = x^8 + \binom{8}{1}x^7y + \binom{8}{2}x^6y^2 + \binom{8}{3}x^5y^3 + \binom{8}{4}x^4y^4 + \binom{8}{5}x^3y^5 + \binom{8}{6}x^2y^6 + \binom{8}{7}xy^7 + y^8$

$= x^8 + 8x^7y + 28x^6y^2 + 56x^5y^3 + 70x^4y^4 + 56x^3y^5 + 28x^2y^6 + 8xy^7 + y^8$

3. $(x-y)^6 = x^6 + \binom{6}{1}(x^5)(-y) + \binom{6}{2}(x^4)(-y^2) + \binom{6}{3}(x^3)(-y)^3 + \binom{6}{4}(x^2)(-y)^4 + \binom{6}{5}(x)(-y)^5 + (-y)^6$

$= x^6 - 6x^5y + 15x^4y^2 - 20x^3y^3 + 15x^2y^4 - 6xy^5 + y^6$

5. $(a+2b)^4 = a^4 + \binom{4}{1}a^3(2b) + \binom{4}{2}a^2(2b)^2 + \binom{4}{3}a(2b)^3 + (2b)^4$

$= a^4 + 8a^3b + 24a^2b^2 + 32ab^3 + 16b^4$

Problem Set 15.6

7. $(x-3y)^5 = x^5 + \binom{5}{1}x^4(-3y) + \binom{5}{2}x^3(-3y)^2 + \binom{5}{3}x^2(-3y)^3 + \binom{5}{4}x(-3y)^4 + (-3y)^5$

$= x^5 - 15x^4y + 90x^3y - 270x^2y^3 + 405xy^4 - 243y^5$

9. $(2a-3b)^4 = (2a)^4 + \binom{4}{1}(2a)^3(-3b) + \binom{4}{2}(2a)^2(-3b)^2 + \binom{4}{3}(2a)(-3b)^3 + (-3b)^4$

$= 16a^4 - 96a^3b + 216a^2b^2 - 216ab^3 + 81b^4$

11. $(x^2+y)^5 = (x^2)^5 + \binom{5}{1}(x^2)^4y + \binom{5}{2}(x^2)^3y^2 + \binom{5}{3}(x^2)^2y^3 + \binom{5}{4}(x^2)y^4 + y^5$

$= x^{10} + 5x^8y + 10x^6y^2 + 10x^4y^3 + 5x^2y^4 + y^5$

13. $(2x^2-y^2)^4 = (2x^2)^4 + \binom{4}{1}(2x^2)^3(-y^2) + \binom{4}{2}(2x^2)^2(-y^2)^2 + \binom{4}{3}(2x^2)(-y^2)^3 + (-y^2)^4$

$= 16x^8 - 32x^6y^2 + 24x^4y^4 - 8x^2y^6 + y^8$

15. $(x+3)^6 = x^6 + \binom{6}{1}x^5(3) + \binom{6}{2}x^4(3)^2 + \binom{6}{3}x^3(3)^3 + \binom{6}{4}x^2(3)^4 + \binom{6}{5}x(3)^5 + (3)^6$

$= x^6 + 18x^5 + 135x^4 + 540x^3 + 1215x^2 + 1458x + 729$

17. $(x-1)^9 = x^9 + \binom{9}{1}x^8(-1) + \binom{9}{2}x^7(-1)^2 + \binom{9}{3}x^6(-1)^3 + \binom{9}{4}x^5(-1)^4 +$

$\binom{9}{5}x^4(-1)^5 + \binom{9}{6}x^3(-1)^6 + \binom{9}{7}x^2(-1)^7 + \binom{9}{8}x(-1)^8 + (-1)^9$

$= x^9 - 9x^8 + 36x^7 - 84x^6 + 126x^5 - 126x^4 + 84x^3 - 36x^2 + 9x - 1$

19. $\left(1+\frac{1}{n}\right)^4 = 1 + \binom{4}{1}\left(\frac{1}{n}\right) + \binom{4}{2}\left(\frac{1}{n}\right)^2 + \binom{4}{3}\left(\frac{1}{n}\right)^3 + \left(\frac{1}{n}\right)^4$

$= 1 + \frac{4}{n} + \frac{6}{n^2} + \frac{4}{n^3} + \frac{1}{n^4}$

21. $\left(a-\frac{1}{n}\right)^6 = a^6 + \binom{6}{1}a^5\left(-\frac{1}{n}\right) + \binom{6}{2}a^4\left(-\frac{1}{n}\right)^2 + \binom{6}{3}a^3\left(-\frac{1}{n}\right)^3 +$

$\binom{6}{4}a^2\left(-\frac{1}{n}\right)^4 + \binom{6}{5}a\left(-\frac{1}{n}\right)^5 + \left(-\frac{1}{n}\right)^6$

$= a^6 - \frac{6a^5}{n} + \frac{15a^4}{n^2} - \frac{20a^3}{n^3} + \frac{15a^2}{n^4} - \frac{6a}{n^5} + \frac{1}{n^6}$

23. $$\left(1+\sqrt{2}\right)^4 = 1+\binom{4}{1}\sqrt{2}+\binom{4}{2}\left(\sqrt{2}\right)^2+\binom{4}{3}\left(\sqrt{2}\right)^3+\left(\sqrt{2}\right)^4$$
$$= 1+4\sqrt{2}+12+8\sqrt{2}+4$$
$$= 17+12\sqrt{2}$$

25. $$\left(3-\sqrt{2}\right)^5 = 3^5+\binom{5}{1}(3)^4\left(-\sqrt{2}\right)+\binom{5}{2}(3)^3\left(-\sqrt{2}\right)^2+$$
$$\binom{5}{3}(3)^2\left(-\sqrt{2}\right)^3+\binom{5}{4}(3)\left(-\sqrt{2}\right)^4+\left(-\sqrt{2}\right)^5$$
$$= 243-405\sqrt{2}+540-180\sqrt{2}+60-4\sqrt{2}$$
$$= 843-589\sqrt{2}$$

27. $(x+y)^{12}$

First four terms

$$x^{12}+\binom{12}{1}x^{11}y+\binom{12}{2}x^{10}y^2+\binom{12}{3}x^9y^3$$
$$x^{12}+12x^{11}y+66x^{10}y^2+220x^9y^3$$

29. $(x-y)^{20}$

First four terms

$$x^{20}+\binom{20}{1}x^{19}(-y)+\binom{20}{2}x^{18}(-y)^2+\binom{20}{3}x^{17}(-y)^3$$
$$x^{20}-20x^{19}y+190x^{18}y^2-1140x^{17}y^3$$

31. $(x^2-2y^3)^{14}$

First four terms

$$(x^2)^{14}+\binom{14}{1}(x^2)^{13}(-2y^3)+\binom{14}{2}(x^2)^{12}(-2y^3)^2+\binom{14}{3}(x^2)^{11}(-2y^3)$$
$$x^{28}-28x^{26}y^3+364x^{24}y^6-2912x^{22}y^9$$

33. $$\left(\mathrm{a}+\frac{1}{\mathrm{n}}\right)^9$$

First four terms

$$a^9+\binom{9}{1}a^8\left(\frac{1}{n}\right)+\binom{9}{2}a^7\left(\frac{1}{n}\right)^2+\binom{9}{3}a^6\left(\frac{1}{n}\right)^3$$
$$a^9+\frac{9a^8}{n}+\frac{36a^7}{n^2}+\frac{84a^6}{n^3}$$

35. $(-x+2y)^{10}$

First four terms

$(-x)^{10} + \binom{10}{1}(-x)^9(2y) + \binom{10}{2}(-x)^8(2y)^2 + \binom{10}{3}(-x)^7(2y)^3$

$x^{10} - 20x^9y + 180x^8y^2 - 960x^7y^3$

37. $\binom{8}{3}x^5y^3 = 56x^5y^3$

39. $\binom{9}{4}x^5(-y)^4 = 126x^5y^4$

41. $\binom{7}{5}(3a)^2b^5 = 189a^2b^5$

43. $\binom{10}{7}(x^2)^3(y^3)^7 = 120x^6y^{21}$

45. $\binom{15}{6}(1)^9\left(-\frac{1}{n}\right)^6 = \frac{5005}{n^6}$

Further Investigations

51. $(2+i)^6$

$= 2^6 + \binom{6}{1}(2)^5(i) + \binom{6}{2}(2)^4(i)^2 + \binom{6}{3}(2)^3(i)^3 + \binom{6}{4}(2)^2(i)^4 + \binom{6}{5}(2)(i)^5 + i^6$

$= 64 + 192i - 240 - 160i + 60 + 12i - 1$

$= -117 + 44i$

53. $(3-2i)^5$

$= 3^5 + \binom{5}{1}(3)^4(-2i) + \binom{5}{2}(3)^3(-2i)^2 + \binom{5}{3}(3)^2(-2i)^3 + \binom{5}{4}(3)(-2i)^4 + (-2i)^5$

$= 243 - 810i - 1080 + 720i + 240 - 32i$

$= -597 - 122i$

CHAPTER 15 Review Problem Set

1. $P(6, 6) = 6! = 720$

2. $\frac{9!}{3!2!1!1!1!1!} = 30{,}240$

3. $5 \times 6 \times 5 = 150$

4. Sitting Arlene next to Carlos
There are 6 possible seats for Arlene.

The number of orders are
$6 \times 1 \times 1 \times P(5,5) = 720$

Similarily, seating Arlene next to Carlos gives $6 \times 1 \times 1 \times P(5,5) = 720$.

So there are $720 + 720 = 1440$ ways to seat Arlene and Carlos next to each other.

5. $C(6, 3) = \frac{P(6, 3)}{3!} = \frac{6!}{3!3!}$
$C(6, 3) = 20$

6. $C(7, 3) \times C(6,2)$
$35 \times 15 = 525$

7. $C(13, 5) = \frac{P(13, 5)}{5!} = \frac{13!}{8!5!}$
$C(13, 5) = 1287$

8. Three digit numbers
$2 \times 4 \times 3 = 24$

Four digit numbers
$5 \times 4 \times 3 \times 2 = 120$

Five digit numbers
$5 \times 4 \times 3 \times 2 \times 1 = 120$

$24 + 120 + 120 = 264$

9. Committees with no man
$C(5, 3) = 10$

All possible 3-person committee
$C(9, 3) = 84$

Committee with at least one man
$84 - 10 = 74$

10. If the 1st person is on the committee and the 2nd person is not on the committee
$C(6, 3) = 20$

If the 2nd person is on the committee and the 1st person is not on the committee
$C(6, 3) = 20$

If neither is on the committee
$C(6, 4) = 15$

$20 + 20 + 15 = 55$

11. Subset with A but not B
$C(6, 3) = 20$

Subset with B but not A
$C(6, 3) = 20$

Subset with A or B but not both
$20 + 20 = 40$

12. $\frac{6!}{4!2!} = 15$

13. $C(3, 2) \times C(4, 1) \times C(5, 1)$
$3 \times 4 \times 5 = 60$

14. $8 \times C(6,2)$
$8 \times 15 = 120$

15. $n(S) = 2^3 = 8$
$n(E) = C(3, 2) = 3$
$P(E) = \frac{n(E)}{n(S)} = \frac{3}{8}$

16. $n(S) = 2^5 = 32$
$n(E) = C(5, 3) = 10$
$P(E) = \frac{10}{32} = \frac{5}{16}$

17. $n(S) = 6 \times 6 = 36$
$n(E) = 5$
$P(E) = \frac{n(E)}{n(S)} = \frac{5}{36}$

18. $n(S) = 6 \times 6 = 36$
$n(E') = 10$
$n(E) = 36 - 10 = 26$
$P(E) = \frac{26}{36} = \frac{13}{18}$

19. $n(S) = P(5, 5) = 120$
If two people are side by side, there are three seats to fill and then double the number of ways for the order of the two side by side and times four for the different seat positions for the side by side people.

$P(3, 3) \times 2 \times 4 = 6(2)(4) = 48$
Therefore, not side by side
is $120 - 48 = 72$

$n(E) = 72$

$$P(E) = \frac{n(E)}{n(S)} = \frac{72}{120} = \frac{3}{5}$$

20. $n(S) = P(7, 7) = 5040$

$n(E) = 4 \times 3 \times 3 \times 2 \times 2 \times 1 \times 1 = 144$

$$P(E) = \frac{n(E)}{n(S)} = \frac{144}{5040} = \frac{1}{35}$$

21. $n(S) = 2^6 = 64$
n(no heads) = 1
n(one head) = $C(6, 1) = 6$
n(one or less heads) = $1 + 6 = 7$
n(at least two heads) = $64 - 7 = 57$

$$P(E) = \frac{n(E)}{n(S)} = \frac{57}{64}$$

22. $$P(J,J) = \frac{4}{52} \cdot \frac{3}{51} = \frac{1}{221}$$

23. $$n(S) = \frac{6!}{3!1!1!1!} = 120$$

$$n(E) = \frac{5!}{3!1!1!} = 20$$

$$P(E) = \frac{n(E)}{n(S)} = \frac{20}{120} = \frac{1}{6}$$

24. $n(S) = C(7, 3) = 35$
$n(E) = C(6, 3) = 20$

$$P(E) = \frac{20}{35} = \frac{4}{7}$$

25. $n(S) = C(8, 4) = 70$
Committees with Alice but not Bob
$C(6, 3) = 20$

Committees with Bob but not Alice
$C(6, 3) = 20$

Committees with Alice or Bob but not both
$n(E) = 20 + 20 = 40$

$$P(E) = \frac{40}{70} = \frac{4}{7}$$

26. $n(S) = C(9, 3) = 84$
$n(E) = C(4, 1) \times C(5, 2) = 40$

$$P(E) = \frac{40}{84} = \frac{10}{21}$$

27. $n(S) = C(13, 4) = 715$
n(no women) = $C(6, 4) = 15$
n(at least one women) = $715 - 15 = 700$

$$P(E) = \frac{n(E)}{n(S)} = \frac{700}{715} = \frac{140}{143}$$

28. $$P(\text{no red}) = \frac{8}{13} \cdot \frac{8}{13} = \frac{64}{169}$$

$$P(\text{at least 1 red}) = 1 - \frac{64}{169} = \frac{105}{169}$$

29. $$P(RB) = \frac{4}{12} \times \frac{3}{12} = \frac{1}{12}$$

$$P(BR) = \frac{3}{12} \times \frac{4}{12} = \frac{1}{12}$$

$$P(\text{one B, other R}) = \frac{1}{12} + \frac{1}{12} = \frac{1}{6}$$

30. $$P(R,B) = \frac{4}{11} \cdot \frac{7}{10} = \frac{14}{55}$$

$$P(B,R) = \frac{7}{11} \cdot \frac{4}{10} = \frac{14}{55}$$

$$P(\text{one B, other R}) = \frac{14}{55} + \frac{14}{55} = \frac{28}{55}$$

31. $$P(\text{no red}) = \frac{4}{7} \times \frac{3}{6} = \frac{2}{7}$$

$$P(\text{at least one red}) = 1 - \frac{2}{7} = \frac{5}{7}$$

32. $n(S) = 4^3 = 64$
$n(E) = 4$

$$P(E) = \frac{4}{64} = \frac{1}{16}$$

33. $P(B|S) = \dfrac{P(B \cap S)}{P(S)} = \dfrac{0.05}{0.10} = \dfrac{1}{2}$

$P(S|B) = \dfrac{P(B \cap S)}{P(B)} = \dfrac{0.05}{0.15} = \dfrac{1}{3}$

34 **a.** $P(M|CD) = \dfrac{P(M \text{ and } CD)}{P(CD)} = \dfrac{0.09}{0.19} = \dfrac{9}{19}$

b. $P(CD|M) = \dfrac{P(M \text{ and } CD)}{P(M)} = \dfrac{0.09}{0.10} = \dfrac{9}{10}$

35a. $P(C|S) = \dfrac{P(C \cap S)}{P(S)} = \dfrac{\frac{200}{1000}}{\frac{700}{1000}} = \dfrac{2}{7}$

b. $P(S|C) = \dfrac{P(C \cap S)}{P(C)} = \dfrac{\frac{200}{1000}}{\frac{450}{1000}} = \dfrac{200}{450} = \dfrac{4}{9}$

36. $(x+2y)^5 = x^5 + \binom{5}{1}x^4(2y) + \binom{5}{1}x^3(2y)^2 + \binom{5}{1}x^2(2y)^3 + \binom{5}{1}x(2y)^4 + (2y)^5$

$= x^5 + 10x^4y + 40x^3y^2 + 80x^2y^3 + 80xy^4 + 32y^5$

37. $(x-y)^8 = x^8 + \binom{8}{1}(x^7)(-y) + \binom{8}{2}(x^6)(-y)^2 + \binom{8}{3}(x^5)(-y)^3 +$

$\binom{8}{4}(x^4)(-y)^4 + \binom{8}{5}(x^3)(-y)^5 + \binom{8}{6}(x^2)(-y)^6 + \binom{8}{7}(x)(-y)^7 + y^8$

$= x^8 - 8x^7y + 28x^6y^2 - 56x^5y^3 + 70x^4y^4 - 56x^3y^5 + 28x^2y^6 - 8xy^7 + y^8$

38. $(a^2 - 3b^3)^4 = (a^2)^4 + \binom{4}{1}(a^2)^3(-3b^3) + \binom{4}{2}(a^2)^2(-3b^3)^2 + \binom{4}{3}(a^2)(-3b^3)^3 + (-3b^3)^4$

$= a^8 - 12a^6b^3 + 54a^4b^6 - 108a^2b^9 + 81b^{12}$

39. $\left(x + \dfrac{1}{n}\right)^6 = x^6 + \binom{6}{1}(x^5)\left(\dfrac{1}{n}\right) + \binom{6}{2}(x^4)\left(\dfrac{1}{n}\right)^2 + \binom{6}{3}(x^3)\left(\dfrac{1}{n}\right)^3$

$+ \binom{6}{4}(x^2)\left(\dfrac{1}{n}\right)^4 + \binom{6}{5}(x)\left(\dfrac{1}{n}\right)^5 + \left(\dfrac{1}{n}\right)^6$

$= x^6 + \dfrac{6x^5}{n} + \dfrac{15x^4}{n^2} + \dfrac{20x^3}{n^3} + \dfrac{15x^2}{n^4} + \dfrac{6x}{n^5} + \dfrac{1}{n^6}$

40. $\left(1-\sqrt{2}\right)^5 = 1^5 + \binom{5}{1}(1)^4\left(-\sqrt{2}\right) + \binom{5}{2}(1)^3\left(-\sqrt{2}\right)^2 + \binom{5}{3}(1)^2\left(-\sqrt{2}\right)^3$

$$+\binom{5}{4}(1)\left(-\sqrt{2}\right)^4 + \left(-\sqrt{2}\right)^5$$
$$= 1 - 5\sqrt{2} + 20 - 20\sqrt{2} + 20 - 4\sqrt{2}$$
$$= 41 - 29\sqrt{2}$$

41. $(-a+b)^3 = (-a)^3 + \binom{3}{1}(-a)^2 b + \binom{3}{2}(-a)b^2 + b^3$

$$= -a^3 + 3a^2b - 3ab^2 + b^3$$

42. $(x-2y)^{12}$

Fourth term $= \binom{12}{3}(x)^9(-2y)^3$

$$= 220(x^9)(-8y^3)$$
$$= -1760x^9y^3$$

43. $(3a+b^2)^{13}$

Fourth term $= \binom{13}{9}(3a)^4(b^2)^9$

$$= 715(81a^4)(b^{18})$$
$$= 57915a^4b^{18}$$

CHAPTER 15 Test

1. $1 \times 3 \times 2 \times 1 = 6$
Since Abdul can have either end
$2(6) = 12$

2. $6 \times 5 \times 4 \times 2 = 240$

3. $6 \cdot 6 \cdot 6 = 216$

4. $6 \times C(10, 2)$
6×45
270

5. $6 \times 6 = 36$ combinations
1 combination gives a sum of 2
2 combinations give a sum of 3
3 combinations give a sum of 4
4 combinations give a sum of 5
So 10 combinations give a sum less than 6. Therefore, there are 26 combinations with a sum of five or greater.

6. $P(6,6) \cdot P(3,3)$
$= 720 \cdot 6$
$= 4320$ ways for Math 1st – Biology 2nd
4320 ways for Biology 1st – Math 2nd
8640 ways

7. If it contains A, then we need to choose 3 elements from the remaining 6 elements.
$C(6, 3) = 20$

8. $C(4,2) \cdot C(4,2) \cdot C(4,1)$
$= 6 \cdot 6 \cdot 4 = 144$

9. There are 4 of the letter S
There are 3 of the letter A
There is 1 letter F
There is 1 letter R
$\frac{9!}{4!3!1!1!} = 2520$

10. $C(7,4) \cdot C(5,3)$
$= 35 \cdot 10 = 350$

11. E = sum less than 9
E′ = sum 9 or more
There is 1 way to get a sum of 12.
There are 2 ways to get a sum of 11.
There are 3 ways to get a sum of 10.
There are 4 ways to get a sum of 9.
$n(E') = 10$
$P(E') = \frac{10}{36} = \frac{5}{18}$
$P(E) = 1 - P(E') = 1 - \frac{5}{18} = \frac{13}{18}$

12. $C(6,3) = 20$
$P(\text{three heads}) = 20\left(\frac{1}{2}\right)^6 = \frac{5}{16}$

13. $n(s) = P(6, 3) = 120$
E = numbers over 200
E′ = numbers less than 200
$n(E') = 1 \times P(5, 2) = 20$
$P(E') = \frac{20}{120} = \frac{1}{6}$
$P(E) = 1 - P(E') = 1 - \frac{1}{6} = \frac{5}{6}$

14. $C(7,4) = 35$ everyone
$C(5, 4) = 5$ excludes Anwar and Bob
$P = \frac{5}{35} = \frac{1}{7}$

15. $n(S) = C(8, 3) = 56$
E = at least one man
E′ = no men (all women)
$n(E') = C(5, 3) = 10$
$P(E') = \frac{10}{56} = \frac{5}{28}$
$P(E) = 1 - P(E') = 1 - \frac{5}{28} = \frac{23}{28}$

16. $P(\text{three defective}) = \frac{11}{12} \cdot \frac{10}{11} \cdot \frac{9}{10} = \frac{3}{4}$

17. There are $C(5, 3) = 10$ ways to get 3 heads and 2 tails.

$$P = 10\left(\frac{1}{2}\right)^5 = \frac{10}{32} = \frac{5}{16}$$

$$E(x) = n \cdot p = 80\left(\frac{5}{16}\right) = 25$$

18. $E_v = \frac{1}{3000}(500) + \frac{1}{3000}(300) + \frac{1}{3000}(100)$

$$E_v = \frac{1}{6} + \frac{1}{10} + \frac{1}{30} = \frac{3}{10} = \$0.30$$

19. $P(W \cap G) = \frac{7}{19}\left(\frac{12}{19}\right) = \frac{84}{361}$

$$P(G \cap W) = \frac{12}{19}\left(\frac{7}{19}\right) = \frac{84}{361}$$

$$P(E) = \frac{84}{361} + \frac{84}{361} = \frac{168}{361}$$

20. $P(G \text{ and } G) = P(G) \cdot P(G|G)$

$$= \frac{5}{15} \cdot \frac{4}{14} = \frac{2}{21}$$

21. $P(A|B) = \dfrac{P(A \cap B)}{P(B)} = \dfrac{\frac{250}{2000}}{\frac{800}{2000}} = \dfrac{250}{800} = \dfrac{5}{16}$

22. $\left(2 - \dfrac{1}{n}\right)^6$

$= 2^6 + \binom{6}{1}(2)^5\left(-\dfrac{1}{n}\right) + \binom{6}{2}(2)^4\left(-\dfrac{1}{n}\right)^2 + \binom{6}{3}(2)^3\left(-\dfrac{1}{n}\right)^3 + \binom{6}{4}(2)^2\left(-\dfrac{1}{n}\right)^4$
$+\binom{6}{5}(2)\left(-\dfrac{1}{n}\right)^5 + \left(-\dfrac{1}{n}\right)^6$

$= 64 - \dfrac{192}{n} + \dfrac{240}{n^2} - \dfrac{160}{n^3} + \dfrac{60}{n^4} - \dfrac{12}{n^5} + \dfrac{1}{n^6}$

23. $(3x + 2y)^5 = (3x)^5 + \binom{5}{1}(3x)^4(2y) + \binom{5}{2}(3x)^3(2y)^2 + \binom{5}{3}(3x)^2(2y)^3 + \binom{5}{4}(3x)(2y)^4 + (2y)^5$

$= 243x^5 + 810x^4y + 1080x^3y^2 + 720x^2y^3 + 240xy^4 + 32y^5$

24. $\left(x - \dfrac{1}{2}\right)^{12}$

Ninth term $= \binom{12}{8}(x)^4\left(-\dfrac{1}{2}\right)^8$

$= \dfrac{495x^4}{256}$

25. $(x + 3y)^7$

Fifth term $= \binom{7}{4}x^3(3y)^4 = 35x^3(81y^4) = 2835x^3y^4$

PROBLEM SET 11.6 Partial Fractions

1. $\dfrac{11x-10}{(x-2)(x+1)}=\dfrac{A}{x-2}+\dfrac{B}{x+1}$

Multiply both sides of the equation by $(x-2)(x+1)$.

$11x-10=A(x+1)+B(x-2)$

Let $x=-1$.

$$\begin{aligned}11(-1)-10&=A(-1+1)+B(-1-2)\\-21&=-3B\\7&=B\end{aligned}$$

Let $x=2$.

$$\begin{aligned}11(2)-10&=A(2+1)+B(2-2)\\12&=3A\\4&=A\end{aligned}$$

$\dfrac{11x-10}{(x-2)(x+1)}=\dfrac{4}{x-2}+\dfrac{7}{x+1}$

3. $\dfrac{-2x-8}{x^2-1}=\dfrac{-2x-8}{(x+1)(x-1)}=\dfrac{A}{x+1}+\dfrac{B}{x-1}$

Multiply both sides of the equation by $(x+1)(x-1)$.

$-2x-8=A(x-1)+B(x+1)$

Let $x=-1$.

$$\begin{aligned}-2(-1)-8&=A(-1-1)+B(-1+1)\\-6&=-2A\\3&=A\end{aligned}$$

Let $x=1$.

$$\begin{aligned}-2(1)-8&=A(1-1)+B(1+1)\\-10&=2B\\-5&=B\end{aligned}$$

$\dfrac{-2x-8}{(x+1)(x-1)}=\dfrac{3}{x+1}+\dfrac{-5}{x-1}$

OPTIONAL Problem Set 11.6

5. $\dfrac{20x-3}{6x^2+7x-3}=\dfrac{20x-3}{(2x+3)(3x-1)}$

$$\frac{20x-3}{(2x+3)(3x-1)}=\frac{A}{2x+3}+\frac{B}{3x-1}$$

Multiply both sides of the equation by $(2x+3)(3x-1)$.

$20x-3=A(3x-1)+B(2x+3)$

Let $x=-\dfrac{3}{2}$.

$$20\left(-\frac{3}{2}\right)-3=A\left[3\left(-\frac{3}{2}\right)-1\right]+B\left[2\left(-\frac{3}{2}\right)+3\right]$$

$$-33=-\frac{11}{2}A$$

$$6=A$$

Let $x=\dfrac{1}{3}$.

$$20\left(\frac{1}{3}\right)-3=A\left[3\left(\frac{1}{3}\right)-1\right]+B\left[2\left(\frac{1}{3}\right)+3\right]$$

$$\frac{11}{3}=\frac{11}{3}B$$

$$1=B$$

$$\frac{20x-3}{6x^2+7x-3}=\frac{6}{2x+3}+\frac{1}{3x-1}$$

7. $\dfrac{x^2-18x+5}{(x-1)(x+2)(x-3)}=\dfrac{A}{x-1}+\dfrac{B}{x+2}+\dfrac{C}{x-3}$

$x^2-18x+5=A(x+2)(x-3)+B(x-1)(x-3)+C(x-1)(x+2)$

Let $x=1$.

$(1)^2-18(1)+5=A(3)(-2)$

$-12=-6A$

$2=A$

Let $x=-2$.

$(-2)^2-18(-2)+5=B(-3)(-5)$

$45=15B$

$3=B$

Let $x=3$.

$(3)^2-18(3)+5=C(2)(5)$

$-40=10C$

$-4=C$

$$\frac{x^2-18x+5}{(x-1)(x+2)(x-3)}=\frac{2}{x-1}+\frac{3}{x+2}+\frac{-4}{x-3}$$

9. $\dfrac{-6x^2+7x+1}{x(2x-1)(4x+1)}=\dfrac{A}{x}+\dfrac{B}{2x-1}+\dfrac{C}{4x+1}$

$-6x^2+7x+1=A(2x-1)(4x+1)+Bx(4x+1)+Cx(2x-1)$

Let $x=0$.

$$\begin{aligned}-6(0)^2+7(0)+1&=A(-1)(1)\\1&=-A\\-1&=A\end{aligned}$$

Let $x=\dfrac{1}{2}$.

$$\begin{aligned}-6\left(\frac{1}{2}\right)^2+7\left(\frac{1}{2}\right)+1&=B\left(\frac{1}{2}\right)(3)\\-\frac{3}{2}+\frac{7}{2}+1&=\frac{3}{2}B\\3&=\frac{3}{2}B\\2&=B\end{aligned}$$

Let $x=-\dfrac{1}{4}$.

$$\begin{aligned}-6\left(-\frac{1}{4}\right)^2+7\left(-\frac{1}{4}\right)+1&=C\left(-\frac{1}{4}\right)\left(-\frac{3}{2}\right)\\-\frac{3}{8}-\frac{7}{4}+1&=\frac{3}{8}C\\-\frac{9}{8}&=\frac{3}{8}C\\-3&=C\end{aligned}$$

$$\frac{-6x^2+7x+1}{x(2x-1)(4x+1)}=\frac{-1}{x}+\frac{2}{2x-1}+\frac{-3}{4x+1}$$

11. $\dfrac{2x+1}{(x-2)^2}=\dfrac{A}{x-2}+\dfrac{B}{(x-2)^2}$

$2x+1=A(x-2)+B$

Let $x=2$.

$$\begin{aligned}2(2)+1&=B\\5&=B\end{aligned}$$

Let $x=0$ and substitute 5 for B.

$$\begin{aligned}2(0)+1&=A(-2)+5\\1&=-2A+5\\-4&=-2A\\2&=A\end{aligned}$$

$$\frac{2x+1}{(x-2)^2}=\frac{2}{x-2}+\frac{5}{(x-2)^2}$$

13. $$\frac{-6x^2+19x+21}{x^2(x+3)} = \frac{A}{x} + \frac{B}{x^2} + \frac{C}{(x+3)}$$

$$-6x^2+19x+21 = Ax^2(x+3) + B(x+3) + Cx^2$$

Let $x = -3$.

$$\begin{aligned} -6(-3)^2+19(-3)+21 &= C(-3)^2 \\ -90 &= 9C \\ -10 &= C \end{aligned}$$

Let $x = 0$.

$$\begin{aligned} -6(0)^2+19(0)+21 &= B(0+3) \\ 21 &= 3B \\ 7 &= B \end{aligned}$$

Let $x = 1$ and substitute -10 for C and 7 for B.

$$\begin{aligned} -6(1)^2+19(1)+21 &= A(1)^2(1+3)+(7)(1+3)+(-10)(1)^2 \\ 34 &= 4A+18 \\ 16 &= 4A \\ 4 &= A \end{aligned}$$

$$\frac{-6x^2+19x+21}{x^2(x+3)} = \frac{4}{x} + \frac{7}{x^2} + \frac{-10}{(x+3)}$$

15. $$\frac{-2x^2-3x+10}{(x^2+1)(x-4)} = \frac{A}{x^2+1} + \frac{B}{x-4}$$

$$-2x^2-3x+10 = A(x-4) + B(x^2+1)$$

Let $x = 4$.

$$\begin{aligned} -2(4)^2-3(4)+10 &= B(4^2+1) \\ -34 &= 17B \\ -2 &= B \end{aligned}$$

Let $x = 0$ and substitute -2 for B.

$$\begin{aligned} -2(0)^2-3(0)+10 &= A(0-4)+(-2)(0^2+1) \\ 10 &= -4A-2 \\ 12 &= -4A \\ -3 &= A \end{aligned}$$

$$\frac{-2x^2-3x+10}{(x^2+1)(x-4)} = \frac{-3}{x^2+1} + \frac{-2}{x-4}$$

17. $\dfrac{3x^2+10x+9}{(x+2)^3}=\dfrac{A}{x+2}+\dfrac{B}{(x+2)^2}+\dfrac{C}{(x+2)^3}$

$3x^2+10x+9=A(x+2)^2+B(x+2)+C$

Let $x=-2$.

$$\begin{aligned}3(-2)^2+10(-2)+9&=C\\ 1&=C\end{aligned}$$

Let $x=0$ and substitute 1 for C.

$$\begin{aligned}3(0)^2+10(0)+9&=A(4)+2B+1\\ 9&=4A+2B+1\\ 8&=4A+2B\\ 4&=2A+B\end{aligned}$$

Let $x=1$ and substitute 1 for C.

$$\begin{aligned}3(1)^2+10(1)+9&=A(9)+B(3)+1\\ 22&=9A+3B+1\\ 21&=9A+3B\\ 7&=3A+B\end{aligned}$$

Solve the system.

$4=2A+B$
$7=3A+B$

Multiply first equation by -1, then add the equations.

$$\begin{aligned}-4&=-2A-B\\ 7&=3A+B\\ \hline 3&=A\end{aligned}$$

$7=3A+B$
$7=3(3)+B$
$7=9+B$
$-2=B$

$\dfrac{3x^2+10x+9}{(x+2)^3}=\dfrac{3}{x+2}+\dfrac{-2}{(x+2)^2}+\dfrac{1}{(x+2)^3}$

19. $\dfrac{5x^2+3x+6}{x(x^2-x+3)}=\dfrac{A}{x}+\dfrac{Bx+C}{x^2-x+3}$

$5x^2+3x+6=A(x^2-x+3)+(Bx+C)(x)$

$5x^2+3x+6=Ax^2-Ax+3A+Bx^2+Cx$
$5x^2+3x+6=(A+B)x^2+(-A+C)x+3A$

$5 = A + B$
$3 = -A + C$
$6 = 3A$
Therefore $A = 2$
Substitute 2 for A in the equation $5 = A + B$.
$5 = 2 + B$
$3 = B$

Substitute 2 for A in the equation $3 = -A + C$.
$3 = -2 + C$
$5 = C$

$$\frac{5x^2 + 3x + 6}{x(x^2 - x + 3)} = \frac{2}{x} + \frac{3x + 5}{x^2 - x + 3}$$

21. $$\frac{2x^3 + x + 3}{(x^2 + 1)^2} = \frac{Ax + B}{(x^2 + 1)} + \frac{Cx + D}{(x^2 + 1)^2}$$

Multiply both sides of the equation by $(x^2 + 1)^2$.
$2x^3 + x + 3 = (Ax + B)(x^2 + 1) + (Cx + D)$
$2x^3 + x + 3 = Ax^3 + Ax + Bx^2 + B + Cx + D$
$2x^3 + x + 3 = Ax^3 + Bx^2 + Ax + Cx + B + D$
$2x^3 + x + 3 = Ax^3 + Bx^2 + (A + C)x + (B + D)$

$2 = A$
$0 = B$
$1 = A + C$
$3 = B + D$

Substitute 2 for A in the equation $1 = A + C$.
$1 = 2 + C$
$-1 = C$

Substitute 0 for B in the equation $3 = B + D$.
$3 = 0 + D$
$3 = D$

$$\frac{2x^3 + x + 3}{(x^2 + 1)^2} = \frac{2x}{(x^2 + 1)} + \frac{-x + 3}{(x^2 + 1)^2}$$

PROBLEM SET **Appendix A**

1. $26 = 2 \cdot 13$

3. $36 = 2 \cdot 2 \cdot 3 \cdot 3$

5. $49 = 7 \cdot 7$

7. $56 = 2 \cdot 2 \cdot 2 \cdot 7$

9. $120 = 2 \cdot 2 \cdot 2 \cdot 3 \cdot 5$

11. $135 = 3 \cdot 3 \cdot 3 \cdot 5$

13. $6 = 2 \cdot 3$
$8 = 2 \cdot 2 \cdot 2$
The least common multiple is $2 \cdot 2 \cdot 2 \cdot 3 = 24$.

15. $12 = 2 \cdot 2 \cdot 3$
$16 = 2 \cdot 2 \cdot 2 \cdot 2$
The least common multiple is $2 \cdot 2 \cdot 2 \cdot 2 \cdot 3 = 48$.

17. $28 = 2 \cdot 2 \cdot 7$
$35 = 5 \cdot 7$
The least common multiple is $2 \cdot 2 \cdot 5 \cdot 7 = 140$.

19. $49 = 7 \cdot 7$
$56 = 2 \cdot 2 \cdot 2 \cdot 7$
The least common multiple is $2 \cdot 2 \cdot 2 \cdot 7 \cdot 7 = 392$.

21. $8 = 2 \cdot 2 \cdot 2$
$12 = 2 \cdot 2 \cdot 3$
$28 = 2 \cdot 2 \cdot 7$
The least common multiple is $2 \cdot 2 \cdot 2 \cdot 3 \cdot 7 = 168$.

23. $9 = 3 \cdot 3$
$15 = 3 \cdot 5$
$18 = 2 \cdot 3 \cdot 3$
The least common multiple is $2 \cdot 3 \cdot 3 \cdot 5 = 90$.

25. $\frac{8}{12} = \frac{2 \cdot 2 \cdot 2}{2 \cdot 2 \cdot 3} = \frac{2}{3}$

27. $\frac{16}{24} = \frac{2 \cdot 2 \cdot 2 \cdot 2}{2 \cdot 2 \cdot 2 \cdot 3} = \frac{2}{3}$

29. $\frac{15}{9} = \frac{3 \cdot 5}{3 \cdot 3} = \frac{5}{3}$

31. $\frac{3}{4} \cdot \frac{5}{7} = \frac{15}{28}$

33. $\frac{2}{7} \div \frac{3}{5} = \frac{2}{7} \cdot \frac{5}{3} = \frac{10}{21}$

35. $\frac{3}{8} \cdot \frac{12}{15} = \frac{\overset{1}{\cancel{3}} \cdot \overset{3}{\cancel{12}}}{\underset{2}{\cancel{8}} \cdot \underset{5}{\cancel{15}}} = \frac{3}{10}$

37. $\frac{3}{4} \cdot \frac{1}{2} = \frac{3}{8}$
To make one-half of the recipe, $\frac{3}{8}$ cup of milk is needed.

39. $\frac{2}{3} \cdot \frac{1}{3} = \frac{2}{9}$
The portion being used by Mark is $\frac{2}{9}$ of the disk space.

41. $\frac{2}{7} + \frac{3}{7} = \frac{5}{7}$

43. $\frac{7}{9} - \frac{2}{9} = \frac{5}{9}$

45. $\frac{3}{4} + \frac{9}{4} = \frac{12}{4} = 3$

47. $\frac{11}{12} - \frac{3}{12} = \frac{8}{12} = \frac{2}{3}$

49. $\frac{5}{24} + \frac{11}{24} = \frac{16}{24} = \frac{2}{3}$

51. $\frac{1}{3}+\frac{1}{5}=\left(\frac{1}{3}\cdot\frac{5}{5}\right)+\left(\frac{1}{5}\cdot\frac{3}{3}\right)=$

$\frac{5}{15}+\frac{3}{15}=\frac{8}{15}$

53. $\frac{15}{16}-\frac{3}{8}=\frac{15}{16}-\left(\frac{3}{8}\cdot\frac{2}{2}\right)=$

$\frac{15}{16}-\frac{6}{16}=\frac{9}{16}$

55. $\frac{7}{10}+\frac{8}{15}=\left(\frac{7}{10}\cdot\frac{3}{3}\right)+\left(\frac{8}{15}\cdot\frac{2}{2}\right)=$

$\frac{21}{30}+\frac{16}{30}=\frac{37}{30}$

57. $\frac{11}{24}+\frac{5}{32}=\left(\frac{11}{24}\cdot\frac{4}{4}\right)+\left(\frac{5}{32}\cdot\frac{3}{3}\right)=$

$\frac{44}{96}+\frac{15}{96}=\frac{59}{96}$

59. $\frac{1}{3}+\frac{1}{4}+\frac{1}{2}=$

$\left(\frac{1}{3}\cdot\frac{4}{4}\right)+\left(\frac{1}{4}\cdot\frac{3}{3}\right)+\left(\frac{1}{2}\cdot\frac{6}{6}\right)=$

$\frac{4}{12}+\frac{3}{12}+\frac{6}{12}=\frac{13}{12}=1\frac{1}{12}$

Rosa has $1\frac{1}{12}$ pounds of berries.

61. $\frac{1}{4}-\frac{3}{8}+\frac{5}{12}-\frac{1}{24}=$

$\left(\frac{1}{4}\cdot\frac{6}{6}\right)-\left(\frac{3}{8}\cdot\frac{3}{3}\right)+\left(\frac{5}{12}\cdot\frac{2}{2}\right)-\frac{1}{24}=$

$\frac{6}{24}-\frac{9}{24}+\frac{10}{24}-\frac{1}{24}=\frac{6}{24}=\frac{1}{4}$

63. $\frac{5}{6}+\frac{2}{3}\cdot\frac{3}{4}-\frac{1}{4}\cdot\frac{2}{5}=$

$\frac{5}{6}+\frac{1}{2}-\frac{1}{10}=$

$\left(\frac{5}{6}\cdot\frac{5}{5}\right)+\left(\frac{1}{2}\cdot\frac{15}{15}\right)-\left(\frac{1}{10}\cdot\frac{3}{3}\right)=$

$\frac{25}{30}+\frac{15}{30}-\frac{3}{30}=\frac{37}{30}$

65. $\frac{3}{4}\cdot\frac{6}{9}-\frac{5}{6}\cdot\frac{8}{10}-\frac{2}{3}\cdot\frac{6}{8}=$

$\frac{1}{2}-\frac{2}{3}+\frac{1}{2}=$

$\left(\frac{1}{2}\cdot\frac{3}{3}\right)-\left(\frac{2}{3}\cdot\frac{2}{2}\right)+\left(\frac{1}{2}\cdot\frac{3}{3}\right)=$

$\frac{3}{6}-\frac{4}{6}+\frac{3}{6}=\frac{2}{6}=\frac{1}{3}$

67. $\frac{7}{13}\left(\frac{2}{3}-\frac{1}{6}\right)=\frac{7}{13}\left(\frac{4}{6}-\frac{1}{6}\right)=$

$\frac{7}{13}\left(\frac{3}{6}\right)=\frac{7}{13}\left(\frac{1}{2}\right)=\frac{7}{26}$

69. Let $x=$ amount of the estate left to the church.

$\frac{1}{4}+\frac{2}{5}+x=1$

$\left(\frac{1}{4}\cdot\frac{5}{5}\right)+\left(\frac{2}{5}\cdot\frac{4}{4}\right)+x=1\cdot\frac{20}{20}$

$\frac{5}{20}+\frac{8}{20}+x=\frac{20}{20}$

$\frac{13}{20}+x=\frac{20}{20}$

$x=\frac{7}{20}$

The church received $\frac{7}{20}$ of the estate.